Lecture Notes in Networks and Systems

Volume 115

The series "Lecture Notes in Networks and Systems" publishes the latest developments in Networks and Systems—quickly, informally and with high quality. Original research reported in proceedings and post-proceedings represents the core of LNNS.

Volumes published in LNNS embrace all aspects and subfields of, as well as new challenges in, Networks and Systems.

The series contains proceedings and edited volumes in systems and networks, spanning the areas of Cyber-Physical Systems, Autonomous Systems, Sensor Networks, Control Systems, Energy Systems, Automotive Systems, Biological Systems, Vehicular Networking and Connected Vehicles, Aerospace Systems, Automation, Manufacturing, Smart Grids, Nonlinear Systems, Power Systems, Robotics, Social Systems, Economic Systems and other. Of particular value to both the contributors and the readership are the short publication timeframe and the world-wide distribution and exposure which enable both a wide and rapid dissemination of research output.

The series covers the theory, applications, and perspectives on the state of the art and future developments relevant to systems and networks, decision making, control, complex processes and related areas, as embedded in the fields of interdisciplinary and applied sciences, engineering, computer science, physics, economics, social, and life sciences, as well as the paradigms and methodologies behind them.

**** Indexing: The books of this series are submitted to ISI Proceedings, SCOPUS, Google Scholar and Springerlink ****

More information about this series at http://www.springer.com/series/15179

Aleksei V. Bogoviz · Yulia V. Ragulina
Editors

Industry Competitiveness: Digitalization, Management, and Integration

Volume 1

Springer

Editors
Aleksei V. Bogoviz
National Research University
Higher School of Economics
Moscow, Russia

Yulia V. Ragulina
RUDN University
Moscow, Russia

ISSN 2367-3370 ISSN 2367-3389 (electronic)
Lecture Notes in Networks and Systems
ISBN 978-3-030-40748-3 ISBN 978-3-030-40749-0 (eBook)
https://doi.org/10.1007/978-3-030-40749-0

This Springer imprint is published by the registered company Springer Nature Switzerland AG
The registered company address is: Gewerbestrasse 11, 6330 Cham, Switzerland

Industry Competitiveness in the New Economy

This timely book presents a remarkable collection of chapters that provides readers with a coherent framework for understanding the factors driving industry competitiveness in contemporary conditions of economic digitalization and the ongoing transition to Industry 4.0. In particular, this volume contains selected papers presented at the Scientific and Practical Forum "Industry. Science. Competences. Integration," which was held at the RUDN University (Moscow, Russia) on November 28, 2019.

Presenting contributions by scientists, engineers, and field experts, the book focuses on using advanced technologies and applications, building innovative and resilient systems in industrial enterprises, developing competitive management systems, creating competence networks, and enhancing integration in order to foster and sustain industry competitiveness. Both qualitative and quantitative studies are included, and this collection of diverse perspective adds to the richness of the volume's insights.

Along with reviewing deep theoretical concepts and innovative approaches, the publication provides practical applications and technological solutions to real-world problems existing in the industry. Recent advances in management theory and practice focused on the forces driving competition in the industry are also extensively covered by the leading scholars and practitioners. Particularly, the book covers the concept of Industry 4.0; management in digital economy; management of competitiveness; artificial intelligence and knowledge management approaches; centers of global technological excellence; competitive and resilient systems in the energy industry; compliance and anti-corruption mechanisms; and integration and competence networks between universities, industry, science, and markets.

The resulting book is of great interest for a wide audience due to highly stimulating discussions on the determinants and patterns of industry competitiveness. This publication is highly valuable for researchers and practitioners interested in applying new concepts, methods, and insights into their work. It is also suitable for graduate students and advanced undergraduates, who are interested in the current state of industrial development, driving forces of competitiveness, innovations, advanced technologies, competence networks, and management systems based on knowledge and artificial intelligence.

Aleksei V. Bogoviz
Yulia V. Ragulina

Contents

Industry 4.0 and Competitiveness: Technologies, Economic Development, and Human Resources

Digital Design at Industrial Enterprises of Russia: State of the Scientific and Methodological Base

Rysta Sh. Tarzhmanova(✉), Anna M. Nasonenko, and Gabriel H. Alberto

Peoples' Friendship University of Russia (RUDN University), Moscow, Russia
rysty88@mail.ru, nasonenko-am@rudn.ru,
alberto.gabriel@yandex.ru

Abstract. The article provides a detailed analysis of the scientific and methodological base of the digital design of industrial enterprises in Russia, namely the standards of a unified system of design documentation (USDD) for working with electronic documents. It is approved at the state level as mandatory for compliance, which includes state standards, guiding technical materials, and methods. According to the authors, the analyzed standards for electronic design documentation are more descriptive than practical. The importance and significance of improving the standards of digital design are based on the fact that they set the vector of industrial development for several years ahead, forcing enterprises to move in a given direction and introduce contemporary high technology. This circumstance will allow bringing the industry of the Russian Federation to the forefront. The results of the study may become the basis for the development of a national standard for digital design.

Keywords: Industry · Enterprise · Production · Risks · Standards · Digital design · Digital double

1 Introduction

The risk in production is a complex of possible circumstances that may cause a decrease in the efficiency (profitability) of production. Moreover, in this situation, considering the question of cost optimization of the organization for the development and production of new products, it should be noted that the cost of production is directly dependent on the number and size of its creation costs. Through the use of the new paradigm of digital design and modeling, it became possible to get away from the traditional situation when the number of product changes due to errors or new, previously unaccounted for information. Accordingly, the increasing costs of their introduction are distributed throughout the entire development life cycle, namely from the design stage to the start of mass production (Khvostikova 2012). It is well known that the company incurs high costs due to such changes. As a result, it becomes fundamentally possible to concentrate the bulk of the changes and costs at the design stage, thereby minimizing the total amount of costs, reducing costs, and ensuring the creation

A. V. Bogoviz and Y. V. Ragulina (Eds.): ISCI 2019, LNNS 115, pp. 3–8, 2020.
https://doi.org/10.1007/978-3-030-40749-0_1

of high-tech products of the new generation in the shortest possible time (Guryanov et al. 2017).

Consequently, the digital approach, the cost of production, and the development of a prototype product are reduced. Since, when conducting virtual tests on a digital twin of the product, many technical and economic characteristics of the product can be improved and made perfect without the cost of producing a physical sample of the product.

At present, in Russia, over 500 leading industrial enterprises (including "Russian Helicopters" Holding, "VSMPO-AVISMA" Corporation, "UEC" JSC, "Uralvagonzavod NPK" JSC, and "AvtoVAZ" PJSC) are engaged in technical re-equipment, in particular, the introduction of innovative computer technologies, CAD/PLM software, as well as unique equipment. Next, we will examine the current situation in the field of digital design at industrial enterprises in Russia and the reasons for the lag behind world leaders (Kupriyanovsky et al. 2017).

2 Materials and Method

Thanks to contemporary technologies and programs, organizations have the opportunity to receive ready-made design documentation after creating a digital double of products, i.e., a virtual image endowed with all the physical and technical and economic properties of the products planned for release and allowing virtual testing of the product.

For a significant increase in the technological potential of an industrial enterprise, a transition to a different level of industrial development, competently drawn up all-Russian and industry design standards, are needed according to which documentation can be drawn up and accepted.

The current situation and the level of digital design at domestic industrial enterprises are represented by five levels of development of design and production processes, successively increasing inefficiency (Corporate Training Department of Moscow School of Management 2017), as shown in Fig. 1.

We will give some explanation on the illustration:

1. Electronic drawing. At the same time, just replacing the shuttle with a computer occurs, but a flat drawing is still a storage medium.
2. Electronic 3D model and drawing. The designer creates a three-dimensional electronic model of the product, but still produces drawings, as a document legalized by USDD. Information is transferred to technological, production, planning, and other units in the form of paper drawings and statements. The computer model is illegitimate, but at the same time, productivity is increased by no more than 30%.
3. The electronic model with technological information is the primary source (legalized medium) of design and technological information. The designer creates the model, applying technological information in the form of three-dimensional annotations on it to prepare the production.
4. The electronic model and electronic composition are no longer applied locally in the departments, but it applies to all services of the enterprise, including warehouse,

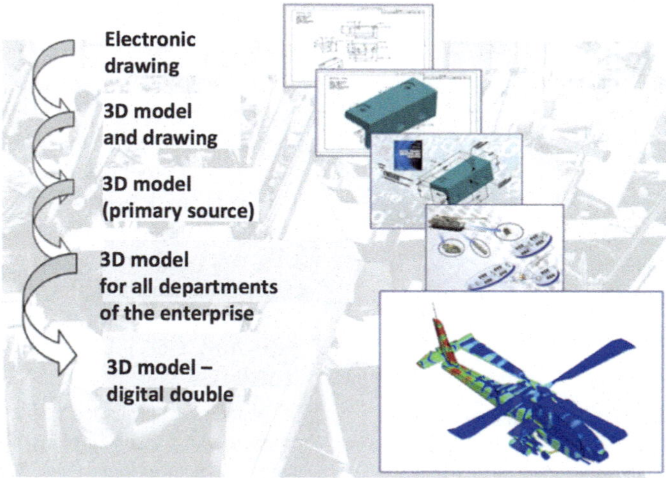

Fig. 1. The levels of development of internal processes of design and production. *Source*: (Corporate Training Department of the Moscow School of Management 2017).

sales, procurement, equipment, maintenance manuals. This significantly saves time for data re-entry and significantly reduces errors.

5. The electronic model is a "digital double" (DD). If at the previous stages, the computer model was only "geometric," then at this stage, it is saturated with physical properties, which allows one to create a "digital double" identical to a real product for virtual calculations and tests, as well as for modeling production processes.

This is a radically different, highly effective level of design and technological preparation of production. The timing of products on the market is shrinking at times, and quality is improving, weight, production costs are decreasing, especially if we are dealing with technically sophisticated products.

To transfer Russian industrial enterprises to a new level of digital design, a scientific and methodological base is needed. It should be approved at the state level as mandatory for compliance, including state standards, guiding technical materials, and methods.

3 Results

We analyzed the current state of the regulatory framework of the Russian Federation in the field of digital design, which is presented in Table 1.

Currently, the USDD regulatory framework for working with electronic documents is represented by the following standards (Bilenko et al. 2017):

According to the results of the analysis, we can say that the high standards for electronic design documentation are more descriptive than practical.

Table 1. The regulatory framework of USDD on work with electronic documents.

Name	Description
GOST 2.102-2013 USDD. Types and completeness of design documents	In this standard, concepts such as an electronic model of a part, an electronic model of an assembly unit, an electronic structure of a product are introduced. These documents are defined as mandatory in the set of necessary design documentation (DD) at the stage of development of working design documentation. Requirements for these documents are presented in GOST 2.056-2014, GOST 2.057-2014, GOST 2.053-2013
GOST 2.051-2013 USDD. Electronic documents	The standard establishes general requirements for the implementation of electronic design documents. A description is given of what parts an electronic document should consist of, what types of electronic documents are. It is declared that, in principle, design documents can be developed, agreed, taken into account, and stored electronically
GOST 2.052-2015 USDD. Electronic product model	The standard establishes general requirements for the implementation of electronic models of the product (parts and assembly units). The basic concepts and terminology of computer modeling, types of electronic models of the product are defined here
GOST 2.053-2013 USDD. Electronic product structure	The standard establishes general requirements for the electronic structure of the product. Based on the ESP, product reports are generated, which are traditionally carried out in paper in the form of specifications and statements
GOST 2.054-2013 USDD. Electronic product description	The standard gives concepts about information technical systems (PLM, PDM, and others), with which we can store, record, and describe electronic design documentation
GOST 2.055-2014 USDD. Electronic specification	The standard defines the concepts of electronic specifications, and summary statements define the storage parameters of the content using PDM systems

(*continued*)

Table 1. (*continued*)

Name	Description
GOST 2.056-2014 USDD. Electronic model of details	The standard establishes some requirements for the implementation and design of models of parts and their presentation using CAD systems, supplementing the requirements of GOST 2.052-2015
GOST 2.057-2014 USDD. Electronic assembly unit model	The standard establishes some requirements for the implementation and design of models of assembly units and their presentation using CAD systems, supplementing the requirements of GOST 2.052-2015
GOST 2.058-2016 USDD. Rules for the implementation of the requisite part of electronic DD	The standard establishes the rules for the implementation of the requisite part of electronic design documents in electronic form. The standard supplements the requirements of GOST 2.056-2014, GOST 2.052-2015, GOST 2.055-2014 in terms of requirements for the requisite part of electronic documents

Source: (Bilenko et al. 2017).

4 Discussion

After analyzing the content of existing standards, we believe that the standards for electronic models GOST 2.052-2015, GOST 2.056-2014, GOST 2.057-2014 need to be supplemented in the first place.

This is necessary so that the specialist, taking the standard, can begin to develop a model in compliance with the requirements of the standard. In the above standards, practical examples are not enough. If one compares the standards for electronic documentation, with the standards for paper, then there are striking differences. For paper standards, various cases of design documentation (DD) are drawn in separate drawings, and they are few in the standards for electronic DD. Based on domestic experience, we can say that the designers at the enterprises draw up the design documentation, and normative inspectors check the design documentation, comparing the images in the design documentation with the images in accordance with GOST. And this is the right approach because one needs to be guided by the requirements of standards during development (Rudskoy et al. 2017).

5 Conclusion

It is obvious that in the framework of the unfolding fourth industrial revolution, it is precisely those enterprises that, being in the global trends of the digital economy, that transfer the focus of their activities to the field of digital design and modeling, computer and supercomputer engineering along with methods of multi-criteria, multi-parameter, multi-disciplinary optimization will become the leaders.

Innovative technologies of computer design, armed with the appropriate scientific and methodological base of digital design, namely, the methods and standards that are mandatory for use, will bring the industry of the Russian Federation to the forefront. First, since the standards are designed to reduce design time by 30–40%, one may expect these design costs reducing by 10–20%, improving the quality of new products. Second, the standards set the vector of industrial development for several years ahead, forcing enterprises to move in a given direction and introduce new technologies.

Acknowledgment. The publication has been prepared with the support of the "RUDN University Program 5-100."

References

Bilenko, P.N., Lysenko, S.L., Zavaleev, I.S., Lysenko, L.V.: A comprehensive assessment of enterprise development as a tool to increase labor productivity. High Technol. **18**(7), 22–31 (2017)

Corporate Training Department of Moscow School of Management: Working Paper "Digital Production: Methods, Ecosystems, Technologies". Skolkovo, Russia (2017)

Guryanov, A.V., Shukalov, A.V., Zakoldaev, D.A., Zharinov, I.O., Kostishin, M.O.: Digitalization of design and production procedures as a tool for automating the design of instrumentation in Industry 4.0. Sci. Tech. Bull. Inf. Technol. Mech. Opt. **6**, 1171–1176 (2017)

Khvostikova, V.A.: Technical and economic features of the organization of production processes for the production of high-tech products. Vestnik Voronezh State Tech. Univ. **8**(8), 85–89 (2012)

Kupriyanovsky, V.P., Sinyagov, S.A., Namiot, D.E., Utkin, N.A., Nikolaev, D.E., Dobrynin, A.P.: Industrial transformation in the digital economy: design and production. Int. J. Open Inf. Technol. **1**, 50–70 (2017)

Rudskoy, A.I., Borovkov, A.I., Romanov, P.I., Kiseleva, K.N.: Engineering Education: World Experience in Training the Intellectual Elite. Polytechnic University Press, St. Petersburg (2017)

Competitiveness of Industrial Production in the Regional Context

Marina V. Moroshkina(✉)

Institute of Economics of the Karelian Scientific Center of the Russian Academy of Sciences, Petrozavodsk, Russia
Maribel74@mail.ru

Abstract. A high level of heterogeneity characterizes the development of the Russian Federation. Russian regions are unequally provided with production and industrial potential, as a result of which an imbalance is observed. The aim of the study is the analysis of inter-regional differentiation according to production potential. The object of the study is the Russian regions, and the subject of the study is the process of interregional differentiation in terms of "industrial production volumes." An analysis of inter-regional differentiation is based on calculating the coefficient of variation. The latter determines the presence of the process of regional convergence. Our analysis demonstrates that in the period from 1990 to 2015, Russian regions tend to diverge in terms of "industrial production volumes." The results of the study can be used to develop strategic and program documents.

Keywords: Interregional differentiation · Russian regions · Industrial production · Leading regions · Lagging regions · δ-convergence

1 Introduction

The Russian Federation is characterized by a heterogeneous economic space, the diversity of which lies in the heterogeneity of the constituent entities of the Russian Federation. Differences in regional entities are formed as a result of heterogeneity of territories in size, economic, and production potential. As a result, the uneven development of the regions is observed. Some regions, possessing economic potential, are successfully developing, while other regions, as a result of the lack of competitive advantages and the potential for economic development, represent stagnation processes. The specifics of regional diversity should be taken into account when drawing up strategic development documents and various areas of government.

The subjects of the Russian Federation have different production potentials and have differences in the level and dynamics of regional development. The spatial heterogeneity of the Russian regions depends on many factors and processes that influence the dynamics of economic growth. The presence of competitive advantages entails unevenness in the pace of socio-economic development and living standards of the population, the level of development of industrial potential, and other economic and social characteristics of the territory (Mau and Yanovsky 2001). One of the significant competitive advantages of regional development issues is the availability of production

A. V. Bogoviz and Y. V. Ragulina (Eds.): ISCI 2019, LNNS 115, pp. 9–18, 2020.
https://doi.org/10.1007/978-3-030-40749-0_2

potential. The factor of the availability and development of the manufacturing sector determines the main directions and formats for developing a territory as a result of the formation of competitive advantages caused by the increase in regional production.

The economic development of the Russian regions is a multi-faceted process, depending on a set of indicators. However, as for most subjects of the world community, for Russian regions, an important area of economic development is the development of production areas, enterprises. The presence of production activities forms a competitive environment for the territory, providing it with the advantages associated with the development of the production sector. In the regional context, the formats of development of the production direction determine and shape the factors of regional growth. As a result of the compilation of strategic development documents, these factors are taken into account, and strategic and forecast documents for the development of the territory are compiled based on their analytical assessment.

As a result, the study is a comparative analysis of the Russian regions in terms of production potential. Following the research goal, the hypothesis regarding a positive impact of production potential on the level and dynamics of economic growth in the regional context is formulated. The object of the research is the Russian regions, and the subject is the process of regional stratification of the Russian regions in terms of "industrial production volumes."

2 Literature Review

Changing business conditions changed the dynamics of the development of Russian regions and affected the level of heterogeneity. An increase in differentiation in Russian regions is observed for most economic and social indicators (Bobkova 2013; Glushchenko 2012).

Inequality in a regional context (in terms of social and economic development) is determined by many reasons affecting a level of regional development. These are such reasons as economic development during market institutions formation, a development level of the industrial sector, investments in the region's economy. Also, such reasons include the level of development of institutional components, the economic and geographical location of the territory, infrastructure development, the dynamics of the development of innovative potential.

The issues of regional development and the identification of causes that influence the dynamics of regional growth are the subject of many scientific works. Since the 50s, the theory of spatial development has been formed as a result of the regional distribution of the manufacturing sector, which affects the dynamics of regional growth (Lesh 1959).

The formation of a regional economy as science begins with an analysis of regional differences, identifying factors affecting the level of growth, determining the competitive advantages of a territory. In the literary direction in the regional economy, issues of regional development can be conditionally divided into two main blocks. In one direction, regional development trends in different time periods are analyzed. In another direction, the analysis of regional inequalities is carried out from the perspective of studying factors of influence on the level and dynamics of economic growth.

It is worth noting that the specifics of regional research has a wide geographical range. Regional development issues may include regions in a territorial definition, and more global territorial entities with more extensive borders, such as the dynamics of regional stratification in different countries (Barro and Sala-i-Martin 1991, 1995; Schumpeter 1908).

Researchers pay special attention to issues of regional inequality, as a result of the consideration of individual areas and identifying the main reasons for the growth of this process (Bogoviz et al. 2019a, b). The definition of the process of regional stratification of the Russian regions includes a set of processes of interregional differentiation observed by various economic and social indicators. The directions of this process can have both positive and negative effects on the territory.

The positive aspects of increasing regional differentiation include the desire for lagging regions to achieve the level of development of leading regions. This desire provides the economic entities of the lagging territories with the potential for possible growth, endows them with the ability to search for new, more efficient forms of economic activity based on the competitive advantages of the territory. The result is the formation and development of the manufacturing sector and the services sector, which contribute to the economic development of the territory. The negative aspects of the increase in regional imbalance include an increase in dissatisfaction with the living conditions and quality of life in the lagging regions, which leads to a decrease in economic activity.

The concept of economic activity is a concept that combines various processes of the economic system in a certain period. The processes that affect the level of economic activity of the territory include various areas of regional development. One of the most significant areas of regional growth is the level of industrial development.

In economic theory, domestic and foreign scientists are actively exploring the level and dynamics of the industrial sector. In the modern economy, there is a direction that studies aspects of the development of industrial production.

Classical approaches include the theory of the industrial approach developed by Schumpeter (1908), in which the author concludes that periodic changes in the industry lead to innovations that continue to change the production system. The innovation process has an impact on all areas of economic activity since its result is an impact on the production process.

Another area that has a large number of formats is the theories of production allocation, which are part of the regional economy.

The process of formation of the theory of localization of production is associated with the scientific work of the German economist J. Thünen entitled "An Isolated State in Its Relation to Agriculture and National Economy" (Thünen et al. 1926; Granberg 2000). In this work, the author revealed patterns of distribution of agricultural format production. According to Thünen's theory, the more efficient the production of an agricultural product there, it should be located closer to the settlement. The result of the theory was a model of the location of agricultural production based on optimal placement.

The theory of the German scientist Launhardt (Pinto 1977) is recognized by a different approach to the placement of production potential. The result of this theory is the process of finding the point of optimal location of an industrial enterprise, according to the ratio of existing sources of raw materials and the market for manufactured products (Launhardt 1885; Limonov 2015).

The work of A. Weber "On the location of the industry: the clean territory of the standard," published in 1909, can be considered an approach to the placement of industrial production (Weber 1929). As a result of the developed methodology, the author considers the general format of the location of a manufacturing enterprise based on an analysis of the influence of accessibility factors and cost calculation.

There are many approaches and techniques within which the geographical location and accommodation play a decisive role in territorial development. One of these theories is the theory of growth poles. In this theory, the process of economic growth is dependent on the development of individual industries. The theory of growth poles is found in the works of such domestic and foreign scientists as Gunnar Myrdal, Albert Hirschman (the "core-periphery" model) (Myrdal 1957). Within the structural components of the model, the uneven development of territories is explained by factors in the development of the manufacturing sector, which ensure economic growth. According to this theory, one of the reasons for the uneven development of regions is the level of agglomeration of production (Drobyshevsky et al. 2005).

At the present stage, there are a significant number of approaches and methods for assessing and analyzing regional heterogeneity. Most researchers conclude that one of the reasons for the unevenness lies in the competitiveness of production potential in the territory. Studies on the inequality of regional development based on production potential will make it possible to assess the level of production potential in the region. During the study, the author used the theory of convergence, which allowed us to analyze the imbalances.

Studies of regional inequalities find application in the work of Barro and Sala-i-Martin, who analyze regional differences (Barro and Sala-i-Martin 1991, 1995).

At the same time, the indicated works did not set the task of assessing convergent or divergent development trends of regions from the standpoint of analyzing socio-economic indicators of industrial production development. Such a study will help to understand the reason for the uneven regional development.

3 Methodology

The statistical information used to study the inequality in the development of the Russian regions is the statistical collections of the Federal State Statistics Service "Regions of Russia" (Federal State Statistics Service 2018). The research sample consists of the indicator "industrial production volumes" for each Russian region. The main object of the study is the regions of the Russian Federation, and the subject of the study is the process of regional differentiation in terms of "industrial production volume." As an indicator determining regional differentiation, the coefficient of variation is used. Based on the calculated coefficient, a regulation on the process of rapprochement or divergence of the Russian regions by the studied indicator is formed.

In statistical analysis, the Coefficient of Variation (CV) is used when it is necessary to evaluate the dispersion process of various quantities.

The coefficient of variation is determined by the following formula:

$$\sigma_t = CV = \frac{\sigma}{x_{\text{cp}}} \tag{1}$$

σ is the standard deviation of a random variable;

$$\sigma = \sqrt{\frac{1}{n}\sum_{i=1}^{n}\left(x_i - x_{\text{cp}}\right)^2}$$

X_{av} is the average value of the estimated development indicator;
x_i is an indicator of the level of development;
i is a region;
$i = 1, \ldots, n$ is the number of regions studied.

For analyzing the dispersion process of the studied indicator in a regional context and determine the convergence process, the concept of convergence is used. In the methodological aspect, two areas of convergence are distinguished, which in different ways explain the process of dispersion of the studied indicators.

In turn, δ-convergence determines the process of reducing the dispersion of indicators in the regional analysis; this suggests a decrease in interregional differentiation. To confirm the presence of δ-convergence, it is necessary to fulfill a condition that determines a decrease in the coefficient of variation in the final period, relative to the initial.

$$\delta_{t+T} < \delta_t \tag{2}$$

Thus, β stands for convergence ("absolute convergence") characterizing the processes of convergence of various objects, comparing the growth rates of the studied economic, social indicators, which can be represented by the following system of equations:

$$y = \beta_0 + \beta_1 x \tag{3}$$

Equation (3) is a regression equation of the dependence of the growth rate of the studied indicator for the studied period to the initial period.

Where,

$$y = \frac{\ln x_{iT} - \ln x_{i0}}{T}$$

$$x = lnx_{i0}$$

T is the number of study intervals.

The calculated coefficient of variation will allow us to estimate the level of regional imbalances in the regional context according to the indicator "industrial production volume."

4 Results

The study analyzed the heterogeneity of Russian regions in terms of industrial production. As a result of the analysis of statistical data on the regions of the Russian Federation ("Industrial Production Volume"), the level of heterogeneity of the regions in this direction was analyzed. More than that, the regions that have leading and lagging positions in this direction are identified.

The analysis, based on the statistics of leading and lagging regions, revealed patterns of getting regions into the leading and lagging groups (Table 1).

Table 1. Leading and lagging regions in terms of "industrial production volumes".

Interval	Lagging regions		Leading regions	
	Features	Regions	Features	Regions
1990–1994	*Low level of production*	Altai Republic, Tyva Republic, Republic of Kalmykia, Republic of Adygea	*The structure of the economy has mining specialization, Financial and banking centers*	Tyumen region, Sverdlovsk region, Moscow, Krasnoyarsk region
1995–1998	*Low level of production*	Altai Republic, Tyva Republic, Republic of Kalmykia, Republic of Adygea	*The structure of the economy has mining specialization, Financial and banking centers The specialization of the economy is engineering*	Tyumen region, Sverdlovsk region, Moscow, Krasnoyarsk region
1999–2002	*Low level of production*	Altai Republic, Tyva Republic, Republic of Kalmykia, The Republic of Ingushetia	*The structure of the economy has mining specialization, Financial and banking centers The specialization of the economy is engineering*	Tyumen region, Sverdlovsk region, Moscow, Samara region
2002–2008	*Low level of production*	Altai Republic, Tyva Republic, Republic of Kalmykia, The Republic of Ingushetia	*The structure of the economy has mining specialization, Financial and banking centers*	Tyumen region, Sverdlovsk region, Moscow, Samara region
2008–2017	*Low level of production*	Altai region, Kabardino-Balkarian Republic, Republic of Kalmykia, The Republic of Ingushetia	*The structure of the economy has mining specialization, Financial and banking centers*	Tyumen region, Moscow region, Moscow, St. Petersburg

**Source:* compiled by the author.
**Source:* (Federal State Statistics Service 2018).

The analysis performed in the regional context allows us to determine the "polar" groups of regions according to the indicator "Volumes of the Industrial Center," making it possible to determine the centers of concentration of production potential.

Another approach to the analysis of the indicator "Volume of Industrial Potential" is to assess the level of differentiation of Russian regions. Historically established industrial formats have a pronounced regional character, as confirmed by data in Table 1. The change in the structure of industrial production does not change significantly during the study period. This circumstance allows us to judge the invariability of regional differentiation in terms of industrial production.

In the framework of this study, the analysis of regional differentiation according to the level of industrial production is carried out. This is based on the calculation of the variation coefficient, which allows one to determine the dispersion process of these values for this indicator. Based on formula (1) and statistical data (Federal State Statistics Service 2009, 2018), the coefficient of variation, allowing to assess the unevenness of the regions of the Russian Federation, was calculated.

As a result of calculating the coefficient of variation by the indicator "Industrial Production Volumes," the dynamics of inter-regional differentiation was determined (Table 2).

The calculation of the coefficient of variation shows a steady increase in regional differentiation. In the period from 1990 to 2015, the value of the coefficient of variation increased from 0.91 to 1.68, i.e., by 51%. Analysis of the data in the table shows the presence of σ-convergence per condition (2).

The graphical representation of the coefficient of variation shown in Fig. 1 shows the increase in differentiation.

The analysis of statistical data and the calculated coefficient of variation (see Fig. 1) allows us to observe the process of divergence of regions according to the studied indicator, namely, "Industrial Production Volume." Failure to meet the condition of δ-convergence (2) allows us to talk about an increase in interregional differentiation. The dynamics of the divergence of the Russian regions, analyzed from 1990 to 2015, shows a relatively stable growth trend of heterogeneity, starting from 1990, when the coefficient was 0.91, and ending in 2015 when the coefficient was 1.68. The results obtained allow us to talk about the stability of the manufacturing sector in the studied period. The industrial heritage of the former Soviet Union inherited by the regions and the available resource potential, which has historically established directions and formats, is used by the regions. In current conditions, production potential operates at the facilities of established production associations.

Table 2. The coefficient of variation in terms of "industrial production volume."

Years	1990	1994	1998	2002	2006	2010	2014	2015
Coefficient of variation	0.91	1.00	1.10	1.35	1.73	1.55	1.62	1.68

**Source: compiled by the author.*

**Data is presented at intervals of 4 years.*

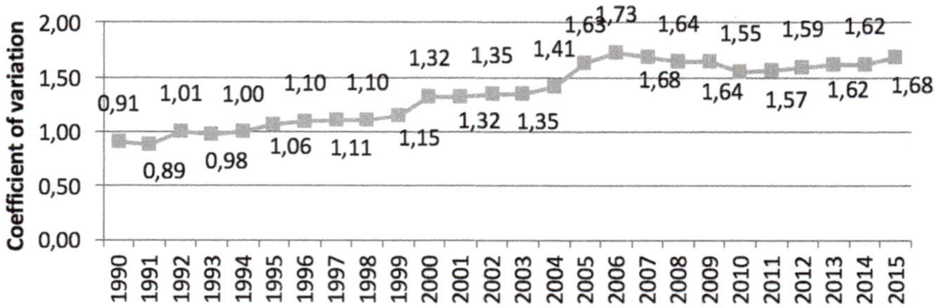

Fig. 1. The coefficient of variation calculated for the Russian regions in terms of GRP per capita.

The level of heterogeneity of the Russian regions in terms of "industrial production," general economic conditions also have an influence, namely structural changes, cyclical fluctuations, crisis manifestations. The aftermath of the 1998 crisis can be observed in the increase in regional differences; a similar process is observed after 2004.

5 Conclusions

In the framework of the study on the indicator "Industrial Production Volume" in the regional context, we conclude that the Russian regions have a significant level of differentiation according to this indicator.

As a result of ranking the regions by the analyzed indicator, specialization of industrial production plays a significant role. Over an extended period, territories, the industry of which is mining, occupy leading positions. As part of this statement, those regions that form the leading group and whose industry has an extractive structure are identified. Transformation processes in the Russian Federation have not changed the raw material orientation of the economy, and extractive industries still occupy leading positions in the manufacturing sector. The leading group also has regions with a high level of development of the banking and financial sector. These regions have the format of metropolitan regions with a high level of development of the infrastructure component, ensuring the development of the entire regional complex. The group of lagging or "subsidized" regions is the opposite in terms of economic development. They have low rates of economic development, specializing in food and agricultural production activities. The lack of large-scale production on the territory does not allow the regions to develop stably. To stimulate existing areas of production, this type of region needs financial support provided in the form of state subsidies and development programs.

Another indicator allows assessing the level of regional differentiation based on the calculation of the coefficient of variation. The behavior of the magnitude of variation in the regional context determined the absence of the process of convergence of the Russian regions by the studied indicator. The convergence theorem based on the calculation of the coefficient of variation and analysis of its behavior in the period under study allows us to draw a conclusion regarding the process of convergence of the studied economic objects. In the framework of this study, an array of Russian regions was examined, which was analyzed for inter-regional differentiation in terms of their industrial production volume. The studied coefficient of variation determined the lack of convergence of the Russian regions; the observed process can be classified as a discrepancy. As a result, throughout the entire research period, the stratification of Russian regions by industrial potential occurs.

The findings of the study suggest a divergence of regions in terms of industrial production development, and the revealed difference between regions does not decrease. As a result, we can talk about building up production capacity and industrial production in leading regions. At the same time, in regions that do not have high production potential, its active growth is not observed.

Stimulating the development of the manufacturing sector is an essential strategic and public objective. The maintenance of existing manufacturing enterprises ensures the development of the economy at the present stage. However, in the future, the process of forming priority areas for economic development should be based on the development of the industrial and production sector in most regions of the Russian Federation.

Acknowledgment. The study is carried out with the financial support of the Russian Foundation for Basic Research as part of the scientific project №19-010-00549 "Interregional Differentiation of the Russian Regions."

References

Barro, R.J., Sala-i-Martin, X.: Convergence across states and regions. Brook. Papers Econ. Act. **1**, 107–158 (1991)

Barro, R.J., Sala-i-Martin, X.: Economic Growth. MIT Press, Cambridge (1995)

Bobkova, A.Yu.: Diagnostics of convergent-divergent processes in the regional fiscal policy. Bus. Inform **11**, 113–118 (2013)

Bogoviz, A.V., Lobova, S.V., Alekseev, A.N., Prokofiev, V.N., Gimelshtein, I.V.: Managing the modernization of regional markets of educational services in the conditions of formation of Industry 4.0. On Horizon **27**(3–4), 187–192 (2019a)

Bogoviz, A.V., Ragulina, J.V., Barcho, M.K.: Influence of innovations on regional socio-economic development. In: Popkova, E. (ed.) The Future of the Global Financial System: Downfall or Harmony. Lecture Notes in Networks and Systems, vol. 57, pp. 1200–1207. Springer, Cham (2019b)

Drobyshevsky, S., et al. (eds.): Factors of Economic Growth in the Regions of the Russian Federation. IET, Moscow (2005)

Federal State Statistics Service: Regions of Russia. Socio-economic indicators. Rosstat, Moscow (2009)

Federal State Statistics Service: Regions of Russia. Socio-economic indicators. Rosstat, Moscow (2018)

Glushchenko, K.P.: Beta convergence myths. J. New Econ. Assoc. **4**(16), 26–44 (2012)

Granberg, A.G.: Fundamentals of the Regional Economy. HSE, Moscow (2000)

Launhardt, W.: Mathematische Begründung der Volkswirtschaftslehre. Verlag von Wilhelm Engelmann, Leipzig (1885)

Lesh, A.: Geographical Location of the Economy. Foreign Literature Publishing House, Moscow (1959)
Limonov, L.E.: Regional Economics and Spatial Development. Yurayt Publishing House, Moscow (2015)
Mau, V., Yanovsky, K.: Political and legal factors of economic growth in the Russian regions. Econ. Quest. **11**, 17–34 (2001)
Myrdal, G.: Economic Theory and Underdeveloped Regions. Gerald Duckworth, London (1957)
Pinto, J.: Launhardt and location theory: a rediscovery of a neglected book. J. Reg. Sci. **17**, 17–29 (1977)
Schumpeter, J.: On the concept of social value. Quart. J. Econ. **23**, 213–232 (1908)
Thünen, J.H., Lassalle, F., Rodbertus-Yagetsov, K.: An Isolated State. Economic Life, Moscow (1926)
Weber, A.: Theory of the Location of Industries. University of Chicago Press, Chicago (1929)

Approaches to the Analysis and Assessment of Environmental Risks in Agriculture Identified Through the Monitoring and Processing of Remote Sensing Data

Victor I. Medennikov[1(✉)], Elena V. Butrova[2], and Tatiana V. Kokuytseva[2]

[1] Institution of Russian Academy of Sciences Dorodnicyn Computing Centre of RAS, Moscow, Russia
dommed@mail.ru
[2] RUDN University, Moscow, Russia
evbutrova@gmail.com, kokuytseva-tv@rudn.ru

Abstract. The article shows a range of approaches to the use of remote sensing technologies for solving problems in various sectors of the economy. The authors present their methodology for assessing agricultural risks in predicting the economic effect of using remote sensing data.

Keywords: Earth · Remote sensing · Risk · Forecast · Agriculture

1 Introduction

The digitalization of the economy and society, which is based on achievements in the field of microelectronics, information and communication technology and in the area of creating new materials, technical means and equipment, has brought the role of spatial data to the most used information resources in the socio-economic, scientific, and social activities of our time, especially in the defense industry. These types of resources more clearly and informatively describe many natural phenomena, the dynamics and location of moving objects, the nature of the state of stationary objects, etc.

The technology for obtaining, storing, transmitting, and processing this data is called Earth's remote sensing (ERS). Earth's remote sensing includes shooting from space, from airplanes, sea and river vessels, from unmanned aerial vehicles (UAVs), ground systems, using various optical-radio-electronic devices. Until recently, only satellite imagery of the earth's surface, which has been in existence for over 70 years, was classified as Earth's remote sensing. Space imagery is well developed in the world currently; it has been widely used in many fields of activity. The range and number of launched types of spacecraft (SC) with a significant change in their technical, organizational, marketing nature is continuously increasing. The number of tasks that use space Earth's remote sensing data is also snowballing. For instance, the well-known research by (Ryabov 2012) mentions 200 problems.

A. V. Bogoviz and Y. V. Ragulina (Eds.): ISCI 2019, LNNS 115, pp. 19–27, 2020.
https://doi.org/10.1007/978-3-030-40749-0_3

Recently, considerable interest has been shown in UAVs because of their ability to shoot in conditions where using spacecraft or manned aircraft is impossible physically, for example, due to weather restrictions (mainly cloud cover), or due to insufficient resolution of the equipment or much more expensive. The increasing use of UAVs for remote sensing is dictated by the following advantages compared with Earth's space remote sensing: conducting surveys at low altitudes, as a result of which centimeter accuracy surveys are obtained (the need for which is great); real-time surveys, a significant reduction in costs for some tasks, including through the acquisition of aircraft by the users themselves. In 2014, about 60% of the entire world UAV market ($6.76 billion) was for military purposes. The U.S. firms dominated the market (65% of the market), of which the leaders are the General Atomics Aeronautical Systems (19%) and the Nortop Grumman Corporation (17%). The Israeli firms also dominated (Elbit Systems, with 8%). The European Union occupied a niche of 6% (*Putin is asked...* 2016). Russia is satisfied with its 2%, mainly for defense purposes.

It is estimated that by 2023, the volume of the global spacecraft-based Earth's remote sensing market will increase by 85% and reach $3.6 billion, of which about $3 billion will be the data obtained using ultra-high-resolution surveys.

Ground-based methods for obtaining heterogeneous information are used in unmanned vehicles. The most significant of them have found application in precision farming, now experiencing a real boom.

All these Earth's remote sensing methods compete in something, supplement each other in something, continually improving. What niche these basic methods can take in order to manage social and economic development can be found shortly, based on the resulting efficiency. For example, three main trends are observed in vehicle control systems. The first trend is that all sensors, data processing tools, a control system are located on a vehicle (V). So far, this option is being tested in different countries. The second trend is that based on the space positioning data of all moving objects, both living and non-living (all must have sensors) and a terrain map, the vehicle is controlled via space satellites. The third trend is that based on sensors installed on various objects along the road and on a vehicle, the vehicle is controlled by a certain electronic police and vehicle control system.

2 Materials and Method

At present, the Earth's remote sensing technologies are becoming one of the necessary elements for the information infrastructure of many countries due to their application for solving many tasks in the field of defense, science, sociology, economics, environmental protection, with a constant expansion of applied areas of application. The growth rates of remote sensing services indicate that they make up about 20% per year. Promising technologies for surveying and processing data are being intensively developed. From 1990 to the present, the number of countries operating an Earth's remote sensing spacecraft has increased from 7 to 20.

Since the Earth's remote sensing technologies began to be actively applied in many sectors of the economy, such as cartography, ecology, forestry and agriculture, land management, geology, logistics, construction, oil and gas transportation systems, weather and climate, oceanology, etc., they should gradually acquire the status of infrastructure technology. In this case, the Earth's remote sensing shortly will have to play in the world economy the same crucial role that railways, telegraph, and telephone communications, as well as electric networks, played in due time.

The current level of Earth's remote sensing technology also allows using the data obtained from various spacecraft, and in the interests of agricultural production. These data already have such a wide range of thematic characteristics that they make it possible to solve a rather significant set of tasks in this area – from designing the borders of fields on maps to forecasting crop yields on areas of entire regions. This is due to the possibility of the wide spatial coverage of space satellite imagery equipment, the emergence of new spectral characteristics of this equipment in the ranges that make it possible to recognize the spectral characteristics of vegetation with the subsequent calculation of various vegetation indices characterizing various state parameters of some plant species. The use of a dynamic series of observations also makes it possible to track trends in changes in the growth and development of plants, the temporal characteristics of agrotechnological measures, carry out monitoring studies of plantings affected by the adverse effects of nature and man, and also carry out many other tasks. For example, Voronina and Mamash (2014) propose a classification of thematic agricultural problems based on MODIS spectroradiometer data. The tasks are divided into the following groups: properties and composition of the soil layer, seasonal dynamic characteristics of vegetation, cartographic parameters, and condition of the land, forecast tasks for the productivity of various crops, atmospheric, and climatic conditions.

3 Results

In Russia, due attention is not paid to the problem of risk assessment and management, even though almost the entire country is located in the area of risk farming. Weighing the choice in the two-criteria problem between profit and risk, Western economists place more weight on the task of risk management rather than profit management.

Risk assessment is a set of methods and models for identifying risks, for determining a set of primary qualitative and quantitative indicators, factors describing the risk situation, affecting the risk level criterion, both individually and in combination. An assessment relies on a systematic approach, considering all systems' activity in conjunction with the environment of its functioning. Also, such an assessment implies analysis and classification (separation) of indicators on external and internal risk factors and indicators of criteria for assessing the level of risk. There are many evaluation methods, both quantitative and qualitative. We will consider the possibility of using the following: statistical methods, method of expert estimates, method of analogs.

The main risk factor for agricultural activities is climatic conditions. This factor is manifested in the instability of plant productivity values, affecting all socio-economic indicators of farms, regions, often the country. Humanity has not yet found ways to completely neutralize the negative impact of weather and climate on the agricultural sector. For example, extensive statistical studies of agricultural production in the Stavropol in the 90s of the last century showed that fluctuations in production volumes by 70% depend on weather and by 30% on economic and technological factors of production. At the same time, economic and technological risks can increase weather risk, and vice versa.

For the production of a certain quality and quantity of agricultural products, it is necessary to observe strict proportions between resources due to technological requirements and the specifics of the products. A change in the quality and quantity of one type of resource entails a change in the structural characteristics of other resources, which ultimately affects the quality and quantity of products. As a result, the factors of production form an integral system of agricultural production. For the analysis of the "resources-output" relationship, the Cobb-Douglas production function, which establishes the dependence of the number of manufactured products on the volumes and proportions of the resources used, is widely used (Medennikov et al. 2017; Ushvitsky et al. 2015). The dependence is sought in the form of a regression equation on a reasonably representative sample of statistical information, as a result of which the parameters of the production function are determined. Rather stringent requirements are imposed on the information entered into the calculation model. The information should reflect the results of either a one-time survey of a sufficient number of homogeneous objects or the results of a survey of the same object for long periods. At the same time, the objects (object) should be in reasonably identical conditions during the time of the survey.

In general, the Cobb-Douglas function is written as follows: $y_j = A_j \prod_{i=1}^{I} x_{ij}^{\alpha_i}$ (power dependence), or $y_j = A_j \sum_{i=1}^{I} x_{ij}^{\alpha_i}$ (linear dependence), where j is the type of product, $j \in J$, $i \in I$, y_j is the effective indicator with respect to which the resource potential is determined (for example, yield), x_{ij} is the volume of the i-th factor used for the j-th type of product, A_j is the technological coefficient, α^i are the elasticity coefficients, showing the growth rate in % of the effective indicator, with an increase in the i-th factor by 1%. Often, it is generally accepted that the relation $\sum_{i=1}^{I} \alpha^i = 1, 0 \leq \alpha^i$ is satisfied. As a result, the Cobb-Douglas function gives an idea of the possible growth factors of the effective indicator (economic growth) and the degree of influence of each source of growth on the effective indicator.

The choice of factors is determined by the purpose of the study. For example, in the 1990s, through the efforts of the Ukrainian Research Institute of Agricultural Economics and the Institute of Cybernetics of the Agro-Industrial Complex, an information system for equally intensive planning of state tasks for farms using the Cobb-Douglas function of the production potential of enterprises and the efficiency of its use for 6 factors was developed:

$$\Pi\Pi = \alpha G^{\alpha 1} Z^{\alpha 2} B^{\alpha 3} T^{\alpha 4} F^{\alpha 5} L^{\alpha 6},$$

where α^i stands for the model parameters, i = (1, 6); G is the amount of water for irrigation; Z is the bonitet arable land; B is the amount of fertilizer applied; T is the number of employees; F is the value of fixed assets; L is feed base volume.

For each type of farm, determined by the main cultivated crop (tomatoes, sugar beets, etc.), the model parameters were determined based on statistical data for 15 years of production and realized costs in 30 farms of the same size and located in the same climatic zones.

To develop a methodology for assessing agricultural risks in predicting the economic effect of the use of Earth's remote sensing data, we will use the following four resources: land, labor, fixed and revolving funds. As a useful indicator, we will take crop yields as the most used for these purposes.

In the economic literature and works on the risk of agricultural production (Benkovskaya 2019), the fluctuation indicator, in the form of the standard deviation, is most often used to measure the magnitude of risk in a quantitative risk assessment. The oscillation index, in the form of the standard deviation, is as follows: $S_{nj} = \sqrt{\sum_{t=1}^{T}(y_{njt} - y_{nj}^{0})^2/T}$ for the jth farm culture n for the observation period T; y_{njt}, respectively, stands for the yields at time t; $t \in T$, y_{nj}^{0} are their average values. Then the variation coefficient $V_{nj} = S_{nj}/y_{nj}^{0}$ can serve as a measure of risk. In entrepreneurial activity, several levels (zones) of risk are distinguished (Benkovskaya 2019), namely minimal (loss up to 25%); increased (loss from 25 to 50%), critical (losses from 50 to 75%), unacceptable (losses from 75 to 100%).

Now, we will consider the case when an enterprise invests in fixed assets using Earth's remote sensing. Based on the Cobb-Douglas function, productivity (efficiency) increases, but the degree of risk does not change. Measuring the degree of risk in innovation is conducted with equations and calculations developed in a way to reflect the prevailing trend most likely. For example, the methodology of A. Kabat and L. Naiken takes into account the autocorrelation coefficient for calculating the indicator characterizing the relative oscillation (Benkovskaya 2019). However, in all these techniques, a sufficiently large number of observations is necessary, which (in principle) is not yet available in Russia due to the short time and uncomplex, fragmented use of Earth's remote sensing data.

For the same reasons, the method of analogs, when one could use the accumulated data in different countries, is not suitable. For example, we consider the state of affairs in Germany and China (Information Agency "Svetich" 2019). German agriculture, through a scientific approach, has been prepared for the introduction of Earth's remote sensing in precision farming technology (PF) (GIS Market Support Association 2019). So, already over 60% of farms of different sizes apply these new technologies. For the search and development of the most PF suitable technologies, the interdisciplinary project "Preagro" has been formed.

The Ministry of Education and Science funds this project, based on the developed concept of PF with the corresponding equipment and technical equipment of agricultural units, implements, and machines. The project is designed to develop new agricultural technologies in crop production, taking into account local characteristics (micro-conditions) of small sections of fields based on Earth's remote sensing data in order to increase the economic efficiency of production. The project involves several scientific, financial, industrial organizations that provide all the necessary funds and resources. Following the developed concept and expert estimates, the use of PF should increase crop yields by 30% with significant resource savings in the amount of 100–150 €/ha. The main area of research and experimentation is the development of fertilizer application technologies, differentiated by small sections of fields, using both Earth's remote sensing data, GPS, and modern GIS technologies. Most farms in Germany have advanced agricultural machinery, computer technology, which provides access to soil map databases, digital data, and Earth's remote sensing images. Service and information and consulting activities to provide the necessary assistance in organizing the implementation of new agricultural technologies, in the collection of soil samples, the formation of digital field maps, and in equipping the equipment with PF equipment are well established.

In China, the first experiments on introducing PF in field cultivation are also conducted. Experiments are taking place near Shanghai in order to bring balanced nutrition of crops to industrial technology. Eleven types of nutrients are tested in 460 test sites. We can already say that the yield of watermelons has increased from 14 to 27%, and the concentration of sugar has tripled, and by 9–13% concerning the collection of rice, and by 18% for wheat.

Taking into account the reduction of agricultural risks and the economic effect of using remote sensing data, the statistical method based on the Cobb-Douglas function needs to be supplemented by expert estimates (Table 1). This approach consists of the competent collection and processing by appropriate mathematical methods of the views of a representative group of experts from specialists in this field of knowledge about risk probabilities and various effects on a given topic. At times, their intuitive opinions, reinforced by proper knowledge and experience, give reasonably accurate estimates. Expert methods are often used in brainstorming technologies when one needs to develop information based on which management decisions are made quickly. This approach is often used when:

(1) The sample of statistical data is small for applying the methods of mathematical statistics;
(2) Statistical methods can not determine the relationship between the studied phenomena due to their belonging to the non-quantitative scale of measurements (qualitative, interval, and ranked nature);
(3) There is a fragmented nature of the information, either in time or due to incompleteness, or even the absence of an explicit connection between the studied phenomena.

An important detail of expert methods is the adequacy of the approach to processing the answers received and interpreting the results.

Table 1. Expert assessments of reducing agricultural risks and the economic effect of the use of Earth's remote sensing data: expert (data), region (name), district (name), economy (name).

Culture	Average yield	Type of Earth's remote sensing	Earth's remote sensing cost per hectare	Yield increase	Reduce the risk of yield loss

As a result of processing the received answers and taking into account the calculations of the Cobb-Douglas production function, the agricultural risk assessment technique for predicting the economic effect of using Earth's remote sensing data is ultimately based on such a mixed method. The calculation results will be presented in Table 2.

Table 2. Final estimates of the reduction of agricultural risks and the economic effect of the use of remote sensing data: region (name), district (name), economy (name).

Crop	Average yield	Risk level (%)	Increase in efficiency (expert)	Growth efficiency (Cobb-Douglas)	Average performance gain	Reduced risk of yield loss (%)	The total risk level (%)

We formalize the calculation results. Let's assume that $r^{\text{э}}_{knjm}$ is the assessment of the k-th expert on reducing the risk of the j-th farm culture n when using the m-th technology of Earth's remote sensing, $r^{\text{оэ}}_{njm}$ is the average assessment of experts, $y^{\text{э}}_{knjm}$ is the assessment of the k-th expert on the increase in productivity (efficiency) of the j-th farm culture n when using the m-th technology of Earth's remote sensing, $y^{\text{оэ}}_{njm}$ respectively, the average score of experts. Then we get a generalized risk assessment $r^{1}_{njm} = V_{nj} \pm r^{\text{оэ}}_{njm}$ and a generalized average yield (efficiency) $y^{1}_{njm} = y^{0}_{nj} \pm y^{\text{оэ}}_{njm}$. Given a sufficient number of experts, we can set the task of finding the choice of the optimal type of Earth's remote sensing technology in the space of risk and effectiveness states.

This shows that the choice of the best Earth's remote sensing technology will be the result of solving an optimization two-criteria problem. In the general case, in the multicriteria optimization procedure, they are guided by the priority Pareto optimality concept, in which the optimal solution is understood as the improvement of some criteria on a specific set of arguments, on which other criteria are not reduced.

We reduce our problem to a form convenient for reasoning in terms of an optimization two-criteria problem. Why we will introduce new expressions.

$f(x)$ is a function of the effectiveness of the xth Earth's remote sensing technology, $z(x)$ is the risk function (risk assessment) of the xth Earth's remote sensing technology, $x \in N$, where N is the discrete set of these technologies.

In Pareto optimization, a particular body (person) making a decision tends to increase the function $f(x)$ and to decrease the function $z(x)$. One can find an acceptable solution via the so-called convolution of criteria by reducing the multicriteria problem to a scalar problem. The convolution of criteria is used in case of comparability of the values of all criteria, i.e., expressed in one measurement unit. Otherwise, the normalization procedure (for example, the logarithm of some criteria) is carried out. We will consider that in our case, such a procedure has been implemented. We will describe the most popular of them and related to our problem statement.

1. *Weighted convolution*. In this case, the general criterion W looks like this: $W = max(\alpha^1 f(x) - \alpha^2 z(x))$ on x for $x \in N^1 \subset N$. That is, the functions $f(x)$ and $z(x)$ enter the general criterion with certain weights α^1 and α^2, while $\alpha^1 + \alpha^2 = 1, 0 \leq \alpha^i, i = 1, 2$.

This convolution is usually used for continuous functions *f(x)* and *z(x)*, although, having introduced some measure of the distance between the Pareto points, this method is also suitable for the discrete case.

2. *The convolution method of imposing restrictions.* In our case, the general criterion W looks like this: $W = max\, f(x)$ on x for $z(x) \leq \beta$ and $x \in N^1 \subset N$. That is, the risk function $z(x)$ should not be greater than some threshold value β.

4 Discussion

The presented approach makes it possible to assess agricultural risks when predicting the economic effect of using the Earth's remote sensing data. This is especially important in the conditions of the formation of market relations in our agricultural sector in the context of the upcoming dramatic changes in production and production management technologies when the contradictions between the effectiveness of the production management process and the effectiveness of risk management are aggravated.

Our managers have to learn to find the necessary balance in this space of decisions, like some Western economists. When making decisions, we have to take into account not only natural but also economic risks, since they are determined by losses associated with additional, unforeseen consumption of resources. Of the many different classification types of risks in this work, based on an assessment of the influence of one of the factors, namely the Earth's remote sensing data on agricultural production, we will consider microeconomic risks, including the risks of individual business units (Benkovskaya 2019). Since only at this level, the factor analysis methods are most developed. On the other hand, at the regional and federal levels, the manifestation of the influence of remote sensing is challenging to track because, at present, experiments with Earth's remote sensing in agriculture are insignificant, uncomplex, and fragmentary. More than that, they only go with individual technology components.

5 Conclusion

The article shows the approaches to the use of Earth's remote sensing technologies for solving problems in various sectors of the economy. The author's methodology for assessing agricultural risks in predicting the economic effect of the use of Earth's remote sensing data is based on the use of the Cobb-Douglas production function. This function determines the dependence of the number of products produced on the volumes and proportions of the resources used, supplemented by the method of expert estimates. As a result, the methodology for assessing risks is reduced to solving an optimization two-criteria problem.

Acknowledgment. The study was financially supported by the Russian Foundation for Basic Research in the framework of the scientific project №19-010-00619 "Development of a Methodology for Applying the Results of the "Digital Earth" Project to Solve the Problems of the National Economy and A Forecast Model for the Economic Effect of Their Application in the Context of Russia's Digitalization."

References

Benkovskaya, L.V.: Improving the statistical analysis of grain production risks and their insurance (2019). http://www.dslib.net/bux-uchet/sovershenstvovanie-statisticheskogo-analiza-riskov-proizvodstva-zerna-i-ih.html

GIS Market Support Association: Precision farming will increase yields (2019). http://www.gisa.ru/55612.html

Information Agency "Svetich": Foreign experience in the application of precision farming technology (2019). http://svetich.info/zarubezhnyi-opyt-primenenija-tehnologii-.html

Medennikov, V.I., Muratova, L.G., Salnikov, S.G., Gorbachev, M.I.: Economic and mathematical modeling of agricultural informatization scenarios. Int. Agric. J. **4**, 23–27 (2017)

Putin is asked to help unmanned aircraft so that it does not repeat the fate of Russian PCs (2016). http://www.cnews.ru/news/top/2016-11-14_putina_prosyat_pomoch_bespilotnoj_aviatsiichtoby

Ryabov, I.V.: Innovative development of the production and marketing of Earth's remote sensing information products. Dissertation of Candidate of Economic Sciences, A. S. Griboedov Institute of International Law and Economics, Moscow, Russia (2012)

Ushvitsky, L.I., Kulagovskaya, T.A., Ter-Grigoryants, A.A.: Risks of agricultural production: the nature, content and methods of assessment. Adv. Mod. Sci. Educ. **1**, 23–30 (2015)

Voronina, P.V., Mamash, E.A.: Classification of thematic tasks of agricultural monitoring using MODIS remote sensing data. Comput. Technol. **19**(3), 76–102 (2014)

Key Parameters of Transport Development in Russia at the Contemporary Stage

Tatiana M. Gainochenko(✉), Anna V. Kurbatova, and Ekaterina S. Kurbatova

State University of Management, Moscow, Russia
{tm_gainochenko, av_kurbatova}@guu.ru,
kurbatova.kate@yandex.ru

Abstract. The entry of humanity into the Chaordic age (in the terminology of D. Hawk) necessitates the introduction of not only systematic methods for assessing the effectiveness of social production but also reinforcing them with non-systemic resources to improve the quality of business and public consciousness. The authors are convinced that, at present, the critical parameters of the country's socio-economic development are the social poverty of Russian society and the low culture of introducing fundamental innovations in the field of the humanities, in particular economics. Relying on the example of the inclusion of individual railway sections in the system of international transport corridors, the possibilities of assessing the influence of the transport factor on social labor productivity at the macroeconomic level using the approach proposed by Professor I. V. Belov back in the late 1980s are described and discussed in the paper.

Keywords: Transport · Complex evaluation · System approach

1 Introduction

Nowadays, a common trend is the search for sources of inertia and threats to the country's socio-economic development in the weakness of leaders, the backwardness of institutions, the strengthening of bureaucracy and corruption, the imposition of economic sanctions against Russia, the acceleration of various kinds of changes in a remarkably complex, uncertain, permeated by mutual relations environment. The information overload of humanity has indeed reached astonishing proportions thanks to the use of modern computer technologies and communications. However, such a phenomenon is not new in the history of humanity. Noteworthy is the Sumerian letter dated to the 4th millennium BC: "Hard times have come. Children ceased to obey their parents, and everyone strives to write a book" (Beshenkov 2006). The emergence of information crises is associated with a violation of the law of necessary diversity in force in the field of management, first formulated by the English cyber scientist U. R. Ashby and later supplemented by the term "performance" (Valuev and Volkova 1991). Knowledge of this law is essential for finding solutions to breaking the information deadlock and moving to a new level of development of methods and technologies for organizing and managing the processes of material and social life. It is

A. V. Bogoviz and Y. V. Ragulina (Eds.): ISCI 2019, LNNS 115, pp. 28–38, 2020.
https://doi.org/10.1007/978-3-030-40749-0_4

known that the minimum complexity of management is achieved by dividing the organization structure into the optimal number of parts without losing the synergistic effect due to the emergence of a new quality of the whole.

However, in practice, it is not always possible to find specific solutions that quickly and without loss overcome the consequences of violating the law of the necessary diversity and performance. For example, the history of the development of the theory and practice of the central problem of economic science, namely, the law of value, indicates a thorny path to finding the objective basis for a fair exchange of material, and, more recently, information products of labor. The discovery of forms of value, ways of measuring money of a different nature using interest on capital, technologies for reducing complicated labor to dull, the development of general indicators that characterize the implementation of the achievements of technology, organization and management of social production on a global and national economies, industries and individual enterprises are certainly significant achievements of human economic thought.

At the same time, the aggravation of global problems actualizes the development and implementation of reliable criteria for assessing the quality of managerial decisions made, based on an understanding of the systemic nature of management objects and reinforcing them with non-systemic resources to improve the quality of business and public consciousness.

The information society is a society in which "high-quality information is circulating in sufficient quantities, and all the necessary means are there for its storage, distribution, and use" (Osipov 2012). Its formation is that an external motivational factor of an objective nature, which creates an opportunity for the Russian Federation to make a "leap into the future" (Glazyev 2018). According to the authors, at the present stage, the most influential internal motivating factors are such order parameters as the introduction of sociocultural synergists that improve the quality of both managerial personnel and the subjects of the mass consciousness of society, as well as methods of a systematic approach to quantifying the system's productive power.

2 Materials and Methods

In the context of economic globalization, the creation of large transnational corporations and the integration of transport of a number of states into the world transport system, an intensive process of formation of international transport corridors (ITC) is underway, providing accelerated promotion of abundant commodity flows between different countries and continents based on the introduction of modern logistics technologies for cargo delivery.

Russia's desire to integrate into the European transport system stimulates the introduction, primarily within the framework of the ITC, of new technologies for the transportation of goods, which can increase the volume of transportation and reduce delivery time, the introduction of a single responsibility for the transportation of goods throughout the transport chain. Such new technologies, for example, include the combined transportation of loaded vehicles, which allows us to move from interspecific competition to cooperation between rail and road transport enterprises. The ITC organization sets the following goals: unification of national legislation, harmonization of transport systems of East and West, creation of an international transport infrastructure that has standard technical parameters and ensures the use of a single transportation technology as the basis for the creation of global logistics systems and the integration of national transport systems in the global transport system. Large cargo flows are concentrated in the zone of gravity to ITC. Also, the coordinated interaction of various modes of transport is carried out there, which provide accelerated and high-quality transportation of goods in these directions with a high level of customer service, on the basis of a single supporting document, a through tariff rate and the full responsibility of the forwarder (operator) for the entire transportation process. At the same time, implementing such a project within the framework of the approach to business management practice being formed over the past two decades and the methods used to evaluate its effectiveness is doubtful.

The approach proposed in this study to the implementation of such breakthrough projects consists of two components. First, there is a methodology for assessing the economic potential of railways, which is described in detail by (Kurbatova and Kurbatova 2015). Second, the use of sociocultural synergies to transform the potential of economic potential into reality with the existing restrictions on labor productivity growth is also necessary to bear in mind.

We give only the main idea that formed the basis of this approach. The productivity of people's production activity is traditionally measured by such economic categories as "labor productivity," "labor intensity," and "productive labor force." The productive labor force is a potential opportunity for living labor, armed with means of labor in accordance with the achieved level of technical and organizational level of production, to provide maximum results per unit time. Labor productivity is the real result of turning opportunities into reality. Increasing labor productivity is limited by the possibilities of its intensification, which is determined by the prevailing social conditions and physiological capabilities of a person. The lower the average per capita income in society, the higher the proportion of time people are forced to spend outside social production to ensure an acceptable standard and quality of life. Therefore, the less is the possibility for the growth of labor intensity, all other things being equal, and the lower is the potential of the productive labor force due to the labor factor. Therefore, in order to take into account the influence of a combination of factors determining the effectiveness of the dissemination of best practices in organizing and managing international transport corridors on the territory of the Russian Federation, a modified version of the indicator "productive labor force" should be used. Professor I. V. Belov proposed this version in the late 1980s. Furthermore, Professor V. A. Persianov actively promoted it to the broader scientific and professional transport community (Gorina and Persianova 2019).

In system objects, productive force is determined by the product of the capacity of one production cycle by the number of cycles. For transport, the capacity of the cycle depends on the duration of the main and auxiliary transport operations, which are determined by the technical and operational indicators of resource use, depending on innovations in the technological, reproductive, and marketing fields. In an extremely general form, the value of the productive power of the system (PP) of any intended purpose can be defined as:

$$\text{ПС} = Q \times N \tag{1}$$

Where, Q is the capacity of one reproduction cycle, natural units of measurement; N is the number of cycles performed over the period T.

Concerning transport, this formula can be converted and presented in an updated cost form:

$$\text{ДПС} = \sum_{i=1}^{n} Q_i \times N_i \times L_i \times d_i \times 10^{-1}, \tag{2}$$

where,

Q_i – is the capacity of the cycle when transporting goods in the *i*-th rolling stock, t;
N_i – is the number of cycles performed by the *i*-th rolling stock for the period T_i;
L_i – is the average range of cargo transportation in the *i*-th rolling stock, km;
d_i – is the average rate of return for the carriage of goods in the *i*-th rolling stock, rubles/10 tkm;
i = 1, 2, 3…n are types of rolling stock.

It should be noted that these conditions are not met when organizing international transport corridors; therefore, the original formula (2), as applied to a specific section of the ITC, should be written as follows:

$$\text{ДПС}_{(\text{МТК})} = \sum_{i=1}^{n} Q_i^{\text{МТК}} \times N_i \times L_i \times d_i \times k_{1i} \times k_{2i} \times k_{3i} \times 10^{-1}, \tag{3}$$

where Q_i^{MTK} is the capacity of the cycle when transporting goods gravitating to a given section of the ITC in the *i*-th rolling stock, *t*; k_{1i} is a coefficient that takes into account the increase in the number of cycles performed by the *i*-th rolling stock when processing it at this section of the ITC; k_{2i} is the same increase in the average transportation distance in the *i*-th rolling stock; k_{3i} is the same increase in the average income rate due to clustering in the transportation of highly tariffed goods in the *i*-th rolling stock.

The capacity of the cycle for the transport of goods using the MTK section is determined by the working fleet of freight cars within the MTK and the average carrying capacity of wagons. The number of cycles that each freight wagon does during the billing period, as well as the average rate of return, is also subject to refinement.

If each ITC passing through the territory of Russia was an independent business entity that has its own (or rented) rolling stock and maintains its internal statistics of traffic, traffic management, etc., then no adjustment of the calculation formula would be required.

The social component of the category "labor productivity" is especially important in transport, since the labor component in the cost structure of transport products is more than 40%. 2725.3 thousand people were engaged in primary, auxiliary, and additional transport activities in transport in 2017. 1926.3 thousand of them were employed in primary activity (Rosstat 2018). Therefore, social policy cannot be separated from economic policy. Moreover, in accordance with the results of the round tables on personnel policies in transport at the State University of Management (State University of Management 2017a, b; Persianov et al. 2017), since the average monthly nominal accrued wages of employees of transport organizations is at the level of 43.5 thousand rubles (the exception is the pipeline and air modes of transport, on which this figure was 75.9 and 97 thousand rubles, respectively) in 2017, the primary barrier to labor productivity growth is the phenomenon of low internal labor motivation (Rosstat 2018).

Analysis of publications on the problem of assessing potential reveals its weak study at the level of individual industry and especially at the enterprise level. Some publications only state the fact that issues of assessing the economic potential of individual enterprises have remained aloof from the attention of researchers.

So, in the period 1860–1920, the production concept of marketing prevailed, and when describing the power capabilities of the enterprise, the emphasis was on the ability of the enterprise to produce goods and services. Estimated indicators were the volume of production, fixed and working capital, the number of employees. With the commodity (1920–1930) and marketing (1930–1950) concepts, the emphasis was on the ability of the enterprise to produce high-quality goods and market them through marketing efforts. The concept of traditional marketing (1960–1980) highlighted the ability of the enterprise to satisfy customer needs. Finally, the socio-ethical concept (1980–1995) and the concept of interaction marketing (from 1995 to the present) emphasize the rational, optimal use of resources and the satisfaction of the interests of society and business partners (*Information portal "market-pages.ru"* n.d.).

So, the term "potential," as applied to an individual enterprise, should characterize not only production capabilities but also the necessary abilities of the enterprise to satisfy consumer demand as much as possible, while rationally using resources and taking into account the interests of the state, society, and business partners.

In turn, the economic potential of transport at the level of a country (region, specific territory) should be understood as the maximum possible quantitative level (volume) of meeting the demand for transportation of certain types of products of material production sectors and people in certain areas, with specific qualitative characteristics (speed, safety, security, etc.) and in a particular (specified) period of time with the best use of available production capacities, namely, transport infrastructure, rolling stock and labor resources.

3 Results

The calculation results of the economic potential of the ITC railway sections according to formula (3), taking into account the conditions of their operation, we will explain with the following options. These options reflected changes in the organization and management of rolling stock, the type of wagons used, and the average profit rate for goods of highly profitable categories and are presented in Table 1.

1. The initial data are as follows: the working fleet of cars served by this railway section before it is included in the ITC system is 50 thousand cars; the average statistical load of a wagon is 55 tons; the average turnaround time of a wagon is 10 days; and the average rate of return on transportation of goods gravitating to a given railway section (import, export, transit, domestic transportation) is 200 rubles/t.

Table 1. Assessment of changes in the economic potential of the railway section before and after inclusion in the ITC system.

An economic potential of the ITC site	Options for introducing technological and managerial innovations			
	Before inclusion in the ITC system (1)	Introduction of a new wagon management technology (2)	A decrease in average wagon static load and an increase in average income rate (3)	An increase in transportation distance and an increase in average income rate (4)
In material terms, million tons	100.375	118.880	107.360	60.800
Change, %	*100.0*	*118.4*	*106.9*	*60.9*
In monetary terms, billion rubles	20.075	26.200	34.360	29.200
Change, %	*100.0*	*130.2*	*171.2*	*141.0*

2. When the site is included in the ITC system, the working wagons fleet served by the site, and the average statistical load remains the same. Thanks to the transition to new technology for managing car flows following international standards, the average turnaround time of a car is reduced to 8.5 days, and the average revenue rate from transportation rises to 220 rubles/t.
3. The inclusion of the site in the ITC system leads to a decrease in the average static load of the car to 50 tons, while maintaining the average turnaround time of the car at the same level (8.5 days). And the inclusion of the site in the ITC system leads to an increase in the average revenue rate from transportation to 320 rubles/ton due to the clustering of production in the ITC gravity zone. The working fleet of wagons within the boundaries of the ITC section under consideration remains at the same level (50 thousand units).

   Despite the decrease in the average statistical load of the wagon, the economic potential of the site in material form, in comparison with the original version, increases slightly (by 6.9%). At the same time, in value-form, it increases significantly (by 71.2%).

   The inclusion of the railway section in the ITC system can lead to an increase in the average time of wagon turnover and a decrease in the economic potential of the section in material form due to an increase in the average transportation distance. However, it can significantly increase in value-form due to an increase in the average revenue rate from transportation.
4. The initial data are as follows: the working fleet of wagons is 50 thousand units; the average turnaround time of a wagon is 15 days; the average statistical load of a car is 50 tons; the average revenue rate from transportation is 480 rubles/t.

Thus, in comparison with the initial version, only due to an increase in the average income rate, even with an increase in turnaround time and a decrease in the average statistical load of a car, the economic potential in monetary terms increased by 41%.

From the above examples, it follows that with the inclusion of railway sections in the ITC system due to higher tariffs for the transportation of cargo mass of clusters, good economic results can be achieved.

The organization of the operation of the ITC railway sections places high demands on the quality of personnel management. In order to identify the dominant motivational trends in labor collectives in transport, sociological surveys of representatives of the senior staff of the personnel services of federal transport authorities in the Russian Federation, transport companies of the state and private sectors of the economy during the round tables on personnel policy in transport were held (State University of Management 2017a, b; Persianov et al. 2017). The methodology for assessing the value orientation, strength, and orientation of the organization's personnel motivation was based on the developed questionnaire for examining the motivation of employees and the technology for its processing, described in the work of Professor A. Ya. Kibanov (Kibanov and Ivanovskaya 1999).

4 Results and Discussions

Summing up the work done, we should conclude that two components determine the level of economic potential. An objective component is a combination of labor, intangible, material and natural resources, involved and not involved for some reason in production and having a real opportunity to participate in it. A subjective component is the ability of workers and collectives to use resources and create the maximum amount of material goods and services and the ability of the administrative apparatus of an enterprise, organization, industry, and an economic system as a whole to make optimal use of available resources.

According to the first component, one should recognize that to date, the ITCs in Russia (apparently in other neighboring states) are not isolated infrastructure complexes[1], but they work in close interconnection with other sections of railways. The ITCs do not have their own organizational and managerial structures. Any independent statistical accounting and reporting are not conducted on them. Today, information on the size and structure of traffic can only be obtained by calculation. At the same time, it is necessary to solve several issues related to establishing the borders of the ITCs and the features of the organization of traffic management.

The external environment for the ITC railway sections is the rest of the transport network that is not within the boundaries of any particular ITC. The same applies to the wagon fleet: it accesses the entire railway network, but when it enters the ITC zones, it must be passed through the sections and processed at the stations in a special (usually accelerated) mode.

The analysis carried out in the work allows us to determine the economic potential of railways at the country level (region, specific territory, section of the ITC) as the maximum possible quantitative level (volume) of satisfying the demand for transportation of certain types of products of material production sectors and people in

[1] The exception is the United Transport and Logistics Company.

certain areas, with specific qualitative characteristics (speed, safety, security, etc.) and in a particular (specified) period of time with the best use of available production capacities, namely transport infrastructure, rolling stock, and labor resources. It allows to quantify the economic potential of the Russian sections of the ITC with a contribution to GDP and in the form of an increase in the cost of goods (mass of goods) as a result of their movement. Also, one can quantitatively determine the impact of new technologies, organization, and management of transportation on the increment of the economic potential of these sections of the railway the network.

Acceleration of wagon turnover (development of additional freight through the use of freed wagons), as well as the creation of innovative industrial centers in these sections, which may be industrial free economic zones (FEZ), will increase the level of the economic potential of the ITC railway sections. In turn, they will provide ITC with a highly tariffed cargo base competitive in international markets.

The clustering of production can significantly enhance the effect of creating industrial free economic zones in these areas. Clusters not only facilitate the entry of regions and countries to the international commodity market with ready-made competitive products but also provide industrial cooperation for enterprises of several countries that require a material basis for goods exchange, namely, a network of international transport corridors that deliver cluster products to neighboring and remote countries.

The idea of clustering, widely spread abroad, is gradually paving the way in Russia, which, in recent years, has been convinced by network methods of development of regions and branches of the Russian economy.

According to the second component of assessing the level of economic potential, it was revealed that the dominant type of motivation in labor collectives in transport is pragmatic orientation with a high level of motivation power. In the direction of motives (motives for achieving, maintaining, and lack of motivation in general), different motives dominate in different groups of personnel. So, for the administrative staff of the middle and lower management, the leading one is the preservation motive, while for the production staff, private incentives to work are absent altogether, and external ones are limited by low wages and below the average level of living conditions and quality of life.

In a business environment, a culture of motivation by fear is widespread; the system of penalties significantly prevails over rewards. The staff turnover rate is high, including among administrative and managerial personnel of middle and lower level. The emerging trend in the free formulation of the reasons for the dismissal of this category of employees, namely the "humiliating situation," is of particular interest. This reason occupies the highest position among the reasons for voluntary dismissals. This confirms the lag of managerial practice from the theoretical achievements of both domestic (Gerchikov 2004) and foreign management schools (Kellerman 2008).

The discovery of the phenomenon of internal motivation and its differentiation according to the types of subordinates places high demands on the managerial staff. Namely, the leader must be able to not only choose the most suitable management style and methods of their manifestation in specific situations but also take into account the types of the internal motivation of people in the organization. Different authors give different names for these types, but the general thing is that only that type of workers reacts positively to negative incentives that have minimal successful motivation.

Consequently, modern management does not use the enormous potential inherent in the following rule: reward for the positive is more effective than punishment for the negative. This leads to significant economic losses, particularly due to an increase in the level of various kinds of diseases caused by a split personality due to the mismatch of values and people's beliefs with the norms of behavior in a particular business environment.

5 Conclusions

Thus, through various methods of enriching the business ethos, the absolute limit of manageability that has historically and managerially developed in our country can be overcome. The theory and practice of managerial thought have developed such methods of enriching business ethos as mutual comparisons and ethical borrowings, the creation of an ideologue of a civilizational level, the formation of an image of a desired future, the priority of legality with a dominance of ethos, deontic techniques, international organizational and legal outsourcing for internal tasks, coding activities and certification of products, organizations and specialists on behalf of and according to the criteria of professional associations, subjectivity of ethos, collective self-organization, etc. At the same time, using an approach to assessing the economic potential of management objects that is adequate to the current level of their development (including railway sections in the ITC system) will allow for the implementation of breakthrough technological solutions in various areas of society, without resorting to the most bloody value in the history of mankind, namely justice. For example, the choice of an effective mode of using the transportation potential can be made by the method of factor analysis (chain substitutions), the essence of which is as follows.

It can be assumed that some complex indicator is analyzed (in our case, this is the economic potential of the ITC) $y = F(x_1, x_2, \ldots x_n)$, which is a function of several independent variables. The base value (for example, planned or design level) is $y_0 = F(a_0, b_0, \ldots m_0)$. Let the actual value of the indicator is $y_1 = F(a_1, b_1, \ldots m_1)$. The successive replacement of the basic values of variables with actual allows us to evaluate their impact on the complex generalizing indicator in the form of the following increments:

- Due to a change in X_1:

$$\Delta y_a = F(a_1, b_0, \ldots m_0) - F(a_0, b_0, \ldots m_0);$$

- Due to a change in X_2:

$$\Delta y_b = F(a_1, b_1, \ldots m_0) - F(a_1, b_0, \ldots m_0);$$

- Due to a change in X_n:

$$\Delta y_n = F(a_1, b_1, \ldots m_1) - F(a_1, b_1, \ldots m_0).$$

In the analysis, and statistical aggregates can be used if they are accumulated over a number of years. These aggregates (time series) can be processed by the method of correlation and regression.

Acknowledgments. The authors are grateful to Doctor of Economics, Professor V. A. Persianov for many years of conscientious and disinterested scientific and pedagogical work, for the full support of our research work, for the invaluable gift of enlightenment that he leads at the Department of Freight Forwarding Services, State University of Management. We thank the departing professors, doctors of economic sciences N. N. Gromova and N. S. Uskova, from whom we learned a lot while studying and working at the first Department of Transport Management organized in our country, which trained managerial staff on general transport problems. The authors express their special gratitude and wish for health and prosperity to Doctor of Economics, Professor V. S. Gorin and colleagues for the opportunity of constructive and fruitful joint work over the years of reforming the higher education system.

References

Beshenkov, S.A.: Computer Science. Systematic Course. BINOM Knowledge Lab, Moscow (2006)

Gerchikov, V.I.: Motivation, Stimulation, and Remuneration. HSE, Moscow (2004)

Glazyev, S.: Jerk into the Future: Russia in New Technological and World Economic Structures. Knizhnyy mir, Moscow (2018)

Gorina, V.S., Persianova, V.A. (eds.): Scientific thought in the development of transport in Russia: historical retrospective, problematic issues and strategic guidelines. "TransLit" Publishing House, Moscow (2019)

Information Portal "market-pages.ru": The evolution of marketing concepts (n.d.). http://market-pages.ru/inmark/3.html

Kellerman, B.: Types of subordinates: what a manager should know. Harvard Bus. Rev. Russia **1**, 64–71 (2008)

Kibanov, A.Ya., Ivanovskaya, L.V. (eds.): Personnel Management. Publishing House "Prior", Moscow (1999)

Kurbatova, A.V., Kurbatova, E.S.: Assessment of the economic potential of railway transport. Univ. Vestnik **11**, 127–134 (2015)

Osipov, A.L.: Information Technology in Management. Novosibirsk State University of Economics and Management, Novosibirsk (2012)

Persianov, V.A., Stepanov, A.A., Gainochenko, T.M.: Research methodology and management on transport: main issues at the final stage of Russian transport sector reforms. Upravlenie **3**, 11–16 (2017)

Rosstat: Transport of Russia. Rosstat, Moscow (2018)

State University of Management: Round table "Personnel policy and improving the quality of transport education". Vestnik Transp. **3**, 2–4 (2017a)

State University of Management: Round table "Personnel policy in transport: transition to engineering and management education". Vestnik Transp. **9**, 2–5 (2017b)

Valuev, S.A., Volkova, V.N. (eds.): System Analysis in Economics and Organization of Production. Politekhnika, Leningrad (1991)

Competitive Advantages of Innovative Development of High-Tech Manufactures Based on the Creation of Special Economic Zones

Helena E. Panfilova[1(✉)], Alexey I. Tikhonov[2], and Alexey V. Savin[1]

[1] State University of Management, Moscow, Russia
helena_panfilova@mail.ru, savinguu@yandex.ru
[2] Moscow Aircraft Institute (NRU), Moscow, Russia
mai512hr@mail.ru

Abstract. The aim of the study was to identify the primary competitive advantages of industrial organizations located in special economic zones and related to high-tech sectors of the economy. When processing the analytical material, statistical methods of summarizing, grouping, comparing, as well as methods of a systematic approach, complex analysis, and calculation of the arithmetic mean were used. The results of the study indicate that the presence of a preferential tax regime, professional experience among the management team of the management company and the provision of "clean rooms" for the organization of high-tech industries are the determining factors in the competitiveness of innovative enterprises being located in a special economic zone. The formation of a favorable innovation climate at growth points contributes to the exit of the Russian economy from the crisis under the conditions of anti-Russian sanctions. The novelty of the results of the study lies in the development of a methodology for rating the attractiveness of special economic zones from the perspective of an integrated approach to host innovative resident companies.

Keywords: Creation criteria · Competitive advantages · Special economic zone · Technology park · Technopolis

1 Introduction

In the context of economic globalization, the activation of foreign investors in the Russian market, as well as the introduction of anti-Russian sanctions, issues of the prospective development of high-tech sectors of the Russian economy came to the fore (Krutikov et al. 2016). Improving the competitiveness of entrepreneurship in small and medium-sized businesses is associated with placement in special economic zones, which provide customs, tax, and other benefits (Gobillion et al. 2012). However, in practice, a number of conflicts arise related to the legal status of the subjects of special economic zones, the peculiarities of interaction in carrying out business activities on the basis of the Agreements (Kuzmin 2013).

A. V. Bogoviz and Y. V. Ragulina (Eds.): ISCI 2019, LNNS 115, pp. 39–47, 2020.
https://doi.org/10.1007/978-3-030-40749-0_5

Over the past five years, the number of technology parks in Moscow has grown from 6 in 2014 to 35 in 2018, which was 28% of the total number of technology parks in Russia (Nadezhin 2010). In the technology parks of Moscow, about 1,700 residents were accommodated, more than 45 thousand jobs were created, and more than 48 billion rubles were invested in infrastructure with the accumulated result (Kolychev 2015).

2 Materials and Method

The Special Economic Zone (SEZ) is interpreted as "a part of the territory of the Russian Federation, which is determined by the Government and within the framework of which there is a special regime for carrying out business activities, and the customs procedure of a free customs zone can be applied" (Federal Law of July 22, 2005 N 116-FZ) (Consultant Plus n.d.).

The main goals of forming SEZs are usually considered such as increasing the flow of foreign investment, increasing the number of joint scientific and technical work, raising the competitiveness of import-substituting products for high-tech sectors of the economy, as well as creating new high-performance jobs and competence development centers for professions of the future.

SEZs are created to assist small and medium-sized businesses through the provision of services for the provision of space for industrial and innovative activities, the provision of transportation and warehousing of goods, etc. (Drueva 2013).

Russian legislation provides certain guarantees to ensure the sustainable operation of SEZ residents. First, it protects from adverse consequences as a result of changes in tax legislation. Second, these are the guarantees for judicial settlement of disputes. In addition, individual tax, customs, and land regimes are created by law (Jensen and Winiarczyk 2014).

On the other hand, a number of requirements are imposed on residents of the SEZ as business entities that carry out industrial-production or technical-innovative activities (Table 1).

Currently, in the Russian Federation, there are more than forty special economic zones, including technical-innovative and industrial-production ones. Practice shows that, in comparison with all-Russian indicators, the costs of investors in implementing projects in these zones are 30% lower on average.

In order to optimize management by the Management Company within the framework of the SEZ, a cluster approach is being implemented (Zimenkov 2012). So, the following 6 clusters are represented in the special economic zone of "Technopolis "Moscow" JSC: "Microelectronics, optics, and electronic equipment," "Energy-efficient technologies," "Medical technologies, equipment, and biopharma," "Information and communication technologies," "Contemporary materials and nanotechnologies," as well as "High-tech industries" (Nekuchaeva 2011).

The implementation of entrepreneurial activity in the SEZ is based on the use of special tax, customs, land regimes. Its effectiveness is directly related to improving the mechanisms of legal regulation of intersectoral interaction (with the participation of the Ministry of Economic Development and other government bodies, a management

Table 1. Requirements presented by the Technopolis "Moscow" JSC to business entities.

No	Type of requirements	Requirements
1	Formal	Registration of organizations in Moscow within the boundaries of the municipality in whose territory the special economic zone is located
		The absence of branches, representative offices, and also separate subdivisions outside the special economic zone of the organization
		Organizations conducting business in a special economic zone are not entitled to transfer their rights and obligations
		Reaching an Agreement on the implementation of industrial-production or technical-innovative activities
		Conducting activities in the territory of the SEZ in accordance with the Agreement (Article 13 of the Federal Law "On the SEZ")
2	Economic	The total amount of investments related to the acquisition, creation, modernization, reconstruction, and preparation for using non-current assets of the applicant (at capital construction projects planned for lease), determined for implementation within no more than 5 calendar years from the date of signing of the agreement and per square meter the total area of capital construction projects planned for lease is at least 50,000 rubles (net of value-added tax)
		The planned average monthly wage of workers employed at the capital construction projects scheduled for lease is at least the size of the average monthly income from labor activity in the city of Moscow (according to Rosstat). This average monthly income is determined for the year in which the completion of the investment is planned (for reference, 59,823 rubles)
		The amount of proceeds from the applicant's planned activity per one square meter of the total area of capital construction projects scheduled for lease is at least 200,000 rubles (excluding value added tax) for the second calendar year from the planned date of completion of the investment

company, strategic investors, and representatives of small and medium-sized businesses).

The Technopolis "Moscow" was created as a special economic zone of the technology-innovative type in 2005 (Ministry of Economic Development of the Russian Federation n.d.). Subsequently, the area of the SEZ was expanded by joining land plots located in Pechatniki, Silino, and Stary Kryukovo.

To organize effective interaction between state authorities, the management company, residents, and investors, a supervisory board of the SEZ is created, which may include representatives of educational and research organizations located within the same municipality as a special economic zone.

As criteria for the creation of the SEZ, competitive advantages were approved based on the geographical location, the availability of infrastructure, and the provision of human resources. In the future, Decree of the Government of the Russian Federation dated May 10, 2017 No. 552 introduced adjustments related to the assessment of the economic efficiency of their functioning "… by indicators of profitability, rentability, and payback period" (Government of the Russian Federation 2017) (Table 2). Also, the requirements for justifying the environmental efficiency of creating the SEZ, and the size of the cost of budget funds per square meter of the total area of the SEZ are legally established. The SEZ management company is responsible for attracting 50% of the planned extra-budgetary investments.

Table 2. Evaluation of the effectiveness of the exclusive economic zone by economic indicators.

Used economic indicator	Calculation method	The standard value of the indicator	Note
Rentability	The ratio of the total investment of residents, tax deductions of residents and the management company to the budgets of all levels, customs payments to the amount of funds of the federal budget, the budget of the subjects and the local budget	Not less than 2	For industrial zone
		Not less than 1 (at the end of the tenth year of project implementation, cumulative)	For technical and innovative zone
Profitability	The ratio of the revenue of residents and the management company to the total amount of the federal budget, the budget of the subjects and the local budget	Not less than 1, 2	For industrial zone
		Not less than 1 (at the end of the tenth year of project implementation, cumulative)	For technical and innovative zone
Payback period	The difference between the size of the revenue of residents, the management company and the means of various budgets when creating the infrastructure of the Russian Federation is taken into account	No more than 8 years	For industrial zone
		No more than 15 years	For technical and innovative zone

3 Result

An analysis of the specialization of technology parks in Moscow showed that among the technology parks, the most widespread area is "Electronics and microelectronics," which covers 70.37%. The specialization "Information and telecommunication technologies" is found in 66.67% of technology parks. Further, the directions "Energy, energy efficiency, and energy-saving" and "Devices and power supplies" are on par since their coverage is 55.56%. 37.04% of technology parks specialize in new materials. Only 3.70% of Moscow's technology parks are focused on vacuum plants, industrial design, and production of oilfield, drilling, and geology prospecting equipment. 96.30% of technology parks have the following specialized facilities: a congress center, meeting rooms, training rooms, and a conference room. Based on the information received, 92.59% have parking. 62.96% of technology parks possess technology centers for common use, equipped with expensive technological and scientific equipment that residents can use. Only 11.11% of the technology parks analyzed have an additive technology center and a customs post, which is important for foreign residents and residents receiving materials or sending their products for export. All studied technology parks rent an office, laboratory, or industrial premises. Also, based on the information received, 92.59% of technology parks provide telecommunication services and services for providing parking spaces for vehicles. Only 7.41% of the studied innovation subjects offer garbage collection and waste disposal services. Technology parks and technopolises also undertake energy, water, and heat supply.

The analysis highlighted the main competitive advantages of the SEZ from the standpoint of their influence on the competitiveness of high-tech sectors of the economy of the region and the country as a whole. So, in our opinion, one of the key advantages is the provision of the necessary innovative technological equipment on a preferential rent basis, which significantly increases the chances of technology parks to attract companies as residents. In the course of the study, Moscow's technology parks and technopolises were evaluated in terms of completeness of the provision of service packages. Each service was evaluated on a 5-point scale by an expert. The maximum number of points for a service package is 99 points. 41% of the analyzed technology parks scored more than half the points of the maximum number, i.e., more than 50 points. According to available information, the Technopolis "Moscow" provides the most substantial package of services, the total score of which is 89. The following technology parks and technopolises were also included in the top 5 leaders: the "Skolkovo" technopolis with 71 points, the "Eleron" technology park and the "Slava" technology park with 65, and the "Kalibr" technology park with 64. Then, they were ranked according to the criterion of the most significant contribution. Table 3 shows an example of listing for the provision of innovative technological equipment.

The next competitive advantage of special economic zones is the creation of new jobs. The problems of job shortages and unemployment are equally important. The Russian economy does not generate enough jobs. The management companies of technology parks and technopolises (and the residents themselves partially) solve this social problem by creating new jobs. The range of the considered indicator for the studied technology parks and technopolises ranged from 3 to 3,000 created jobs (Table 4). Thus, the "Skolkovo" technopolis contributed to the solution of this urgent problem by creating 3,000 jobs, and the MSU Science Park, along with the "Moscow" technopolis, created 2,500 each. Technopolises and technology parks are able to reduce the percentage of unemployment if they create jobs by universal forces. In this way, qualifications and education can be more fully utilized, informal employment is to be reduced, and other issues related to the labor market should be addressed. 63% of technology parks created less than 1,000 jobs.

Table 3. Ranking of technology parks owning innovative technological equipment.

Rank group	No	Name of technology parks	Total points
Group 1	1	Moscow Technopolis	202.0
	2	Skolkovo Tesnopolis	182.0
	3	Slava	153.0
	4	Mosgormash	145.0
Group 2	5	Kalibr	137.0
	6	TechnoSpark	128.8
	7	MSU Science Park	128.1
	8	Polyus	126.8
	9	Pulsar	125.2
	10	Fiztechpark	120.8
	11	Eleron	119.4
	12	Strogino	116.0
	13	Sapfir	113.9
	14	Svyaz Inzhiniring	101.8
Group 3	15	VTI	100.5
	16	Temp	98.2
	17	Fotonika	96.4
	18	Nagatino	96.0
	19	Vizbas	94.1
	20	Modul	89.0
	21	Otradnoye	86.0
	22	TSNIITMASH	86.0
	23	Vodnyy stadion	83.5
	24	Ricor High-Tech Innovation Center	80.0
	25	Elma	73.8
	26	ITELMA	64.8
	27	Tisnum	54.1

The competitive advantage of the SEZ is also the presence of business incubators, logistics centers, laboratories, customs posts, and clean rooms. The presence of the listed infrastructure is estimated at the maximum number of points. The customs post at the territory of the technology park/technopolis is especially attractive for foreign residents and residents who send products for export or purchase raw materials and materials for production abroad. Such an infrastructure in the special economic zone allows residents to receive customs preferences in the form of exemption from import and export duties on raw materials and equipment, as well as exemption from VAT. A resident can declare goods without leaving the territory of the technology park. And custom clearance, which can last six months, now takes half an hour (Usmanov 2017). Customs posts and warehouses of temporary storage are present in the following technology parks: "Moscow" Technopolis, "Skolkovo" Technopolis, and "Temp" Technology Park. Having clean rooms is also a huge plus. Such rooms, made according

Table 4. The number of jobs created in the SEZ.

No	Name of technology parks	Number of jobs created	Points
1	Skolkovo Technopolis	3000	3
2	MSU Science Park	2500	3
3	Moscow Technopolis	2500	3
4	Mosgormash	1600	2
5	Otradnoye	1528	2
6	Elma	1509	2
7	Slava	1450	2
8	Strogino	1200	2
9	Sapfir	1134	2
10	Svyaz Inzhiniring	771	1
11	Nagatino	710	1
12	Pulsar	694	1
13	Fiztehpark	600	1
14	Ricor High-tech Innovation Center	575	1
15	Fotonika	550	1
16	ITELMA	500	1
17	Modul	410	1
18	Kalibr	367	1
19	VTI	366	1
20	Tisnum	364	1
21	Vodnyy stadion	330	1
22	Eleron	231	1
23	TechnoSpark	168	1
24	Vizbas	3	1
25	Polyus	–	0
26	Temp	–	0
27	TSNIITMASH	–	0

to cleanliness standards, for example, from IOS 5 class to IOS 7 class, will be in demand by companies engaged in research, biopharmaceuticals, as well as serial production of microelectronics, etc. The "Skolkovo" technopolis and "Moscow" technopolis, as well as such technology parks as "Polyus," "TechnoSpark," "Slava," "Pulsar," "Sapfir," "Fotonika," and "Tisnum," rent clean rooms. The logistics center has loaders and other technical equipment, provides packaging, picking, and delivery services. So, residents have the opportunity to use the services of a logistics center company, which is selected by the management company. Accordingly, the quality is guaranteed and verified. Only 15% of Moscow's technology parks equipped these centers on their territory. Business incubators are also a valuable infrastructure for the technology park.

4 Discussion

As a result of the analysis of the activity of the main economic zones, from creation to a stable position in the market, as well as on the basis of the study of scientific and normative reference books, the following key competitive advantages of the SEZs are highlighted (in terms of their impact on the competitiveness of high-tech industries, regional economy, and the country as a whole): the presence of individual tax, customs, land regimes; providing the necessary innovative technological equipment on a preferential rental basis; creation of new jobs; the presence of business incubators, logistics centers, laboratories, customs posts, and clean rooms. Ultimately, all this contributes to the creation of favorable conditions for the development of innovative high-tech industries. The criteria for creation and effectiveness also allowed us to develop a methodology for rating the attractiveness of special economic zones from the perspective of an integrated approach to host innovative resident companies.

5 Conclusion

On the example of the special economic zone (Technopolis "Moscow" JSC), the formal and economic requirements for business entities engaged in industrial-production or technical-innovative activities in the territory of the SEZ are formulated.

Based on the analysis of the regulatory framework, including Government Decisions, codes, and law enforcement practice, criteria for creating and evaluating the effectiveness of the functioning of the SEZ, the structure of governing bodies was established.

An analysis of the specialization of technology parks in Moscow showed that, as a rule, the activities of residents are concentrated in innovative and high-tech fields. From these positions, the study conducted a rating of special economic zones of Moscow for critical competitive advantages and showed the attractiveness and prospects of their further development.

Acknowledgment. The work was carried out as part of the initiative research "Development of Business Systems in the Russian Economy."

References

Consultant Plus: Legal base (n.d.). https://consultant.ru

Drueva, A.A.: The legal status of residents of special economic zones in the Russian Federation. Lawyer **22**, 11–16 (2013)

Gobillion, L., Magnac, T., Laurent, H.: Do unemployed workers benefit from enterprise zones? The French experience. J. Public Econ. **7**, 88–92 (2012)

Government of the Russian Federation: Decree "On Amending the Criteria for Creating a Special Economic Zone", 10 May 2017, No. 552 (2017)

Jensen, C., Winiarczyk, M.: Special economic zones – twenty years later. CASE Network Studies and Analyses, 467 (2014)

Kolychev, A.M.: Economic (Business) Law. NORMA INFRA-M, Moscow (2015)

Krutikov, V.K., Dorozhkina, T.V., Tyutin, D.V., Hudy-Hyski, D., Yakunina, M.V.: Special Economic Zones. Streltsov I.A., Kaluga (2016)

Kuzmin, D.V.: Problematic issues of evaluation of the innovative development of the Russian Federation. State Audit. Right Econ. **2**, 73–79 (2013)

Ministry of Economic Development of the Russian Federation: SEZ "Technopolis Moscow" (n. d.). http://www.ved.gov.ru/files/images/SEZ%20Technopolis%20Moscow.pdf

Nadezhin, N.N.: Entrepreneurial activity as a sphere of the interaction of private law and public law interests and means of regulation. Sci. Gazette Philos. Sociol. Law **20**(91), 168–172 (2010)

Nekuchaeva, M.Yu.: Prospects for the Development of Special Economic Zones in Russia. MAKS Press, Moscow (2011)

Usmanov, R.M.: Creation of "special zones" as a measure to improve the standard of living of the population of a municipality. Law Pract. **3**, 13–20 (2017)

Zimenkov, R.I.: Free Economic Zones. Unity-Dana, Moscow (2012)

The Efficiency of Russian Enterprises Based on the Assessment of Their Competitiveness

Victor P. Ivanitsky, Elena G. Knyazeva, and Elena A. Razumovskaya(✉)

Ural State Economic University, Yekaterinburg, Russia
rasumovskaya.pochta@gmail.com

Abstract. The study determines the comparative effectiveness of Russian enterprises from different sectors of the national economy. The authors focus on finding arguments in favor of restructuring and diversifying its structure or refute such a need. It is generally accepted that the raw material orientation is a risk factor since it creates a country's dependence on the cost of hydrocarbons and the national currency. Such an opinion is justified, especially in the long run. However, the authors set a goal to practically substantiate the state of affairs by evaluating the effectiveness based on an in-depth analysis of the competitiveness of enterprises in the raw materials sector and several others, from the category of primary industries. The main research approach was a dynamic one due to its versatility and flexibility, which ensured a comparative assessment of the competitiveness of enterprises from different sectoral and quantitative samples. The results obtained during the study can be applied to develop measures to increase the competitiveness and investment attractiveness of individual companies and industries, as well as to optimize the tools of state regulation of the economy.

Keywords: Economic efficiency · Competitiveness assessment · Dynamic method · Economic sectors

1 Introduction

Those countries with well-diversified market economies have the advantage of being resistant to the volatility of domestic and global demand for manufactured industrial products. The approaches to assessing the performance of industrial enterprises are different, but the main enlarged groups of comparative analysis are considered to be the state and private and industry format groups (Stiglitz 2008). The role of the state in a developed economy is such that it does not act as a principal owner, confining itself to its inherent functions that determine the inefficiency of management in the industrial sector, namely: the need for electoral opinion, the obligation to finance meritorious benefits, the inevitable bureaucracy in regulation and control while ensuring social guarantees, and others (Porter and Heppelmann 2017). The structure of the Russian economy, both sectoral and in the institutional aspect concerning property, has features that predetermined the direction of this study. Given the share of the public sector (about 72%) and the predominance of the commodity sector in the category of

A. V. Bogoviz and Y. V. Ragulina (Eds.): ISCI 2019, LNNS 115, pp. 48–58, 2020.
https://doi.org/10.1007/978-3-030-40749-0_6

systemically critical national economies of Russia (68%), the need to verify the hypothesis about the low efficiency of commodity companies relative to others, in particular, the metallurgical and manufacturing industries, seems promising for research. Tasks for assessing the effectiveness and competitiveness of Russian enterprises are among the leading Russian experts in the field of strategic management (Gelvanovsky 2012; Kovalenko 2013).

In this regard, the authors consider it essential to assess the effectiveness of leading Russian companies based on an analytical study using the dynamic method of comparing competitiveness indicators for a sample of the largest domestic companies from the primary (mining) and manufacturing (metallurgical) sectors of the economy. Differentiation by ownership was not deliberate.

The relevance of the chosen research format is due to a resonant discussion in society on the planned privatization of part of state-owned companies from different sectors until 2024.

2 Materials and Method

The spectrum of its indicators determines the choice of a methodology for assessing the efficiency of industrial enterprises based on competitiveness; using absolute indicators of revenue or profit for this is not entirely incorrect and uninformative. The dynamics of stock prices on the stock market can be used as one of the tools for such an assessment. However, fluctuations in stock prices often do not reflect changes in the efficiency of issuing enterprises since the volatility of quotations is often determined by market factors, namely geopolitical, speculative, and others. Also, some domestic enterprises (Nizhnekamsk-Neftekhim, Metalloinvest, Sibur, Uralkali, etc.) do not place their securities on the stock exchange; therefore, it is impossible to analyze the quotes of their securities.

Another standard tool for assessing the effectiveness of enterprises is the ratings generated by leading information and analytical agencies (world agencies are Forbes Global-2000, Fortune Global-500, Financial Times Global-l500; and domestic agencies are Forbes-200, RBC-500, Expert-600). However, the ratings reflect the ranking in terms of revenue or market capitalization, and operating efficiency (profitability) or financial stability in the ratings is extremely rare; therefore, the positions of companies in such ratings and their actual effectiveness may diverge. So, according to the results of 2014, in the domestic ratings, among the most productive enterprises were such companies as Mechel, SU-155, and Transaero. Their resonant bankruptcies in 2015 allow us to conclude that the published ratings do not take into account many of the resulting parameters, for example, a competitive status.

Evaluation of effectiveness in terms of competitiveness is one of the significant elements of a system for assessing the combined effect of the financial and economic activities of enterprises. A wide range of existing competitiveness assessment methods makes it possible to classify them as follows:

- Matrix methods based on the assessment of the company's product portfolio;
- Product methods based on assessing the competitiveness of company products;

- Operational methods based on the assessment of various components of a company's operations;
- Valuation methods based on the market value of the company;
- Dynamic methods based on the assessment of key economic indicators of the company in dynamics (Voronov et al. 2016).

The essence and tools of these methods are described in detail in particular scientific sources, the analysis of the applied aspects of their application (Voronov 2015) allows us to conclude that most of them are aimed at assessing the competitiveness of an individual enterprise. When there is a need to conduct a comparative assessment of the competitiveness of several enterprises in one or several industries, then the solution of such a problem for most methods is hugely complicated and time-consuming; an exception is a dynamic approach based on the assessment of the company's key economic indicators in dynamics.

The dynamic method allows one to evaluate the competitiveness of both an individual enterprise and in comparison, with others, which allows one to get an idea of the situation in the industry. The essence of the dynamic method reflects the efficient use of economic resources by the enterprise; operational efficiency, strategic positioning, and financial condition reflect this (Ansoff 1965; Andrews 1971). The key indicators of these sources are the profitability of economic activity, the dynamics of sales volumes, as well as the degree of liquidity (Hamel and Prahalad 1996). A quantitative assessment of each of the sources of competitiveness is carried out by comparing the specified indicators of the analyzed company with the corresponding indicators of competitors.

Using the indicated indicators for quantifying competitiveness sources, such as operating efficiency (K_R), strategic positioning (K_I), and financial condition (K_L) ratios, we obtain a single integrated indicator reflecting the assessment of the competitiveness of the analyzed company (K) (Rasumovskaya et al. 2017, 2018):

$$K = K_R \cdot K_I \cdot K_L, \tag{1}$$

The operating efficiency ratio is calculated as follows:

$$K_R = \frac{R_A}{R_S}, \tag{2}$$

where K_R is the coefficient of operational efficiency; R_A is the operational efficiency of the company for the reporting period; R_S is the operational efficiency of the sample for the reporting period.

The authors understand the sample as a set of competitors, in comparison with which the competitiveness of the studied company is assessed. Aggregated indicators for the sample are determined by summing the corresponding indicators of companies from the sample.

$$R_A = \frac{S_A}{E_A}, \tag{3}$$

where R_A is the operational efficiency of the company for the reporting period; S_A is the revenue (sales) of the analyzed company for the reporting period; E_A is the costs (expenses) of the analyzed company for the reporting period.

$$R_S = \frac{S_S}{E_S}, \tag{4}$$

where R_S is the operational efficiency of the sample for the reporting period; S_S is the revenue from the sample for the reporting period; E_S is the sampling costs for the reporting period.

In the general case, the composition of the costs must satisfy the following condition: the net profit of the enterprise is the revenue reduced by costs. Depending on the objectives of the analysis and openness of the data, the composition of costs may vary. Operating costs are acceptable to use as costs to analyze the competitiveness of core activities.

Strategic positioning ratio:

$$K_I = \frac{I_A}{I_S}, \tag{5}$$

where KI is a strategic positioning coefficient; IA is the index of changes in the company's revenue for the reporting period; IS is the index of the change in revenue for the sample for the reporting period.

$$I_A = \frac{S_A}{S_{0A}}, \tag{6}$$

where IA is the index of changes in the company's revenue for the reporting period; SA is the revenue of the analyzed company for the reporting period; S0 A is the revenue of the analyzed company in the previous period.

$$I_S = \frac{S_S}{S_{0S}}, \tag{7}$$

where I_S is the index of the change in revenue for the sample for the reporting period; S_S is the revenue from the sample for the reporting period; S_{0S} is the revenue from the sample in the previous period.

Financial ratio:

$$K_L = \frac{L_A}{L_S}, \tag{8}$$

where K_L is the financial ratio; L_A is the liquidity of the company at the end of the reporting period; L_S is the ample liquidity at the end of the reporting period.

$$L_A = \sqrt{\frac{CA_A}{CL_A}}, \tag{9}$$

where: L_A is the liquidity of the analyzed company at the end of the reporting period; CA_A is the current assets of the company at the end of the reporting period; CL_A is the company's current liabilities at the end of the reporting period.

$$L_S = \sqrt{\frac{CA_S}{CL_S}}, \quad (10)$$

where L_S is the liquidity for the sample at the end of the reporting period; CA_S are the current assets sampled at the end of the reporting period; CL_S – a short-term sampling obligation at the end of the reporting period.

The criterial value coefficient of competitiveness:

1. $0 < K < 1$ means low competitiveness of the studied enterprise in relation to the sample;
2. $K = 1$ means a comparable competitive position of the company with the sample;
3. $K > 1$ means higher competitiveness of the studied enterprise in comparison with the sample.

In general, the higher the value of K, the analyzed company is more competitive in relation to sample-forming companies.

3 Results

The integrated competitiveness coefficient is extremely rarely taking values of less than 0.4 and more than 1.8. In accordance with the value of the competitiveness coefficient, the gradation of the competitive status of enterprises is presented in Table 1.

Without taking into account the dynamics, the value of the indicator does not allow forming an exhaustive idea of the object of study, the analysis of indicators in dynamics makes it possible to obtain complete information about the state of the enterprise and form a forecast for its development. Within the framework of the methodology of the dynamic approach, dynamic series is formed to ensure the representativeness of the data array and to increase the reliability of the results significantly.

Table 1. Scale for assessing the competitiveness of enterprises (Stiglitz 2008).

K values	Interpretation of the competitiveness of the enterprise under study regarding sampling
$K < 0.60$	Extremely low
$0.60 \leq K < 0.80$	Very low
$0.80 \leq K < 0.95$	Low
$0.95 \leq K < 1.00$	Moderately low
$K = 1.00$	Satisfactory (equal to competitors)
$1.00 < K \leq 1.05$	Moderately high
$1.05 < K \leq 1.2$	High
$1.2 < K \leq 1.4$	Very high
$1.40 < K$	Extremely high

The selection of research and comparison objects includes 32 Russian enterprises from the raw materials and manufacturing industries. The basis for selection as objects of the study was the criteria for the industry sector of the largest enterprises in terms of revenue.

Based on expressions (1)–(10), the calculation of the competitiveness indicators of all enterprises in the sample was carried out for 2014–2018. The source of information was financial reporting data published by enterprises on Internet sites. Costs (E) used in the calculations were calculated by the difference in excess of revenue over net profit. The calculation results are presented in Table 2; all enterprises are ranked in the order of calculated competitiveness according to the results of 2018.

The authors' attempt to compare the enterprises of the raw materials and manufacturing industries allowed us to formulate the following main conclusions: 11 out of 33 enterprises in the sample (33.3%) belong to the raw materials sector, the remaining 22 (66.7%) belong to the chemical industry, metallurgy, and mechanical engineering.

Table 2. The competitiveness of Russian enterprises.

No.	Enterprises	Industry	Competitiveness indicators (*K*)				
			2014	2015	2016	2017	2018
1	Polyus	Precious metals mining	1.065	6.985	4.621	4.614	3.988
2	ALROSA	Diamond mining	1.119	1.418	4.296	4.2981	3.679
3	VSMPO-AVISMA	Non-ferrous metallurgy	1.353	1.643	2.041	2.038	2.001
4	Polymetal	Precious metals mining	0.637	1.299	2.014	2.008	1.897
5	Metalloinvest	Ferrous metallurgy	1.159	1.115	1.999	1.974	1.886
6	Nizhnekamsk-neftekhim	Chemical industry	1.124	1.697	1.654	1.642	1.632
7	Transneft	Oil transportation	1.066	1.506	1.566	1.553	1.457
8	Sibur	Chemical industry	0.823	1.255	1.547	1.521	1.487
9	Norilsk Nickel	Non-ferrous metallurgy	1.794	1.406	1.546	1.491	1.482
10	Slavneft	Hydrocarbon Production and Processing	1.614	1.572	1.551	1.533	1.477
11	Uralkali	Chemical industry	1.387	0.884	1.512	1.487	1.422
12	PhosAgro	Chemical industry	0.881	1.766	1.507	1.488	1.412
13	NLMK	Ferrous metallurgy	1.358	1.692	1.486	1.474	1.314
14	KamAZ	Engineering	0.799	0.775	1.424	1.422	1.386
15	Severstal	Ferrous metallurgy	0.893	1.453	1.377	1.364	1.359
16	MMK	Ferrous metallurgy	0.981	1.091	1.373	1.371	1.313
17	Gazpromneft	Hydrocarbon Production and Processing	1.070	1.286	1.471	1.306	1.287
18	Rusal	Non-ferrous metallurgy	0.966	1.227	1.135	1.129	1.112
19	United Aircraft Corporation	Engineering	1.045	0.747	1.119	1.117	1.008
20	Tatneft	Hydrocarbon Production and Processing	1.511	1.661	1.152	1.112	0.988

(*continued*)

Table 2. (*continued*)

No.	Enterprises	Industry	Competitiveness indicators (*K*)				
			2014	2015	2016	2017	2018
21	Transmashholding	Engineering	0.860	0.545	1.081	1.072	0.897
22	Surgutneftegas	Oil production	7.416	2.482	1.278	1.076	0.889
23	TMK	Ferrous metallurgy	0.857	0.790	1.004	0.978	0.875
24	Akron	Chemical industry	0.830	1.798	0.975	0.966	0.818
25	Evraz	Ferrous metallurgy	0.787	0.845	0.927	0.921	0.801
26	Eurochem	Chemical industry	0.721	1.265	0.929	0.922	0.789
27	Lukoil	Hydrocarbon Production and Processing	1.191	1.032	0.937	0.912	0.744
28	GAZ	Engineering	0.402	0.618	0.907	0.899	0.731
29	SUEK	Hydrocarbon Production and Processing				0.799	0.712
30	Rosneft	Hydrocarbon Production and Processing	0.886	0.799	0.772	0.695	0.668
31	ChTPZ	Ferrous metallurgy	0.874	0.955	0.573	0.551	0.547
32	AvtoVAZ	Engineering	0.517	0.297	0.484	0.448	0.434
33	Mechel	Ferrous metallurgy	0.150	0.164	0.315	0.303	0.291

At the end of 2018, the leading domestic enterprises with the highest competitiveness indicators include the following: Polyus (raw materials sector), ALROSA (raw materials sector), VSMPO-AVISMA (processing industry), and Metalloinvest (processing industry). In total, seven out of the ten most competitive enterprises belong to the category of processing, while the enterprises of the commodity sector have the highest rates (Polyus and ALROSA).

The reason for the leadership of Polyus and ALROSA companies, which are significantly ahead of all other companies, is the positive impact of the devaluation effect of 2014 and stable world prices for precious metals and diamonds in 2015–2017. In 2018, the situation did not change significantly; on the contrary, short-term and multidirectional changes in the foreign market strengthened the existing trends.

Contrary to forecasts, the devaluation did not provide competitive advantages to companies in the metallurgical sector, such as MMK (No. 16), Rusal (No. 18), ChTPZ (No. 31), and Mechel (No. 33), as well as engineering companies such as UAC (No. 19) and AvtoVAZ (No. 32) for a combination of reasons, namely, due to the fall in world prices for metals and the pool of Western countries' economic sanctions against Russian companies.

The AvtoVAZ company demonstrates low competitiveness despite state support. Throughout the study period, a deterioration in this indicator is visible, the peak of which occurred in 2015.

Low competitiveness over a long time horizon is highly likely to lead to the failure of the enterprise, so the measures currently being implemented concerning the Mechel are quite logical. The authors believe that only state support allows the AvtoVAZ company to maintain production. Should it be concluded that the non-primary sector, according to preliminary estimates, is the least competitive? The question is somewhat complicated since many manufacturing and other industrial enterprises are oriented towards export in production, namely metallurgy, and some enterprises are focused on the domestic market, for example, consumer engineering. Therefore, the reasons for the low competitiveness should not be sought in the manufacturing companies themselves but the lack of consumer demand.

The analysis is enlarged and preliminary. The results of assessing the competitiveness of enterprises in one industry should be considered more reliable. It is such an analysis that will ensure compliance with the research logic. Therefore, the authors conducted a grouping of companies by industry. Then, they assessed the competitiveness of each industry group.

The following groups became such groups from the formed sample: the raw materials industry and the processing industry (mainly metallurgy and engineering). The authors did not include other industries in the sample, such as telecommunications, electric power, construction, and trade since they differ significantly in scale, and their condition is determined by a spectrum of other reasons than the selected enterprises.

First, let's review the raw materials sector. This industry group is of particular interest because it makes a significant contribution to filling the state budget and GDP, and also because the state of this industry affects the state of the monetary policy of the entire national economy of Russia. The competitiveness indicators of the raw materials sector enterprises (Table 3) show that Polyus and with a small margin ALROSA are leaders (K values 2.086 and 2.012, respectively) in 2018. Such enterprises as Surgutneftegas, Lukoil, SUEK, Rosneft have the lowest competitiveness score in 2018 (K = 0.717).

Table 3. Indicators of competitiveness of enterprises of the commodity sector (Stiglitz 2008).

Enterprises	2014	2015	2016	2017	2018
Polyus	0.991	0.880	1.091	2.123	2.086
ALROSA	1.132	1.094	1.082	1.878	2.012
Polymetal	1.112	1.065	0.987	1.245	1.963
Transneft	0.985	1.146	1.213	1.086	1.714
Slavneft	0.825	1.111	1.512	1.507	1.511
Gazpromneft	0.863	1.221	1.578	1.470	1.483
Tatneft	1.281	1.545	1.312	1.152	1.158
Surgutneftegas	6.654	2.358	1.866	1.120	1.131
Lukoil	0.987	0.941	0.952	0.937	1.344
SUEK	0.524	0.623	0.648	0.827	0.912
Rosneft	0.675	0.688	0.672	0.661	0.717

Second, there is the engineering sector. The competitiveness indicators of engineering enterprises are presented in Table 4.

Table 4. Indicators of competitiveness of enterprises of the engineering complex.

Enterprises	2014	2015	2016	2017	2018
KamAZ	1.143	1.272	1.301	1.477	1.511
UAC	1.761	1.515	1.644	1.298	1.388
Transmashholding	1.241	0.892	0.973	1.088	1.122
GAZ	0.448	1.081	0.817	0.914	0.921
AvtoVAZ	0.628	0.423	0.434	0.422	0.412

KamAZ (K = 1.511) and UAC (United Aircraft Corporation) (K = 1.388) are the obvious industry leaders; Transmashholding is slightly behind the leaders but also demonstrates a high level of competitiveness. (K = 1,122); also, GAZ has low competitiveness (K = 0.921). AvtoVAZ has the lowest competitiveness indicators in the industry group (K = 0.412), which even devaluation processes did not allow to strengthen in the domestic market.

4 Discussion

The results of the study on the presented industry groups allow us to conclude that in the fuel and energy complex, there is an absolute parity of competitiveness of state and private companies; in telecommunications, private companies take the lead; state-owned enterprises are more competitive in mechanical engineering and the electric power industry. Thus, in the study of the competitive status of various industry groups, it was found that there are no unequivocal leaders according to the criterion of private or state ownership.

However, this conclusion requires a more complete and comprehensive justification, for which the authors conducted a consolidated assessment of the competitiveness of enterprises in the sample from Table 2. The consolidated assessment required, through the previously presented mathematical tools, calculations for the following two aggregated groups of companies: the public and private sectors. The results are presented in Table 5.

Table 5. Comparative dynamics of indicators of competitiveness of enterprises in the raw materials and engineering sectors.

Industries	2014	2015	2016	2017	2018
Raw materials	0.869	0.987	0.837	0.985	1.032
Engineering	1.216	1.028	1.194	1.015	0.989

The results of the calculations show that according to the results of 2018, enterprises in the commodity sector are more competitive; however, their advantage about machine-building enterprises is very insignificant (difference K = 0.043). This is probably due not so much to fundamental industry advantages as to better market conditions in the global hydrocarbon market.

There is a strong impact of challenging macroeconomic conditions in which the Russian national economy and all its economic actors are located. The question of the reasons for this influence may have a multidimensional answer, for which it is necessary to conduct several narrower studies in related fields of economic and financial science.

5 Conclusion

The results obtained in the course of the study allow us to conclude that the hypothesis of the unequivocal inefficiency of enterprises in the commodity sector did not find confirmation. It is important that the study was conducted taking into account the conditions of the Russian economy for a limited period, during which the economy was affected by the crisis, the consequences of which affected the situation of all sectors and economic entities.

Nevertheless, the result obtained is valuable for private practice, since it can, to a certain extent, serve as the basis for increasing state financing of large raw materials enterprises as part of anti-sanctions support and other projects. Also, another interpretation of the results should not be ruled out, namely: higher competitiveness (and, therefore, efficiency) of the raw materials industry may be caused not by its growing effectiveness, but by a decrease in the performance of the high value-added sector in the production process. In turn, this is due to a significant deviation from the economic "normal". Namely, one could consider unstable economic conditions that do not allow the dynamic development of technology sectors. This once again confirms the need for structural reforms in the Russian economy (Mau 2016), the result of which should be the creation of similar economic conditions for all subjects under which access to capital and the market for the sale of products will become genuinely transparent.

Acknowledgment. The authors are grateful to the leadership of the Technical University of the Ural Mining and Metallurgical Company (Verkhnyaya Pyshma), personally to the rector of UMMC Technical University Lapin, Vyacheslav Aleksandrovich and the head of the Department of Applied Economics Voronov, Dmitry Sergeyevich for their great assistance in preparing the study.

References

Voronov, D.S.: The relationship between the competitiveness of the enterprise and thecompetitiveness of its products. J. Mod. Competition **9**(1), 39–53 (2015)

Voronov, D.S., Korsunov, P.P., Krivorotov, V.V., Matveeva, T.V.: Assessment and Planning of the Competitiveness of the Enterprises of the Fuel and Energy Complex. UrFU, Yekaterinburg (2016)

Gelvanovsky, M.I.: Methodological approaches to ensure the competitiveness of international integration groups in the context of globalization. Eurasian Econ. Integr. **1**, 44–58 (2012)

Kovalenko, A.I.: Theoretical and methodological aspects of the use of the concept of "competitiveness" in scientific research. Mod. Compet. **6**, 65–79 (2013)

Mau, V.A.: Anti-crisis measures or structural reforms: Russia's economic policy in 2015. Econ. Issues **2**, 5–33 (2016)

Razumovskaya, E.A., Voronov, D.S., Pridvizhkin, S.V.: The use of the dynamic method to assessing the competitiveness of metallurgical companies. Econ. Entrep. **8–1**(85–1), 521–525 (2017)

Razumovskaya, E.A., Voronov, D.S., Pridvizhkin, S.V.: Comparative assessment of the competitiveness of companies in the private and public sectors of the Russian economy based on operational efficiency. Russian Entrep. **19**(6), 1847–1866 (2018). https://doi.org/10.18334/rp.19.6.39156

Stiglitz, J.: Public Sector Economics. Infra-M, Moscow (2008)

Andrews, K.R.: The Concept of Corporate Strategy. Dow Jones-Irwin, New York (1971)

Ansoff, H.I.: Checklist for competitive and competence profiles. In: Ansoff, H.I. (ed.) Corporate Strategies, pp. 89–99. McGraw-Hill, New York (1965)

Hamel, G., Prahalad, C.K.: Competing for the Future. HBS Press, Boston (1996)

Porter, M.E., Heppelmann, J.E.: Why every organization needs an augmented reality strategy. Harv. Bus. Rev. **95**(6), 46–57 (2017)

Application of the Blue Ocean Strategy in the Assessment of the Feasibility of Space Projects

Mikhail N. Plakhotnik(✉), Andrey A. Zybrichev, and Valeriy M. Smirnov

Peoples' Friendship, University of Russia, (RUDN University), Moscow, Russia
plami@bk.ru

Abstract. The article uses the blue ocean strategy as a methodology for assessing the feasibility of a space project. We consider the approach and construction of a methodology for the analysis and assessment of the feasibility of space projects, relying on the methods of economic and mathematical modeling. We chose the Kohonen self-organizing card method as the leading one. An analysis of the implemented space projects was carried out using this method. The paper concludes with final remarks about the analysis and its possible further application.

Keywords: Blue ocean · Space project · Methodology · Kohonen neural networks · Space project feasibility assessment · Space project efficiency

1 Introduction

In scientific literature, the term "blue ocean" is new. Based on the world-famous work "The Blue Ocean Strategy" by V. Chan Kim and Rene Moborn (Chan Kim and Moborn 2018), one can note that the blue ocean does not exist but is being created. Creating something new has a high risk of failure. However, if successful, a breakthrough in the research area and obtaining superprofits will be provided. The blue ocean can arise in all areas of human activity, in our case, in the space industry. The blue ocean strategy carries the essence of reducing search risk (rethinking boundaries), planning risk (focusing on the situation as a whole, and not on numbers) and scale risk (where the following reverse system works supply-demand). Space activity is associated with high risk, as well as obtaining superprofits from the creation of a fundamentally new device (complex, etc.). Additionally, a high level of competitiveness in the manufacturing country with other countries is remarkable.

In our opinion, one of the possibilities of applying the blue ocean strategy is the development of a methodology for assessing the implementation of space projects. This technique will minimize the risks of project failure, maximize profits, and create unique equipment (apparatus, etc.).

A. V. Bogoviz and Y. V. Ragulina (Eds.): ISCI 2019, LNNS 115, pp. 59–66, 2020.
https://doi.org/10.1007/978-3-030-40749-0_7

The feasibility of a space project is evaluated in the following cases:

- When selecting space projects that have not yet begun;
- When a space project is already underway, and it is necessary to optimize management;
- When a space project has already been implemented, to analyze the results obtained.

The tasks of evaluating the feasibility of space projects provide for the comparison of many different, conflicting properties, which gives reason to classify these tasks as multicriteria. It is known that the solution of multicriteria problems is problematic due to the almost complete absence of formalized methods.

The article discusses the possibility of using the Kohonen self-organizing maps (Kohonen 2008) (SOM) to classify space projects that are effective in implementation. Also, we present the methodology of using such a network to organize samples at different stages of the project life cycle.

2 Materials and Method

2.1 General Characteristics of the Methodology

The task of assessing the effectiveness of scientific space projects is a significant problem, from the successful solution of which the reliability of the conclusions on the scientific significance of the results, on the social and economic efficiency of the planned and completed works depends. The following tasks are solved when one uses this technique:

- Forming a system of criteria and indicators of the effectiveness of space activities (concerning the evaluation of scientific space projects);
- Formalized analytical and qualitative assessment of scientific space projects based on a set of quality criteria;
- Visualized representation of the project assessment (building a project profile).

Particular forms of presenting complex information by implementing organizations were developed to analyze the effectiveness of scientific space projects. Besides, at the request of experts, terms of reference, estimates, schedules, reports, and other necessary materials for the project being evaluated are made available to them.

A standardized questionnaire was developed to conduct the examination of projects in the field of space research and receive expert responses. Forms and surveys are the framework documents; their indicators can be modified, deleted, or added in each case. The capabilities of the developed methodology allow a comprehensive assessment of the effectiveness of projects with several quality indicators from a few to several tens. The technique is not applicable in those cases when qualitative indicators are not fundamentally reduced to quantitative (numerical) values and cannot be measured on scales.

The technique cannot be used without additional modification if there are criteria that can take only discrete (integer) values.

2.2 Summary of the Methodology

The tasks are solved by a multicriteria method with obtaining initial data by expert estimates. Questionnaires on rating scales form expert information for the formation of a system of criteria and for the assessment of particular standards for the effectiveness of space projects. The data array is used as source material for calculating analytical quality assessments, as well as for constructing a graphic image of the project's effectiveness ("project profile") and obtaining qualitative characteristics.

Multicriteria assessment is carried out both according to the classical nonlinear compromise scheme (Piganov and Tyulevin 2009) and by the method of embedded scalar convolutions of the vector criterion (Piganovet al. 2015). In the latter case, projects are analyzed and evaluated according to aggregated standards, which makes the analysis convex and contrasting.

2.3 Software Implementation of the Methodology

As a software tool for building the Kohonen's self-organizing maps, we chose the analytical platform "Deductor Academic 5.1" (Deductor n.d.).

The program consists of the following blocks:

1. Project Information;
2. Organization of the examination procedure;
3. Processing of examination results;
4. Conclusion and visualization of results.

The program is controlled using the keyboard or mouse and consists of entering data and commands for performing calculation procedures in the corresponding information fields.

The "Information about the project" block contains the following:

- Fields for entering the project name and the implementing organization, data on timing, the amount of funding;
- A standard form for the presentation of information completed by the project executors;
- A technical task, a report, another project information;

The block "Organization of the examination procedure" includes the following:

- A list of expert experts and information about them;
- Examination form;
- Characteristics of the criteria system.

The block "Processing the results of examination" accumulates the answers of experts, generates an array of data for subsequent processing, and also calculates various categories of assessments of project efficiency based on formulas.

The block "Conclusion and visualization of results" provides the results of the calculation of various categories of assessments of project effectiveness, and it also builds a "project profile."

2.4 Kohonen Networks

The Kohonen networks are classified as self-organizing neural networks; they allow us to identify groups of input data vectors that are characterized by similar properties (Kohonen 2008). The Kohonen network is a single-layer network built from WTA neurons (Khaikin 2006) (Winner Takes All). The structure of such a network is shown in Fig. 1.

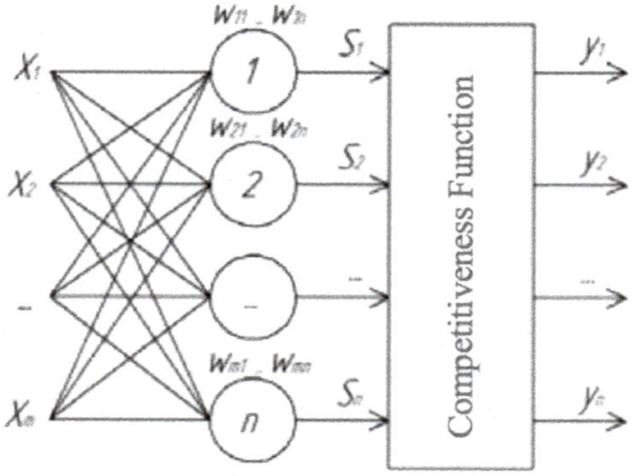

Fig. 1. A Kohonen network structure.

The network structure is as follows. Each neuron is associated with each component of the input data x1, x2,…, xm (input vector), which is a set of informative parameters for each instance of the sample in this case. Each neuron is a linear weighted adder:

$$S_j = b_j + \sum_{i=1}^{m} w_{ij}x_i,$$

where S_j – is the result at the output of the adder; j – is the number of the neuron; i – is the number of the informative parameter; b_j – is the threshold; w_{ij} – is the weight of the i - th parameter of the j - th neuron. From the output of each neuron, the result goes to the competition function, which determines the neuron with a maximum value of the result at the output. It assigns the value of 1. The output signals are set to zero.

For the winning neuron, the following condition is satisfied:

$$d(x, w_j) = \min_{1 \leq i \leq n} d(x, w_i),$$

where $d(x, w)$ – is the distance between the input data vectors and the synaptic weight vector, j – is the winner neuron number. A neuron whose synaptic weight vector w_i is the least different from the input vector becomes a winning neuron. In the case of determining the maximum result at the outputs of several neurons, a result that is equal to one is assigned to one of them, and zero is assigned to the outputs of the remaining neurons. The winner neuron determines a group to which the input data vector is closest.

Weights are adjusted every training cycle:

$$w_j^{(k+1)} = w_j^{(k)} + \eta_j^{(k)} \left[X - w_j^{(k)} \right],$$

where k – is the number of the learning cycle; X – is an input vector; $\eta_{\rm j}^{(k)}$ – is the coefficient of the learning speed of the j-th neuron in the k-th cycle (learning rate). Thus, a neuron whose synaptic balance vector was closer to the input vector is corrected and becomes even closer. The weights of the closest neurons to the winning neuron are also adjusted. Thus, the probability of this neuron becoming a winner neuron when applying a closer data vector to the input increases.

Training of the Kohonen network consists of selecting such values of weights at which errors are minimized if the input data is replaced in a sense with closer data to these weights (Khaikin 2006).

2.5 Assessment Methodology

The assessment technique based on the Kohonen self-organizing maps is as follows:

- Determining the vector of input data from the results of a training experiment;
- Defining network building parameters;
- Analysis of the results with the determination of the classification accuracy.

The "Canopus-V" project is selected as the subject of research. The leading indicators affecting the assessment of the implementation of this space project are the following: the amount of financing, the success of launches of the created equipment, deadlines, as well as presence/absence of errors in the implementation of the project. The sample size is 20. The project financing is used as informative parameters.

Organizing a network requires determining the following construction parameters:

- The size of the training and test set;
- Network size (number of cells);
- Example recognition criterion (recognition error value);
- Speed and radius of training;
- Number of clusters.

Since the sample size is small, the training set should be as large and sufficient as possible for the correct organization of the network. At the same time, the test set should be sufficient to test this network with 100% traffic. Based on the conditions set, the size of the training set was 15 copies (90% of the sample), the size of the test set was five copies (10% of the sample).

The network size is chosen in such a way as to exclude the appearance of "dead" neurons, i.e., of such neurons whose synaptic weight vector is significantly removed from the input data vector. Such neurons turn out to be inactive when training the network because they cannot win the competition with the nearest neurons. This situation leads to the fact that the input data are interpreted by a smaller number of neurons than was originally set, which introduces distortions in the final result. To study the presented samples, networks of 10 × 10 cells were selected. The cell view is selected hexagonal since it more correctly displays the Cartesian distance between the centers of the cells. The remaining parameters of the construction of the Kohonen network are presented in Table 1 (Kohonen 2008).

Table 1. Parameters for building a Kohonen network.

Network parameter	Specified criterion
Recognition error rate	An example is considered recognized if the recognition error is less than 0.05
Number of eras	200
Learning speed *n*	At the beginning of training was 0.3 At the end of the training was 0.005
Training radius	At the beginning of training was 2 At the end of training was 0.2
Clustering	The fixed number of clusters is 2

From the whole set of network options constructed, the only option was chosen in accordance with the given parameters for each sample. The minimum average recognition error for the training set was achieved with 100% recognition of both the training and test sets. Since a fixed number of clusters (equal to two) is given, it is not difficult to determine to which class of products the constructed network relates each instance from the classification table.

3 Results

3.1 Results of Network Construction

Figures 2, 3, 4 and 5 show that the map is divided into regions. In this case, the red area represents the maximum efficiency of the project, and the dark blue area represents the minimum efficiency. As can be seen in Fig. 2, from the moment of creation, the Canopus-V space project was in the zone of maximum efficiency (the marker is located in the red area). Figures 3 and 4 show that the marker first moved to light green, and then to the green area, which indicates a decrease in project efficiency. And in Fig. 5, we show that the marker is in the blue area. This means that without critical interventions by the Roscosmos State Corporation in the implementation process, the project's effectiveness would soon be 0, leading to its premature completion.

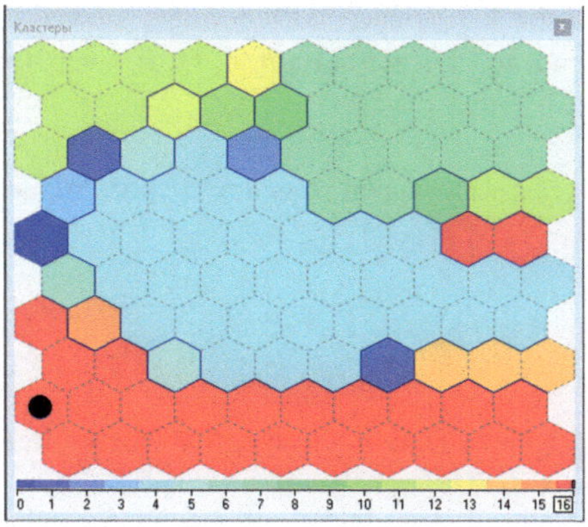

Fig. 2. Maximum efficiency.

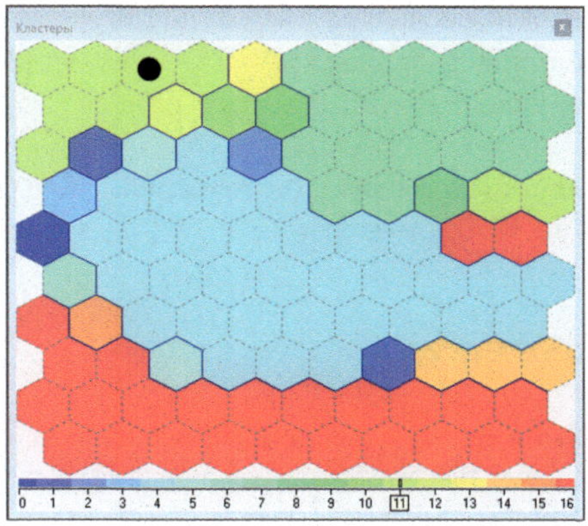

Fig. 3. Reduced efficiency.

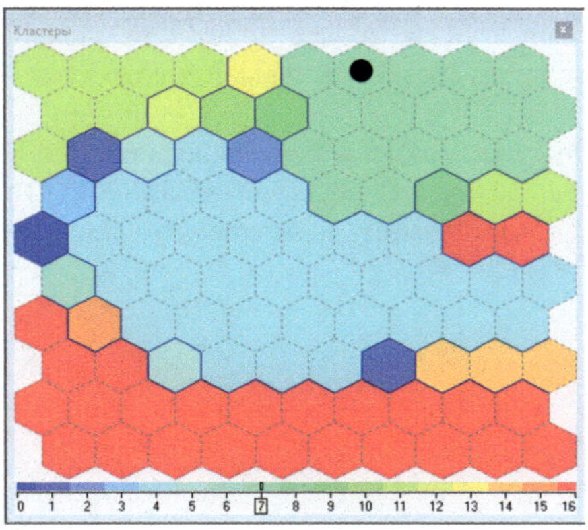

Fig. 4. Restricted efficiency.

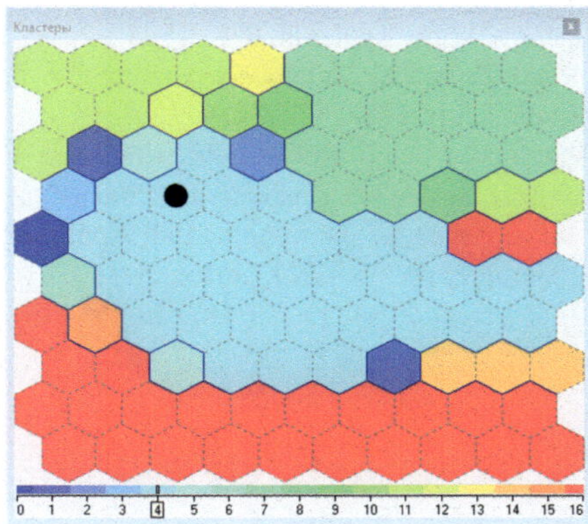

Fig. 5. Minimum efficiency.

4 Conclusion

The results are explained by the fact that a small sample was used as input data, leading to small training sets. The network size was also limited since the network size increases, and the "dead" neurons appear (which should be avoided).

The advantages of the developed methodology include the fact that the trained network is able to recognize any stage of the space project implementation that is not included in the sample, which cluster analysis algorithms could not do by definition. Thus, the developed methodology can be recommended both as a method for verifying the implementation of a space project, and as a methodology for assessing the implementation of a space project.

In sum, it is necessary to note that the creation of a methodology for assessing the feasibility of a space project will provide a breakthrough in the field of creation and implementation of space projects. Using this technique, minimization of organizational risk will occur, i.e., overcoming organizational problems and shortcomings, as well as minimizing managerial risk. This would lead to the maximum control over the implementation of the space project.

Acknowledgements. The research has been conducted with support from the "5-100" Program at the RUDN University.

References

Chan Kim, W., Moborn, R.: Blue Ocean Strategy. How to Find or Create a Market Free from Other Players. Mann, Ivanov and Ferber, Moscow (2018)

Deductor: Advanced analytics without programming (n.d.). https://basegroup.ru/deductor/description

Khaikin, S.: Neural Networks: A Full Course. Williams, Moscow (2006)

Kohonen, T.: Self-Organizing Maps. BINOM. Knowledge lab, Moscow, (2008)

Piganov, M.N., Tyulevin, S.V.: Prediction of the reliability of electronic equipment. In: Scientific and Technical Statements of SPbSPU. Series: "Informatics Telecommunications Control", vol. 1, pp. 175–182 (2009)

Piganov, M.N., Tyulevin, S.V., Erantseva E.S.: Individual prognosis of quality indicators of space equipment elements. In: Proceeding XIII International Conference CADSM 2015: The Experience of Designing and Application of CAD Systems in Microelectronics, Lviv, Ukraine, pp. 367–371 (2015)

Digitalization as a Direction of Neo-Industrial Transformation of Strategic Urban Development in the Ural Macroregion

Ivan A. Antipin(✉) and Olga Yu. Ivanova

Ural State University of Economics, Yekaterinburg, Russia
aia87@mail.ru, trtskmiusl7@mail.ru

Abstract. The purpose of this article is to study digitalization as a direction of neo-industrial transformation of the strategic development priorities of the cities of the Urals macroregion, as well as to identify barriers to managing digitalization processes. The paper methodologically is based on the theoretical principles of innovative economics, strategic management, regional and spatial economics. Assessing the processes of strategic guidelines transformation for urban development, occurring due to the development of digital technologies, was carried out using a combination of the following methods: logical, dialectical, causal, content analysis, calculation of statistical indicators, etc. The scientific novelty of the presented research s that the authors identify the barriers to the management of digitalization processes. More than that, they analyze and discuss the prospects for developing strategic planning in the formation of priorities for scientific, technical, and innovative development of cities.

Keywords: Strategy · Strategic planning · Strategic management · Largest cities · Digitalization

1 Introduction

The strategic goal of Russia approved in 2008 is "to achieve a level of economic and social development that corresponds to the status of Russia as a leading world power of the 21st century, which occupies leading positions in global economic competition and reliably ensures national security and the realization of the constitutional rights of citizens" (Government of the Russian Federation 2008). Achieving this goal is impossible without conducting systematic work on the development of the national economy, the introduction of information, telecommunication, network, and other competitive digital technologies. As President of the Russian Federation V. V. Putin notes in his message to the Federal Assembly, "It is necessary to focus on areas where the powerful technological potential of the future is accumulating, a large-scale systemic program for developing the economy of a new technological generation, the so-called digital economy, should be launched" (Presidential Executive Office 2016). This statement preceded the development of the national program "Digital Economy of the Russian Federation" (Government of the Russian Federation 2017) and a series of priority federal projects, in each of which there are digitalization aspects.

A. V. Bogoviz and Y. V. Ragulina (Eds.): ISCI 2019, LNNS 115, pp. 67–75, 2020.
https://doi.org/10.1007/978-3-030-40749-0_8

The Government of Russia identifies the crucial task in the long-term socio-economic development of the Ural Federal District, which is the formation of a modern information and telecommunication infrastructure, ensuring a high level of accessibility and providing quality services on its basis (Government of the Russian Federation 2011). The development of digitalization in the Urals is supposed to be in line with the following areas:

- Formation of a single information space;
- Transition to digital broadcasting in most of the county;
- Eliminating "digital inequality" between different regions;
- Implementation and widespread use in the district of navigation and information systems based on the GLONASS global navigation satellite system;
- Creation of a universal logistics and information operator based on the national postal operator;
- 4th generation technology development, etc.

The increasing importance of digitalization processes and the growing attention to their development in strategic planning documents are the result of the completion of the fourth technological structure and the deployment of the fifth in the economic space of the Urals. At the end of the 20th and beginning of the 21st centuries, on the basis of innovative development and a change in the role of technology, the trends of active modernization of traditional industries taking place in the context of the formation of new sectors of the economy (called "a new industrialization") began to attract more and more attention of researchers. These trends are widely covered in scientific publications, including publications by Ural scientists (Silin et al. 2017; Tatarkin and Romanova 2014).

The growth of high-tech science-intensive industries in the economic space of the Urals, including the creation of high-tech electric trains, unmanned aerial vehicles, the invention of polymer materials with unique properties, new electronic and optical equipment, is not possible without the involvement of digital and information technologies. They are deeply integrated into the economy and social life, determining the emergence of the digital economy.

Given the fact that the Urals is a traditionally industrial region, the potential points of technological growth of the economy are identified in the strategy for the long-term socio-economic development of the Ural Federal District. Namely, the strategy addresses metallurgical, machine-building, chemical-pharmaceutical, and timber industry complexes. They have signs of competitive industries that meet the criteria of technological prospects, including in new markets (Romanova 2018). In particular, thanks to the Ural enterprises, the concept of "white metallurgy" was introduced into everyday life. The concept implies using the ultra-modern metallurgical production based on a combination of high (including digital) technologies.

The large and most significant cities of the macroregion, which are the drivers of industrial and scientific-technical development, are forced to impose even more stringent requirements on the formation of strategic documents for socio-economic development and the elaboration of strategic management mechanisms. Forming the future image of modern cities, both developers and implementers of strategic documents have to work with large amounts of information, as a rule. In this regard, in the development strategies of the largest cities, such terms as "digital economy," "digital and cognitive technologies," "big data," "databases," etc. must be present. Issues related to forecasting, strategic planning, and strategic management of digitalization processes are becoming increasingly important (Antipin 2018; Silin et al. 2018; Surnina and Shishkina 2013).

In light of the above aspects, the purpose of this article is to combine the following three aspects:

- Digitalization research as a direction for the transformation of strategic priorities of the largest cities of the Ural macroregion;
- Definitions of digitalization development process and control barriers;
- Justification of the prospects for developing strategic planning in the field of prioritization of scientific, technical, and innovative development of cities.

In this study, the analysis of strategies for socio-economic development of cities is carried out according to the stages of the cycle of strategic planning and strategic management, namely: environmental analysis, goal-setting, strategy construction (structure), mechanisms for implementing the strategy, a system for monitoring and managing the implementation of the strategy.

2 Materials and Method

Systemic transformations in the strategic development of the country are impossible without the implementation of similar, coordinated processes in the regions and cities located on its territory. In June 2014, in the Russian Federation, Federal Law No. 172-FZ *On Strategic Planning in the Russian Federation* (State Duma of the Russian Federation 2014) was approved. It does not contain a general definition of the term "strategy," although interpretations of more specific terms are presented, including "strategic planning document," "Strategy for the Socio-Economic Development of the Russian Federation," "Strategy for the Spatial Development of the Russian Federation," "Strategy for the Socio-Economic Development of the Subject of the Russian Federation," "Strategy for the Socio-Economic Development of the Municipality." The lack of legislatively approved language is compensated by a wide range of definitions of the term "strategy" that appear in scientific sources (Animitsa and Bochko 2003; Surnina 2004; Fatkhutdinov 2005).

The need for a clearer understanding of the new requirements for strategic documents in order to increase their relevance to current realities is argued by a combination of the following interrelated circumstances:

First, the development of the modern world is based on progress, including in the scientific, scientific, technological, and innovative fields. This process is characterized by significant dynamics, the speed of changes that are generated by the increasing globalization of economic, political, and socio-cultural life. Achievements and potential of this progress should correlate with the tasks stated in strategic documents;

Second, in a rapidly and continuously changing environment, the most significant cities have to clearly determine the priorities of their strategic development and the ways to achieve them. Not just the development of development strategies, including the development of the digital economy, but the formation of the image of the future is of particular importance (Romanova 2018). The importance of the problem of shaping the image of the future is noted in the National Technological Initiative, which emphasizes the need to plan work from the future (the so-called "preferred reality") to the present.

The experience of strategic planning of the largest cities in the world (such as Amsterdam, Barcelona, London, New York, Singapore, Stockholm, Rio de Janeiro, Chicago, and others) indicates that more and more attention is paid to the role of information and communication technologies in improving the functionality of urban systems, the use of digital technologies, including telemedicine, distance learning, online commerce, 3D modeling, etc. The most important global trend accompanying the process of new industrialization is the emergence of industry 4.0 and the digital economy.

Industry 4.0 involves fundamental changes in the automation and digitalization of technology, and the transition from the Internet of People to the Internet of Things (Keshelava 2017). The Internet of Things provides the ability to monitor, control, and manage devices remotely, and also allows one to generate new ideas, collect and process information from big data streams in real-time (Atzori et al. 2010).

The concept of the "digital economy" is relatively new. The consequence of this is the widespread use of the terms "digital economy" and "digitalization" without a clear distinction and uniform interpretation. Nevertheless, in the scientific literature, there are many approaches to the disclosure of the term "digital economy." The authors are close to the position expressed by Dr. R. V. Meshcheryakov, who believes that there are two approaches to the term "digital economy." The first approach is "classic." According to this approach, the digital economy is an economy based on digital technologies; moreover, it is correct to characterize so exclusively the field of electronic goods and services (telemedicine, distance learning, sale of medical content, etc.). The second approach is an expanded one, according to which the "digital economy" is an economic production using digital technologies (Urmantseva 2017).

Thus, we consider digitalization as one of the critical areas of the neo-industrial transformation of the strategic development of cities.

3 Results

An analysis of the socio-economic development strategies of the largest cities of the Ural macro-region, which was carried out under the prism of the goal set in this study, allowed us to obtain the following results.

First of all, the strategic priorities of the largest cities of the Ural macroregion include the development of an innovation-oriented economy, the creation of favorable conditions for the development and application of high (including digital) technologies.

So, for example, one of the seven strategic directions for the development of Yekaterinburg is "Yekaterinburg is an inter-regional innovation-oriented industrial and financial center". The tasks of the above strategic direction include developing the digital economy with a focus on improving the efficiency of interaction between business, government, and society.

The Chelyabinsk Strategy is based on an innovative scenario. The main goal is to create the status of a modern metropolis, with the appropriate appearance, infrastructure development, level of services, quality of life. In the field of technology development, the following three-level task is set: “to switch from technology imports to technology development and to the independent creation of new technologies.”

The strategic development goal of Perm is to improve the quality of life of the population based on the innovative development of the city’s economy. Its implementation is expected due to the development of a diversified economy based on dynamically developing and innovative sectors, the creation of conditions for the development of a knowledge economy, and the transition of the urban economy to the sixth technological order.

The strategic goal of the development of Orenburg is to create conditions for improving the quality of life of the population through the development of the city’s economy, combining the modernization of traditional industries and the emergence of new leaders in innovative economic development. According to the developers of the strategy, this is possible under the condition of restructuring the economy.

The development of an innovative economy is not considered as a general goal of the city of Kurgan. However, the objectives for the sub-goal “Development of Economic Potential” include creating the conditions for increasing the innovative development of the economy of the city of Kurgan and forming the innovative and technological center of the Kurgan region.

The priority of Ufa is to become a high-tech center. As indicated in the strategy of the city, a “strong economy” is achieved on the basis of “promoting sustainable development of industry and innovation”.

The triune goal of the city of Izhevsk also includes transition to an innovative economy, which requires work in the following areas: entrepreneurial (production), investment activity, small and medium enterprises, labor market regulation, development of the economic basis of the city.

Second, despite the strategic priorities identified, based on the content analysis (its brief results are given in Table 1), we can conclude that most of the strategies of the cities studied practically do not contain concepts related to the concept of the digital economy and describe the mechanisms of its formation and development.

Table 1. The use of “new” terms in the development strategies of cities of the Ural macro-region, the number of times.

Terms	Yekaterinburg	Chelyabinsk	Perm	Orenburg	Kurgan	Ufa	Izhevsk
Information (processes, technologies, systems, etc.)	36	18	30	1	4	2	1
Digital (economics, technology)	6	1	–	–	–	8	–
Big data	–	–	–	–	–	–	–
Smart city	2	–	–	–	–	2	–
Internet of things	1	–	–	–	–	1	–
Industry 4.0	–	–	–	–	–	–	–
New industrialization	–	–	–	–	–	–	–

Source: Compiled by the authors on the approved strategic planning documents.

Third, the specially developed programs (projects) for the development of the digital economy, in general, and individual digital technologies, in particular, are reflected only in the strategies of two of the seven cities studied.

In particular, the "Digital Yekaterinburg" strategic program is highlighted in the development strategy of Yekaterinburg. In the development strategy of Ufa, the project "Digital Platform for Supporting the Production and Marketing Activities of SMSP " is provided as part of the implementation of the priority direction "Creating Favorable Conditions for the Development of Small and Medium-Sized Enterprises in Priority areas of the city's economy" and in the framework of the priority direction "Effective and Open Power."

Fourth, In the development strategies of the studied cities, there is no block for the strategic management of digitalization development processes and an analysis of the directions of using their potential in the urban economy.

The development of this analysis faces several barriers that can be divided into two groups ("information" and "methodology").

Information barriers include a number of critical aspects. There is no comprehensive approach to monitoring digital transformation, including the measurement of the main directions and factors for the effective use of information and communication technologies (ICT). It should be emphasized that in the statistical collections "Regions of Russia: Main Socio-Economic Indicators of Cities," information on the development of ICT and digital technologies in the largest cities of the country is not presented. In the statistical collections "Informations Society in the Russian Federation," there is no urban slice (only regional). The latest statistical compilation, "Information Society: Development Trends in the Subjects of the Russian Federation," was published in 2015, and there was also no urban analysis slice in it. More than that, the current forms of general statistical observation do not reflect modern processes of digital transformation; there is no correlation with international standards. This problem does not allow taking into account the specifics of the digital transformation of state and municipal government, health care, culture, education, science, etc., and making correct international comparisons (Shaposhnik 2018).

Methodological barriers include the following: lack of methodological tools for assessing the readiness of the urban environment for digitalization and adaptability of digital technologies in the urban economic space, assessing the socio-economic effects of the use of digital technologies. The areas of development of the digital transformation assessment methodology should include such areas as ICT infrastructure, human capital, business environment, government policy and regulation, information security, and trust.

4 Discussion

Digitalization affects all key sectors and spheres of modern life, which are concentrated in cities while being integral parts of ensuring their livelihoods. However, various cities of our country are at different stages of the spread and adaptation of digital technologies. There are two levels of digitalization: the primary level implies the availability, quality, and availability of digital infrastructure; and the secondary level involves the intensity and skills of using the existing infrastructure, the nature of the services used, i.e., the presence of “digital competencies” (Higher School of Economics 2018).

Most Russian cities are at the first level of digitalization, their residents make purchases online, receive electronic government services, communicate on social networks, etc. And only a few of the largest cities in Russia move to the second level of digitalization. The second level involves the large-scale introduction of artificial intelligence technologies, cybersecurity technologies, the “Internet of things” (industrial Internet), large databases, and solutions developed for “smart” cities. So, the Research Institute of Technology and Communications (Moscow) identified the Top 10 most “smart” cities in Russia based on an assessment of a set of reasonable indicators in 2017. It included (as the integral value decreases) such cities as Moscow, St. Petersburg, Kazan, Yekaterinburg, Krasnoyarsk, Novosibirsk, Ufa, Sochi, Perm, Rostov-on-Don (National Research Institute of Technology and Communications 2017). This list includes only three cities of the Ural macroregion, in the strategies of which digitalization is named among the priorities of strategic development.

There are studies demonstrating that the development strategies of an increasing number of foreign cities (Amsterdam, Barcelona, London, New York, Singapore, Stockholm, Rio de Janeiro, Chicago, and others) emphasize the role of information and communication technologies in improving the functionality of urban systems (Angelidou 2017).

Among all the materials considered in the study, the socio-economic development strategy of Yekaterinburg has the most strong structure of strategic directions. The target vector, main objectives, strategic vision, as well as indicators for monitoring and evaluating effectiveness, without which it is impossible to achieve city development priorities, are presented in detail within each of the strategic directions. The city’s strategy presents a mature program for the development of digitalization and the use of its potential in the urban economy – “Digital Yekaterinburg.” Within the framework of this strategic program, an orderly transition to the development of intelligent urban infrastructure has been worked out, and ensuring information security in the context of digitalization growth is provided. The mechanisms for implementing the strategic program are the projects “Modern Information Environment” and “IT Cluster.” A system of performance indicators for the strategic direction has been developed; their values are predicted.

In our opinion, just such an approach to strategic planning, tested in Yekaterinburg, should be taken as a basis in the formation of priorities for scientific, technical, and innovative development of other cities in the macroregion.

5 Conclusion

Based on the analysis, we can conclude that, on the one hand, the need to use digital technologies, and in the long run the transition to a digital economy, is becoming a conscious area of socio-economic development of the Ural cities, which is reflected in the processes of transformation of strategic planning and management. Nevertheless, on the other hand, the theoretical, methodological, and informational basis for this is poorly developed; therefore, the development strategies of the largest cities do not fully contain the tools for solving the tasks set in them and should be updated to more closely correspond to modern realities and global challenges. Best practices, including the experience of Yekaterinburg, should become a guideline for further work in the field of updating the programs of socio-economic development of cities. Over time, the quintessence of these practices can be the basis for the formation of unified mechanisms for strategic planning and strategic management aimed at developing digitalization and using its benefits in the city economy.

Acknowledgment. This article was prepared with the financial support of the Russian Foundation for Basic Research as part of the scientific project No. 18-010-00833 "Neo-Industrialization in the Space of the Macroregion in the Context of a Cyclical Wave Methodology (On the Example of the Urals)."

References

Angelidou, M.: The role of smart city characteristics in the plans of fifteen cities. J. Urban Technol. **24**(4), 3–28 (2017)

Animitsa, E.G., Bochko, V.S.: The Development Strategy of the Largest City: A Look into the Future (Scientific and Methodological Approach). Uralskiy Rabochiy Publishing House, Yekaterinburg (2003)

Antipin, I.A.: Digitalization as a tool for transforming the strategic development of the territory. Theor. Pract. Soc. Dev. **12**, 90–94 (2018)

Atzori, L., Iera, A., Morabito, G.: The internet of things: a survey. Comput. Netw. **54**(15), 2787–2805 (2010)

Fatkhutdinov, R.A.: Strategic Management. Delo Publishing House, Moscow (2005)

Government of the Russian Federation. Order "On approval of the Concept of long-term socio-economic development of the Russian Federation for the period until 2020" (November 17, 2008 No. 1662-r) (2008). https://www.audar-info.ru/na/editSection/index/type_id/3/doc_id/5823/release_id/49752/

Government of the Russian Federation. Order "On approval of the Strategy for socio-economic development of the Ural Federal District until 2020" (October 6, 2011 N 1757-r) (2011). http://www.consultant.ru/document/cons_doc_LAW_120672/

Government of the Russian Federation. Order "On approval of the program "Digital Economy of the Russian Federation"" (July 28, 2017 No. 1632-r) (2017). http://www.consultant.ru/document/cons_doc_LAW_221756/

Higher School of Economics. Digitalization in small and medium-sized cities of Russia (2018). https://urban.hse.ru/data/2018/06/06/1149766040/2018-06-GSU-HSE_pres_v6.pdf

Keshelava, A.V.: Introduction to the "Digital" Economy. VNIIGeosystem Publishing House, Moscow (2017)

National Research Institute of Technology and Communications. Indicators of smart cities NIITS 2017 (2017). http://niitc.ru/projects/indikatory-umnykh-gorodov-niits-2017

Presidential Executive Office. Message from the President of the Russian Federation to the Federal Assembly of the Russian Federation (December 1, 2016) (2016). http://www.consultant.ru/document/cons_doc_LAW_207978/

Romanova, O.A.: Strategies for the socio-economic development of the Russian regions in the context of the challenges of the digital economy. Izv. Ural State Econ. Univ. **19**(3), 55–68 (2018)

Shaposhnik, S.B.: Measuring the digital economy: international standards and Russian statistics. In: Scientific conference "Lomonosovskiye Chteniya – 2018", Moscow State University, Moscow, Russia (2018). https://digital.msu.ru/wp-content/uploads/2018-04-19-Шапошник.pdf

Silin, Y.P., Animitsa, E.G., Novikova, N.V.: Regional aspects of new industrialization. Reg. Econ. **13**(3), 684–696 (2017)

Silin, Y.P., Dvoryadkina, E.B., Antipin, I.A.: A study of the priorities of the strategic development of a new industrial city. Manager **9**(6), 2–16 (2018)

State Duma of the Russian Federation. Federal Law "On strategic planning in the Russian Federation" (June 28, 2014 No. 172-FZ) (2014). http://www.consultant.ru/document/cons_doc_LAW_164841/

Surnina, N.M.: Strategic analysis and planning methods at the national, regional and corporate levels (case study). Ural State Economic University Publishing House, Yekaterinburg (2004)

Surnina, N.M., Shishkina, E.A.: Development of a methodology for regional strategic planning: improving coherence and effectiveness. Manager **1**(41), 56–63 (2013)

Tatarkin, A.I., Romanova, O.A.: Theoretical foundations and stages of the new industrialization of the Russian economy. Grand Altai Res. Educ. **1**, 60–70 (2014)

Urmantseva, A.: Digital economy: how professionals understand this term (2017). https://ria.ru/20170616/1496663946.html

Main Directions in the Innovative Environment Formation at Knowledge-Intensive Enterprises Operating in the Field of Mechanical Engineering

Mikhail N. Petrov(✉)

FSUE TsNIIMash, Moscow, Russia
mnp0973@gmail.ru

Abstract. This article is devoted to the development of a project management paradigm based on synergetic methods. The system analysis of the project as an open, nonlinear dissipative structure is carried out. Based on the analysis, the paper outlines the main directions for developing an innovative environment at an enterprise of the knowledge-intensive segment of mechanical engineering necessary for the effective use of modern design practices. The paper concludes with highly appreciated insights and strong recommendations.

Keywords: Project management paradigm · Innovative environment · Synergetics · Self-developing fractal structures

1 Introduction

Increasing the efficiency of implementing high-tech projects in the knowledge-intensive segment of mechanical engineering is one of the essential conditions for the further dynamic development of the economy of the Russian Federation. However, the current state of many projects in the knowledge-intensive industries indicates a significant lag in terms, quality, and excess of cost parameters in relation to the planned indicators. This tendency is mainly due to the existing methodological basis of project activities, which was formed based on the general scientific paradigm of the first half of the twentieth century.

The formation of a new scientific paradigm in the second half of the 20th and the beginning of the 21st century determines the need to find new design practices that would form the basis of a promising project management paradigm and a new 21st-century scientific management paradigm. This paradigm implies a transition from determinism and reductionism to holistic principles, the widespread use of ontological models based on new interpretations of quantum mechanics, the application of the principles of synergetics and theory of phase transitions.

A. V. Bogoviz and Y. V. Ragulina (Eds.): ISCI 2019, LNNS 115, pp. 76–82, 2020.
https://doi.org/10.1007/978-3-030-40749-0_9

2 Materials and Method

New approaches to the formation of a promising project management paradigm require the comprehension of project activities based on modern scientific tools that have shown their effectiveness in related fields of knowledge.

First of all, we are talking about synergetics, which appeared based on the development of system analysis, modern mathematical apparatus (fractal geometry, the theory of determinate chaos and dissipative structures), natural-scientific knowledge. More than that, one also should note a superdisciplinary gradient that determines scientific knowledge in the field of nonlinearity, chaos, and complexity, forming the holistic paradigm of the modern picture of the world (Knyazeva and Kurdyumov 2014).

Exactly synergetics, according to academician V. S. Stepin, a leading scientist in the field of philosophy of science, is destined to become the basis of a promising scientific paradigm of the 21st century (Mainzer 2009).

For the analysis of the project as a complex socio-technical system, from the position of the system synergetic approach, it is advisable to use such fundamental categories of synergetics as bifurcation, fractality, temporality, emergence, and evolution. Besides, it is necessary to take into account the role of such a category as chaos in the self-organization of complex nonlinear systems and the mechanisms of formation of multiple spectra of discrete structures-attractors.

We will reveal these concepts in detail, applying them to the processes of management of high-tech programs and projects.

The project activity, from the position of synergetic approaches, is a set of stable spatio-temporal relationships created in the course of the project and forming new environments with non-linear properties.

In this case, the project can be considered as a set of all possible spatio-temporal relationships that affect the process of project implementation as a complex socio-technical system.

During the implementation of Lorentz's experiments, which later found development in the scientific schools of Kurdyumov and Samara, there was an understanding of the presence of the called structures-attractors. In accordance with this phenomenon, any project environment has a "filling" not implemented at the moment (potential) structures-attractors, forming due to their fields of attraction scenarios of the project – the only possible areas of implementation.

At the same time, as a result of choosing a set of possible evolutionary paths of the project environment, all other scenarios of implementation are closed.

In turn, the transformation of the internal properties of the non-linear project environment will reformat the entire probabilistic spectrum of the ways of its scenario implementation.

Thus, it is necessary to ensure high flexibility of organizational changes, to possess the causal mechanism of occurrences and development of attractors, without creating models unusual for them for design forms.

Considering strange attractors as the formed paths of scenario development of the project environment, it is possible to determine the areas and processes occurring in them, which subsequently would proceed in the same way in the upcoming phases of the project, or processes that have already taken place earlier and are taken as the basis for the current processes of the project implementation.

One can explain this notable phenomenon using the fact that the considered attractors, as potential ways of developing a non-linear project environment, are "described by invariant-group solutions in which space and time are not free but in a certain way connected" (Knyazeva and Kurdyumov 2014) (Fig. 1).

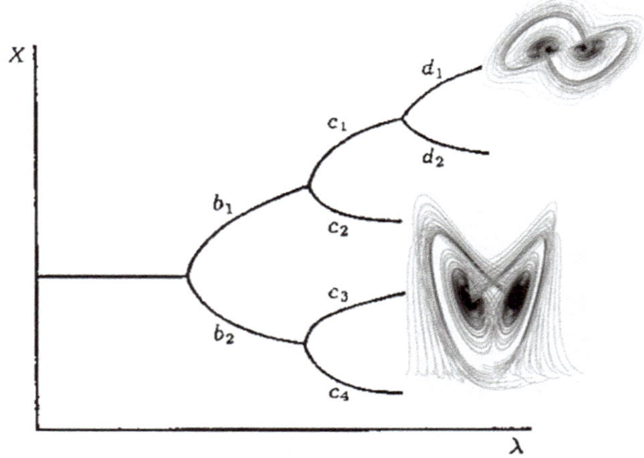

Fig. 1. The evolution of the phases of development of project activities.

To ensure successful working conditions with attractors, which at the current moment cannot be determined and localized with high accuracy, iterative sequences are necessary. Some corrections are made, along with a planned correction of the strategy for implementing project activities without excessive formalization of constraining organizational forms.

This approach determines the need for the formation of an innovative environment and the feasibility of applying modern design practices during the project implementation – flexible and extreme.

It is also essential to take into account such a crucial synergetic phenomenon as temporality, i.e., the compression (extension) of the processes of the project environment in time. In accordance with this property, the management of project activities is advisable to carry out as a result of topologically correctly organized fluctuation effects at the points of bifurcations of the project – the space-time invariants of the open, nonlinear medium. The project environment enters into a promising spectrum at such points, including attractors that are currently undeveloped as potential states of the future system.

Also, it is necessary to take into account such a primary category of a non-linear dissipative environment as emergence. In particular, the properties of the new system are higher than the sum of the properties of its constituent parts, which is also called the systemic or synergistic effect.

Thus, we can conclude that cooperation, in relation to the goals of the project, is beneficial in all its forms and leads to significant savings in resources (both material and intellectual) of participants in network communities.

Representing the evolution of the project environment as a sequence of bifurcations with access to one or many potential attractors, one of the most significant tasks of project management can be formulated: the formation of a minimally costly, from the point of view of resource support, "soft" control is based on point fluctuation disturbances, topologically organized in the zone of bifurcation points of the design environment.

For the practical implementation of this approach, it is necessary to estimate the coordinates of phase transitions (spatio-temporal invariants) and to model potential attractors that are optimal for project purposes, which in turn determines the relevance of search forecasts functionality for project activities.

Chaos is given high importance to the processes of self-organization of non-linear dissipative systems. It is a universal principle of creation, which contains in its potential the whole spectrum of possible development paths (attractors) of all aspects of society and nature.

Periodic immersion in chaos is necessary for the life of a complex system and is a way to update its organization (Knyazeva and Kurdyumov 2014).

Interpretation of the role of chaos as the primary tool for the evolution of an open non-linear dissipative environment of a project determines for those organizations belonging to the knowledge-intensive segment of mechanical engineering the task of choosing organizational management structures and models of organizational behavior with a high level of flexibility and adaptive abilities (Kochetkova 2012).

The basis for solving these organizational problems can be the theory of self-organizing systems or fractal management, which is focused on launching and further supporting self-developing trends in conditions of chaos and uncertainty.

Within the framework of the main provisions of this theory, "…in the conditions of the chaotically changing external environment, the most stability and viability will have the systems constructed on a model and similarity of a fractal" (an element that has the property of self-similarity) (Kochetkova 2012).

The functioning of various non-linear systems of human society is an object of study of fractal management, the successful evolution of which "… turns out to be directly dependent on the optimal interaction between network, hierarchical, and in some cases also market structures" (Oleskin 2017).

Network structures in conditions of uncertainty and chaos are characterized by high flexibility and stability, which is their fundamental difference from hierarchies (in the system plan, it is a degenerate network). At the same time, they are characterized by high organizational complexity, low controllability, high creation and maintenance costs, low regularity development.

The heterochrony of network structures (or asynchronization according to the rhythms of participants in network communities), together with its self-reference (lack of external control), is not only a powerful tool to increase the innovative potential of the network participants but also the basis for the formation of non-trivial ideas, "… ahead of their time and exceeding what was achieved at the bureaucratic science level" (Oleskin 2017).

Also, the non-linear nature of interconnection is a severe tool to increase the personal initiative of participants in network communities, forms the necessary creative environment, and ultimately contributes to the creation of innovative ideas (Chuchkevich 1999).

It should be noted that organizational management structures implemented in the form of hierarchies have many significant drawbacks that reduce the effectiveness of their use in conditions of high environmental uncertainty (chaos). Disadvantages include a high number of management levels, bureaucracy, dependence on the personal qualities of managers, low adaptive ability. However, there are many indisputable advantages, such as primarily high system capacity, a transparent communication system, and quick response. They determine the inappropriateness of their use in solving complex non-linear tasks in the implementation of project activities and simultaneously make it acceptable to use when implementing standard tasks with a low level of uncertainty (Chuchkevich 1999).

3 Results

The carried-out analysis demonstrates the need for parallel use in the knowledge-intensive segment of the economic complex of the two organizational subsystems discussed above, which need to interact in the format of the general management systems of organizations, namely (a) hierarchies (providing functions and a production unit) and (b) heterarchies (innovative activity and development).

This conclusion, in the context of the existing specifics of high-tech mechanical engineering, requires the organizational separation of the structural divisions of (a) design and development activities and (b) research and development work with the allocation of current activities that are unusual for these structures, support functions: material and technical support, economic and contractual.

In the context of this model, the project council of organizations will play the role of corporate structures-adapters. It consists of the head office of project management and project offices, accumulating communication flows of competence centers, organizational units of the supporting activity ("back office"), and structures that carry out research and development work.

Collegial councils (scientific and technical councils, as well as portfolio, investment, and tender committees, which are characterized by high flexibility and adaptability due to their specificity) are called upon to ensure the symbiosis of the hierarchy and hierarchy and the implementation of this organizational model as corporate "meta-managers."

The formation of corporate networks should be implemented in the form of project fractals, a set of separate group units groups of similar projects, for the effective functioning and the formation of their creative potential in terms of innovation generation (network emergence). More than that, its systematic complication and development of communication interaction of network participants are highly necessary (Fig. 2).

The formation of network idea generators ("factories of thoughts") should be implemented on the basis of further development of the population-communicative

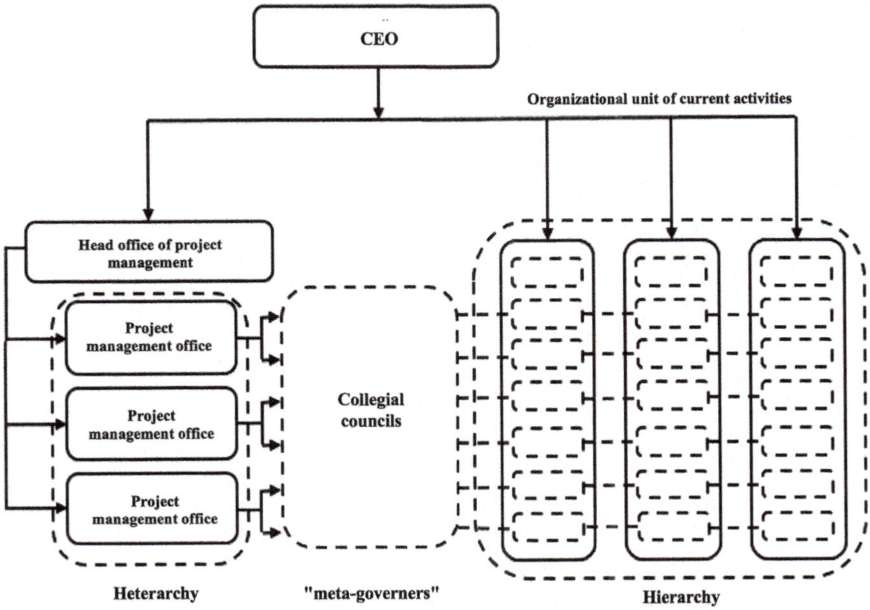

Fig. 2. Organizational approaches to the formation of an innovative environment.

network paradigm (outstripping development in the creation of promising concepts) in the form of network sectoral and intersectoral communities of scientists and practitioners built in the format of virtual communities based on a single information environment (as in Linkedin).

4 Discussion

Based on the research conducted, we can draw the following conclusions:

1. Representation of project activities in the form of a set of space-time invariants, forming a different hierarchy of new environments with non-linear properties conducted in this study, determines one of the priorities of project management, which is the formation of the most "soft" (least expensive in terms of project resources) management of the project environment as a result of topological fluctuations (impacts) in the areas of bifurcation points of options of project scenarios. This approach requires assessing the coordinates of the phase transition points (for the formation of the optimal goals of the project structures-attractors), which determines the importance of the forecast as one of the key functions of the project activities.
2. Emergence determines the high importance and effectiveness of external and internal cooperation for saving resources and realizing project goals.
3. As a tool for the evolution of the project environment, chaos is an effective means of deriving the spectrum of variations of the project development paths (at

bifurcation points, where the instability zone of the dissipative project environment is high) to one of the desirable structures for project activity-attractors. This fact, in turn, determines the need for the reengineering of business processes and organizational modeling for the formation of the organizational environment in the companies belonging to the knowledge-intensive segment of mechanical engineering to improve the efficiency of modern methods of project management.

4. One of the significant conclusions of this work is that it is necessary to use two organizational subsystems in knowledge-intensive organizations in the mechanical engineering segment in parallel, interacting in the format of a common economic system of corporate governance, namely hierarchies (functions and production) and hierarchies (managing the development of innovative products).

5 Conclusion

Synergetic methods, as the foundations of a promising 21st-century scientific paradigm, are one of the most promising tools for building a project management paradigm that meets the challenges of the present and solves many critical tasks related to improving the efficiency of high-tech knowledge-intensive mechanical engineering projects.

References

Chuchkevich, M.M.: Bases of Management of the Network Organizations. Institute of Sociology RAS, Moscow (1999)

Knyazeva, E.N., Kurdyumov, S.P.: Synergetics: Nonlinearity of Time and Landscapes of Co-evolution. KomKniga, Moscow (2014)

Kochetkova, A.I.: Fundamentals of Control in Chaos. Rid Group, Moscow (2012)

Mainzer, K.: Complex System Thinking: Matter, Mind, Humanity. New Synthesis. Book House "LIBROCOM", Moscow (2009)

Oleskin, A.V.: Network Structures in Biosystems and Human Society. Book House "LIBROCOM", Moscow (2017)

Digital Platforms in Hospitality Business

Marina Y. Sheresheva[1(✉)], Igor I. Savelyev[1], Nelli N. Nadzharyan[1], and Alexey V. Kudryavtsev[2]

[1] Lomonosov Moscow State University, Moscow, Russia
m.sheresheva@gmail.com, sii-33@mail.ru, nellinorikov-na@gmail.com
[2] Vladimir Branch of RANEPA, Vladimir, Russia
nb-33@inbox.ru

Abstract. The article presents the results of desk research and a case study on the importance of digitalization in hospitality. The authors highlight five stages of the hospitality industry evolution in terms of digital transformation, discuss the role of digital platforms in the hospitality business, and present the results of the hotel business operations analysis. The two automated systems of integration to optimize internal processes and improve the quality of data processing demonstrate that digital technologies in the hotel business are essential. They allow using traditional management schemes on a new basis, leading to a higher level of customer understanding and service customization. Russian hoteliers need to use digital platforms more intensively and more competently to maintain and develop their competitiveness.

Keywords: Hospitality business · Digitalization · Digital platforms · Competitiveness

1 Introduction

Nowadays, tourism and hospitality are not only a rapidly growing but also a highly profitable industry, an essential part of the global services market. According to the World Tourism and Travel Council (WTTC) statistics, the tourism industry contribution to the world GDP is 10.4%, with over 313 million jobs created worldwide. In particular, the hospitality industry is now ready to reach a 170 billion dollars record in the total booking (JLL 2018). Therefore, this sector determines the conditions for the sustainable development of many national economies.

The hospitality industry is highly dependent on providing a high level of service (Chathoth et al. 2016). At the same time, an individual approach is of particular value to contemporary customers (Radojevic et al. 2015). In this regard, the hospitality companies are paying increasing attention to customer relationships and customization of services (Prachayakupt et al. 2017), and to collaborative innovations (Marasco et al. 2018).

Customer relationship management requires specific competencies, as well as the ability to use customer information in order to increase competitiveness (Melián-González and Bulchand-Gidumal 2016; Sigala 2016, 2018). Hoteliers have quite easy

A. V. Bogoviz and Y. V. Ragulina (Eds.): ISCI 2019, LNNS 115, pp. 83–90, 2020.
https://doi.org/10.1007/978-3-030-40749-0_10

access to customer data. After check-in at the hotel, the customer deals with several points of contact, such as the reception, restaurant, or gift shop. All hotels maintain their individual and independent customer records.

In Russia, most hoteliers currently have a significant, sometimes excessive amount of information about their customers. Still, access to data at different points of contact with hotel guests is usually limited; and effective data processing is insufficient. Many industry actors lack the necessary competences to analyze data and make informed management decisions.

Many hotels develop their solutions based on unique metrics and methodologies to satisfy the technological requirements of serving modern guests. This has led to a diverse range of approaches that, in essence, aim to provide a similar set of services. Although these approaches may work well within the framework of a single hotel, there is often a lack of potential for the customer-oriented combination of services. This imposes restrictions on the usability of guest preference profiles on a broader scale, in particular, using modern digital platforms.

According to Accenture, platform business models represent a fast-increasing proportion of the growth of the digital economy, with the top 15 public platform companies already representing $ 2.6 trillion in market capitalization (Accenture 2018). However, as mentioned above, there is no standard data processing platform in the hospitality industry, which can lead to loss of potential income in a highly competitive environment.

The need for such a platform in the hospitality industry remains open. Many researchers and practitioners argue that the availability of alternative digital platforms does not impede and even contributes to the ability to provide personalized services to hotel guests (Brousseau and Penard 2007). At the same time, experts agree that hotel business needs to develop and use solutions that would ensure productive interaction with customers using advanced digital service platforms that are becoming widespread in all areas of society (Loonam et al. 2018).

When developing management decisions for emerging markets, including Russia, it is necessary to take into account the specifics of these markets, the key characteristics of which are significantly different compared to the "traditional" developed markets (Engardio 2006; Sheth 2011). Thus, there is a need to adapt traditional approaches and hotel management practices to Russian conditions.

2 Materials and Method

The hospitality industry has undergone significant changes due to the emergence and development of digital technologies. The main trends are as follows:

- Growing user independence and awareness, as well as and the possibility of online booking, lead to changes in customer behavior (Horner and Swarbrooke 2016);
- Digital "arms race" among hotels due to the rising importance of technological equipment to provide a positive guest experience (Šeric et al. 2016; Subramanian et al. 2016);

- Integrating the 'customer journey' (Yachin 2018) in and outside the hotel: the user's technology and platform (smartphone) are integrated with the services provided by the hotels (Wang et al. 2016).

One can divide the evolution of the hospitality industry in terms of digital transformation into five stages:

- *Stage 1* (1980s): Attempts to use the emerging new infrastructure to improve user experience. At this stage, many innovations are not yet widely spread.
- *Stage 2* (1990s): The extensive use of the Internet in the hospitality industry, the provision of online booking services. At this stage, the industry is more likely to master new opportunities; technologies serve more to improve the inside processes such as booking.
- *Stage 3* (2000–2010): Several significant changes occur at once. New services provide users with aggregated information and allow more confident choices based on data. The fast spread of Wi-Fi and smartphones creates an environment in which an equal interaction between the hotel and the visitor has become possible. Hotels start to provide users with access to Wi-Fi. The competition among hotels becomes tough. Moreover, the emergence of Airbnb in 2008 fundamentally changes the industry; all hotels face growing competition from individuals that offer apartments for short-time rent. Independent travel becomes widespread with booking. Customer experience outside of hotels changes significantly. Booking through applications, increase in booking speed, access to a large amount of data - all this changes consumer behavior, reduces the importance of some factors (good location), and increases the importance of other factors (price, technological equipment, post-sales user experience).
- *Stage 4* (2010–2017): The penetration of digital transformation from the "outside world" into the hotel internal environment. The user experience inside the hotel is changing, and new service standards are being set, competition is moving into the area of technological equipment since 2/3 of hotel guests consider hotel investments in technologies that improve the customer experience as "extremely important" or "important" (Oracle 2017).
- *Stage 5* (2018–present): digital transformation will increasingly affect the industry, both in terms of increasing competition and integration of the external and internal experience of hotel guests. Personal devices outside the hotel will manage the customer experience in the hotel (synchronization of digital keys with the smartphone, number management through applications, etc.), thus smoothing the general perception of the entire customer journey.

For the sustainable growth of the hospitality business, experts consider it necessary to adapt to the changing consumption and behavior of a new 'technophiles' generation, i.e., people who are interested in modern technologies and enjoy using them. Digitalization of services is mandatory for 'technophiles'; they are attracted by online booking, location-based services, and personalized communication, as well as integration with social networks. The goal of digitizing services is to move to a digital business model in providing hospitality services for customers (Leung 2019).

Summarizing different definitions, we can say that the digital platform is a business model with technology that simplifies exchanges between several groups - for example, end-users and manufacturers. Digital platforms can mediate the flow of information and thus enable the interconnection of products and services, data, and flows between different actors (Ruutu et al. 2017). In hospitality, a digital platform gives guests the ability to view, plan, and select events, thereby facilitating the smooth integration of technology into their travel experience (Lin et al. 2018). Digital platforms depend on communities since a critical point for their development is to generate trust with clear general conditions regarding intellectual property and data ownership. It also helps consumers and suppliers trust each other online through evaluation mechanisms (de Reuver et al. 2018).

The main advantages of the digital platform include revenue generation, cost reduction, and the ability to adapt to innovations for new products and services. Interaction with a digital platform implies a complete revision of the service packages and technological base currently used in the hotel industry. The purpose of these improvements is to digitalize services and personalize the experience (Langford 2016).

Nowadays, many third-party applications provide digital services that guests are familiar with and rely on. The reduction in revenue from these third-party applications and services is a growing concern for hoteliers (Mohanty et al. 2016). To compete with these third-party applications, hoteliers develop their applications that provide services both inside and outside the hotel. Thanks to special incentives such as coupons and bonuses, guests can be encouraged to use their applications. Providing digital services of the same quality as third-party application services requires a reliable technological infrastructure base with specialized computing and communication capabilities.

In the Russian hospitality industry, the level of most hotels' technological equipment is quite low, and business processes are mostly conservative (Sheresheva and Kopiski 2016; Sheresheva et al. 2016). At the same time, there is a significantly increased interest in the application of digital technologies, due to the understanding that it is possible to transfer habitual communication mechanisms to the digital environment.

Taking into account this trend towards digitalization in the Russian hospitality industry, we conducted a case study, intending to show that the use of new technologies helps to build a potentially more profitable business model.

The four-star business hotel "Mandarin," opened in November 2008, was chosen as the object of the case study.

The methods used in the case study included the thorough analysis of all business processes in the hotel (described in the BPWIN program, AllFusion Process Modeler 7), in-depth interviews with hotel managers, statistical analysis, and the survey conducted in February–March 2019 (734 hotel guests interviewed within 18 days).

3 Results and Discussion

In the Mandarin hotel, all departments work on the automated control system *Edelweiss* developed by Edelink, a company that provides a wide range of functional modules for hotels and cooperates with more than 500 hotels in Russia. Another company (hiinc) started to cooperate with the Mandarin hotel in 2018. This company provides *handy* smartphones that help the hotel to collect complex big data analytics. The *handy* system functions as follows: to unlock the phone, a guest has to type his or her last name, first name, middle name, date of birth, and room number. Thus, the hotel can receive personal guest data immediately after unlocking the smartphone.

Still, the analysis has shown that *handy* and *Edelweiss* functioned separately. Each system, *handy* and *Edelweiss* had the ability to personalize; each of them provided a guest profile collected, but they did not support the cross-analysis of data. As a result, this has led to the fact that each system had its database. Isolated pieces of information hindered the exchange of knowledge about guests' preferences and behavior.

To improve the processing of information about guests and launch the personalization process, the suggested solution was the integration of the *handy* system with the *Edelweiss* system.

The cost of the project design and implementation amounted to 73,955 rubles, which represents the initial investment (capital costs). CFS (critical success factor) was the leading indicator (Jusoh et al. 2018). For the project, CFS is sales, since it directly depends on the use of data from *handy* integration. Over the next month (April 2109), sales increased by 5%.

CFS before the stage of project implementation was calculated as the ratio of the Mandarin hotel profit in 2018 (11,621,000 rubles) to the number of employees (79):

CSF before automation = 11,621,000/79 = 147,101 rubles per year.
CSF after project implementation was as follows:
CSF after automation = 147,101 * (1 + 0.05) = 154,456 rubles per year.

Therefore, CSF increase will be equal to the ratio of CSF before and after project implementation:

Δ CSF = 154,456 – 147,101 = 7,355 rubles per year.
CSF per month = 154,456/12 = 12,871 rubles per month.

To assess the economic effect of the project implementation on the Mandarin hotel total result, we used the investment analysis method and calculated the indirect effect. The calculation of the projected costs after the integration of work in the new system is presented in Table 1.

The total cost of ownership (TCO) amounted to 57,000 rubles. In this case, the TCO suggests a successful outcome of project implementation (CSF increases). For evaluation, the Cash Flow (CF) in rubles and Discounted Cash Flow (DCF) were calculated:

CF = Revenue in rubles per month − Expenses in rubles by month = 23,497.
DSF = Cash Flow Multiplied by Discount Factor = 95,494.

Table 1. The total cost of the project ownership.

No	Costs amount per year (RUB)	Number per year	Cost per month (RUB)*	Amount per year (RUB)
1	Update	2	4,500	9,000
2	Training	3	2,000	6,000
3	System administration	12	3,500	42,000
4	**Total**			**57,000**

*Data from the previous experience of introducing technologies in the hotel.

For a complete picture of the indirect economic effect of the project implementation, we calculated the return on investment index (ROI), internal rate of return (IRR) and net present value (NPV). The inflation for 2018 was 4.3%.

ROI = The ratio of the amount of cash flow for the year and capital expenditures.

IRR - calculated using the standard function IRR (Microsoft Excel).

NPV - when calculating this indicator, the total DCF value minus capital costs is used.

The result was the following indicators:

$$ROI = 95{,}494/73{,}955 = 1.3$$
$$IRR = IRR\ (DCF) = 4\%.$$
$$NPV = 95{,}494 - 73{,}955 = 21{,}539 \text{ rubles.}$$

The obtained values allow us to conclude that the project is efficient, since the value of NPV is >0, and ROI is >1 (for every ruble of investment, 30 kopecks in profit).

4 Conclusion

The case study confirmed that the introduction of digital technologies has a positive impact on hotel competitiveness. The developed model for optimizing information processing in the Mandarin hotel benefits from the application of digital technologies and helps to build a potentially more profitable business model and thus to reach a new level of communication with customers.

One can conclude that the use of digital technologies in the hotel business allows the use of traditional management schemes on a new basis. This leads to a higher level of customer understanding and service customization.

Therefore, the use of digital technologies is essential for the hospitality industry. Hotels that do not take full advantage of these new opportunities may lose their competitiveness and market share. Russian hotels need to be more flexible and active in optimizing and improving business processes, including using the capabilities of digital platforms.

References

Accenture: Digital platforms will define the winners and losers in the new economy (2018). https://www.accenture.com/_acnmedia/PDF-85/Accenture-Digital-Platforms-Pov.pdf

Brousseau, E., Penard, T.: The economics of digital business models: a framework for analyzing the economics of platforms. Rev. Netw. Econ. **6**(2), 81–114 (2007)

Chathoth, P.K., Ungson, G.R., Harrington, R.J., Chan, E.S.: Co-creation and higher order customer engagement in hospitality and tourism services: a critical review. Int. J. Contemp. Hosp. Manag. **28**(2), 222–245 (2016)

de Reuver, M., Sørensen, C., Basole, R.C.: The digital platform: a research agenda. J. Inf. Technol. **33**(2), 124–135 (2018)

Engardio, P.: How China and India Are Revolutionizing Global Business. McGraw-Hill, New York (2006)

Horner, S., Swarbrooke, J.: Consumer Behaviour in Tourism. Routledge, London (2016)

JLL: Global Hospitality Research (2018). Retrieved from jll.ru

Jusoh, A., Mardani, A., Omar, R., Štreimikienė, D., Khalifah, Z., Sharifara, A.: Application of MCDM approach to evaluate the critical success factors of total quality management in the hospitality industry. J. Bus. Econ. Manag. **19**(2), 399–416 (2018)

Langford, G.: Travel and hospitality industry outlook. J. Qual. Assur. Hosp. Tour. **1**(3), 65–83 (2016)

Leung, R.: Smart hospitality: Taiwan hotel stakeholder perspectives. Tour. Rev. **74**(1), 50–62 (2019)

Lin, P.M., Tung, V.W.S., Qiu Zhang, H., Gu, Q.: Tourist experience on memorable hospitality services. J. China Tour. Res. **14**(2), 123–145 (2018)

Loonam, J., Eaves, S., Kumar, V., Parry, G.: Towards digital transformation: lessons learned from traditional organizations. Strateg. Change **27**(2), 101–109 (2018)

Marasco, A., De Martino, M., Magnotti, F., Morvillo, A.: Collaborative innovation in tourism and hospitality: a systematic review of the literature. Int. J. Contemp. Hosp. Manag. **30**(6), 2364–2395 (2018)

Melián-González, S., Bulchand-Gidumal, J.: A model that connects information technology and hotel performance. Tour. Manag. **53**, 30–37 (2016)

Mohanty, S.P., Choppali, U., Kougianos, E.: Everything you wanted to know about smart cities: the internet of things is the backbone. IEEE Consum. Electron. Mag. **5**(3), 60–70 (2016)

Oracle: Creating the Coveted Hotel Guest Experience. An Oracle Hospitality-Phocuswright Study (2017). https://www.oracle.com/webfolder/s/delivery_production/docs/FY16h1/doc35/Guest-Experience-Report-2016-V7.pdf

Prachayakupt, S., O'Mahony, B., Sillitoe, J.F.: An integrated customer relationship model to improve retention in hospitality services. In: Dhinam, M.C. (ed.) Opportunities and Challenges for Tourism and Hospitality in the BRIC Nations, pp. 95–116. IGI Global, Hershey (2017)

Radojevic, T., Stanisic, N., Stanic, N.: Ensuring positive feedback: factors that influence customer satisfaction in the contemporary hospitality industry. Tour. Manag. **51**, 13–21 (2015)

Ruutu, S., Casey, T., Kotovirta, V.: Development and competition of digital service platforms: a system dynamics approach. Technol. Forecast. Soc. Chang. **117**, 119–130 (2017)

Šeric, M., Gil-Saura, I., Mollá-Descals, A.: Can advanced technology affect customer-based brand equity in service firms? An empirical study in upscale hotels. J. Serv. Theory Pract. **26**(1), 2–27 (2016)

Sheresheva, M., Kopiski, J.: The main trends, challenges and success factors in the Russian hospitality and tourism market. Worldw. Hosp. Tour. Themes **8**(3), 260–272 (2016)

Sheresheva, M., Polyanskaya, E., Matveev, N.: The Russian hospitality market: what is the basis for success in a challenging environment? Worldw. Hosp. Tour. Themes **8**(3), 310–326 (2016)

Sheth, J.N.: Impact of emerging markets on marketing: rethinking existing perspectives and practices. J. Mark. **75**(4), 166–182 (2011)

Sigala, M.: Social CRM capabilities and readiness: findings from Greek tourism firms. In: Inversini, A., Schegg, R. (eds.) Information and Communication Technologies in Tourism 2016, pp. 309–322. Springer, Cham (2016)

Sigala, M.: Implementing social customer relationship management: a process framework and implications in tourism and hospitality. Int. J. Contemp. Hosp. Manag. **30**(7), 2698–2726 (2018)

Subramanian, N., Gunasekaran, A., Gao, Y.: Innovative service satisfaction and customer promotion behaviour in the Chinese budget hotel: an empirical study. Int. J. Prod. Econ. **171**, 201–210 (2016)

Wang, D., Xiang, Z., Law, R., Ki, T.P.: Assessing hotel-related smartphone apps using online reviews. J. Hosp. Mark. Manag. **25**(3), 291–313 (2016)

Yachin, J.M.: The 'customer journey': learning from customers in tourism experience encounters. Tour. Manag. Perspect. **28**, 201–210 (2018)

The Philosophy of Digital Transport and Logistics

Victor S. Gorin, Alexei A. Stepanov, Olga A. Tetssoeva, and Artem O. Merenkov(✉)

State University of Management, Moscow, Russia
auto_guu@mail.ru

Abstract. Recently, Russia has embarked on the digital economy development. The scholars' opinion regarding this term interpretation diverges currently. An aggregate interpretation is provided by Academician Kleiner, who defines the digital economy as the environment state. The digital economy has a frame nature, its key characteristics includes forecasting the future state of the environment. At the same time, digitalization in specific sectors of the economy remains an important issue. "Digital" transport and logistics require a different management approach. The basis of digitalization is to consider the requirements of the end-user, the basis of the "digital" service of the 21st century. The purpose of this research is to understand the phenomenon of "digital" economy concept in Russia. The paper focuses on the transport system efficiency and considers the "digital" transport service quality indicators.

Keywords: Digital transport and logistics · Customer service · Intelligent transport systems · Transport management · Mobility

1 Introduction

The history of the human civilization development picks up the stream (Table 1). The nature of the intensity of changes dictates the evolution pace of scientific, technical, intellectual (Government of Russia 2017), economic, and information development. Each next period has its own driver in the high technology production (Marx 2001).

In the 19th century, the combustion engine made a true revolution in human life and production. In the 20th century, it was the introduction information technology in industry and the emergence of the Internet as the World Wide Web. Such changes are called "intermittent equilibrium," a situation where the world (system), jumping from one state to another, never comes back (Fig. 1). At the same time, progress goes along the path of dualism, i.e. along technical innovations; more than that, the consciousness and human psychology change, forming social demands.

In our opinion, in the 21st century, the digitalization of the real economy sector should become such a progress engine, creating what could correspond to the characteristics of artificial intelligence by the middle of the century.

A. V. Bogoviz and Y. V. Ragulina (Eds.): ISCI 2019, LNNS 115, pp. 91–99, 2020.
https://doi.org/10.1007/978-3-030-40749-0_11

Table 1. The evolution stages in transport progress

Ways (routes)	Facilities
Water	Using the first floating transportation means, as well as the natural abilities of humans
Roads	The wheel invention, the use of live traction force
Ranged roads	Inventing the internal combustion engine, using the motorized technology, developing the railway transport
Outer space	Constructing the carrier rocket, launching the first satellite, the first human space flight
Digital transportation systems	Using automation tools to accumulate large data amounts and develop the decision-support tools, the transport digitization
Intelligent transport system	Using artificial intelligence and digital technology to achieve transport work automation of all types

Source: Developed by the authors.

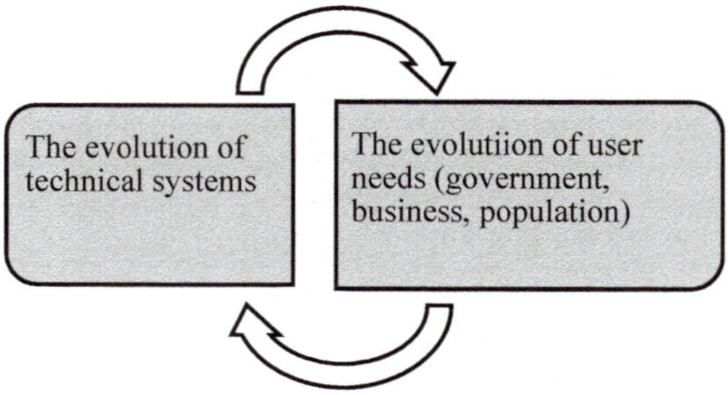

Fig. 1. The dualism in the evolution of transport systems.

2 Materials and Methods

The above facts from the economics and society history and the desire to answer philosophical questions in the field of transport economics determine the scientific relevance of our research.

The purpose of this study is to understand the "digital" transport and logistics phenomenon as the component of the wider concept of the Russian "digital" economy.

We state the following research objectives:

- Determining a set of effective operation indicators of the transport complex in the digital economy;
- Analyzing the "digital" service quality;
- Determining the supply and demand factors in "digital" transport and logistics;
- Studying the role of information and communications environment in building intelligent transport systems (ITS);
- Studying the foreign experience of transport digitization based on the comparison of "digital" projects in Russia and abroad.

The scientific novelty of our research includes the following:

- Clarification of the transport service concept in the context of economic digitalization;
- Determining the quality cycle of "digital" transport and logistics;
- Determining the basic study principles of the individual components and the whole system structure of "digital" Russian transport and logistics;
- Developing an algorithm for the "digital" transport and logistics network;
- Providing recommendations on the further development of the Russian "digital" transport and logistics.

According to Greenberg (2018), one of the main elements containing the potential for the Russian economy development is the necessity to link the countries' territory into a single space. On the one hand, it is a tendency to space compressing. In essence, we are talking about a tendency to compress space, on the one hand, and to create the prerequisites for creating a single information and communication space, on the other. In this regard, the role of transport is greater than ever.

To accomplish this ambitious task, the transport system (TS) should be significantly transformed, acquiring the innovative network form that could literally "envelope" the territory. At the same time, our development efforts should be focused not on quantitative, but on qualitative measures. Avoiding traditional development methods (roads construction and reconstruction) would break the vicious transport demand and supply cycle, creating a situation where construction and reconstruction costs more and more. The new infrastructure generates a new transport demand, to satisfy which an unlimited budget for construction and reconstruction is needed.

In this regard, the philosophy of automation, intellectualization, speed increase, transportation quality improvement, losses liquidation, and cost cutting should be integrated into the "digital" transport and logistics basis (Fig. 2). In the "digital" economy such quality cycles are individualized and based on customer preferences, which are automatically determined by the system. At the same time, the economic growth acceleration should take place not only due to the digitalization, but mainly thanks to an increase in knowledge, experience, and training, which are necessary for the digital technology implementation.

This initiative should be developed due to intersection of different interests (Fig. 3). The project for the development of "digital" transport and logistics in Russia has already started. The digital transport complex platform in Russia exists as an integral part of the Russian "Digital Economy" project.

The companies Platon, RZD, and GLONASS are interacting and cooperating in the digital transport development in 2019. The program "Digital Economy of the Russian Federation" is currently implemented by the Government of the Russian Federation decree (Government of Russia 2017). The basic directions for the "digital" economy development are presented in Fig. 3. One of the most important factors in the formation of the "digital" transport and logistics system implies taking into account the interests of various groups of users (Fig. 4).

The legislative activity should result in approving the "digital" transport and logistics concept, as well as in developing a set of principles for creating a multi-level

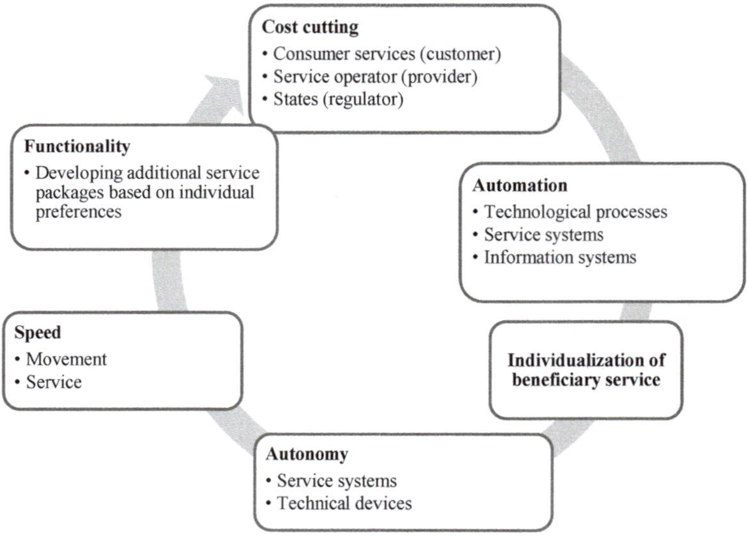

Fig. 2. The quality cycle of "digital" transport and logistics.

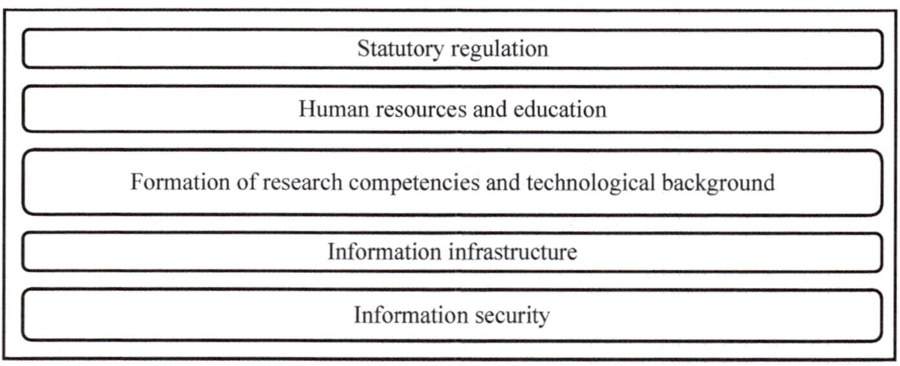

Fig. 3. The basic directions of developing the system of "digital" transport and logistics in Russia.

national transport model of transport macro-planning in a single information space while maintaining end-to-end control and common policy.

From the practical point of view, solving the issue is not limited to the interactions between large domestic companies in the field of innovative transport solutions. The "Caravan" project development is an example of a successful approach to digitalization (Comnews 2017). The project purpose is to create a domestic corporate intellectual transport system (ITS), involving the concept of an integrated digital roads infrastructure and the key system components deployment development.

The integrated ITS deployed along the route includes: a high-precision vehicle positioning system; a system for measuring flow intensity changes, an incident

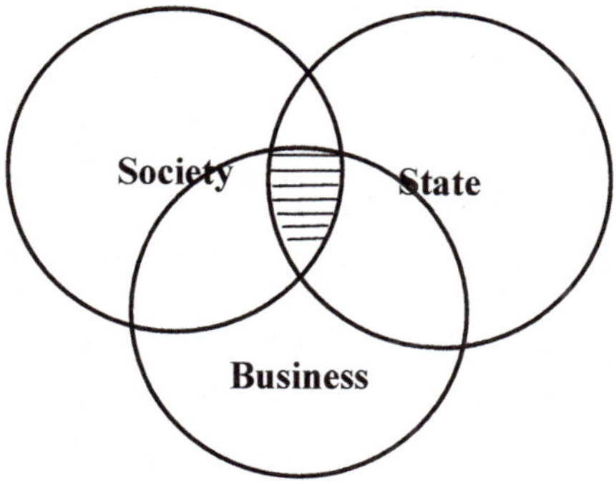

Fig. 4. The groups interested in the implementation of "digital" transport and logistics.

detection and recording system. The project offers a number of exceptional services: building a digital road model providing cyber security and data transmission protection systems; a modular platform for the ITS and road users' interaction (Bueno et al. 2015). The supporting systems include LTE cellular communication systems and engineering communications.

Developing the national "digital" transport and logistics system is the major trend in all advanced countries. Successful projects have been implemented in several European countries. In particular, the Transport Integrated Model for Europe (Tri-mode), which has a number of functions (EU transport systems coverage, passengers and goods movement monitoring, scanning traffic and energy consumption streams, considering economic factors of transport), can serve as an example of such global models (Cao and Cao 2017). The system's tasks include such functions as planning, control, and goal setting. Such transport models are the mathematical algorithms for the social, economic, and transport data in a specialized development environment. Digital transport infrastructure databases – the Transmodel standard (there are counterparts in the Czech Republic and the UK) – are also actively developing (de Oña and de Oña 2015).

3 Results

Intelligent transport is based on the big data management. The required data amount already exists. According to the Moscow Deputy Mayor for Transport Issues Maxim Leksutov, 160,000 traffic surveillance cameras are functioning on a daily basis in Moscow (Unified Transport Portal 2019). All the gathered information flows to the traffic organization center in real time, where transport management in Moscow operates. Some functions are also automated. However, even with large amounts of data, there is a need for a manager making decision.

As the analysis of the Russian transport systems in cities (including Moscow, Kazan, and Sochi) shows, creating a unified information transport complex system, developing management tools for transport decision-making, as well as moving from the transport flow modeling to the process of assessing the transport and economic impact (taking into account the impact on the environment) can be an effective management option. The work with these data should be carried out relying on the data fusion principle, i.e. processing and combining data from different institutions in order to increase their accuracy.

4 Results and Discussion

Currently, the Russian transport complex is characterized by a partial platformization. There are no genuine "digital" transport and logistics nowadays. It is necessary to form a single data space (information and communication space) for more integrated solutions development. Developing a similar environment based on the blockchain technology (distributed registry, network) is efficient, especially when taking into account modern technologies.

The "digital" transport and logistics system should be a multi-level one. The area topography is the basis. The basis of the area topography has over itself a plurality of platforms (regional, national, local, and international) that function as a single transport and communication space-form, the functions of which are determined by a stable set of peoples' movements and interaction. Intersection platforms are the next level. Meetings and interactions are the opportunity to accumulate a communicative potential. At the same time, the distribution of energy potential should resemble a distributed network in its shape, which is distinguished by speed, flexibility, and adaptation to the changing environmental conditions. The value of such a network is located on its circumference. The state's task is to build such a "digital" transport and logistics network, in which users play the key role in a decentralized process, because they generate added value.

This network model assumes the presence of multitude consumer services centered on aggregator platforms (like Uber, Trans.edu, etc.). The operator (provider) is responsible for the quality of service provided to the user, which solves the following tasks:

- Providing access to the data;
- Harnessing the potential of network scale effect;
- Creating a single user IP;
- Regulating payment methods and systems;
- Interacting with service providers;
- Solving property right issues.

An additional network effect is in the implementation of the "preferred attachment" principle: the greater the binding potential, connecting digital objects to each other, the higher the network potential of the system as a whole. More than that, the network and the connection process should function in *instant availability conditions*. The algorithm for developing a transport network has the following order (Fig. 5).

The impetus for the network model development (including economic) is convenient services and the information availability presentation. Moreover, information competition for a passenger has recently become more acute. The service quality problem is largely determined by the lack of necessary information, including

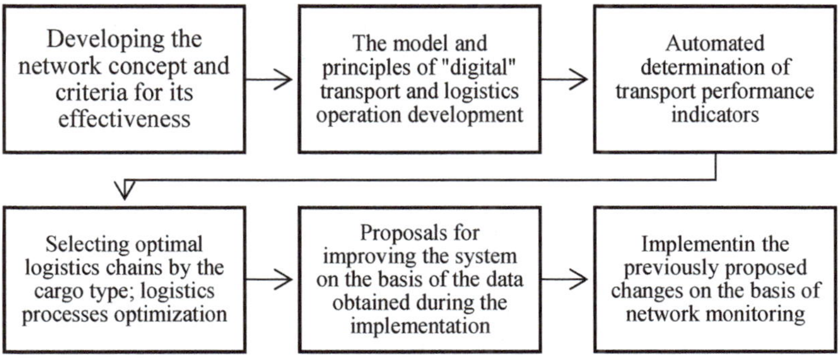

Fig. 5. Algorithm for developing a "digital" transport and logistics network.

regulatory and reference character, which is the industry information space core, which includes a set of dictionaries, classifiers, and reference books, standards, and regulations used in all aspects of its activities.

Decentralization, the network nature of the economy, information (open data) are the basis for the ITS development, contributing to the transport mobility growth, the cargo traffic intermodality, the improved quality reporting, cost reductions, and the international cooperation development. Ring signatures, homomorphic encryptions, and zero-knowledge proof provide confidentiality, allowing users to store data in such a way that some of their constituents can be checked and calculated without disclosing personal information.

The research network communications allows us to say that users (passengers, transport companies) are what they find themselves connected to today. The network speed is ahead of the management decisions. The network intelligence will become faster and more effective than the human mind and an experienced manager, because a successful decisions history database is formed with mutual connection.

The Airbnb and Uber platforms, which are implemented in the decentralized organizations format, are the examples of network services. The companies are not only a successful combination of digital (new) and material (old) worlds. They are global and cross-border services, where user interaction is conducted through a web interface. They put in place unused benefits, free homes and cars, and provide access to the market. This is a sharp historic increase in supply. Management of such companies requires an understanding of modern computer technology, as well as the principles of traditional sectorial management, including transport and logistics.

Another example is the "La'Zooz" app for drivers (Cassano 2015). It creates a new business paradigm, in which the services cost is exchanged for crypto currency in combined forms (virtual and physical). Thus, the cost becomes of a closed loop nature, as the service provider receives payment in exchange for information that serves as the primary basis for making a deal. The relationship between the seller and the consumer has a peer-to-peer network (without an intermediary) character, which creates prerequisites for the local mini-economies emergence. A customers network is formed around suppliers that operate under perfect competition, as can be seen from Fig. 6. One chooses a supplier on the basis of price policy and service quality.

Characteristics explaining the increasing communication speed						
Instant access to services	Transport system accessibility (barriers elimination)	Transport communication reliability	Increasing number of people using transport services	Increasing time spent by passengers in transport	Reduced tickets prices	Growing trip marketing component

Fig. 6. Distinctive features of the "digital" model of the transport service.

Nowadays, it seems reasonable to consider the transport business as a customized model, i.e. on the basis of the customers' needs. "Digital" transport and logistics are the transport systems driven by use cases. It is such a transport service system in which the choice of configuration of transport service packages is predetermined based on the user's needs digitization. The "digital" transport service model has a number of characteristics (Fig. 6).

5 Conclusion

As can be seen from Fig. 6, one of the basic principles of the modern transport system is the accessibility, which is expressed in eliminating the barriers. Such "barriers" can be economic (flexible pricing system) and information (working with electronic services, navigation ease) ones, as well as of the physical/infrastructure nature (low-floor rolling stock, passengers with additional needs adaptation to the transport, etc.) It must be emphasized that the "digital" transport and logistics erase the difference in classes and the clients' economic viability (another barrier). It is advisable to carry passengers in a single class, probably for free. The economic difference would be noticeable only in additional services packages that a particular passenger can afford. A similar model is implemented in catering restaurants such as McDonald's and Starbucks coffee houses. In short, attendance is free in such places, and the income is generated by creating a comfortable atmosphere, which encourages a client to visit such places more often. A longer consumer access time increases chances to offer the visitor additional payed services.

The concept of services individualization is realized in practice due to the existence of service packages strategy, i.e. when a passenger (of a train) chooses a menu, a film, or a music album on the basis of his/her personal desires. "Digital" technologies allow predicting such requests, particularly due to the analysis of "Big Data," access to which is provided by the clients themselves by installing the carrier's application. The range of customized services is set due to the fact that the client exists in several dimensions. First of all, it happens the real world and the virtual one, in which transport (business, company) works with a "digital" human clone (digitized needs). A clone-intellectual model of an individual's consumer behavior is based on retrospective analysis, predicting future behaviors. A clone-dynamic (evolving, complemented) model exists in open space (probably with permission, and sometimes even without human approval). This allows assessing an individual by the various groups of interests (sports, medicine, displacement, work, leisure, etc.). Interacting with each other through such platforms, companies update a client model, thus helping both themselves and other market players. The combination of electronic clones and the digitization of needs allow us to form a networked society, which would be the key to improving the living standards.

Acknowledgments. We express gratitude towards the Scientific School on Transport Management of the State University of Management for their invaluable assistance in organizing the research.

References

Greenberg, R.S.: Thoughts on the imperatives of economic modernization in Russia. Economicheskoye Vozrozhdeniye Rossii **2**(56), 41–46 (2018)

Marx, K.: Capital. AST Publishers, Moscow (2001)

Bueno, P.C., Vassallo, J.M., Cheung, K.: Sustainability assessment of transport infrastructure projects: a review of existing tools and methods. Transp. Rev. **35**(5), 622–649 (2015)

Cao, J., Cao, X.: Comparing importance-performance analysis and three-factor theory in assessing rider satisfaction with transit. J. Transp. Land Use **10**, 65–82 (2017)

De Oña, J., de Oña, R.: Quality of service in public transport based on customer satisfaction surveys: a review and assessment of methodological approaches. Transp. Sci. **49**(3), 605–622 (2015)

Government of the Russian Federation: Order on the approval of the program "Digital Economy of the Russian Federation" (28 July 2017, no. 1632-p, Moscow, Russia) (2017). http://base.garant.ru/71734878/

Unified Transport Portal: Digitalization of the Moscow transport system (2019). http://www.report2010-2017.transport.mos.ru/pdf/ar/ru/mega-projects_digitalization.pdf

Comnews: "Caravan" goes to the second stage (2017). https://www.comnews.ru/content/110735/2017-12-04/karavan-idet-na-vtoroy-etap

Cassano: Could La'Zooz be the ride-sharing app we've been waiting for? (2015). https://www.fastcompany.com/3041403/could-lazooz-be-the-ride-sharing-app-weve-been-waiting-for

Transformation of the Concept of Population Transport Service Quality in the Environment of Urban Digital Agglomerations

Elizaveta B. Bednyakova(✉), Yana S. Ignatova, and Elena V. Kuptsova

State University of Management, Moscow, Russia
bettyvin@mail.ru, sakuptsov@mail.ru, ysignatova@gmail.com

Abstract. The article presents the retrospective view on the urban environment formation in terms of organizing public transportation and the social attitude towards the concept of transport services quality for the population. The research goals are the following: (1) to develop the concepts of the client as a consumer of transport services and the quality of the transport service process; (2) to formulate a new paradigm of transport services for the population of urban agglomerations and basic conditions for the functioning of the passenger transport and logistics infrastructure. The paper devotes sufficient attention to such issues of legal regulation of the development of large urban agglomerations and their transport, as well as the place of a human in the transport system of the city. The negative impact issues of vehicles and excessive workload of transport infrastructure on the city ecology, the population life and health are also reviewed. The article argues that significant changes are made to the contour of the transport process quality when changing the society development paradigms.

Keywords: Urban environment · Transport services · City development paradigm · Transport services quality · Large city ecology

1 Introduction

The formation of the policy focused on urban agglomerations transport based on the sustainable development principles requires reconsidering a number of categories related to transport services. This is concerned with understanding the issues of interaction between the state and society on the regulation of transport service processes, development of a new customer understanding and, accordingly, the new quality of transport services, as well as changes in the transport service paradigm of the population itself. Along with the special order of state regulation in this area, it is necessary to change priorities in the development strategy of urban agglomerations.

The purpose of the study is to develop the following concepts of the client as a transport services consumer and the quality of the transport services process. Also, the paper aims to formulate a new paradigm of transport services for the population of

A. V. Bogoviz and Y. V. Ragulina (Eds.): ISCI 2019, LNNS 115, pp. 100–107, 2020.
https://doi.org/10.1007/978-3-030-40749-0_12

urban agglomerations and basic terms for the functioning of passenger transport and logistics infrastructure.

The analysis of definition analysis of the process and transport services quality as a scientific category leads to the conclusion that they do not reflect current trends in the urban environment formation.

2 Materials and Method

The study used the Russian regulatory framework materials in the field of state regulation of transport activities in cities, state statistics, as well as the official research results on the development of urban transport systems published in the licensed publications.

The following research methods were used:

- Collection of secondary information on the transport services organization, built on the principles of sustainable urban agglomerations development in the media and consulting agencies publications;
- Analysis of norms in the field of regulating the processes of organizing transport services and the transportation quality, operating both in the territory of Russia and the EU;
- Analysis of trends analysis in the following areas: developing passenger transport technologies and urban transport systems management; identifying the relationship of transport policy with the cultural and mental characteristics of transport services consumers; conducting a retrospective analysis of the evolution of ideological views on the formation of transport policy related to the priority of preferential movement;
- Proving logically (using the systems approach) the stated hypothesis about the need to change the paradigm of transport services for the population of urban agglomerations.

3 Results and Discussion

A retrospective look at the role of public transport in the Russian cities allows distinguishing a total of three periods, taking into account the perspective of the preferential movement of public transport, a person, and a personal car. Changes in society attitudes to the preferential movement priority and the public transport role are closely correlated with the population automobilization level (Fig. 1).

Until the mid-1980s, there was a priority in the movement for public transport in the urban environment, while the person as a passenger was in second place. In the early 1990s, the situation began to change due to the formation of the domestic car market. As a result, there was a sharp increase in the number of cars. A personal car came out on top in the priority movement syste = m. The public passenger transport (PPT) went second. However, the street and road networks in the 1960s and 1980s were completely unsuitable for the high car traffic. It is well-known that the Soviet town-planning norms proceeded from the following number: 60 cars per 1000 citizens.

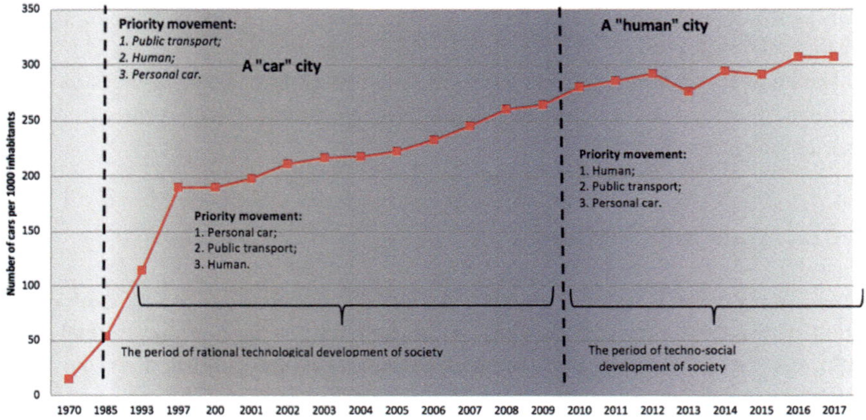

Fig. 1. The scheme of the transport service paradigm, depending on the number of cars among the population of Moscow. Source: (Autostat 2017).

However, this number was increased by the largest cities in three times by the middle of the 1990s. (Blinkin and Reshetova 2013). The country as a whole passed this mark by the end of the century. Unfortunately, during the process of developing cities and road networks, a new reality associated with an increase in the number of cars was almost completely ignored. Such trends and policies for urban development have resulted in traffic congestion and increased environmental pollution, leading to an increase in the average travel time, in the road accidents (RTA), and the population morbidity. Thus, the growing number of cars among the Moscow population in 2000 amounted to almost 200% of the 1991 level. This decade is characterized by the most intensive growth. In 2010, it was approximately 130% of the 2000 level, and 103% in 2017 of the 2010 level. Arguably, the second period (1985–2010) in the Moscow urban transport system development is characterized by the "rational manufacturability" principle.

In response to the negative impact of vehicles on the city ecology and the growing social awareness of the "green technologies" priority, Moscow has changed the transport strategy, giving public transport a priority on the roads. First of all, it was intended to reduce traffic congestion and dismiss traffic jams. The second priority was to contribute to improving the environmental situation in the city.

Since 2010, the public transport share in the total trips number has increased by 10%. Also, the cost of upgrading the rolling stock fleet seriously increased in Moscow. Their share in the city budget has grown from 12 to 18.5%, accounting for a quarter of all state programs in Moscow expenses. Moscow will spend 3.2 trillion rubles from 2012 to 2020 for the program "Development of the Transport System" (McKinsey & Company 2018a). According to that research, "high efficiency, convenience, and relative accessibility of public transport" in Moscow can be considered as achievements.

According to the Center for the Study of Negative Problems of Megapolises (Institute for Transport Economics and Transport Policy Studies, Higher School of Economics), the traffic flow in Moscow is distributed as follows: private car – 27%, mass public transport – 71% (more than 70% are subway trips), others – 2% (McKinsey & Company 2018b).

The third urban development period begins from 2010–2012. This period can be called the “Peoples’ City.” The main focus is on a human, their need for mobility due to the preferential development of public transport over personal transport. In this new system of public transport services, a person is considered not only as a passenger, but as a pedestrian, a passenger, and a car enthusiast. It seems fair to say that this period is connected with the change of the transport services paradigm for the city’s population. In particular, it is a certain move from the “city for car” approach to the “city for peoples” concept.

Meeting the population needs and improving the quality and efficiency of transport services is fully possible only in a digital environment using urban information technology, i.e. as a part of the “smart city” system creation. This would allow the city authorities to interact with the large cities’ population directly and control the urban development in order to improve the quality of life. Currently, certain initiatives are already underway in this direction at the state level. For example, the Ministry of Construction, Housing and Utilities of the Russian Federation has approved the standard of the federal project “Smart City” in February 2019 (Minstroy 2019). However, the list of mandatory requirements for “smart cities” does not even mention the public transport system.

In turn, the change in the transport services paradigm is a consequence of major changes in the social mentality. The latter is expressed in the movement from the principle of technogenic rationality in the social development towards techno-socio-natural development.

The traditional contour of the quality of transport process is changing under new conditions. In the approach proposed by the authors on the formation of the quality contour of the transport process, the “quality” is focused not only on meeting the needs of direct consumers (passenger), but also of indirect consumers pedestrians, car enthusiasts, or even society as a whole. The traditional quality contour takes into account only the process characteristics of moving passengers in space – accessibility and comfort. Environmental requirements of the service standard are associated exclusively with the requirements for the rolling stock power units of carriers. Also, access to urban highways of a large number of private cars, in addition to the increasing load on transport infrastructure and the natural environment, carries the risk of (a) increasing the population morbidity, (b) reducing their ability to work and people life expectancy from harmful emissions into the atmosphere, as well as (c) increasing the number of RTAs.

As practice shows, the number of accidents involving buses (Table 1) is also growing every year in Russia (Astapenko 2018). The table shows that while reducing the mortality rate by 34%, there is an increasing number of the injured by 1.3%.

Table 1. Accidents on roads with the participation of buses in the intra-city transport service for 2016–2018.

Years	Accidents	Injured	Dead
2016	5290	7264	257
2017	5608	7700	324
2018	5252	7820	206

The bus users and injured persons of road accidents and their consequences often lose their health, habitual way of life, work opportunities, as well as require care from relatives, etc. (Ignatov 2014). And this is an additional loss in the city budget income. Compulsory insurance of civil liability for causing harm to life, health, or passengers' property (acquired by the transport companies and individual entrepreneurs) solves these problems only partly. Most of the damage to peoples' health is compensated within 30% of the sum insured, i.e. 600 thousand rubles. Clearly, it does not compensate all the treatment costs.

In the context of transport services, the "quality for passengers" is measured and evaluated by the "minimum transport standard" (MTS) indicators. The scientific community puts forward various proposals for its content. Thus, representatives of the Scientific School of the Department of Management of Forwarding Services of the State University of Management (Moscow, Russia) presented the layouts of the Target Social Standard for Bus Services, namely:

1. (Merov 2009) proposed a system of MTS indicators for Moscow, which included the population transport mobility; modal split; urban transport reliability; transport discrimination level of the population; transportation comfort level; specific lost fund of free time per day; the public transport share in total environmental pollution; the accidents level caused by public transport; urban passenger transport efficiency.
2. (Martynov 2013) proposed a target social standard of service quality for basic bus transportation services, which has two components the Operational Performance Standard (ODS) and the Service Delivery Standard (SDS). The ODS includes such indicators of the transport services quality as reliability, load, and security. The SDS includes accessibility, integration, and information. All indicators are a system relevant to the technological approach to the urban transport development. Information on compliance with standards comes from the regional information transport system.
3. (Tettsoeva 2009) proposed the following version of the system of quality indicators for international bus traffic, with their further inclusion into the national standard (Tettsoeva 2009): safety indicators; environmental friendliness indicators (rationing by CO2 level and noise); reliability indicators; availability indicators; ergonomic indicators. It was proposed to calculate the generalized specific quality indicators through individual estimates averaging.

The current situation requires clarifying the approach to assessing the transport services quality in general and the Social Standard content in particular.

Evaluation of the transport services quality should be focused on meeting the expectations of direct customers, while taking into account the society and the carrier interests (municipal or commercial passenger transport company). The direct customers' interests are reflected in the Social Standard. The authors share the view on the Social Standard structure as a combination of ODS and SDS, as proposed earlier by (Martynov 2013). At the same time, they consider it necessary to supplement the substantive part of SDS with the requirement to change the rolling stock fleet structure of commercial organizations performing passenger transportation in favor of electric cars.

In the context of adopting a new paradigm of technogenic socio-natural society development, the traditional quality contour of the transport process ris to be changed. We believe that regarding the transport policy implementation on the sustainable development principles, the transport services quality should be one of the elements of the system for ensuring the population's quality of life in the urban environment. It should be evaluated according to the following model:

$$Q = f(T, S, N),$$

where T – innovative technology and technology of the transport industry, S – society interests, N – natural environment.

This model emerges from the hypothesis put forward earlier in the text by the authors about changes in the paradigm of developing urban transport systems, which is also to be conducted on the basis of the PPT movement priorities.

Thus, the quality of transport services by PPT is a complex indicator that reflects the interests of direct and indirect consumers and is characterized by their physical indicators and cost indicators (operational, capital, or disconnected) for their implementation. The direct consumers' interests are expressed in natural terms of the consumer value of passengers: (a) implementation costs for the state and commercial carriers, (b) the level of the transport infrastructure and the system friendliness as a whole for the motorists, and (c) the minimum level of harmful effects on the environment for the society.

The complex quality indicator assessment can be given on the basis of the standard parametric model:

$$\mathrm{K} = \mathrm{f}\left(\mathrm{T}, \mathrm{A}_{\mathrm{fis,fin}}, \mathrm{E}, \mathrm{C}, \mathrm{S}_{\mathrm{d}}, \mathrm{S}_{\mathrm{a}}\right),$$

where T is the innovation level of the applied vehicles and technologies in passenger transport; $\mathrm{A}_{\mathrm{fis,fin}}$ – the public transport physical and financial accessibility; E – the effectiveness of the city transport system components (public and private transport); C – the service comfort in the urban transport system (transportation and transport service, including services digitalization); S_{d} – city sustainable development index; S_{a} – citizens social activity in the transport policy formation.

4 Conclusion

Considering all the consumers of the urban transport infrastructure as passenger, pedestrians, and car enthusiasts at the same time, building a new transport environment requires not only the implementation of a set of initiatives on the technical PPT system reforms, but also the positive image creation of the PPT in the people minds.

In this regard, there is a necessity in the active joint work of the city administration and active citizens to promote the sustainable development principles in the society. Also, it is necessary to integrate various types of urban public transport into a single transport process. Another problem is the disposition of contradictions between the transport infrastructure and the urban environment on the basis of reconstruction and development of transport and transfer nodes with transport and pedestrian traffic streamlining, parking lots, cultural and community facilities saturation. Among other necessary measures we would like to note a few more: (1) protecting and improving the city environment; (2) improving transport accessibility conditions in the direct residence places, as well as the quality of service and comfort in the PPT for each direct and potential passengers; (3) providing transport services customization (car sharing).

These measures will allow to attract those customers who have the opportunity to choose the way to travel around the city. It would also contribute to the promotion of decisions in the field of city transport policy in favor of public transport. Another benefit is to the opportunity to combine individual and social aspirations into a single effort aimed at the sustainable cities development (UITP 2017).

We also would like to note in the context of our research that the state, society, and business need an image of the future with a balanced economic policy in relation to the agglomerations and transport development (Bednyakova 2010). In our perspective, this state can be expanded as follows:

- The socially oriented state (according to the Constitution of the Russian Federation) must comply with the socially oriented urban systems having a developed transport infrastructure;
- There is a need to make a better use of the existing urban transport production capacities, their modernization and development based on their own strengths and resources (economic, scientific, technical, and spiritual-moral potential), with the steady state support;
- Strengthening the city and its transport system integrity through a rational combination of government regulation, relying on the market mechanisms and the support for integration processes (as a synergy source);
- Strictly observing the interests of the state, carriers, cities, and agglomerations residents in projects and development programs of the urban transport systems.

Acknowledgment. The authors express their gratitude to V. A. Persianov, V. S. Gorin, A. A. Stepanov, and M. Ya. Blinkin, who are all prominent representatives of the transport industry dealing with the public transport development, for their significant support to the study.

References

Astapenko, E.: Accident statistics in Russia continue to spoil the faulty cars and bus drivers (2018). https://www.kolesa.ru/news/statistiku-avarijnosti-v-rossii-prodolzhayut-portit-neispravnye-mashiny-i-voditeli-avtobusov/

Autostat: Rating of Russian million cities by car availability (2017). https://www.autostat.ru/news/29680/

Bednyakova, E.B.: Ideology in the settlement systems and urban planning. J. Univ. Bull. Vestnik Univ. **24**, 45–57 (2010)

Blinkin, M.Ya., Reshetova, E.M.: Road safety: background, international experience, basic institutions. Publishing House of the Ed. home of the Higher School of Economics, Russia (2013)

Ignatov, Ya.S.: Compulsory insurance of civil liability of the carrier to passengers: how it works. Transp. Bull. **6**, 34–46 (2014)

Martynov, D.A.: Regulation of the quality of public services provided by bus in the region (Thesis for Candidate of Economic Sciences). State University of Management, Moscow (2013)

McKinsey & Company: Transport systems of 24 cities of the world: components of success (2018b). https://www.mckinsey.com/ru/~/media/McKinsey/Business%20Functions/Sustainability/Our%20Insights/Elements%20of%20success%20Urban%20transportation%20systems%20of%2024%20global%20cities/Urban-transportation-systems_rus_e-version.ashx/

McKinsey & Company: McKinsey called the Moscow transport system one of the most efficient in the world (2018a). https://www.vedomosti.ru/realty/articles/2018/07/19/775823-mckinsey-transportnuyu-sistemu-moskvi/

Merov, Yu.M.: Urban passenger transport under conditions of state regulation and the operation of market mechanisms (using the example of the Moscow megacity) (Candidate of Sciences Dissertation). Publishing House GUU, Moscow (2009)

Minstroy: The digitalization project of the urban environment "Smart City" (2019). http://www.minstroyrf.ru/trades/gorodskaya-sreda/proekt-tsifrovizatsii-gorodskogo-khozyaystva-umnyy-gorod/

Tettsoeva, E.M.: Improving the quality of international bus transport (Candidate of Sciences Dissertation). Publishing House GUU, Moscow (2009)

UITR: Quality management training in public transport (2017). https://zapdoc.site/trening-po-upravleniyu-kachestvom-v-sfere-obshestvennogo-tra.html/

The Main Areas of Integrated Digitalization of the Strategic Management of Competitiveness of Enterprises from the Russian Industrial High-Tech Branches

Natalia V. Kandybko[1(✉)], Alexander A. Ryazanov[2], and Suren A. Khachaturian[3]

[1] Military University of the Ministry of Defense of the Russian Federation, Moscow, Russia
nataliya_v_@mail.ru
[2] Moscow Witte University, Moscow, Russia
alekryazanov@yandex.ru
[3] Russian Presidential Academy of National Economy and Public Administration, Moscow, Russia
surel311@gmail.com

Abstract. The paper focuses on the results of the study of the main directions and possibilities of the digital transformation of the strategic management of the competitiveness of enterprises in high-tech industries. The authors analyzed the essence and content of the process of strategic management of the enterprise's competitiveness, presented their own definitions of the enterprise's competitiveness, competitive potential, competitive advantage. More than that, the authors proposed a structure of the enterprise's competitiveness system, identified, and substantiated the main directions of integrated digitalization of the strategic management of the competitiveness of high-tech enterprises of Russian industry. In addition, technological solutions used in the process of digital transformation are also discussed in the paper.

Keywords: Management · Enterprise · Competitive potential · Competitive advantage · Competitive weakness · Digitalization

1 Introduction

One of the key priorities of the socio-economic development of the Russian Federation is to increase the competitiveness of socio-economic systems of various levels, including enterprises in high-tech sectors of the Russian industry. This problem is solved both through the realization of their competitive advantages and through forming or acquisitioning new competitive advantages, which include the creation of management systems for their competitiveness. In the context of the ongoing digital transformation of the Russian economy, the fundamentals for such systems should be founded on advanced concepts and methods of strategic data management, innovative

A. V. Bogoviz and Y. V. Ragulina (Eds.): ISCI 2019, LNNS 115, pp. 108–114, 2020.
https://doi.org/10.1007/978-3-030-40749-0_13

business models, promising information and telecommunication technologies, modern digital devices, systems, and platforms. At the same time, in enterprises of high-tech industries, a much greater effect is achieved by forced integrated rather than sequential "patchwork" digitalization of competitiveness management.

The research topic is relevant due to both the insufficient theoretical development of questions about the nature and content of the management of the enterprise competitiveness, and the need for scientific understanding of the role and possibilities of digitalization in the field of competitiveness management of high-tech enterprises, as well as the practical relevance of research related to the development and justification of promising ways for increasing the competitiveness of enterprises in the Russian industrial high-tech sectors of.

The need to increase the competitiveness of enterprises in high-tech sectors of the Russian industry in the context of their digital transformation necessitates solving a scientific problem consisting of the development of theoretical and methodological provisions for the integrated digitalization of strategic management of their competitiveness. The aim of the research is to solve the scientific problem of developing theoretical and methodological provisions for the integrated digitalization of strategic management of the competitiveness of enterprises in the Russian industrial high-tech sectors. To achieve this goal, there is a need in solving a set of interrelated tasks: (1) to study the nature and content of competitiveness management at enterprises; (2) to identify certain specifics of the digitalization impact on this process; (3) to develop theoretical and methodological foundations for integrated digitalization of strategic competitiveness management of enterprises in high-tech sectors of Russian industry.

Theoretical and methodological foundations of the digital transformation of socio-economic systems were laid and developed in the works of many foreign and domestic experts. Among them were (Bogoviz et al. 2018), (Ponomareva et al. 2018), (Khachaturyan et al. 2018), (Lobova et al. 2019), (Ragulina et al. 2019), and others. However, the issues concerning digitalization of the strategic management of enterprise competitiveness in high-tech industries has not yet been adequately reflected in scientific research.

2 Materials and Methods

In the process of solving the main tasks of the research, such methods of studying social and economic phenomena as dialectic, systemic, historical and logical, comparative analytical, economic and mathematical modeling, expert estimates, as well as the methods of scientific abstraction, induction and deduction, normative and positive analysis, and synthesis were used.

3 Results and Discussion

According to the authors, enterprise competitiveness management is carried out at strategic, functional, and operational levels. Its basis is strategic management, which essence and content are presented in Table 1 and Fig. 1.

Table 1. Stages of the strategic management of enterprise competitiveness.

No	Functions	Stages	Description
1.	Goal setting	Strategic goal setting regarding enterprise competitiveness	Development and refinement of the mission elements, strategic vision, and other elements of the strategic target system
2.	Data collection	Strategic monitoring of the competitive environment	The periodic collection, pre-processing, systematization, and accumulation of data on the state and development trends of the main elements of the external and internal environment of the enterprise
3.	Data analysis	The strategic competitive analysis	Data transformation on the state and development trends of the main elements in external and internal environments into analytical conclusions on promising areas for competition
4.	Managerial decision making	The selection of a competitive enterprise strategy	Assessing the identified promising areas for competition and ensuring a reasonable choice of the best option (or thereof combination)
5.	Planning	The development of a competitive enterprise strategy	Development of a long-term plan for conducting competition: detailing the process of increasing competitiveness in the chosen direction at stages and events, distributing enterprise resources between them, calculating intermediate results, determining responsible executors
6.	Organization, stimulation, coordination of staff	Implementation of the enterprise's competitive strategy	Implementation of a long-term competition plan
7.	Control and regulation	The strategic control of the enterprise competitiveness	Comparing the achieved competitive position of the enterprise with the target settings, adjusting the content of other stages of the strategic management of the enterprise competitiveness according to the results of this comparison

The object of this process is enterprise competitiveness, which can be defined as its dynamic manageable complex property, (a) determined by the competitive potential not inferior to the competitors and the possibility of its use, (b) expressed in the ability to redistribute the system created in the socio-economic system (market, industry) for a

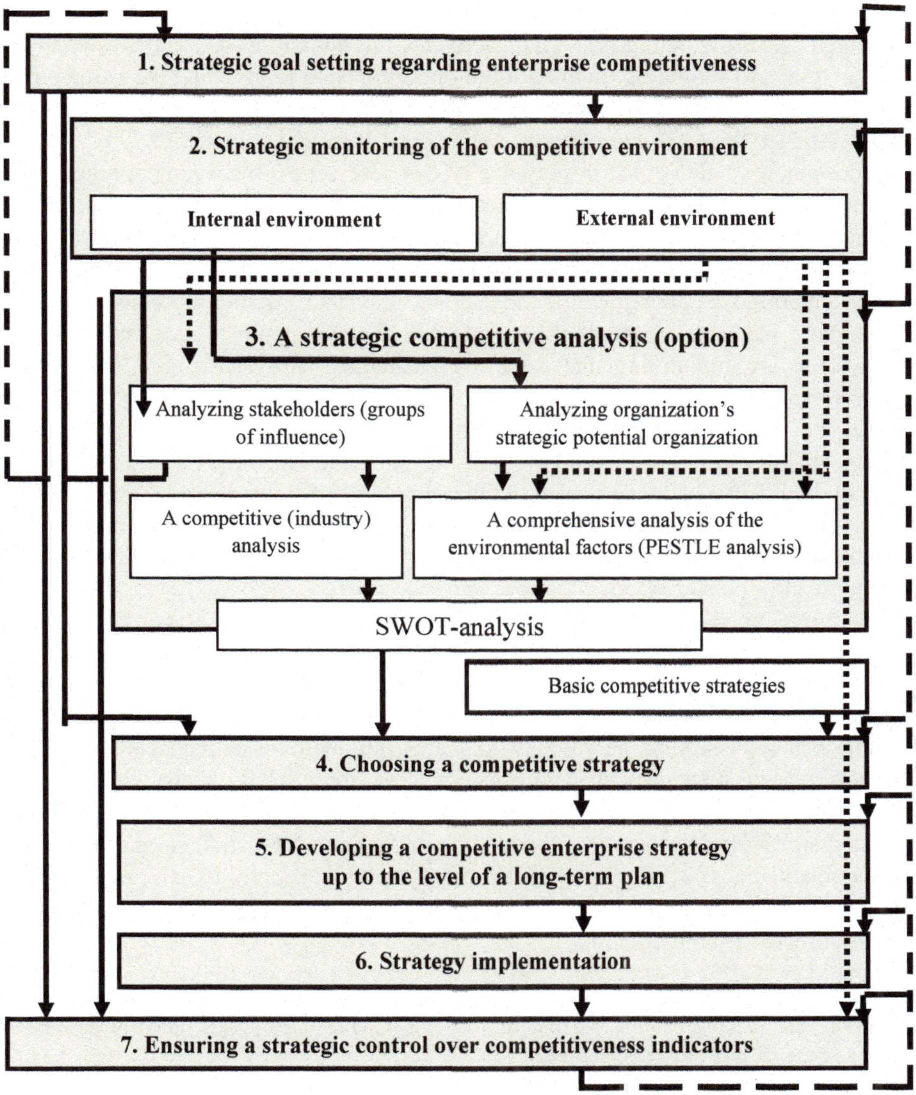

Fig. 1. Stages in the strategic management of enterprise competitiveness.

particular value period in their favor, and (c) recognized by the external environment. The most important element of enterprise competitiveness is its competitive potential, which is a unique combination of competitive advantages and weaknesses of the enterprise. In turn, the enterprise's competitive advantage is a temporal (dependent and manageable) set of its characteristics that exceeds the set of competitors' characteristics and provides the company with the opportunity to redistribute the value created (in the industry) for a certain period in their favor, as well as to achieve better financial and economic indicators. Finally, the competitive weakness of the enterprise is the antithesis of the competitive advantage, subject to the temporal dynamics and a managed set of the enterprise's characteristics, inferior to the set of characteristics of competitors and providing them with the opportunity to redistribute the value created on the sales market (in the industry) for a certain period in their favor, as well as to achieve better financial and economic indicators, than this enterprise.

The authors believe that the essence of competitiveness management comes down to identifying and overcoming its competitive weaknesses, as well as identifying, forming, acquiring, accumulating, maintaining, and realizing its competitive advantages. The material and technical basis of this process, according to the authors, should be the enterprise's competitiveness management system, structurally consisting of the managing, managed, targeted, supporting, and functional subsystems, as well as the subsystem for monitoring the status of managed subsystems and the external environment.

In the context of digital transformations of socio-economic systems at various levels, digitalization of competitiveness management mechanisms at enterprises in Russia should also be implemented. At the same time, the comparative analysis carried out during the study allows us to conclude that at the enterprises of high-tech industries, this transformation should be involved, rather than focal, which would allow obtaining a significantly greater effect.

The integrated digitalization will combine the capabilities of both ERP and CRM systems integrated on the basis of a single platform, database management and mathematical modeling systems, innovative technologies of crowdsourcing, blockchain, the Internet of Things, business analysis of large amounts of data, cloud computing, machine learning, neural networks, artificial intelligence, management decision support, cloud management, and modern data sources. Its main directions and opportunities, identified during the study, are presented in Table 2.

It should be noted that the integrated digitalization of the strategic management of the competitiveness of a high-tech enterprise is a prerequisite for transforming the latter into an intelligent enterprise, the development of the concept of which is a further area of the research by the authors.

Table 2. The main directions and opportunities for the integrated digitalization of the strategic management of enterprise competitiveness.

Stages in strategic management of enterprise competitiveness	Digitalization directions, technological solutions	Expected effects
Strategic goal setting regarding competitiveness	Crowdsourcing	Increasing the objectivity of assessing competitive positions and management ambitions
Strategic monitoring over the competitive environment	Electronic media, organizations' websites, social networks, instant messengers, technologies of the Internet of Things,	Increasing the managers' situational awareness, reliability, completeness, timeliness of information; ensuring the

(continued)

Table 2. (*continued*)

Stages in strategic management of enterprise competitiveness	Digitalization directions, technological solutions	Expected effects
	blockchain, integration of information systems, CRM, database management	uniformity of the information environment; releasing the personnel from routine operations, increasing labor productivity
Strategic competitive analysis	Intelligent data processing based on business analysis technologies for big data, cloud computing, machine learning methods, neural networks, artificial intelligence	Increasing the objectivity of analytical results, identifying a greater number of competitive advantages, weaknesses, opportunities, and threats of the competitive environment; releasing the personnel from routine operations, increasing labor productivity
Selecting a competitive enterprise strategy	The technology support for management decision-making	Increasing the objectivity of evaluating alternatives, reducing the risk of error, and accelerating decision making
Developing a competitive enterprise strategy	The mathematical modeling system using the digital twin of the enterprise	Releasing the personnel from routine operations, increasing labor productivity
Implementing an enterprise competitive strategy	The cloud management technologies, comprehensive automation of business processes, the ERP system	Increasing the speed of interactions between business units, increasing labor productivity
Strategic control over enterprise competitiveness	ERP & CRM systems	Increasing the managers' situational awareness, reliability, completeness, timeliness of obtaining information, objectivity of assessing a competitive position of the enterprise, releasing the personnel from routine operations, increasing labor productivity

4 Conclusion

The authors believe that digital transformations of the key mechanisms competitiveness management of enterprises in high-tech sectors of the Russian economy will provide the opportunity for the practical implementation of an integrated approach for solving the urgent tasks in the current geopolitical and socio-economic conditions of increasing their competitiveness, as well as improving the quality of the data used in the information environment of enterprises, situational management awareness, decisions objectivity, and labor productivity. The basic functions of management determine the key directions in the integrated digitalization of the strategic management of the competitiveness of a high-tech enterprise and involve using a large number of modern systems, technological solutions, methods, and devices integrated into a single digital platform – the prototype of an intelligent enterprise.

References

Bogoviz, A.V., Lobova, S.V., Alekseev, A.N., Koryagina, I.A., Aleksashina, T.V.: Digitization and internetization of the Russian economy: achievements and failures. In: Advances in Intelligent Systems and Computing, vol. 622, pp. 609–616 (2018)

Khachaturyan, A.A., Ponomareva, S.V., Melnikova, A.S.: Long-term planning of assets and scientific and technological development of the economy of branch markets in digitalization conditions. St. Petersb. State Polytech. Univ. J. Econ. **32**, 396–428 (2018). https://doi.org/10.18720/IEP/2018.4/17

Lobova, S.V., Ragulina, Y.V., Averin, A.V., Simonov, S.Y., Semenova, E.I.: Methods of digitization of the Russian economy with the help of new internet technologies. In: Advances in Intelligent Systems and Computing, vol. 726, pp. 221–228 (2019)

Ponomareva, S.V., Khachaturyan, A.A., Melnikova, A.S.: Impact of digitalization and industrialization on asset planning and scientific and technological development of the production and economic system of the Russian Federation. St. Petersb. State Polytech. Univ. J. Econ. **11**(4), 60–69 (2018). https://doi.org/10.18721/je.11404

Ragulina, Y.V., Shkodinsky, S.V., Mishchenko, V.V., Romanova, Y.A.: Scenarios of development of INDUSTRY 4.0 in the conditions of knowledge economy formation and their consequences for modern economic systems. Stud. Syst. Decis. Control **169**, 227–234 (2019)

Prospects for Developing a Competitive Position of the Russian Aerospace Industry in the World Market in Conditions of Economic Digitalization

Igor V. Chistov[1(✉)], Sergey E. Zakutnev[1], and Aleksandr A. Ryazanov[2]

[1] Military University of the Ministry of Defense of the Russian Federation, Moscow, Russia
ivchistov@mail.ru, szakutnev@yandex.ru
[2] Moscow Witte University, Moscow, Russia
alekryazanov@yandex.ru

Abstract. The paper focuses on the results of a comprehensive study conducted by the authors on the main modern directions of domestic space activity and the competitive position of the Russian aerospace industry in the world space market, which revealed the loss of the leadership in almost all areas of space exploration and the mismatch of the market share occupied by Russia with the status of a leading world space power. The paper analyzes the impact that global economic digitalization has on the competitive position of the Russian aerospace industry in the global space market, discusses the need for targeted management of industry competitiveness, and considers the structure of this process. In so doing, special attention is paid to the formation of the industry information ecosystem. The author's definitions of the world market of space, industry competitiveness, and digitalization are also provided in the paper and rigorously discussed.

Keywords: Space activities · Aerospace industry · World space market · Digitalization · Industry competitiveness · Industry competitiveness management · Information technology development · Industry information ecosystem

1 Introduction

Modern space activity is an essential, integral component of science, economics, ensuring military security and international cooperation. Since its inception, it has belonged to those areas where Russia had the leading position. However, the collapse of the USSR, de-industrialization of the post-Soviet space in the 1990s, economic crises, anti-Russian policy of the United States and its allies, and serious miscalculations on behalf of the leadership in the domestic aerospace industry has led to the loss of Russian leadership in almost all areas of space exploration (Table 1). The most crucial factor that has a significant impact on all spheres of public life at present is

A. V. Bogoviz and Y. V. Ragulina (Eds.): ISCI 2019, LNNS 115, pp. 115–122, 2020.
https://doi.org/10.1007/978-3-030-40749-0_14

digitalization, which is both a threat and an opportunity for the development of space aeronautics in Russia.

Thus, the research topic is relevant due the need for developing a scientific understanding of the role and place of digitalization in managing competitiveness of socio-economic systems. More than that, there are still demanding questions on the insufficient theoretical understanding of the nature and content of digitalization, as well as the practical significance of the research related to the development and justification of promising ways to increase the competitiveness of leading branches of the Russian economy.

The need to fundamentally improve the competitive position of the Russian aerospace industry in the global market of space under the ongoing digitalization of global economy necessitates solving a scientific problem consisting in the development of theoretical and methodological provisions for the formation of the industry information ecosystem, which is an element of the competitiveness management system of the industry. The purpose of the study is to solve the scientific problem of developing theoretical and methodological provisions for the formation of the industry information ecosystem of the aerospace industry of the Russian Federation. The solution to this scientific problem is concretized in the form of a set of interrelated tasks: to investigate the essence of the world space market, analyze industry competitiveness and digitalization, discuss the current state and prospects for the development of the space activities of Russia. Another set of questions is related to the competitive position of the Russian space industry in the global space market in the context of digitalization. There is also a strong need to identify the influence features that digitalization has on the competitive position of the industry. Developing theoretical and methodological foundations for the formation of the industry information ecosystem of the aerospace industry of Russia as an element of the competitiveness management system of the industry.

Theoretical and methodological foundations of the digitalization of socio-economic systems have been substantiated and developed in the work of many foreign and domestic scholars, such as (Bogoviz et al. 2018); (Ponomareva et al. 2018), (Lobova et al. 2019), (Ragulina et al. 2019), and others. However, there are not enough comprehensive studies devoted to the analysis of the impact of digitalization on the competitiveness of socio-economic systems.

2 Materials and Methods

In the process of solving a scientific problem, such methods of studying socio-economic phenomena as dialectic, systemic, comparative analytical, historical and logical, economic-statistical, and expert assessments, as well as the methods of scientific abstraction, induction and deduction, normative and positive analysis and synthesis were used.

Table 1. Main directions of space activity of the Russian Federation.

Main directions of space activity		Position of the Russian Federation	Indicators
Manned space exploration	Human-crewed spacecraft launches	World leadership	Russia is one of 2 countries operating the GAC, absolute leadership in the number of launches of human-crewed spacecraft
	Transport spacecraft launches	Lost: world leadership, parity with the USA	Russia is one of 4 countries launching transport spacecraft; it has unstable leadership in the number of the launches, and it lags in this field behind the USA and Japan
	Construction and operation of inhabited orbital stations	Lost: world leadership, parity with the USA	Technology transfer to the USA; key participation in the project of International Space Station (ISS)
	Preparation of a human-crewed flight to the Moon and Mars	Lagging behind the USA	Leading the "Lunar Program" after NASA astronaut's landing on the Moon; necessary technological capacity is present
Launching space carrier rockets (SCR)		Lost: world leadership and parity with the USA	Russia has lost its technological leadership and leadership in the number of SCR launches; a high accident rate of SCR; absence of a new SCR generation
Operating the ground-based infrastructure of space activities		Lagging behind the USA and China	Launching the *Vostochny* spaceport in 2016; absence of the infrastructure there to launch a human-crewed spacecraft; gaining control over the project of the spaceport *Morskoy Start* in 2016
Operating the orbit groups of uncrewed space vehicle (USV)		Lagging behind the USA, EU, and China	Russia has lost the third place in terms of the number of USVs in the orbit group; there is a technological lag of Russian space instrument engineering from the USA and the EU; a high share of imported electronic component base; 50% of domestic USVs are not meeting the world level; a high accident rate of the Russian USV; a fail to satisfy part of the needs of the national economy with the orbit group of USV

(*continued*)

Table 1. (*continued*)

Main directions of space activity	Position of the Russian Federation	Indicators
Deep space exploration	Lagging behind the USA, EU, China, Japan, and India	Not a single Russian-made USV was withdrawn from the near-earth orbit in 1990–2018; one Russian-European USV was withdrawn

Sources: (Makarov and Payson 2009), (Bzhilianskaya 1997), (Gingerich et al. 2015), (Iliopoulos and Esteban 2020), (Grimard 2012), (Borowitz and Battat 2016).

3 Results and Discussion

The institutionalization of the commercial area of space activities has led to the formation and development of the world space market (WSM), which the authors consider possible to define as a system of domestic and international goods/money relationship arising from the implementation of space activities and the use of its results in the fields of economy, defense, politics, science, education, and culture. Currently, experts identify the segments of launch services, spacecraft production (SC), production of ground-based space equipment, and satellite services in the structure of the WSM (Iliopoulos and Esteban 2020).

The analysis of the competitive position of the enterprises of the Russian aerospace industry in these WSM segments allows us to conclude that it does not correspond to the status of the profile industry of the leading world space power. The domestic companies are practically not represented in the segments of satellite services and the production of ground equipment. The insignificant (2%) presence of Russian enterprises in the spacecraft production segment of the world space market is due, first of all, to the technological lag in Russian space instrument engineering. The largest market share (10%) of the Russian aerospace industry is in the launch services segment of the WSM. However, the primary sources of revenue of *Roscosmos Group of Companies* in this segment (services for preparing, delivering to the ISS and returning NASA astronauts to the Earth, deliveries of *Soyuz-ST* SCR to the French company *Arianespace* for launching at the Kourou spaceport) cannot be considered stable in modern geopolitical conditions even in the medium term (Erygina and Serdyuk 2014; Bauer et al. 2012; Roscosmos 2019).

The most important factor that has a significant impact on all spheres of public life at present is digitalization, which, according to the authors, represents a significant socio-economic transformation based on strategic data management, mass introduction, and development of new ideas, processes, models, and technological solutions related to digitalization of data. Under the influence of the digitalization of the global economy, the WSM configuration changes significantly. The analysis carried out by the authors revealed the stagnation of its traditional segments (spacecraft production and the provision of launch services) and the dynamic growth of revenue of ground equipment manufacturers and satellite service providers.

Since the Russian space industry is currently represented only in the traditional WSM segments, and its main competitive weakness is the technological lag behind the US and the EU, the main challenges to digitalization for the industry's competitive position are its further reduction in the traditional WSM segments, up to the complete crowding out of the Russian aerospace technology from the international arena. One more challenge is the emergence of technological entry barriers, that are insurmountable for the industry, to more profitable segments of production ground space equipment and the provision of satellite services (Table 2).

Table 2. Main challenges of digitalization of the world economy for the competitive position of the aerospace industry of Russia on the WSM.

Main challenges of digitalization of the global economy for the industry	Prospects for the development of a competitive position in the industry
An increased technological lag behind the US and the EU; a reduced quality level; an increased accident rate; Russian aerospace technologies loosing competitiveness, along with space launches and satellite services	Medium-term: Reduction of the market share occupied by the industry in the launch services and SC production; emergence of insurmountable technological entry barriers for the industry in the segments of the production of ground-based equipment, and satellite services Long-term: Ousting of the domestic aerospace technology from the market; emergence of insurmountable technological entry barriers for the industry in the emerging segments of orbital production, maintenance, and repair of spacecraft; expansion of foreign manufacturers into the domestic market

On the other hand, the accelerated digitalization of the aerospace industry of the Russian Federation, according to the authors, will help to fundamentally improve the competitive position of the industry in the WSM. At the same time, the process of digitalization of the industry should be carried out as part of the targeted management of its competitiveness. The authors consider it possible to define the competitiveness of the industry as a multidimensional concept characterizing the dynamic, manageable property of the industry, and determined by the competitive potential (a unique combination of competitive advantages and competitive weaknesses of the industry) that is not inferior to similar industries of other countries and other sectors of the national economy. There is a possibility to use competitive advantages to redistribute the value created for a certain period in a particular segment of the world and national economy, in their favor, as well as achieving better development indicators than other sectors of financial and economic.

Table 3. Prospects for developing a competitive position of the aerospace industry of Russia on the WSM under the influence of the industry information ecosystem.

Main directions in the formation of the industry information ecosystem	Main expected beneficial effects	Prospects for developing a competitive position in the industry
Implementing a distributed integrated management system (DIM-system)	Improving the quality of aerospace technology (AT), launch and satellite services, reducing their cost	Medium-term: Increasing the market share of the industry in launching services and spacecraft production segments, entering segments of the production of ground-based space equipment and satellite services, developing orbital production segments, maintenance, and repair of spacecraft Long term: World leadership on a new technological basis
Digitalization of production and life cycle of aerospace technology (implementation of PLM-system)	Reducing the launching time for a new samples of AT at the WSM; improving the quality of AT, launch and satellite services; creating conditions for import substitution of electronic components, compensating for the technological lag of space instrumentation, technological breakthrough in the production of ground-based space equipment, orbital production, orbital maintenance and repair of spacecraft	
Implementing a mathematical modeling system	Reducing the launching time of new samples of AT at the WSM	
Using the "Big Data" technology	Increasing the speed of managerial decision making, creating the conditions for a technological breakthrough in the field of competitiveness management of the industry	
Implementing a new service paradigm	Improving the quality of launching and satellite services, creating conditions for the implementation of pilot business projects	
Using technology for industry data analysis based on artificial intelligence	Business processes optimization, industrial robotics development, creating conditions for a technological breakthrough in the field of managing competitiveness of the industry, orbital production, orbital maintenance and repair of spacecraft	

(*continued*)

Table 3. (*continued*)

Main directions in the formation of the industry information ecosystem	Main expected beneficial effects	Prospects for developing a competitive position in the industry
Implementing a system for supporting management decision-making	Improving the quality of managerial decisions, including those in the field of industry competitiveness management	
Developing a corporate infrastructure for information technology	Ensuring the effective interaction of industry processes and ecosystems	
Supporting the personnel working at the Roscosmos Group of Companies	Full integration into the industry information ecosystem	

The authors believe that the strategic component (strategic competitive analysis, strategic goal-setting, development, implementation, and control of the industry's competitive strategy) should become the basis for managing industrial competitiveness. The leading directions in competitiveness management at the functional level are, first of all, marketing and resource directions (financial, material, technical, scientific, technological, innovative, personnel, etc.), which necessitates the development and implementation of relevant industry functional strategies. At the same time, in the conditions of digitalization, a sectoral strategy for information technology development must be developed and implemented, aimed at forming the industry information ecosystem and fundamentally improving its competitive position in the WSM (Table 3).

4 Conclusion

The authors believe that the digitalization of the world economy poses serious challenges to the competitive position of the Russian aerospace industry in the global space market. In the medium term, it can manifest itself in crowding out the domestic aerospace technology created on the basis of new technological solutions, as well as in creating conditions for the expansion of foreign spacecraft manufacturers into the domestic space market.

At the same time, the forced digitalization of the industry carried out as part of its comprehensive technological modernization will fundamentally improve the competitive position of the Russian aerospace industry and create conditions for restoring Russia's world leadership in the field of space research and exploration.

References

Bauer, V.P., Kovkov, J.V., Moskva, A.M., Senchagov, V.K.: The state and mechanisms of development of the rocket and space industry in Russia. RAS Institute of Economics, Moscow (2012)

Borowitz, M., Battat, J.: Multidisciplinary evaluation of next steps for human space exploration: technical and strategic analysis of options. Space Policy **35**, 33–42 (2016)

Bogoviz, A.V., Lobova, S.V., Alekseev, A.N., Koryagina, I.A., Aleksashina, T.V.: Digitization and internetization of the Russian economy: achievements and failures. In: Advances in Intelligent Systems and Computing, vol. 622, pp. 609–616 (2018)

Bzhilianskaya, L.: Russian launch vehicles on the world market: a case-study of International joint ventures. Space Policy **13**(4), 323–338 (1997)

Erygina, L.V., Serdyuk, R.S.: Growth trends of Russian space industry. Siberian J. Sci. Technol. **1**(53), 207–211 (2014)

Gingerich, D.E., Forrest, J.S., Abiin, A.A., Maricle-Fitzpatrick, T.M.: The Russian R-16 Nedelin disaster: an historical analysis of failed safety management. J. Space Saf. Eng. **2**(2), 65–73 (2015)

Grimard, M.: Will the US remain the real leader of human space exploration? A comparative assessment of space exploration policies. Acta Astronaut. **75**, 1–14 (2012)

Iliopoulos, N., Esteban, M.: Sustainable space exploration and its relevance to the privatization of space ventures. Acta Astronaut. **167**, 85–92 (2020)

Lobova, S.V., Ragulina, Y.V., Averin, A.V., Simonov, S.Y., Semenova, E.I.: Methods of digitization of the Russian economy with the help of new internet technologies. In: Advances in Intelligent Systems and Computing, vol. 726, pp. 221–228 (2019)

Makarov, Y., Payson, D.: Russian space programmes and industry: defining the new institutions for new conditions. Space Policy **25**(2), 90–98 (2009)

Ponomareva, S.V., Khachaturyan, A.A., Melnikova, A.S.: Impact of digitalization and industrialization on asset planning and scientific and technological development of the production and economic system of the Russian Federation. St. Petersb. State Polytech. Univ. J. Econ. **11**(4), 60–69 (2018). https://doi.org/10.18721/je.11404

Ragulina, Y.V., Shkodinsky, S.V., Mishchenko, V.V., Romanova, Y.A.: Scenarios of development of Industry 4.0 in the conditions of knowledge economy formation and their consequences for modern economic systems. In: Studies in Systems, Decision, and Control, vol. 169, pp. 227–234 (2019)

Roscosmos: Year report – 2018 (2019). https://www.roscosmos.ru/media/img/2019/august/godovoi.otcet.goskorporazii.roskosmos.2018.g.pdf

The Platform of Information and Economic Security of the Russian Industrial Enterprises

Harutyun H. Khachaturian[1(✉)], Svetlana V. Ponomareva[2], and Anna S. Melnikova[2]

[1] Market Economy Institute of the Russian Academy of Sciences, Moscow, Russia
karutyun@yandex.ru
[2] Perm National Research Polytechnic University, Perm, Russia
psvpon@mail.ru, dtw@yandex.ru

Abstract. This paper is devoted to creating an information and economic security platform of Russian industrial enterprises by taking into account the digital development vector. The authors discuss such topical problems as (1) the lack of basic terms in-country legislation and (2) the need to create an information and economic security platform of industrial enterprises. This new scientific research is represented by two significant positions (1) the model of a sustainable Russian security development formed by five elements (economic, social and environmental reliability, with the addition of information and digital security, as the IT industry, artificial intelligence, nanobiotechnology, research and development of Russian scientists theorists, and practitioners currently being implemented) was built; (2) the platform of Russian information and economic security was prepared by taking into account the negative determinants of the country's development digital vector. As the study results reveal, the authors concluded that it would be reasonable to clarify the terms "social security" and "digital security" at the legislative level. They proposed to introduce the concept of the sustainable Russian security development model, with the addition of the innovative and the security, digital and informational, components to the traditional ones (economic, sociological, and environmental safety).

Keywords: Industrial enterprises · Security · Model · Digitalization · Industrialization

1 Introduction

The retrospective origins of the sustainable development concept have got their start in the 1970s and are based on economic, sociological, and environmental elements. The research rationale is connected with the transition of the Russian Federation to Industry 4.0 and the processes of digitalization. The study object is Russian industrial enterprises; the subject is the information and economic security of industrial facilities. The study purpose is to build a model of sustainable Russian security development, based on the analysis of domestic regulatory literature, and formed by five elements

A. V. Bogoviz and Y. V. Ragulina (Eds.): ISCI 2019, LNNS 115, pp. 123–129, 2020.
https://doi.org/10.1007/978-3-030-40749-0_15

(economic, social and environmental reliability, to which information and digital security are added), and to prepare a platform of Russian information and economic security, taking into account the negative determinants of the country's development digital vector. It makes sense noting that, during the transformation processes in scientific and technological development, socio-economic systems digitalization, and information processes globalization, domestic and foreign scientists are faced with the issue of changing the sustainable development concept, taking into account the trends in the security actualization of all the infrastructures of the society, industry, and the state.

2 Materials and Methods

The authors used general scientific methods of scientific cognition, as well as the hypothetical-deductive method of scientific cognition and modeling, to build a model of sustainable Russian security development, formed by five elements (economic, social, and environmental reliability, with the addition of the information and the digital security, as the IT industry, artificial intelligence, nanobiotechnology, research and development of Russian scientists theorists, and practitioners are currently being implemented). Recently, the research topic has been popular with scholars. The following foreign and domestic scientists have significantly contributed to the topic of the research: (Khachaturyan et al. 2018); (Alekseev et al. 2018); (Ogurtsov et al. 2016); (Meissner et al. 2017); (Herter and Ovtcharova 2016); (Geissbauer et al. n.d.); (Schlaepfer et al. 2017); (Ponomareva and Zheleznova 2018); (Bresnahan and Yin 2017); (Chen and Kamal 2016). The research authors have repeatedly addressed the problems of industrialization (Khachaturyan et al. 2018; Ponomareva et al. 2018), informatization (Melnikova 2018), and digitalization (Khachaturyan and Ponomareva 2018).

3 Results and Discussion

The authors propose to introduce the concept of the model of the Russian security sustainable development, with the innovative and security (digital and informational) components, presented in more detail in Fig. 1, in addition to the traditional ones (economic, sociological, and environmental safety).

The extensive time period of the polemical review of the periodic and monographic literature allows concluding on the research problems, the essence of which is to update the data and the theoretical scientific work component due to the transition of social, economic, technological, and other processes to the digital progress vector. The model of the sustainable Russian security development is formed by five elements: economic, social, and environmental reliability, as well as information and digital security in connection with the current implementation of the IT industry, artificial intelligence, nanobiotechnology, and research and development conducted by domestic scientists, theorists, and practitioners in the studied sphere.

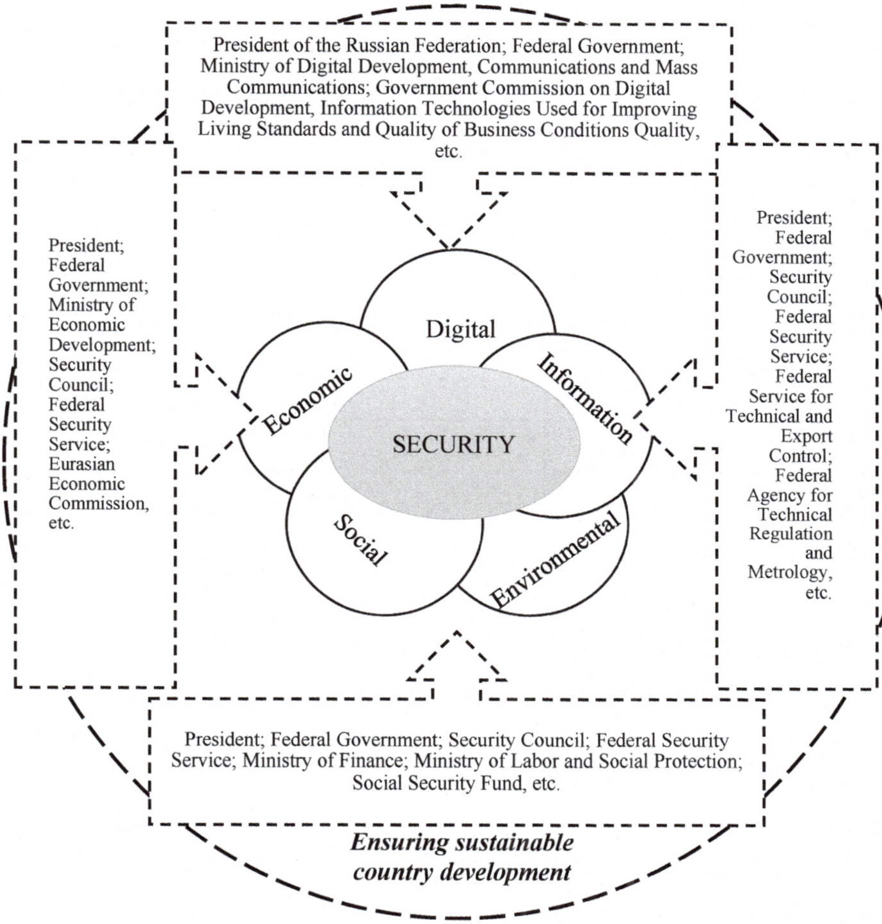

Fig. 1. The model of the sustainable security development in Russia. Source: Developed by the authors as a result of the study.

The model relies on regulators for each type of dedicated security of sustainable development, namely: digital, economic, social, and information protection, which forms a holistic and compact information infrastructure of high-tech industrial enterprises' information and economic security. A further step in the study of selected research issues could be the expansion of the controlling bodies of the industrial entities' environmental-economic component.

After studying the sustainable Russian security development using the hypothetical-deductive scientific knowledge method, there is a necessity to proceed to consideration of the constructed model components (Fig. 1), presenting the Russian information security platform (Fig. 2), taking into account the detection of country's digital progress negative factors.

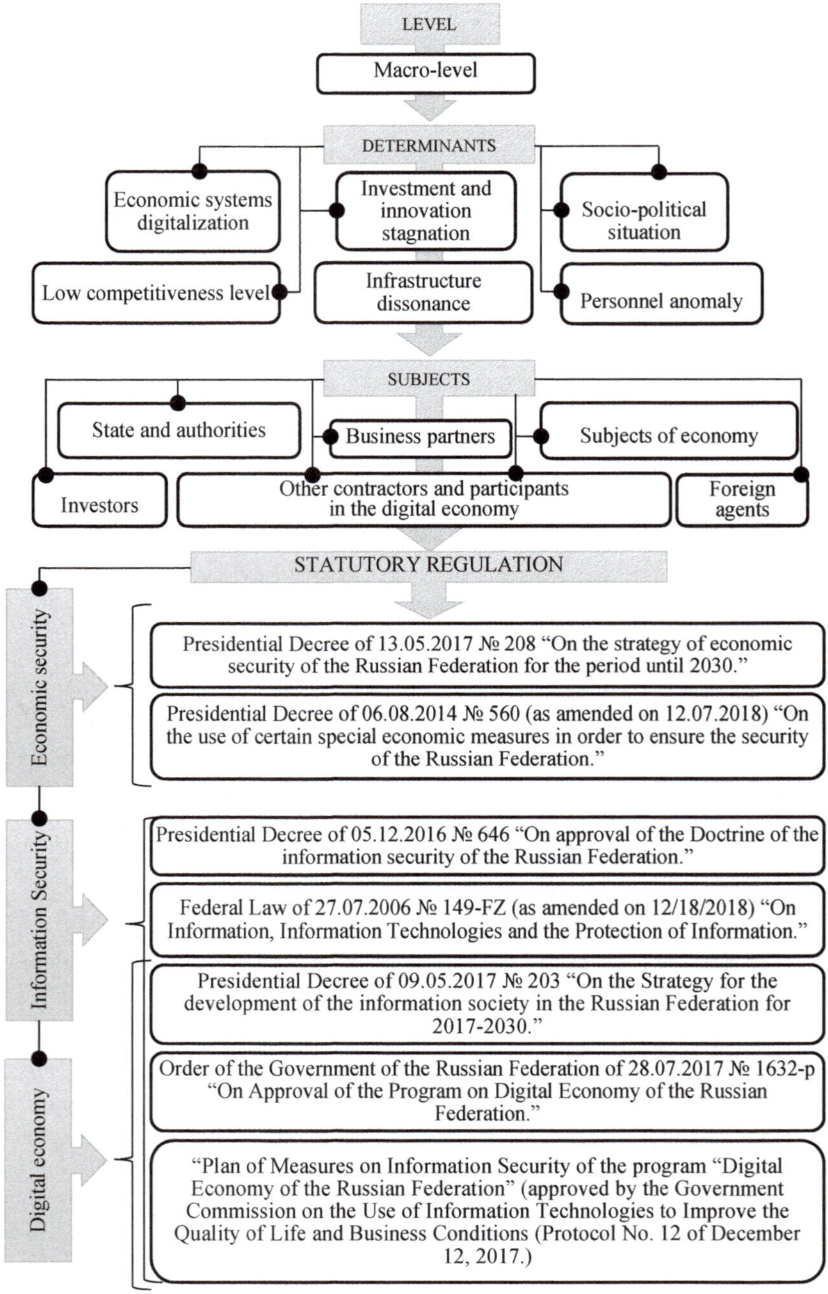

Fig. 2. The information security platform developed in Russia.

The macroeconomic level is considered in the platform model. The identification of problematic factors was based on the economic security strategy presented in May 2017 in the decree of the president of Russia. Threats to economic security include six determinants: economic systems digitalization; investment and innovation stagnation; infrastructure dissonance; low competitiveness; social and political situation; and personnel anomaly. Participants of the information and communication security platform are the state and municipal authorities, high-tech industrial enterprises and other real economy entities, investors and business partners, foreign agents and other counterparties, participants in the country's information, and economic security. Regulations are divided into three segments: economic security, information security, and digital security (digital economy).

Based on the information-economic security platform presented in Fig. 2 with the digitalization trend and socio-economic globalization, the main challenge that Russian scientists and the Russian government face is clarifying the terms "social security" and "digital security" at the legislative level.

4 Discussion

The processes of collaboration among international agencies is understood to be part of economic security, but they are not applied to information security. Only the national standards of the Russian Federation's doctrine of information protection have been developed. The renewal of the term "digital economy" by Russian and foreign scientists also implies the economic security available at the regulatory level in digital development, although the author's platform revealed the extent of information security in digital developments and the absence of the all-important economic security component. There exist such concepts as "social support," "social security," and "social insurance," and the law does not consider the term "social security" as a category. Moreover, no separate regulations can be found in the Russian regulatory legal acts, which requires further attention from scholars and researchers.

5 Conclusion

Taking into account the significant shortcomings identified in this paper, the authors presented a host of issues related to Russian information and economic security, taking into account the digital transformations in industry. The scientific novelty is represented by two significant positions: (1) the model of the sustainable Russian security development, formed by five elements (economic, social and environmental reliability, with the addition of information and digital security, as the IT industry, artificial intelligence, nanobiotechnology, research and development of Russian scientist theorists, and practitioners are currently being implemented) was built; (2) the platform of Russian information and economic security was prepared taking into account the negative determinants of the country's development digital vector.

References

Alekseev, A.N., Evdokimov, S.Y., Khachaturyan, K.S., Tarasova, A.Y., Khachaturyan, A.A.: The financial strategy of development of industry 40 in the countries with a developing economy. Espacios **39**(12), 1 (2018)

Bresnahan, T., Yin, P.-L.: Adoption of new information and communications technologies in the workplace today. Innov. Policy Econ. **17**(1), 95–124 (2017)

Chen, W., Kamal, F.: The impact of information and communication technology adoption on multinational firm boundary decisions. J. Int. Bus. Stud. **47**(5), 563–576 (2016)

Geissbauer, R., Schrauf, S., Koch, V.: Industry 4.0 – Opportunities and Challenges of the Industrial Internet assessment, pp. 5–8. PricewaterhouseCoopers (n.d.). https://www.pwc.nl/en/documents/pwcindustrie-4–0.pdf

Government of the Russian Federation: Order "On Approval of the Digital Economy of the Russian Federation Program" (July 7, 2017 No. 1632-p) (2017). http://base.garant.ru/71734878/

Government of the Russian Federation: The plan of activities in the direction of "Information Security" of the program "Digital Economy of the Russian Federation" (approved by the Government Commission on the use of information technology to improve the quality of life and business conditions (protocol №. 12 of December 12, 2017)) (2017). http://www.consultant.ru/

Herter, J., Ovtcharova, A.: Model-based visualization framework for cross-discipline collaboration in industry 4.0 scenarios. Procedia CIRP **57**, 398–403 (2016)

Khachaturyan, A.A., Khachaturyan, K.S., Abdulkadyrov, A.S.: Model of innovational development of the modern Russian industry. In: Advances in Intelligent Systems and Computing, vol. 622, pp. 44–50 (2018)

Khachaturyan, A.A., Ponomareva, S.V., Melnikova, A.S.: Long-term planning of assets and scientific and technological development of the economy of branch markets in digitalization conditions. St. Petersb. State Polytech. Univ. J. Econ. **32**, 396–428 (2018). https://doi.org/10.18720/IEP/2018.4/17

Khachaturyan, A.A., Ponomareva, S.V.: Scientific and technical development of a high-tech company and the digital economy development. In: SHS Web of Conferences, vol. 55, p. 01020 (2018). https://doi.org/10.1051/shsconf/20185501020

Meissner, H., Ilsena, R., Auricha, J.: Analysis of control architectures in the context of Industry 4.0. Procedia CIRP **62**, 165–169 (2017)

Melnikova, A.S.: Tools for the formation of the information infrastructure of high-tech industrial enterprises. News Ural State Univ. Econ. **19**(6), 123–134 (2018). https://doi.org/10.29141/2073-1019-2018-19-6-10

Ogurtsov, E.S., Rylov, A.A., Durdyeva, D.A., Lebedev, N.A., Khachaturyan, K.S., Safyan, A. M., Rochev…, K.V., Tsareva, G.R.: The role of business incubators in supporting economic growth and advancement of small business of the Russian Federation. Indian J. Sci. Technol. **9**(29), 89339 (2016)

Ponomareva, S.V., Zheleznova, I.V.: Intra firm planning and mathematical modeling of owner's equity in industrial enterprises. In: IOP Conference Series: Journal of Physics: Conference Series, vol. 1015 (2018). https://doi.org/10.1088/1742-6596/1015/3/032106

Ponomareva, S.V., Khachaturyan, A.A., Melnikova, A.S.: Impact of digitalization and industrialization on asset planning and scientific and technological development of the production and economic system of the Russian Federation. St. Petersb. State Polytech. Univ. J. Econ. **11**(4), 60–69 (2018). https://doi.org/10.18721/je.11404

Presidential Executive Office: Decree "On the application of certain special economic measures in order to ensure the security of the Russian Federation" (August 6, 2014 No. 560, as amended July 12, 2018) (2014). https://bazanpa.ru/prezident-rf-ukaz-n560-ot06082014-h2342454/

Presidential Executive Office: Decree "On the approval of the Doctrine of Information Security of the Russian Federation" (December 5, 2016 No. 646) (2016). https://bazanpa.ru/prezident-rf-ukaz-n646-ot05122016-h2878924/

Presidential Executive Office: Decree "On the Strategy for the Development of the Information Society in the Russian Federation for 2017-2030" (May 9, 2017 No. 203) (2017). http://base.garant.ru/71670570/

Presidential Executive Office: Decree "On the strategy of economic security of the Russian Federation for the period up to 2030" (May 13, 2017 No. 208) (2017). https://bazanpa.ru/prezident-rf-ukaz-n208-ot13052017-h2988663/

Schlaepfer, R.C., Koch, M., Merkofer, P.: Challenges and solutions for the digital transformation and use of exponential technologies. Deloitte, pp. 3–9 (2017). http://deloitte.com/content/dam/Deloitte/ch/Documents/manufacturing/ch-en/manufacturingindustry-4-0-24102014.pdf

State Duma of the Russian Federation: Federal Law "On Information, Information Technologies and Information Protection" (July 27, 2006, No. 149-FZ, red. December 18, 2018) (2006). http://www.consultant.ru/document/

The Resource Selection Model for Meeting the Information Needs of Users in the Sphere of Small Agribusiness

Mariya S. Kudasheva(✉)

Penza State Technological University, Penza, Russia
msi1958.20@mail.ru

Abstract. The study rationale is in the fact that the digitalization of the economy requires new approaches to organizing information services for the most vulnerable collective users in the field of agricultural production. The purpose of the study is to consider the process of selecting an information resource to meet the information needs of representatives of small agribusiness. The leading methods of the study are an abstract-logical method, the "5W + 1H" formula set, as well as information theories, probabilities, and statistical methods. The paper proposes a multi-criteria model for selecting an information resource to meet the information needs of the representatives of small agribusiness within the framework of a specific request, depending on the characteristics of the formula "5W + 1H." The processes of assessing particular criterial functions and compiling the objective function of the model are also considered. The use of a multicriterial selection model is demonstrated on the example of the informational need of an agricultural agrotechnical consumer cooperative for updating the material and technical resources. The significance of the research topic is determined by the fact that in most cases for small agribusiness, information is the only resource (capital) at the initial stage of operation. More than that, the search process is hampered by the lack of the necessary level of qualification and information culture. The multi-criteria model of selecting an information resource can find its application both in scientific research of information needs and in modeling and development of information technologies for agribusiness.

Keywords: Information need · Information request · Multi-criteria model · Function · Criterion · Agribusiness

1 Introduction

The creation of the final product value in agriculture, as in any production process, is accompanied by not only material and financial flows but informational flows, as well. With the modern realities of business operation within the digital economy, the information flow is becoming a key resource for competitiveness and business performance. Furthermore, considering the sector of smallholder farming in agricultural production, in many cases, information is the only agribusiness capital (Msoffe and Ngulube 2016).

A. V. Bogoviz and Y. V. Ragulina (Eds.): ISCI 2019, LNNS 115, pp. 130–138, 2020.
https://doi.org/10.1007/978-3-030-40749-0_16

The analysis of the structure of agricultural production in Russia reveals a tendency towards a significant contribution of smallholder farming to the agricultural production of the country, the stability of rural territories, and the implementation of the strategy of import substitution (Lyamin 2016). At the same time, small forms of managing experience difficulties providing financial resources, marketing products, and supplying the activities of the economy with adequate state support. These difficulties are interconnected with the inefficiency of the information flows and the incompetence of the representatives of small agribusiness in information exchange (Gandasari et al. 2018). At the same time, there is a system of information support for small business forms that do not meet the demands of the digitalization of the economy. One of the problems faced by the representatives of small agribusiness in the course of their activities is the difficulty in choosing a relevant and effective information source.

This situation has persisted for a long period and is exacerbated by the transition to a digital economy. The transition requires the search for new ways to increase the efficiency of the already created elements of the agricultural consulting system. It is proposed to consider the process of information support of small agribusiness from the concept of information needs as a new direction. Moreover, at the first stage of this scientific search, it is proposed to pay attention to the choice of information resources to meet the information needs of users in the field of small agribusiness.

2 Materials and Method

The methodology of the study is based on the abstract-logical method used for constructing the theory and including various techniques and operations from which analysis, synthesis, deduction and induction, formal logic, and hypothetical assumption will be used. When considering definitions and basic concepts in the study, the monographic and evolutionary methods were used. In order to describe the model of the user's information needs in agribusiness, the Kipling method is used (formula "5W + 1H").

The creation of a multiple-criteria model for selecting an information resource is carried out using the tools of set theory, information theory, probability theory, and methods of mathematical statistics. In order to evaluate the parameters and calculate particular criterion functions, the expert assessment is used. The representatives of small agribusiness, namely the chairmen of the agricultural consumer cooperatives in the field of agrotechnical services, acted as the experts.

3 Results

The need for a business entity for information is called information needs. In agribusiness, smallholder farming (which includes farms, personal farmsteads, cooperatives, and individual entrepreneurs) can be considered collective users of information.

The information value is determined by its ability to contribute to the achievement of any goals. This concept can be considered from the point of view of the place of the representative of agribusiness in the process of satisfying information needs. A smallholder farming can act in two roles: as a producer of information and as the consumer of the information.

The primary information need of the information producer is all the available information about the distribution channels for services (products) (Flores et al. 2019). As an information consumer, an agribusiness representative makes an information request reflecting his need for information on state support and opportunities to attract financial resources, on new equipment and technologies in agricultural production, and information on upcoming events (seminars, fairs, conferences), etc.

As a rule, the subject receives a set of data on the issues of interest from different sources. The entire set of available data can be converted into the information during the processing and into knowledge during the application. The availability of information, accumulated experience, and knowledge is often the only resource (capital) of smallholder farming in agricultural production.

The information behavior of small agribusiness is determined by the presence of a problem of access to reliable and relevant information and the complicated process of selecting information sources in accordance with an information request. The information request is formed depending on the model of the information needs of the user. It is proposed to use the formula "5W + 1H" as a toolkit for describing the model (Jia et al. 2016). This technique consists of posing six questions that determine the principles for satisfying the user's information needs (Barahnin and Fedotov 2008): where, who, why, what, when, and how.

With the use of this formula, it is possible to determine the essence of the informational interest of a particular source of informational needs, taking into account their subjective nature. It will allow one to formalize an information request and increase the efficiency of the process of meeting the information needs of the representatives of small agribusinesses by, for example, modeling the choice of information resource.

A description of the choice of an information resource to satisfy a specific information request of a user depending on the formula characteristics "5W + 1H" can be implemented using the multi-criterial selection model. It is proposed to use the objective selection method using sequential data streams (Salnikov 2016). The formation of a model for information resource selection consists of the implementation of the following processes:

1. The set of fundamental characteristics of the information needs of the subject w_i are isolated, and their maximum values $\gamma_{i\ are\ determined}$:

$$w_i^M = \{w_1, w_2, \ldots w_n | \gamma_1, \gamma_2, \ldots \gamma_n\}.$$

2. The parameters $w_{i\ are\ normalized}$ by the maximum value of γ_i, which will allow the obtaining of information needs coefficients $K_i = w_i/\gamma_i$.
3. The analytical expressions of the particular criteria functions $f^M(K_i)$ are formed. Among all the normalized parameters, the coefficients of the significant parameters K_i^s is singled out. They have a significant influence on the choice of an information resource to satisfy a specific information request of the user. Their criteria functions are denoted as $f^{(M)}(K_i^S)$.

4. The objective function of the selection model $J^{(M)}_{(x)}$ is formed as a dependence of the criterion functions:

$$J^{(M)}(x) = J^{(M)}\left\{\sum_{i=1}^{m} \alpha_i * f^{(M)}(K_i), f^{(M)}(K_i^S)\right\},$$

where α_i – the weighting factor of the criterion functions; m – the number of private criterion functions.
Weighting factors are considered as indicators of the relative significance of the initial parameters. If there are significantly different coefficients, the identification of their priority can be difficult, so in this case, it is set the same. The criterion functions of the essential parameters $f^{(M)}(K_i^S)$ are taken into account in the target function without weight coefficients; that is, they have a direct impact on its behavior.
5. Following the range of the values of the particular criterion functions, the values of the analytical model of the objective function are divided into intervals or zones R_i. Each zone is characterized according to the information needs. It is proposed to divide the values of the analytical model of the objective function into three zones: R1 – zone of low information content, R2 – zone of medium information content, and R3 – zone of high information content.
6. Many alternative options for information resources are being clarified to satisfy the information request within the framework of the information need, depending on the characteristics of the formula "5W + 1H":

$$X^{(M)} = \{X_1, X_2, \ldots, X_n\}.$$

7. Following the proposed set of alternative options for the information resources $X^{(M)}$, particular criteria functions are calculated, and the objective function value of the selection model $J^{(M)}_{(x)}$ is determined for each X.
8. The decision-making on the choice of an information resource under certain conditions is formed as follows: "Hitting the value of the analytical model of the objective function in the interval of values will determine the information resource belonging to one of the informative zones. The preferred resource is an information resource from the high information content zone."

When using the model of a multi-criterial selection of an information resource, it is proposed to use the following information properties as the initial characteristics to satisfy the information needs of small agribusiness:

- relevance R – the information ability to correspond to the information request;
- completeness P – the information property to comprehensively characterize the response to any information request;
- relevance A – the ability of information to maintain its usefulness (value) for satisfaction over time.
- Availability Ds – the information property to characterize the possibility of its receipt by the subject of the information needs;
- reliability Dt – the information property of not having hidden errors.

In order to assess these characteristics, the information and probability theories, and methods of mathematical statistics are used as tools. Since the maximum value of each characteristic is equal to one, then the stage of normalizing the characteristics is skipped, and many coefficients of information needs are obtained:

$$K_i = \{K_R, K_P, K_A, K_{Ds}, K_{Dt}\}$$

K_R, K_P, K_{Dt} were chosen as the essential parameters, as the most influential in meeting the information need. For the selected parameters, the use of the following particular criteria functions is proposed:

1. Relevance:

$$f^{(M)}(K_R^S) = \frac{N_P}{N_0},$$

where N_p – the amount of relevant information; N_0 – the total amount of the analyzed information.
In the analysis, the total number of sources viewed per request is meant by N_0, and the number of relevant sources among the total volume is meant by N_p (Chipiga 2014).
2. Completeness. In order to construct a particular criterion function of the parameter K_P, it is necessary to fill in the object-characteristic table of each source for satisfying the information need. The object-characteristic table is a two-dimensional matrix D, which columns give a list of the names of objects, processes or phenomena that are included in the circle of the informational needs of the subject, and the rows indicate the names of their characteristics (parameters), the values of which are necessary to satisfy the informational needs. A set of the object-characteristic tables for a specific model of information needs is described in accordance with the formula "5W + 1H" as information needs inventory. Next, let us denote by $d_{\mu\nu}$ an element located in the μ row and the νth column of the matrix D of the component under consideration of the corresponding information inventory, and:

$$d_{\mu\nu} = \begin{cases} 1, \text{if information for this item is available} \\ 0, \text{if information for this item is not available} \end{cases}$$

Then, as a particular criteria function for evaluating the completeness of information, we take the following formula:

$$f^{(M)}(K_P^S) = \frac{\sum\limits_{\forall\mu}\sum\limits_{\forall\nu} d_{\mu\nu}}{mn}$$

where m – the number of rows, n - the number of columns in the considered object-characteristic table (Chipiga 2014).

3. Credibility:

$$f^{(M)}(K^{S}_{Dt}) = 1 - \frac{d}{Q},$$

where d – the number of distortions; Q –the amount of information.
The partial criterion functions of parameters that are not included in the essential group have the following form:

1. Relevance:

$$f^{(M)}(K_A) = \frac{Z_t}{Z_{to}},$$

where Z_t - the usefulness of the information in the time of its use, Z_{t0} – the usefulness of information at the time it was received.

2. Availability:

$$f^{(M)}(K_{Ds}) = \frac{1}{2E}\sum_{e=1}^{D}(M_e),$$

where E – the objects of the possible number of the access to the information for satisfying the information needs; M_e – the object of the access to the information for satisfying the information needs; e – the number of the source of access to information.

In order to evaluate the parameters and calculate particular criteria functions, the use of an expert assessment is proposed. The representatives of small agribusiness were chosen as experts.

The objective function of the model will be considered as the following expression:

$$J^{(M)}(x) = f^{(M)}(K^{S}_{R}) + f^{(M)}(K^{S}_{P}) + f^{(M)}(K^{S}_{Dt}) + 0.5 * f^{(M)}(K_A) + 0.5f^{(M)}(K_{Ds}).$$

The weighting coefficients before the partial criterion functions of the parameters K_A and K_{Ds} take into account their relative importance when choosing an information resource to satisfy an information request within the informational needs of a representative of small agribusiness.

Since the range of the particular criteria functions varies from 0 to 1, the value of the objective function falls into the range 0 … 4. The interval of the values of the objective function is divided into three zones: (1) low informative zone (0–1.5); (2) medium informative zone (1,5–3); (3) high informative zone (3–4). The information resources falling into the zone of low information content will be considered unsatisfactory in relation to the information needs of a representative of small agribusiness. The information resources, which function value belongs to the zone of high information content, are considered a priority for meeting the information request within the information needs of the representatives of small agribusiness. The information resources, which value can be attributed to the medium informative zone, can be considered permissible to satisfy the information need.

4 Discussion

Let us demonstrate the application of the proposed multiple-criteria model for selecting an information resource in a particular case of the information need of a representative of small agribusiness. At the first stage, the information need is described using the formula "5W + 1H" (Table 1).

Table 1. The model of information needs.

Part of the formula "5W + 1H"	Characteristic
Where?	Agrotechnical works
Who?	Chairperson of the Agricultural Consumer Cooperative
Why?	Searching for the sources of the material and technical resources supply (new equipment acquisition)
What?	Manufacturers and suppliers of the material and technical resources, their products, prices, and working conditions Current information
When?	Immediately
How?	In the text or tabular form; electronically; in its original form

Source: Developed by the author.

Let us clarify the alternative options for implementing the selection model, depending on the characteristics of the formula "5W + 1H":

X_1 – the website of the manufacturer (supplier) of the material and technical resources;

X_2 – the website with the reviews of material and technical resources and their manufacturers (suppliers);

X_3 – a fair or an exhibition of the material and technical resources;

X_4 – direct contact with the manufacturer (supplier) of the material and technical resources;

X_5 - the official websites of the government agencies.

Let us estimate the model parameters. For this, the chairmen of agricultural agrotechnical consumer cooperatives located in the Penza region (Russian Federation) were involved as experts. The evaluation results are as follows:

1. Relevance: when assessing the relevance, let us proceed from the assumption that the total amount of information, that the user will get acquainted with, and the determination of its relevant part will be carried out in the number of sources. The expert assessment allowed to determine the following data: for X_1 – $N_o = 100$, $N_p = 15$; for X_2 – $N_o = 100$, $N_p = 20$; for X_3 – $N_o = 40$, $N_p = 8$; for X_4 – $N_o = 10$, $N_p = 8$; for X_3 – $N_o = 5$, $N_p = 2$.

2. Completeness: to evaluate this parameter, the object-characteristic table for each information source is composed. The processes or phenomena that pertain to the informational needs of the subject are indicated as objects: (1) data on the material and technical resource and (2) data on the maintenance of the material and technical resource. Their characteristics include (1) price, (2) the reliability of the supplier (manufacturer), (3) the reputation of the supplier (manufacturer), (4) quality, and (5) additional costs. Accordingly, matrices for each of the sources X are obtained.
3. Reliability: in assessing the reliability, we will proceed from the assumption that the determination of the amount of information and the number of distortions will be carried out in signs. The expert assessment allowed to determine the following data: for $X_1 - Q = 10\ 000$, $d = 120$; for $X_2 - Q = 10\ 000$, $d = 250$; for $X_3 - Q = 5\ 000$, $d = 100$; for $X_4 - Q = 4\ 000$, $d = 20$; for $X_5 - Q = 10\ 000$, $d = 120$.
4. Relevance: when assessing the relevance, let us proceed from the assumption that the determination of the usefulness of the information at the time of its use and at the time of its receipt will be carried out in percent. The expert assessment allowed to determine the following data: for $X_1 - Z_{to} = 90$, $Z_t = 85$; for $X_2 - Z_{to} = 90$, $Z_t = 85$; for $X_3 - Z_{to} = 90$, $Z_t = 65$; for $X_4 - Z_{to} = 95$, $Z_t = 93$; for $X_5 - Z_{to} = 60$, $Z_t = 10$.
5. Accessibility: after analyzing the description of the selection model depending on the characteristics of the "5W + 1H" formula, we came to the conclusion that there are six objects of access to information ($E = 6$): a computer, a phone (smartphone), a tablet, an electronic information board (stand) with the electronic search and the access to the owner's web-site, and a personal contact and a print edition. If we imagine a sequence of objects as a vector consisting of 1 - if the access to the information through the object is possible and 0 - if it is impossible. The following data is obtained: for $X_1 - M_1 = [1, 1, 1, 1, 0, 0]$, for $X_2 - M_2 = [1, 1, 1, 0, 0, 0]$, for $X_3 - M_3 = [0, 0, 0, 1, 1, 1]$; for $X_4 - M_4 = [0, 1, 0, 1, 1, 1]$; for $X_5 - M_5 = [1, 1, 1, 1, 0, 0]$.

The implementation of the model of multi-criteria selection of an information resource was carried out using the mathematical editor MathCAD.

As a result of the analysis, it was determined that, first of all, it is worth choosing the source X_4, direct contact with the manufacturer (supplier) of the MTR, as the information resource for satisfying the information need proposed for modeling. Based on the simulation results, the following rating of information resources can be compiled:

X_4 – direct contact with the manufacturer (supplier) of the MTR ($J^{(M)}(X_4) = 3{,}251$);

X_2 – the website with the reviews of material and technical resources and their manufacturers (suppliers) ($J^{(M)}(X_2) = 2{,}572$);

X_3 – a fair or an exhibition of the material and technical resources ($J^{(M)}(X_3) = 2{,}266$);

X_1 – the website of the manufacturer (supplier) of the material and technical resources ($J^{(M)}(X_1) = 2{,}077$);

X_5 – the official websites of the government agencies ($J^{(M)}(X_5) = 1{,}467$).

The study of the informational interest of a specific source of the informational needs within the framework of small agribusiness will allow the informational diagnostics of all processes in the information-consulting system, which involves information modeling of an object based on the known information about it, using special methods, techniques, and technologies to assess its condition and identify trend development.

5 Conclusion

The paper focuses on a multi-criteria model for selecting an information resource to meet the user's information needs in the field of agribusiness. The scope of the proposed model extends to the entire set of possibilities for modeling the information needs of collective users in the field of small agribusiness. Scholars, researchers, representatives of the organizations providing agribusiness infrastructure, authorities, and local governments, as well as other people interested in information and consulting services for agribusiness, can use this research. It is possible to adapt the model to a broader area of research of the information needs of collective users in various fields of activity by describing the information need using the "5W + 1H" formula and selecting the appropriate characteristics of the information need. The approach to the consideration of the information services for agribusiness through the selection of the resources with high information content in the frames of the information request can be used in the development of information resources by organizations in the market of information and communication technologies.

References

Barakhnin, V.B., Fedotov, A.M.: A study of the information needs of the scientific community to build an information model for describing its activities. Bull. Novosibirsk State Univ. (Ser. Inf. Technol.) **6**(3), 48–59 (2008)

Chipiga, A.F.: Information Security Basics. Publishing House of SKFU, Stavropol (2014)

Flores, H., Villalobos, J.R., Ahumada, O., Uchanski, M., Meneses, C., Sanchez, O.: Use of supply chain planning tools for efficiently placing small farmers into high-value, vegetable markets. Comput. Electron. Agric. **157**, 205–217 (2019)

Gandasari, D., Sarwoprasodjo, S., Ginting, B., Susanto, D.: Orchid consortium communication network in Indonesia. Pertanika J. Soc. Sci. Humanit. **26**(3), 1797–1814 (2018)

Jia, C., Yu, Y.T., Cai, Y., Tse, T.H.: 5W+1H pattern: a perspective of systematic mapping studies and a case study on cloud software testing. J. Syst. Softw. **116**, 206–219 (2016)

Lyamin, B.M.: On the role of small business forms in increasing the stability of rural territories. Russ. Electron. Sci. J. **3**(21), 65–78 (2016)

Msoffe, G.E.P., Ngulube, P.: Agricultural information dissemination in rural areas of developing countries: a proposed model for Tanzania. Afr. J. Libr. Arch. Inf. Sci. **26**(2), 169–187 (2016)

Salnikov, I.I.: Methods of Multicriterial Selection of Means for Implementing the Designed Information Technology System Under Heterogeneous Conditions. Privolzhskiy Dom znaniy, Penza (2016)

The Platform Format of Technological Development as a Manifestation of NBIC Convergence Configuration

Yuri A. Dmitriev[1], Tatyana A. Lachinina[2], Maksim A. Chirkov[3], and Maksim S. Chistyakov[1(✉)]

[1] Institute of Economics and Management, Vladimir State University named after A. G. and N. D. Stoletovs, Vladimir, Russia
managvlgu33@mail.ru, shreyamax@mail.ru
[2] Russian Engineering Academy, Moscow, Russia
t_lachinina@mail.ru
[3] Political Economy Department, Lomonosov Moscow State University, Moscow, Russia
mospil@mail.ru

Abstract. A wide range of involved areas of knowledge indicates a transdisciplinary focus on the formation of the fundamental basis of the NBIC convergence and, in turn, generates a particular sequence of problem layers of a worldview, philosophical, and ethical nature. The scientific community and politicians are faced with the question of moral and humanistic character and social philosophy, connected with the influence of the so-called "big technological four" on the transformational restructuring of the civilization, and on distinction changes that will affect the life outlook of every person. Considering the evolutionary formation of NBIC convergence from the point of view of the dialectical significance of the technological component in the chain of perception of the model of world civilization, we can draw an analogy with the formation of technological platforms for promising evolutionary development, which in synergetic and emergent interaction will create a configuration of favorable living conditions and socio-economic background. In turn, this trend will determine the trajectory of significant aspects of society, making it possible to verify the shape of the influence of convergent technologies on them.

Keywords: NBIC convergence · Technology · Technological wave · Industrial revolution · Economic platform · Cyber-physical systems

1 Introduction

The world has entered an active race for new knowledge, allowing us to provide many new breakthroughs in technological development. Biotechnology, genetic engineering, alternative energy, nanotechnology, and additive, cognitive, and social technologies are all actively developing, building on traditional machine technologies. A transition to hybrid technologies is developing, where various combinations of machine and non-machine technologies, together with information technologies, are used as instruments

A. V. Bogoviz and Y. V. Ragulina (Eds.): ISCI 2019, LNNS 115, pp. 139–148, 2020.
https://doi.org/10.1007/978-3-030-40749-0_17

for regulating and directing natural processes to achieve human goals, thereby providing an open door for new technological revolutions (Bodrunov 2018).

The uniqueness of the fifth and sixth technological waves was the practical application of fundamental scientific knowledge. The concept of a "knowledge-based economy" was formed. From an economic point of view, it means that scientific knowledge is taking an increasing share in the added value. The term "innovation" means not just any improvement, but improvement created through the applied development of scientific knowledge (Bodrunov et al. 2013).

The basic scientific achievements of the VI industrial-technological wave, presented in the format of the NBIC convergence, are the synergistic quintessence of fundamental scientific principles and the results of the search for strategic technological solutions.

2 Materials and Methods

The theoretical and methodological basis for the discussion was the conceptual, scientific provisions and research results of domestic and foreign scientists on the evolution of technological waves. The development of modern society viewed through the prism of the coherent quintessence of NBIC technologies makes up the platform foundations of the VI technological wave. The general scientific methods of cognition (i.e., dialectical, complex, systemic, integrated, and cluster approaches or the method of comparative analysis) were used as research methods.

3 Results

In this perspective, the discussed topics will address the need for the verification of the convergence definition. Convergence is the process of merging and rapprochement, which leads to a stable equilibrium development. Given this interpretation, it is necessary to distinguish the semantic component of identically close concepts from a scientific point of view, such as diversification, which is the formation of new directions associated with existing ones, and diffusion, which can be positioned as intersubjective penetration of subject areas on a mutual conceptual basis. It is implied that convergence contributes to its conservation subject areas seeking rapprochement and interpenetration of scientific areas (directions) on the platform of priority and succession, as well as the parity of views and approaches, which, in a single symbiosis, provides the practical realization and importance of its transformation into technological solutions. The current reality implies the diversification not only of mutual penetration but also of symmetric reciprocal technology transfer, leading to changes in the shape of technological boundaries and also leading to certain qualitative changes in the complex interdisciplinary interaction at the crossroads of various scientific fields and industries.

The conceptual provisions of convergent technologies were formed relatively recently, at the beginning of the twenty-first century. Manuel Castells, one of the ideologists of this concept, noted the increasing role of the processes of convergence of certain technologies in highly integrated format systems, while the existing obsolete technological trajectories become less visible and significant (Castells 2006). This concept was based on the synergistic interaction of nanotechnology (N), biotechnology (B), information technology (I), and cognitive technology (C). In 2001, the US National Science Center proposed the acronym NBIC; the NBIC-convergence terminology was put forward in 2002 by Michael Roco (M.C. Roco) and William Bainbridge (W. Sims Bainbridge) (Kiuwer Academic Publishers, 2003). The authors presented the terminology in the report "Converging Technologies for Improving Human Performance," prepared at the World Technology Evaluation Center (WTEC) (Pride and Medvedev 2008).

The NBIC convergence is characterized by the following features that arise in the process of the convergence of the vectors of scientific and technological development: the interaction of intensive format in scientific and technological evolution; the extended range of sampling (coverage) of subject areas considered, analyzed and subjected to reactive influence; a significant synergy effect; and variational perspectives of qualitative dynamics of large-scale technological solutions.

In turn, NBIC economics has close points of contact with the theory of technological waves formulated by academicians Glazyev and Lvov (1986), which explains the mechanism of formation of the differentiation of territories; the basis is the level of social and technological development. From their point of view, the "technological wave" is a system of interconnected industries (including technological chains dependent on each other) with different technological levels, which can be considered as a subsystem of a more general economic system (Bodrunov and Galbraith 2017). In connection with scientific and technological progress, there is a transition from lower structures to higher, more progressive ones (Glazyev et al. 2009).

According to the theory of technological waves, consistent with the theory of long waves developed by N.D. Kondratyev, the state in its evolutionary development goes through a sequential series of economic cycles (about 40–60 years long), characterized by a specific production. At the same time, different scientists determine the change of cycles differently: according to G. Mensh, it is the transition to the next technological path; according to K. Freeman, it is a change in technological paradigms (systems); and according to V. Polterovich, it is the sequence of development from one innovative pause to another.

In 1975, the West German economist G. Mensch, in his book *Technological Path*, updated the idea of J. Schumpeter on the correlation of the intermittent introduction of clusters of large-scale basic technological innovations with long waves of economic dynamics.

According to E. Laszlo, human civilization is in a state of bifurcation, which can be interpreted as a branched orientation of the evolution of modern society. Such a critical historical stage requires the verification of optimal development strategies that allow the generation of vector directions for correcting the situations of numerous modern threats and the crisis of globalization development toward sustainable evolution, and the leveling out of the aggravations of environmental, energy, economic, and sociocultural problems.

Converged technologies penetrating everyday human reality will contribute to the corrective replacement of previous technological platforms (i.e., the transition to the next technological wave). Technological transition through NBIC convergence can be described as strategically critical. It will make changes to the vertical labor division, which in many ways, will complicate professional graduation in the vertical labor plane. Also, a shift in the social aspects of the development of civilization poses a certain threat.

As an example, let us cite the informational side of convergence, namely, Big Data technologies. These technologies enable the analysis of significant volumes of data with different variable scenarios, which allows for predicting the course of many social and economic processes with a relatively high degree of probability. With the development and complexity of these technologies, the forecasting can develop into a simulation of the socioeconomic reality's processes. It should be noted that a distinctive feature of these technologies is the ability to process information arrays, which are difficult to analyze in the traditional ways using the currently relevant tools. The development of convergent information technologies will allow civilization to analyze processes of a social nature, for example, crises. The virtual visualization of crisis scenarios will become a tool for leveling their possible manifestations. The reverse side of convergent information technologies is the so-called global gamification, which is fraught with the threat of their possible application to the correction of political processes and the design of a change in the political system. If we direct the vector of the development of information technologies in a positive direction, then it can contribute to the modeling of possible negative variations in political life, socioeconomic development, and society (in the general sense), which can arise under the influence of various economic, political, environmental, and technological factors.

The "digital component" of the NBIC convergence is capable of replacing both an object and a biological being through the formation of the so-called "digital image". Thereby, both the object of material nature and the saturation of information will be of value. Digital immortality acquires quite real outlines from the perspective of development of man and civilization (Pride and Medvedev 2008).

The formation of a new technological structure on the converged technology platform will be facilitated by the institutional conditions for inter-organizational interaction and the generation of the configuration of its network, created based on cluster technologies. In particular, J. Schummer points out that one of the fundamental reasons for the high competitiveness of successful regions is the presence of spatially closely concentrated inter-organizational network interactions (Schummer 2008). The contour (boundary) of the inter-organizational network as a system forms a high level of the interaction intensity between the intranet residents (elements), while this intensity significantly exceeds the one between them and external elements.

The evolutionary development and economic expansion of the NBIC convergence presuppose technological and infrastructural interconnection between companies, which, in turn, are important elemental links, including the formation of high-tech cluster structures. This relationship becomes the so-called trigger impulse in the development of a cluster formation on a different platform for the technological development of the NBIC economy.

In this context, the Klein-Rosenberg model of the innovative development vector (after the names of scientists S. Klein and N. Rosenberg) is of interest, based on the central line of innovation, which is a certain sequence of stages with a single focus—

from generating an idea to its implementation and output to the market. Also, the stages have not only direct connections but also inverse ones, allowing corrective actions of the results of the intermediate stages of the process. The resulting feedback loops between the knowledge generation, a technological product, and the anticipated market demand lead to the formation of an adaptive nature of innovative development based on a problem-oriented format for translating innovative ideas (Abuzyarova 2015).

Thus, the formation of the "triple helix" model—representing the organizational platform of innovative scientific and technical development on the fundamental quintessence of the interaction between scientific, academic institutes and experimental design bureaus, industrial potential, and the government administrative offices of power resources—is quite natural. H. Itskowitz and L. Leidesdorf proposed this model of partnership in the form of a hybrid social configuration. It has a classic, linear character of ideas regarding innovative development, based on scientific potential and technological products, which are conductors in the implementation of the achievements of fundamental scientific thought.

Following the new industrial and technological revolution, a change in the social structure and the whole face of society will inevitably follow. Society and the technologies upon which it relies will create incomparably higher opportunities for satisfying human needs. The wide distribution of knowledge—intensive material production, accompanied by a progressive decrease in material costs and vigorous innovations growth—will soon be able to satisfy the needs for the material benefits that people need (Bodrunov 2017).

Keeping in mind the urgent tasks, we must also remember the **strategic perspective**. Where will the flow of the industrial stream of the new generation lead us? Will it lead us to a type of production in which the technological development of new knowledge will play the main role? At one time, J. K. Galbraith asked a similar question. We are also obliged to reflect on this, considering the changes that occurred half a century after the release of his book.

4 Discussion

According to most scholars, the fifth technological wave, which dominates the structure of the economy of the industrially developed states, has reached threshold values for its growth, which determines the final stage of its evolutionary cycle. Also, from an economic point of view, it has exhausted reserves as a platform basis for further economic development. At this stage, there is also the parallel design of the reproductive system of the sixth technological wave, which predetermines the variability of the scenario and the dynamics of the evolution of civilizational development for the next 30–40 years. Various approaches to embedding states in the outline of a future model of transition to a new economic wave are proposed (Polterovich 2009).

Addressing the Federal Assembly on March 1, 2018, V.V. Putin said: "The one who uses this technological wave will break far ahead. Those who cannot do this, it, this wave, will simply overwhelm, drown" (Kornienko 2018).

Press Secretary of the President of the Russian Federation D. Peskov, drawing the journalists' attention to these words, said the following: "I urge you to pay attention to the main idea of the message—this is an idea that the President emphasized at the very beginning—the idea of a technological breakthrough that will come around the world. Moreover, in this case, the task of Russia is to have time to make this breakthrough, we have everything necessary for it because the world is on a jump-like development path and the main thing for us now is to saddle this technological wave" (Peskov 2018).

According to J. Schumpeter, the so-called radical innovations in the system form a different (new) quality of the technological foundation, generating waves of changes in the structure of the entire social development model.

The NBIC-convergence is a kind of trigger for the large-scale generation and implementation of high-tech startups ("radical" technologies) of the platform economy, the core of which is the sixth TW, contributing to the formation of the reproductive potential of the new technological structure. When combined, the latest "breakthrough" technologies can potentiate existing technological products, creating fundamentally different outlines of the reproductive capabilities of the technological framework, leveling the reduction in the return on production factors, which will lead to the stimulation and growth of economic development.

Earlier, the areas of scientific knowledge and the sphere of practical activity, which seemed diverse and multidirectional, with the advent of NBIC technologies, acquired a new quality of vector development on a single platform of synergetic and emergent technological interaction, which contributed to the formation of the concept of convergence of high technologies on a platform basis. It is the evolution of NBIC technologies in the convergence plane of innovative platform interaction that sets a new quality of impulse knowledge of significant civilizational development. Also, the prospects that open up with the development of qualitative manifestations of the quintessence of the properties of NBIC technologies require philosophical reflection. Indeed, the results of the comprehensive development and implementation of NBIC convergence can be stunning transformational changes in everyday reality. Therefore, scientific thought is faced with the task of various aspects of understanding the advantages and consequences of the new era of convergent technologies, including those from the worldview perspective. The NBIC convergence, due to its transdisciplinarity and trans technology, forms a new scientific direction and generates value guidelines for civilizational development, which may differ from existing ones.

The paradigm of the development of science vector dramatically changed at the end of the twentieth century, from research and studies of world structure to civilization development based on platform technologies, where scientific achievements themselves can create elements of the world's structure. Therefore, the human mind has become able to a certain extent to create the surrounding reality and everyday reality. It is safe to say that science has reached the point where the latest human achievements should harmoniously flow into the production sphere and everyday life, thereby creating favorable conditions for the creative generation of the latest breakthrough technological products, creating new market niches while improving the existing ones.

Knowledge-intensive material product is a new quality of the main resource and the result of a new industrial twenty-first-century economy. It possesses features of both informational and ordinary material products. From the first, a knowledge-intensive product inherits the information component and many of its properties and problems. From the second, it receives objective utility for the reproduction of both material production itself and the human qualities necessary for it (Bodrunov 2016).

The so-called breakthrough technologies, which form the basis of the VI technological wave, are the driving force of the fourth industrial revolution and transform the existing production models. Also, in the short term, they can modify existing sectors of the economy to an unusual form for the modern layman. Thus, a potential threat is generated for those states that fail to join the wave of technological re-equipment and adaptation of the economy to a new stage of civilizational development. This factual reality of the impending near future potentiates the possible manifestations of breakthrough technologies in the most unexpected areas, including in combinatorial combination with already existing technological solutions.

As breakthrough technologies develop, they become a kind of catalytic accelerator for approximating foreseeable technologies.

The concept of the fourth industrial revolution was first introduced in 2011 (Industry 4.0, 2011). In full accordance with this concept, the nearest future development will be based on the introduction of breakthrough technologies, including the use of cyber-physical systems (CPS) in production activities, as well as their use in order to meet the needs of society in various aspects of life. It is the breakthrough technologies that make up the core of the VI technological structure that will make available the implementation of projects that were previously (at the current level of scientific thought evolution) considered unrealizable in the existing technological plane.

It should be noted that the introduction of "cyber-physical systems" as a key driver of the IV industrial revolution, along with the positive dynamics of the technological civilizational development, can become a catalyst for adverse changes in the existing model of employment and division of labor. A tendency toward fundamental changes in labor relations has already been outlined. The nature and content of labor are being transformed. This stage for certain types of the professional activity can become a catastrophe due to modifications of the organizational component of the functioning of the labor market and, as a result, the emergence of anew formal filling of professions, new requirements for qualifications and education, the emergence of new professions, and the "extinction" of existing professions.

According to the American scholar E. Toffler, traditional ideas about the quality of life change with the development of information society. They are increasingly dependent on information (digital) products, on accessibility to those. The World Wide Web is converged into everyday reality so strongly that this resource is more and more transformed from a purely domestic one into a strategic factor influencing world civilization. This factor provides an opportunity for the middle class to initiate as a subject in innovative economic development. Scholars M. Castells and R. Florida, analyzing the changes in the existing characteristic features of the middle class, introduced the concepts of "creative class" and "information producers" (Castells 2006; Florida 2011). It is in the conditions of the current stage of the evolution of information society that the middle class, and especially its creative layer, becomes not only a mass user but also a manufacturer of innovative high-tech products. However, in this context, we recall the opinion of some representatives of the expert community that not all entities of the middle class will be able to adapt to the development pace and confront the challenges of the fourth industrial revolution. Moreover, its scale in different states can change significantly.

5 Conclusion

The NBIC convergence of technologies within the framework of the sixth technical wave assumes the implementation of a policy to translate the fourth industrial revolution into modern reality, including the platform of cyber-physical production systems that may have the configuration of neuronet and social networks. The outlines of convergence of the main technological groups are more clearly traced:

1. Aggregation technologies for large volumes of data (including sensors for collecting and transmitting information); The Internet of things, as well as the inter-machine interaction built on its basis; the development of the energy-efficient wide-area networks; the development of the so-called secondary supporting technologies necessary for the data centralization and remote decision-making visualization;
2. Artificial intelligence technologies; machine learning and machine intelligence;
3. Smart Technologies;
4. Technological solutions for visualization of the graphical and touch user interface of virtual reality;
5. Additive manufacturing technologies (lower cost of 3D printers and printing); expanding the range of materials for 3D printing;
6. Next-Generation Robotics Technologies in Synergy with the Artificial Intelligence potential;
7. The latest technologies for the generation of clean energy and with a minimum low transmission and storage coefficient.

The focus of the sixth technical wave will be the development of nanotechnology, which will open new horizons for the development of optical nanomaterials, materials with special nanocoating, nanophotonics, nanobiotechnology, thermonuclear and green energy, etc. In turn, the implemented technologies will make it possible to implement the following scientific areas of thought in the aspect of practical application:

- The foresight of the design thought will allow the development of parts and materials with predefined properties and characteristics;
- Material, resource, and energy efficiency at the household and production levels; environmentally friendly energy generation technologies at a lower cost;
- Non-waste production and cycling of household waste on a resource recycling platform;
- The so-called “smart” technologies, which will become the platform basis for almost all spheres of human life, will significantly improve and transfer the quality of life to a new plane, increase labor productivity, and proportionally reduce the cost of manufacturing the high-tech products for everyday and industrial use;
- a new stage in the evolution of medical and pharmaceutical science;
- the multidisciplinary effect of fundamental scientific research based on the sixth technological wave (TW);

- the finite element method in convergence with the visualization technologies will make it possible to switch to the universal virtual design of products and objects before their validation and launch into mass production, which will enable foresight regarding characteristics, properties, and competitive advantages.

The current stage of civilizational evolution is scientific and high-tech, including the involvement of cluster formats of the NBIC convergence. Due to its "breakthrough" orientation, convergent technologies transform society through the emergent and synergetic influence. Based on the platform approach, the NBIC convergence generates a fundamentally different environment whose basis is the highly innovative integration of civilizational genesis, including improving sociocultural values.

A philosophical understanding of the genesis of the fourth industrial revolution leads to an understanding of its essence based on the convergence of the so-called three components of evolution: technological (material), digital (virtual), and biological. An active initiative on the "Industry 4.0" platform allows the realization of the huge intellectual potential of a person, which induces creative reserves for reformatting into a different quality of the social environment and contributes to the more finely adjusted tuning of the scenarios of civilization development.

References

Abuzyarova, M.I.: The essence of open innovation management as a modern model of innovation management. Econ. Law **85**, 98–102 (2015)

Bodrunov, S.D.: The Future. New Industrial Society: Reboot. Publishing house of the S. Yu. Witte Institute of New Industrial Development, St. Petersburg (2016)

Bodrunov, S.D.: The return of industrialization – the return of Galbraith: from NIO.2 to noospheric civilization. Econ. Revival Russ. **2**(52), 17–21 (2017)

Bodrunov, S.D.: Noonomics. Publishing house of the S. Yu. Witte Institute of New Industrial Development, St. Petersburg (2018)

Bodrunov, S.D., Galbraith, J.K.: The New Industrial Revolution and the Problems of Inequality. Plekhanov Russian University of Economics, Moscow (2017)

Bodrunov, S.D., Greenberg, S.D., Sorokin, D.E.: Reindustrialization of the Russian economy: imperatives, potential, risks. Econ. Revival Russ. **1**(35), 19–49 (2013)

Castells, M.: The Information Age: Economics, Society, and Culture. Higher School of Economics, Moscow (2006)

Florida, R.: Creative Class: People Changing the Future. Publishing House "Classic-XXI", Moscow (2011)

Glazyev, S.Yu., Dementiev, V.E., Elkin, S.V., Kryanev, A.V., Rostovsky, N.S., Firstov, Yu.P., Kharitonov, V.V., et al.: Nanotechnology as a Key Factor in the New Technological Structure in the Economy. Trovant, Moscow (2009)

Kornienko, E.: Decisive breakthrough: Vladimir Putin outlined the main danger to the Russians (2018). https://iz.ru/715088/ekaterina-korinenko/reshitelnyi-proryv

Lvov, D.S., Glazyev, S.Yu.: Theoretical and applied aspects of NTP management. Econ. Math. Methods **5**, 793–804 (1986)

Polterovich, V.M.: Innovation pause hypothesis and modernization strategy. Econ. Issues **6**, 4–22 (2009)

Pride, V., Medvedev, D.: The phenomenon of the NBIC convergence: reality and expectations. Philos. Sci. **1**, 97–117 (2008)
Schummerm, J.: From nano-convergence to NBIC-convergence (2008). https://pdfs.semanticscholar.org/7614/e7933fcb14c3daf674ebc2abab43129ad343.pdf
Smirnov, S.: Peskov called the real meaning of Putin's message (2018). https://www.vedomosti.ru/politics/articles/2018/03/02/752591-peskov-putina

Digital Economy and Management: Creating Strategic Capabilities

A Model of Effective and Sustainable Innovative Development of a Company: Analyzing Foreign Experience

Janusz Sobon[1(✉)], Ecaterina D. Velixar[2], and Marcin Prokopiuk[3]

[1] Department of Entrepreneurship and Innovation, The Jacob of Paradies University in Gorzow Wielkopolski, Gorzów Wielkopolski, Poland
sj@list.pl
[2] MGIMO MIEP, Moscow, Russia
iamunikum@mail.ru
[3] Warsaw School of Economics, Warsaw, Poland
marcin2pro2@o2.pl

Abstract. The article analyzes the foreign practice of selecting and using key performance indicators (KPI) of innovation. It also reviews methodological principles and features of their justification. The system of key performance indicators (KPI) of innovation is defined using the example of Eni, one of the leading oil and gas companies. The authors explore and summarize the best practices for developing and implementing the system of key performance indicators (KPI) in management.

Keywords: Innovative activity · Methodology · Key indicators · Efficiency · Oil and gas complex · KPI justification methodology · R&D

1 Introduction

Today, the innovative activity of any company is a crucial factor in its success. If a company claims to be a leader in the global market, it needs to carry out innovative activities always. Thanks to innovative developments, companies can reduce production costs, reduce or eliminate harmful environmental impacts, reduce the number of injuries at work, and develop fundamentally new technologies. An essential role in the implementation of innovations belongs to startups as trendsetters of innovation (Stepanov and Veselovsky 2017).

The problem of evaluating the effectiveness of innovation in energy companies is currently one of the most critical tasks. The explanation for this lies in the fact that the last decade is the decade of the consistent introduction of new progressive and highly efficient technologies and innovations in oil and gas companies. Today, the oil and gas industry is one of the most high-tech sectors of the economy.

However, to date, there is no adequate toolkit for evaluating the effectiveness of innovation in companies in the fuel and energy sector in both domestic and foreign science and practice.

A. V. Bogoviz and Y. V. Ragulina (Eds.): ISCI 2019, LNNS 115, pp. 151–158, 2020.
https://doi.org/10.1007/978-3-030-40749-0_18

In this regard, the challenge is the formation of reliable tools for managing innovative activities based on the use of reliable and adequate information about its effectiveness and, in particular, using KPI indicators (Cimino et al. 2016).

2 Materials and Methods

The scientific works of scholars from different countries of the world on the problem of the formation and use of key indicators of the effectiveness of innovation, and in oil and gas companies, made up the methodological and methodical basis of the study.

The article uses theoretical and methodological works, recommendations of various research institutions.

In preparing the article in the research process, we used monographic, abstract-logical methods, factor and system analysis, as well as graphic, expert, economic-statistical, and other methods of analysis.

3 Results

International experience shows that cost reduction in the oil and gas complex can be ensured as a result of introducing innovations based on the latest achievements of scientific and technological progress. These innovations include, first of all, managerial innovations based on the use of new concepts of creative management of innovation performance related to using the system of key performance indicators (KPI), which are successfully used in the practice of managing innovative processes in foreign companies (Savina et al. 2015).

Today, most energy companies solve problems of increasing the efficiency of innovation through research centers, innovation committees, cooperation agreements with universities. For constant and continuous control over the effectiveness of change, companies use various teaching methods and indicators. One such useful tool is a system of key performance indicators (KPI) (Stepanov et al. 2019).

Positive experience in the formation of a system of key performance indicators (KPI) of innovative activity was developed at the Eni fuel and energy company.

Eni is a vertically integrated company operating in 71 countries. About thirty-two thousand employees are employed in it.

Eni operates in four key areas.

The First Direction is Exploration and Production. Exploration and production form a crucial activity of Eni in more than 40 countries. This direction is aimed at the search and production of oil and gas. The Eni's strategy is to ensure organic production growth based on a high-quality asset portfolio and long-term relationships with host countries.

The Second Direction is Gas and Electricity. This direction is focused on the value chain, namely the supply, trading operations, and marketing of gas and electricity. The company exports about 60% of its gas outside Italy to the European market. The company provides a leading position in the gas market in Europe through competitive advantages, including an international approach, long-term access to natural resources and infrastructure, market knowledge, product diversity, and a reliable customer base (Eni 2017).

The Third Area is Oil Refining and Marketing (Environmentally Friendly Chemistry). Oil refining and marketing divisions deal with the issues of refining and entering the fuel market. Today, Eni is the largest oil refining company in Italy. This company holds a 25% share in the retail marketing market in the fuel and energy sector. The strategy of the oil refining and marketing divisions is to reduce costs and increase margins and return on profitability.

The fourth area is corporate projects for the development of renewable energy sources, which are designed for the long term.

At Eni, much attention is paid to the formation of a system of key performance indicators (KPI) of innovation. Such a system of indicators was developed directly in the departments of the company involved in R&D. At the same time, the development of KPI innovation was based on the following points:

- A strategic vision of the main activities of the company based on the implementation of the latest achievements of scientific and technological progress and innovation;
- Features of the organizational structure of innovative activities of the company;
- Maximizing the possibility of a comparable analysis of KPIs used in R&D of a company and KPIs used in operations;

The strategic vision of the main activities of the company based on the implementation of the latest achievements of scientific and technological progress and innovation.

Thanks to ongoing innovation, the vertically integrated company Eni is the foremost oil and gas company in the world. The company exercises a significant influence on innovative development in all areas of activity and introduces new technological developments over the years. The company carries out many groundbreaking discoveries and developments in its center for research and development of renewable energy and the environment. The company also collaborates with leading technological institutes. The company has registered more than five hundred patents.

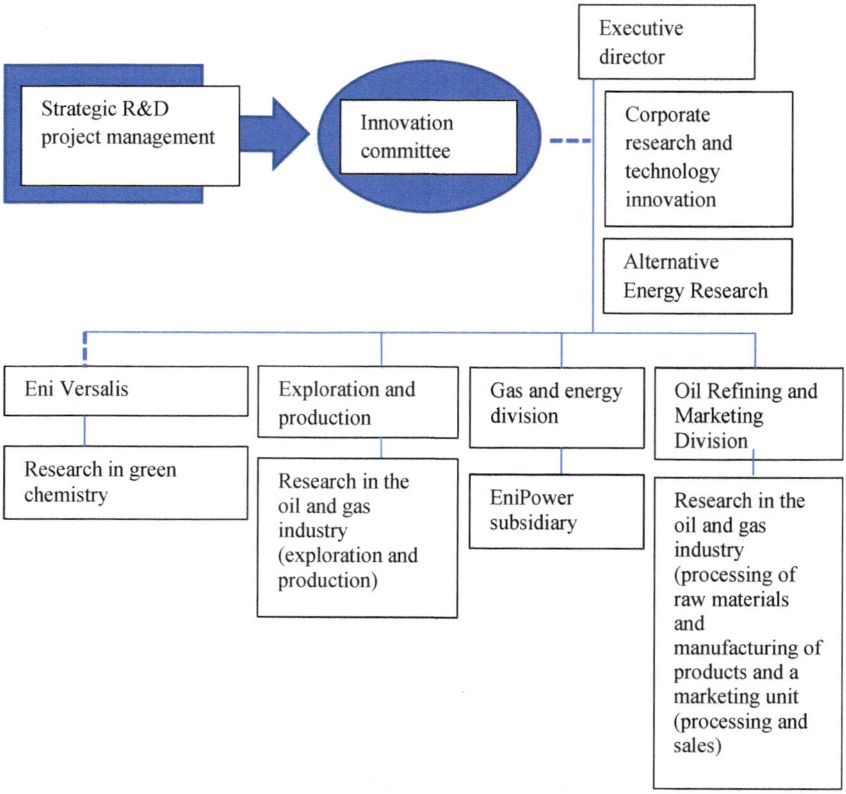

Fig. 1. The organizational structure of the research and development in the Eni company. *Source:* Compiled by the authors.

3.1 Features of the Organizational Structure of Innovative Activities of the Company

The organizational structure of the innovative activities of the Eni company is presented in Fig. 1.

For the organizational structure of the innovation activity of the company Eni, the non-standard subordination of departments involved in R&D is characteristic. At the same time, it has its specific features. First of all, it is an atmosphere of creative freedom of innovation from the bureaucratic pressure of the company's apparatus. So, the innovation committee, whose goal is the strategic management of R&D projects, is not directly subordinate to any unit in the company. This organ is independent and responsible only to the executive director, which allows the committee not to limit innovation to market interests and needs of other divisions of the company.

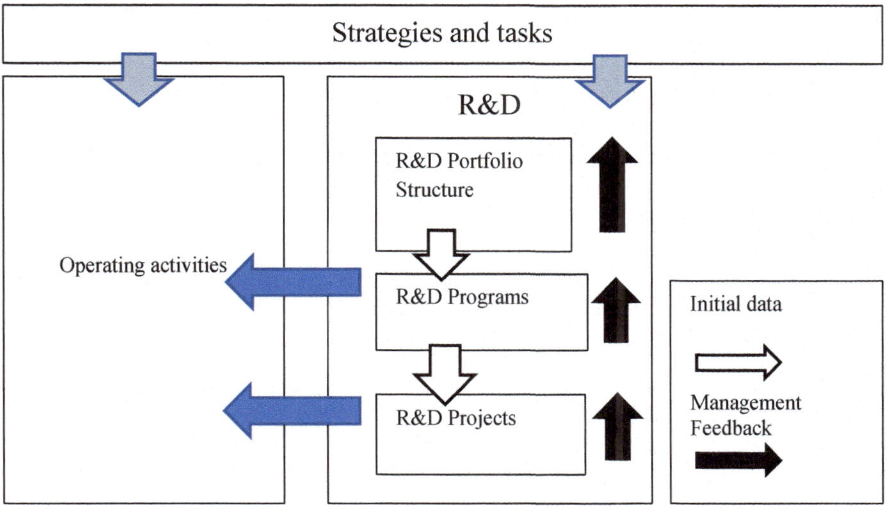

Fig. 2. R&D management at Eni: the relationship between strategy and operations.

The organizational structure also has a Corporate Research and Technological Innovation department, as well as an independent research center for alternative energy sources (RES).

The R&D management structure at Eni reflects the real relationship between the strategy and the operations of the company, namely: management determines the key strategy of the company and assigns it to tasks. Strategic tasks from the top management of the company go to the unit engaged in innovative activities. They are implemented in the structure of the R&D portfolio, where they are further distributed between the respective programs and project groups. Also, the results obtained are reported to the management. The direct adjustments are made to the operating activities of the company (Fig. 2).

3.2 Maximizing the Possibility of a Comparable Analysis of KPIs Used in R&D of a Company and KPIs Used in Operations

The development of key performance indicators (KPIs) for a comparable analysis of the company's KPIs and the measurement of key performance indicators in R&D was carried out based on scientifically based requirements adopted by the Eni innovation committee:

- Focusing on R&D priority strategic areas;
- To substantiate such indicators that best characterize the positive trends and the results of innovation in solving strategic (long-term, medium-term) tasks in each of the areas;
- Assigning to each KPI indicator a parametric, normative expression that reflects the goals and objectives of the innovation strategy in a concentrated form (Stepanov and Veselovsky 2017).

The following requirements for measuring key performance indicators in R&D were:

- Measuring the economic benefits of R&D in all departments;
- Determination of the intangible component of the cost of the development of R&D;
- Measuring the effectiveness of the operational level of R&D;
- A key performance indicator system should not use too many human resources to identify all the variables;
- A review of the effectiveness of R&D should be carried out regularly.

At the same time, the methodology identified the following difficulties in the analysis of the aggregate R&D portfolio indicator:

- Time frame. Often, the value of R&D builds up and becomes apparent after the completion of the R&D project;
- Integration with other production functions, for example, administrative function and business management, is required;
- The immeasurability of know-how;
- Innovation cycles differ in different departments of the company;
- A limited number of examples can be found in open sources.

The system of indicators KPI of operating activities and KPI used in the research and development of Eni company are presented in Fig. 3.

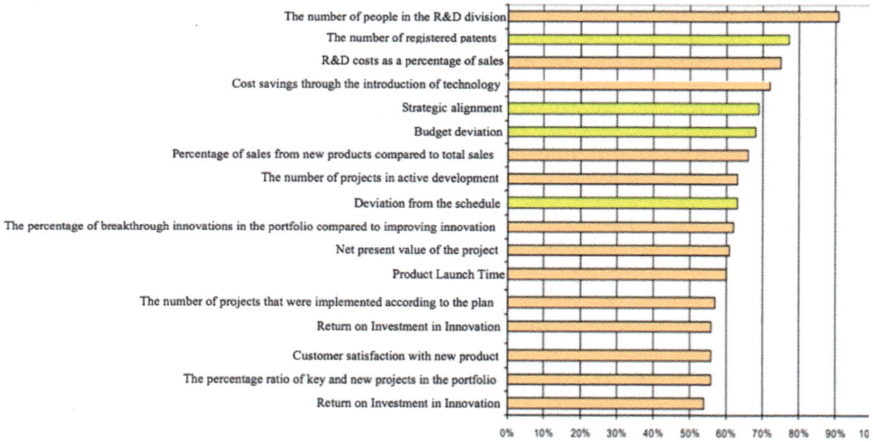

Fig. 3. A comparable analysis of Eni KPIs and frequently used KPIs in R&D.

4 Discussion

As a result, the following system of key performance indicators (KPI) of innovation was highlighted at Eni. Economic benefits are measured by analyzing the results of technology implementation in each division of the company, purchasing licenses by third parties, and reinforcing new areas of activity.

In the exploration and production unit, economic benefits include the following:

- Saving on capital costs;
- Savings in operating activities;
- Increased oil and gas production;
- Increase in net present value;
- Increase in initial stocks (Eni 2017).

In the oil refining and marketing and petrochemical technology divisions, economic benefits include the following:

- Marginal profit from innovative products;
- Operating margin from innovation;
- License fees.

In the division of renewable energy sources and sources in the early stages of technology development, economic benefits include the following:

- The net present value of new business units;
- The actual results of operations are used to measure the economic effect, with the exception of the net present value. The latter is used to capture the value of innovative technologies, demonstrating a long application cycle, mainly in exploration and production and renewable energy sources.

Comparing the most used key performance indicators in R&D in the largest companies and Eni, the following KPI indicators used in the most significant companies were recommended:

- A total number of employees in the R&D department;
- A number of projects in active development;
- The percentage of R&D expenses in relation to sales;
- The percentage of the portfolio of key and new projects;
- The net present value of the entire new product and service portfolio;
- Number of patents;
- Eni KPI measurement;
- Cost reduction relative to R&D investment;
- Budget deviation;
- Deviation from the schedule;
- Percentage of sales of the current year concerning new products and services;
- A number of products sold in the last n-th number of years.

An indicator of registered patents was recognized as one of the key indicators of the effectiveness of innovative activities of the company. At the same time, the analysis for the period 2013–2018 showed that the peak in the number of registered patents in the company was in 2014 and amounted to 50 licenses. At the same time, already in the next year (2015), they registered the minimum number for the five years, namely 22 patents.

5 Conclusion

Despite fluctuations in the financial performance of Eni, investment in R&D showed steady dynamics. Based on indicators, criteria were identified, on the basis of which it is possible to analyze key indicators of innovation activity efficiency.

Eni also has its own performance measurement system. Based on the criteria identified by Eni, as well as the criteria for measuring the impact of R&D on the company's operations, an analysis of the effectiveness of investing in R&D was carried out. This analysis showed that, despite the reduction in net sales from Eni's operations, the company continues to invest in R&D. The ratio of R&D costs to net sales from operating activities shows a positive investment trend. The company focuses on long-term prospects. Therefore, the innovative technologies developed by it will pay off in the next ten years.

References

Cimino, R., Siciliano, L., Stella, A.: R&D portfolio metrics: setting up a key performance indicators system for measuring R&D performance (2016). http://www.eni.com

Eni: The Eni Report: Integrated-Annual-Report-2017 (2017). https://www.Эни.com/en_IT/documentations.page

Savina, M.V., Solodukha, P.V., Stepanov, I.A., Khominich, I.P., Novichkov, A.V.: Essence, place, and role of creative economy in the innovation development paradigm. Rev. Eur. Stud. **7**(6), 77–85 (2015)

Stepanov, A.A., Veselovsky, M.Ya.: The development of innovative startups in Russia: the regional aspect. Acad. Strateg. Manag. J. **16**(1) (2017). http://www.abacademies.org/articles/the-development-of-innovative-startups-in-russia-the-regional-aspect-1939-6104-16-SI-1-117.pdf

Stepanov, I.A., Korniyko, Yu.V., Scherbakova, M.A.: Key indicators of the effectiveness of regional innovation development. UEcS **3**(121), 17–26 (2019)

A Mechanism for Preventively Assessing the Competitiveness of High-Tech Products in Conditions of Digital Transformations

Lidia S. Zelentsova, Nataliya G. Shalamova, and Yulia V. Vorontsova(✉)

State University of Management, Moscow, Russia
zelentsovals@yandex.ru, ngshalamova@yandex.ru, jvms2008@yandex.ru

Abstract. The article considers the problem of determining the competitiveness of the designed products at the stage of the formation of the technical task, relevant for high-tech industries. The study aimed to develop an original approach to assessing the competitive advantages of high-tech products based on the disclosure of the triune nature of competitiveness. On this basis, it further determines an optimal ratio of technical excellence and price efficiency at the product design stage. An analysis of the methods for assessing the competitiveness of high-tech products shows that, for all their diversity, they suffer from the lack of methodologically developed issues of determining the technical and economic parameters of a new product that precedes the design process. In this regard, the authors hypothesized the need for a preventive assessment of competitiveness, which would avoid errors in the formation of the technical and economic appearance of the product. Accordingly, it reduces design costs and predicts optimal costs for the production and operation of high-tech products. In order to ensure the identification of the competitive advantages of high-tech products at the early stages of the formation of the technical and economic appearance of the product, the authors have developed a preventive competitiveness assessment mechanism focused on using digital technologies. This should be considered as the scientific result of the study. The proposed mechanism can find application in the design and technological bureaus of the aviation industry, in particular, when designing helicopters, and in other branches of engineering and industry as a whole.

Keywords: High tech products · Preventive assessment mechanism · Online monitoring · Design features · Technical and economic appearance · Technical excellence · Competitiveness · Cost-effectiveness

1 Introduction

The relevance of the study is due to the fact that, at present, increasing the competitiveness of manufacturers of high-tech products (HTPs) in the world and domestic markets is determined by their ability to timely develop and produce new generations and modifications of HTPs that most satisfy the requirements of consumers and ensure

A. V. Bogoviz and Y. V. Ragulina (Eds.): ISCI 2019, LNNS 115, pp. 159–167, 2020.
https://doi.org/10.1007/978-3-030-40749-0_19

the return on investment life cycle investments (Chursin 2017). The process of designing an HTP plays a decisive role in creating a competitive HTP both in terms of ensuring the technical excellence of the product and in terms of the cost of its creation (Dal Pont and Azzaro-Pantel 2014). Studies of the famous consulting company McKinsey show that about 80% of the cost of a product is formed at the stage of its development (Gold 2016).

In the context of digital transformation, opportunities for implementing a fundamentally new approach to ensuring the competitiveness of HTP appear. This is supported by a survey of the rapidly developing high-tech Russian companies in 2016, according to which 50% of companies noted the technology of "digital design and modeling" as the most relevant for creating new products or giving new properties to existing positions (Borovkov et al. 2018). Currently, foreign countries are actively implementing projects aimed at the development of the digital economy (Thoben et al. 2017). Domestic industrial enterprises are at the beginning of this path. The costs of 55% of Russian industrial enterprises for digitalization and the development of IT infrastructure do not exceed 1% of their budget (Balenko 2018).

At the same time, it is essential to emphasize that the use of digital production technologies is critically important for enterprises producing complex HTPs. The use of digital technologies in the industry can lead to an increase in labor productivity by 3–5%, a reduction in the time taken to bring products to market by 20–50%, and costs for quality assurance by 10–20% (Khachirov and Khubulova 2018). However, the realization of the potential of digital technologies is possible provided that the HTP design management system is transformed, first of all, by creating effective mechanisms to ensure the development of competitive products (Chursin and Makarov 2015).

The following domestic and foreign authors have made a significant contribution to the coverage of various aspects of competitiveness management: G. L. Azoev, D. Barney, I. S. Golubev, R. Grant, V. I. Grushenko, F. Kotler, Jean-Jacques Lambin, M. Porter, C. Prahalad, A. A. Thompson, R. A., Fathutdinov, G. Hamel, A. A. Chursin, and others.

Currently, many methods and techniques for assessing the competitiveness of products have been developed. They differ in the selected criteria and methods for determining the assessment results. However, as a rule, they are not focused on a proactive assessment of competitiveness in the early stages of product development and are not brought to the level of an effective mechanism (Timofeev and Strigul 2016). Today, the development of a mechanism for assessing the competitiveness of HTPs at the design stage is becoming not only an urgent task but also really achievable on the basis of digital production.

The aim of the study is the formation of a mechanism for the preventive assessment of competitiveness (MPAC), which ensures the determination of the competitive advantages of HTPs at the design stage.

Achieving the goal involves solving the following tasks:

- Identifying features of the design process in terms of ensuring the HTP competitiveness;
- Determining the main components of the HTP competitiveness and approaches to their assessment at the early stages of the formation of the technical and economic appearance of the product;

- Determining the structure and content of MPAC HTPs;
- Formation of an approach to designing a HTP, based on a preventive assessment of its competitiveness.

2 Materials and Method

At the heart of the proposed approach to designing a HTP on the basis of a preventive assessment of its competitiveness, the idea is that the competitiveness of products has the following triune essence: (1) comparative (competitiveness as a phenomenon arises in the presence of competition), (2) value (products must have a specific value that distinguishes it from similar products in order to satisfy the needs of consumers), and (3) economic (products must be sold and bring an income to the organization) (Demchenko 2015). Based on this, the main components of the HTP competitiveness are technical excellence (TE), which characterizes value competitiveness, and price efficiency (PE), which determines economic competitiveness (Golubev and Levochkin 2010).

The technical perfection of the HTP is the superiority of its technical characteristics concerning similar products, which is achieved through scientific and technological innovations. In a certain period, the results of the STP determine the maximum possible level of quality of the HTP, which can be achieved using advanced developments, in other words, it sets the upper value of the TE. At the same time, the implementation of scientific and technical progress in a particular organization depends on several factors and conditions such as access to markets for new products, the availability of competent specialists, the ability to attract the necessary financial resources, etc. Existing restrictions determine the lower value of the TE indicator. It should also be noted that the achievement of the highest value of the TE indicator may be impractical because some scientific and technical developments can be significantly ahead of time, the social need for them may not yet be formed.

An essential component of the HTP competitiveness is its cost-effectiveness. The price of the HTP should meet the expectations of consumers and be profitable for its developer (manufacturer), ensure cost recovery, and increase its profitability. To determine the market price for the design of HTPs in the context of incomplete and uncertain information, the concept of boundary pricing seems useful. According to this concept, the price is set between the upper maximum permissible limit (upper limit) and the lower objectively necessary border (lower limit) of the price of the newly created HTP. The price of the upper limit is determined from the condition of equal profitability, i.e., equality of indicators of technical and economic efficiency of new and replaced products. This price determines the limit of the cost of the product, above which it becomes unprofitable for the consumer to purchase it. The determination of the upper price limit is based on the processing of statistical data on prices and sales volumes of products similar in characteristics to the new HTP. The price of the lower limit is determined based on the conditions of profitability for the developer (manufacturer) of a new product. This price level is dictated by the costs that are necessary for the development (manufacture) of the sample. To determine the prices of the lower limit, the methods focused on the full accounting of the costs necessary for the development and manufacture of the product are used (Golubev and Levochkin 2010).

A preventive assessment of the HTP competitiveness at the design stage is carried out using online monitoring of data on the technical and cost characteristics of competitors' products, the formation of a database of competitive advantages, and their predictive analysis using digital technologies.

The proposed approach to improving the design process of HTP on the basis of a preventive assessment of its competitiveness provides for the following stages of work.

1. Carrying out a comparative analysis of the designed product with competitors' products and with the ideal model, the characteristics of which are determined based on the advanced achievements of scientific and technical progress, to identify problem areas in the HTP design.
2. Identifying the key areas of technical design improvement to ensure the HTP competitiveness.
3. Determinung the possible ratios of the competitiveness components of the HTP ("technical excellence/price efficiency") for each key area.
4. Choosing the rational directions for the formation of the technical and economic appearance of the HTP.

3 Results

The HTP design process has several features that have a significant impact on its competitiveness (Gupta et al. 2016). These include the following:

- The high complexity of the design object, namely a large number of components and assemblies included in the HTP, of possible design solutions;
- The need to use specialized scientific and technical knowledge to form the technical appearance of the product, accumulated in the process of fundamental and applied research;
- The long term development of HTP due to the following factors: the complexity of the products, the large amount of new knowledge that must be acquired and applied when creating models of the new generation of HTPs, and the considerable time spent on coordinating the work of specialized institutes and enterprises involved in the development; large volume of tests and control operations;
- High requirements for the competencies of personnel performing a full cycle of work on the HTP design of HTP using digital technologies;
- A significant level of design costs due to the high complexity of the work performed, the use of expensive resources;
- Increased risks associated with the possibility of developing a non-competitive HTP in the absence of complete and accurate information about potential demand and future requirements for the technical and economic characteristics of the designed product.

Improving the efficiency of the design process is achieved by creating a mechanism for preventively assessing competitiveness (MPAC). The purpose of this mechanism is to determine and form the competitive advantages of the HTP at the early stages of the formation of the technical and economic appearance of the product. The basis for the implementation of MPAC is the integrated use of modern information technologies, knowledge bases, heuristic techniques, and informal methods of synthesis and analysis (Sebo et al. 2011).

The MPAC HTP proposed by the authors provides a process of preventive assessment of the HTP competitiveness through the interaction of functional blocks (Fig. 1).

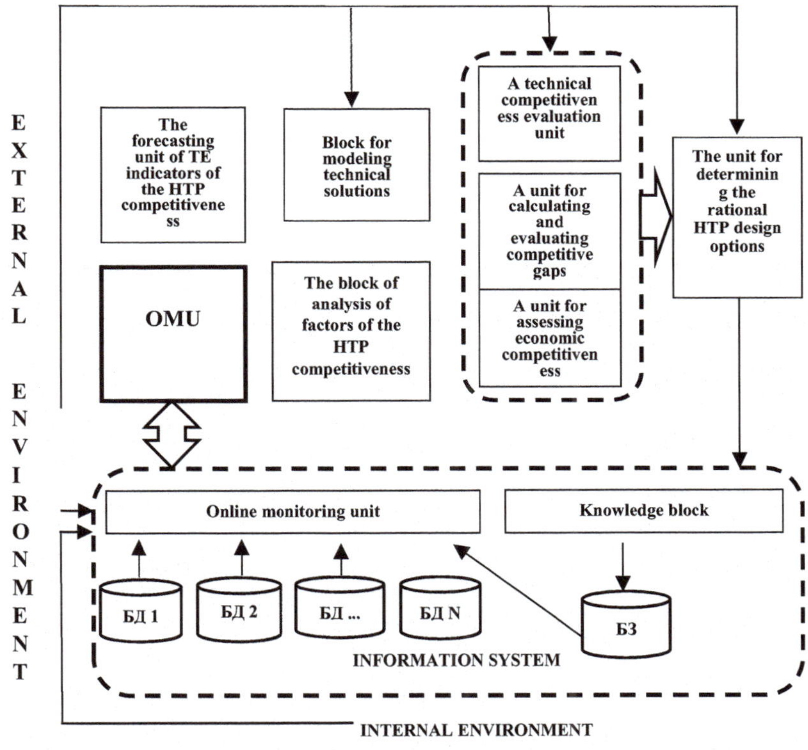

Fig. 1. Logical scheme, MPAC HTP. Source: Developed by the authors

The MPAC HTP analysis unit is designed to identify trends in the development of key external and internal factors of HTP competitiveness. Based on a retrospective data analysis, predictive models are built, and their verification and validation are carried out. Those models that have successfully passed verification and validation are used to determine the predicted values of technical and economic (T-E) indicators of HTP competitiveness.

The block of modeling of technical solutions is designed to determine the key areas of technical improvement of the design. In this block, various options for changing the problem areas of the HTP design are formed in order to improve the technical product characteristics. For this, a computer-aided design system, a database on the composition of the HTP (and its technological environment), a knowledge base on competitive advantages are used (Chang 2013).

The technical competitiveness assessment unit of the HTP is used for calculating the values of the normalized indicators of the competitiveness of the designed product and competitors' products. This is done according to technical characteristics in relation to the characteristics of an ideal model.

In the unit for assessing the HTP economic competitiveness, the calculation of the development costs, production, and operation of the product for the initial and advanced design options are carried out. Based on it, the lower limit of the price of the product and the cost of its operation are determined, as well as the values of normalized indicators of competitiveness by economic characteristics concerning the characteristics of an ideal model are calculated (Table 1).

Table 1. Values of normalized private indicators of competitiveness in terms of technical and economic characteristics and their weight coefficients for three HTP design options and competitor products.

Indicators: *T - technical*, *E - economic*	The weight coefficient of the normalized private indicator of product competitiveness	The value of the standardized private indicator of competitiveness			
		Designed product			Competitor products
		Option 1	Option 2	Option 3	
T 1	0.073	0.920	0.920	0.942	0.846
T 2	0.052	0.904	0.904	0.912	0.967
T 3	0.219	0.863	0.889	0.908	0.938
T 4	0.156	0.725	0.755	0.785	0.897
E 1	0.167	0.956	0.948	0.934	0.818
E 2	0.333	0.842	0.866	0.874	0.914

The unit for calculating and evaluating competitive gaps is aimed at identifying the problem areas of the created design of HTP. For this, a comparative analysis of the designed product with competitors' products according to technical and economic characteristics is carried out, as a result of which the gaps in the competitiveness of the designed HTP relative to the competitor's product, taken as the basis for comparison, are determined.

The value of the positive gap in the competitiveness of the designed product relative to the competitor's product (the amount of lead) R^+ is calculated by the following formula:

$$R^+ = \sqrt{\sum_{i\in I^+} R_i^{\,2} \cdot b_i} = \sqrt{\sum_{i\in I^+} (P_i - P_{i_6})^2 \cdot b_i}\,;\ \mathrm{I}^+ = \{\mathrm{i} \in \mathrm{I}\colon \mathrm{R_i} > 0\} \tag{1}$$

The negative gap in the competitiveness of the designed product relative to the competitor's product (lag value) R^{-} is calculated by the following formula:

$$R^{-} = \sqrt{\sum_{i \in I^{-}} R_i^{2} \cdot b_i} = \sqrt{\sum_{i \in I^{-}} (P_i - P_{i_6})^2 \cdot b_i}\,;\ \mathrm{I}^{-} = \{\mathrm{i} \in \mathrm{I}: \mathrm{R_i} < 0\} \quad (2)$$

where R_i is the gap value according to the private standardized indicator of competitiveness by the i-th characteristic of the designed HTP relative to the competitor's product, taken as a comparison base; P_i (P_{i_6}) is the value of the private standardized indicator of competitiveness according to the i-th characteristic of the designed product (the product of a competitor taken as a comparison base); b_i is the weight coefficient of the private standardized indicator of product competitiveness by the i-th characteristic; moreover, the following condition is satisfied $\sum_{i=1}^{n} b_i = 1$; I is the set of indices of technical and economic characteristics of the HTP.

Table 2 shows the results of calculations of the values of competitive discontinuities for the three HTP design options relative to the competitor.

Table 2. The results of the calculation of the values of competitive gaps for the three design options HTP.

Indicator	Option 1	Option 2	Option 3
Positive gap	0.059834	0.056763	0.054036
Negative gap	0.088202	0.068154	0.053330
The ratio of the positive gap to the negative gap	0.678372	0.83286	**1.013248**

In the block for determining the rational directions of the formation of the technical and economic HTP appearance, one compares the values of competitive gaps calculated for the considered options of the product design. Rational options with the best (maximum) values of the ratio of the positive gap to the negative gap, which have a high ($\mathrm{R}^{+}/\mathrm{R}^{-} \geq 1{,}25$) or a sufficient degree of stability ($1{,}0 \leq \mathrm{R}^{+}/\mathrm{R}^{-} < 1{,}25$), are selected. In this example, the third design of the HTP satisfies these conditions.

The information system contains a database of HTP; a database of the technological environment, a database of scientific and technological developments, a knowledge base about the HTP competitive advantages, and the methods and techniques for assessing the GTP competitiveness, online monitoring, and knowledge building blocks.

The online monitoring unit provides continuous real-time monitoring of the critical factors of the external and internal environment that affect the competitiveness of the HTP. Intelligent sensors in equipment, "cloud technologies," mobile monitoring technologies, etc. can be used for this.

The knowledge accumulation unit collects and processes the information on trends and advanced achievements in the development of high-tech products and technologies; on production management methods; about the HTP competitive advantages and the methods and techniques for assessing the HTP competitiveness, etc.

The purpose of the organizational and managerial unit (OMU) is to ensure the coordinated operation of all the MPAC HTP blocks. It includes the principles, goals, criteria, methods, and resources for managing the process of proactive assessment of the HTP competitiveness.

Using the proposed MPAC HTP at the design stage will allow for a directed search for design options for the designed product, ensuring the best ratio of indicators of "technical excellence/market efficiency."

4 Discussion

The scientific novelty of the presented work is indisputable:

- Justification of the integrated approach to assessing competitiveness, based on its triune nature, allowing to take into account the requirements for the designed HTP both on the part of the consumer and the developer;
- Modeling the technical and economic characteristics of the appearance of HTP by calculating the positive and negative competitive gaps relative to similar products of competitors;
- Development of MPAC HTP, the originality of which consists in predicting technical and economic indicators, positioning on determining design options that have superiority over competitors' products.

5 Conclusion

The article substantiates the relevance, scientific and practical significance of the problem of determining the parameters of design competitiveness of products and the possibility of solving it at a fundamentally new level in digital production by developing a mechanism for the preventive assessment of the competitiveness of high-tech products (MPAC HTP).

The proposed concept of the triune essence of competitiveness made it possible to distinguish such indicators as technical excellence and market efficiency as the main components of the HTP competitiveness. In the developed MPAC HTP, it is proposed to calculate gaps (lags) in its technical and economic characteristics from similar characteristics of a competitor's product in order to identify critical areas of technical design improvement. Modeling of technical solutions and calculating the required costs for the development and production of various design options for high-tech products allows one to choose the most rational areas for creating the technical and economic appearance of the HTP.

The developed mechanism is a set of interacting blocks that implement the proposed approach to the design of HTP based on a proactive assessment of its competitiveness using digital technologies.

Acknowledgment. The authors are grateful to the State University of Management and the Peoples' Friendship University of Russia for the opportunity to participate in research on the topic "Methods of Optimizing the Economic Efficiency of a Complex Technical System: Design at a Given Cost and Competitiveness in the Market."

References

Balenko, E.: The Ministry of Industry and Trade appreciated the readiness of Russian enterprises for digitalization (2018). https://www.rbc.ru/technology_and_media/03/07/2018/5b3a26a89a794785abc9f304

Borovkov, A.I., Ryabov, Yu.A., Maruseva, V.M.: A new paradigm of digital design and modeling of globally competitive products of the new generation (2018). http://fea.ru/news/6721

Chang, K.-H.: Product manufacturing and cost estimating using CAD/CAE (2013). http://dx.doi.org/10.1016/B978-0-12-401745-0.00001-0

Chursin, A.A.: Management of Competitiveness in the Conditions of Innovative Development of the Economy, 1st edn. Economy, Moscow (2017)

Chursin, A., Makarov, Y.: Management of Competitiveness Theory and Practice, 1st edn. Springer, Cham (2015)

Dal Pont, J.-P., Azzaro-Pantel, C.: New Approaches in the Process Industries: The Manufacturing Plant of the Future, 1st edn. Wiley, Croydon (2014)

Demchenko, S.G.: Improving the competitiveness of industrial products - an adequate response to Western sanctions: a methodological approach. Actual Probl. Econ. Law **1**(33), 10–16 (2015)

Gold, S.: The Power of Going Digital: Q&A with Siemens' Raj Batra (2016). https://www.mapi.net/blog/2016/04/power-going-digital-qa-siemens-raj-batra

Golubev, I.S., Levochkin, S.B.: The content basis of the design competitiveness of aircraft. Bull. Mosc. Aviat. Inst. **17**(1), 88–96 (2010)

Gupta, Sh., Kota, S., Mishra, R.P.: Modeling and evaluation of product quality at the conceptual design stage. Int. J. Syst. Assur. Eng. Manag. **7**, 163–177 (2016)

Khachirov, A.D., Khubulova, V.V.: Industry in the context of the digital economy. Bull. Acad. Knowl. **2**(25), 226–232 (2018)

Sebo, D., Petricova, Z., Jancus, T.: Using a multicriteria evaluation to assess the competitiveness of products. J. Prod. Eng. **14**(1), 43–46 (2011)

Thoben, K.D., Wiesner, S.A., Wuest, T.: "Industry 4.0" and smart manufacturing–a review of research issues and application examples. Int. J. Autom. Technol. **11**(1), 4–16 (2017)

Timofeev, D.V., Strigul, L.S.: Analysis of methodological approaches to assessing the competitiveness of technology at the design stage. Bull. Polotsk State Univ. **6**, 51–58 (2016)

Self-organization of Business Processes in the Environment of Web-Economic Relations

Vladlena V. Preobrazhenskaya(✉)

Bauman Moscow State Technical University, Moscow, Russia
preo-v@bmstu.ru, preovet42@yandex.ru

Abstract. Digital technologies of network communication have given rise to a rapidly evolving web relations environment in which business processes begin to transform in terms of organization and management methods. The relevance of developing new management approaches is determined by the competitive advantage provided by organizational capital in the conditions of innovative production and the digital economy. The purpose of the study is to highlight such forms of business process organization that have a particular development perspective in the web economic environment. For this, the following methods were used: the methodology of epistemology and philosophy of science, provisions of economic theory and theory of self-organization, production organization approaches, and project management methods. The main result of the research is the concept of self-organization of business processes in the web economic environment. This provision is based on the following conclusions. In network communications, means of validation of subjects of economic relations are developed, which provides legal consolidation of responsibility for obligations assumed. The range of available national and international business resources in commodity form (finance, production processes, professional functions) is expanding, which reduces the need for own (endogenous) means of production. A local target integration into a standard business process with a fixed end goal and a limited scope of production activity becomes the basis of self-organization as a new form of organization and management in the future digital economy.

Keywords: Digital economy · Web-relations · Self-organization · Business process · Project management · Organization · Management

1 Introduction

The rapid development of digital technologies and their dissemination in the field of economic relations has given rise to the mainstream of the digital economy, namely, a persistent combination of empirical and theoretical discourses, which are based on a definition that is simplified to a complete distortion of its content. Everyone understands that the production, distribution, exchange and consumption of numbers has never been and never will be the goal of economic relations. Digitalization is associated with the expectation of economic growth while the economic growth rate is an

A. V. Bogoviz and Y. V. Ragulina (Eds.): ISCI 2019, LNNS 115, pp. 168–176, 2020.
https://doi.org/10.1007/978-3-030-40749-0_20

indicator of the rate of entry into the target area, but not an identifier of the goal of economic activity. In a strict scientific understanding, a digital (as well as analog) form of transmission of audio and visual information is not even an economic resource, which, for many, does not serve as an obstacle to expect an economic miracle from digital technologies. Digital technologies can become a means of accelerating business processes but in no way an end in itself, the ultimate dream of economic entities (Ragulina et al. 2018). The goal of the present (and future!) economy remains the same is an expanded reproduction of material goods without the constant consumption of which knowledge, information, creativity, and other attributes of post-industrial forms of the economy become an aimless empty sound.

From theoretical and practical perspectives, such a direction of research, in which the capabilities of digital technologies are considered in the context of their influence on the emergence of new and development of traditional forms of production relations, seems relevant (Chursin et al. 2018; Veselovsky et al. 2018). Since the capitalist commodity form of these relations has received its worldwide political consolidation for the foreseeable future, the subject of the proposed study is limited to the analysis of new forms of relations within business processes, i.e., microeconomics. This approach allows one to focus on the study of new forms of organization of production without undue enthusiasm for the digital form of coding and decoding information.

The purpose of the proposed study is to connect the traditional ways of organizing business processes with the emerging opportunities of web-based economic relations based on digital technologies, to reveal the main directions of development of project management systems in the web-based economic environment.

Achieving this goal requires highlighting digital technologies that are most able to influence future economic relations. Also, while focusing this goal, one must do the following: (a) considering the essence of business processes from economic and organizational positions; (b) highlighting the specifics of the transformation of economic resources into means and factors of production activity; (c) finding forms of organizing business processes that have a particular development perspective in a web-based economic environment; (d) highlighting the development trends (dialectics) of the ways of organizing business processes; (e) finding and describing new organizational forms for the implementation of joint business activities.

2 Materials and Method

The research material is the project management system as one of the forms of business process organization. We rely on the following research tools: a general scientific methodology of economic theory, self-organization theory, provisions of epistemology and philosophy of science, as well as applied methods for studying the processes of production organization and project management.

3 Results

In economic relations, the "digital economy" means the widespread use of information and analytical technologies designed to increase the share of industrial production through its innovative development (Bodrunov 2016). For the innovative development path, increased demand for intellectual resources is characteristic, which leads to an even higher concentration of intellectual capital in the sectors of material production. The modern concept considers intellectual capital as a unity of various parts, in which structural capital is separately allocated. The structural capital of a corporation includes knowledge crystallized in the form of a management structure, production processes, communication culture in a corporate social environment (Glossary n.d.). The independent value of structural capital lies in the additional competitive advantage that it provides to the corporation and which is reflected in the increase in its market value. It is important to emphasize here that in the process of innovative development of industrial production, management organization methods and methods of business communication in the business environment become elements of productive forces that have independent value.

Communication in any social environment includes the ability to exchange information (communication) and personal interaction (interactivity). In the corporate social environment, these abilities are realized in the production process, are subordinate to the goals of the performance of their duties by personnel, are organized and regulated by the corporate management system. The corporate organization of business processes is characterized by the strict nature of business relations, determined by job descriptions, ethical standards, and rules of business etiquette, which are imperative. This means a particular limitation of communicativeness and interactivity, which, being the basic abilities of production personnel, are capitalized and become elements of structural capital, means of achieving a corporate goal. Thus, the corporate form of organization of business processes concentrates economic resources in an individual space, subordinates them to corporate goals, transforms them into production factors.

In March 1989, with the advent of the WWW (World Wide Web), the monopoly of corporate forms of business communication came to an end (Slobodyan 2014). The original meaning that Tim Berners-Lee put into the web technology (WWW) lies in the matrix topology of the "each with each" relationship. The web matrix laid the foundation for direct interaction (web relationships), in which there is virtually no need for intermediaries and intermediates. The affiliation of web relations to digital technologies can be justified only by the fact that information exchange is carried out using digital electronic devices that were invented and used long before WWW. Nevertheless, the electronic-digital basis of the technical base of web relations (Internet) has given rise to simplify these relations themselves to the level of digital.

The advantages of web relations have been strengthened by the rapid development of the technical base for their implementation, namely the Internet. Currently, the advantages of the web environment are the following:

Availability: Any subject of relations, having technical means of accessing the Internet, gets access to global web resources. This can be done around the clock, regardless of geographical location, via wired and wireless (including satellite) communication channels;

Independence, i.e., the absence to date of significant restrictions and external control over the activities of subjects of relations and objects of their interaction.

Self-organization: A relatively independent evolution of the web environment under the influence of the needs of the subjects of relations (network users).

Efficiency: Web relations can be carried out in real-time in individual and collective forms of communication.

Low imputed costs: An insignificant amount of expenses for technical means and time of communication, as well as constant trends towards their reduction.

The social effectiveness of network relationships has contributed to the emergence of web-based economic relationships. Their distinguishing feature was the direct exchange not only of information and knowledge but also of economic resources. Their goal was not an idle pastime, but obtaining material income. Their means were not ways of entertainment, but actually the means of implementing business plans. Payment systems and banks became pioneers in this area of relations (Martin and Harper 2018). Currently, financial and many commodity transactions are carried out via the Internet, which has spawned new ways of organizing payments, trade, and attracting economic resources. Web-based economic relations stimulate the emergence of new and the development of well-known forms of organizing business processes.

In social philosophy, there is a concept of economic singularity, according to which economic relations arose at the moment of separation (separation in time) of the production process (food production) from consumption. Consequently, the end of financial and economic relations in the future is forecasted due to the irreversible convergence of acts of production and consumption. It is assumed that the era of the predominance of production of specific individual products to order a specific consumer is to come. In other words, the production will not be mediated by market factors of consumption (Boldachev 2016; Bogoviz et al. 2019). To establish such post-market relations, new forms of organization of production, providing the necessary speed and flexibility of business processes, are required. One such form is considered an "electronic enterprise."

In the modern concept of "Industry 4.0," the concept of "digital enterprise" is interpreted as an organizational form that integrates vertical and horizontal value chains (PwC 2016). Vertical integration covers internal corporate – endogenous production processes. Horizontal integration includes external non-corporate – exogenous participants in the business process. One of the most essential characteristics of this concept is digital business models, which are understood as fundamentally new services. The revolution of these models is associated with comprehensive, personalized customer service based on integrated digital platforms – services. It is emphasized that, in the future, the diversification of production is aimed at expanding the range of individual services. In the industrial sectors of the economy, this means an increase in the share of production forms for a single individual order and a decrease in mass-production for statistical market needs.

In a web relations environment, the production process itself becomes a commodity, transforms into a production service. Namely, one should consider the production of individual products with specified characteristics or the performance of separate production functions with a given legal responsibility. In the process of developing this environment, the form of the electronic enterprise should inevitably change, as digital business models are increasingly associated with personalized customer service and the exogenous environment continuously expand. On the other hand, the rapid development of integrated digital platforms, namely cloud services, eliminates the importance of vertical integration. Many endogenous manufacturing processes will not require exclusively corporate forms of robust business relationships. This leads to the emergence and development of new non-corporate ways of organizing business processes that would coexist along with the development of corporate forms of production. Therefore, it is advisable to consider the non-corporate organization of business processes, because this is due to their flexibility and competitive advantage. Relatively low overhead costs give them an additional development perspective in future web economic relations, in the digital economy of the future. The prototype of the non-corporate form of business process organization is the project management system.

4 Discussion

A project is a time-limited business event, the purpose of which is to create a single product (result). As a specific form of organizing joint activities, the project is sufficient for the production of products that do not have analogs (producing unique products). The organization of a business process in the form of a project (project organization) has several features that distinguish it from serial-mass (process organization) production activities. The main difference between project management from functional process forms of organization is a priori established limitation of goals and means. Corporations use the design organization of business processes as a tool to achieve strategic objectives related to the diversification of activities and innovative development. However, one should pay attention to the following main advantage of the project management system: a project is a modern analog of the future production of specific individual products on order.

From a theoretical point of view, the project belongs to the class of control tasks with a fixed final state. In practice, this means obtaining a result without cyclic repetition of an operating activity (production process). The specifics of the formulation and solution of the problem with a fixed goal lies in the constancy and limitation of production factors in space and time[1]. Strict teleological certainty (finiteness of the goal) leads to a fixation of the time of direct transformation of economic resources into means of achieving it in time, which means a one-time nature of the action of production factors. Reusable resources become funds on time, and as part of the project[2]. After achieving the goal, they undergo a reverse transformation. Namely, they again become public economic resources.

[1] Factors of production are the production resources which become an element of the production process, i.e., used for the production of goods and services (Lopatnikov 2003).

[2] Only due to the virtue of being related to some goal, i.e. by virtue of a teleological definition, they acquire the meaning of means (Konstantinov 1960–1970).

The inverse transformation effect inherent in the design organization of business processes is of great practical importance because project implementation can be based on temporary (rental) production relations. This reasonably well-known property of the project management system in the environment of web-economic relations gives rise to a new form of non-corporate organization of commercial activities based on rental project management. The formulation of the project management task, based on the concept of temporary use of resources, allows minimizing capital costs. This is effective with the venture nature of new financing and is especially useful in future production business processes under a single individual order, with a decrease in mass-production under the statistical needs of the market. The concept of rental management is limited by the business idea of the project and the cost of materials that have the nature of one-time use (productive consumption), which are almost impossible to return back after use. We will take a closer look at the prospect of the temporary use of production resources in a web-based economic environment.

Labor capital. In conditions of increasing competition in the labor market, traditional corporate forms of labor activity are less and less able to ensure full employment of the working population, which provides the necessary income level. Non-standard (flexible) forms of employment become the response of the device (Razumova and Artamonova 2017). In the digital economy, the development of non-corporate forms of employment alternatives to traditional corporate forms of employment is projected. These include freelance, i.e., work remote from the employer, performing one-time tasks by crowd workers (freelancers working via the Internet). Corporations are increasingly engaging in temporary participation in their projects of individual crowd workers who can work on a flexible schedule and anywhere in the world.

Outstaffing represents a transition from using the labor of individual crowdfunders to the rental of working capital in the form of entire labor collectives. Outstaffing enterprises provide formally registered employees to the client corporation on the terms of a contract that perform work for it.

Outsourcing is distinguished by the fact that separately allocated production functions are rented here. The cumulative economic effect of using outsourcing is integral, i.e., it consists of direct and indirect benefits that the corporation receives from outsourcing its own functions (Kotlyarov 2017).

The above analysis indicates the priority in web economic relations of design methods for organizing business processes over process, rental forms of using resources over purchased. It remains to highlight one significant phenomenon that arises in the environment of web-economic relations. The concept of setting the problem of rental project management gives rise to self-organizing forms of business processes. We will consider this in more detail.

The basis of the digital economy is digital enterprises, the operational activity of which is the personalized customer service on order. An order is the same (as a project) time-limited business process with a fixed goal, which is to obtain a single result (income). In the environment of web-based economic relations, business processes themselves take on a marketable appearance. Along with raw materials, materials, serial production, separately allocated production functions become free-market resources, such as production of goods with specified characteristics (processing); management functions of various components of the business process (management);

monitoring the rational use of resources (controlling); design, planning, and construction (engineering); rational provision of resources (logistics), etc. Almost all types of business actions or their final results are offered to the customer in the form of separately provided commercial services.

The international division of labor, the specialization of digital enterprises, and the expansion of the competitive market contribute to lowering the cost of commercial services. For the one-time production of specific products on order, temporarily using (renting) exogenous market resources instead of their own (endogenous) funds becomes economically feasible. This establishes the prevalence of horizontal (non-corporate) value chains in economic relations. The feature of horizontal ties lies in the flexibility of economic ties between business process entities. The last advantage gives rise to its continuation, namely the lack of an organizational component in the factors of production, which is in abundance in the case of a predominance of vertical (corporate) value chains. In these conditions, an alternative form, namely self-organization of the business process, is born and develops.

From the position of the general theory of systems, the main factor in system formation, in general, and self-organization, in particular, is the final result (Popov 2008). In a commercial application, the result is the primary goal of the business process. As a rule, this is income from commercial activities, since the global goal is identified by the customer, so long as the participants in the execution of the order to pursue individual local commercial goals. This is achieved by solving local problems by the subjects of the business process and obtaining the expected income as a result. Therefore, the interconnection of local tasks (the goal tree) defines the structure of the organization of the business process. Not a rigidly determined corporate vertical chain of value creation but a flexible, probabilistic horizontal chain of online connections leads to the self-organization of subjects of economic relations into a typical business process. The flexibility of economic relations is mainly due to their temporary nature, as a result of which these relations can be further strengthened or weakened by changing the target tree in various business processes in order.

Thus, the development of web-economic relations and the principle of rental (temporary) project management formulated above lead to fundamentally different ways of coordinating commercial activities, to forms of self-organization of business processes. The development and strengthening of non-corporate (temporary) labor relations in the limit eliminate the need for a corporate (permanent) form of organizing business processes and replacing its activities based on temporary (rental) project management for many types of production. Online economics expands the possibilities of system decomposition of the project towards a more flexible hierarchy of sustainable elements (types of work, costs, the nomenclature of resources, etc.). Free market conditions make it possible to rationalize the nomenclature of resources regardless of their territorial, national, or corporate affiliation, which contributes to the reduction of funds for estimating the costs of their delivery at the right time and place. Free web access to the labor resources offered on the open market allows one to carry out all project management in the form of outsourcing and outstaffing. The possibilities of web-based economic relations increase the flexibility of organizing production and management functions to the level of self-organization.

Thus, the goal of the proposed study is achieved. The concept of setting the problem of rental project management and analysis of the development of online economic conditions have revealed the effect of internal self-organization of future business processes.

5 Conclusion

In conclusion, it should be noted that the development of web-based economic relations expands the arsenal of ways of flexible organization of production and management, one of which is the self-organization of business processes. This form of business process organization becomes indispensable for obtaining individual results, namely: innovative products, services, technologies, knowledge in the digital economy. The territorial and interstate distribution and functional specialization of digital production in the conditions of the online economy and the free international market expand the resource field of business processes. This gives rise to self-organizing forms of commercial activity, the basis of which may be further researched with respect to the ways of developing rental project management.

References

Bodrunov, S.D.: The new industrial society of the second generation: man, production, development. Soc. Econ. **9**, 5–21 (2016)

Bogoviz, A.V., Lobova, S.V., Alekseev, A.N., Prokofiev, V.N., Gimelshtein, I.V.: Managing the modernization of regional markets of educational services in the conditions of formation of Industry 4.0. Horiz. **27**(3-4), 187–192 (2019)

Boldachev, A.: Finita la history - political-cultural-economic singularity as an absolute crisis of civilization. In: Frumkin, K.G. (ed.) Singularity. Images of "Posthumanity", pp. 7–37. Algorithm, Moscow (2016)

Chursin, A.A., Kashirin, A.I., Strenalyuk, V.V., Ostrovskaya, A.A., Kokuytseva, T.V.: The approach to detection and application of the company's technological competences to form a business-model. In: IOP Conference Series: Materials Science and Engineering, vol. 312, p. 012003 (2018)

Glossary of terms of normative and technical documentation (n.d.). Structural capital. https://normative_reference_dictionary.academic.ru/75458/

Konstantinov, F.V. (ed.).: Philosophical Encyclopedia, vols. 1–4. Sovetskaya entsiklopediya, Moscow (1960–1970)

Kotlyarov, I.D.: Problems of evaluating the effectiveness of outsourcing. Bull. Inst. Econ. Russ. Acad. Sci. **6**, 87–99 (2017)

Lopatnikov, L.I.: Economics and Mathematics Dictionary: Dictionary of Modern Economics. Delo, Moscow (2003)

Martin, T., Harper, H.: Robot for the client. How Russian banks will change in the near future (2018). https://www.forbes.ru/finansy-i-investicii/355517-robot-dlya-klienta-kak-izmenyatsya-rossiyskie-banki-v-nedalekom

Popov, E.I.: Organization and self-organization as tools to increase the efficiency of industrial production. Prod. Organ. **2**, 9–11 (2008)

PwC: "Industry 4.0": creating a digital enterprise (2016). https://www.pwc.ru/ru/technology/assets/global_industry-2016_rus.pdf

Ragulina, Y.V., Bogoviz, A.V., Alekseev, A.N.: Modeling the management of innovational processes in regional economy. In: Advances in Intelligent Systems and Computing, vol. 622, pp. 568–573 (2018)

Razumova, T.O., Artamonova, M.V.: Flexible forms of employment in the system of criteria of the concept "Future in the world of work". Sci. Stud. Fac. Econ. Electron. J. **9**(4), 7–27 (2017)

Slobodyan, E.: What is the World Wide Web and who invented it? (2014). http://www.aif.ru/dontknows/file/1123050

Veselovsky, M.Y., Izmailova, M.A., Bogoviz, A.V., Lobova, S.V., Ragulina, Y.V.: System approach to achieving new quality of corporate governance in the context of innovation development. Quality - Access Success **19**(163), 30–36 (2018)

Innovations in Banks: Areas and Challenges

Anna I. Serebrennikova(✉), Yuliya V. Kuvayeva, and Tatyana A. Chilimova

Ural State University of Economics, Yekaterinburg, Russia
serebrennikova_ai@usue.ru, ykuvaeva1974@mail.ru, tachil77@mail.ru

Abstract. Currently, the competitiveness and sustainable development of banks are mainly dependent on innovation. The research goal is to study and systematize those reasons responsible for the failure of innovation in banks. When conducting the research, the following methods and approaches were used: a systematic approach, structuring, comparison, generalization, description, observation, induction, deduction, abstraction methods. Based on the analysis of approaches to the definition of the concept of banking innovation, areas of innovation in the bank identified. An approach to the analysis of the problems of introducing changes in the context of specific areas is proposed, fundamental differences and possible commonality of the identified problems in different areas are also shown. The authors conclude that there is the possibility of applying the approach to the analysis of issues for concretizing places and the reasons for their occurrence, selecting introduced innovations, and developing an innovative bank development strategy.

Keywords: Bank · Innovation · Banking innovation · Innovation areas · Innovation implementation problems · Banking product

1 Introduction

Recently, innovations in all sectors of the economy are firmly associated with trends in the sustainable and active progressive development, as well as with increased competitiveness. Banks and banking activities are no exception. The use of innovations in banking is not only justified but also necessary. This is a consequence of the functions of the central financial intermediaries performed by banks, the state, and activity of which should be as close as possible to the level of economic development. Innovative development of the banking sector is seen as a condition for sustainable economic growth in the long term, including its national security (Anikina et al. 2016; Thalassinos et al. 2013, and others).

It is noted that the absence or insufficiency of the use of innovations in the activities of banks leads to stagnation and the possibility of losing the integrity of the bank as a system (Thalassinos et al. 2012).

As a rule, authors emphasize the positive result of innovative activity, expressed in increasing the efficiency of the bank, mobilizing and distributing resources, creating additional value for the bank's customers, increasing satisfaction with banking products and services, etc.

A. V. Bogoviz and Y. V. Ragulina (Eds.): ISCI 2019, LNNS 115, pp. 177–184, 2020.
https://doi.org/10.1007/978-3-030-40749-0_21

Along with this, there is a fallacy and danger of considering only the positive effect of innovation, because they not only carry additional ones, including risks that are not specific to the bank, but can also be virtual or imitative, worsening the situation of the bank and its customers (Rusanov 2011). Besides, even the positive nature of innovation does not guarantee the success of their implementation for several reasons, including related to the Russian features of innovation, the state of the banking system and the financial market.

Based on the preceding, the purpose of this study is to deepen the essential understanding of the content of innovations in banks, which is achieved by analyzing implementation problems in the context of areas of innovation.

To achieve the goal of the study, the authors to solve the following problems:

- Systematization of views on approaches to defining the concepts of innovation and banking innovation;
- Identification of directions for the implementation of innovations in the bank and their characteristics;
- Formulation and analysis of the problems of introducing banking innovations in the context of areas of innovation.

2 Materials and Method

Based on the goals and objectives, the starting point of the study is the analysis of the concepts of innovation and banking or financial innovation, which allows us to identify the main areas of innovation and clarify their essence. Until now, there is no consensus on the nature of banking innovation, which gives rise to many definitions and the need to analyze existing approaches (Gribanov et al. 2017; Andrushchuk 2018; Gadisova 2011; Ivanova 2010; Mitrokhin and Gribanov 2018; Yakovenko and Timchenko 2013). In addition to the above works, in determining the directions of innovations and their characteristics, the already established research in this direction was used (Murshudli 2018; Savinova 2018; Chernykh 2009), focusing on specific areas of innovation.

The identification and systematization of the main problems and difficulties in the use of innovations and innovative activities were carried out based on previously conducted theoretical studies and practical observations. (Ashba 2019; Batalova 2013; Masalimova and Streltsov 2018; Mitrokhin and Gribanov 2018; Shikova and Borisoglebskaya 2016).

This study is based on the use of a wide range of methods of scientific knowledge, namely: a systematic approach and structuring of the views presented in literary sources, methods of comparison and generalization in the characterization of areas of innovation, a method of observation in identifying problems of the introduction of innovations and their interconnection, as well as such general scientific methods such as description, induction, deduction, and abstraction.

3 Results

As noted above, there are a large number of definitions of the concepts of innovation and banking innovation. The systematization of definitions allowed researchers to identify the main approaches to the definition of a concept.

Approaches to the definition of financial (banking) innovation:

- New financial products or services;
- Automation and informatization of financial processes, new methodology and methods of financial operations;
- A new organization of business processes of a credit organization;
- New institutional forms of banking business organization (Yakovenko and Timchenko 2013).

Another point of view is quite close to the one presented, according to which systemic, productive, instrumental, and processing approaches are distinguished (Gribanov et al. 2017). Subsequently, the number of approaches is reduced to the following three: instrumental, processing, and effective (Mitrokhin and Gribanov 2018). Another point of view recognizes the following approaches: product, technological, and business processing to determine the essence of innovation. At the same time, the business processing approach includes "new institutional forms of banking business organization" (Andrushchuk 2018).

According to the author, the absence of contradictions between the different approaches identified by different scientists is noteworthy, because none of them deny the presence of a fundamentally new or improved product, including methods of its implementation, as a result of the application of innovations. The main difference is the focus on techniques or tools, which is a natural consequence of the fact that the result of any company's activity is a product. Orientation towards the result is of particular importance in the case of banks as a consequence of the characteristic properties of a banking product. In particular, the lack of the possibility of accumulation of a banking product indicates its mobility, which means that it is highly dependent on the current level of bank development. Consequently, the author considers it appropriate to highlight the main areas of innovation in banks, which, in one way or another, can affect the banking product. The directions of innovations in the bank and their manifestations are given in Table 1. Note that the author's view is in many respects consistent with Russian legislation and the theory that defines the list of typical changes faced by the production manager (Schumpeter 1982).

We will give a more detailed description of the directions of innovation and consider the possible problems of their implementation. Within the framework of this article, we will focus on the first two directions, which most clearly reflect the specifics of banking activities. We will start with the product line. The expansion and updating of the product line directly demonstrate the bank's function in providing consumers with banking products. The reason for innovation is a change in the economic and market situation, the competitive environment, customer needs, and the transition to another business model.

Table 1. Directions of innovative development in a bank.

Direction of innovation	Manifestations of innovation
Product	• Expanding the product line by improving existing products and developing new ones; • Diversification of activities
Technological	• Automation and informatization of operations and processes; • Improving and developing new technologies for financial transactions; • Using modern information technologies (including the creation of a single technological platform and environment)
Business processing	• Improving existing business processes and developing new ones; • Using new achievements of business process management
Institutional	• Reorganization of the organizational structure and management of the company; • Creation of corporate structures; • Mergers and Acquisitions

Source: Compiled by the author.

Competition in the Russian market has several features. First, competition is more at the level of additional service than at the level of the products themselves, the set of which is almost standard (Zemlyakov et al. 2017). Second, although experts acknowledge the high level of competition in the banking sector, its unevenness should be taken into account. The largest banks with state participation hold the bulk of the market; they set the direction for the formation of the product line and technological innovations. Third, banks are forced to compete with non-banking institutions offering alternative products of banking products.

In this area of innovation, the author notes the possibility of the following problems:

- *Secondary innovation.* Focus on existing products, which cannot give significant results in the long run. Innovation refers to modifying or combinatorial;
- *Conservativeness.* Adherence to traditional lines of business and products may be a likely consequence of the bank's conservatism. Customers can be conservative through skepticism about brand new products;
- *Excessive diversification.* The pursuit of expanding the product line leads to the dispersal of the bank's resources, the impossibility of a precise study of all products, the decrease in the economic effect of their use, and the loss of competitive position concerning non-banking companies that sell one or more products. The inclusion of non-standard products that are not characteristic of banking activities may lead to a change in priority areas of business with the erosion of banking specifics;
- *Product line imbalance.* Excessive enthusiasm for product development, especially with the use of digital service channels and information technologies, may lead to the predominance of high-tech products with enhanced consumer qualities mainly for one generation. Ongoing research, including devoted to banks, show significant

differences in the preferences of technologies and services for each subsequent generation (Ashba 2019). A continuation of this problem is the following;

- *Unbalanced customer base.* The crowding out of existing products and access channels to more modern ones can lead certain groups and generations of customers from the "comfort zone," which in turn could reduce their loyalty to the bank. In particular, a similar situation is developing now, when there is an apparent lack of products for older people, despite their quantity and financial potential. Of course, product improvement is necessary because only it can give real competitive advantages, but in this case, the bank must find a middle ground and carry out transformations gradually;
- *Excessive product improvement.* Here, at first glance, there is a contradiction. Only a fundamentally new (breakthrough) product can bring maximum and long-term effects. Still, its introduction must be carefully prepared because the consumer can not understand the product.

Moreover, in addition to the client's readiness for technological changes, banks should take into account the level of their financial literacy and the formation of needs. Unfortunately, the optimal combination of financial and information literacy in Russia is rare. As practice shows, the demand for investment products through remote access channels is minimal and does not exceed 2% of all operations (Andrushchuk 2018).

The technological direction of innovations in the bank is inextricably linked with the grocery. The trend is associated with automation, informatization, and digitalization. On the one hand, innovations are related to the provision of products and lead to their qualitative, fundamental change under certain conditions. On the other hand, the scope for innovation within the bank is broad. Increasing the efficiency of the entire bank, in the end, they favorably affect the process of bringing the product to the client. The author shares the view that these innovations are based on the achievements of other industries and are external in relation to the bank (Ivanova 2010).

We have formulated the following main problems of the technological area of innovation:

- *The forced character of innovation.* In terms of maintaining competitiveness, the forced nature of innovation is natural. However, we need to consider that, once involved in the digital information race, the bank becomes dependent "on the constant updating of technologies and the need for innovation" (Masalimova and Streltsov 2018). To maximize the effectiveness of introducing innovations, the bank should have a strategy based on a vision of prospects and make decisions based on it. The problem is most significant for medium and small banks, which are forced to be content with compromise solutions with limited resources;
- *Secondariness of innovation.* It is determined by the external nature of technological innovations based on existing technologies and software, which should be adapted to the specifics of the bank. The problem can be mitigated by conducting its development, inaccessible to most banks. "It should be noted that the association "Fintech," whose goal is to introduce technological solutions to the financial market, includes only the largest banks, such as Sberbank, VTB, Gazprombank, Alfa Bank, Tinkoff Bank, Raiffeisen Bank, Otkrytiye Bank" (Pertseva 2018). Most banks are doomed to follow leaders;

- *Costs.* Many authors emphasize this problem. Costs accompany the entire cycle of innovation, namely the development (acquisition), commissioning, including adaptation to specific conditions, and operation. The question of the justification of the costs of innovation (Batalova 2013) and the assessment of their effectiveness are becoming highly relevant. The largest banks are again in the most advantageous position;
- *Personnel.* The problem of personnel is complex and multifaceted. The first facet is a change in the set of specialties and staff structure, a decrease in the need for specialists in the field of finance and banking. The second one implies changing a set of competencies of specialists and maintaining their sufficient level. So, the ability to set tasks for a further solution by technical means becomes a significant competency for a bank employee. The third one is the innovative potential of employees, i.e., both their openness to new ideas and ability to generate and introduce innovations. The fourth aspect is the need to optimize the staff motivation system to develop their competencies and skills.

The above list of problems is not exhaustive, but it demonstrates the difference between the issues of different directions.

4 Discussion

The systematization of problems and the identification of their causes shows the difference in the issues related to a particular area of innovative activity, indirectly confirming the correctness of the chosen approach to analysis. The division of innovations in the directions is due to the complexity of the banking business. Note that the complexity of innovation processes and their relationship within the banking business give rise to a multiplicity of definitions of the concept of banking innovation.

The author considers it necessary to emphasize the close connection of problems from different groups, which is caused by the orientation of all processes in the bank towards the creation of the final product. So, at present, the banking product targeted at product innovations, and the channels for bringing it to the consumer, transforming under the influence of technological innovations, are difficult to separate.

The following circumstance is noteworthy: a problem with one name (secondariness) is present in both groups of issues, but it would seem to differ in content. However, regardless of the line of business, the cause of these problems is the use of existing innovations and imitation.

It poses two crucial questions. The first key question: "When does an innovation from a relevant and useful event, which is filled with meaning and gives an impetus to the further development of the bank, turning into a mere replication of existing achievements, which can only help to "stay afloat" for some time?" Obviously, in solving this issue, the time horizon and frequency of application of innovation should be taken into account. Regional distribution of innovation will also be essential. It is no secret that the level of bank innovations in those countries with developing economies is much lower than the level of innovation in international banking (Murshudli 2018).

Our second question: "Where is the line between innovation and imitation? Consideration of the content of innovation and the problems associated with its implementation will help answer this question". Perhaps getting rid of imitation will be a step towards overcoming uniformity in the banking products market and developing fundamentally breakthrough products.

5 Conclusion

The current state of development of the country's banking system is characterized by increased competition and crisis phenomena, highlighting increased competitiveness and the need for a transition to a sustainable development paradigm. The use of innovations can provide banks with a reliable basis for strengthening competitive positions in the market and sustainable development.

At the same time, in combination with external conditions, innovations can lead to severe transformations in banks, including diversification of activities and a change in the business model. These transformations can change the vector of the bank's development, up to the abandonment of specialization in banking. The increased susceptibility of banks to innovations in such a situation makes them as vulnerable as possible to unsuccessful innovations and problems arising from their implementation.

In the framework of the study, the authors identify the problems of introducing innovations in the context of banks' areas of innovative activity. This approach combines theoretical developments and practical observations, which can be an impetus for the development of the theory and deepening the essential understanding of banking innovations. The systematic application of the proposed approach will help in identifying specific places of occurrence of problems and increase the likelihood of their timely elimination, contribute to the adoption of informed decisions about the need for innovation, as well as the development of an innovative bank development strategy. In turn, this will have a positive impact on strengthening the bank's competitive position in the market and increasing its stability.

References

Andrushchuk, V.V.: Innovative aspects of increasing the competitiveness of Russian banks. Acad. Vestn. **2**, 79–86 (2018)

Anikina, I.D., Gukova, V.A., Golodova, A.A., Chekalkina, A.F.: Methodological aspects of prioritization of financial tools for the stimulation of innovative activities. Eur. Res. Stud. J. **19**(2), 100–112 (2016)

Ashba, A.A.: Features of the competitive behavior of commercial banks in the digitalization of the economy (on the example of Sberbank of Russia). Theory Pract. Soc. Dev. **1**(131), 67–71 (2019)

Batalova, N.T.: Problems of forming an innovative environment in the banking sector and potential solutions. Vestn. USUES Sci. Educ. Econ. Ser. Econ. **4**(6), 65–68 (2013)

Chernykh, S.I.: Russian Banks in a Competitive Environment. IE RAS, Moscow (2009)

Gadisova, T.M.: Financial innovations as a strategic factor in the effective and sustainable development of commercial banks in Russia. Actual Issues Econ. Sci. **1**(21), 95–99 (2011)

Gribanov, A.V., Mitrokhin, V.V., Vilkova, M.V.: Innovation as the main factor for the sustainable development of a commercial bank. Eur. Res. Stud. J. **XX**(4A), 408–416 (2017)
Ivanova, O.V.: About the essence of banking innovation. FES Finan. Econ. Strategy **3**, 21–26 (2010)
Masalimova, E.V., Streltsov, M.A.: Development of innovative products and services as the main factor in the competitiveness of the Russian banking sector. In: Koschegulova, I.R. (ed.) Collection of Scientific Papers on Management. Economic Analysis. Finance, pp. 271–276. Ufa State Aviation Technical University, Ufa (2018)
Mitrokhin, V.V., Gribanov, A.V.: Innovation as a tool to ensure sustainable development of the banking sector. Bus. Educ. Law Vestn. Volgograd Inst. Bus. **3**(44), 127–134 (2018)
Murshudli, F.F.: Innovative trends in international banking business (on the example of Azerbaijan). Vestn. MGIMO Univ. **1**(58), 186–212 (2018)
Pertseva, S.Yu.: Digital transformation of the financial sector. Manag. Innov. **18**, 48–52 (2018)
Rusanov, Yu.Yu.: Innovative methods and tools of banking risk management. Financ. Money Invest. **1**(37), 25–26 (2011)
Savinova, E.A.: Features of the innovative development of the banking sector in Russia. Sci. J. “Econ. Sociol. Law” **1**(9), 65–70 (2018)
Schumpeter, J.A.: Theory of Economic Development. Progress, Moscow (1982)
Shikova, E.I., Borisoglebskaya, L.N.: Development of banking innovations in modern conditions. Univ. Vestn. **10**, 176–179 (2016)
Thalassinos, E.I., Maditinos, D., Paschalidis, A.: Observing evidence of insider trading in the Athens stock exchange. J. Econ. Struct. **1**(1), 1–15 (2012)
Thalassinos, E.I., Venediktova, B., Staneva-Petkova, D.: Way of banking development abroad: branches or subsidiaries. Int. J. Econ. Bus. Adm. **1**(3), 69–78 (2013)
Yakovenko, S.N., Timchenko, A.A.: Financial innovations in the activities of commercial banks: theoretical and economic aspects. Mod. Prob. Sci. Educ. **2**, 335–341 (2013)
Zemlyakov, Yu.D., Rudneva, N.M., Mamaeva, I.L.: The need and problems of improving the banking products market. Vestn. Int. Acad. Syst. Res. Inform. Ecol. Econ. **20**(2), 54–57 (2017)

The Role and Specificity of Human Capital in the Formation of a Digital Economy

Svetlana A. Korshunova(✉)

Voronezh State Technical University, Voronezh, Russia
skorsh5@gmail.com

Abstract. The article analyzes the process of transformation of human capital in the digital economy and reveals the current specifics of human capital, formed under the influence of digital technologies. Using systemic and synergetic research methods allowed us to consider the modern digital economy as a dynamic, self-developing system in which human capital is the primary source of innovative transformations. The authors discuss the mechanism of the dynamics of current social processes that influenced the formation of new qualitative content of human capital. They highlight the following global trends in the development of modern human capital: the transformation of human capital into an independent factor of production, the increasing role of intellectual capital in the digital economy, the emergence of new requirements that modern employers place on human capital. It is proved that, at present, human capital takes the form of intellectual capital associated with the heuristic nature of new knowledge. The authors conclude that the young generation is becoming the bearer of modern intellectual capital, in which unique values, a new type of motivation, new forms of communication, and new skills (including digital) are actively being formed.

Keywords: Digital economy · Human capital · Intellectual capital · Generation Z · Digital skills

1 Introduction

Currently, most developed countries in the world are undergoing a civilization transition associated with the introduction of information technology in all spheres of human life. The emergence of the global Internet network has become a real digital-information revolution, which determined the new specifics of modern reality, namely the formation of a digital society. Digital society should be understood as a system of economic, social, and cultural relations based on the use of digital technologies.

Thanks to the development and implementation of information and communication technologies, all aspects of human activity and, above all, the economy of modern society have transformed. The information has become the leading economic resource of humanity, capable of ensuring fast and efficient business conduct using online trading, electronic payments, online advertising, and online banking. Thus, the modern economy, directly related to the use of digital computer technologies, has turned into a

A. V. Bogoviz and Y. V. Ragulina (Eds.): ISCI 2019, LNNS 115, pp. 185–192, 2020.
https://doi.org/10.1007/978-3-030-40749-0_22

digital economy, and this process can be considered a definite global trend of the last decade.

The digital economy is an economic activity that is characterized by the active implementation and practical use of digital technologies for the collection, storage, processing, and transmission of information. Worldwide, the adoption of the digital economy is seen as a fundamental transformation of the traditional economic model.

The emergence of the term "digital economy" is associated with the names of two researchers, namely Don Tapscott and Nicholas Negroponte. In 1995, the renowned Canadian researcher Don Tapscott released the book "Digital Economy," in which he first described the new economic reality shaped by network technologies (Tapscott 1995). Tapscott predicted many changes that Western society should face, including predicted the rapid growth of digital technologies and their impact on the development of the economy, public administration, the nature of social relations.

In the scientific literature, the author of the term "digital economy" is considered to be the American scientist Nicholas Negroponte from Massachusetts University. In 1995, he published a non-fiction book called "Being Digital" (Negroponte 1995). Negroponte gave his characterization of the genesis of the digital economy and analyzed the benefits that the new economy and society as a whole receive with the development of Internet technologies. First of all, this concerned new opportunities that opened up for the economic sphere, because the formation of a new paradigm of economic relations made it possible for business entities to communicate and exchange information in real-time and from around the world.

The dominant economic and social potential brought by new information and communication technologies was quickly appreciated in all developed countries of the world. Since the 2000s, governments in many countries have begun developing strategies for the transition to a digital economy.

As for the prospects for the digital development of the Russian economy, they are quite realistic and are perceived by the country's leadership as an essential task of the next decade. The Digital Evolution Index study conducted by Mastercard and the Fletcher School of Law and Diplomacy at Tufts University in 2017, as well as the analysis of the Digital Society Index, which was carried out by the Oxford Economics in conjunction with the Dentsu Aegis Network in 2018, showed fairly good competitiveness and high digital development potential of Russia. In many respects, Russia is included in the group of promising countries, demonstrating a steady rate of digitalization of the economy.

In 2017, the Government of the Russian Federation approved the program "Digital Economy of the Russian Federation," the purpose of which is to launch in Russia a large-scale system program for the development of the economy of a new technological generation. The critical areas in the formation of the digital society of Russia were named personnel and education. The importance of these areas is determined by their relationship with such a resource of social and economic development as human capital.

Most contemporary researchers consider human capital as the main prerequisite for the innovative activity of the economy (Kapelyushnikov 2012). The accumulation of human capital is a critical factor in the economic and social well-being of modern society. This means that it can become a driver of technological change, or, conversely, act as a brake on the process of digitalization of society.

The purpose of this study is to analyze the features of the development of human capital in the context of the formation of a new digital reality and determine the current specifics of human capital, which he received during the implementation of information and communication technologies in all spheres of human life.

2 Materials and Method

The presented work is interdisciplinary. It was carried out at the intersection of economic and humanities, which allowed us to expand the problem field of research significantly.

In the analysis of global processes of transformation of modern society, affecting the formation of a new specificity of human capital, a scientific methodology was used, including the following research methods: functional analysis, a system method, induction and deduction, as well as synergetic method. From the perspective of a systematic approach, the economic sphere of production and consumption is one of the elements of the social system. Therefore, the economy experiences all the consequences of the transformation of a post-industrial society, which is gradually acquiring the character of a digital society. At the same time, human capital acts as a complex system object of socio-economic research.

The synergetic methodology, which has successfully proved itself in the framework of many interdisciplinary studies, allows us to consider economic systems as dynamic self-developing systems that exist in a state of uncertainty characteristic of the whole modern society. The active development of human capital is determined by the combined impact of factors such as institutional, demographic, economic, socio-political, cultural, environmental, etc.

3 Results

In economics, it is customary to understand human capital as all acquired knowledge, skills, abilities, motivations, professional experience, and health that people possess and which they (people) can use to produce goods and services.

The creators of the theory of human capital T. Schulz and G. Becker began to consider human abilities from an economic point of view since they can bring economic benefits. Along with other forms of capital, human capital is in demand in the labor market and has a particular value (Becker 1993). Its owner possesses this capital, i.e., a human himself who can maintain his knowledge, skills, abilities even after their sale. If human capital acts as an asset, along with financial or physical assets, then it also needs investments. Therefore, all economists involved in the theory of human capital emphasized the need for investments that must be invested in people in order to stimulate the growth of human capital (Schultz 1971).

The development of new technologies and globalization processes have led to an increase in the role of human capital in the global economic system. Throughout the

20th century, the relationship between human and physical capital has consistently evolved in favor of human capital as a result of which it has become the main factor in the economic well-being of developed countries.

Currently, the problem of the specifics of human capital, which is being formed within the framework of the new digital reality, is relevant. Contemporary society faces previously unknown challenges. These challenges include the following: the intensification of information flows and the associated acceleration of the rhythm of life; the introduction of new digital technologies in all areas of the global economy; the emergence of virtual reality as a new form of human life. The answer to these challenges is the transformation of the internal content of human capital. The specifics of human capital begin to change under new requirements for personal characteristics. Moreover, such a request is made by both employers and the whole of modern society.

Thus, the need for scientific research capable of identifying a qualitatively new content of human capital, which is emerging in the modern digital information environment, arises.

When considering the structure of human capital, one should bear in mind that almost all of its elements do not have a constant character. Knowledge, skills, and even health are acquired and accumulated by a person in the process of life. Like physical capital, human capital is subject to depreciation and is dependent on investment. Other factors are affecting the specifics of human capital. Recently, studies have emerged that note the dependence of human capital on the transformation of traditional social institutions in modern society (Dyatlov and Dobrokhotov 2018). A more in-depth analysis shows that the modernization processes taking place in the institutional environment, in turn, are the results of the dynamics of the entire post-industrial society. Therefore, the general trends in the development of human capital manifested in world practice as a whole should be highlighted.

One of the main trends in modern society is a specific growing imbalance in the development of the main types of human capital. In the digital economy, intellectual capital gains a competitive advantage, which means that the organization that most successfully uses its intellectual capital and the intellectual capital of its individual employees is highly competitive in the market.

The concept of intellectual capital (IC) has been actively developed in Western and domestic science since the 90s of the XX century. Thomas Stewart, an American journalist and a well-known columnist for economic issues, is the author of the term "intellectual capital." In his book "Intellectual capital: A new source of wealth for organizations" (Stewart 1997), he first drew attention to the specific impact that the information revolution had on the activities and assets of modern companies. As a result of the introduction of information technology in the manufacturing sector, a new type of organization arises, namely, an intelligent company, which differs from a traditional industrial company in its intangible assets. And for this reason, the financial structure of such a company is not amenable to conventional economic analysis (Ruus et al. 2010).

Intellectual capital is a combination of human knowledge, skills, abilities, mobility, and creativity, providing the ability to create a product in the process of the movement of this capital as part of human capital (Suprun 2006). However, this definition does not reflect the specific feature of intellectual capital, which lies in the nature of knowledge that forms its basis. If human capital is based on reproducible knowledge, then intellectual capital is formed only due to new knowledge of a heuristic nature.

Intellectual capital can be individual and social. Knowledge, abilities, skills, and intellectual potentials form intellectual capital. Organizational intellectual capital is the qualification, knowledge, experience, technology that employees of an organization possess. Aggregate intellectual capital, as one of the types of human capital, carries a clearly expressed synergetic principle. The total intellectual capital is formed not just by adding up the individual capital of employees but is the result of their interaction. The synergistic effect is achieved in this way; it many times increases the efficiency of the organization. Before our eyes, intellectual capital, as one of the forms of human capital, is beginning to transform the world radically.

In such conditions, the new requirements that the digital economy and digital society impose on human capital are being formed. We are talking primarily about those requirements that must meet the knowledge, abilities, skills, and professional experience of modern humans.

4 Discussion

At the end of the 20th century, already in the era of the establishment of post-industrial society, the adaptive capabilities of a person as a carrier of human (intellectual) capital began to appear when a demand for certain personal qualities of a person appeared, such as individualism, creativity, originality, creative courage, etc. When selecting employees, the leaders of modern companies with a large volume of intellectual assets choose employees who are free from professional stereotypes and patterns of thinking, who can make creative decisions and produce new ideas. The employees themselves, in addition to material interest, demonstrate a new type of professional motivation associated with the possibility of personal growth, self-realization, and the semantic content of their work. A few years ago, this only concerned employees of high-tech companies. Recently, entire sectors of the economy have appeared, the production success of which directly depends on the unique abilities of workers in these sectors.

Economic theories of human capital, continuing to consider a person only from the perspective of his production functions, significantly narrow the idea of the intellectual capabilities of modern humans. It is necessary to expand the context of scientific research, including in the analysis, not only economic but also sociocultural, psychological, and worldview factors that determine the motivations of modern humans. First of all, this applies to representatives of the last two generations of youth who received in sociology the well-known names "Generation Y" and "Generation Z." Representatives of these two generations are experiencing the most significant influence of digital technologies, as they constitute the most active audience of users of electronic gadgets, social networks, and Internet resources.

Philosophers, sociologists, and psychologists record the emergence of new spiritual meanings and value orientations among the young generation. The formation of a digital society and digital economy provides a person with new opportunities, namely: forms a new value system for him, new consumer practices, and also offers people new means and forms of communication. The globalization of the information space contributes to the spread of these trends around the world. For example, more and more young people in the world choose self-development and self-improvement as their primary life strategy.

Studies show that cash compensation is no longer perceived as the primary motivation for working in developed countries. A famous American author Daniel Pink argues that cash incentives for company employees work only if the employee performs tasks based on the simple decision algorithm. However, when solving complex, non-standard tasks, monetary motivation does not work. Currently, this applies to 70% of people in developed countries whose professional activities are associated with elements of creativity and analysis (Bogoviz et al. 2019a, b).

For such creative employees, we can offer another type of motivation, which is based on the following three central values: freedom of choice, mastery, and having a worthy goal. These values are intrinsic to all spiritually and intellectually developed people. Therefore, company leaders must create in their organizations an atmosphere conducive to the implementation of these values in labor and professional activities.

At first glance, these changes affect only creative people in developed countries. However, research on "Generation Z" notes that modern youth have the same requirements for their work (Stillman and Stillman 2017). Representatives of "Generation Z" (digital youth) prefer socially useful activities, and 93% of youth say that the company's attitude to society affects the decision to work with it or to work in it.

Thus, common carriers of human capital demonstrate its new qualities, which are manifested in new values, new types of motivation, special (intellectual) knowledge, and specific skills. Skills are understood as the ability to use theoretical knowledge in solving practical professional problems. In addition to the communication skills demanded in modern society, the so-called digital skills come first. Digital skills include the following: general ICT skills allowing each employee to use digital technology in their daily work; professional ICT skills of system design, programming, application development, data and network management; problem-oriented skills, complementary skills, and using digital economy services (Sukhomlin et al. 2017).

In 2016, under the auspices of the Organization for Economic Cooperation and Development (OECD), an international organization that includes developed countries, several studies were conducted on how digital technologies create a demand for new skills adapted to these technologies (OECD 2016). Studies prove that the need for common and advanced ICT skills exists among workers in various sectors of the economy. Another conclusion is that the spread of digital technology is changing the way we do work, increasing the need for such additional skills as the ability to process information, carry out independent management, and solve the problems of communication between employees and customers.

Successful possession of these skills ensures the inclusion of workers in a specific information and communication environment and contributes to the development of globalization processes in society.

One of the forms of cultural globalization of a person becomes his ability to global mobility, that is, a skill that allows an employee to change his place of work quickly, the sphere of activity, and even his place of residence. An employee beginning his career understands well that the list of disappearing professions and even industries are continually growing, so it would be a mistake to dwell on a narrow professional sphere.

Labor mobility includes a flexible work schedule, the ability to work remotely, and moving to related professional areas.

The presence of personal qualities, such as creativity, independence, and a balance between work and own life, forces the contemporary employers and business structures to develop new principles for labor relations. The already appearing regulatory documents reinforce the possibility of flexible employment. In Russian law, the Institute of Distance Labor has been formed (Gimpelson and Kapelyushnikov 2006). This allowed some companies (Sberbank, Yandex, etc.) to transfer employees to partially remote work. Following the example of Western companies, rotation programs are being introduced, which are based on the policy of compulsory movement of workers between regional branches of companies, including the change of functional roles and positions.

5 Conclusion

Summing up, it can be argued that the mechanism of modern dynamic processes affects all areas of modern society and determines the role and specificity of human capital. Particularly, it includes the carriers of intellectual capital, who possess unique knowledge and skills and create new digital technologies. Penetrating into all spheres of society, such technologies change social structures and processes, giving them a digital character. At the same time, human capital receives a unique role in a changing society. Acting as a stock of specific knowledge, it can become a source of modern innovation and one of the leading causes of global economic and social transformations.

The digital society and specifically the digital economy are beginning to present qualitatively new demands on the content of modern human capital. As a result, human capital significantly changes its internal content, receiving all the signs of intellectual capital. The specificity of intellectual capital lies in the fact that it is formed at the expense of new knowledge of a heuristic nature. The modern young generation (digital youth) is gradually becoming the carrier of intellectual capital, which focuses on its system of values based on creativity, freedom of choice, and self-development. Such workers have a new type of motivation (having a worthy goal); new professional skills are being formed, including digital skills, communication skills, and global mobility skills.

References

Becker, G.: Human Capital: A Theoretical and Empirical Analysis, with Special Reference to Education, 3rd edn. The University of Chicago Press, Chicago (1993)

Bogoviz, A.V., Lobova, S.V., Ragulina, J.V.: Shift of the global investment flows in the conditions of formation of digital economy. Lecture Notes in Networks and Systems, vol. 57, pp. 1216–1223 (2019a)

Bogoviz, A.V., Lobova, S.V., Ragulina, J.V.: The cost and value of human capital in the modern digital economy. Lecture Notes in Networks and Systems, vol. 57, pp. 1224–1230 (2019b)

Digital evolution index 2017 (n.d.). https://www.mastercard.us/en-us/governments/insights-research/digital-evolution-index.html
Digital society index 2018 (n.d.). https://www.oxfordeconomics.com/recent-releases/the-digital-society-index-2018
Dyatlov, S.A., Dobrokhotov, M.A.: Forms of implementation of human capital in the digital economy. News St. Petersburg State Univ. Econ. Period. Sci. J. **4**(112), 25–29 (2018)
Gimpelson, V.E., Kapelyushnikov, R.I.: Precarious Employment in the Russian Economy. Publishing House HSE, Moscow (2006)
Kapelyushnikov, R.I.: How Much is the Human Capital of Russia? Publishing House of Higher School of Economics, Moscow (2012)
Negroponte, N.: Being Digital. Alfred A. Knopf, New York (1995)
OECD: Skills for a Digital World: 2016 Ministerial Meeting on the Digital Economy Background Report. OECD Digital Economy Papers, vol. 250, pp. 1–56 (2016). https://doi.org/10.1787/5jlwz83z3wnw-en
Ruus, I., Pike, C., Fernsten, L.: Intellectual Capital: Management Practice. Graduate School of Management, St. Petersburg State University, St. Petersburg (2010)
Schultz, T.W.: Investment in Human Capital: The Role of Education and Research. Free Press, New York (1971)
Stewart, T.A.: Intellectual Capital. The New Wealth of Organizations. Currency Doubleday, New York (1997)
Stillman, Д., Stillman, J.: Gen Z @ Work: How the Next Generation Is Transforming the Workplace. Harper Business, New York (2017)
Sukhomlin, V.A., Zubareva, E.V., Yakushin, A.V.: Methodological aspects of the digital skills concept. Mod. Inf. Technol. IT Educ. **13**(2), 146–152 (2017)
Suprun, V.A.: Intellectual Capital: The Main Factor of Economic Competitiveness in the XXI Century. Kombook, Moscow (2006)
Tapscott, D.: The Digital Economy: Promise and Peril in the Age of Networked Intelligence. McGraw-Hill, New York (1995)

Cognitive Production Factors in the Digital Economy

Vladlena V. Preobrazhenskaya(✉) and Evgeniya N. Gorlacheva

Bauman Moscow State Technical University, Moscow, Russia
preo-v@bmstu.ru, gorlacheva@yandex.ru

Abstract. In the historical context, the digital economy coincides with the processes of the re-industrialization of the national economy in many countries, among which Russia is far from last. In these processes, intellectual capital acquires significant importance, which receives a powerful tool for its amplification in digital technologies. The capitalization of a person's mental abilities is accompanied by business growth, which could not be ensured by the behavioral factors of human capital, characteristic of the industrial era of economic development. Restructuring of production factors associated with a decrease in the value of behavioral factors and an increase in the role of mental-cognitive functions in production processes occurs. The purpose of our research is to study the restructuring processes and highlight the main factors of production as business resources, transformed into means of innovative development under the influence of digital technologies. Provisions of the philosophy of science, economic theory, political economy were used for research purposes. The concept of the cognitive factor of production (CFP), as the main means of innovative development in the digital economy, can be considered the main result.

Keywords: Digital economy · Intellectual capital · Human capital · Cognition · Cognitive factors of production

1 Introduction

The development of digital technology has historically combined with the era of post-industrial economies. In the concept of post-industrialism[1], the main resource of production activity, a significant factor in production[2] is information. The prevalence of information and knowledge in productive activities gave reason to proclaim the onset of a more advanced era of development, namely the information society based on the increased concentration of intellectual capital. Excessive enthusiasm for the

[1] Post-industrialism is a theory of a new stage in the development of society, which follows an industrial society. The leading role in such a society is given to the service sector, science, education, and *the main object of activity is the sale of information* (Encyclopedic Dictionary. 2009).

[2] Factors of production focus on the production of resources that become an element of the production process, i.e., used for the production of goods and services (primary resources) such as labor, capital, land. And such factors as the concentration of wealth, combination, and distribution of labor have a significant impact on production efficiency (Lopatnikov 2003).

A. V. Bogoviz and Y. V. Ragulina (Eds.): ISCI 2019, LNNS 115, pp. 193–200, 2020.
https://doi.org/10.1007/978-3-030-40749-0_23

reorganization of economic relations based on information technologies and the creation of a peripheral economy have led to a weakening of the material foundation of its development in some advanced countries. Reindustrialization, i.e., restoration of the national sector of large material production, has become an alternative. This concept began to receive government support in the USA, where large material production should revive as the central sector of the national economy.

The restoration and further development of material production take place in the context of the expansion of the global system of digital technologies, which gives rise to a new, neo-industrial way of production relations (Bodrunov 2018). It is distinguished by the prevalence of web-based economic ties, the growing innovative component in the results of production activities (Bogoviz et al. 2019). Such trends necessitate a more concentrated use of intellectual capital in the material production sectors. At the same time, the value of the human component of intellectual capital is enhanced by a person's ability to transform information into knowledge, knowledge into action, action into a creative product.

Economic development trends indicate a diminution in the future of the role of the behavioral abilities of human capital through their replacement with fundamentally new forms of mental skills reinforced by digital technologies (Bogoviz et al. 2019). This cannot but lead to a consistent restructuring of production factors in the process of developing communication tools, the emergence of new numerical methods in digital technologies.

The purpose of our research is to highlight the mental abilities of human capital, which under the influence of digital technologies, become no less significant factors of innovative products along with material means.

Achieving the stated goal requires to identify areas of human capital use in the digitalization of economic relations and to highlight the main component of the mental abilities of human capital, which has a unique market value in the digital economy. It is also required to reveal the mechanisms of mental abilities transformation into a factor of production; to study the tools of the influence of human intellectual abilities on the competitive position of a business; to reveal the prospects for the effective use of selected factors of production in the development of web-economic relations.

2 Materials and Method

Cognitive activity, as a function of cognition, producing knowledge in various forms of its existence, is the material of research. The research is based on (a) provisions of the epistemology and philosophy of science, (b) a descriptive methodology of economic theory, (c) insights form advances in political economy, (d) applied methods of studying the processes of cognition, (e) methods of scientific organization and production management.

3 Results

The concept of the "digital economy" means the widespread use of information and analytical technologies designed to increase the share of industrial production through its innovative development. For the innovative development path, increased demand for intellectual resources is characteristic, which leads to an even higher concentration of intellectual capital in the material production sectors. The modern concept considers intellectual capital as a unity of various parts, in which human capital is recognized as an active component of intelligence, which is a carrier of knowledge, skills, and experience. Thus, the market value of human capital is determined by the active side of intelligence, which allows one to transform information into knowledge, knowledge into action, action into a fundamentally new, innovative commodity product.

One should note that any expedient activity requires the ability to think, i.e., based on intelligence, even if it is artificial. The difference between human capital in a market understanding is that here, a person's mental abilities acquire a separate commodity value. The exceptional property of human capital to produce non-obvious (creative) solutions becomes the basis of competitive advantage in high-tech, innovative production. A clear tendency to change the content of the labor process as a whole is revealed. The creative activity begins to dominate the reproductive labor characteristic of the past era of industrial development (Buzgalin and Kolganov 2019). These processes give reason to researchers to predict the onset of a creative revolution that will make the creative worker a self-sufficient participant in the process of capitalist production (Khubiev 2018).

Considering human capital as a production resource, we will interpret the concept of "intelligence" as the ability of human thinking to transform information into knowledge, analyze a priori knowledge in the process of their development. We will understand the capitalization of mental skills as the use of knowledge, labor skills, professional experience of corporate personnel for production activities. In the digital economy, digital skills related to the ability to establish and maintain web-based economic relationships stand out separately (Carretero et al. 2017).

Capitalization of thinking means the transformation of a single human resource into a corporate means of achieving the goal of production activity. In this regard, the value of human capital is determined not so much by the potential of knowledge, skills, and experience as by the results of its use, which are crystallized in the final products of production. Hence, the need for a commodity form of these abilities becomes apparent, and this form ensures the ability transformation from a resource into a means of achieving commercial goals; the prevalence of these agents transforms them into factors, i.e., necessary production conditions.

Labor skills are transformed into production activities through their understanding, deliberate dismemberment and integration into a system[3]. This is done by expediently

[3] Skills are actions that, as a result of prolonged repetition, become automatic, i.e., not in need of element-wise conscious regulation and control. Only those skills that are meaningful, deliberately dissected, and integrated into systems as their highest form enable a person to maintain conscious control over their actions. It is relatively easy to rebuild them (Philosophical Encyclopedic Dictionary 2010).

processing the primary (automatic) work forms. Professional experience is a unity of labor skills, knowledge, and abilities. Skills are the ability to apply acquired knowledge to industrial practice, i.e., turn knowledge into action[4]. From the perspective of epistemology, knowledge is understood as a form of the existence of consciousness in the way of a minimized activity scheme[5].

A comparison of these concepts allows us to establish the origin of the mental abilities of the human capital of a commodity form. It arises from the type of mental activity that connects knowledge and action with the power of reasonable necessity. Unlike behavioral action, a motor reaction to an event prevailing in industrial work, this form of mental action uses the analytical and synthetic power of the mind. This creative form of mental faculties is *cognition*. Through indicators of individual perception of a person, the competencies of specialists in the selection of production personnel are evaluated. Various models and calculation methods representing and measuring integral indicators of individual cognitiveness of applicants for the performance of separately allocated production functions are used for this (Azhmukhamedov 2011). The theory of cognition has found practical application in various methods of increasing the effectiveness of human capital management. In particular, a cognitive approach has been formed; it serves to overcome poorly structured problems in the management of complex systems (Avdeeva et al. 2007). The interpretation of the cognitive process as a purposeful transition from ignorance to understanding allows us to single out cognition as the most developed way of knowledge to exist. Separation of cognitive abilities, which have an independent commodity value in social processes, reveals the need for their separate accounting in human capital.

The productive power of cognition lies in the ability to think to transform the primary form of knowledge (information) into a higher active form (expedient activity), which is a minimized action scheme. The present and future values of human capital lie in its cognitive ability, i.e., an ability to transform information into knowledge, knowledge into productive activity, activity into a commodity product. Cognitive activity transforms knowledge from a resource into a means of production by imparting new meaning and content to information by linking it to a commercial goal, i.e., by establishing teleological certainty. The above considerations allow us to single out a single cognitive basis of new production activity, which includes knowledge, experience, skills.

4 Discussion

For the production process, the cognitive nature of human capital makes it possible to carry out high-tech productive activities, i.e., it is a means to an end. In the form of expedient action, cognitive technologies become a necessary condition for obtaining a

[4] Skills are the ability to quickly, accurately, and knowingly perform certain actions based on acquired knowledge and acquired skills. This ability is regularly associated with applying knowledge in practice, as well as in the process of training and production activities (Vishnyakova 1999).

[5] Knowledge is a form of social and individual memory, a reduced pattern of activity and communication (Kasavin 2009).

result, i.e., they are transformed from a resource into a production factor (Maksimov 2001). Thus, the cognitive factor becomes an element of the production process. Its quantitative side determines the necessary level of mental ability of human capital, which corresponds to the needs of a particular production, and the level of its cognitive capacity. The necessary sufficiency can be formulated as a condition for the *cognitive correspondence* of human capital to the specifics of production activity.

The practical side of cognition is manifested through the use of the mental functions of the human mind in terms of its ability to turn information perceptions into knowledge, followed by critical analysis. This specificity gives rise to a phenomenon that can be called the *cognitive growth effect*. The essence of this phenomenon is that in the process of production activity, the cognitive functions of transforming perception into knowledge are supplemented by inverse ones, namely, the functions of converting knowledge into perceptions. This explains the process of growth of labor competencies in the process of complicating production activities. From the standpoint of the philosophy of science, no substantive and informational activity is possible without such an expanded (productive) action of cognition (the ability to turn knowledge into perceptions through tools). In this expanded functional state, cognition exists in the form of a production resource, from an individual "ability to think," it is transformed into a collective ability to produce.

For further research, the fact that the specificity of using cognition as a means of production leads to the concept of knowledge that is meaningful through an understanding of its productive purpose seems important. In this interpretation, the concept of "cognition" becomes not a synonym for knowledge, but an alloy of knowledge with the ability to put it into practice in the material-objective activity. As applied to the production sphere, "cognition" is a single concept of knowledge and methods of their industrial application; the concept of knowledge, which constantly grows under the influence of cognitive production processes, becomes verified and fixed by practice. It is evident that the production process – in its constancy and innovative development – contains an equally constant process of cognitive growth. This process is a continuous change in the quantity and quality of "cognitions" – the knowledge and skills (firmly mastered and fixed by labor activity) are necessary for the continuous development and improvement of modern and future production processes.

We will call the necessary level of saturation "cognitions" of human capital the cognitive capacity of the labor process. Cognitiveness becomes a factor not only because of the causality and conditionality of its relationship to production[6]. Strict teleological certainty (the concreteness of the goal) leads to the transformation of cognitive, economic resources into means of achieving it, turning into a factor of production. In the process of completing the goal (production activity), they undergo quantitative and qualitative growth as publicly available market resources.

Attention should be paid to the following circumstance: the consequence of the existence of cognition in the form of a factor of production is its transformation into a measure of human capital. This happens by establishing the unity of the qualitative and

[6] Factors of production - production resources that become an element of the production process, i.e., used for the production of goods and services (Lopatnikov 2003).

quantitative sides of this means of production[7]. The qualitative side of cognition is cognitive polymorphism, expressed in a triad of experience, knowledge, creativity, which in different proportions are conditions for the practical application of this resource to technology, organization, and production management. In the digital economy, the importance of creative labor, digital competencies in the elements of the cognitive structure of human capital is enhanced. There is an evolutionary restructuring of production activities. The principle of structural conformity leads to a corresponding change in the cognitive structure of human capital.

The quantitative side is associated with the selection of such a numerical characteristic of cognition, which changes due to the number, size, and order of communication of elements, i.e., structural changes. The presence of a quantitative characteristic becomes a necessary side of the cognitive measure, a condition for converting it from a means into a controllable production factor. The development of information technology, numerical optimization tools in the form of cloud services significantly increases the cognitive capacity of production activities due to the complexity of labor. Adequate assessment of the magnitude of quantitative changes becomes the basis for correct wages in the digital economy. Otherwise, there will be a demotivation of personnel and the associated risks of lost profits from the rational use of human capital.

Thus, the goal of the study has been achieved, namely: cognition is highlighted as the primary mental ability of a person, which, under the influence of the development of digital technologies, becomes a significant factor in innovative production.

5 Conclusion

In conclusion, it must be emphasized once again that in the context of the digitalization of economic relations, the main direction of the use of human capital is the mental abilities of labor resources in the commodity form of their existence. The main component of mental ability is cognition, which has a separate market value in the digital economy. Linking cognition to the goals of commercial (corporate) activity transforms this form of mental (intellectual) ability into a cognitive factor of production. Through cognition, human intellectual abilities are transformed by capital into their competitive advantage by using knowledge, skills, creativity to produce innovative products to create new (innovative) needs. In the process of developing web-based economic relations, the effectiveness of the functioning of human capital can be significantly improved through the use of cognition as a measure of the intellectual abilities of labor, necessary and sufficient for productive activities.

[7] Measure is a philosophical category expressing the dialectical unity of the qualities and quantitative characteristics of an object. The quality of any object is organically linked to a certain quantity. As part of this measure, quantitative characteristics can change due to changes in the number, size, order of communication of elements, speed, degree of development, etc. A measure is a kind of zone within which a given quality can be modified while maintaining its essential characteristics (Ilyichev et al. 1983).

References

Avdeeva, Z.K., Kovriga, S.V., Makarenko, D.I., Maksimov, V.I.: Cognitive management approach. Manag. Probl. **3**, 2–8 (2007)

Azhmukhamedov, I.M.: A fuzzy cognitive model of specialist competency assessment. Vestnik Astrakhan State Tech. Univ. (Ser. Manag. Comput. Sci. Inform.) **2**, 186–190 (2011)

Bodrunov, S.D.: Noonomics. Cultural Revolution, Moscow (2018)

Bogoviz, A.V., Lobova, S.V., Ragulina, J.V.: Distortions in the theory of costs in the conditions of digital economy. Lecture Notes in Networks and Systems, vol. 57, pp. 1231–1237 (2019)

Bogoviz, A.V., Ragulina, J.V., Barcho, M.K.: Influence of innovations on regional socio-economic development. Lecture Notes in Networks and Systems, vol. 57, pp. 1200–1207 (2019)

Boldachev, A.: Finita la history - political-cultural-economic singularity as an absolute crisis of civilization. In: Frumkin, K.G. (ed.) Singularity. Images of "Posthumanity", pp. 7–37. Algorithm, Moscow (2016). https://dom-knig.com/read_191204-3

Buzgalin, A.V., Kolganov, A.I.: Transformations of the social structure of late capitalism: from the proletariat and the bourgeoisie to the precariat and the creative class? Sociol. Stud. **1**, 18–28 (2019)

Carretero, S., Vuorikari, R., Punie, Y.: Digcomp 2.1: The digital competence framework for citizens. Publication Office of the European Union, Luxembourg (2017). http://publications.jrc.ec.europa.eu/repository/bitstream/jrc106281/web-digcomp2.1pdf_(online).pdf

Encyclopedic Dictionary (2009). http://niv.ru/doc/dictionary/encyclopedic/index.htm

Glossary of terms of normative and technical documentation: Structural capital (n.d.). https://normative_reference_dictionary.academic.ru/75458/%d1%81%d1%82%d1%80%d1%83%d0%ba%d1%82%d1%83%d1%80%d0%bd%d1%8b%d0%b9_%d0%ba%d0%b0%d0%bf%d0%b8%d1%82%d0%b0%d0%bb

Haken, G.: Information and Self-organization: A Macroscopic Approach to Complex Systems. Kom Kniga, Moscow (2005)

Ilyichev, L.F., Fedoseev, P.N., Kovalev, S.M., Panov, V.G.: Philosophical Encyclopedic Dictionary. Soviet Encyclopedia, Moscow (1983)

Kasavin, I.T.: Encyclopedia of Epistemology and Philosophy of Science. "Kanon +"; ROOI "Reabilitatsiya", Moscow (2009). https://www.twirpx.com/file/748736/

Khubiev, K.A.: Innovative economy versus hired labor. Polit. Econ. Issues **1**, 55–61 (2018)

Kotlyarov, I.D.: Problems of evaluating the effectiveness of outsourcing. Bull. Inst. Econ. Russ. Acad. Sci. **6**, 87–99 (2017)

Lopatnikov, L.I.: Economics and Mathematics Dictionary: A Dictionary of Modern Economics. Delo, Moscow (2003)

Maksimov, V.I.: Cognitive technology - from ignorance to understanding. In: Proceedings from CASC 2001: The 1st International Conference on Cognitive Analysis and Situation Management, vol. 1, pp. 4–41. IPU RAS, Moscow (2001)

Martin, T., Harper, H.: Robot for the client. How Russian banks will change in the near future (2018). https://www.forbes.ru/finansy-i-investicii/355517-robot-dlya-klienta-kak-izmenyatsya-rossiyskie-banki-v-nedalekom

Philosophical Encyclopedic Dictionary (2010). http://philosophy.niv.ru/doc/dictionary/philosophy/index.htm

Popov, E.I.: Organization and self-organization as tools to increase the efficiency of industrial production. Prod. Organizer **2**, 9–11 (2008)

PWC: "Industry 4.0": Creating a digital enterprise (2016). https://www.pwc.ru/ru/technology/assets/global_industry-2016_rus.pdf

Razumova, T.O., Artamonova, M.V.: Flexible forms of employment in the system of criteria of the concept “Future in the world of work.” Sci. Stud. Fac. Econ. Electron. J. **9**(4), 7–27 (2017)
Slobodyan, E.: What is the worldwide web, and who invented it? (2014). http://www.aif.ru/dontknows/file/1123050
Vishnyakova, S.M.: Vocational Education: Dictionary. Key Concepts, Terms, Relevant Vocabulary. Nmts SPO, Moscow (1999)

Theoretical Justification and Practice of Risk Management for Economic Security of the Dealer Networks of Carmakers

Tatyana Yu. Serebryakova[1(✉)] and Olga Yu. Kurtaeva[2]

[1] Cheboksary Cooperative Institute (Branch) of the Russian University of Cooperation, Cheboksary, Russia
tserebryakova@rucoop.ru

[2] Peoples' Friendship University of Russia, Moscow, Russia
olkurt@list.ru

Abstract. With the development of a risk-based approach to management, there is a need for its use in many management processes. Economic safety is no exception. Economic security is closely related to the control function of management. The whole essence of economic security is to prevent risks. The control based on risk is an element of the economic security system. The area of interaction is risk management. Therefore, account and risk analysis can be used at the creation of a system of economic security. Risks should be identified for each level of management and process. The progress of the processes is reflected in the accounting. Information from accounting should be fully used for the tasks of the economic security of a business. It is proposed to use the idea of "triple bookkeeping" by Yuji Ijiri, expanding it to "quadro-accounting." The proposed method is implemented in the environment of automated information systems since the necessary detail can be obtained only with the use of computer technology. The essence is that the change in the capital is analyzed by the difference in the balance sheet data, and then by income and expenses. The results are presented in a special form according to factor analysis of capital due to changes in property and liabilities, as well as the impact of changes in the components of profit. They give the grounds for identification and primary risk assessment.

Keywords: Economic security · Accounting records · Analysis · Risks · Control · Management

1 Introduction

With the growth of the enterprise economy, there is a need to use a risk-based approach to many management processes. Economic security, which in terms of working with risks is closely related to control, is no exception. The control based on risk is based on risk management. There are risk management models, for example, COSO ERM[1],

[1] Available at: http://www.pqm-online.com/assets/files/lib/std/iso_31000-2009%28r%29.pdf.

A. V. Bogoviz and Y. V. Ragulina (Eds.): ISCI 2019, LNNS 115, pp. 201–213, 2020.
https://doi.org/10.1007/978-3-030-40749-0_24

ferma[2], ISO 31000[3] and others. They describe risk management processes to varying degrees. None of them contain methods for using the data of accounting and management accounting. At the same time, it is accounting that contains the most comprehensive information base on the economic activities of organizations, which can serve as the basis for any control actions, including risk control in the interests of business security.

The work connected with the identification of risks in many respects is based on economic analysis. It is based on financial and management accounting data. Therefore, the opportunity of studying accounting and economic analysis for identifying and assessing economic security risks seems interesting.

The issues of risk management are the subject of research by many scientists. Risks are considered from various positions, including to ensure the security of IT systems embedded in ISO 31000 (Barafort et al. 2017), risks and their minimization associated with the activities of managers to embellish the position of the organization if their income depends on profit (Wang et al. 2018), business risks arising in connection with social networks (Demek et al. 2018; Williams and Hausmann 2017). Many authors consider risks from the quantitative side for various technical non-standard situations; these issues are also related to management problems (Goerlandt et al. 2017; Pasman et al. 2017; Khastgir et al. 2017).

Some authors investigate the nature of risks, its tools, and efficiency. And on this basis, form methods of using risks for business management (Aven and Zio 2014; Lathrop and Ezell 2017; Cinzia and Giovanni 2017; Olechowski et al. 2016; Thekdi and Aven 2016). In a paper devoted to risk analysis, (Hansson and Aven 2014) brought up very topical issues about whether the doctrine about risk analysis is scientific. Does this area of knowledge possess attributes inherent in any science: subject, method, own field of research, methodology, social needs. The authors make an affirmative conclusion, linking it with socially demanded phenomena, holistic concepts, methods, and applied tools. In general, it is possible to agree with a similar outcome. However, it should be noted that risks, especially in economic processes, are very subjective at the stage of identification and forecast. In this connection, the theory of risks should have reliable methods of assessment and analysis. Besides, the objective methods of identification, evaluation, and risk analysis allow using methods for predicting them automated control systems for the calculation of key indicators of economic security.

One should remember that there is an entirely reliable and objective information system – accounting. These data can be used for economic security purposes. Since accounting has sufficient objectivity and the degree of automation, it can become the basis for the construction of objective and relatively unambiguous, universal methods of working with risks based on IT-systems. In the first instance, we are talking about financial and economic risks of an internal nature, which signal the presence of dangerous situations, and in turn, affect the safety of the business. Automated information

[2] Available at: http://www.ferma.eu/app/uploads/2011/11/a-risk-management-standard-russian-version.pdf.

[3] Available at: http://www.valtars.ru/files/upload/Actual_info/coso_upravlenie_riskami_organizacii_integrirovannaya_model.pdf.

systems allow one to identify the following risks: marketing risks, foreign exchange, management risks, and many others that should be investigated by the organization's economic security service.

The purpose of the study is the development and justification of the methodology for using accounting systems to identify, assess, and analyze business risks in the interest of providing economic information. Research tasks come down to studying opportunities of using the reporting data for identification of risks of safety, creation of a basic model of identification of risks according to reporting data, and development of the main directions of risk analysis.

2 Materials and Method

There are two possible fundamental principles of accounting, which can be identified in contemporary conditions as a result of accounting in the scientific and practical activities of specialists - Unigraphics and digraphic.

The Unigraphics principle refers to accounting in the form of a regular recording record in the form of a list. Digraphic means accounting, which is based on double entry.

In recent decades, a new paradigm has emerged in accounting theory, the so-called triple recording, the ideologist of which is Y. Ijiri (Ijiry 1986; Ijiri 1988). Ijiri believes that "double-entry bookkeeping is not an absolute system, and can be logically expanded" (Ijiry 1986).

The accounting principle of triple recording is that all accounts are divided into static, dynamic, and impulse. Assets and liabilities are recorded on static accounts, for this purpose, a familiar account, but the records are made on the debit side. Income and expenses are recorded on dynamic accounts, and the records are made on loan. On impulse accounts – the rate of change in income and expenditure in a certain context. The type of accounting proposed by Yuji Ijiry is significant for managers since it clearly and intelligibly reveals the capital gain for the period. More than that, this is demonstrated in comparison with the increase in property. Thus, it becomes clear in what types of assets the incremental capital is invested, as well as due to what factors it has increased. Similarly, it is possible to understand the situation regarding the loss.

Like any new theory, the concept of triple recording has its limitations; among the main and most important limitation is the lack of universality in the set and types of factors that affect capital with a positive or negative effect. Essentially, accounting on tepbit accounts, that is, in the third dimension, from the standpoint of impulse strength, is very conditional, since the result of the analysis of the influence of capital growth (decrease) factors is taken into account. Many forces can not be determined from the accounting data because the change in the capital is influenced by numerous external and internal factors. The same factors that can be arithmetically calculated and recorded on tepbit reflect the impact of market conditions on the growth of capital, through the manifestations in the change in the composition of assets, sales returns.

On the one hand, the set of indicators that can be disclosed using information from accounting accounts is limited. On the other hand, it is not universal, and conception is subjective and depends on the professional training of personnel involved in such accounting. Many accounting researchers in their work pointed to all these and other shortcomings (Fraser 1993; Karelskaya and Suga 2015; Samusenko 2013; Sen and Bala 1999; Wojtowich 2002).

We believe that the technique described by Yu. Ijiry cannot be recognized as accounting. It can be used to build analytical calculations because, in essence, we are talking more about the form of presentation of credentials and the analysis resulting from it. It is entirely possible to agree with S.A. Samusenko, S. N. Karelskaya, and E. I. Suga (Karelskaya and Suga 2015; Samusenko 2013) is that it is convenient to take into account the influence of factors on tepbit, this will reveal the values and reasons for the change in capital, i.e., the most important information for the accounting model based on profit management. Moreover, in our opinion, this model should be recognized not so much as an accounting model, but as an analytical one. In other words, this is a model for the presentation and disclosure of analytical information according to accounting data. In this connection, we support the conclusions made by (Karelskaya and Suga 2015) contained on the availability of additional accounting and informational capabilities of triple bookkeeping along with significant limitations.

Among the shortcomings should be noted one more. Those records offered for tepbit influence the factors on dynamic indicators of activity of the organization. They characterize the change in capital. Factors changing "static" data on assets and liabilities remain aloof. (For example, in a situation where there is a revaluation of assets with its value attributed to additional capital.) Capital increases but does not form a financial result of activities. Besides, changes in the structure of resources of the organization can have an impact on the subsequent events influencing modifications directly in the organization's capital in the future. Such changes can only be judged by the changes in the balances of assets and liabilities. It can be said in another way, using the terms of Y. Ijiry. A change in the structure of resources through a change in the values of certain assets and liabilities is a consequence of the manifestation of impulses and forces. They can be measured in the same way. For the organization, it is important to analyze their sizes and structure. The additional features that the Ijiry triple accounting model provides, in our opinion, are that it fits well with automated accounting systems. Accounting and management accounting is a subsystem of an organization's automated information system. This creates the opportunity to build special reporting forms that provide users with the necessary information in an automatic mode. In this case, without changing the essence of accounting and standard financial statements, you can get the format of the balance sheet, with emphasis on the indicator "capital", the format of the report on financial results (profit and loss) with an indication of capital gain and a breakdown of the main factors influencing it. This allows management to take a fresh look at familiar data. Once one established standard operations to determine the influence of the most objective factors, it would be possible to provide the necessary unification, which can be further detailed to the required details.

Such a logical layout is convenient in many cases as a "starting point." We believe that in this way, it is possible to organize accounting in order to obtain information for risk management and their control in the implementation of the tasks of ensuring economic security. At the same time, we believe that such information is primarily analytical. It does not contradict the idea about the accounting of risks by the results of the analysis as identification of risks is in many respects based on the economic review.

3 Results

First, factor analysis should be used to identify economic security risks. Factor analysis is a method of economic analysis. It is very important because it is a control tool that allows defining values of key indicators that need to be controlled for the understanding of economic threats and degree of economic security (danger). The main directions of internal and external economic threats can be defined using critical indicators (among which, of course, the factors changing profit). Second, control is inextricably linked with risk management, because in our opinion, risk management and analysis are a control tool. By studying the components of the financial result from the standpoint of the dynamics and causes of changes in capital, assets, liabilities, income, expenses, it is possible to obtain objective information to assess the risks of economic security in conjunction with control.

Depending on a model of presenting the financial position of the organization (at historical or fair value) used as the accounting one, the change in the capital will reflect the increment of funds (resources) of the organization based on their actual mass (historical value) or at the fair value date of presentation. In any case, the change in the capital will be a crucial indicator characterizing the organization's activities at a particular time (at reporting date). All possible deviations of financial results from the planned to a lesser extent affect the amount of capital and its dynamics in a negative sense. This is an object of attention to economic security services, as it carries financial threats. If one wanted to understand the existing risks of not achieving capital gains in the established amounts (calculated at historical cost or fair value), it would be necessary to analyze the impact of various factors on its value.

In our opinion, based on the idea of triple recording Y. Ijiry, it can be improved in the framework of the automated information system. In our opinion, it is more appropriate to build a quatro-model, which can be successfully used as an automated analytical technique for risk management, established from the results of accounting and financial reporting indicators. In this context, one considers accounting records as the generalization in the system of accounts on the principle of double record. In turn, accounting reports are understood as the reporting made according to business accounts in any format, including accounting (financial) and management.

The model of such an analysis may look as follows. The *first format* characterizes capital gains due to changes in resources. For all asset accounting accounts, growth is determined in monetary terms. The increase can be expressed as a positive value or negative if there is a decrease in the asset in total terms. Similarly, the amount of change in receivables is determined. Liabilities also participate in the calculations, but their changes should be taken into account with the opposite sign. For example, an increase in liabilities will have a negative value, and their decrease will be taken into account with a "+" sign. Thus, as a result of the first calculation, the capital gain will be determined for the analyzed (reporting) period.

The *second format* characterizes capital gains due to changes in the use of resources. It is based on information on the accounts of sales, profits, and losses. The positive values are income and benefits, and the negative are expenses and damages. The resulting value is the financial result of the reporting (analyzed) period.

The *third format* characterizes the influence of factors on the changes that have occurred on the accounts of property and liabilities. Two objective factors: quantity and price, affect changes in property accounts. The influence of these factors should be described for each type of property. Thus, the risk management will have information on the reasons for the decrease (increase) in assets for use in risk analysis. It is necessary to determine the factors affecting the value of the change in obligations. Here, objective factors can be recognized as the turnover in comparing the initial and final costs for understanding the trend, and other factors are not excluded.

The *fourth format* reveals the change in capital under the influence of various reasons. Since capital change is the financial result of the reporting (analyzed) period, the influence of factors can be achieved through the usual formula of balance sheet linking. However, for control and risk management purposes, it is preferable to have calculations related to assessing the influence of factors on sales, cost, other income, and expenses. It is possible to use a multiplicative formula that describes the dependence of the main indicator on several factors. However, in this case, it is possible to use the idea of Y. Ijiry. To summarize sales in the context of price/quantity factors for enlarged groups, specific products (types of work), cost of sales in the context of cost elements, and within them, for more specific factors. The main thing is that the change in the capital should be fully disclosed in the context of the factors that affected its value.

Such accounting cannot be carried out manually, it is meaningless without the use of automated accounting systems, due to the high complexity, which will not be justified by the value of the information received, if done manually. As a result, we get automated calculations based on accounting data, a turnover balance. The most preferred model scheme will look as follows (Fig. 1).

Let's consider both formats 1 and 2, which are calculated according to the results of the work of the dealership of the automaker "X" in 2018, and presented below in Tables 1 and 2.

According to format 1, the risks from the perspective of economic security are the presence of value under "value-added tax" (No. 5), its growth, as well as the growth of loans (No. 10) and accounts payable (No. 11). Key indicators of economic security, such as growth in stocks (No. 4) and a decrease in fixed assets (No. 1), are subject to further study.

The key indicator for format 2 (for specialists of the economic security service) is the presence of other expenses (No. 11) in a significant amount and interest expenses (No. 9). Derivative key alarm indicators demand additional studying of economic security in format 2, such as a low gross income (which takes less than 10% of revenue), heavy expenses for sale, and management (which practically do not become covered by a gross profit).

A source of information	
Balance sheet report	Report on financial results

Format 1	Format 3	Format 2	Format 4
Growth of assets and liabilities for the period	The impact of factors on asset growth	Total income for the period (by type of business and type of product)	The influence of factors on the total income of the period (by the variety of business factors quantity/price)
Addition:			
Stock change (by type) for the period	Influence of factors of quantity and price	Return on sales	Option 1: Influence of price and quantity factors; Option 2: Influence of factors of profitability of sales and profits
Change in accounts receivable for the period	Effect of turnover factors	Cost of sales (minus)	Influence of factors depending on the calculation items (quantity and price of material costs, number, and productivity of personnel costs, etc.)
Change in monetary assets for the period	Effect of turnover factors	Business expenses (minus)	
Subtraction:		Management expenses (minus)	
Change in liability	Effect of turnover factors	Other expenses	Factors of an interest rate, unplanned income
		Other expenses (minus)	Factors of an interest rate, unplanned income
		Tax (minus)	Factors of change of taxable basis
		Changes in the capital which did not enter the financial result of the period (+/-)	Factors affecting the appearance of such income/expenses
Capital growth for the period	Capital growth for the period	Capital growth for the period	Capital growth for the period

Final document			
Report on changes in assets and liabilities	Report on the reasons for changes in assets and liabilities	Report on total capital for the period	Report on the impact on total capital

Fig. 1. A conceptual diagram of a quadro-model for identifying economic security risks.

Table 1. Format 1: analysis of the impact of property and liabilities on the capital.

No.	Indicators	According to the report on a financial position (balance sheet)			
		December 31, 2018	December 31, 2018	Change: + increase; − decrease	Impact: + positive; − negative
	Capital	87788	69160	+18654	18654
1.	Fixed assets other non-current assets	52723	52732	−9	−9
2.	Financial investments	37	–	+37	+37
3.	Deferred tax assets	20097	22932	−2835	−2835
4.	Stocks	177795	127382	+50413	+50413
5.	Value-added tax	4439	1310	+3129	+3129
6.	Accounts receivable	20524	24017	−3493	−3493
7.	Monetary resources and equivalents	44783	35995	+8788	+8788
8.	Total assets	320398	264368	+56030	+56030
9.	Deferred tax liabilities	733	759	−26	+26
10.	Short-term loans	90447	90009	+438	−438
11.	Accounts payable	136910	100230	+36680	−36680
12.	Estimated liabilities	5253	4969	+284	−284
13.	Total liabilities	233343	195967	+37376	−37376

4 Discussion

Further work with the results of calculations to identify and assess the risks of economic security, in this case, have the following focus.

Key signaling indicators require an immediate assessment and decision on the existence of a threat to economic security. In our case, the presence of the balance under the item "value-added tax" should be studied from the standpoint of the absence of violations of tax legislation, which could lead to failure to provide tax refunds of this amount shortly. Besides, it is necessary to establish whether the diversion of the organization's financial resources due to the late receipt of value-added tax from the budget is justified for the organization's business purposes. This is one of the threats to economic security.

In the "Report on Changes of Assets and Liabilities," there is information on the growth of stocks of goods in stock for specific car models, and the increase is due to a quantitative factor, as can be seen from "Report on reasons for changes in assets and liabilities." Understanding the degree of danger of the situation requires studying the data of "The Report on the Cumulative Capital." It is clear whether a particular type of

Table 2. Format 2: Analysis of the impact of income and expenses on the capital of the organization.

No.	Indicators, their impact on capital (+ positive; − negative)	For the 2018 year	Impact
	Capital gains	+18654	
1.	Revenue	+1894563	From the increasing price and quantity of the sold goods, according to the form activity of car sales and decrease in repairs
2.	Cost of sales	−1712502	From the increase in cost and the number of cars sold
3.	Gross profit	+182061	From the increase in the trade margin and the relative reduction in the cost of sales of cars
4.	Business expenses	−87936	The influence of factors depending on the cost of sales and management, including the quantity and price of material costs, the number, and productivity of personnel, etc.
5.	Management expenses	−47123	
6.	Sales profit	+47002	
7.	Income from participation in other organizations	+6	Profitability of a subsidiary (dependent) company
8.	Interest receivable	+22	Rates of interest
9.	Interest expenses	−16662	Interest rates and cash reserves
10.	Other income	+8556	Factors of an unplanned income
11.	Other expenses	−17922	Factors of unplanned expenses
12.	Profit before tax	+21002	
13.	Tax payments and deferred taxes	−4496	20% of profit before tax
14.	Net profit	−16506	Capital increase
15	The actions directed on the rise in capital	+2654	
16.	Dividends paid	−506	

finished product was sold, in what volume, and at what cost. Data from the “Report on Influence on the Cumulative Capital,” in comparison with the previous period, show what the trend in sales of this product is. At the same time, it is necessary to study plans of the organization from which you can see that the increase in stocks of goods is due to overall sales growth in the reporting period and the plan for the future period. At the same time, it is necessary to study the portfolio of orders for this product, as well as the market situation. These studies will provide the basis for action to limit the consequences of a possible unjustified increase in stocks. It is possible that such measures were not needed if threats to the sale of cars would not be identified shortly.

There may be another situation when the reports on the cumulative capital and on influence on the cumulative capital, in comparison with the data of the same reports for the previous period, show that:

- The share of sales of a certain product is reduced;
- The sale price of this product is reduced;
- The number of a certain product is reduced.

The observed external market risk can be minimized by reducing the production of this product. For a full assessment of the situation by THE "Report on Changes of Assets and Liabilities," one can see what stocks of this product are in warehouses, as well as how the cost of purchased goods behaves in comparison with the selling price of this product on the market. This information must be evaluated to understand the quantitative and qualitative characteristics of the risk, as well as to develop management tactics in relation to the identified risk: not to take restrictive actions, establish measures to reduce the risk to an acceptable level, and others.

Growth of liabilities of the organization, including for loans, is one more key alarm indicator for economic security. The first reason for this is a well-known situation in which an increase in debt means deteriorating solvency, in addition to objective reasons. Each case must be studied. In our case, there should be no reason for concern, since we see that the organization has resources that can cover debts to suppliers (money in the first place). In addition, for further events in January 2019, it is clear that the liabilities are being repaid. The second reason why the growth of liabilities is dangerous for the organization's economy is the likelihood of uncontrolled accumulation of debts that are not secured by assets due to insufficiently profitable activities. The signal of which is an increase in loans. For the economic security service, it is essential to establish how much the interest rate on loans corresponds to the profitability of sales in both relative and absolute terms. In our case, this ratio is unsatisfactory, since the interest paid is slightly less than half of the profit from sales. In this case, it is necessary to check the financial plans and budgets to understand the risk appetite and the possible change in strategy.

Of course, the signal indicator is the presence of other expenses, unless this indicator is associated with the sale of other assets. In any case, it requires the study and assessment of risks, including the irrational use of the organization's funds.

Signal critical indicators from the perspective of capital gains are presented in Table 3:

Final documents help identification of various risks of the organization. The reports on changes of assets and liabilities, on the reasons for changes of assets and liabilities, and on the cumulative capital for the period" more likely to identify market risks. The report on influence on the cumulative capital (especially in comparison to the same report for the last period) helps to identify financial risks.

The proposed final documents should be formed in electronic form with the ability to print the necessary fragments on key indicators that have fallen into the risk zone due to the identified negative deviations. In addition, signal documents should be automatically generated on the occurrence of risk situations, for example, at a decrease in sales with a simultaneous increase in stocks and similar cases. Formation of the list of situations demonstrating risks is a long and labor-intensive process. Signal documents

Table 3. Crucial signal risk indicators of the organization for their capital for economic security.

No.	Signal key indicator	A source
1.	The presence of residues on the line "value-added tax"	Balance
2.	The presence of growth in the line "value-added tax"	Format 1
3.	The presence of an increase in the line "Stocks" while a decrease in the line "Income"	Balance and format 2
4.	Decrease in an indicator on the line "Fixed assets" at simultaneous decrease on the line "Income"	Balance and format 2
5.	The presence of an increase in the line "Stocks" with a simultaneous increase in the line "Short-term loans" and (or) "Lenders"	Format 1
6.	The ratio of capital gains and the indicator on the line "Income"	Format 1 and format 2

should be used as primary documents for registering risks. This should be the impetus for their registration by the internal control department.

It is necessary to understand which accounting reports are expedient to apply for the purposes of the quadro-model. According to the standards of financial reporting, various reserves are formed in accounting, the need for this is due to the uniformity of the reflection of the financial result by periods. For an accounting of risks, it is necessary to use the reporting made at historical cost. Such reports contain a financial result that is not distorted by additional leveling procedures.

Separately, one can examine statements that reflect fair value. The value of the proposed model lies in the fact that it can be used both at historical cost and at fair value. In the second case, equity growth will not always correlate with the financial result, since bringing the value of a tangible asset to fair value may not always be related to the financial result (Serebryakova et al. 2016). This point is taken into account in the proposed model by adding the line "changes in equity not included in the financial result of the period." Besides, the proposed model considers the growth over the period in the value of total capital, and not net profit.

The format of the quadro-model is of particular interest, constructed based on the "report on cash flow." Such a report provides information on the financial result of the paid implementation. In comparison to the reports on shipment, the financial risks caused by gaps between delivery and payment, and the risks of project financing can be identified.

Further research should be aimed at a thorough study of the signal indicators for the statement of the identified risks on the account in the automatic mode.

5 Conclusion

Risk management and its control for use by economic security services are becoming a necessity for organizations operating with significant information flows. The monitoring of economic security indicators in such conditions is possible only with the use of automation. Moreover, it is possible to talk about the use of a risk-based approach in matters of economic security only in the conditions of applying information technologies.

The automated systems of accounting of risks demand objectively created algorithms based on the concrete data collected in the automated information system. Accounting systems are one of the sources of such data.

Economic analysis is a method of identifying risks in economic security. Relying on analytical methods, these risks are assessed and controlled after they are identified. The analysis in the process of economic security differs from the usual analysis at the stages of setting the task and drawing conclusions.

To identify the risks of economic security based on accounting, it is necessary as a report to obtain forms of documents containing the result of analytical calculations in comparison with normative or optimal values. The beginning can be the most straightforward analysis of indicators that form capital.

The capital increase is the primary goal of the business. In this regard, it is necessary to determine the risks and threats for this purpose. The model for the presentation of capital from financial statements is the most reliable and reasonable. Accordingly, the identified risks of economic security will be objective. This is the essence of the quadro-model for determining risks. Any risks carry threats, and threats are an object of study by economic security services.

References

Aven, T., Zio, E.: Foundational issues in risk assessment and risk management. Risk Anal. **34**(7), 1164–1172 (2014)

Barafort, B., Mesquida, A.-L., Mas, A.: Integrating risk management in IT settings from ISO standards and management systems perspectives. Comput. Stand. Interfaces **54**(3), 176–185 (2017)

Demek, K.C., Raschke, R.L., Janvrin, D.J., Dilla, W.N.: Do organizations use a formalized risk management process to address social media risk? Int. J. Account. Inf. Syst. **28**, 31–44 (2018)

Fraser, I.A.M.: Triple-entry bookkeeping. Account. Bus. Res. **23**(90), 151–158 (1993)

Goerlandt, F., Khakzad, N., Reniers, G.: Validity and validation of safety-related quantitative risk analysis: a review. Saf. Sci. **99**(B), 127–139 (2017)

Hansson, S.O., Aven, T.: Is risk analysis scientific? Risk Anal. **34**(7), 1173–1183 (2014). https://doi.org/10.1111/risa.12230

Ijiri, Y.: Accounting matrices and three-dimensional arrays. Account. Educ. **3**(2), 270–285 (1988)

Ijiry, Y.: Framework for triple-entry bookkeeping. Account. Rev. **61**(4), 745–759 (1986)

Karelskaya, S.N., Suga, E.I.: Triple-entry bookkeeping J. Ijiri. Audit Statements **8**, 62–74 (2015)

Khastgir, S., Birrell, S., Dhadyalla, G., Sivencrona, H., Jennings, P.: Towards increased reliability by the objectification of Hazard Analysis and Risk Assessment (HARA) of automated automotive systems. Saf. Sci. **99**(Part B), 166–177 (2017)

Lathrop, J., Ezell, B.: A systems approach to risk analysis validation for risk management. Saf. Sci. **99**(Part B), 187–195 (2017)

Cinzia, M., Giovanni, V.: Project risk management: a deterministic quantitative technique for assessment and mitigation. Int. J. Proj. Manag. **35**(3), 320–340 (2017)

Olechowski, A., Oehmen, J., Seering, W., Ben-Daya, M.: The professionalization of risk management: What role can the ISO 31000 risk management principles play? Int. J. Proj. Manag. **34**(8), 1568–1578 (2016)

Pasman, H.J., Rogers, W.J., Mannan, M.S.: Risk assessment: what is it worth? Shall we just do away with it, or can it do a better job? Saf. Sci. **99**(Part B), 140–155 (2017)
Samusenko, S.A.: Triple accounting system of Yuji Ijiri and prospects of its application. Int. Account. **28**, 44–54 (2013)
Sen, D.K., Bala, S.K.: Triple-entry bookkeeping - an extension of double-entry bookkeeping: a brief analysis. Tribhuvan Univ. J. **XXI**(1), 1–21 (1999)
Serebryakova, T.Y., Klychova, G.S., Fedorova, L.P., Gordeyeva, O.G.: Revisiting the fair value and historical cost and their impact on the financial indicators of the company. Int. Bus. Manag. **10**(23), 5663–5668 (2016)
Thekdi, Sh., Aven, T.: An enhanced data-analytic framework for integrating risk management and performance management. Reliab. Eng. Syst. Saf. **156**, 277–287 (2016)
Wang, T.-Sh., Lin, Y.-M., Werner, E.M., Chang, H.: The relationship between external financing activities and earnings management: evidence from enterprise risk management. Int. Rev. Econ. Financ. (2018). https://doi.org/10.1016/j.iref.2018.04.003. (in press, accepted manuscript)
Williams, S.P., Hausmann, V.: Categorizing the business risks of social media. In: Proceedings from CENTERIS 2017: International Conference on ENTERprise Information Systems; ProjMAN 2017: International Conference on Project MANagement; HCist 2017: International Conference on Health and Social Care Information Systems and Technologies (2017). Sci. Direct Procedia Comput. Sci. **121**, 266–273
Wojtowich, P.: Theoretical and practical problems of the system troj sizes. Zeszyty Teoretyczne Rady Naukowej **65**, 132–153 (2002)

Model Estimates of Internal Economic Risks in Innovative Processes

Gabriel H. Alberto(✉), Fode M. Sissoko, and Mateus A. Muatchavua

Peoples' Friendship University of Russia (RUDN University), Moscow, Russia
alberto.gabriel@yandex.ru, Fode0334@gmail.com,
muatchavua14@hotmail.com

Abstract. The article formulates some approaches to the construction of an internal risk assessment model for the innovation processes. These principles should contain methods for analyzing and managing risks that can be involved in all stages of an innovative project. Risk assessment and analysis models are traditionally based on different principles: constructing and analyzing the risk matrices, developing the integrated indicators to assess the risk, constructing models based on dynamic equations, etc. The main components of the analysis, with which one can implement risk management, are risk assessment and development of effective anti-risk measures.

Keywords: Algorithm · Innovative project implementation · Mathematical model

1 Introduction

Risk management involves constant analysis, which consists in assessing the level of risk caused by the presence of any adverse factors and developing a plan of anti-risk measures (Alekseev and Kurilova 2015; Chursin et al. 2016).

The risk analysis carried out at the enterprises implementing innovative projects can be represented by the following scheme (Fig. 1):

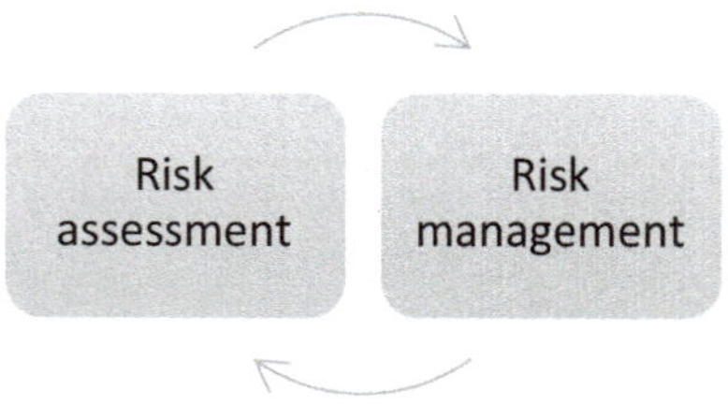

Fig. 1. Scheme of risk analysis.

At each stage of the innovative project implementation, thanks to the offered scheme of risk analysis, it is possible to reveal (based on risk assessment) results from

A. V. Bogoviz and Y. V. Ragulina (Eds.): ISCI 2019, LNNS 115, pp. 214–220, 2020.
https://doi.org/10.1007/978-3-030-40749-0_25

adverse factors which can lead to additional financial expenses, non-compliance with the schedule of implementation of the project, decline in quality of the developed products or service (Damodaran2010; Karimov 2015; Koshelev 2011; Milykh 2008; Sekerin et al. 2008; Sholomitsky 2005).

The following describes a model for assessing the internal economic risks arising from the implementation of innovative projects in the enterprise.

2 Methods

In the theory of risk analysis, one of the standard approaches to assessing the level of risk is an approach based on assessing the probability of project failure, due to the lack of necessary resources, including analytical information (a sufficient database of big data) and competencies necessary for implementing a knowledge-intensive project. As is known, innovatively active enterprises cooperate with many enterprises – suppliers of components, centers of competence, including with foreign companies. The problems which arose at other enterprises can lead to a disruption in the supply of necessary components, and lead to a failure to meet the schedule for the implementation of an innovative project.

With the purpose of assessing the level of risk associated with the adverse effect of such factors, first and foremost, consider the overall function of the performance of the innovation project, expressing the value of the integral performance indicator (efficiency) for the implementation of the innovative project through the composition and volumes of the resources used:

$$Y = F(X_1, X_2, \ldots, X_n)$$

In this formula, Y is an integral indicator of efficiency (effectiveness) (which can be as dimensionless or accepting a dimensional value depending on the economic setting of task; for example, prime costs of innovative products), and the values $X_1, X_2, \ldots, X_n$ express the volumes of n types of various resources necessary for the implementation of an innovative project. As a basis for constructing concrete of the type of functional dependence, we take the approach used in constructing the production function. In economic theory, many examples have been proposed to construct adequate models of production functions that describe production processes at the enterprise level, and even at the level of its production links, so we would assume that this model describes the process of creating innovative products in the enterprise. Suppose that the expected effectiveness of the project on the development of innovative products is equal to Y^0, for which necessary resources in volumes, respectively $X_1^0, X_2^0, \ldots, X_n^0$, i.e.

$$Y^0 = F(X_1^0, X_2^0, \ldots, X_n^0).$$

One of the most important and widely used models of production functions in economic theory is the following model of production functions (a special case of a generalized *CES* function):

$$Y = AX_1^{\alpha_1} X_2^{\alpha_2} \ldots X_n^{\alpha_n},$$

where $A > 0$ is a technological coefficient, $\alpha_1, \alpha_2, \ldots, \alpha_n$ are positive numbers, such that $\alpha_1 + \alpha_2 + \ldots + \alpha_n = 1$. Suppose that the resource support for the development of innovative products has its level of internal economic risk for each type of resource. Thus, the values $X_1, X_2, \ldots, X_n$ can be considered as random variables. The expected volumes of resources of each type, directed to the development of innovative products, taking into account risk, are equal respectively:

$$M(X_j) = X_j^0(1 - R_j), \quad j = 1, 2, \ldots, n. \tag{1}$$

here, for the variable R_j marked the level of risk associated with ensuring the development of innovative products resource X_j.

In this case, as a quantitative assessment of the risk level associated with the resource provision of the innovation project, we would use the probable value of providing the resource X_j, concerning the planned value of providing this resource X_j^0:

$$R_j = M\left((X_j^0 - X_j)/X_j^0\right), \tag{2}$$

where X_j^0 is a planned provision of a resource, X_j is probable value of providing with a resource, taking into account risk factors. That is, when X_j is fully provided with a resource, the risk level associated with the use of this resource becomes equal to zero. It would appear reasonable that the level of risk associated (for example, with resources attracted from outside) is higher than the level of risk associated with the provision of own resources of the enterprise. Under the assumptions made, the integral indicator of the effectiveness of the implementation of the innovative project Y is also a random variable. In the case of pairwise independence of $X_1, X_2, \ldots, X_n$, we obtain the equality:

$$M(Y) = A(M(X_1))^{\alpha_1}(M(X_2))^{\alpha_2} \ldots (M(X_n))^{\alpha_n}. \tag{3}$$

Then from (1) and (2) it follows:

$$\begin{aligned} M(Y) =& A\left(X_1^0\right)^{a_1}\left(X_2^0\right)^{a_2} \ldots \left(X_n^0\right)^{a_n}(1 - R_1)^{\alpha_1}(1 - R_2)^{\alpha_2} \ldots (1 - R_n)^{\alpha_n} \\ =& Y^0(1 - R_1)^{\alpha_1}(1 - R_2)^{\alpha_2} \ldots (1 - R_n)^{\alpha_n} \end{aligned}$$

Then, considering that $E(Y) = Y^0(1 - R)$, where R is the level of risk associated with the neediness of components, we can write down the equality:

$$(1 - R) = (1 - R_1)^{\alpha_1}(1 - R_2)^{\alpha_2} \ldots (1 - R_n)^{\alpha_n}.$$

Hence, since $\alpha_1 + \alpha_2 + \ldots + \alpha_n = 1$ inequality follows

$$(1 - R) \geq \min\{(1 - R_j)\}.$$

Then

$$R \leq \max(R_j), \tag{4}$$

The estimate (3) is exact, i.e., at specific values of the parameters $\alpha_1, \alpha_2, \ldots, \alpha_n$ (3) goes into exact equality. Besides, the estimate (3) does not depend on concrete parameters of the production function, i.e., on real values of the parameters $\alpha_1, \alpha_2, \ldots, \alpha_n$, and coefficient A.

Evaluation of internal risks of the innovation project is carried out as follows. As the risk level R_Y of not achieving the planned value of the performance indicator Y, we will take the following value:

$$R_Y = \max(R_j) \tag{5}$$

Using the values $X_1, X_2, \ldots, X_n$ and $X_1^0, X_2^0, \ldots, X_n^0$, formula (5) can be written in the following form:

$$R_Y = 1 - M\left(\frac{Y}{Y_0}\right) = 1 - M\left(\min\{X_1/X_1^0, X_2/X_2^0, \ldots, X_n/X_n^0\}\right) \tag{6}$$

In practice, risk assessment is carried out by statistical or expert methods.

We propose a mathematical model for the quantitative analysis of the risk level associated with insufficient resource security. This model will be based on the results of expert evaluation.

According to the formula (6), the task of finding the risk assessment R_Y is reduced to the task of finding the expected value $B = \min\{X_1/X_1^0, X_2/X_2^0, \ldots, X_n/X_n^0\}$, which is a function of random variables $X_1, X_2, \ldots, X_n$. As we previously assumed, the values of $X_1, X_2, \ldots, X_n$ are independent.

The basis of the algorithm is the procedure of expert survey, using the method of pairwise comparisons and the Monte Carlo method of statistical testing. The value of the random variable X_i (according to the Monte Carlo method) is calculated by the formula:

$$X_i = b_{j-1}^i + (b_j^i - b_{j-1}^i)\frac{(\xi - \sum_{l=1}^{j-1} p_l^i)}{p_j^i}. \tag{7}$$

Here $[b_{j-1}, b_j], j = 1, 2, \ldots, k$, the segments in which, according to the expert, would be the actual value of the volume of the provision of resources X_i, and p_j^i is the corresponding probability for this segment. To calculate this probability, one can use the hierarchy analysis method, which involves pairwise comparisons. Value ξ is a random number having a uniform distribution over the interval (0, 1).

To obtain the estimated distribution of the random variable X_i, it is necessary, following the algorithm, to play a given number of tests. In practice, it is enough to conduct several thousand tests (3–5 thous.), although the computing power of modern computers allows millions of such tests, which increases the accuracy of calculations.

To estimate each random variable X_i, its mathematical expectation is calculated based on the resulting estimated distribution.

Thus, the use of the described technique allows us to determine the estimate of the mathematical expectation of a random variable B, the value of which is necessary to find the level of risk associated with the resource supply of the enterprise.

We give a model example of the practical use of the described mathematical tools to assess the level of the internal economic risk of the implementation of an innovative project.

Suppose that the following resources are needed to implement an innovation project: rare-earth metals (X_1), information resources (X_2), unique equipment (X_3). As a result of the expert survey, estimates of the volume of money (in bn. rub.), which can be allocated for the purchase of these resources for the implementation of the innovation project, and the corresponding probability.

So, for each resource, the possible intervals were determined in which the amount of money funds needed to acquire each of the resources can be found.

Further, using the method of analysis of hierarchies and the methods of expert evaluation, it is necessary to determine the probability with which the volume of the necessary funds will fall into the selected intervals. For this, the simplest hierarchical structure is considered, and a matrix of pairwise comparisons is compiled. Then, by processing the matrix of pairwise comparisons according to the method proposed by T. Saati, a probability vector is obtained. The hierarchical structure corresponding to this task is shown in Fig. 2 (Table 1):

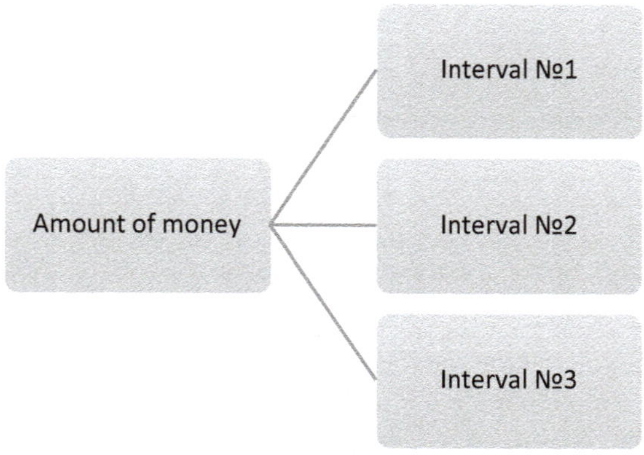

Fig. 2. The hierarchical structure for determining the probabilities of falling volumes of money at appropriate intervals.

Table 1. Example of analyzing the risk level of resource neediness of the innovative project.

Component	Variable	Interval expert estimates (bn. rub.)		
Rare-earth metals	X_1	[1.67, 1.77]	[1.77, 1.83]	[1.83, 1.92]
	p^1	0.067	0.148	0.785
Informational resources	X_2	[0.043, 0.05]	[0.05, 0.057]	[0.057, 0.06]
	p^2	0.096	0.105	0.799
Unique equipment	X_3	[0.046, 0.05]	[0.05, 0.055]	[0.055, 0.061]
	p^3	0.082	0.158	0.761

Source: a group of authors.

Applying the methodology for estimating the expectation of a random variable B (3,000 tests) described above, and the formula (6), we obtain that the level of risk in this case is:

$$R = 0.067.$$

The obtained value serves as a guideline when performing anti-risk measures and monitoring the level of risk caused by the action of the described factors. Ongoing anti-risk measures should be aimed at reducing the risk level, that is, at approaching the risk level to zero value:

$$R \to 0.$$

This toolkit can be used to assess the level of risk for several targets of an innovative project. For this, it is necessary to assess risk levels for each of the target indicators.

3 Conclusion

The paper proposes a step-by-step toolkit for assessing risks in organizations during the planning of their financial and economic activities. The developed methodological guidance can become the basis for the development of regulations and standards within organizations that ensure the integration of a risk management system in terms of identifying, assessing, and ranking risks in existing systems for short-term, medium-term, and long-term planning of enterprises.

Acknowledgments. This study has been supported by the RUDN 5-100 Program.

References

Alekseev, A.N., Kurilova, A.A.: Risk management as a financial risk management system of the company. Econ. Soc. **2–1**(15), 154–163 (2015)

Chursin, A., Tyulin, A., Yudin, A.: The model of risk assessment in the management of the company's competitiveness. J. Appl. Econ. Sci. **11**(8), 1781–1790 (2016)

Damodaran, A.: Strategic Risk Management. Williams Publishing House, Moscow (2010)

Karimov, S.M.: Fundamentals of modern risk management and risk management techniques. Econ. Entrep. **11–1**(64), 850–858 (2015)

Koshelev, V.A.: Management of innovative risks in housing construction (Cand. of Sci. (Economic) dissertation). Samara, Russia (2011)

Milykh, F.G.: The model of management of own risk the production level of the enterprises of light industry. Entrepreneurship **2**, 45–42 (2008). https://elibrary.ru/item.asp?id=9950840

Sekerin, A.B., Stroyev, S.P., Selyutin, V.D.: The fuzzy set model management of the risk of economic insolvency of a manufacturing enterprise. Risk Manag. **2**(46), 28–35 (2008). https://elibrary.ru/item.asp?id=11706821

Sholomitsky, A.G.: Risk theory: the choice at uncertainty and modeling of risk. High School of Economics, Moscow, Russia (2005)

High-Tech CluFigrs as a Basis for Improving the Russian Defense Industrial Complex with Respect to Digitalization

Arsen S. Abdulkadyrov[1(✉)], Vladimir Yu. Hakhalev[2], and Svetlana S. Hakhaleva[2]

[1] Institute of Socio-Political Research, Russian Academy of Sciences, Moscow, Russia
ars.rggu@mail.ru

[2] Military University of the Ministry of Defense of the Russian Federation, Moscow, Russia
vkhakhalev@rambler.ru, skhakhaleva@yandex.ru

Abstract. The paper examines topical issues of the high-tech cluster structures formation in the Russian advanced defense-industrial complex, taking into account a given digital development vector. As results of the research conducted, the authors identified the following problem points: (1) there are numerous problems existing in the military-industrial complex development, including the current situation in the state defense order (SDO) and military construction in general; (2) lack of theoretical research and practical developments, as well as theoretical and methodological support, on how high-tech defense-industrial clusters in the current economic conditions should be considered. The scientific novelty of our paper is represented by three significant provisions. Based on the scientific analysis of "cluster formation" as a concept, a new scientific term "high-tech defense-industrial cluster" is formulated by the authors. A high-tech defense-industrial cluster is understood as a complex multifunctional economic system, formed initially on the basis of closely interconnected industrial and defense-industrial complex enterprises, organized on co-operative lines, with institutional and innovation environments giving a strong impetus for further sustainable development and competitive advantages based on the national economy digitization. Second, the author's systematization of key factors for creating high-tech defense-industrial clusters was prepared and proposed, their distinctive system properties were highlighted, and an algorithm for the high-tech defense-industrial clusters formation, taking into account their industry specification, was presented. The research result is the conclusion that creating high-tech clusters in the defense-industry complex is a critical step in digitizing the industry. This process would strengthen the defense-industrial complex, its organizational production and structure research potential in line with the tasks of ensuring the Russian national security.

Keywords: High-Tech defense industry cluster · National security · Defense capability · Digitalization · Innovation · Competitiveness

A. V. Bogoviz and Y. V. Ragulina (Eds.): ISCI 2019, LNNS 115, pp. 221–226, 2020.
https://doi.org/10.1007/978-3-030-40749-0_26

1 Introduction

The conceptual cluster formation basis and the effective formation of the cluster policy were laid in the foreign researchers' works. The relevance of the research topic is connected with the need to improve the competitiveness and efficiency of the Russian defense-industrial complex in the national economic system digitalization conditions. The research object is the Russian defense-industrial complex, considered as the national economy specific economic-production system. The research subject is a specific military-economic relation arising between federal executive bodies, defense enterprises, and the national innovation system's links. The research goal is to study the conceptual foundations in the formation of the high-tech defense-industrial clusters (HTDICs) as one of the digital economy's key areas, based on the clustering theory and methodology, specific algorithms, models and strategies for its implementation in the Russian defense-industrial complex in order to increase its competitiveness and long-term performance. The conceptual clustering foundations are considered by the authors as a strategic imperative of creating high-tech integrated structures in the defense industry complex.

2 Materials and Methods

In the paper, the authors used such traditional research methods as dialectical, logical, scientific generalizations, tabular and graphical methods to review, process, and analyze factual material, which ensured the validity of theoretical conclusions and practical suggestions. Methods of structuring goals, expert assessments, and economic modeling were also used as a methodological toolkit. In present-day conditions, when Russia joined the WTO, it implements many federal targeted development programs. In this situation, the problem of forming different innovative clusters becomes particularly relevant; therefore, this topic of scientific research has recently become quite popular.

The following authors, among foreign and domestic scientists, should be mentioned for their contribution: (Alekseev et al. 2018), (Abdulkadyrov et al. 2017), (Ogurtsov et al. 2016) and (Chen and Kamal 2016). The authors of this paper have repeatedly addressed the issues of the effective economic growth strategy grounded in the high-tech, innovative products manufacturing (Khachaturyan et al. 2018a, b), as well as investment and innovative development in the medium term on the basis of digital technologies (Chen and Kamal 2016).

3 Results and Discussion

Based on the analysis of "cluster formation" as a scientific concept, a new term "high-tech defense-industrial cluster" is formulated by the authors. A high-tech defense-industrial cluster (HTDIC) is understood as a complex multifunctional economic system, originally formed on the basis of closely interconnected industrial and defense-industrial complex enterprises, organized on co-operative lines, with institutional and innovation environments serving a strong impetus for sustainable development and increasing competitive advantages based on the national economy digitization. Such a cluster operates on the basis of digitalization, centers for the scientific knowledge generation, and business ideas, as well as the highly qualified personnel, which should produce dual-use products with long-term competitive advantages. The authors also highlight the high-tech defense-industrial cluster's system properties, which are presented in Fig. 1.

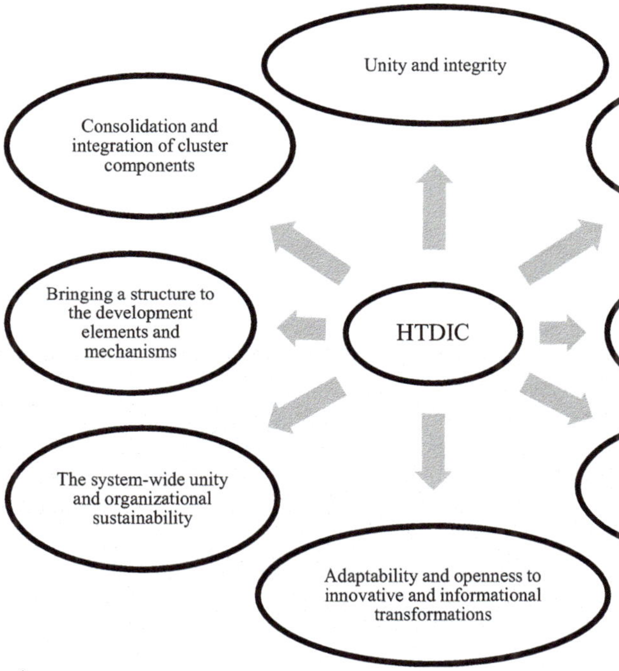

Fig. 1. The structure of system characteristics inherited in the high-tech military-industrial cluster. *Source*: Developed by the authors as a result of the study.

The scientific analysis and the polemical review of periodicals and monographic literature led to the conclusion about the current problems in this sector of the national economy. The authors believe that high-tech defense-industrial cluster is to be considered with respect to integrity, continuity, organizational stability, self-organization, and self-development. This entity should be based on using the latest digital achievements, leading to a synergistic effect of the basic cluster structures functioning in it. Thus, according to the authors, if the key parameters of the development phases coincided (at least for two future cluster system's elements,) this would lead to the self-organization of the rest and bring the entire system to a qualitatively new development level.

The examples of successful implementation of reformatting existing integrated systems in the defense production field into cluster structures are the Rocket and Space Corporation "Energia," "Almaz-Antey," "Sukhoi Civil Aircraft," and others.

These cluster forms have achieved success primarily due to the high degree of research centers, universities, and digital modern technologies participation. After reviewing and comprehensively analyzing the structural system of the proposed high-tech cluster characteristics, using the system-based approach to the cluster basic using the hypothetic-deductive method of scientific knowledge, we switched to consider the high-tech defense industrial cluster functional network structure.

The high-tech clusters formation in the Russian defense industry is fully correlated with the Order of Russian Government of 28.07.2017 No. 1632-p (approved the Program "Digital Economy of the Russian Federation") (Government of the Russian Federation 2017). It is a key factor in the formation of the national economy's competitive advantages and the public-private partnership promotion (Russian Federation 2012; Russian Federation 2015). More than that, this factor is one of the most important mechanisms for the military economy development, taking into account the vector of the entire country's national economy digital development (Bogoviz et al. 2019).

The model of the Russian defense and security integrated is formed by five elements: economic, directly military and social, and components to which information and digital security are added, in connection with the IT industry introduction, artificial intelligence, digital and nano-bio technologies, Russian scientists and practitioners' current researches and developments. In a generalized form, the algorithm for creating a high-tech defense-industrial cluster can be represented as a universal gradation of actions for the organization of each cluster subsystems formations in the form of five stages:

1. Determining cluster development priorities, taking into account national interests;
2. The existing scientific and technological potential and the assessment of cluster-forming structural units;
3. Identifying the constituent components of critical technologies emerging in a cluster structure;
4. Creating and identifying specific requirements and standards for all substructures of a high-tech cluster;
5. Determining a set planned activities, projects, and vectors of introducing and implementing tools for the organizational, economic, and innovative support of the created structure elements.

Based on the proposals considered in this paragraph and accordance with the economic systems and industries' digitalization trend, as well as economic processes globalization and integration, it is necessary to identify "sore spots" in the field of the defense-industrial complex clusterization facing the Russian scientific community and Russian government bodies.

4 Discussion

The existing development problems in the "knowledge generation" sphere of the innovation and technological lag of the Russian national economy, taking into account its digital development vector, including the defense-industrial complex (one of its basic generators), are mostly objective. First of all, it is connected with the disproportions in the economic structure and institutional environment difficulties. As such, the creation and development of the high-tech cluster in the defense industry system is the most important direction for solving the accumulated problems of science, technology, and innovation. It serves as the basis for improving the complex, taking into account the national economic system digitization. However, a large number of theoretical and practical clustering issues in the country's defense industry has not been resolved yet. There are a lot of reasons for that, which requires further attention, study, and analysis by the authors.

5 Conclusion

Taking into account the identified significant shortcomings, the authors presented a further study range of issues related to the high-tech defense-industrial clusters formation in Russia, considering digital transformations in the defense-industrial complex and the Digital Russian Economy program approved by the Russian Government. In total, two provisions represent the scientific novelty. First of all, the authors coined the definition of a "high-tech defense industry cluster." Its most significant properties are integrity, consistency, using digital technologies, self-organization and self-development, leading to a functioning cluster structures synergistic effect. Second, the authors summarized the key factors for creating high-tech defense-industrial clusters, highlight their distinctive system properties, and provided an algorithm for the formation of high-tech defense-industrial clusters, taking into account the country's digital development trends.

References

Abdulkadyrov, A.S., Ryzhov, I.V., Strokov, A.I., Kamzolov, Y.V.: Actual aspects of improving the high-tech products organization. Bull. High. Educ. Inst.: Technol. Text. Ind. **5**(371), 18–27 (2017)

Alekseev, A.N., Evdokimov, S.Y., Khachaturyan, K.S., Tarasova, A.Y., Khachaturyan, A.A.: The financial strategy of development of industry 4.0 in the countries with a developing economy. Espacios **39**(12), 1 (2018)

Bogoviz, A.V., Lobova, S.V., Ragulina, J.V.: Shift of the global investment flows in the conditions of formation of digital economy. In: LNNS, vol. 57, pp. 1216–1223 (2019)

Chen, W., Kamal, F.: The impact of information and communication technology adoption on multinational firm boundary decisions. J. Int. Bus. Stud. **47**(5), 563–576 (2016)

Government of the Russian Federation: Order "on approval of the digital economy of the russian federation program", 28 July 28 2017 No. 1632-p (2017). http://base.garant.ru/71734878/

Khachaturyan, A.A., Khachaturyan, K.S., Abdulkadyrov, A.S.: Model of innovational development of the modern Russian industry. In: Advances in Intelligent Systems and Computing, vol. 622, pp. 44–50 (2018a)

Khachaturyan, A.A., Ponomareva, S.V., Melnikova, A.S.: Long-term planning of assets and scientific and technological development of the economy of branch markets in digitalization conditions. St. Petersburg State Polytechnical Univ. J. Econ. **32**, 396–428 (2018b)

Ogurtsov, E.S., Rylov, A.A., Durdyeva, D.A., Lebedev, N.A., Khachaturyan, K.S., Safyan, A. M., Rochev, K.V., Tsareva, G.R.: The role of business incubators in supporting economic growth and advancement of small business of the Russian Federation. Indian J. Sci. Technol. **9**(29), 89339 (2016)

Russian Federation: Federal Law "On State Defense Order" 29 December 2012 No. 275-FZ. Coll. legislation grew. Federation, 53 (2012)

Russian Federation: The Federal Law "On Public-Private Partnership, Municipal-Private Partnership in the Russian Federation and Amendments to Certain Legislative Acts of the Russian Federation" 13 July 2015 No. 224-FZ. Rossiyskaya Gazeta, 6727 (2015)

Financial Instruments for Reproduction of the Real Sector of the Economy

Salihbek G. Abdulmanapov[1(✉)], Vezirhan K. Giraev[2], and Khadizhat M. Khadzhalova[3]

[1] Research Institute "Management, Economics, Politics, and Sociology", Dagestan State University of National Economy, Makhachkala, Russia
salih@mail.ru

[2] Dagestan State Technical University, Makhachkala, Russia
vgaaf@rambler.ru

[3] Institute of Social and Economic Research, Dagestan Scientific Center, Russian Academy of Sciences, Makhachkala, Russia
chaha@mail.ru

Abstract. Implementing the strategic objective of the breakthrough development of Russia is impossible without a combination of market and state regulation mechanisms. Full relevance of the latter is growing due to the increased dependence of national economies on the contradictory nature of the market economy in the context of the world economy's globalization. First of all, it concerns the impact of fiscal and tax policies on the reproduction processes in the real sector of the economy in the context of new realities. An important issue is the development of theoretical provisions on the use of fiscal instruments of state regulation of expanded reproduction in the real sector of the economy, serving as the basis for the industrialization of the national economy. The authors state that reconsideration of fiscal instruments for the state regulation of the reproduction of enterprise capital is needed. The scientific novelty of the study lies in the theoretical justification of the methods of financial support for the reproduction of capital of enterprises using the fiscal instruments to stimulate innovative entrepreneurship in the context of globalization of the world economy. The shift in emphasis of state regulation of reproduction processes towards fiscal policy makes it possible to differentiate the instruments used, depending on the degree of their influence on the processes in the real sector of the national economy.

Keywords: Finance · Budget · Fiscal policy · Economic growth · Government regulation · Capital reproduction

1 Introduction

The strategy and tactics of modern industrial production are based on an organic combination of market and state regulators with the widespread use of the latest scientific discoveries and achievements of technology and digital technologies, computer science, management, and other areas of material and spiritual activity. One of the effective state regulation levers of the reproduction processes is fiscal policy (Andrianov 1996).

A. V. Bogoviz and Y. V. Ragulina (Eds.): ISCI 2019, LNNS 115, pp. 227–232, 2020.
https://doi.org/10.1007/978-3-030-40749-0_27

Hence, the increased attention to the problem of justifying the need to reorient the state tax and budget policy to stimulate the development of the real sector of the economy and increase the investment in human capital. The purpose of the study is to develop theoretical principles and substantiations of practical recommendations on the use of fiscal instruments of state regulation of expanded reproduction in the real sector of the economy as the basis for the industrialization of the national economy.

2 Materials and Methods

Such general methods of scientific knowledge as the method of system analysis, scientific abstraction, comparative analysis, logical, and historical methods were used in the course of the study. The problems of using fiscal instruments in a market transformation are reflected in the works of such well-known scholars as Keynes, J., Samuelson, P., Seligman, E., Stiglitz, D., and also such Russian scholars as Goncharenko, L. I., Gorsky, I. V., Karavaeva, I. V., Kirina, L. S., Lykova, L. N., Maiburova, I. A., Panskova, V. G., Pinskaya, M. R., and others. At the same time, in the context of globalization and intensification of international competition (Chursin et al. 2016; Ragulina et al. 2020), the formation of the effective state regulation policy of reproduction processes as the basis for national wealth is becoming increasingly important.

The enthusiasm for the service economy in the concept of a post-industrial society, which relies on a change in the structure of employment through the recognition of the production nature of trade and consumer services (Gershuny and Miles 1983; Sassen 1996), has somewhat diminished the industry role. Recently, there has come the understanding that the basis for society and state wealth, including the success of the service economy, is based on the "technical" capabilities of the real economy sector.

Erhard, the author of the post-war Germany revival reforms and its subsequent industrialization, noted: "… economic prosperity is closely linked to the state fate, and, vice versa, the recognition of each government and each state is directly related to success or failure of their economic policies" (Erhard 1991). The state's economic policy is implemented using appropriate methods of influence on the socio-economic development of the country, which are usually divided into administrative and economic ones, according to the organizational and institutional criteria. There is a close interconnection between these methods.

The implementation of the financial influence of the state on economic entities is carried out using the financial system of the state. First of all, through the interconnectedness and interdependence of such components as state and local finance, finance of enterprises and organizations, household finance, and the financial market. Consequently, the financial policy of the state should be considered as a set of legislatively regulated and mutually agreed measures of financial influence, which are implemented through the financial system, aimed at the processes of mobilization and use of financial resources in order to ensure sustainable macroeconomic development of the country.

Monetary methods of influence on the development of the economy act as means of its stabilizing and stimulating. Applying the methods and tools of monetary policy as one of the stimulators of economic growth should be subordinated to ensure the transition from the macroeconomic stabilization tactics to the strategy of active state growth policy. However, as practice shows, monetary approaches do not eliminate the deep sources of inflationary processes (Kabanov et al. 2018). The unilateral application of only one method of state influence on the economy (although it is effective) without combining it with others is not enough to stimulate the reproduction and industrialization of the real sector of the economy.

The modern methods of the state's influence on the economy are not limited to monetary methods. First of all, it concerns fiscal policy. Among the scholars, there prevail two approaches to determining the nature and purpose of fiscal policy. So, the first group of scholars tend to identify fiscal policy with budgetary policy, the others consider it as a combination of tax and budgetary policies. In other words, they focus on tax and expenditure policy. The proponents of the first approach are several scholars, considering the fiscal policy as "… a budget policy aimed at changing the aggregate demand by changing government spending and taxes" (Kabanov et al. 2018).

Unification of the concepts of fiscal and budgetary policy is also peculiar to P. Heine. He argues that fiscal policy is a budgetary policy aimed at regulating and preventing undesirable changes in the aggregate demand by implementing planned changes in government spending and taxes (Heine 1997).

The leading place in the fiscal regulation belongs to the tax policy instruments. This is led by the comprehensive nature of the taxation system. It is located at the intersection of the state fiscal interests and the taxpayers' economic interests. The influence of the tax policy on the quantitative and qualitative parameters of the economic system, in fact, occurs through the realization of the internal potential of the taxation system, which is displayed in two functions – fiscal and regulatory.

3 Results and Discussion

Ensuring an effective tax policy is based on the optimal, for a specific period, combination of fiscal and regulatory properties of taxes following the priorities of the socio-economic development of the state. Tax cuts and tax breaks are the most widely used tools for tax regulation. At the same time, lowering base rates does not provide a targeted impact on specific areas and activities, improving only the business climate as a whole. Tax benefits provide unique opportunities for solving a variety of economic and social problems, which cannot be achieved by reducing the tax rate (Fisher et al. 1993). In the last quarter of the 20th century, they have become an important factor in international competition for foreign direct investment (Morisset 2003). Investment incentives are important to stimulate innovation and R&D. Thus, tax policy instruments should be economically justified, applied, taking into account specific conditions and factors (Bogoviz et al. 2019).

In the fiscal regulation system, a significant role is played by the budget, which, in present conditions, has turned from a balance of monetary incomes and state expenses into a powerful macroeconomic regulator of social reproduction, stimulating economic growth, increasing employment, and activating other socio-economic processes. Thus, budget expenditures are aimed at ensuring sustainable rates of economic growth, ensuring defense capability and social protection of the population (Pozdnyakova et al. 2019).

The tools for implementing the state functions in the sphere of the structural and economic growth policies (besides the administrative and tax) include, first of all, a direct financing from the budget of infrastructure projects. Second, there is a direct state financial support for enterprises, cheaper loans by subsidizing part of interest rates at the expense of budget funds, providing public goods at a price lower than market prices. Third, providing state guarantees for the types of activities that ensure the cumulative effect of economic growth and structural reforms (Dolan and Lindsay 1994).

These instruments can be considered as the financial privileges of a selective action or subsidies in a broad sense. These also include budget investments, government loans, guarantees, etc. The provision of subsidies to enterprises does not make a positive regulatory impact on the economic activity of market agents if their mechanism does not meet the principles of transparency of distribution and use; the certainty of goals, conditions of delivery and expected results; legal support of responsibility and effective use; subsidy urgency. Following the abovementioned principles, it is customary to divide subsidies for enterprises support into direct (budget loans, state loan guarantees, budget investments) and hidden loans (concessional loans, deferral of principal and interest payments, loans write-off, tax restructuring debt.)

According to the IMF experts, the so-called quasi-fiscal operations are operations that are hidden and are financed by the national bank and other state financial institutions that are not reflected in the budget, but affect the budget deficit size (preferential loans to government agencies, state corporations, and private entities; making unsecured or contingent liabilities, etc.). In our opinion, hidden subsidies aimed at solving the current problems of inefficient enterprises cannot provide the appropriate regulatory impact, since they do not help to eliminate the root causes of the inefficiency of their economic activities. Given the significant destructive impact of this type of subsidies on the economy, its use should be minimal or should be dismissed altogether (Silvestrov et al. 2018). The principal difference between hidden and direct subsidies to enterprises in the real sector is that the mechanism of the latter potentially has regulatory properties in order to ensure the efficient use of limited public resources.

Budget investments are an important tool to stimulate expanded reproduction. At the present stage, for optimizing and improving the efficiency of budget investments, the public-private partnership practice is widely used. This approach is justified and allows to distribute risks in the context of limited budgetary resources.

State guarantees and loans as the instruments for economic regulation, provide easier access to the credit resources to enterprises that, under the conditions of the general market, are not able to use bank loans. As a rule, they are provided to state corporations, strategic enterprises for the purpose of financing investment programs, technical production re-equipment, new technologies introduction, etc. (Ivanter and Panfilov 2011; Ignatieva and Mariev 2009).

Targeted state loans allow the provision of enterprise funds from the budget on the terms of payment, maturity and repayment, which increases the responsibility for their effective use.

An important tool for the regulation of economic processes is the state order (government procurement). For example, in the US and the UK, government procurement is almost 20% of GDP. The positive impact of the state procurement is realized only when such conditions as the competition among potential suppliers, holding state tenders for the selection of the most acceptable counterparties; real consumer protection in contracts between customers and performers; strict compliance with contract terms are met.

4 Discussion

The tasks and fiscal instruments of state regulation of the reproduction of capital of enterprises are defined as (a) improving the quality of the production means through public investment, financing R&D and introducing their achievements into production, tax incentives for the scientific and technical progress development and enterprises investment and innovation activities; (b) improving the quality of the workforce by directing public investment in the formation of human capital and stimulating private investment in its development.

The basic financial concepts were further developed. *The financial policy of the state* – a set of legislatively regulated and mutually agreed measures of financial influence through the financial system aimed at mobilizing and using financial resources for ensuring sustainable macroeconomic development of the country. The *fiscal policy* is a system of economically sound measures to form and improve the state tax system and change the structure of budget expenditures to ensure stable economic growth.

The contradictory effect of tax and budget policy instruments on capital reproduction is revealed: deterioration of reproduction conditions as a result of taxation of individual and corporate incomes, and, accordingly, the reduction of private savings and investments; the improvement of reproduction conditions as a result of the provision of tax benefits, state subsidies, soft loans, but distorting economic decisions and weakening competition.

5 Conclusion

Thus, provided that the instruments of fiscal policy are justified and adequate to the functions and macroeconomic conditions of the state, they are able to provide a positive effect on economic growth and reproduction processes in the real economy sector. In modern conditions, they are used by all countries of the world for macroeconomic regulation. The difference lies only in the significance of their impact on economic development, which is determined by the degree of the state redistribution of the country's GDP.

References

Andrianov, V.: State regulation and self-regulation mechanisms in a market economy (world experience and Russia). Voprosy Ekonomiki **9**, 28–36 (1996)

Bogoviz, A.V., Zhuravleva, T.A., Popkova, E.G., Zarudneva, A.I., Alpidovskaya, M.L.: Effective tax policy of the state: conceptual foundations and methodology of evaluation. Stud. Syst. Decis. Control **182**, 91–98 (2019)
Chursin, A., Tyulin, A., Yudin, A.: The model of risk assessment in the management of company's competitiveness. J. Appl. Econ. Sci. **11**(8), 1781–1790 (2016)
Dolan, E.J., Lindsay, D.E.: Macroeconomics (Trans. From English). Litera Plus, St. Petersburg (1994)
Erhard, L.: Welfare for all (Trans. From it.). Nachalo-Press, Moscow (1991)
Fisher, S., Dornbusch, R., Schmalenzi, R.: Economy. Delo, Moscow (1993)
Gershuny, J., Miles, I.: The New Service Economy: The Transformation of Employment in Industrial Societies. Praeger, New York (1983)
Heine, P.: The Economic Way of Thinking (Trans. from English.). Catalaxia, Moscow (1997)
Ignatieva, E.D., Mariev, O.S.: Methodological foundations and tools for assessing the potential of self-development of regional socio-economic systems. Econ. Reg. **2**, 116–129 (2009)
Ivanter, V.V., Panfilov, B.C.: The end to growth economy or a change of the development paradigm? Bull. Russ. Acad. Sci. **7**, 594–598 (2011)
Kabanov, V.N., Zhilnikov, A.Yu., Zablotskaya, T.Yu.: Financial guarantees of the regional budget - the potential for growth of investments in fixed assets. Econ. Reg. **14**(1), 315–325 (2018)
Morisset, J.: Tax incentives: using tax incentives to attract foreign direct investment/world bank group, private sector and infrastructure network, public policy for the private sector. Note number 253, 4 p (2003)
Pozdnyakova, U.A., Bogoviz, A.V., Lobova, S.V., Ragulina, J.V., Popova, E.V.: The model of well-balanced taxation for overcoming the shadow economy in modern Russia. Stud. Syst. Decis. Control **182**, 207–215 (2019)
Ragulina, J.V., Bogoviz, A.V., Lobova, S.V., Alekseev, A.N., Pyatanova, V.I.: Strategy of increasing the global competitiveness of Russia's economy and Russia's becoming a new growth vector of the global economy. Lect. Notes Netw. Syst. **73**, 203–210 (2020)
Sassen, S.: Transnational Economies and National Migration Policies. Institute for Migration and Ethnic Studies. University of Amsterdam, Amsterdam (1996)
Silvestrov, S.N., Bauer, V.P., Eremin, V.V.: Evaluation of the dependence of the investment multiplier on changes in the structure of the regional economy. Econ. Reg. **14**(4), 1463–1476 (2018)

The Influence of the Asymmetry of the Banking System of the Russian Federation on Interbank Competition in the Conditions of Digitalization of Banking Activities

Igor E. Risin, Elena F. Sysoeva, and Maria R. Kostyanskaya(✉)

Voronezh State University, Voronezh, Russia
risin@mail.ru, selfin@mail.ru, krivkina.maria@yandex.ru

Abstract. The purpose of the study is to assess the level of the asymmetry of the Russian banking system, as well as to identify the factors influencing the state of the interbank competition in the context of the digitalization of banking activities. The statistical data for 2011–2019 of the Bank of Russia were used as an information base. The study allowed to evaluate the contemporary domestic market of banking services as an oligopoly market with a competitive environment and concentration in the federal center, which negatively affects the development of interbank competition in the country. The revealed external and internal factors affecting the state of interbank competition allow to identify possible directions for improving the regulatory framework and structure of the Russian banking system in the context of digitalization of the economy as a whole and banking activity in particular, especially in the direction of developing interbank competition for the purpose of improving banking services for the real sector of the economy. The authors argue that the decentralization of cash flows of credit organizations, taking into account the regional component, is to be made.

Keywords: Interbank competition · Asymmetry of the banking system · Banking digitalization

1 Introduction

In the context of the globalization of financial markets, the limited access of Russian commercial banks to the foreign capital market, and the tendency for digitalization of banking activity, one of the pressing issues regarding the functioning of the Russian banking system is to increase its competitiveness.

The most powerful participant in the financial market is the state, which has a significant impact on the competitors in the financial and banking sector. The three largest banks with state participation in their capital can be distinguished. They are Sberbank PJSC, VTB Bank PJSC, and Rosselkhozbank PJSC. They initially receive significant competitive advantages over the banks without state participation, which deprives the system of competitive relationships.

A. V. Bogoviz and Y. V. Ragulina (Eds.): ISCI 2019, LNNS 115, pp. 233–242, 2020.
https://doi.org/10.1007/978-3-030-40749-0_28

One of the problems in the development of interbank competition in Russia is that modern approaches to regulating relations between agents of the Russian banking market regarding competition are restrictive rather than stimulating. In essence, this approach is based on a system of formal criteria in the form of special norms and requirements, which in practice turn into specific "barriers" of normal functioning for most Russian banks, except banks with state participation (Magomedov and Suleymanova 2008).

The policy of regulation of competition in Russia requires determining the characteristics of the Russian banking system and developing measures for regulating banking activity from the standpoint of competitive relations. Equal competition and market discipline require clear, general rules of the road and an effective system for responding to their violation.

2 Materials and Method

The information base of the study was the statistical data of the Central Bank of Russia (Bank of Russia) on the state of the Russian banking system (Central Bank of the Russian Federation 2019a, b, c), methodological recommendations on participation in the selection of Russian credit organizations as authorized banks under the soft loan program for SMEs (JSC "Corporation SME" 2019)

A systematic approach using methods of comparative analysis, synthesis, classification, grouping, and observation were used to achieve the objective of the study. These methods allowed us to identify the quantitative parameters of the asymmetry of the Russian banking system, the composition of the factors influencing the development of interbank competition, and, in particular, the digitalization of banking activities.

3 Results

Until June 1, 2017, there were no official differences in the regulation of the activities of commercial banks. The distinction of this activity and, therefore, its regulation was determined by a set of appropriate licenses for banking operations.

On September 12, 2010, the Basel Committee on Banking Supervision approved a global reform designed to increase the financial stability of the global banking (and financial as a whole) systems, primarily by increasing the bank liquidity reserves and improving their quality, which was called "*Basel III*" (Basel III Capital Accord).

In Russia, the introduction of new standards on the requirement for the structure of assets and capital of the banks began in January 2013 and was fully completed by January 2015 and by January 1, 2019, on the structure of the reserves.

In order to ensure compliance with the requirements of *Basel III*, the Russian Federation began to use a three-tier regulatory system for commercial banks from June 1, 2017. The commercial banks are currently divided into:

- Systemically significant (Central Bank of the Russian Federation 2015), which are subject to short-term liquidity requirements and additional requirements for capital adequacy following *Basel III*;

- Universal licensed banks that are required to comply with all mandatory economic standards, including the standards of the stable net funding and financial leverage;
- Banks with a basic license, which are restricted to the performance of several banking operations and have only five economic standards established. These restrictions are associated with the need to reduce the risks of active operations of medium and small banks, as well as with their preservation in the context of tightening regulatory requirements.

In the last decade, the works of economists devoted to the development of regional banking systems came to a consensus: the key problems of regional banks are the strict requirements for the amount of capital (Glezman and Zike 2013). Indeed, the small size of banks often prevents them from mobilizing resources to invest in large projects. Due to the lack of attracted resources and to their proximity to the market, regional banks, instead of becoming a key link in the banking system of the region, specialize in the settlement and cash services and work with a relatively small circle of clients, including non-banking businesses of the owners (Andieva 2013).

Partially, this problem should be solved by "softer" requirements for the banks with a basic license; the reporting procedure should be simplified and the requirements for complying with mandatory standards relaxed, including the terms of capital adequacy and liquidity standards (Central Bank of the Russian Federation 2018a).

Out of the desire to increase the return on assets and capital, banks often resorted to illegal activities, servicing shadow cash flows, and distorting official financial statements, which led to large-scale license reviews.

As a result, the number of existing credit organizations with the right to carry out banking operations has decreased over the past five years, from 923 on January 1, 2014, to 484 on January 1, 2019 (Information about the banking system, 2014–2019), or by 47.6%.

One of the factors of interbank competition is its regional distribution. The distribution of banks across the Russian Federation shows a pronounced asymmetry. Most of the registered banks are located in the Central Federal District, mainly in the Moscow region (Table 1).

Table 1. Distribution of the banks in Russia.

Year	Total in Russia	Region								
		CFD		NWFD	SFD	NCFD	VFD	UFD	SFD	FEFD
		Total	Incl. Moscow and Moscow region							
2011	1012	585	525	71	47	57	118	51	56	27
2012	978	572	512	69	45	56	111	45	54	26
2013	956	564	506	70	46	50	106	44	53	23
2014	923	547	498	70	46	43	102	42	51	22
2015	834	504	459	64	45	28	92	35	44	22
2016	733	434	392	60	42	22	85	32	41	17

(continued)

Table 1. *(continued)*

Year	Total in Russia	Region								
		CFD		NWFD	SFD	NCFD	VFD	UFD	SFD	FEFD
		Total	Incl. Moscow and Moscow region							
2017	623	358	321	49	38	17	77	29	37	18
2018	561	319	284	43	35	17	71	26	32	18
2019	484	272	239	41	25	12	67	23	28	16

Source: (Central Bank of the Russian Federation 2019a).

As of February 2019, there were 147 banks with a basic license operating in the Russian Federation. Their distribution in the country is as uneven as in the banking system as a whole. More than a third of such banks are in the Central Federal District, 70% of which are located in the Moscow region (Table 2).

Table 2. The distribution of banks with a basic license for the Russian Federation for February 2019.

Region	The number of banks with a basic license registered in the region
Central Federal District	66
Including in Moscow and Moscow region	46
Northwestern Federal District	13
Southern Federal District	10
North Caucasus Federal District	9
Volga Federal District	20
Ural Federal District	8
Siberian Federal District	16
Far Eastern Federal District	5

Source: (Central Bank of the Russian Federation 2019c).

A similar situation can be seen in the volume of lending: more than half of the loans issued to legal entities in Russia are in the Moscow region, 80% of which are issued by the 30 largest banks. It is important to note that of the 30 largest banks, 24 are registered in the Central Federal District, and 23 of them are in Moscow (Central Bank of the Russian Federation 2019b) (Table 3).

Table 3. The information on the volume of lending to legal entities as of 01.01.2019.

Region	Total, mln, ₽	30 largest banks, mln, ₽	The share of the 30 largest banks in lending, %
Russian Federation	45 005 060	39 672 895	88%
Central Federal District	27 145 782	24 498 180	90%
Share in total lending	60.32%		
Including Moscow and Moscow region	24 486 270	19 513 526	80%
Share in total lending	54.41%		
Northwestern Federal District	4 508 165	3 941 160	87%
Share in total lending	10.02%		
Southern Federal District	2 695 761	2 273 783	84%
Share in total lending	5.99%		
North Caucasus Federal District	268 413	239 091	89%
Share in total lending	0.60%		
Volga Federal District	4 535 966	3 656 283	81%
Share in total lending	10.08%		
Ural Federal District	2 455 323	2 104 510	86%
Share in total lending	5.46%		
Siberian Federal District	2 296 100	2 037 839	89%
Share in total lending	5.10%		
Far Eastern Federal District	1 099 549	922 049	84%
Share in total lending	2.44%		

Source: (Central Bank of the Russian Federation 2019b).

As follows from Tables 1, 2 and 3, the asymmetry of the banking system of the Russian Federation contributes to the outflow of financial resources to the federal center, which slows down the economic growth of the regions.

4 Discussion

It was assumed that the banks with a basic license would work with the target audience, which is small and medium-sized businesses, and their activity should ensure the availability of banking services in the regions, which will allow their economic growth.

Domestic scholars also made this assumption. In particular, Afanasyeva O. N., and Marushkin E. O. (Afanaseva and Marushkin 2017) believe that "the use of proportional banking regulation … can ensure the creation of the necessary regulatory framework to maintain competition, as well as to stabilize and make the entire banking system more stable. Small and medium-sized credit organizations are the most optimal basis for developing the availability of credit instruments, financial resources for individuals, and small and medium-sized businesses in various constituent entities of the Russian Federation."

However, practice shows that the assumed role of banks with a basic license was not implemented.

In particular, in order to maintain the role of banks with a basic license in lending to small and medium-sized enterprises (hereinafter referred to as SMEs), the rules for granting subsidies to JSC "Corporation SMEs" for lending to such organizations have been amended. According to the amendments, banks with a basic license can participate in the selection of authorized banks under the soft loan program, while for large banks, a threshold value of the minimum allowable capital amount of ₽ 100 billion has been introduced. At the same time, the applicant bank must have experience in lending to SMEs; its structure should include a structural unit that provides and monitors loans issued to such entities. In addition, for banks with a capital of less than ₽ 50 billion, which currently includes all registered banks with a basic license, an additional requirement that the share of loans to SMEs should be at least 50% or at least ₽ 10 billion is imposed (JSC Corporation "SME" 2019). This condition casts doubt on the possibility of participation in the program for banks with a basic license, since the amount of net loan debt in the assets of the bank with the maximum capital among banks with such a license amounted to ₽ 928,461 thousand in the first three quarters of 2018 (Central Bank of the Russian Federation 2018b).

Similar doubts about the increasing role of banks with a basic license in the lending process for SMEs are also presented by the statistics. According to the Ministry of Economic Development of Russia, the total number of accredited banks is 70, and only 11 of them have a basic license (Ministry of Economic Development of the Russian Federation 2019).

Also, the largest share of loans granted to SMEs is accounted for by 30 of the largest banks; the national average is 73.1%. Sberbank PJSC, VTB Bank PJSC, and Raiffeisenbank JSC remain market leaders in terms of the volume of SME loans issued in 2018.

The distribution of the volume of loans granted to SMEs by federal districts is presented in Table 4.

Table 4. The volume of loans granted to SMEs as of 01.01.2019.

Region	Total, mln, ₽	30 largest banks, mln, ₽	The share of the 30 largest banks in lending, %
The Russian Federation	6 816 081	5 042 009	73.97
Central Federal District	2 769 879	2 143 649	77.39
Share in total lending	40.64%		
Including Moscow and Moscow region	2 063 748	1 564 625	75.81
Share in total lending	30.28%		

(*continued*)

Table 4. (*continued*)

Region	Total, mln, ₽	30 largest banks, mln, ₽	The share of the 30 largest banks in lending, %
Northwestern Federal District	737 220	563 635	76.45
Share in total lending	10.82%		
Southern Federal District	572 091	321 543	56.20
Share in total lending	8.39%		
North Caucasus Federal District	123 044	102 474	83.28
Share in total lending	1.81%		
Volga Federal district	113 4157	807 413	71.19
Share in total lending	16.64%		
Ural Federal District	495 566	365 428	73.74
Share in total lending	7.27%		
Siberian Federal District	673 198	534 698	79.43
Share in total lending	9.88%		
Far Eastern Federal District	310 927	203 168	65.34
Share in total lending	4.56%		

Source: (Central Bank of the Russian Federation 2019b).

The results of the analysis show that the banks with a basic license still cannot be called significantly competitive not only in the country as a whole but also in the context of a region. Serious competition, caused by the prevailing opinion that a large bank is more reliable, does not allow the banks with a basic license to take their market share, and, therefore, their existence is called into question.

The risk of the loss of competitiveness, along with the risk of the loss of the liquidity, market risks, regulatory risks, and reputational risks, has a direct impact on the financial situation of a particular bank and poses a threat to the banking system as a whole. The number of financial resources held by the bank in the market is directly proportional to the number of risks it takes. Based on the available data, it can be concluded that the most significant risk is borne by the largest banks, which account for 88% of the lending market, and this figure jeopardizes the functioning of the entire economy in case of a crisis in at least one of these banks. This situation indicates the need for the decentralization of the cash flows of credit organizations, taking into account the regional component.

At present, the most important factor in the interbank competition is the digitalization of banking. More than a third of the revenues of the commercial bank comes through digital channels. Digitalization today can be divided into two areas. Firstly, it is a model of remote service in which the bank is mobile, and all services are available through smartphone applications. Secondly, these are technologies hidden from customers—petabytes of information and statistical models (Bankir 2018). These areas can be implemented only by banks with significant financial resources. In order to compete under the conditions of the digital economy, regional banks need staff with the appropriate competencies and a modern technical base to competently manage project portfolios and analyze large amounts of data in order to make management decisions permanently. Large-scale funds are also required for protection against cyberattacks, unauthorized access to information, and the growth of cybercrime as a result of the spread of the mobile services. Consequently, in the medium term, only the most technologically advanced and efficient banks will flourish.

Nevertheless, in the context of the digitalization of banking activities, there are certain advantages for regional banks with a basic license. Thus, the Bank of Russia has announced the size of the commission (0.34% per annum of the amount of the attracted deposit), which banks will pay for attracted depositors through its marketplace. For large banks with a wide branch network, such a commission is quite high since attracting a depositor is cheaper for them. However, for small regional banks that wish to provide representation in a new segment of the financial market or another region or that wish to increase the volume of deposits quickly, such a commission may be acceptable. The marketplace platform reduces the cost of the distribution of financial products for businesses that cannot afford extensive advertising campaigns (Kommersant 2019), which can help strengthen the position of the banks with a basic license, taking into account the digital component of their activities.

However, savings on commissions cannot significantly affect the trend of reduction in the commercial banks that can survive in the digitalization of banking.

One of the factors influencing interbank competition in the Russian Federation is also the introduction of the economic sanctions against systemically important Russian banks (Sberbank PJSC, VTB Bank PJSC, Rosselkhozbank PJSC, Gazprombank JSC) and other major banks (Vnesheconombank of the Russian Federation, JSC "AB" Russia", JSC "Sobinbank", etc.). The sanctions provoked the reorientation of large banks to fund in the Russian financial market and at the same time, reduced the ability of medium and small banks to compete for depositors' funds and in the market of interbank loans, first of all, refinancing loans of the Bank of Russia. This situation ensured an increase in the cost of borrowing in the domestic financial market, especially for the long term, since the bulk of "long-term financing" in previous years was provided to Russian commercial banks with a universal license on the foreign market. In addition to the direct increase in interest rates for Russian commercial banks and their borrowers, an additional negative factor is the increased exchange rate risk for the loans in foreign currency.

5 Conclusion

The institutional asymmetry of the Russian banking system, expressed in the concentration of the vast majority of commercial banks in Moscow and the Moscow region; the asymmetry of cash flow, manifested in the scale of operations of the systemically important banks, and banks with universal and basic licenses, have a significant impact on the state of interbank competition.

The competitive deterioration of medium and small banks (with a basic license) was influenced by such internal and external factors as:

- There was a significant decrease in profit and profitability (assets and equity) in the banking sector, which limited the opportunities for organizational growth of commercial banks and the equity buildup through profits refinancing;
- The toughening of regulatory requirements in connection with the *Basel III* implementation;
- In the context of systemic risks caused by sanctions pressure on the Russian economy and increased volatility of financial markets, there was a significant outflow of customer funds from small and medium-sized banks to banks with state participation, despite their lower funding interest rates;
- The transformation of the financial product markets into digital platform markets, which dramatically changes the role of financial intermediaries. The digitalization of the interaction between clients and banks, the use of digital tools to support sales, the creation of a system for Big Data management and the adaptation of IT architecture to the new requirements, which are the most effective areas of digital transformation for commercial banks, require large expenses that only banks with significant financial resources can afford.

The intended role of banks with a basic license to work with a target audience, for example, with clients related to small and medium-sized businesses, was not realized. Thus, the SME Corporation selection system and the technology for interacting with selected banks were made for the large banks and did not take into account the reorientation of the Program 6.5 to the banks with a basic license and regional banks.

All this allows for evaluating the modern domestic banking services market as an oligopoly market with a competitive environment and concentration in the Moscow region, which negatively affects the development of interbank competition, banking services, and lending to customers in the regions.

References

Afanasyeva, O.N., Marushkin, E.O.: New directions of regulation by the bank of Russia of the activities of commercial banks. Vestnik of Volzhsky University after V.N. Tatischev, 4–1, 3 (2017)

Andieva, E.V.: Region banks: functioning and development. Terra Economicus 3–3 (2013). https://cyberleninka.ru/article/n/banki-regiona-funktsionirovanie-i-razvitie

Bankir, Ru.: Banking of the future: digital as a competitive advantage (2018). https://bankir.ru/publikacii/2018092

Kommersant: Banks will pay for attracting deposits through the marketplace (2019). https://www.kommersant.ru/doc/3908026

Central Bank of Russian Federation: Content. Bank of Russia Bulletin 98(2052) (2018a). https://www.cbr.ru/Queries/XsltBlock/File/48359?fileId=-1&scope=2052

Central Bank of the Russian Federation: on approval of the list of systemically important credit organizations (2015). http://www.cbr.ru/press/PR/?file=20102015_100129ik2015-10-20T10_01_03.html

Central Bank of the Russian Federation: Balance sheet (published form) to 01.10.2018 Bank "Aspect"(2018b). http://www.cbr.ru/credit/main/coinfo/f806/?regnum=608&dt=201810

Central Bank of the Russian Federation: Information about the banking system of the Russian Federation (2019a). https://www.cbr.ru/statistics/?PrtId=lic

Central Bank of the Russian Federation: Information on placed and attracted funds (2019b). https://www.cbr.ru/statistics/pdko/sors/retro/

Central Bank of the Russian Federation: Reference book of credit institutions (2019c). http://www.cbr.ru/credit/main.asp

Glezman, L.V., Zike, R.V.: The role of regional banks enhancement in the economic development of Russia. Russ. J. Entrep. **16**, 64–70 (2013)

JSC Corporation "SME": Guidelines for participation in the selection of Russian credit institutions as authorized banks under the program of concessional lending to SMEs (2019). http://economy.gov.ru/wps/wcm/connect/21282de1-2a51-4f14-99d0-dbdda3a76512/metodik.pdf

Magomedov, G.I., Suleymanova, A.M.: On some issues of development of competitive relations in the banking sector. Russ. J. Entrep. **2**(106), 37–42 (2008)

Ministry of Economic Development of the Russian Federation: The Ministry of Economic Development of the Russian Federation approved 70 banks for concessional lending to SMEs at a rate of 8.5% (2019). http://economy.gov.ru/minec/press/news/2019022501

The Algorithm for Rapid Estimation of the Production Costs of Multidimensional Products of an Industrial Enterprise

Danil A. Tolstobrov(✉), Nadegda A. Tolstobrova, and Sergey A. Fedoseev

Perm National Research Polytechnic University, Perm, Russia
2told@mail.ru, tolstobrova@mail.ru, fsa@gelicon.biz

Abstract. The introduction of modern software allows industrial enterprises to rationalize the use of working hours of employees, as well as to reduce labor and material consumption of products. This paper focuses on the algorithm for the operational forecasting of the production costs for an industrial enterprise operating in the perfectly competitive market in the context of the customers providing limited information. The use of the software available on the market to optimize the nesting cards for sheet material during the process of preliminary calculation of a client's application is justified only for the oligopoly market (or with a guaranteed order). However, this exceeds the cost of the work for its detailed processing dozens of times. The author's algorithm for the rapid estimation of the production costs is based on the ABC method and the "Mueller and Son" nesting design techniques, which are simple to implement. It allows for editing the application for one or a whole group of multidimensional products at the request of the customer at the calculating stage, for ensuring the optimal balance between the price expected by the client and the profitability laid down by the industrial enterprise. According to the algorithm for a non-standard application, a multidimensional product is presented as a rectangular parallelepiped. The production costs for each layer of the used material are considered separately. A graphical study of the change in the rate of sheet material consumption depending on the installed overall dimensions of a multidimensional product and the useful width of the roll made it possible to find their ratio, which leads to the peak values of material overruns. In order to avoid them during the order processing, an algorithm has been developed and tested to find the optimal material consumption. The application of these recommendations will allow industrial enterprises to increase the value of products and the enterprise itself for each participant of the relationship. This will also become the basis for the growing competitiveness of the domestic products.

Keywords: Industrial enterprise · Lean manufacturing · Computer Assisted Design System · Algorithm · Production costs · Nesting card · Multidimensional product

A. V. Bogoviz and Y. V. Ragulina (Eds.): ISCI 2019, LNNS 115, pp. 243–252, 2020.
https://doi.org/10.1007/978-3-030-40749-0_29

1 Introduction

Modern digital technologies expand the range of possible activities of industrial enterprises (IE), allowing them, depending on the nature and complexity of the work performed by the staff, to simplify it, partially or fully automate it, giving employees the opportunity to act as a supervisor (operator) of the tasks performed by the software (Kablov and Finogeev 2016). The subject of the study is the mechanisms for predicting the consumption of rolled (sheet) materials at the harmonization phase between the client and the IE engaged in serial production by the individual requirements of the client. The development of this study is of strategic importance in accordance with the order No. 1907 "On approval of Recommendations on the application of the principles of lean manufacturing in various industries," entered into force on June 20, 2017.

The following circumstances became the basis for recommendations on the implementation of the lean manufacturing management system (LMMS): reducing the production cycle; increasing the efficiency of IE, regardless of the production scale; and increasing the value of products and the enterprise as a whole for each internal and external participant of the relationship.

The study aims to develop an algorithm for calculating material costs based on an operational assessment of the parameters of multidimensional products of the IE to justify the preliminary cost of the order in conditions of limited information.

2 Materials and Method

The IE, performing a sheet metal cutting during the production process to fulfill the requirements of the LMMS, should use the technologies for efficient optimization of sheet metal nesting cards that reduce the amount of unused materials and the working hours for their formation, while maintaining the consumer value of the products (Smirnov and Kremlova 2017).

The loss of the consumer value of the product may be caused by the placement of connecting seams on the surface or structural load-bearing nodes. As a result, the placement of dividing seams into its parts is carried out only after the justification at the technological, functional, economic, and consumer levels. The cutting is carried out according to the group of patterns corresponding to the product.

The software, existing on the market—CADS (Computer Assisted Design System) solving the problem of optimizing nesting sheets differs in the form of reporting, the system of integration with the equipment, the range of steps covered from the beginning of product design to the formation of accounting documentation (Lovigin 2016).

This paper focuses on the methods that allow the rapid assessment of the production costs of a limited series of products according to the individual requirements of the customer. In the research process, preference was given to the smallest CAD in terms of the performed functionality or its separate independently functioning blocks, which are convenient for integration into the workflow and specialize only in the creation and correction of product details that form cutting cards and a set of reporting documentation.

The introduction of CAD software allows one to increase the efficiency of the use of sheet material in the formation of nesting cards in comparison with the manual selection from 5% to 12% and to reduce the working time of employees in drawing up a nesting card to several minutes instead of no less than half an hour. Also, modern CAD software, depending on the group of patterns that are being processed, can offer several different visualized options for cutting maps for selection and subsequent editing (Murzakaev and Lyalin 2013).

CAD, depending on its configuration, can provide an employee with additional opportunities at the stage of preparation for calculating material nesting cards, grouping patterns according to their correspondence to the final product. An additional advantage of CAD is the generated reports on the planned material consumption and the compiled design specifications simplifying the stages of accounting, design, coordination, and final assembly of products.

The quality of such calculations performed by CAD directly depends on the completeness of the information about the final product provided to it. Such information saturation of the program requires significant time costs for the employees (operators) for the initial processing and insertion (Kablov and Finogeev 2016).

This approach to the design and creation of complex products allows us to achieve high-performance indicators for a serial or large-scale production when the customer has already confirmed the order, or during the calculating stage, when the application is higher than the cost of working time for employees to process it more than ten times.

It can be argued that it is characteristic for an IE that adheres to the principle of minimizing production costs and possible future risks for a non-standard or large-scale order to have a long-term stage of settlements and coordination with the client of the configuration and order value. Such a strategy corresponds to the oligopoly, where the market share occupied by each participant has already been determined. The client is associated with such a manufacturer by contractual, temporary restrictions, or territorial proximity. Placing an order with another large software company with a production schedule planned for a long period will force the client to agree to increased lead time.

It is explained that such IEs are characterized by the gradual development of their production capabilities. Each introduction of new production facilities goes through a planning and economic feasibility study over an extended period. Increasing the IE intensity without an appropriate system of internal interaction and intermediate control will lead to a future loss in the quality of products.

The IEs with small-scale production mainly correspond to the perfectly competitive market, consisting of a large number of active participants dynamically developing their client portfolio. Such IEs can create a monopoly or oligopoly market for themselves—only if they sensitively respond to the individual expectations of their client.

Being in the perfectly competitive market, the IE cannot afford to deal with the calculation of the client's application for a long time, since the period for determining the potential executor of the order, allocated by the client, takes from three to five working days. An extension of the term for providing a response to a client can only happen if the client expects an exclusive product that has no analogs.

Often, the condition in the mandatory promptness of the response to the client request is supplemented by incomplete information. This forces the IE to make calculations without having an accurate idea of the future production costs associated with it. The situation in which all the information about the software order is known corresponds to a repeated serial order in most cases. However, in this case, the influence of competitors without any additional restrictions will force the IE to reduce the profitability to the maximum permissible minimum payback on the production costs.

The limited information provided by the client in the initial application is explained by several reasons: commercial secret, the client's reluctance to provide detailed information about his activities to his competitors.

The fulfillment of a product or a group thereof deviating from the standard order may lead to unforeseen costs for the software. It is established that the preliminary calculation of the cost of order fulfillment based on incomplete information is rarely amenable to increase. The only argument for the recalculation is the dramatic effect of the missing information on the production process. Being aware of this, at the stage of calculating the application, the IE will try to increase the estimated profitability of the order optimally. However, here, we should take into account competitors' influence on the potential choice of the client.

Therefore, in order to maintain its position in the market, software should be guided in the direction of conducting non-standard orders using fast algorithms with a high-reliability degree of assessment.

Each non-standard order is a multidimensional product, described by a variety of consumer properties such as $X = (X_1, X_2, \ldots, X_N)$. The order can be adjusted with the client during the coordination stage, following the established external conditions while maintaining its general structural diagram. With this approach, the IE can maintain the optimal value of the forecasted profitability and adjust—in agreement with the client—the product parameters to achieve the expected price.

The author's algorithm can be considered one of these methods, based on a combination of the ABC method and the "Mueller and Son" nesting design techniques. In order to simplify the calculation system, each non-standard product can be represented as consisting of one or more figures of the shape of a rectangular parallelepiped (FSRP). Then, each of these FSRPs can be described by a combination of its overall dimensions (x, y, z), the presence of each of its faces (A, B, C), as well as a combination of layers of materials used for the top and bottom coatings and filler (Tolstobrov 2019) (Fig. 1).

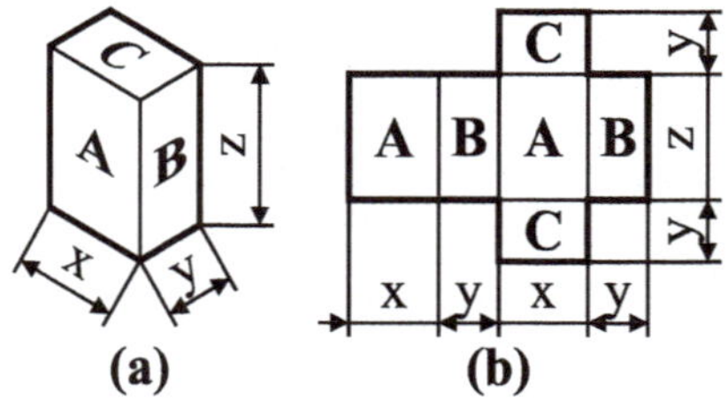

Fig. 1. FSRP product and its surface.

Let us assume the condition that on the entire involved surface of a separate figure of the FSRP, one type of material is used. If the FSRP is a product consisting of several layers (top, bottom, and filler), then for the calculation period, it is divided into a group where only one material corresponds to each. This paper discusses products made from textile materials. They are characterized by a significant excess of the total length of the roll to its usable width and each of the overall dimensions of the FSRP, then take the roll length to infinity and rely only on its useful width (W) in the calculations (Fayzrakhmanov 2011; Mukhacheva 1984).

In the production process at the cutting site of materials, their scrap is accumulated, which requires sorting and storage. The work on the placement of individual parts of the product (patterns) on the resulting trim increases the working time, and its further demand is difficult to predict. The product cost should include additional costs for the resulting trim from the used materials (Maslov and Shutkevich 2013).

Each FSRP product can be considered as a group of patterns representing a separate face or a combination of several. To simplify the calculation system, we accept the conditions that each of the patterns is a figure close to a rectangular shape, having two overall dimensions, and also all the patterns are not tied to the direction of the fabric pattern in the roll—technical fabrics do not have a clear pattern direction. Therefore, the pattern, placed on a roll with a useful width *W*, has two overall dimensions: *Xw*—across the roll and *Lw* along with the roll. Figure 2 shows a graphical representation of two placement options: when the pattern fits completely into the roll – Fig. 2a, and when the material for the pattern must be combined because it exceeds the roll width – Fig. 2b. The *delt* value is defined as the remainder of dividing *Xw* by *W*, provided that $Xw > W$.

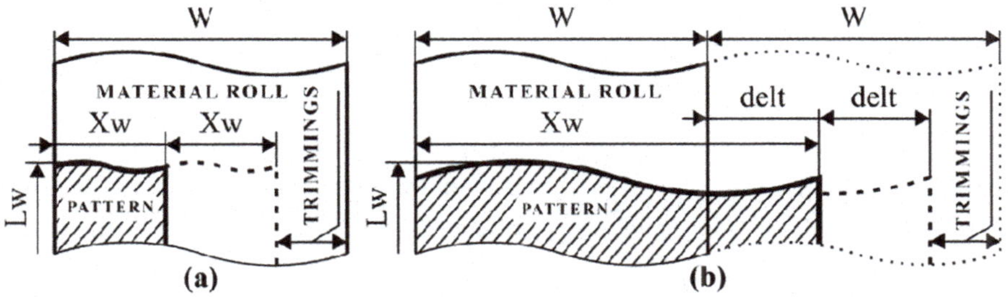

Fig. 2. Graphic examples of placing a part on a roll of material

The algorithm for calculating the distribution coefficient of the material consumption for the manufacture of patterns with a width *Xw* across a roll with a useful width *W* will be calculated according to the system of logical Eq. 1:

$$\begin{cases} Step\ 1: \ If\ W \times Xw = 0,\ then\ k = 0,\ else \\ Step\ 2: \ If\ Xw \leq W,\ then\ k = W/(Xw \times [W/Xw]),\ else \\ Step\ 3: \ If\ Rz = 0,\ then\ k = 0,\ else \\ Step\ 4: \ If\ (W \mod Xw) = 0,\ then\ k = 1,\ else \\ Step\ 5: \ k = (W \times [Xw/W] + W/[W/(Xw \mod W)])/Xw \end{cases} \quad (1)$$

where:

W – the useful roll width;
Xw – overall pattern size across the roll;

Rz – a permission to combine the material in patterns (0 – No, 1 – Yes);
k – the coefficient of the material distribution;
[ ] – the operation to determine the integer part of the division;
mod – the operation to determine the division remainder.

Equation 1. The algorithm of the linear coefficient of distribution of material consumption.

The calculation of the coefficient *k* consists of the following steps:

Step 1: *checking for zero values among W and Xw;*
Step 2: *calculating k if the size of Xw does not exceed W;*
Step 3: *verification of the permission to combine the material;*
Step 4: *definition of k if the Xw is a multiple of W;*
Step 5: *final calculation of k.*

Based on this algorithm for determining the distribution coefficient of the material consumption *k*, we trace its dependence on the set values *Xw* and *W*. According to this calculation method, we can set:
If $Xw \leq W$ then $k^{Rz=0} = k^{Rz=1} \geq 1$. The combination of the material within the boundaries of the pattern is not considered the principle of non-violation of the consumer value of the final product and not the creation of additional work for employees;
If $Xw > W$ then $k^{Rz=0} = 0$, a $k^{Rz=1} \geq 1$. The creation of the pattern is possible only by combining the material.

Let us analyze a chart of the dependence of the distribution coefficient of the material consumption *k* on the *Xw*/*W* ratio with the current condition of the permission for combining the material within the boundaries of the piece *Rz* = 1 – Fig. 3.

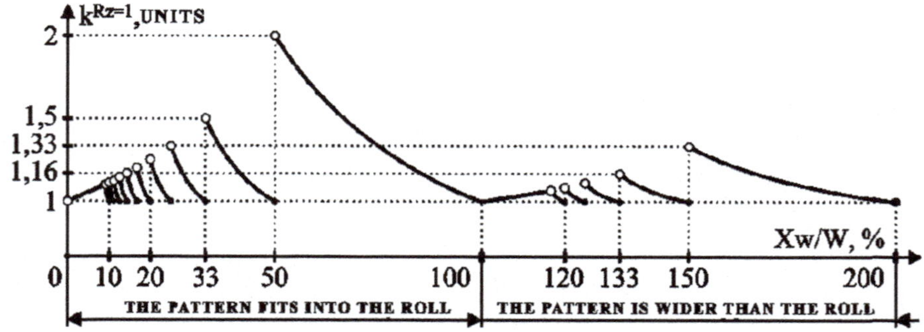

Fig. 3. The dependence chart of $k^{Rz=1}$ to Xw/W ratio

From the chart $k^{Rz=1}(Xw/W)$, it can be seen that the distribution coefficient of the material flow is non-linear and takes values equal to unity for the following points:

- The multiple ratios of the roll width W to pattern size Xw ($W\,mod\,Xw = 0$, at $0 < Xw \leq W$);
- The multiple ratios of the width of the roll W to the remainder of the division of the pattern exceeding the W roll ($W\,mod(Xw\,mod\,W) = 0$, at $Xw > W$).

These points of the multiple ratios correspond to the optimal material consumption. Within a gradual increase in the size Xw and passing through the multiple combination point, a sharp increase in the value of k occurs as a result of the appearance of a scrap of the material unsuitable for the use with this type of pattern. A further increase in Xw leads to a gradual approach to the next point of optimal material consumption. The highest value of k close to 2 is observed when Xw passes the value of W/2.

Based on the accepted statements about the shape of the patterns close to the rectangular and the lack of reference to their position on the roll, a reduction in the material consumption should be achieved by considering each of the overall dimensions of the patterns to the width of the roll separately. Let us rename the overall dimensions of the pattern as Xw_1 and Xw_2. Equation 2 should determine the minimum distribution coefficient of the material area consumption:

$$k_S^{Rz}(Xw_1, Xw_2, W) = \min\left(k^{Rz}(Xw_1, W), k^{Rz}(Xw_2, W)\right) \tag{2}$$

Equation 2. The equation for the minimum coefficient of the distribution of the flow material area.

Based on Eq. 2, it is possible to construct a chart K_S of the distribution coefficient of the consumption of the area of the material (Fig. 4). For the convenience of

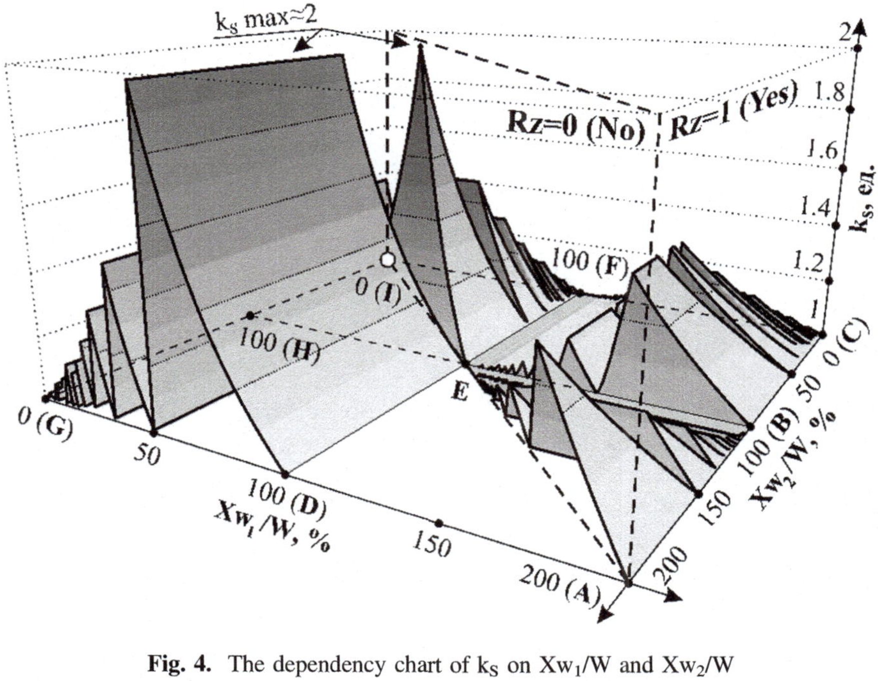

Fig. 4. The dependency chart of k_S on Xw_1/W and Xw_2/W

comparison, the diagonal connection of the charts is presented with the prohibition and permission to combine the material within the borders of the pattern.

The coordinate plane enclosed between the axes Xw_1/W и Xw_2/W is marked by the points A, B, C, D, E, F, G, H и I. The ACI figure reflects the area where the material alignment is allowed $Rz = 1(Yes)$; the AGI figure represents an area with a ban on the combination of material $Rz = 0(No)$. Such a combination of two charts is permissible since it does not interfere with the analysis of K_S indicators for a two-dimensional figure whose dimensions can be interchanged without distorting the results.

The chart shows the preservation of the connection of K_S with the multiplicity: W to Xw_1 – figure DFIG; W to Xw_2 – figure BHIC; W to $(Xw_1\, mod\, W)$ – figure ACFD; W to $(Xw_2\, mod\, W)$ – figure AGHB.

The EFIH area reflects the values at $0 < Xw_1\, and\, Xw_2 \leq W$, and the pattern does not exceed the useful roll width.

3 Results

For the optimal selection of the material consumption during the study, 14 FSRP assembly options were classified that combined 191 different combinations of placing patterns and their groups on the material. The search for 1,540 non-repeating combinations of sizes x, y, and z of the FSRP figure, carried out according to this algorithm with the moving over of each of the sizes starting from 0.05 W to W with a step of 0.05 W and the sifting out of options with the indicator $k_S > 2$, made it possible to create a chart (Fig. 5).

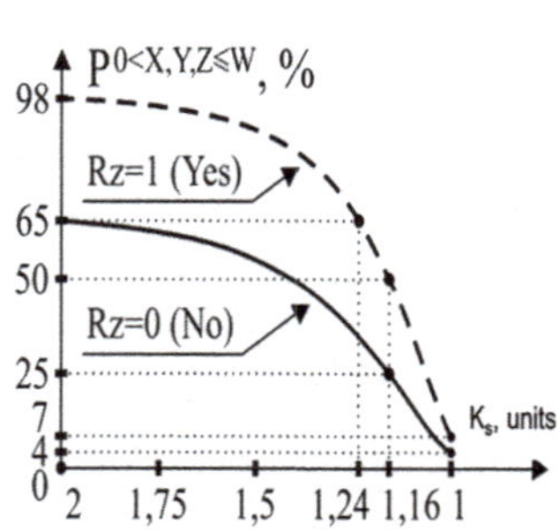

Fig. 5. The chart of the probability of order P fulfillment from k_S at $0 < x, y, z \leq W$

This chart reflects the results of calculations using the conditions for permission and the prohibition of crushing material within the boundaries of the pattern. According to this chart, the coefficient is $kS = 1,16$. According to this chart, the coefficient x corresponds to a probability of 25% at $Rz = 0$ and 50% at $Rz = 1$. It suggests that when choosing the optimal material consumption, a separate analysis should also be carried out to resolve the combination of material in the boundaries of the patterns for each layer: upper, lower, and filler.

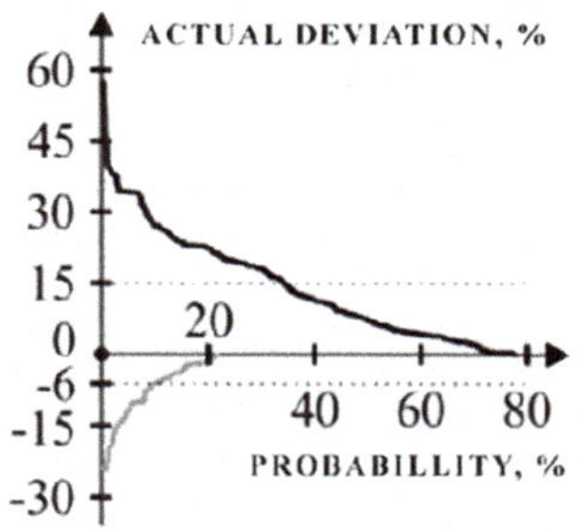

Fig. 6. The probability chart of deviation of actual data from the forecast calculations

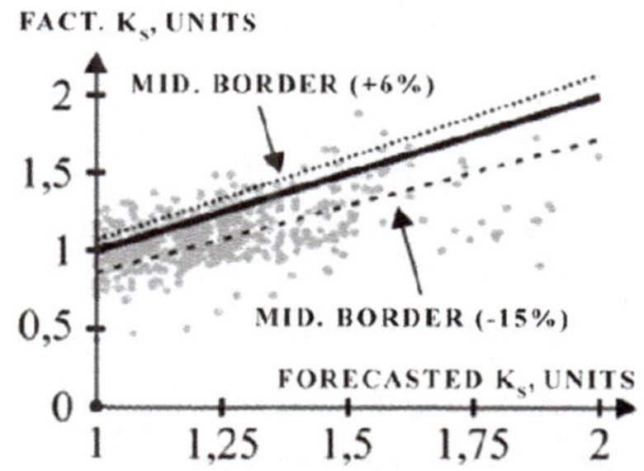

Fig. 7. The compliance chart of the actual material consumption with the forecast calculations

The analysis of more than 3.3 thousand separate reports on the material consumption allowed us to build charts of the probability of deviation of the actual material consumption and the correspondence of the actual data on the material consumption to the forecast calculations – Figs. 6 and 7.

From the figures above, it is seen that the average excess of the material consumption in 6% is a deviation in 10% of cases, and more than 30% of cases correspond to the economic consumption of material amounting to 15%. Also, Fig. 7 shows the points with the indicator $k_S < 1$, which means that the company actively applies the lean manufacturing method using the material residues.

4 Conclusion

This scientific work is of great importance for the application of quick calculations in conditions of limited order information. The simplicity of the algorithm and the convenience for the implementation, without tight binding with the software, makes it possible to use it on the software of various scales, creating the condition for implementing the concept of marketing interaction with the client.

Acknowledgment. The author would like to express his gratitude to the industrial enterprise LLC NPO *PermNefteGaz* for the opportunity to carry out this scientific research and to his supervisor Dr. Sc. Fedoseeva S. A. for the help in the scientific work.

References

Fayzrakhmanov, R.I.: Optimization of the process of cutting industrial materials according to the criterion of minimum material losses in the presence of technological limitations Ph.D. thesis, Ufa State Aviation Technical University, Ufa, Russia (2011)

Kablov, E.V., Finogeev, A.G.: Automation of calculations for cutting and rationing of materials by means of a system of technological processes integrated into CAD. Basic Res. **5**(2), 252–259 (2016)

Lovigin, A.: Overview of the Russian CAM market for 2016 (2017). https://sapr.ru/article/25359

Maslov, V.I., Shutkevich, L.N.: Improving the quality of processes for accounting and distribution of business waste metal sheets. Mod. Eng. Sci. Educ. **3**, 491–500 (2013)

Mukhacheva, E.A.: Rational cutting of industrial materials. Application in ACS. USSR: Engineering, Moscow (1984)

Murzakaev, R.T., Lyalin, D.A.: The compaction algorithm for a nesting map based on a two-dimensional gravitational simulation model. Mod. Sci. Actual Probl. Theory Pract. (Ser. Nat. Tech. Sci.) **9**(10), 34–41 (2013)

Smirnov, A.A., Kremlova, N.A.: Organizational and economic aspects of lean manufacturing at a machine-building enterprise. Sci. Tech. Statements SPbSPU (Econ.) **10**(3), 171–182 (2017)

Tolstobrov, D.A., Tolstobrova, N.A., Fedoseev, S.A.: The economic and mathematical model for calculation of preliminary production costs of multidimensional products for objects of protection having a complex geometric shape. In: IOP Conference Series: Materials Science and Engineering, vol. 483, p. 012065, 1 July 2019

Improving the System of State (Municipal) Procurement as a Factor for Stimulating Competition in the Digital Economy

Tatiana A. Guseva(✉), Elena A. Demenina, and Elena E. Smetanina

Orel State University named after I.S. Turgenev, Orel, Russia
guseva2002@rambler.ru, aegis-orel@yandex.ru, eleyna@list.ru

Abstract. The purpose of the study is to identify the mechanisms for improving procurement for state and municipal needs, serving as a driver of economic development. During the research, various approaches and methods were used, their application made it possible to research the object under the study in interdependencies and relationships with other objects, identify trends, generalize facts, and phenomena, and draw conclusions. The result of the study is the analysis of the current contract system and the proposals for improving existing legislation. The authors conclude that the proposed measures for the further improvement of the procurement procedures are designed to ensure competition, counteract corruption, and ultimately lead to the efficient spending of budget funds and the improvement of public administration.

Keywords: Competition · State (municipal) needs · Budget cost efficiency · Contract system

1 Introduction

Procurement to meet state (municipal) needs have been the focus of legislative attention for more than eight years. The improvement of the legal norms governing this area is carried out as a part of a large-scale reform of the public finance management system. One of the key areas for improvement is the provision of a competitive and transparent procurement system, including the use of digital technologies when concluding state and municipal contracts (Guseva et al. 2017; Guseva and Khovanskaya 2018).

The volume of state and municipal procurement carried out following the Federal Law of April 5, 2013 No. 44-FZ, "On the contract system in the field of procurement of goods, work, services to meet state and municipal needs" (State Duma of the Russian Federation 2011) increased in 2018. During the I–III quarters of 2018, 2.53 million contracts were concluded (in the I–III quarters of 2017 – 2.49 million contracts), which is 1.6% higher than in 2017. The total volume of the contracts concluded in 1–3 quarters of 2018 amounted to ₽ 4.7 trillion (in I–III quarters of 2017 – ₽ 4.02 trillion), which is 17% higher than the same indicator in 2017. The average price of the contract in the I–III quarters of 2018 amounted to about ₽ 1.85 million, which is 14.2% higher

A. V. Bogoviz and Y. V. Ragulina (Eds.): ISCI 2019, LNNS 115, pp. 253–258, 2020.
https://doi.org/10.1007/978-3-030-40749-0_30

than the same indicator in 2017 (₽ 1.62 million) (Ministry of Finance of the Russian Federation 2018). Given the above, anti-corruption management and competition stimulation are urgent tasks.

The use of digital technologies in public and municipal procurement will be an important step in improving the efficiency of the budget expenditures, which, as in all previous years, will be implemented, including the use of the best international practices. Further improvement of public finance management, taking into account international best practices, is associated with the following:

- the implementation and ongoing review of budget expenditures;
- the participation of the citizen in the budget process;
- the development of a system of long-term (strategic) planning of budget investments;
- the increase in the completeness and quality of financial reporting of the public sector;
- the digitalization of the public administration system and the budget process (using blockchain technology, smart contracts). Given that in current conditions, the circulation of information takes on increasingly high-tech forms, and for the secure transfer of large amounts of data, the most suitable means is blockchain technology.

The relevance of creating and implementing measures to improve the efficiency of the public finance management system is dictated by the desire to achieve those goals that are indicated in the strategic state planning documents (Government of the Russian Federation 2019).

2 Materials and Method

Both general and private scientific methods of cognition were used as the methodological basis of the study. The systematic approach is used as an interdisciplinary, general, scientific, methodological concept. When applying this methodology, the system properties of the studied object are identified. The systematic approach and the general systems theory were used, which take into account the subject, scope, and interdisciplinary nature of the study, as well as the possible limits of their application for studying legal phenomena and, in particular, legislation.

The study also used the method of analyzing legal documents, particularly regulatory acts issued by various government bodies of the Russian Federation. The application of this method made it possible to obtain information on legal regulation in the research field formed by public authorities related to both the legislative and executive branches.

With the use of the modeling method, the proposals have been developed to improve the use of the catalog of goods, works, and services in the Unified Information System in procurement. The use of this method was expressed in the formation of ideas about the role and importance of the catalog, taking into account the conditions and characteristics proposed by the authors.

3 Results

From 2017 to 2018, the Russian Federation made a transition to a new stage in the development of the contract system, whose key features follow:

- in order to ensure maximum openness, all-state (municipal) purchases have been electronically transmitted;
- only those electronic sites that meet the established requirements (primary and secondary) are functioning, including those relating to the minimum volume of purchases at these sites;
- procurement information is unified by using the catalog of goods, works, and services, which is placed in a single information system in the field of procurement;
- access to procurement for small businesses is simplified;
- a single register of procurement participants is maintained in a single procurement information system.

4 Discussion

The accumulated experience in the functioning of the contract system allows the distinction of the following areas of its further development.

1. First of all, the procedures of procurement themselves need to be improved: there are too many of them, and they are excessively complicated. Some requirements for the content of applications are redundant. Technical errors are possible during their preparation. This allows the customer to act in bad faith, thereby affecting the procurement outcome. In addition, individual procurement methods help the customer to determine subjective criteria for evaluating tenders. These techniques allow us to get away from the competitive procedures and conclude a contract with a single supplier.

The competitive procedures themselves, which in some cases have lost significant distinguishing features, need further improvements.

2. A significant share of purchases (up to 30%) still come from single suppliers (service providers, contractors), which makes it impossible for small businesses to offer purchases at lower prices (Ministry of Finance of the Russian Federation 2018).

The Unified Trade Aggregator "Birch" is intended to reduce the number of contracts concluded with a single supplier (RT-Project Technologies Joint-Stock Company 2019). It operates on the principle of an electronic store. The use of this electronic resource allows establishing suppliers (contractors, service providers) on a competitive basis by preplacing procurement participants with applications, offers, and the ability to quickly choose between their offers with minimal prices.

3. To exclude abuses on the part of customers, which are fitting the procurement subject to specific suppliers, it is advisable to replenish the catalog of goods, works, and services, which contains a description of typical objects of procurement, in the Unified Information System for Procurement. It will allow us to offer interchangeable goods (works, services) in the supply and solve the indicated problem of abuse.

In addition, maintaining the catalog and its timely replenishment will allow the automatic control and analysis of the procurement information, namely: to evaluate and compare the procurement activities of customers, to calculate the relative prices of comparable goods, works, and services, to identify cases of description of the procurement object for unreasonable restrictions on access to procurement, and, accordingly, corruption.

When using the catalog in the field of procurement, it is advisable to compare the information from the catalog with the other information systems, including the state industry information system and a single structured reference-classification of drugs. It seems important to ensure the possibility of the formation and use by the participants of the contract information system on the comparable purchase prices, as a result of the application of the catalog of goods, works, and services, and, also, to establish cases of phased prohibitions on the purchase of goods, works, and services in excess of a certain price level and (or) characteristics, stipulated by the regulation rules.

4. An important area of development is the improvement of standardization rules, which provide requirements for the qualitative and quantitative characteristics of goods (works, services), excluding their purchase with excessive consumer properties. The current rules do not apply to all purchases for state (municipal) needs, they are not integrated with other procurement institutions, there is no universal set of parameters to be standardized (characteristics, price, volume), and there is no list of the mandatory standardization of the maximum price. These problems lead to procurement cases in terms of budget spending efficiency that are very curious and difficult to explain.

The proposals for improving the standardization institute differ in diametrically opposite points of view: from the need to establish a single list of goods, works, and services to be standardized; to link it to the catalog of goods, works, and services; to the fact that the standardization institution imposes redundant functions on the customer (Ermakova 2016).

5. The improvement of anti-corruption effectiveness in the procurement sector is provided by the Federal Antimonopoly Service (hereinafter, the FAS Russia).

Nowadays, there are various options for special software modules that allow automatic submission of the price at electronic auctions on behalf of the auction participant up to the price limit set by the participant. The use of such robots is an absolutely legal practice. Recently, however, the FAS Russia has increasingly come across both the use of legal auction robots for illegal purposes—to implement anti-competitive agreements—and the development and use of illegal auction robots; that is, software tools deliberately created to implement a cartel or other anticompetitive agreement on bidding. The markets that have been embraced by the "digital" conspiracies are completely different. These include medicines with consumables, repair of heating network sections, sale of property seized in pursuance of judicial acts or acts of other bodies, and removal of municipal solid waste (Tenishev 2019).

FAS Russia has given explanations related to the functionality features of such programs and containing recommendations on the identification and suppression of their illegal use by bidders. So, in 2017, the central office of FAS Russia identified a cartel of participants for the procurement of medical supplies, which used auction robots to maintain the maximum price automatically, while the total amount of initial prices amounted to more than ₽ 145 million (FAS Russia 2017).

About six cases of maintaining prices at the tenders using automated auction robots have also been considered in the territorial bodies of FAS Russia (Perm, Yaroslavl, Leningrad, Bashkortostan, Murmansk, and Udmurt Offices of FAS Russia) to date.

The above has led to the development of the so-called *big digital cat*—a set of software tools based on blockchain technology, which allows the identification of signs of anticompetitive agreements at the tenders and helps form an evidence base confirming the conclusion and implementation of the relevant conspiracy, using both open and closed information sources (FAS Russia 2018).

The corresponding program provides systematic and automatic identification of cartels and other anticompetitive agreements online; allows more efficient use of the databases available at various federal executive bodies (including the Federal Tax Service, the Federal Treasury, the Ministry of Digital Development, Telecommunications and Mass Media of the Russian Federation, the Federal Financial Monitoring Service and others) without additional budgetary expenses; and reduces the latency of cartels and other anticompetitive agreements.

In the long term, as predicted by the FAS Russia, there should be a decrease in the absolute number of cartels subject to the active development of the antitrust compliance institution and the adoption of laws aimed at increasing the effectiveness of the fight against cartels, as well as increasing the interaction rate with law enforcement.

Smart contracts seem to play an essential role in achieving the goal of increasing transparency in the procedures of public and municipal procurements. The use of digital technologies in concluding, securing, and executing state and municipal contracts will minimize unreasonable refusal to fulfill contractual obligations and accelerate the exercise of rights and obligations of the parties, bypassing the stage of human decision-making, and avoiding certain actions depending on the will of the parties to the contract. Before talking about the legal content of the *smart contract*, this concept should be defined. It is proposed to regard legal contracts or elements of legal contracts concluded in electronic form as smart contracts when the fulfillment of obligations is automated and provided by a computer program (Yurasov and Pozdnyakov 2017). An example of such a performance, which may be in demand in the system of state and municipal procurement, is the terms of the contract described using computer code. The code allows the program to take action, i.e., to write off money from the party's account to a contract when information on the performance of an obligation by another party is received.

Thus, the digitalization of public (municipal) procurement is not only a way to objectify the activities of the contract parties, which increases the opposition to this sphere of various types of offenses, but also an important determining factor in economic development.

5 Conclusion

Therefore, such measures as:

- the optimization of the competitive procedures themselves;
- the transmission of purchases, even of small volumes, into digital format (the bidder aggregator "Berezka");

- the creation of a single cycle - coordination of procurement plans with the catalog of goods, works, and services, standard contracts;
- the conclusion of smart contracts;
- the creation of appropriate programs providing systematic and automatic identification of cartels and other anticompetitive agreements online, and several other measures designed to increase the efficiency and transparency of the procurement, increase competition, and reduce corruption in the economy.

References

Ermakova, A.V. (ed.): Proceedings from: XI All-Russian Practical Conference-Seminar "State and Municipal Procurement". PRINT PRO LLC., Moscow (2016)

FAS Russia: The case of a cartel implemented with the help of auction robots (2017). https://fas.gov.ru/news/23274

FAS Russia: FAS creates a web service "big digital cat" (2018). https://fas.gov.ru/news/26154

Government of the Russian Federation: Order "On approval of the concept of increasing the efficiency of budget expenditures in 2019–2024 (2019)". http://government.ru/docs/35598/. Accessed 31 Jan 2019 No. 117-r

Guseva, T.A., Khovanskaya, E.A. (eds.): Proceedings from CIEDR 2018: The International Scientific and Practical Conference "Contemporary Issues of Economic Development of Russia: Challenges and Opportunities" (2018). https://dx.doi.org/10.15405/epsbs.2019.04.36

Guseva, T.A., Astafichev, P.A., Sizov, V.E.: Legal aspects of the protection of national interests when executing state (municipal) procurements in Russia. Information (Japan) **20**(10), 7307–7318 (2017)

Joint Stock Company: RT-Design Technologies. Rules for the operation of a single trade aggregator (2019). https://agregatoreat.ru/materials/regulations

Ministry of Finance of the Russian Federation: Statistical indicators characterizing the results of procurement to meet state and municipal needs according to the results of I-III quarters of 2018 (2018). https://www.minfin.ru/ru/perfomance/contracts/purchases/

State Duma of the Russian Federation: Federal Law On Procurement of Goods, Work, Services by Certain Types of Legal Entities (2011). http://www.consultant.ru/cons/cgi/online.cgi?req=doc&base=LAW&n=312202&fld=134&dst=1000000001,0&rnd=0.1772995394005079#04230990184269232. Accessed 8 July 2011 No. 223-FZ

Tenishev, A.P.: Anticartel investigations: the main results and work plans of the FAS Russia for 2019 (2019). https://www.consultant.ru/law/interview/tenishev5/

Yurasov, M.Yu., Pozdnyakov, D.A.: Smart contract and the prospects for its legal regulation in the era of blockchain technology (2017). https://zakon.ru/blog/2017/10/9/smart-kontrakt_i_perspektivy_ego_pravovogo_regulirovaniya_v_epohu_tehnologii_blokchejn#_ftn6

Target Costing System as a Tool for Improving the Competitiveness of an Economic Entity

Marina I. Samogorodskaya, Irina A. Bejnar, and Tatiana S. Narolina(✉)

Voronezh State Technical University, Voronezh, Russia
marta17@yandex.ru, beinar@mail.ru, narolina@inbox.ru

Abstract. The paper focuses on the main features of the target costing concept, its differences from the traditional system of cost accounting, and the advantages and disadvantages that appear during its practical application. The authors established that the use of the target costing system provides an increase in the profitability of a manufactured product due to the formation of target costs. The conducted studies have shown that this method is most effective at the stage of the development and design of an innovative product. The authors calculated the target cost based on the target price and the target profit.

Keywords: Cost management · Cost accounting · Target costs · Target price · Target profit

1 Introduction

Since the 1990s, the priority task of the Russian economy has been to increase the competitiveness of domestic producers. Over the years, a lot of mechanisms, methods, and tools have been developed in order to solve this problem. Individual enterprises have achieved some success in this area. However, despite the constant and comprehensive attention paid to this problem, the current state of the domestic national economy irrefutably indicates that it is still far from being resolved. The severity of the current situation is also confirmed by a significant number of publications and discussions on these topics.

In our opinion, the main reason for difficulties lies in the wrong choice of priorities by Russian manufacturers. The main goal of economic activity in a market economy is to obtain an acceptable profit. At the same time, we will abandon the widespread idea of obtaining the maximum possible profit because qualitative assessments are a relative concept. Today, the maximum profit corresponds to one level, while the next day, it may change due to environmental factors. The desire to earn as much as possible "here and now" often leads manufacturers to pursue their commercial interests in opposition to the interests of consumers. The typical problems that Russian consumers face are inadequate quality, difficulties of after-sales service, and excessive pricing.

At the same time, considerable experience has been accumulated from the efficient functioning of manufacturers without a detriment to consumers. In particular, the practice of Japanese enterprises indicates that the most competitive and, as a result,

A. V. Bogoviz and Y. V. Ragulina (Eds.): ISCI 2019, LNNS 115, pp. 259–266, 2020.
https://doi.org/10.1007/978-3-030-40749-0_31

successful enterprises are those where consumer interests are paramount. Whereas in Western Europe, the USA, and Russia, manufacturers solve the problem of optimizing the costs of remedial work by conscientiously dooming certain consumers to purchase low-quality products, in Japan, most enterprises operate on the "zero defects" principle. The statistics are inexorable evidence of the effectiveness of the Japanese strategy. The level of quality cost in the production cost in the United States averages to 25%. In Japan, it is 13% (Samogorodskaya 2014, 2011).

In addition to the product quality criteria in consumer choice, the price factor is also of key importance. In Russia, the pricing system for goods and services is cost-based. The costs to the manufacturer for the production and sale of products are the priority when setting prices, which can lead to the point of absurdity. Losses from defective products are included in the actual cost and are laid down in the product's price. In other words, manufacturers in Russia pass all their problems on consumers. Trying to achieve competitiveness with such an attitude toward consumers is Sisyphean labor. In the conditions of the Russian market, where manufacturers use an analytical method of calculating the cost that is regulated by the domestic accounting system, the price competition is minimized. For the production of the same-type products, the standard types of raw materials, fuel, and energy resources are used, which are purchased at approximately the same price from the suppliers. The tariff wage system used in most Russian enterprises also provides approximately the same level of wage costs.

As a result, prices for finished products do not differ much from one manufacturer to another. So, the cost-based pricing system is convenient for domestic entrepreneurs and fully meets their interests. However, consumer interests here are infringed. Therefore, we believe that in order to develop competition among Russian manufacturers, it is necessary to change the approach to pricing and, as a consequence, to a cost management system for production and sales of products.

2 Materials and Method

The extensive experience in the application of various cost management systems is amassed in the world practice. Among them are both well known, variously modified, and used around the world, as well as recently developed innovative systems that provide the ability to influence cost accounting at different pricing stages (Beynar 2009). In the course of the study of the leading cost management systems (Schegoleva and Khvostikova 2014; Egorova and Yudanova 2015; Sarzhevskaya and Peretyatko 2016; Kazachenko 2015; Ermakova 2015), the authors formulated priority characteristics that make it possible to assess the applicability of these systems in various conditions (Table 1).

Table 1. The Key features of leading cost management systems.

Characteristic	Standard costing	Absorption costing	Direct costing	Activity-based costing	Target costing	Kaizen costing	Costs – volume - profit	Benchmarking	Cost-killing	Life cycle cost	Chain of consumer values
Application universality	+		+			+					
Possibility of application in a specific financial and economic situation								+	+	+	
Availability of standard information requirements	+	+	+								
Additional information requirements								+		+	
The need for employee motivation						+			+		
Entering new markets					+		+	+		+	+
Consolidation in existing markets						+		+			
Short-term profit growth				+					+		
Long-term profit growth					+	+					
Improving the competitiveness of products due to low costs and low prices					+			+			+
Launching new products and introducing them to a new or developed market					+					+	
Cost reduction				+			+		+		
Assuring the targeting costs level					+			+			+
Ensuring the excess of income over costs in the long run				+	+					+	
Efficient use of the available resources	+									+	+
Cost management as a priority in enterprise management					+	+				+	+
Cost management plays a supporting role	+	+	+	+			+	+	+		
Cost management is considered along with other management objects	+	+	+	+		+	+	+	+		

A comparative analysis allows us to identify several areas of the development and application of these systems:

(1) universal (normative in the history of domestic production) characterized by applicability at enterprises of any type with standard information requirements (Standard Costing, Absorption Costing, Direct Costing);
(2) specialized, providing the possibility of application in non-standard situations with additional requirements for information support and cost reduction in the short-term (Benchmarking, Cost-Killing, Life Cycle Cost);
(3) innovative, providing not only the development of new markets due to the priority management of the target costs level, but also the orientation of the enterprise on a long-term profit (Target Costing, Kaizen Costing, Chain of Consumer Values).

Of all of the analyzed systems, only the Target Costing system satisfies the conditions necessary for increasing the competitiveness of the enterprise and products (the price factor, the launch of innovative products in new markets, ensuring long-term profit growth).

The Target Costing concept, developed by Japanese specialists in production management and management accounting in the 1960s, and successfully used in practice for almost 40 years, is aimed at the development and implementation of a target pricing oriented to market realities. It is represented by a holistic multi-aspect corporate strategy, including, along with the planning of innovative production, control over the formation of production costs and the reduction of total costs (Sakurai 1989).

3 Results

Innovative products, as shown by recently changed business conditions, become a determining factor in the success of enterprise competitiveness. Thus, the functioning in a changing situation of modern markets and progressively developing technologies makes a change in the course of design (construction, development) wholly justified. The main stages of designing new products as a part of the application of the Target Costing concept take on the reverse order, from calculating costs to finding technical (technological) solutions (Kerimov 2007):

(1) the market analysis for identifying consumer preferences;
(2) the analysis of the possibilities of creating (changing) products following i.1;
(3) the establishment of the upper maximum allowable price and marginal cost;
(4) the identification of the ways to minimize costs;
(5) the search for technical (technological, organizational) solutions.

The price-formation of the toothpaste is proposed as an example of the use of the Target Costing system in the context of Russian realities.

In 2015, the company "SPLAT-Cosmetics," located in the village of Okulovka, the Novgorod region, launched the innovative toothpaste brand *Biomed*. The priority task was to develop the most harmless product that does not contain the components typical of most kinds of toothpaste (fluorine, triclosan, chlorhexidine). Also, the new toothpaste should be focused on caries prevention due to the enamel remineralization and strengthening.

Since Russia was experiencing the aftereffects of the economic crisis and a significant increase in prices in 2015, it was possible to offer new products to a reasonably saturated market only at relatively low prices. In the course of marketing research preceding the development of the innovation, the priority consumer requirements for a new product and an acceptable price level for natural toothpaste for consumers were identified.

After systematizing and analyzing the information collected during the interviewing of potential buyers, the target price for Biomed Superwhite toothpaste was set at ₽ 70 per tube. In order to achieve the goals set by SPLAT-Cosmetics in terms of conquering a particular market segment, the profitability of the new product was initially set at 12%.

The consumer preferences identified in the course of the marketing research are presented in Table 2, taking into account their importance for the formation of the target price of innovation.

Table 2. The influence of consumer preferences on the target price of the product.

Preferences	Significance	Target cost, ₽.
Natural composition	0.28	17.5
Gentle whitening	0.21	13.12
Caries protection	0.19	11.88
Fresh breath	0.14	8.75
Environmentally friendly packaging	0.12	7.5
Fast tooth cleansing	0.06	3.75
Total	1.0	62.5

The design cost for the production and sale of the Biomed Superwhite toothpaste, calculated at the stage of research and development, was ₽ 75.4. The projected sales volume of the new product amounted to 3 million units. Thus, the potential proceeds from the sale based on the cost of the project should be ₽ 253,344,000, with a potential profit of ₽ 27,244,000.

The structure of cost elements tied to the established preferences of consumers is presented in Table 3.

Table 3. The structure of the design cost.

Preferences	Target cost, ₽.	Cost elements	Design cost, ₽.	Deviation, ₽.
Natural composition	17.5	Emulsifier Polysorbate 80	12.24	5.26
Gentle whitening	13.12	Abrasive components: silicic acid, soda, dicalcium phosphate dihydrate Plant Enzymes: Papain, Bromelain	22.33	−9.53
Caries protection	11.88	Calcium hydroxyapatite, cinnamon extracts, plantain and birch leaves, coconut oil, thyme essential oil	7.93	3.95
Fresh breath	8.75	L-Arginine; essential oils and antiseptics, vitamins E, K, PP, carotene	11.42	−2.67
Environmentally friendly packaging	7.5	Biodegradable plastic tube	14.97	−7.47
Fast tooth cleansing	3.75	Sodium bicarbonate and other salts	6.51	−2.76
Total	62.5		75.4	

Next, we need to compare the target cost with the design, and highlight those consumer preferences for which the design cost exceeds the target. The negative cost difference for "gentle whitening" is almost fully offset by savings in "natural composition" and "caries protection." Since the criterion of "quick tooth cleansing" is not significantly important for consumers, technologists have proposed to reduce the specific gravity of salts used in the composition of the product. It was decided to adjust the composition in favor of the use of cheaper essential oils and antiseptics to reduce costs by the criterion "fresh breath." The biggest problem was the negative cost difference by the "environmentally friendly packaging" criterion. As a result of the analysis of the supplier market, a manufacturer of biodegradable plastic tubes, which agreed to a packaging price of ₽ 8 if the order is increased by 40%, was found. After conducting repeated marketing research, the management of SPLAT-Cosmetics decided to take this step, since there was a chance of an increase in demand for the new product due to its cost-cutting.

As a result of the recalculation and approvals, the adjusted target cost of the Biomed Superwhite toothpaste was established (Table 4).

Table 4. The calculation of the adjusted target cost.

Preferences	Target cost, ₽.	Design cost, ₽.	Adjusted target cost, ₽.
Natural composition	17.5	12.24	12.24
Gentle whitening	13.12	22.33	22.33
Caries protection	11.88	7.93	7.93
Fresh breath	8.75	11.42	8.75
Environmentally friendly packaging	7.5	14.97	8.0
Fast tooth cleansing	3.75	6.51	3.75
Total	62.5	75.4	63.0

A slight excess of the initially set target cost, in this case, is not critical. The amount of revenue from the sale of the product at the adjusted target price will be ₽ 296,352,000, and the potential profit will be ₽ 31,752,000. Thus, the value of the gain in profit from the product sale as a result of changes in the system of assessing the production costs and products sale will be ₽ 4,608,000.

The above example clearly illustrates the benefits of the Target Costing concept over the traditional costing approach.

4 Discussion

Among the undoubted advantages of the Target Costing method is the possibility of the transition to strategic planning and the regulation of the processes of generating costs and profits of the enterprise, which is especially important for high-tech industries, as well as adjusting the process of developing new products following consumer requests.

However, the introduction of a new cost accounting system requires a significant restructuring of intra-organizational priorities because there will be a necessity not only for interconnecting the long-term development strategies and short-term marketing goals but also, most likely, for radically changing the entire management system of an enterprise (Beynar 2015; Pakhomova and Narolina 2017).

The analysis of the practice of applying the Target Costing method in foreign enterprises (Ansari et al. 1997; Lisitsin and Zozulya 2014; Kuvshinov 2014) shows that the constant monitoring of market conditions and prospects for demand for developed products should be taken as a priority criteria. At the same time, one should be quite critical of the continuity of the design process, because it is not always possible to find the optimal ratio between the subjective needs of consumers, specific market conditions, and objective opportunities for product development; all of this must be taken into account.

It seems, however, that the most controversial requirement for the mentality of a Russian entrepreneur is the need to establish close interaction between various structural divisions and employees. In other words, a successful enterprise should be a single system and work as a whole, with permanent accounting and cost control adopted in the everyday practice of business processes, a sufficient level of employee motivation without undue pressure, and an increase in labor productivity during the development process.

5 Conclusion

The task of increasing competitiveness is a known quantity for any business entity. However, even though at present much attention is paid to this problem in domestic enterprises, this does not always lead to the achievement of the desired results. Increased competitiveness can be gained by achieving certain competitive advantages. One of these advantages is provided by the "reasonable price" strategy, which follows the Target Costing concept. The application of this concept in the practice of an enterprise initially sets up the production system for the manufacturing of products with optimal costs. The result is a "reasonable" price for the consumer, substantial revenue, and significant financial results. The efforts spent on finding ways and means to reduce the project cost to bring it to the "target" level will pay off later due to the quick sale of the finished products. The accelerated launch of a new product to the market, the lack of overstocking of warehouses, the teamwork of all departments of the enterprise, and optimal costs are the obvious benefits of using Target Costing in the enterprise. It should be noted that this system of cost management, in fact, does not contradict the domestic accounting system, which makes it most acceptable for domestic producers, unlike other cost management systems under consideration.

References

Ansari, S.L., Bell, J., Cypher, J., Dears, P., Dutton, J., Ferguson, M., Hallin, K., Marx, C., Ross, C., Zampino, P.: Target Costing: The Next Frontier in Strategic Cost Management. McGraw-Hill, New York (1997)

Beynar, I.A.: The main directions for reducing the cost of information security. Inf. Secur. **12**(2), 297–300 (2009)

Beynar, I.A.: Methods for assessing the feasibility of the cost of providing information security systems. Compet. Innov. Finance **2**, 44–48 (2015)

Egorova, S.E., Yudanova, L.A.: Comparative analysis of new methods and systems for cost accounting and calculation of production costs. Bull. Pskov State Univ. **2**, 94–106 (2015)

Ermakova, N.V.: Methods to stimulate cost reduction: theoretical and practical aspects. Tavrichesky Sci. Obs. **2**, 156–158 (2015). http://elibrary.ru/download/95980050.pdf

Kazachenko, A.N.: The possibilities of cost management methods and conditions for their use (2015). www.cfin.ru/management/finance/cost/comparison.shtml

Kerimov, V.E.: Target - costing as a tool for determining the target cost. Finance: Theory Pract. **4**, 70–77 (2007)

Kuvshinov, M.S.: Analysis of the conformity of management methods to current management tasks. Int. Account. **23**, 14–23 (2014)

Lisitsin, V.G., Zozulya, D.M.: Concepts of cost management in a strategic aspect. Sci. Educ. Econ. Econ. Entrepreneurship Law Manag. **5**(48), 26–32 (2014)

Pakhomova, Yu.V., Narolina, T.S.: The concept of innovation management of enterprise development. In: Proceedings of the Seventh International Scientific and Practical Conference "Management of the Socio-Economic Development of Regions: Problems and Their Solutions". University Book Closed Joint-Stock Company, Kursk, Russia (2017)

Sakurai, M.: Target costing and how to use it. J. Manag. Summer, 39–50 (1989). https://maaw.info/JournalofCostManagement-2.htm

Samogorodskaya, M.I.: Economic tools for quality management processes. Bull. Voronezh State Tech. Univ. **7**(11–3), 90–94 (2011)

Samogorodskaya, M.I.: Economic aspects of a quality management system. Voronezh State Technical University, Voronezh, Russia (2014)

Sarzhevskaya, V.V., Peretyatko, E.S.: Comparative characteristics of methods for calculating the cost of finished products: abc – costing and target - costing. New Sci.: Curr. State Ways Dev. **12–1**, 183–187 (2016)

Schegolev, A.S., Khvostikova, V.A.: Modern methods of accounting and cost management in industrial enterprises. Econominfo **22**, 37–39 (2014)

Khromova, I.I., Akhidzhak, S.A., Matsko, A.S.: Methodical bases of management accounting in the conditions of application of modern calculation systems. Naukovedenie **9**, 1 (2017). http://naukovedenie.ru/PDF/51EVN117.pdfFloatPlacement>

Management and Competitiveness: Theoretical Foundations and Practical Applications

A Balanced Scorecard in the Company Management Strategy in Contemporary Conditions

Darina A. Sizova(✉), Tatiana V. Sizova, and Elena S. Adulova

Plekhanov Russian University of Economics, Moscow, Russia
Darina3@yandex.ru, Stv-60@mail.ru,
E.S.Adulova@rostec.ru

Abstract. The relevance of the presented research is due to the development of strategic goals and the search for tools to ensure their achievement by building a balanced scorecard. Today, a balanced scorecard is a contemporary tool for the strategic management of a particular enterprise, consisting of an optimally selected set of financial and non-financial indicators, which eventually allows evaluating its effectiveness in all aspects of the activity. This system is closely related to the long-term strategic goals of the company, and it is aimed at making management decisions for their effective achievement that ensure business management integrity. The goal of the study set for the authors was to study the feasibility of creating and implementing a balanced scorecard in the organization for effective company management. In the process of studying the issue, the authors of this article used such methods of scientific knowledge as the method of generalizations and comparisons, statistical and factor analysis, and others. The authors focused on the problems associated with the introduction of a balanced scorecard in organizations. According to the results of the analysis, the authors obtained the following conclusions. The authors define the essence and purpose of the balanced scorecard in the strategic management of the company are defined. The relationship between the balanced scorecard and the strategic goals of the company is also established.

Keywords: Balanced scorecard · Strategic management · Financial performance

1 Introduction

In contemporary reality, there are several designs for setting the company's strategy, and one of them is a balanced scorecard. This system covers the following main areas in which the strategic goal setting is possible: finance, customers, business processes, employees.

The effectiveness of using a balanced scorecard has been proved with the case of companies already implementing it. At the same time, not every company manages to develop and implement this system at home. They face complex operations during development and many structural elements (Chursin et al. 2017). Building a balanced

A. V. Bogoviz and Y. V. Ragulina (Eds.): ISCI 2019, LNNS 115, pp. 269–278, 2020.
https://doi.org/10.1007/978-3-030-40749-0_32

scorecard in a company requires to conduct a scientific analysis of the company's business processes on an ongoing basis and to develop practical models. Today, methodologies for building a balanced scorecard are continually being improved and continue to evolve.

2 Materials and Method

In the process of researching the issue of creating a balanced scorecard for effective company management, such methods of scientific knowledge are used as the method of generalizations and comparisons, statistical and factor analysis. The paper relies on general provisions of the specialists in the field of economics and project experience. When analyzing the literature on the subject, we analyzed a large number of research papers. The foundations of a balanced scorecard are described in the works of its creators, namely Norton, D. and Kaplan, R. Also, the works of the following domestic experts were used: A. V. Sorokina, D. A. Sizova, T. V. Sizova, V. G. Kandalintseva, E. M. Mikhailova, and others.

3 Results

With the advent of the information age (20th century), the ability to manage the scale and volume of production, quickly and economically introduce new technologies, went by the wayside (Ragulina et al. 2019). In this period, for success, companies must have the ability to correctly and fully use their intangible assets. They will help the company to develop relations with existing customers and establish contact with new ones, to introduce an information system, to develop a motivation program for employees and mobilize them, to determine the target segment and introduce new products and services for it, as well as to produce goods and services of higher quality and lower cost.

The activities of companies in the information age are mainly focused on the following aspects (Fig. 1):

- Customer focus;
- Business process integration;
- Interaction with suppliers and customers;
- Innovation;
- Global-scale;
- Intellectual workers (Caplan and Norton 2017).

In Russian realities, with rapidly changing market conditions and the entry of foreign companies into the market, many enterprises, even having strategic goals, try to control and achieve only operational indicators set for the short term.

Many companies prescribe strategic goals in their accounting policies only as one of the necessary items. As a result, many middle managers do not assume that the company has long-term plans. For them, the priority is only indicators of operating activities, and there is no connection between the short-term and long-term goals of the company.

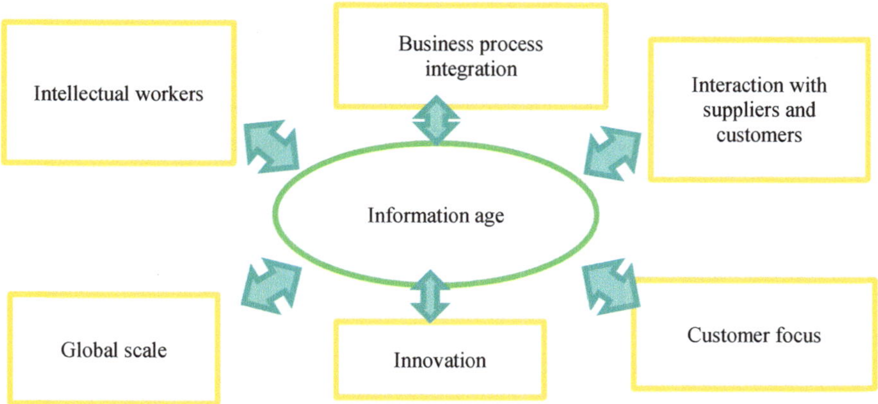

Fig. 1. Aspects of the information age. *Source*: (Kandalintseva 2015).

If a company seeks long-term work and development, strategic goals become essential. Contemporary top managers try to increase the market value of their enterprises, but it consists of the achievements of long-term plans, i.e., the obtained indicators should reflect the degree of achievement of strategic goals. Many companies use only financial indicators as the main criteria for evaluating performance. Even the performance assessment of employees is assessed in "key performance indicators" (KPIs), which are financial in the very end. These indicators may include the following: planned sales figures, setting an upper limit on the sum of costs of operating expenses, and others. It is incorrect to use exclusively financial indicators as a method of evaluating the activities of a company during the information age (Sizova et al. 2017). The information era in the economy and activities of the organization is aimed at increasing its value by investing in its personnel, suppliers, and customers of the company, to create a positive business reputation. Financial indicators are aimed at analyzing the past period and identifying problem areas or sectors. Nevertheless, they do not show the direction of the company and do not tell what needs to be changed in the business process or company employees to improve performance and increase productivity. The target values of financial indicators that the company wants to achieve at a certain point in time can be predicted and set, but it is impossible to predict the activities of the company on the basis of financial indicators and in what position it will be at a particular point in time (Sizova and Sizova 2017a).

In today's highly competitive environment, management needs an extensive and structured set of tools to manage a large and complex company. These indicators should fully reflect the state of the "environment" and the "climate within the company." Under current conditions, a company needs to understand the goals that must be set for the enterprise and translate them into tasks for each structural unit. The methods to achieve and fulfill them require attention. One of the tools that reflect the effectiveness of the company may be a balanced scorecard (BSC).

A balanced scorecard (BSC) is a system for the strategic management of an enterprise, consisting of an optimally selected set of financial and non-financial indicators, which allows evaluating its effectiveness in all aspects of activity (Loshilina 2007). A balanced scorecard is built based on long-term strategic goals and reflects the stages of their achievement. It is continually modified and adjusted to the needs and individual requirements of the company. However, its structure, development requirements, and strict relationship with the strategic goals of the company remain unchanged.

After research conducted by N. Kaplan and D. Norton, the introduction and development of a balanced scorecard among American companies began. The successful spread of a balanced scorecard in the U.S. companies and a positive response about its work (Xerox, Metro Bank, GM) was due to the expansion of management capabilities in the strategic direction, as well as convenient control of crucial indicators and assessment of the trajectory of movement towards the goals.

In continuation of the development of the system, the following sample sequence of drawing up a management plan for the enterprise was developed:

1. Clear articulation of the company's strategy in the form of strategic objectives;
2. Formation and maintenance of the relationship between strategic goals and indicators of their achievement;
3. Development of a company employee information system;
4. Planning and setting goals and strategic initiatives;
5. Expanding strategic feedback and information.

Currently, the work of many companies concentrates on the interests of customers and the quality of internal and external business processes. This forces managers to analyze customer needs, incoming claims, the reasons for supply disruptions, and the proportion of identified product defects and to communicate the results of this analysis to staff. For this, financial and non-financial performance indicators are used in the preparation of reports for top management. However, one can consider this set of indicators a full-fledged BSC, especially if it does not meet specific rules. According to Caplan and Norton (2017), "Multifaceted indicators in a properly constructed system should consist of interconnected goals and evaluation criteria for their achievement, consistent and complementary" (p. 56). The relationship between strategy and system is the relationship between goals and criteria for their evaluation, and this can be controlled. Causal relationships are fundamental in building a balanced scorecard. It should be reflected in each goal and associated performance indicators, both financial and non-financial.

A balanced scorecard is a set of interrelated indicators that help interpret the company's strategy for management (Sorokina 2011). It allows one to transform the goals of the strategy into indicators, which are grouped in the following four blocks: financial, client, internal business processes, and staff training and development. Thus, indicators are transformed into an understandable language for employees, informing them of the fundamental factors of success in the present and future.

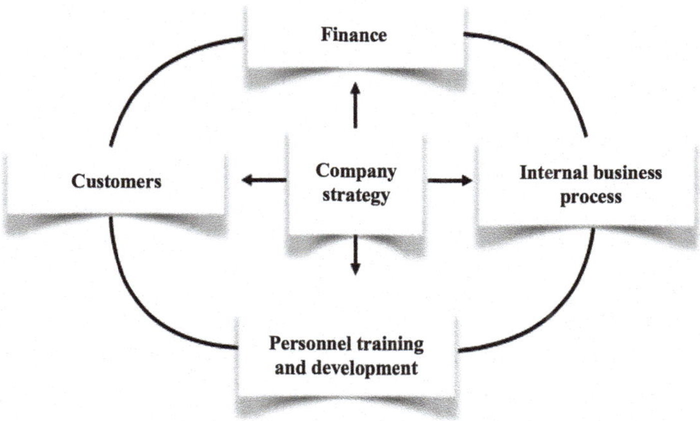

Fig. 2. Structure of a balanced scorecard. *Source*: (Sorokina 2011).

Four blocks of BSC should achieve a balance between themselves and between long-term and short-term goals. Each block of a balanced scorecard, with its own internal strategy and factors, has an impact on the next component. Thus, a balanced scorecard gets a hierarchical look. At the same time, each perspective of the system links the result and the factors of its achievement and is aimed at implementing a single strategy (Fig. 3).

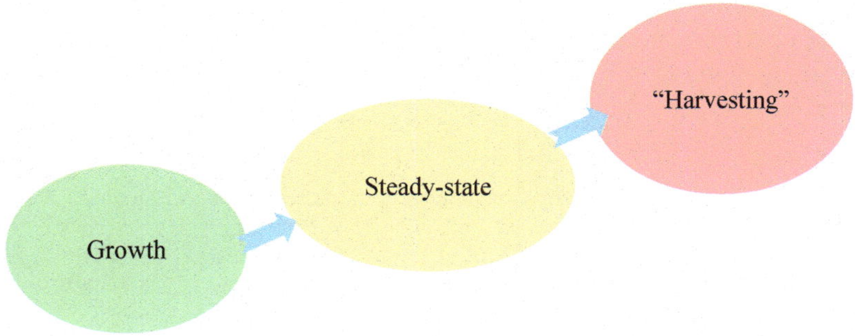

Fig. 3. Stages of strategic development. *Source*: (Porter 2005).

Each stage is timed to a specific stage in the development of the company. At the same time, the same indicators can have different meanings, to which the company seeks. At the growth stage, cash flow may be negative, and profitability at the lowest level, but this does not mean that the company does not achieve its goals.

At the stage of sustainable growth, the company still needs to invest, but it is assumed that the company is increasing its market share. At this stage, the selected indicators are aimed at covering weaknesses to increase the company's profitability.

At the harvest stage, the main objective of the company is to increase the cash flow from the core business and reduce the need for working capital.

After choosing a strategy, the top management of the company must determine the strategic goals (directions) that will lead to its implementation and translate these goals into tasks that are required to achieve it.

For each stage of the strategy, we can identify strategic financial directions that will lead to the achievement of the enterprise strategy:

- Revenue growth and expansion of the structure of activities;
- Reduce costs and increase productivity;
- Asset use and investment strategy (Caplan and Norton 2017).

After setting the strategy and determining strategic goals, a selection of indicators occurs that will reflect the effectiveness of the enterprise and the achievement of goals. The selection of critical indicators depends on strategic goals at each stage of the company's development (Sizova and Sizova 2017b). Following the strategic direction, a system of indicators, between which there are causal relationships, is being developed.

With an increase in the number of offers of products or services, with an increase in the occupied market share, development of products with the creation of higher added value and price revision, the first strategic goal is achieved (revenue growth and expansion of the structure of activities). Cost reduction and productivity increase are characterized by lower direct and indirect costs and the use of available resources with other business units (Sizova et al. 2018). The strategic orientation of the use of assets and investment strategy is associated with a reduction in working capital, which is used to maintain production and expand the activities of the enterprise. For the third focus, the optimal use of fixed assets, the activation of previously unused resources, and their direction on the development of the enterprise are also characteristic.

All selected indicators are structured into a matrix of stages of business processes. In our example, the matrix will consist of 3 × 3 blocks and have the following form (Table 1).

Table 1. The matrix of indicators of strategic financial goals.

Stages	Strategic directions		
	Revenue growth and expansion	Cost reduction	Asset use
"Harvesting"	Sales growth indicators	Revenue/Personnel	Investments
	Percentage of new product sales revenue		Research and development
Steady-state	Target customer shares	Own costs versus competitors	Working capital liquidity ratio
	Cross sales	Cost reduction	ROCE by major asset categories
	Product and customer profitability	Indirect costs	Asset utilization ratio
Growth	Product and customer profitability	Unit cost	Payback
	Percentage of nonprofit customers		Performance

Source: (Caplan and Norton 2017).

A balanced scorecard includes several essential elements for each stage of strategic planning.

When drawing up a balanced scorecard, the following are considered: its components, strategic goals, cause-effect relationships, key performance indicators, and target values. The interconnection of all stages is consistent and is shown in Fig. 4.

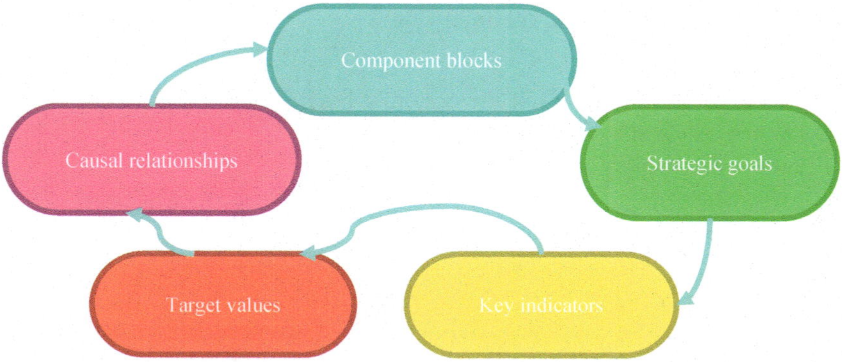

Fig. 4. Elements of a balanced scorecard. *Source:* (Zaitseva 2012).

A balanced scorecard helps to formulate and relate the financial indicators of all company divisions.

Let us consider individually each block involved in the construction of a balanced scorecard (see Fig. 2).

Block 1. Financial component

The financial unit of the system is aimed not only at the formation of long-term indicators, but it also allows one to control and evaluate short-term factors that help to achieve long-term goals. A balanced scorecard helps to articulate financial criteria that reflect the long-term goals of the company. Using a balanced system, we can establish the relationship between financial performance and all departments of the enterprise at various stages of business processes (Balishev and Protopopova 2014). When developing a system, we need to take into account that all goals and indicators of its other components should be related and influence the result of one or several goals of the financial component. The performance of each component of the system is a link to achieving financial goals, which are the most critical areas of a business strategy.

Block 2. Client component

Before the information age, enterprises paid great attention to production and innovation, not focusing on customer needs enough. Currently, most companies concentrate on meeting customer needs and increasing their market share (Kandalintseva 2015).

At the beginning of the BSC construction, a company needs to determine its target market segment, after which goals and indicators are formulated. According to statistics, for companies that have already implemented a balanced scorecard, two sets of characteristics can be distinguished: general (key) indicators and factors for achieving results (Gordan et al. 2018).

Key indicators are present in the BSC of all enterprises, and factors help determine what needs to be done to achieve a result.

The key indicators include the following:

- Company market share;
- Preservation and expansion of the customer base;
- Customer satisfaction rate;
- Customer profitability.

These indicators are universal for any activity, but each company should consider them according to individual client groups.

Block 3. Internal business processes

This block (developed when preparing a balanced scorecard) is associated with the company's business processes, the fullness of which is determined by its type of activity. The formation of indicators in this area is determined by the selected indicators in the financial and client components. When compiling the system, it is recommended to determine the complete chain of internal business processes in value terms (from innovation to after-sales service) (Zaitseva 2012).

Block 4. Training and staff development

The strategic goals of this block are factors that help to achieve the goals of the first three components of the system. By the examples of many companies that have already implemented a balanced scorecard, one can distinguish the following areas of this component: job opportunities and information systems, motivation, a delegation of duties and compliance of personal goals of employees with corporate goals. The introduction of BSC at the employee level can be brought into a system of indicators such as (Mihailova 2008):

1. Percentage of top managers involved in the system;
2. Percentage of employees involved in the system;
3. Percentage of top managers whose personal goals are aligned with those of the BSC;
4. Percentage of employees whose personal goals and objectives are aligned with the goals of the BSC;
5. Percentage of employees who achieved their personal goals.

In addition to the indicators mentioned above, we can also highlight the following three key indicators that assess the personnel situation in the company: employee satisfaction, maintaining the personnel base, and employee efficiency.

4 Discussion

Each block considered by us has a relationship with other components of a balanced scorecard. Nevertheless, it has the following hierarchical form: personnel qualifications affect the process of production of goods and the provision of services, which in turn affects customer loyalty and market share. Moreover, this leads to different financial results of the enterprise and affects its value (Bricheeva 2010).

5 Conclusion

When drawing up a balanced system, it is necessary to adhere to the following specific stages of its development: setting strategic goals, choosing key indicators that reflect achievement, drawing up a strategic map, setting target values for the coefficients, developing a motivating system for employees, developing strategic measures aimed at implementing the strategy, communicating the existence of a balanced scorecard and its essence to each unit and employee of the enterprise.

Acknowledgments. The authors thank their colleagues for their help.

References

Balishev, N.B., Protopopova, M.N.: On the issue of assessing the probability of bankruptcy of an enterprise. Vestn. Tula Branch Financ. Univ. **1**, 5–7 (2014)

Bricheeva, N.N.: Application of the hierarchy analysis method in building a balanced scorecard. Izv. South. Fed. Univ. Tech. Sci. **2**, 214–220 (2010)

Caplan, R., Norton, D.: Balanced scorecard: From strategy to action. Olimp-Biznes, Moscow, Russia (2017)

Chursin, A.A., Shamin, R.V., Fedorova, L.A.: The mathematical model of the law on the correlation of unique competencies with the emergence of new consumer markets. Eur. Res. Stud. J. **20**(3), 39–56 (2017)

Kandalintseva, V.G.: Innovative business: The use of a balanced scorecard. RANEPA, Moscow, Russia (2015)

Loshilina, I.V.: Modern management technologies (2007). https://www.businessstudio.ru/procedures/strategi/bsc_strategy_formula/

Mihailova, E.M.: The basic principles of developing a balanced scorecard (BSC) for enterprises (2008). https://cyberleninka.ru/article/n/osnovnye-printsipy-razrabotki-sistemy-sbalansirovannyh-pokazateley-ssp-dlya-predpriyatiy

Nils-Gordan, O., Jan, R., Magnus, V.: Balanced scorecard: Practical guide to use. Williams, Moscow, Russia (2018)

Porter, M.: Competitive strategy: Methodology for analysis of industries and competitors. Alpina Business Books, Moscow, Russia (2005)

Ragulina, J.V., Alekseev, A.N., Bogoviz, A.V., Lobova, S.V.: Automatization of the labor resources market in the age of the internet of things: conceptual substantiation and risk management. Studies in Computational Intelligence, vol. 826, pp. 909–915 (2019)

Sizova, T.V., Sizova, D.A.: Methodological approaches to the diagnosis of the financial crisis of the company. J. Legal Econ. Stud. **2**, 254–260 (2017b)

Sizova, T.V., Sizova, D.A.: Making managerial decisions based on the results of the diagnosis of the crisis financial condition of the company. J. Legal Econ. Stud. **3**, 135–141 (2017a)

Sizova, T.V., Sizova, D.A., Volobuev, N.A.: Diagnostics of the level of economic security of organizations in contemporary conditions of economic development. In: Materials of the International Scientific and Practical Internet Conference "Public and Private Finance in an Innovative Economy", pp. 205–214. Plekhanov Russian University of Economics, Moscow (2017)

Sizova, T.V., Sizova, D.A., Adulova, E.S., Kalinkina, K.E.: Transaction costs and the ways to optimize them (2018). https://sgemsocial.org/ssgemlib/spip.php?article6091

Sorokina, A.V.: The mechanism for implementing an effective company strategy using a balanced scorecard. MIIT, Moscow, Russia (2011)

Zaitseva, N.A.: Necessary conditions and obstacles for the implementation of a balanced scorecard. In: Proceedings of the Free Economic Society of Russia, vol. 166, p. 158 (2012)

Risk Analysis in the Implementation of High-Tech Projects

Gabriel H. Alberto[1(✉)] and Igor V. Kiryakov[2,3]

[1] Department of Applied Economy of TsUOP, RUDN University, Moscow, Russia
alberto.gabriel@yandex.ru
[2] Meditsinskiye Resheniya LLC, Moscow, Russia
ivkiryakov@gmail.com
[3] Department of Mathematical Modeling and Information Technologies, RUDN University, Moscow, Russia

Abstract. The article presents a method of risk analysis in the implementation of high-tech projects based on sensitivity assessment. The authors state that risk analysis contributes to the success of its implementation and reduce losses occurring during its implementation. The hypothesis of this study is that the timely and complete identification of risks in high-tech projects contributes to the success of its implementation and reduces losses that occur during its implementation. Practical value is the search for risk analysis methods that maximize the accuracy of their identification in high-tech enterprises. The novelty of the study is determined by the possibility of applying the presented risk analysis model in other high-tech projects.

Keywords: Risk · Knowledge-intensive projects · Chelators · Efficiency · Net discounted income · Loss · Reduction

1 Introduction

The orientation of the economy to high-tech projects requires objective tools for assessing the future effectiveness of a project. Usually, the intuition and personal impressions of the investor or his representative act as such tools. Furthermore, often, science-intensive projects and government development institutions invest tens of millions of rubles in a young team of enterprises, assessing the risks of non-return of money based on the opinions of just one or two experts. If a slightly larger number of specialists is involved in the project assessment and an attempt is made to summarize their opinions somehow objectively, then experts are at best offered to give the project simple estimates, from which they then calculate the usual arithmetic mean (Zubarev and Menshikov 2014; Ragulina et al. 2019; Bogoviz 2019; Chursin and Vlasov 2016).

Meanwhile, there are accurate and objective mathematical methods that allow working with fuzzy systems for assessing quality and choice. In our opinion, one of these methods can be objectively based on methods for assessing the sensitivity of projects to external factors affecting them (Vlasov 2011). Especially, this has become relevant in recent years, when partial pressure on Russia does not allow accurate

A. V. Bogoviz and Y. V. Ragulina (Eds.): ISCI 2019, LNNS 115, pp. 279–285, 2020.
https://doi.org/10.1007/978-3-030-40749-0_33

forecasting of the ruble exchange rate, and energy prices in the world. In such conditions, the implementation of science-intensive projects should be based on the search for domestic suppliers, given the possibility of closing part of foreign sales markets.

One of the high-tech industries is the chemical industry. In recent years, this industry began to develop most actively through the introduction of new types of equipment in production, the launch of new products on the market. The development of each project in this industry requires a large number of scientific studies (Khokhlov 2014).

2 Materials and Methods

The study is conducted on the project for the production of adsorbent based on hydrolysis lignin based on the production site of "Vostok" LLC. Feed safety is one of the most critical factors for ensuring both health and high productivity of animals and birds. One of the acute problems of animal husbandry and poultry farming is the contamination of feeds by mold metabolism products. Almost all households and private farmsteads face this problem. Effective adsorbents can solve this problem.

The development and implementation of the production of enterosorbents based on hydrolysis is an essential major scientific, economic and social problem, the solution of which would not eliminate the deficiency in enterosorbents, reduce the final product cost, solve the problem of the qualified use of hydrolysis of "Vostok" LLC, and reduce environmental pollution environment.

3 Results and Discussion

The main indicators of the project are provided in Table 1.

Table 1. Project performance indicators.

Indicator	Value
Payback period	36 months
Discounted payback period	60 months
Average rate of return	14,89%
NPV	2028 thousand rubles
Profitability index	1,04
Internal rate of return	1,86%
Modified internal rate of return	0,27%
Duration	3,35 years

To determine the riskiness of this high-tech project, in our opinion, it is advisable to turn to its sensitivity method to external factors.

In this case, the external factors that have a negative impact on the project include the following:

- The possibility of banning the purchase of products by Western countries (it was initially assumed that Germany would be a major buyer of products);
- Closing access to cheap raw materials from Central Asia as a result of natural disasters;
- The instability of the ruble may adversely affect revenue from the sale of products to foreign countries.

Table 2 shows the data serving as an information basis for determining the level of risks affecting project performance.

Table 2. Information basis for determining the level of risks affecting project performance.

Year	Sales (tons per year)	Variable costs per ton of products, thousand rubles	Fixed costs for the planned production output, thousand rubles	Cost of the planned output	Selling price of one ton of products	Planned revenue, thousand rubles	Revenue from sales, thousand rubles	Discount coefficient (at a capital price of 10%)	Discounted net cash flow, thousand rubles
1-st	500	3	500	2,000	4.5	2,250	250	0.909	227
2-nd	1,000	3	500	3,500	4.5	4,500	1,000	0.824	824
3-rd	1,200	3	500	4,100	4.5	5,400	1,300	0.751	976
Total									2,028

To analyze the project sensitivity, a set of key risk assessment parameters was identified, among which are highlighted:

- The physical volume of sales;
- Variable costs;
- Fixed costs;
- Selling price;
- Capital price.

Subject to the risk of prohibiting the acquisition of products by Western countries (Germany), project performance may decline by 10%. We will present the calculation of the parameters of the specified risk in Table 3.

Thus, the loss of income as a result of the risk of imposing sanctions on the purchase of sorbents in Germany will amount to 2028 − 1713 = 315 thousand rubles.

Next, we assess the implementation of natural disaster risk, which would block access to cheap raw materials. At the same time, the company's expenses will increase by 10%. The calculated indicators are shown in Table 4.

The specified risk will lead the project to reduce revenues by the following amount: 2028 − 1136 = 892 thousand rubles.

Next, we will evaluate the risk of changes in the exchange rate of the ruble. However, given the negative dynamics of the ruble and the fact that the primary sales are in foreign currency, a depreciation of the ruble may positively affect the results of the project. The calculated indicators are shown in Table 5.

Table 3. Project indicators subject to the risk of a ban on the acquisition of products by Western countries (market loss of 10%).

Year	Sales (tons per year)	Cost of the planned output	Selling price of one ton of products	Planned revenue, thousand rubles	Revenue from sales, thousand rubles	Discount coefficient (at a capital price of 10%)	Discounted net cash flow, thousand rubles
1	2	3	4	5	6	7	8
1	550	2,150	4.5	2,475	325	0.909	295
	450	1,850		2,025	175		159
2	1,100	3,800	4.5	4,950	1,150	0.824	948
	900	3,200		4,050	850		700
3	1,320	4,460	4.5	5,940	1,480	0.751	1,111
	1,091	3,773		4,909	1,136		853
Total							2,355
							1,713

Table 4. Sensitivity assessments of project indicators for disaster risk management.

Year	Sales (tons per year)	Variable costs per ton of products, thousand rubles	Fixed costs for the planned production output, thousand rubles	Cost of the planned output	The selling price of one ton of products	Planned revenue, thousand rubles	Revenue from sales, thousand rubles	Discount coefficient (at a capital price of 10%)	Discounted net cash flow, thousand rubles
1	500	3	500	2,000	4.95	2,475	475	0.909	432
					4.09	2,045	45		41
2	1,000	3	500	3,500	4.95	4,950	1,450	0.824	1,195
					4.09	4,091	591		487
3	1,200	3	500	4,100	4.95	5,940	1,840	0.751	1,382
					4.09	4,909	809		608
Total									3,008
									1,136

Table 5. Estimates of the sensitivity of project indicators in the implementation of currency risk.

Year	Sales (tons per year)	Variable costs per ton of products, thousand rubles	Fixed costs for the planned production output, thousand rubles	Cost of the planned output	Selling price of one ton of products	The planned revenue, thousand rub.	Revenue from sales, thousand rub.	Discount coefficient (at a capital price of 10%)	Discounted net cash flow, thousand rubles
1	500	3,3	500	2150	4,5	2250	100	0.909	91
		2,7		1850			400		364
2	1000	3,3	500	3800	4,5	4500	700	0.824	577
		2,7		3200			1300		1071
3	1200	3,3	500	4460	4,5	5400	940	0.751	706
		2,7		3740			1660		1247
Total									1374
									2681

This risk will have a positive effect on the project, with a decrease in the ruble exchange rate, ruble revenue for the project will increase by 2681 − 2028 = 653 thousand rubles.

Thus, the most severe risks to which the project is exposed are:

- The possibility of a ban on the acquisition of products by Western countries (it was initially assumed that Germany would be a major buyer of products);
- Disabling access to cheap raw materials from Central Asia as a result of natural disasters.

Next, we will compile Table 6 to determine the coefficient of elasticity, as well as indicators of sensitivity and predictability of project variables.

Table 6. Calculation of indicators of project sensitivity to risks.

Indicator	The percentage change in the variable	Percent Change NPV	Sensitivity indicator (elasticity)	Variable rank	Sensitivity level	Predictability level
The volume of production, units (Q)	−10	84.5%	8.45	3	High	High
Unit variable costs, rub. (VAC)	10	67.7%	6.77	4	Average	Average
Fixed costs, rub. (FC)	10	93.9%	9.39	2	High	High
Unit sales price (P)	−10	56.0%	5.60	5	Average	Low
Capital price, % (d)	10	98.0%	9.80	1	High	High

Based on specific data, it is possible to compile a matrix of project sensitivity to risks and their predictability.

All variables were in the risk zone II. That is, all identified risks require close attention to changes.

Table 7. Matrix of project sensitivity to risks and their predictability.

Variable predictability	Variable sensitivity		
	High	Average	Low
Low	I	I	II
Average	I	II VAC,	III P
High	II Q, FC,	III d	III

Further, as possible options for optimizing the risks of a science-intensive project, it is advisable to build at least three of the following standard scenarios: pessimistic, optimistic, and most likely. The indicated scenarios are given in Tables 8.

We will summarize the results in Table 10.

Next, we need to determine the average expected value of the net present value, the quadratic deviation, and the coefficient of variation for certain risks (Table 11).

Table 8. The best scenario.

Year	Sales (tons per year)	Cost of the planned output	Selling price of one ton of products	Planned revenue, thousand rubles	Revenue from sales, thousand rubles.	Discount coefficient (at a capital price of 10%)	Discounted net cash flow, thousand rubles	Year	Realization (tons per year)
1-st	550	2.7	450	1,935	4.95	2,723	788	0.917	722
2-nd	1,100	2.7	450	3,420	4.95	5,445	2,025	0.842	1,705
3-rd	1,320	2.7	450	4,014	4.95	6,534	2,520	0.772	1,945
Total									4,373

Table 9. The worst scenario.

Year	Implementation (tons per year)	Cost of the planned output	The selling price of one ton of products	Planned revenue, thousand rubles	Revenue from sales, thousand rubles	Discount coefficient (at a capital price of 10%)	Discounted net cash flow, thousand rubles	Year	Realization (tons per year)
1-st	450	3.3	550	2,035	4.09	1,841	−195	0.901	−175
2-nd	900	3.3	550	3,520	4.09	3,681	161	0.812	131
3-rd	1,091	3.3	550	4,150	4.09	4,462	312	0.731	228
Total									183

Table 10. The probability of the impact of risks on the project in different development scenarios.

Scenario	The probability of the result	NPV value
The worst	0.25	183
Main	0.5	2028
The best	0.25	4373

Table 11. Indicators of a knowledge-based project.

Indicator	Project		
	Pessimistic	Expected	Optimistic
NPV	183	2,028	4,373
Probability	0.25	0.5	0.25
NPV given probability	45.75	1,014	1,093.25
Total NPV subject to probability M (NPV)	–		2,153
Deviation from the average	−1970	−125	2,220
Squared deviation from mean	3,880,900	15,625	4,928,400
Dispersion			8,824,925
Standard deviation(s)	–		2,970.7

To assess the risk in this methodology, the standard deviation indicator is used, and the expectation indicator is M (NPV).

In accordance with the "three-sigma" rule, the value of a random variable, NPV in this case, with a probability close to 1 is in the interval [M − 3s; M + 3s].

Thus, the total amount of losses resulting from the occurrence of risk events during the implementation of the investment project is 2970.7 thousand rubles (which suggests a high degree of risk of this project).

In other words, the risk price of the investigated project is 2970.7 thousand rubles of possible losses, i.e., adoption of the project may entail the possibility of losses for not more than 2970.7 thousand rubles.

The risk of loss of suppliers can have the most significant impact on both the profitability and investment attractiveness of the project. At the same time, the net present value will change by 50.6%.

With the loss of suppliers of cheap raw materials, the net present value will decrease by 56.0%.

Thus, in order to reduce the risks of this project, it is advisable to look for new markets in countries in which agriculture is developed. The Middle Eastern countries (Egypt in particular) may be of great interest.

At the same time, in order to get protected from losses due to natural disasters in the country of cheap raw materials, it is advisable to study the possibility of procurement in other regions, or the possibility of replacing this raw material with its substitutes.

4 Conclusion

In conclusion, the hypothesis was confirmed that the timely and complete identification of risks in high-tech projects contributes to the success of its implementation and the reduction of losses occurring during its implementation. In this case, losses could exceed the NPV as a whole and lead to losses for the project.

Acknowledgements. The study has been supported by the Program "5-100" of the RUDN University (Moscow, Russia).

References

Bogoviz, A.V.: Industry 4.0 as a new vector of growth and development of knowledge economy. Studies in Systems, Decision and Control, vol. 169, pp. 85–91 (2019)

Chursin, A., Vlasov, Y.: Innovation as a Basis for Competitiveness: Theory and Practice. Springer, Berlin (2016)

Khokhlov, N.V.: Risk management. UNITY DANA, Moscow, Russia (2014)

Ragulina, J.V., Alekseev, A.N., Bogoviz, A.V., Lobova, S.V.: Automatization of the labor resources market in the age of the internet of things: conceptual substantiation and risk management. Studies in Computational Intelligence, vol. 826, pp. 909–915 (2019)

Vlasov, A.N.: The main aspects of the investment analysis of real estate development projects. Transp. Bus. Russ. **10**, 24–25 (2011)

Zubarev, A.M., Menshikov, V.V.: Management of a high-tech project for the development and application of an information system in a chemical enterprise. Adv. Chem. Chem. Technol. **28**(10), 125–128 (2014)

Effectiveness Management

Victor G. Antonov(✉), Elena V. Kuptsova, and Michael V. Samosudov

State University of Management, Moscow, Russia
v.antonov1949@yandex.ru, sakuptsov@mail.ru, samosudov@mail.ru

Abstract. Effectiveness is a common term, but it is not sufficiently defined. Effectiveness is often confused with performance and not considered as the basis for the systemic sustainability of organizations. The paper focuses on the essence of effectiveness as an economical category through the prism of implementing managerial activities in the current and long-term periods of organizations. The genesis of various approaches to assessing effectiveness is examined, and various methodological recommendations for assessing the level of effectiveness are analyzed. The special attention is paid to the methods and tools of managerial influences that ensure the effective functioning and development of organizations. The authors define the category of "effectiveness" in a broad and narrow sense, in the tactical and strategic aspects of the activity of socio-economic systems. The authors state that successful long-term commercial activities of the organizations are based on the balance between the economic and social interests of its stakeholders. A model for management activities to ensure the effectiveness of the socio-economic system is proposed.

Keywords: Effectiveness · Effect · Organization · Management · Owner · Top manager · Efficiency indicators · Measurement · Results · Management subject · Management object

1 Introduction

The following logical chain can describe the activities of any entrepreneur: to create a business – to how to manage a business – to develop a business. At the same time, the entrepreneur seeks to obtain economic benefits, compensating his invested funds, and stimulating business development. It dictates the need to objectively measure and evaluate business economic conditions, which is not obtained by everyone and not every time. What is the reason for obtaining wrong estimates?

We are used to assessing the economic viability of a business through the category of effectiveness as a ratio of "cost – result." In this case, the term "result" is considered synonymous with the term "effect". With this approach, the mechanism for obtaining the desired effect is reduced exclusively to the calculated actions, and activities for its formation are not disclosed. The business practice shows that there is no direct connection between resource availability and the effect achieved. Therefore, effectiveness is the result of an activity that needed to be managed.

A. V. Bogoviz and Y. V. Ragulina (Eds.): ISCI 2019, LNNS 115, pp. 286–293, 2020.
https://doi.org/10.1007/978-3-030-40749-0_34

At the same time, in the most general view, effectiveness (*Latin – effective, productive, yielding results*) is an indispensable condition for the organization's existence, acts as the development indicator, and its most important incentive. Besides, effectiveness may vary depending on what goals, their corresponding effects, and costs of those resources we consider.

2 Materials and Methods

In order to manage effectiveness of a socio-economic system, it is necessary to determine the object and the management system, their interaction order. If we consider effectiveness as a result of the activity of the socio-economic system in the primary stakeholders' interests, it is necessary to define the areas of responsibility of all key stakeholders. It is also necessary to define what needs to be managed in order to influence the effectiveness level. In our opinion, it is necessary to manage, first of all, the processes of goal-setting; second, evaluate all business participants' behavioral patterns, based (including) on the creation of proper business ethics; third, focus processes; finally, the material and informational flows. All these management aspects should be integrated into one subsystem of "Efficiency Management" (Fig. 1).

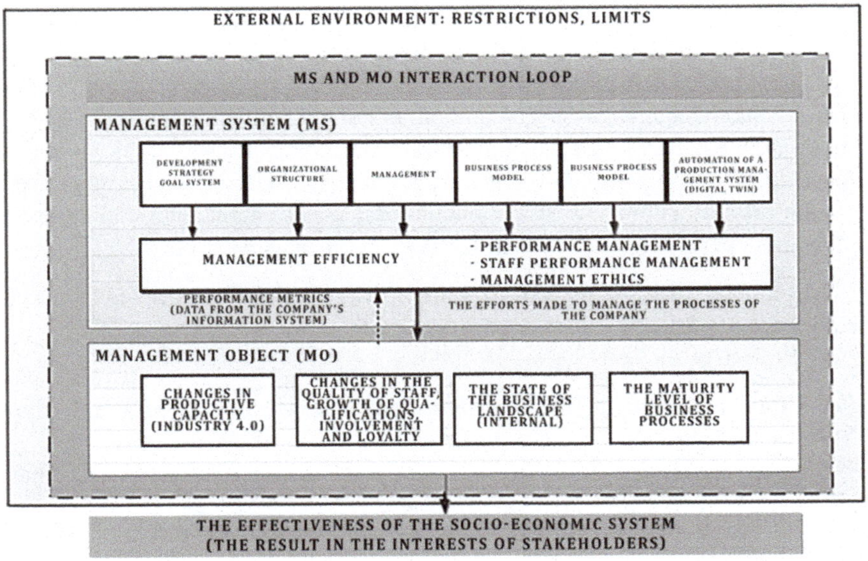

Fig. 1. The scheme of the company management for achieving the target effectiveness.

The interaction of control and production systems is carried out through the activities exchange (Tikhomirova 1984). The first one is expressed in the efforts made (managerial influences) to manage the processes of the organization, which leads to specific changes in the production personnel labor and finds expression in the created final product. It is important to note that these managerial influences should be applied timely and reflect the result in the interests of stakeholders. The way the control object responds to these control actions can be evaluated through efficiency metrics. These estimates must not be "posthumous." Creating digital models of the production facilities makes it possible to simulate these reactions and manage them in real-time. The firm view on many domestic entrepreneurs on the staff as a resource is not economically justified. The staff in a modern business is not just one of the stakeholders but an investor (bringing the ability to work to the business). If the staff does not share the owner's goals and due to a number of reasons and is not loyal to the company, then achieving the target result becomes quite problematic. There is a gap in the interaction of the subject and object of management at the level of interests of the main stakeholders (Antonov and Maslennikov 2013). It is necessary to understand and systematically manage activities of a socio-economic system to achieve its target effectiveness.

The above makes it necessary to develop new approaches to the formation and evaluation of effectiveness and allows us to put forward the following scientific hypothesis:

- Six components determine the economic viability of any socio-economic system: a target result and a period during which the result must be achieved; available resources; system participants' behavior; restrictions, limits and influence forces; a system ability to upgrade; consistency of all previous components;
- A balance of economic and social interests of the stakeholders, in which investments in the organization's activities are justified in terms of achieving the desired result for each party, ensures a successful long-term commercial activity;
- Effectiveness requires management, both in current activities and in the long term;
- Conditions for developing the target effectiveness of a socio-economic system is the availability of competencies in the field of strategic leadership and competitiveness; an operational know-how based on innovations in engineering and technology; creating an enabling environment for staff and management ethics.

The purpose of the study is to develop methodological approaches to managing effectiveness of various socio-economic systems.

The objectives of the study are as follows:

- Identifying and justifying the need for introducing two components of effectiveness (tactical and strategic);
- Revealing the essence of the organization of the system efficiency management.

In a historical retrospective, the view on the category of effectiveness underwent some changes: through the definition of the "effectiveness" category (Petty 2012; Quesnay 2008; Ricardo 2007) to its correlation with management (Cyert and March 1992; Meyer 2004; Drucker 2003; Kozlova 1983), and further to the formation of a set of factors that ensure the required effectiveness level, taking into account the organization's strategy, mission and corresponding objectives (Sink 1989; Kono 1991; Kleiner 2007; Popkova 2007; Otenko 2004; Vertakova et al. 2010).

This paper is the result of an initiative study of issues of formation and evaluation of the effectiveness of socio-economic systems. Both theoretical and empirical methods were used in the course of the study. (1) Empirical research: monitoring trends in the management of Russian companies based on publications in the open press and on the Internet. (2) Theoretical research: collecting information on the evolution in approaches and views on the category of the "effectiveness" in the works of domestic and foreign scholars and practitioners; evaluating companies' efficiency (for example, UBER, Delikatnyy pereyezd, and a number of small and medium-sized enterprises). In this study, the authors also rely on their own experience in company management and consulting activities.

3 Results

The results of the study testify in favor of the previously put forward scientific hypothesis about the need for new approaches to the formation and evaluation of effectiveness. The effectiveness should be considered in two dimensions: as a management activity (in the broad sense) and as an evaluative indicator of the result obtained (narrow).

Figure 2 shows the formation chain of a system of generalized efficiency indicators, depending on the target management function chosen by the owner. This chain was developed by the authors taking into account the methodological materials presented by (Yazovsky 2015).

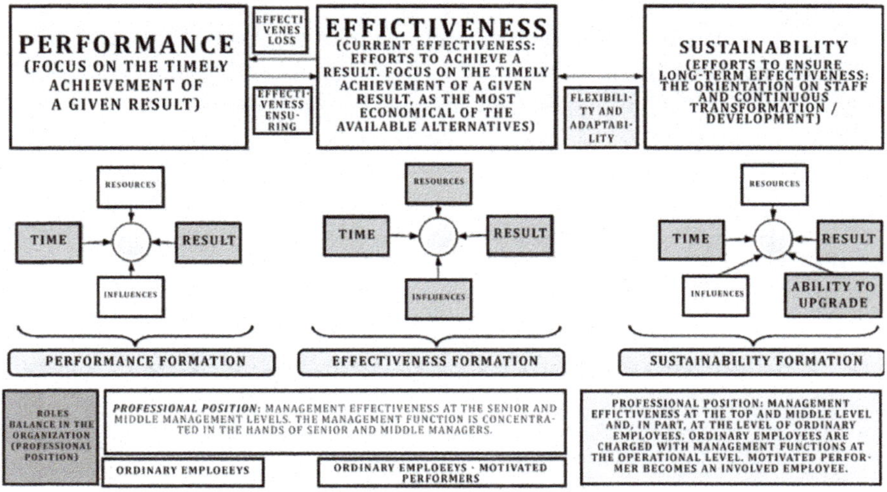

Fig. 2. The formation chain of a system of generalized efficiency indicators, depending on the target management function chosen by the owner.

Each stage in the chain is the result of the consistency of a specific previously named set of components. The timely achievement of a result gives effectiveness (productivity). At the same time, there is a danger that a quick result, as a rule, is obtained with unreasonably high costs of resources. The reason is that the development stage is practically absent in the process of making management decisions (development, adoption, implementation). Orientation to the timely achievement of a given result as the most economical of the available alternatives ensures current effectiveness. Economic feasibility of the result is given by the ratio of the result to the number of resources spent on its achievement, which takes into account the economic interests of all interacting parties (Yazovsky 2015). It is expressed in terms of net profit and profitability. Correlation of the result with influences and restrictions (legislation, public opinion, etc.) sets the social justification for this result.

Attention to the current efficiency without taking into account functional (systemic) sustainability can lead to the loss of sustainability and leading, as a result, to the termination of organization's activities. Effectiveness and sustainability are inversely related. Therefore, carried away by the struggle for the effectiveness, it is easy to lose stability. According to foreign publications (Journal of Marketing Management, Forbes), the organizations' life expectancy over the past 25 years has decreased to the level of 20 years, and the period of maintaining competitiveness has decreased to less than ten years. So, summarizing the above, any functionally stable organization is effective, but not any effective organization is stable.

That is why the coefficient of functional stability (C_{fs}), which characterizes the systemic stability of a business entity from the standpoint of functions effectiveness, is a more reliable indicator of the economic viability of the business in the long term than the efficiency indicator: C_{fs} = f (controllability; resource base; behavior of active elements of the system).

Along with that, effectiveness is an indispensable condition for the existence of the organization and its stability. i.e., absence of a definite difference between the incoming cash flow and costs for a sufficiently long period leads the organization to lose the function ability. Therefore, a business manager must understand what to do and how to manage more effectively.

The authors propose to assess the processes in the formation of performance, effectiveness, and system stability of the organization based on the interconnected regulatory parametric management models *f(m)* and models of operational processes *f(o)* (Fig. 3).

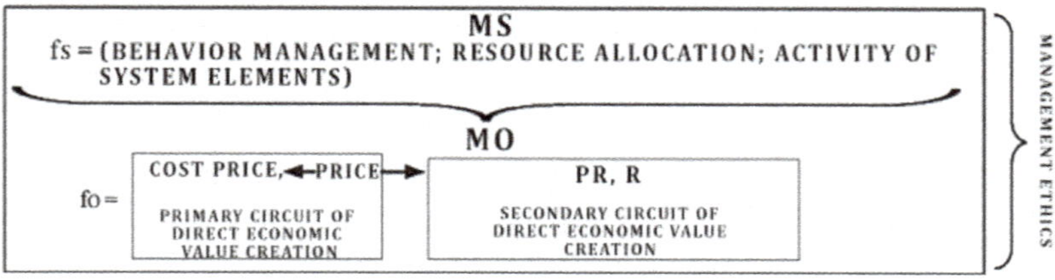

Fig. 3. The management model for the socio-economic system effectiveness.

The management model *f(m)* reflects the interaction of a subject and an object within an elementary social system. Each participant has specific characteristics, a behavior vector, available and expected resources, and an active system element.

In response to managerial influences from a management subject, a model of operational processes *f(o)* reflects the processes of creating and distributing profits/direct economic value (the business model determines the indicator).

The authors believe that there are no single universal criteria for assessing the organizations' operational effectiveness since they all differ in type, industry sector, the maturity level of business processes, and goals. However, in any case, the operational effectiveness associated with the management object, in terms of creating profits, will be associated with the following groups of indicators reflecting the effectiveness of the labor (labor resources), fixed assets, working capital and capital investments; intangible assets; technical and technological innovation, and sales effectiveness.

On this basis, system effectiveness is evaluated through a set of indicators reflecting each of these groups compared with the normative (given by a control system) values. Therefore, to manage operational effectiveness, each owner/manager should have a panel of metrics implemented in a digital environment and reflecting these aspects of the operation of this particular organization.

To assess the organizations' operating efficiency in terms of profit distribution, we can rely on three indicators that are suitable for most organizations (return on equity (RoE), return on the capital employed (ROCE), earnings per share (EpS), and EpS growth. Figure 2 clearly shows that the process of formation of sustainability gives an impetus to the transformation of the existing system of managing the organization's activities towards the management of effectiveness.

Moreover, the depth and scope of these changes can be quite different and even vary over a relatively short time interval. It is worth noting that this transformation is achieved by changing the balance of the roles of rights and responsibility for making management decisions. New responsibility areas are being formed in the conditions of time shortage and high requirements to the professional competencies; the line personnel is involved in making managerial decisions to improve executable business processes leading, ultimately, to ensure the economic viability of the business.

Thus, the result of the study is the definition of "effectiveness" notions as:

- In tactical activities (in the narrow sense), effectiveness is the result of the response of the socio-economic system to managerial influences on its elements in the interests of stakeholders. Wherefrom all the existing alternatives to get the desired result, the most economical one is chosen considering resource costs, time necessary to obtain the result, taking into account the influences of the business environment, and public institutions in the system development. This result is expressed in the achieved level of targeted financial and operational indicators that form the effectiveness model. This definition is valid for the assessment of current activities. However, ensuring the economic viability over a long period requires constant renewal, flexibility.
- In the strategic activities (in the broad sense), effectiveness is determined the systemic sustainability of the business, understood as the ability of the socio-economic system to be updated in a timely manner in response to the changes in the external and internal environment, based on the synergy of the productive actions of the owner, management, and employees. The update refers to a new state of the socio-

economic system, which ensures its reliable functioning and development ahead of the disturbing effects of the environment.

Further studies, in the framework of the presented topics, the authors associate with: (a) clarifying the conceptual apparatus (as the material accumulates); (b) developing a guidelines for the formation of the effectiveness management systems in small and medium businesses; (c) conducting a more in-depth study based on the proposed concept, highlighting individual groups: owners, top managers, and personnel, as well as analyzing the features of their interaction at various stages of the organization's "life cycle," substantiate key aspects of effectiveness in individual development stages.

4 Conclusion

The current economic state of organizations is characterized by an ever-increasing instability and searching for the adequate measures of the obtained results. This is due, among other things, to the fact that the results of the economic activity reflect multiple and often conflicting goals. Each of the goals may reflect the interests of the individual groups of stakeholders. In order to reduce conflict, it is necessary to use integrated indicators to ensure a certain balance level of various groups' interests in assessing the effectiveness of the organization's socio-economic development. At the same time, it is very important to be able to establish the relationship between the result obtained, as well as the set of actions and the conditions in which this result was obtained. In our opinion, the effectiveness should be considered in two dimensions. In a broad sense, this is a management activity. In a narrow sense, one considers the efforts made to manage the organization's processes in the interests of stakeholders.

"Effectiveness" as an economic category should be fully involved in management activities, as one of the management subsystems. Within this subsystem, tactical and strategic effectiveness is determined, and effectiveness management should employ the normative multifactor models.

References

Antonov, V.G., Maslennikov, V.V.: Navigator of the owner – the owner of an effective business. Manag. Sci. Sci.-Pract. J. **3**, 26–36 (2013)

Cyert, R.M., March, J.G.: A Behavioral Theory of the Firm, 2nd edn. Wiley-Blackwell, Hoboken (1992)

Drucker, P.: Management development in the 21st century. Williams Publishing House, Moscow, Russia (2003)

Kleiner, G.A.: The effectiveness of the mesoeconomic systems of the transition period. Problems of theory and practice of management. Sci. Pract. J. **4**, 24–30 (2007)

Kono, T.: The strategy and structure of Japanese corporations. Economy, Moscow, Russia (1991)

Kozlova, O.V.: Theory of management of socialist production. Economy, Moscow, USSR (1983)

Meyer, M.W.: Evaluation of business effectiveness, Korsunskoy, A.O. (Trans. from English). Vershina LLC, Moscow, Russia (2004)

Otenko, I.P.: Methodological foundations of enterprise potential management. Scientific publication. Kharkiv National University of Economics, Kharkiv, Ukraine (2004)
Petty, V.: Economic and statistical work. Book on Demand, Moscow, Russia (2012)
Popkova, V.V. (ed.): Sustainable economic development in the context of globalization and knowledge economy: conceptual foundations of management theory and practice. Economy, Moscow, Russia (2007)
Quesnay, F.: Selected economic works. Eksmo, Moscow, Russia (2008)
Ricardo, D.: The beginnings of political economy and taxation. Eksmo, Moscow, Russia (2007)
Samosudov, M.V.: Corporate governance: Theory of corporate interaction: Training module. Russian Foreign Trade Academy, Moscow, Russia (2006)
Sink, D.S.: Effectiveness management: planning, measuring and evaluating, monitoring, and enhancing. Progress, Moscow, USSR (1989)
Tikhomirova, A.V.: Evaluation of production management. Economy, Moscow, USSR (1984)
Vertakova, Yu.V., Kharchenko, E.V., Zheleznyakov, S.S.: Integration of approaches to the management of the modern organization. Publishing House of the South-West State University, Kursk, Russia (2010)
Yazovsky, A.: How to improve effectiveness management at the level of performers (2015). http://www.e-xecutive.ru/management/practices/1086065-kak-uluchshit-effektivnost-upravleniya-na-urovne-ispolnitelei/

Competitiveness Management of Medical Industry Enterprises in Integrated Structures

Yuri G. Gertsik(✉)

Bauman Moscow State Technical University, Moscow, Russia
gerzik@bmstu.ru

Abstract. The paper considers the possibilities of cluster structures to increase the competitiveness of participants and cluster products. The authors note the importance of the formation and implementation of the synergistic effect by increasing the efficiency of interaction in the cluster, the use of innovative management. The concept of a logistics-oriented approach and controlling technologies, the development of international certification methods in the Russian Federation have been developed.

Keywords: Competitiveness · Cluster · Controlling · Management · Synergy · Medical industry enterprises

1 Introduction

Currently, most industrial enterprises in Russia (and the medical industry enterprises (MIEs) in particular) are faced with the acute problem of increasing the competitiveness of their products to a level that meets or exceeds international standards (Federal Law No. 488-FZ of December 31, 2014). We consider he MIEs as those enterprises and separate structures of enterprises whose mission is the development and production of medical devices (MDs). The MIE products are classified as high-tech products (Decree of the Government of the Russian Federation No. 305 dated April 15 2014). Accordingly, the solution to the problem of the efficiency and competitiveness of MIE and their products are becoming critical not only for healthcare, but also for practical research in the technological development of industry, the economy, and strengthening national security (Osipov 2012).

An analysis of the published works in this area shows that the efficiency of production and the competitiveness of its products are largely determined by the effectiveness of enterprise management, -innovation, including the management of the competitiveness of enterprises and their products (Mescon et al. 2006; Hauschildt and Salomo 2011). One should note that in addition to the strategic and tactical management of production processes and the effectiveness of MIE management, the efficiency of the process of formation of organizational and economic sustainability (OES) is significant for increasing the competitiveness of MDs and MIEs (Omelchenko and Mirotin 2011).

Taking into account the analysis of the effectiveness of high-tech innovative industries, the work assumes that for enterprises and organizations that have common

A. V. Bogoviz and Y. V. Ragulina (Eds.): ISCI 2019, LNNS 115, pp. 294–300, 2020.
https://doi.org/10.1007/978-3-030-40749-0_35

innovative goals of developing competitive products, the cluster form of interaction is most effective, including for the healthcare sector (Gertsik and Omelchenko 2016; Porter 1991, 2010; Balashov 2011; Agafonov 2017).

This provision also requires improving the efficiency of managing the competitiveness of the MIE in the cluster, mainly due to strengthening its innovative component by, for example, increasing the regulatory impact on the organization of production, the novelty of which is protected by intellectual property protection documents. These problems determine the relevance of research in the field of the possibility of increasing the effectiveness of managing the competitiveness of MIE and its products, namely MDs, in an integrated cluster structure, taking into account the possible use of a synergistic effect (Vetrova et al. 2017).

2 Materials and Methods

The paper used the information and analytical reviews, materials of scientific works aimed at improving the competitiveness and economic development of enterprises, retraining, and advanced training of specialists in the field of the medical industry, healthcare, and economic sciences. Also, we relied on the regulatory documents of state bodies, reference, and informational materials of the websites of manufacturers of medical devices, healthcare institutions, social protection, and labor protection bodies. Besides, articles on the history and dynamics of cluster structures in the pharmaceutical industry and healthcare, materials on theoretical issues of management, competitiveness, and quality control of enterprises and their products were also reviewed.

3 Results and Discussion

The purpose was to analyze the main principles of the methodology for the OES formation and management and the MIE competitiveness, taking into account the developed concept, the logistics-oriented approach, and institutional theory to ensure a synergistic effect in an integrated cluster-type system. The author considers the following tasks as the main ones:

- A comparative analysis of the socio-economic foundations for developing the MIEs' efficiency when integrating them in cluster systems, including healthcare institutions;
- Analysis of the prospects for the implementation of management of economic and organizational integrated cluster systems for the combined solution of the problems of increasing the management effectiveness of health care facilities (HCF) of the healthcare system and MIE, adoption of necessary management strategies that are consistent with the common goals of participants to increase the competitiveness of HCF, MIE, their services and products;
- Justification of the feasibility of forming and analyzing the economic efficiency of the supply system (SS) (Omelchenko et al. 2011), the most appropriate solution to

the main task of both healthcare and the medical industry is to maintain and strengthen human health.

Solving these problems requires creating a system that can be formed from separate structures with their own missions, tasks. In turn, this system should have a uniting, socially significant goal. In general, such a system can be characterized by the classification of economic and social indicators reflecting the system competitiveness, which ensures the efficiency of operation by optimizing costs to obtain desired results (Table 1). The primary role of the state in the effectiveness of the functioning of health care institutions (HCF and MIE) in such a cluster structure is to create new and improve the existing legislative regulations at the level of federal regulatory documents in the field of MD circulation. The Federal Law No. 323 *On the Basics of Protecting the Health of Citizens in the Russian Federation* should be taken as a basis for determining indicators of the efficiency of MD circulation in the Russian Federation.

In accordance with institutional theory (Yerznkyan 2017; Yerznkyan et al. 2017), the management system forms sustainable management mechanisms of interaction in the process of joint activities to achieve a common goal. The initially financially and functionally independent participants in cluster structures regulate their actions by harmonizing formal and informal norms and rules.

According to the author, the classification presented in Table 1, reflects some indicators of the effectiveness of the management of cluster systems in the medical industry and healthcare significant for improving the quality of medical care. To a large extent, these indicators characterize the competitiveness of MIE and MD. Moreover, based on (Komarova 1990), the MIE competitiveness is understood as "the struggle between producers for more favorable conditions for the production and sale of goods, for obtaining the highest profit." And competitiveness for MD is understood as the "rivalry, struggle to achieve better results …" (in this case, in their production and operation).

In a market economy, these concepts are interconnected in most cases. In a planned economy, products may be uncompetitive despite the high competitiveness of the enterprise. They are subsidized by the state and focused on the production of more significant products under government orders. This led to a sharp increase in imports, especially high-tech MD, requiring substantial resources and production costs.

When analyzing the effectiveness of management decisions on the formation of competitiveness of cluster structures and their products, it is advisable to apply project financing methods (PF). In studies, it is advisable to use the CAPM model (Capital Asset Pricing Mode – the Sharp and Lintner model for accounting for capital assets) and the model underlying financial arbitration (Arbitrage Pricing Theory – APT model). The application of the CAPM model is based on the assumption that the systematic risk of financial assets depends on how sensitive they are to fluctuations in the level of profitability of the market portfolio of investments (or the industry average profitability rate). This issue is especially relevant for domestic MIE due to the significant transformation of the financial and economic component during the transition of the country's economy to market conditions of operation and bulk purchases of the imported MD and the resulting competition from foreign manufacturers, respectively. With this in mind, issues of innovative management development, with the maximum use of PF, are of particular importance, given the tendency to equip health facilities with high-tech and high-quality products (Gertsik 2016).

Table 1. Classification of economic and social indicators of the effectiveness of integrated economic structures in the field of healthcare and the medical industry.

Indicators	Medical industry enterprise (MIE)	Health care facility (HCF)
Economic	1. Compliance of economic and industrial activities with regulatory documents in the sphere of MD circulation 2. Improving the MIE and MD competitiveness 3. Expansion of production of innovative MIE products	1. Compliance of economic and medical-preventive activities with regulatory documents in the field of circulation of medical devices 2. Improving the HCF competitiveness and medical technology 3. Development of innovative medical technologies
Social	1. Increasing OES MIE 2. MIE brand promotion 3. Improving the efficiency of MD operating conditions	1. Enhancing OES HCF 2. HCF brand improvement 3. Improving the quality of patient care

We note that quality control, reducing the risks of late deliveries of products and raw materials, is ensured by using the standards of the Supply Chain Management system (SCM) and the Manufacturing Execution Systems (MES). The purpose of applying these standards is to manage the supply and sales processes of an enterprise and control the distribution of goods and production processes from the perspective of a logistics-oriented approach.

The creation of these quality control systems was determined by the requirements of production. Their application and development have own history. So, the term "SCM" (Supply Chain Management) was coined by Oliver and Weber in 1982. SCM is designed to implement and control the purchase of raw materials, production, and sales of products, and it is a logical use of an integrated approach to logistics. This is especially important in the formation of adaptive management systems for modern high-tech production, aimed at improving the competitiveness of high-tech products of industrial enterprises (Chursin 2017).

At the same time, the SCM-system can be divided into two subsystems.

First, Supply Chain Planning (SCP) is the planning of "supply chains." SCP is based on regulatory documents for planning and scheduling product development and sales schedules. SCP also includes systems for the joint development of forecasts taking into account the potential risks of exposure to the internal and external environment of the enterprise. The concept of "supply chain management" defines the application of logistics operations throughout the product life cycle. At this stage, in the cluster, for the formation of adaptive management systems for modern high-tech production (Chursin 2017), it is advisable to use controlling technologies that ensure the adoption of preventive measures to optimize production, orienting to the output of competitive products (Falko 2017).

Second, Supply Chain Execution (SCE) is the execution of "supply chains" in real-time. The use of the SCE system allows both the MIE and the medical and technical cluster as a whole to perform the following actions: (a) to forecast daily/weekly sales of products; (b) to manage product inventories taking into account the selected management model and risks of overproduction or insufficient production volume; (c) to manage replenishment of stocks; (d) to optimize supply planning within the enterprise's logistics network, taking into account the risk of non-fulfillment of the sales plan, the risk of insufficient deliveries of materials and components, the presence of residues, transportation capacities, as well as the risk of various restrictions on the supply conditions and business rules. The latter is especially relevant in the delivery of innovative high-tech MD to various HCF with different staffing, using different diagnostic and therapeutic technologies, having a heterogeneous patient population, but using the same MD produced by MIE. This may lead to changes in the volumes of MD deliveries of a particular MIE with a significant change in the contingent of patients and the used technologies of therapy and diagnosis.

It is also worth emphasizing that in order for the Russian MIEs to compete on equal terms with foreign MD manufacturing companies, fulfilling the requirements for import substitution, it is necessary to organize the work of the MIE in the cluster structure, taking into account the global rules of competition. One of the generally accepted standards of competitiveness is the rule that is typical for many foreign MIEs. This rule implies confirming the level of their development and production of MD by the presence of quality certificates. It should be noted that the "Quality Mark" marked earlier, high-quality products of domestic enterprises but limited competition led to the devaluation of this quality indicator of domestic products. However, recently, taking into account the requirements of a market economy, the "Russian quality system" is being revived in the Russian Federation (Government of the Russian Federation 2015).

The presence of a quality certificate allows the organization to compete with similar organizations both in the domestic and foreign markets, find new partners and enter into additional contracts, including state ones, receive preferential investments, loans, etc. Concerning the field of medicine and the production of medical equipment, quality issues become more relevant. This is due to the fact that not only the company's success in the market, but the life and health of a large number of people with disabilities in a number of important body systems (designed to control and regulate external influences on a person) depend on the proper provision of product quality, quality after-sales service.

In particular, the application of 9,000 family standards, based on the use of a process approach in assessing potential risks, is aimed at achieving the goal of increasing competitiveness. These standards can be applied (and are applied) in the management of innovative MIEs producing high-tech MD. The ISO/IEC 27001 standard is designed to create, implement, monitor, maintain, and improve the information security management system and reduce the risks of its violation. Based on the current market conditions for the development of the country's economy, it is advisable to introduce international certification methods for MD and MIE more widely. Unfortunately, at this stage, more attention is paid to the possibility of conducting international certification of domestic MD and MIE based on international certification centers, referring to the insufficient equipment and qualified staff of Russian certification centers. At the same time, the investment of foreign certification centers by paying for their services in the long term is financially comparable with the creation of a domestic, international center with characteristics exceeding overseas centers, including in terms of interaction with manufacturers of the Eurasian Economic Union (EAEU) and other countries. Cooperation with these countries continues despite sanctions. Russia has many years of experience in conducting technical tests based on the multifunctional medical testing center VNIIMT/NPO "Ekran." Full-scale studies are also conducted in Rostest-Moscow divisions.

With this in mind, the creation of an international testing center in Russia is justified both technologically and economically.

4 Conclusions

These measures provide competitive advantages for the MIEs in integrated structures:

1. The main innovative solutions in the management of enterprises and the cluster should be focused on ensuring the implementation of the competitiveness of participants and products of the cluster, based on the use of the most significant advantages of integration of both enterprises and medical institutions in the cluster.
2. When assessing the effectiveness of enterprises in cluster systems in the medical industry and healthcare, it is advisable to apply project financing methods, including analysis and consideration of the possible dynamics of the risks of exposure to negative factors from the internal and external environment using controlling technologies.
3. In order to increase the efficiency of both the cluster and its participants, management decisions on certification of both products and the enterprise itself in the cluster are possible and appropriate. At the same time, it is necessary to take into account international quality standards based on research and testing laboratories of the Russian Federation, which can increase the competitiveness of both MIE and its products.

References

Agafonov, V.A.: Regional development and cluster strategy. In: Kleiner, G.B. (ed.) Strategic Planning and Enterprise Development. Materials of the Eighteenth All-Russian Symposium, pp. 562–565. Central Economics and Mathematics Institute of the RAS, Moscow, Russia (2017)

Balashov, A.I.: Formation of innovative pharmaceutical clusters. Natl. Interests: Priorities Secur. **13**(106), 29–35 (2011)

Chursin, A.A.: Formation of adaptive management systems for modern high-tech production. Econ. Manag.: Probl. Solut. **4**(5–1), 27–33 (2017)

Falko, S.G.: Controlling in the process of implementation and optimization of production systems. Controlling **1**(63), 2–5 (2017)

Gertsik, Yu.G.: Prospects of cluster initiatives in Russia and the worldwide. Health Soc. Care J. **2**(3), 9–23 (2016)

Gertsik, Yu.G., Omelchenko, I.N.: Organizational and economic stability and competitiveness of the medical industry: The concept and methodology of formation. Publishing House of MSTU named after N.E. Bauman, Moscow, Russia (2016)

Government of the Russian Federation: Decree of the Government of the Russian Federation "On approval of the state program of the Russian Federation "Development of the pharmaceutical and medical industry" for 2013 – 2020" (April 15, 2014 No. 305), Moscow, Russia (2014)

Government of the Russian Federation: Order of the Government of the Russian Federation "On the establishment of an autonomous non-profit organization 'Russian quality system'" (April 30, 2015. No. 780-r) (2015). http://government.ru/docs/17902/

Hauschildt, J., Salomo, S.: Innovations-Management, 5 Auflage. Vahlen Verlag, Munchen (2011)

Komarova, L.N. (ed.): Dictionary of foreign words. "Russkiy yazyk" Publishing House, Moscow, Russia (1990)

Mescon, M., Albert, M., Hedouri, F.: Fundamentals of management. Williams, Moscow, Russia (2006)

Mirotin, L.B., Omelchenko, I.N.: Engineering logistics: Logistic-oriented product life cycle management. Goryachaya liniya-Telekom, Moscow, Russia (2011)

Omelchenko, I.N., Sheresheva, M.Yu., Gorlacheva, E.N., Kolesnik, N.A.: Application of a stochastic approach to analysis of revenue in industrial networks. Sci. Educ. **2**, 1–30 (2011). http://technomag.edu.ru/doc/165985.html

Osipov, V.S.: Clusters as an instrument of state economic policy. Vestn. Econ. RAS **6**, 92–97 (2012)

Porter, M.E.: Towards a dynamic theory of strategy. Strateg. Manag. J. **12**, 95–117 (1991). (Special Issue: Fundamental Research Issues in Strategy and Economics)

Porter, M.E.: What is value in health care? N. Engl. J. Med. **363**, 2477–2481 (2010)

Presidential Executive Office: Federal Law "On industrial policy in the Russian Federation" (December 31, 2014 No. 488-FZ), Moscow, Russia (2014). http://kremlin.ru/acts/bank/39299

Vetrova, N.A., Gladkov, N.Yu., Gorlacheva, E.N., Gudkov, A.G., Omelchenko, I.N., Naraikin, O.S.: Assessment of the synergistic effect arising in the process of intercompany interaction. Nanotechnol.: Dev. Appl. - XXI Century **9**(2), 3–12 (2017)

Yerznkyan, B.A.: Institutional strengthening: three types of relationships. Inst. Stud. J. **9**(1), 27–38 (2017)

Yerznkyan, B., Gassner, L., Kara, A.: Culture, institutions, and economic performance. Montenegrin J. Econ. **13**(2), 71–80 (2017)

Possibilities of Applying Foreign Talent Management Experience in Russian Companies

Marina V. Savelyeva, Tatyana N. Kurina, and Yulia S. Zykova(✉)

Department of Management, Institute of Business Studies, Russian Presidential Academy of National Economy and Public Administration, Moscow, Russia
marina.savelyeva@gmail.com, tk19@rambler.ru, volna21@mail.ru

Abstract. This study examined the strategies, methods and tools of talent management in foreign and Russian companies. The following countries were selected to analyze foreign experience: (a) European countries (Germany, Switzerland, UK); (b) Asian countries (Japan, Singapore, Hong Kong); and (c) North America (USA). The aim of the study was to analyze and identify the most relevant and effective talent management practices of highly qualified specialists and employees holding senior positions, as well as to compare the effectiveness of different approaches and methods of talent management in different countries of the world. As part of the study, a synthesis of the experience of foreign countries in terms of talent management and a comparison with the talent management methods that are used in Russia are presented. In conclusion, we propose the most effective methods and practices practiced abroad that are of value to Russian companies.

Keywords: Management · Company management · Personnel management · Talent management · Top management · Coaching · HR · HR management · Training · Personnel partnership · Relationships · Leadership · Talents development · Personal development · Leadership training

1 Introduction

Today, one of the key tasks facing the leadership of both Russian and many foreign companies is to overcome the shortage of qualified employees. At the same time, the assessment of the significance of this task by specialists exceeds the significance of the problems associated with corruption, the need to adjust the organization's activities in a crisis, or when changes in expectations and behavior of customers (PWC 2016). At the same time, the growing need in organizations for qualified personnel is accompanied by the decreasing working population in Russia. Implementation in *talent management* organizations contributes to the solution of this problem. Talent management refers to a set of personnel management tools that ensure the attraction, development, effective use, and retention of employees who make a significant contribution to the development of the organization.

A. V. Bogoviz and Y. V. Ragulina (Eds.): ISCI 2019, LNNS 115, pp. 301–310, 2020.
https://doi.org/10.1007/978-3-030-40749-0_36

In other words, talent management is an area of HR management, including attracting qualified employees, adapting new employees, as well as retaining staff to meet current and future business goals. In organizations that implement talent management, this practice is included in the sphere of activity not only of the *HR* department but also of all managers.

Talent management also includes training, use, and development of talented employees who are able to solve complex business problems effectively and can continue to occupy leadership positions in the company. This practice makes it possible to identify and retain talented people in key positions, motivate them to achieve high individual results. At the same time, the purpose of talent management is to identify the talents of employees and use them in the best way for the employee and company.

The specifics of talent management is manifested in a change in the approach to personnel management, i.e., the principle of equal treatment of all employees is replaced by the principle of differentiation of employees, the allocation of talented people from the total number for the development of which unique investments are allocated. At the same time, in managing the company as a whole, it is necessary to envisage the selection of the most promising specialists, their career planning, using tools for their development, and their retention, of course.

Thus, the implementation of talent management implies not only the development of an appropriate approach to personnel management but also the formation of an appropriate philosophy in the context of managing the company as a whole. This practice allows not only to improve the quality of staff work but also to increase the activity of talented specialists who can become a sustainable competitive advantage of the organization and increase its competitiveness.

At the same time, the leaders of Russian companies imagine talent management in a slightly different way than their colleagues from abroad (Fig. 1).

It is interesting that in foreign companies in talent management, the emphasis is on preparing future leaders, the culture and behavior of a person in the workplace, and performance management. While in our country, the critical aspects of talent management are constant or variable remuneration, benefits, searching for the right skills, and human resource flexibility. Based on this understanding, talent management methods and tools are selected and used.

2 Measurement Approaches

According to a ranking based on the Global Talent Competitiveness Index *(GTCI)*, published by the INSEAD International Business School in 2018 and covering 125 countries, Switzerland is a leader in the talent competitiveness in the world. Swiss companies achieve a high level of attraction and retention of talented professionals with a combination of an ideal environment that provides regulation through market and business and labor conditions. The ratings of individual countries discussed below are shown in Table 1.

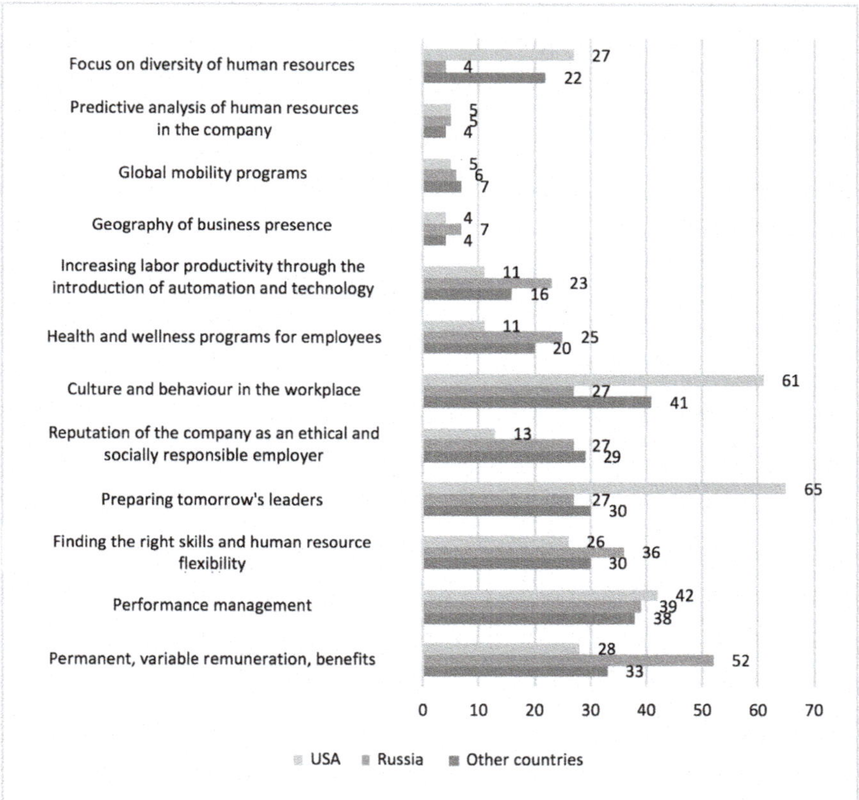

Fig. 1. Representations of talent management in Russia, the USA, and other countries, %. Source: (PWC 2016).

Table 1. Rating of countries according to the quality of talent management and the quality of skills of employees in organizations (Lanvin and Monteiro 2019).

Country	Rating GTCI	The quality of the company's talent management			Employee skills quality	
		Attraction	Development	Retention	Vocational skills	Erudition and horizon
Switzerland	1	5	2	1	1	4
Singapore	2	1	11	26	7	1
USA	3	14	1	13	2	3
United Kingdom	9	9	9	11	27	5
Germany	14	20	13	10	3	23
Japan	22	45	20	20	23	26
China/Hong Kong	45	76	24	62	67	43
Russia	49	109	44	52	47	28

The Right Management (Manpower Group) company conducted a study on the talent management of employees, including medium and high-rank specialists.

According to a company study conducted with executives from 2,221 companies from 13 countries (Right Management (ManpowerGroup) n.d.), the following key talent management initiatives can be noted:

- Development and improvement of leadership skills;
- Executive coaching;
- Assessment of talent to fill a vacant position.

Key talent management initiatives in the context of individual countries are shown in the following tables (Tables 2, 3 and 4).

Table 2. Development and improvement of leadership skills (Right Management (ManpowerGroup) n.d.).

Country	%*
USA	56%
Japan	55%
United Kingdom	54%
Norway	41%

** percentage of the total number of company executives participating in the survey.*

Table 3. Executive Coaching (Right Management (Manpower Group) n.d.).

Country	%*
China/Hong Kong	59%
Germany	41%

** percentage of the total number of company executives participating in the survey.*

Table 4. Talent assessment to fill a vacant position (Right Management (ManpowerGroup) n.d.).

Country	%*
Singapore	44%

** percentage of the total number of company executives participating in the survey.*

Thus, on a global scale, *leadership development* has been identified as the main focus of talent management investments. 46% of organizations plan to invest in it. Organizations also plan to develop targeted skills critical to the organization (39%) and invest in skills assessment (33%).

In the Asia-Pacific region, 39% of respondents said they would focus on executive coaching, in comparison with 22% in the Americas and Europe. Team building is also considered one of the top priorities in the Asia-Pacific region. 44% of respondents plan to invest in this area, in comparison with 26% in Europe and 25% in the Americas.

Most successful companies see employee engagement as a way of thinking, rather than a program that starts and stops. Not only do they measure engagement, they actively listen to what employees are saying and take a holistic approach to action to maximize results and optimize impact.

Given that most organizations around the world plan to invest more in leadership development and skills development at all levels, it is not surprising that 48% said they *would expand their employee engagement programs on complex and substantial projects* to ensure that these investments would benefit. This is true for all regions, while the most significant response was in the Asia-Pacific region, where 59% of respondents said they plan to implement this initiative. In this region, companies from China/Hong Kong (85%) and Singapore (57%) are most interested in implementing an employee engagement strategy.

In the Asia-Pacific region, there is a significant difference in the levels of market maturity and economic growth between the countries that participated in the study. Country results vary widely. For example, given the aggressive growth of labor markets in China/Hong Kong, 85% of respondents plan to expand their HR programs, in comparison to 49% in Germany.

While most organizations (73%) around the world rely on a combination of internal talent development and external recruitment, very few (only 13%) are confident in the strength of their leadership. Confidence in their current management channel is particularly weak in some countries, including Singapore, Germany, and the United States.

The Right Management experts believe that since most companies seem to be looking for new leaders in their organization, further development of the management channel is a prerequisite.

According to the survey, the leaders of companies in several countries believe that their enterprises have a well-developed system for building a management channel that can almost wholly satisfy the company's needs for highly qualified specialists and employees of a managerial level (Table 5).

Table 5. The presence of a well-developed management channel (Right Management (ManpowerGroup) n.d.).

Country	%*
Japan	23%
China/Hong Kong	15%
United Kingdom	14%
USA	12%
Germany	12%
Singapore	12%

** percentage of the total number of company executives participating in the survey.*

According to a study conducted by *Right Management,* one can conclude that most countries focus on developing leadership and management skills for talent.

The development of the management channel system involves promoting the leader on the "ladder of excellence": "self-management – management of others – managers management – a functional method – business management – group management – enterprise management."

The leadership channel is considered a useful tool since it meets several needs of talented and highly efficient employees, namely the need for personal and career growth, involvement, expansion of job responsibilities, and areas of personal influence and leadership in the company.

In more detail, the strategy of building an effective leadership "pipeline" used in foreign companies is as follows (Decision Dynamics n.d.):

- Creating success profiles for key positions;
- Identifying the strengths and talents of the leader and colleagues;
- Measuring and evaluating employee talent and engagement, both in terms of productivity and potential for current and future roles;
- Identifying the strengths required for critical positions and provide attractive career paths;
- Attracting and developing talents with the help of motivating and challenging tasks, projects, and career experiences;
- Realization and disclosure of potential talents in complex tasks, roles, and projects;
- Talent development to increase productivity in current and future positions;
- Preservation of critical talents through the development of motivating and individual development plans.

According to experts at *The Boston Consulting Group*, many companies devote a significant amount of time to top stars, namely future leaders (Bhalla et al. 2018). As a rule, they spend much less time supporting and developing leaders who manage client-oriented functions or functions that directly affect growth, profitability, or other critical performance indicators. These leaders work on value creation; they are also the key to implementing the company's goals throughout the organization, which, according to a study by *The Boston Consulting Group*, directly affects efficiency. In a digital world where creating and managing customer experience is the foundation of building relationships, leaders at the forefront are crucial to success.

The path of an individual employee from a specialist to a leader is the most significant change in the role of any employee; however, in many companies, these newly made "leaders" receive little or no useful support at all. Only the "Welcome to leadership" token training or some control modules that should be addressed when multitasking is held for many of them.

At this stage, company management must be directly involved in the training of this management cohort. Such training may be selective, but this is often enough to bring the leadership and managerial qualities of a young leader to a new level.

Top leaders help provide leaders at the forefront with practical tools for solving the problems they face. During this process, leaders at the forefront build their capabilities in the real work that they perform every day, which entails the necessary changes in the organization and helps attract more interested teams that increase productivity and provide the best experience for customers.

3 New Technologies and Methods

A study by McKinsey (Kerr and Toriello 2012) emphasizes the importance of evaluating departmental databases using new analytical technologies. The survey said, "HR has a lot of information. It could be a bottomless database for a company." The more accurately a company can measure its impact, the more useful this new scenario will be. This trend is known as *"People Analytics,"* i.e., the process of collecting, organizing, and analyzing data on employee behavior in order to use this information to make decisions in the company by predicting trends and improving strategies.

McKinsey also seeks to include all employees in the strategic planning of organizations. This is a trend that focuses on reducing hierarchies within and outside companies, and also makes them more flexible, in addition to exchange relationships that focus on common goals. Socially conditioned evolution in the changing "employer/employee" relationship creates new problems and potential solutions that differ from those that the previous generations of leaders worked with. A "new organization" requires the independent, experienced, globally-oriented teams, young leaders, and a new talent management model.

Although top-level positions remain in companies, leadership is now becoming a "team sport," where leaders must inspire and unite teams, and be able to connect with other groups. Besides, they should share and maximize the use of people and their abilities. Companies must invest even more in leadership development and automate transactional and repetitive functions, allowing leaders to devote themselves to nurturing and caring for the development of culture and talent.

An instrument such as training, in the context of talent management, is usually applicable to young professionals. However, large companies such as *The Boston Consulting Group* (Bhalla et al. 2018) offer a wide range of training programs, including in-depth courses to develop the essential skills needed at every stage of career growth – from a junior consultant to a senior partner.

In this regard, from Russian companies, it is worth noting Sberbank, which conducts training and professional development programs for employees of any level so that they have the opportunity to listen to lectures by colleagues and invited speakers, learn bank products and learn how to better interact with colleagues and clients.

Sberbank specially built an educational center to train its employees. Foreign experts and bank managers teach there. Sberbank is collaborating with *Cambridge English Language Assessment* to help employees learn foreign languages. Studying books from the corporate library helps in the professional and career development of employees.

The Metro Cash & Carry company has separate training programs for executives and managers. For example, a company offers a business course that is prepared according to the standards of leading business administration programs (MBAs). Part of the training takes place at the *Metro* Education Center in France. Also, employee training is not only paid as working time, but employees also receive a supplement for successful training. One of the most attractive tools for attracting, retaining, and developing talent for employees and candidates is international rotation.

The next important tool for the preservation and development of talent is the personnel reserve.

For example, at MTS, a talent pool program is called "Talent Management." In total, 85% of leadership positions are "closed" by internal candidates thanks to the program. Employees holding various positions in the company, from department heads to leading specialists and experts, can enter a two-year program. As part of the "Talent Management" program, MTS personnel is divided into the following three categories: successors (candidates from among the leaders for managerial positions of the next level), line employees with high development potential (*Hi-Po*), and narrow specialists with any unique expert knowledge or skills, or high efficiency (*Hi-Pro*) (Sereda 2018).

Heads of departments nominate candidates for the program during the annual performance evaluation and discuss at the annual collegiate session with the participation of MTS vice-presidents. If an employee has a consistently high assessment of performance, high potential, and interest in career advancement, then his chances of getting into the program are high. After selection, the most critical thing begins, namely the development of a comprehensive plan for further training for each category. The plan includes participation in professional conferences, internships, target programs of the Academy of Success and the Academy of Leadership, programs for the development of strategic thinking, emotional intelligence, managerial skills, and much more.

Speaking about the application of foreign talent management strategies in Russia, it is necessary to take into account the peculiarities of the Russian mentality, the current economic situation, and the labor market itself. For example, for Russia, such a factor is characteristic as low mobility of candidates, i.e., a small (in comparison with the countries of Europe and North America) percentage of applicants is ready to move to another region where a suitable vacancy has opened. According to a study by the Russian personnel holding "Ankor," there is a high demand for mobile personnel from employers in Russia, on the one hand, and a low level of labor mobility in the labor market from the candidate field, on the other (Gordeev 2019). In this case, the employer should provide an appropriate compensation package to interest the candidate.

The involvement of candidates in exciting and large-scale projects continues to be one of the key factors allowing both to retain a talented employee and to guarantee the further development of talent.

In Russia, the leadership approaches of most companies to talent management are still significantly different from the key strategies used by leading countries in this area.

In addition, one should note that in Russia, the approach to forming a pool of candidates varies significantly depending on the size of the company.

For comparison, the following answers were provided as survey outcomes *(Right Management (ManpowerGroup)* n.d.*)* about talent management methods from the U.S. company executives:

- Leadership development (e.g. coaching, employee development with high potential, succession management, productivity management, multicultural competencies).
- Talent acquisition (e.g., hiring, identifying sources of hiring, selection, and onboarding).[1]
- Employee engagement (e.g., engagement and retention strategies, healthcare, performance optimization).
- Individual and team development (e.g., competency modeling, organizational, team, and individual assessment).
- Organizational effectiveness (e.g. implementing development strategies, change management).
- Outplacement and change of position (e.g. support for changes in a career, career management, and development).

4 Conclusion

- In the *United States and Europe*, talent management focuses on intangible promotion. Recently, one of the problems of large companies is the retention of key employees, whose interest can be less and less explained by real bonuses. (Often, top employees themselves call financial compensation "golden cuffs," and new strategies need to be developed to maintain employee interest in continuing to work inside the company). Here, attracting staff to work on complex tasks and projects, expanding the scope of responsibility, providing more freedom of action, and the ability to make decisions independently come to the fore. In the case of high-level talent management, the practice is often when a new project or company line of business is opened in order to retain a valuable employee.
- For *Asian countrie*s, high dynamics of increasing the pool of employees and the development of their talents are characteristic. Since in countries such as Singapore and Hong Kong, the share of expats attracted to critical positions is traditionally high, there is a shift towards the development of their human resources and management. This trend is dictated primarily by financial requirements. In the context of the global unstable economic situation, companies operating in relatively young markets are more likely to opt for leadership development within their own company, rather than hiring executives from North America or Europe.
- In *Russia,* the aggregate material remuneration remains the dominant factor. This approach of company managers does not meet the requirements of the modern market and requires a shift in emphasis on strategies that are applied by world leaders in this field. In particular, the use of such a useful talent management tool as the "development of a management channel" would make it possible to come closer to solving the personnel crisis problem that many Russian companies face.

[1] Onboarding is an acquaintance of new employees with labor activity, adaptation to the social environment of the organization, and their successful inclusion.

References

ANCOR: HR Digest, Issue IV (2015). https://ancor.ru/media/press/digest/pdf/ANCOR_Digest 2015_VI.pdf

Bhalla, V., Caye, J.-M., Lovich, D., Tollman, P.: CEO's Guide to Talent Management Today (2018). https://www.bcg.com/en-ru/publications/2018/ceo-guide-talent-management-today.aspx

Decision Dynamics: Research-Based Talent Management (n.d.). http://www.decisiondynamics.se/en/talent-management

Gordeev, V.: The World Bank lowered its forecast for economic growth in Russia and in the world (2019). https://www.rbc.ru/economics/09/01/2019/5c353d059a7947dfebf839cd

Kerr, J., Toriello, M. (eds.): Research Report. The State of Human Capital 2012. The False Summit (2012). https://www.mckinsey.com/~/media/McKinsey/dotcom/client_service/Organization/PDFs/State_of_human_capital_2012.ashx

Lanvin, B., Monteiro, F. (eds.): INSEAD (2019): The Global Talent Competitiveness Index 2019, Fontainebleau, France (2019). https://gtcistudy.com/the-gtci-index/#gtci-rankings-table

PWC: Russian edition of the 19th Annual survey of executives of the largest companies in the world 2016 Chayaniya of Russian business (2016). http://hr-media.ru/wp-content/uploads/2016/02/CHayaniya-rossijskogo-biznesa-opros-107-CEO-rukovoditelej-krupnejshih-kompanij-iz-Rossii.pdf

Right Management (ManpowerGroup): Talent Management Accelerating Business Performance. Global Trends, Challenges and Priorities (n.d.). https://www.right.com/wps/wcm/connect/5d228cd7-296f-439d-9fa0-a443564f29b5/rm_wp_lo-res_6-27-14.pdf?MOD=AJPERES

Sereda, A.A.: Talent management: creating a talent management development model, Master's dissertation. Russia, Yekaterinburg: Ural Federal University named after the First President of Russia B. N. Yeltsin (2018). http://elar.urfu.ru/bitstream/10995/66208/1/m_th_e.a.sereda_2018.pdf

The Impact of Talent Management on Company Competitiveness

Marina V. Savelyeva[1](✉) and Tatiana N. Kurina[2]

[1] Department of Management, IBDA RANEPA, Moscow, Russia
Marina.savelyeva@gmail.com
[2] IBDA RANEPA, Moscow, Russia
tkl9@rambler.ru

Abstract. The paper focuses on the impact of talent management on company competitiveness. The research methods are mainly theoretical, specifically: analysis, synthesis, comparison, generalization, and identification of signs, features, and properties of the phenomena. The scientific novelty is determined by comparing the performance indicators of the talent management system and the competitiveness indicators of the enterprise, as well as studying the influence of human resources on the formation and development of competitive advantages of the mesoscale. The main results of the study are as follows: human resources are one of the competitive resource advantages of the enterprise; there is a connection between the performance indicators of the implementation of the talent management system and the competitiveness indicators of the enterprise; talent management affects the competitiveness of the enterprise through the selection, selection, hiring and development of the most qualified, experienced and educated personnel. In the course of the study, it was concluded that in the process of managing human resources in general and talents in particular at the level of managerial and production personnel, the enterprise is being managed, developed, and competitively formed.

Keywords: Competitiveness · Enterprise competitiveness · Competitive advantages · Talent management

1 Introduction

Most scholars working on the issues of talent management are inclined to believe that talent management is a tool of the entire management system of an organization and is aimed at improving its performance indicators. Talent management, in this way, increases its competitiveness. Therefore, the hypothesis of the study is as follows: "Talent management affects company competitiveness". This study will attempt not only to prove the hypothesis but also to find out how talent management affects the competitiveness of the company.

A. V. Bogoviz and Y. V. Ragulina (Eds.): ISCI 2019, LNNS 115, pp. 311–317, 2020.
https://doi.org/10.1007/978-3-030-40749-0_37

2 Materials and Methods

The materials used in this study are mainly scientific and educational publications on the topic. The methods used in this article are mainly theoretical. These include a theoretical analysis of the educational literature consisting of the allocation and consideration of individual parties, signs, features, and phenomena properties, as well as the synthesis, comparison, and generalization of the above information.

3 Results

Before proceeding to the main part of the study, let us introduce such definitions as *talent management*, *competitiveness*, and *organization competitiveness*.

Talent management is the concept that is most often considered as a set of tools for personnel management, providing the ability of the organization to attract, effectively use, and retain employees who make a significant contribution to the development of the organization (Digital Encyclopedia "Wikipedia" n.d.).

In other words, talent management is an area of HR management engaged in attracting qualified employees, integrating new employees, and retaining staff for meeting current and future business goals (Community HR Managers "hr-portal" n.d.). In the companies adhering to a talent management strategy, the responsibility of the workforce lies not only in the HR Department, but also on all managers working in the organization.

Talent management is also defined as the purposeful activity of the company to create, use, and develop talented employees who can effectively solve complex business problems and are able to take the leadership positions in the company in the future (File archive for students "StudFiles" n.d.).

Talent management makes it possible to identify and retain key employees in key positions, as well as motivate them to achieve high individual results. The purpose of talent management is to identify the talents of the employees and use them in the best way for the employee and the company.

The key idea of talent management is to change the approach to personnel management through the transition from an ideology of equality to the idea of differentiation, that is, the allocation of talented people from the total mass, to bet on them and invest the company's resources. At the same time, the comprehensive management system in the company must correspond to this ideology and be reduced to work, not only on the selection of the most promising employees but also on planning their designation and retainment, using development tools.

Thus, based on the definitions, we can conclude that talent management is a combination of various tools that are used not only in the framework of personnel management in particular but also in the context of managing the organization as a whole. The quality of work that moves the organization forward, increasing its competitiveness in the market, depends on talent management. In order to understand how this happens, it is necessary to determine what both competitiveness, as a general term, and competitiveness of a company, in particular, are.

Thus, competitiveness, which is based on the concept of competition, is a key element of the market economy, the engine of economic growth and improvement of product quality, acceleration of scientific and technological progress, and reduction in prices and production costs (Gorfinkel and Shvandar 1997). As a factor of economic security, it occupies an important place in the conceptual apparatus of competitive relations (Vorozhbit and Zaitseva 2008).

Competitiveness, in its most general form, is defined as a general indicator of the effectiveness of the subject's functioning. It is a combination of its properties and characteristics to surpass other entities and meet all the requirements at different levels (country, region, industry, enterprise, product) under the same conditions. The ability to survive and develop depends on the achieved level of competitiveness (Potudanskaya et al. 2013). Also, the competitiveness is interpreted and formed at different levels – the macro level, meso level, and micro level. The enterprise level, considered in this study, is otherwise referred to as the meso level.

Nowadays, there are many definitions of the concept of *enterprise competitiveness*. Let us review the most common ones.

A. D. Gryaznova and A. Yu. Yudanov see enterprise competitiveness as its ability to be successful and demanded in the market, to compete with firms' competitors, and to receive more economic benefits compared to the companies that supply similar products (Ilyicheva 2010).

According to Ibraeva, enterprise competitiveness can be defined as the ability of entrepreneurs to produce and sell goods on the national and world markets that are more attractive in price and quality than the goods of national or foreign competitors (Ibraeva n.d.).

Irbe insists that the competitiveness of the subject should be understood as its ability to best use its resources and the formed internal and existing external institutional support for creating competitive advantages. It makes it possible to be more attractive to consumers, investors, and suppliers compared to counterparts while maintaining this financial stability and the positive dynamics of business value (Irbe 2012).

Different scholars identify the concept of competitiveness with the properties of an object, a means of a goal-achieving, or a means of analysis. However, the definition given by M. S. Irbe deserves the greatest attention in the framework of this study since it touches an important aspect that the company creates its competitiveness by using the resources available to it (competitive advantages).

The competitive advantages of the meso-level are as follows:

- resource competitive advantages (raw materials, labor, capital);
- strategic competitive advantages (technology and marketing strategies);
- operational competitive advantages (production and marketing efficiency).

It can be seen that labor resources belong to the competitive resource advantages underlying the competitiveness of the enterprise. From this, it can be concluded that talent management, which is a special case of human resource management, has an impact on the competitiveness of the enterprise since it allows us to manage its competitive advantages and improve them.

In order to understand how talent management affects the competitiveness of the enterprise, it is advisable to compare the performance indicators of the implementation of talent management system with the indicators of competitiveness of the enterprise, stipulating that the competitiveness indicators are a set of competitiveness criteria, which, in turn, are interpreted as qualitative or quantitative characteristics used to determine competitiveness (Tables 1 and 2).

Table 1. The performance indicators of a talent management system.

Indicator	Calculation procedure	Explanations
1	2	3
A number of completed projects by one employee, units	The calculation is based on the number of completed projects	The indicators of the number of projects implemented by a particular employee during the reporting period
Profit earned on one project, ₽	Profit earned from the implementation of all projects for the year/the number of completed projects	It allows us to identify the average profitability of one project
Profit attributable to one worker and one project, ₽/person	Profit per project/number of participants in the implementation of the projects	It allows evaluating the average profit indicator brought by one employee during the implementation of projects for the reporting period
Profitability of the project, %	Net profit received from the project/expenses for its implementation * 100%	It shows the profitability of the completed project as a percentage
Profitability of all implemented projects, %	Net profit received from all projects/expenses for its implementation * 100%	Shows the profitability of the completed project as a percentage
Projects cost, ₽	Costs associated with the implementation of projects/income received	Shows the expenses in the total amount of income received from the project

Source: Developed by the authors.

As can be seen from Table 1, the method under consideration provides a relatively complete assessment of the participation of talented employees in specific activities that bring profit to the company. However, it is worth clarifying that the project activity is far from being carried out by all modern organizations, which does not allow them to use the project method for evaluating talent management. Nevertheless, this is enough to compare the indicators examined here with the most frequently used indicators of the enterprise's competitiveness (Table 2).

Table 2 shows that the performance indicators of the implementation of the talent management system overlap with the enterprise competitiveness indicators. For example, the profitability of the project and the profitability of all implemented projects can be considered as a special case of profitability. The cost intensity of projects, being the cost per ruble, partially characterizes the total cost of production.

Table 2. The indicators of enterprise competitiveness.

Indicator	Calculation Procedure	Explanations
1	2	3
Production costs, ₽	Basic expenses + Overhead expenses = Materials (raw materials) + Depreciation deductions + Remuneration of labor + Other expenses (Internet portal for HR and accountants "okbuh.ru" 2018)	Production costs are the costs associated with the production process (from the start of the production to shipment of the finished goods to the warehouse) (Encyclopedia of the Economist "grandars.ru" n.d.)
Production profitability, %	It is calculated from the volume of production profit/production costs * 100% (Internet portal "Russian Start-Up" n.d.)	Profitability of production is the most important indicator that determines and shows the effectiveness of the enterprise
The share of exported products in total production, %	The volume of the exported products/Total volume of products * 100%	It reflects international competitiveness as a result of comparison with advanced enterprises of developed countries, which is achievable only when the country meets the requirements arising from the globalization process (Encyclopedia of the Economist "grandars.ru" n.d.)

Source: Developed by the authors.

A competitive enterprise is the one that can carry out its activities in market conditions while receiving a profit sufficient to maintain a competitive level of human capital, product quality, and means of production (Financial management center "center-yf.ru" n.d.). Such indicators of the effectiveness of the talent management system as profit per project and profit per person and per one project are closely related to the profit of the enterprise as a whole, and, therefore, affect the competitiveness of the enterprise through the influence on its indicator (which is also true for profitability and cost).

Having obtained even more empirical evidence of the impact of talent management on the competitiveness of a company by identifying the relationship between the performance indicators of implementing a talent management system and the indicators of enterprise competitiveness, it is necessary to study the key role of human capital (especially talents) and its management in the process of increasing enterprise competitiveness.

So, the human resources (namely talents) of the company affect its competitiveness as follows.

(1) It is talents (the most qualified and educated employees) that allow the company to create and introduce innovations modernize production. It is innovations that are one of the most important factors in the competitiveness of the enterprise, country, and products.

(2) The most qualified personnel allow the company to use, develop, and create such competitive advantages of the meso-level as operational (production and marketing efficiency) in a better way. Since it is human resources that are engaged in production and marketing activities at the enterprise, their effectiveness depends on the level of their experience and qualification, which suggests that it is necessary to hire the best of them and further develop them at the enterprise.
(3) The strategic competitive advantages represented by technological and marketing strategies, which are another type of competitive advantages of meso-level, also largely depend on human resources, since people create them.
(4) Furthermore, competitive advantages at the meso level such as resources, which, in addition to labor, are also represented by the capital and raw materials, also depend on human resources since they are under their influence and management.
(5) Also, the competitiveness indicators of the company (profitability, cost, the share of exported products in the total volume of production, etc.) are created and managed by people in different ways, which leads to the conclusion that the quality of human resources (experience, qualifications, education, etc.) affects the competitiveness of the company.

4 Discussion

Talent management affects the competitiveness of the enterprise through the compilation, selection, hiring, and development of the most qualified, experienced, and educated employees, which, in turn, create, develop, and manage the competitive advantages of the company that underlie its competitiveness.

In this context, talent management extends not only to managerial personnel (managers, specialists, and technical executives) but also to production personnel (main and auxiliary workers) since the quality of the products largely depends on them. In turn, the quality of the products is often referred to in determining the enterprise competitiveness and, along with price and marketing, is one of the competitive advantages of the micro level (product level).

Moreover, human resources are one of the competitive resource advantages of the meso level; therefore, competent management of the most qualitative of them (that is, talents) can improve the competitiveness of the enterprise.

The study also provided empirical evidence for the influence of talent management on the competitiveness of the company by identifying the relationship between the performance indicators of the implementation of the talent management system and the enterprise competitiveness.

5 Conclusion

In the course of the study, the hypothesis that talent management has an impact on the competitiveness of the company has been thoroughly proven. The following facts support this:

- human resources are one of the competitive resource advantages of the meso-level (enterprise-level), affecting its competitiveness;
- there is a connection between the performance indicators of the implementation of the talent management system and the competitiveness of the enterprise, indicating the fact that talent management is a particular case of managing the competitiveness of the enterprise as a whole;
- talent management affects the competitiveness of the enterprise through the selection, recruitment, and development of the most qualified, experienced, and educated personnel, who, in turn, create, develop, and manage the competitive advantages of the company, which are the basis of its competitiveness.

Thus, in the process of managing human resources (especially talents as their best part) at the managerial and production personnel level, there occurs management and formation of the enterprise competitiveness.

References

Community HR Managers “hr-portal”: Talent management (n.d.). https://hr-portal.ru/varticle/upravlenie-talantami

Digital Encyclopedia “Wikipedia”: Talent management (n.d.). https://ru.wikipedia.org/wiki/Управление_талантами

Encyclopedia of the Economist “grandars.ru”: Production price (n.d.). http://www.grandars.ru/college/ekonomika-firmy/sebestoimost-produkcii.html

File archive for students “StudFiles”: Organization talent management system (n.d.). https://studfiles.net/preview/3795995/page:15/

Financial management center “center-yf.ru”: Competitiveness indicators (n.d.). https://center-yf.ru/data/Marketologu/Pokazateli-konkurentosposobnosti.php

Gorfinkel, G.Ya., Shvandar, V.A. (eds.): Entrepreneurship Course. Textbook. Finance, Moscow (1997)

Ibraeva, A.E.: Methodological approaches to assessing competitiveness and competitive advantages (n.d.). http://www.konspekt.biz/index.php?text=1557

Ilyicheva, I.V.: Marketing: Teaching Aid. UlSTU, Ulyanovsk, Russia (2010)

Internet portal “Russian Start-Up”: Production profitability formula (n.d.). https://www.russtartup.ru/kak-sozdat-svoj-biznes/buhgalteriya-2/formula-rentabelnosti-proizvodstva-kak-opredelit.html

Internet portal for hr and accountants “okbuh.ru”: Product costing (2018). https://okbuh.ru/sebestoimost/raschet-sebestoimosti

Irbe, M.S.: Improving the competitiveness of an industrial enterprise based on an innovative development model. Competitiveness of the Russian economy: problems and ways of increasing. Collect. Sci. Art. **12**, 123–129 (2012)

Potudanskaya, V.F., Novikova, T.V., Vakhovsky, V.V.: Labor mobility as an indicator of staff competitiveness. Russ. Bus. **14**(15), 85–95 (2013)

Vorozhbit, OYu., Zaitseva, A.A.: Competitive advantages in the system of competitive relations of the organization. Econ. Sci. **40**, 258–261 (2008)

Problems of Transformation Management Business Models in Organizations

Roman R. Chugumbaev[1(✉)] and Nina N. Chugumbaeva[2]

[1] Moscow State University of Humanities and Economics, Moscow, Russia
romanry@ya.ru
[2] MIREA – Russian Technological University, Moscow, Russia
nina-ch2005@mail.ru

Abstract. In a relatively short time, the transformation of relations between business, society, and information turned into a force that changed the direction and speed of development of new management tools of the organization. One of these new areas has become innovative processes in the field of business modeling. The business model has become the basis of the competitiveness of modern economic entities. This study proposes to ensure proactive management of changes in the business model on a systematic basis. The paper proves that the model would protect against threats arising from the spread of new and more effective business models. Also, this business model would ensure the use of opportunities for better business functioning.

Keywords: Business model · Stakeholder theory · Management of change · Integrated reporting · Lean manufacturing

1 Introduction

Currently, the long-term success of an economic entity is less and less dependent on the technical and technological security of the organization, and more and more on the ability of the organization to adequately reflect on changing external and internal functioning conditions. Strengthening the globalization of the economy and the widespread penetration of digital technologies have significantly expanded the scope of competition. Among the most significant consequences in the context of the functioning and development of economic entities is that the circle of critical stakeholders is expanding, and their requirements are growing.

The company's competitiveness is now determined not only and not so much by technological superiority but by a wide range of characteristics that reflect the requirements of such key stakeholder groups. The increased attention and influence of various groups of stakeholders on the functioning of new economic entities contribute to a more active initiation of changes at the level of their business model. Business modeling is increasingly becoming an object of attention in scientific research, as well as in the management practice of companies focused on leadership. Even though the business model of the company is a relatively new modern concept of strategic management, it has already managed to gain the attention of economists around the world and has begun to be applied in competition. The attention coming from a wide range of

A. V. Bogoviz and Y. V. Ragulina (Eds.): ISCI 2019, LNNS 115, pp. 318–325, 2020.
https://doi.org/10.1007/978-3-030-40749-0_38

stakeholders to the company's business model was reflected in the content of international management and reporting standards. In scientific communities, interest in business models has appeared relatively recently (about a couple of decades ago) and has not waned to the present. In the works of the late 1990s and early 2000s, the authors often associate the construction of a business model with the operational aspect of the economic activity of organizations.

Thus, the business model of the enterprise is defined as a representation of the actual existing or planned business in the future, which is done in a form that can clearly demonstrate all the significant properties and characteristics of the enterprise associated with its ability to make a profit and be competitive (Kotelnikov 2007). The Harvard publication also directs the term "business model" to the plane of understanding the operational essence of the business, since it represents this category as a set of defining business solutions and also concessions and compromises used by the company for the generation of profit (Sahlman et al. 1999).

Later, in publications of periodical literature of a professional and scientific nature, other definitions of a business model were identified that were associated with the value nature of the economic entity. A business model is presented as a description of how the organization creates, delivers to customers, and acquires value, which allows the formation of a business value management system (Vashakmadze 2012) and profiting from it (Soolyatte 2010). Researchers from the Financial University under the Government of the Russian Federation consistently adhere to the understanding of the business model as an internal logic of creating added value in business segments on the basis of achieving organizational efficiency and the subsequent distribution of this value between stakeholders, ensuring their satisfaction (Berdnikov and Gavel 2011). Some researchers take into account the global digitalization of social and economic processes and consider a group of "platform" business models as special (Dawson et al. 2016; Konopatov and Salienko 2018).

In the standard on integrated reporting, the business model of the organization is represented as the system of transformation of resources through its commercial activity, in products and results, directed to the achievement of strategic goals of the organization and value creation throughout the short-term, medium-term and long-term periods. The business model plays the role of the core of the company and is based on different capitals, understood as resources, and thanks to its commercial activities, converts them into products (goods, services, by-products, and waste). The organization's activities and its products lead to results in terms of impact on capital. The ability of a business model to adapt to changes can influence the long-term work of the organization (Adams 2015). If we use a generalized understanding of the business model as representing critical aspects of the business that create value following the needs of stakeholders, then the increased activity of modern companies in initiating changes in the business model becomes logical. This is becoming a critical factor in competition and long-term business sustainability.

Therefore, this research is focused on problems of scientifically based management of the transformation of a business model of the modern enterprise.

2 Materials and Method

The study begins with an understanding of the business model as an object of transformational impact. Further, taking into account the modern environment for the functioning of the business, the study analyzes the approaches to the transformation. It proves that it should be carried out on a systematic basis using high-quality management tools. To ensure a sound management system, change the business model identifies critical areas of scientific and practical development, and identifies problems of their elements. The role of the stakeholder approach is determined in the qualitative construction of the proposed management system. To ensure the scientific validity of the creation of such a system, we analyze the reserves of business analysis development and the application of the concepts "lean manufacturing" and "Six Sigma." More than that, the acceptance of the elimination of the stakeholders' requirement is analyzed. The role of the quality of information and analytical support in the management system is illustrated.

3 Results

The flexibility of the business model provides a critical competitive advantage in the 21st-century economy, as its stakeholder orientation maximizes value creation. The approach focused on stakeholders has recently become the most progressive paradigm of business functioning and development (Freeman et al. 2018). The digitalization of society and the economy has ensured a high level of transparency of business processes, which made it very difficult for economic entities to obtain competitive advantages through previous solutions that provide innovations in production technology and increase operational efficiency. Previously, this was possible because only financial information was publicly disclosed to external users who have a direct and indirect financial interest. Other information, mostly, carried a trade secret and was protected by an information security system. Currently, more and more difficult to keep secrets from competitors' information on the technology, strategies, techniques, etc., as all this can be copied. And a continually transforming business model is challenging to replicate. When changes reorganize all processes in the company in such a way as to create maximum value for stakeholders, then unique internal competencies are designed to meet the changing requirements of stakeholders. As noted earlier, the business model of the company is subject to disclosure, according to the international standards of integrated reporting. However, the speed of changing the context of the company's operation is such that the business model is forced to transform continuously, which eliminates the competitors' possibility to "blindly" copy it.

All this suggests that organizations for successful functioning are forced to implement changes in the business model. The introduction of such changes can be carried out as (1) reactive, i.e., changes in the business model are controlled by the mechanisms of adaptation (adjustment) to the difference in the context of the functioning of the organization. Such an approach could well justify itself in the recent past when the conditions for the operation of economic entities were relatively stable in short-term and some aspects in the long-term period. Currently, this approach of implementing the changes is understood as a response to the threats of the external environment, and the solution is aimed at not losing the company's positions. (2) Then, the introduction of such changes can be proactive, i.e., systematic management of the difference in the business model. Such changes occur as a result of targeted managerial influences. This approach is preferable in current conditions as it allows us to use possibilities of the external environment and to initiate decisions in due time. Consequently, this will enable promoting to a more excellent level of functioning.

Building a management system to change the business model of the organization requires Management Sciences to develop similar approaches, methods, and techniques. At the same time, the theory of change management and the theory of stakeholders should become a theoretical basis. This makes it possible to manage changes in the business model, taking into account the understanding of the priorities of its functioning and development from stakeholders. Also, the business model is directed to the forecasting of future changes of requirements of stakeholders that allows taking necessary steps to its transformation (Integrated Reporting Committee n.d.). The orientation of the business model to meet the stakeholders' needs creates a competitive advantage for the company, as this ensures the construction of long-term relationships with stakeholders. Global digitalization opens up new opportunities for companies to meet the needs of stakeholders.

Global digitalization opens up new opportunities for companies to meet the needs of stakeholders.

- Timely initiating changes to the business model or its elements;
- Determining the real needs of the organization in the transformation of the business model;
- Providing a reasonable assessment and choice of solutions for the transformation of a business model;
- Controlling the effective implementation of transformational solutions;
- Reasonably adjusting the strategy for changing the organization's business model.

Timely initiation of changes to the business model or its elements ensures the proactive nature of the management system. This task can be successfully completed ensuring that critical groups of business stakeholders are identified and a monitoring and analysis system for their requirements is functioning (Jones et al. 2017). Transformation processes in the organization should begin with the identification of strategically essential groups of stakeholders for the enterprise, whose requirements will be decisive for the subsequent changes in the business model. The methods of strategic analysis determine the structure of key stakeholder groups. In the rapidly changing conditions of the functioning of modern business, the group of stakeholders is expanding more and more, and the structure of crucial groups is also growing. The principal groups include those on whose interaction the competitiveness, functioning, and development of the company depend. For effective communication with representatives of stakeholders' key groups, the control system of the organization needs high-quality information and analytical providing requirements of stakeholders. The company needs a business model that provides value creation for stakeholders and a competitive level of implementation of their obligations. At the same time, the content of the business model should take into account the degree of priority of each group of stakeholders.

In the case of a large-scale delay in the actual state of fulfillment of requirements from the level necessary to ensure the competitiveness of the organization, there is a need to initiate changes in the business model. The lag scale in implementing the requirements of stakeholders can be insignificant, which can demand irrelevant transformational influences. There are situations where changes can only affect a particular aspect of enterprise management (Marinov and Angelov 2018). The applied analytical tools should determine the actual requirements of stakeholders, among many different in the priority, importance, and validity of provisions. This will determine the real needs of changing the business model of the company and successfully developing a methodology for analyzing stakeholder requirements to manage the effectiveness of the initiation, development, and implementation of IT solutions (Davis 2013).

However, the transformation of the business model is carried out not only by IT solutions, so the question of the development of the scientific foundations of requirements analysis in the broader direction is acute. In current conditions when a variety of activities of the organization increases, and the pool of the realized business processes grows, the business model of the organization has to be formed on the principles of careful use of resources and maximizing value for stakeholder. Integration of "lean manufacturing" and "Six Sigma" concepts into the analysis system of stakeholders' requirements provide scientific justification for determining the necessary changes. At the same time, using the concept of "lean manufacturing," the analysis will determine the most rational complex of processes, and with the help of "Six Sigma" to control their performance (George et al. 2005).

Determining the most optimal solution for changing the business model requires evaluating the alternatives by the expected change in the potential value generated for stakeholders. Due to the broad structure of the key stakeholder groups, a reasonable choice need to consider that if the decision to change the business model would offer better value for some stakeholders, then for others, it can reduce the level of implementation of requirements. It is necessary to reveal such consequences of updating of a business model. However, it is possible to face the shortcomings of analytical support, which raises questions for representatives of the information and analytical sciences. The lack of analytical tools is a significant drawback that does not allow a comprehensive assessment of the impact of innovative solutions on changing stakeholder satisfaction. In our opinion, reserves for the solution to this problem are in the development of elimination methods. This opens up opportunities for identifying reserves of value growth for certain stakeholders while taking into account the fact that the functioning of the company, in reality, affects the fulfillment of the requirements of many other stakeholders. All stakeholders, whose requirements have been eliminated, provide a context for comparing different decisions to change the business model. A unique role in the development of approaches for analyzing the impact of innovative solutions on improving the level of fulfillment of stakeholder requirements is played by the development of new areas of economic analysis such as strategic analysis (Zenkina 2013), business analysis (Barilenko 2016; Malcolm 2014).

When assessing the solution for the transformation of the business model, it is necessary to determine what the result will be in terms of changing the position in the competitive environment, namely: to keep the occupied competitive areas or to move higher and be among the leaders. In the first case, a reasonable choice will be associated with a preliminary assessment of the resource intensity of the solutions. In the second case, assessing the effect of moving to a higher level of functioning is to be considered.

The transformation management system in the business model should contain a justified change strategy. To improve such a strategy, the adjustment should be carried out every time after the implementation of the transformational solutions. After implementing the solution, one should analyze how the actual change in the level of stakeholder requirements corresponds to the planned changes before the implementation of the solution.

The system of management of transformation in business models, built with the above principles will allow:

- Revealing the real needs for change of a business model;
- Evaluating and justifying the choice of solutions for the transformation of the business model;
- Monitoring the effective implementation of the solution;
- Specifying and improving the strategy of changes;
- Providing the continuity of developing a business model.

4 Discussion

Despite the high degree of importance of the proposed management system in ensuring long-term sustainability and strategic success, management science needs to significantly update approaches to solving previous problems in the new conditions, as well as to develop approaches that arise in the new context of the functioning of modern business. As already noted, analytical justification of management decisions requires further development of the application of provisions of the theory of sustainable development, as well as the theories of stakeholder and management of change. Methodological support for stakeholder requirements analysis deserves special attention. Such an analysis is presented as a key task. The validity of developing a program of innovative changes in the business model and the adequacy of assessing the predicted effectiveness of its implementation depend on the quality of the solution.

The systemic management of changing the business model, proposed in the study, also has several issues that are beyond the scope of this article, but also subject to careful scientific analysis. Among the severe problems of building a quality management system for changing the business model are the following issues of organizational nature: issues of staff motivation, the use of IT-solutions for effective automation of management processes, and others.

5 Conclusion

Thus, as a result of the study, it was found that to increase the competitiveness of modern economic entities, it is necessary to ensure the timely transformation of the business model. The paper shows that in current conditions, the proactive nature of the transformation is more preferable; however, it requires a process management system for transforming the business model. Building the proposed system implies that management science should actively develop such areas that would analytically justify the initiation, development, and implementation of transformational solutions. Reserves of development of these directions seem in the adaptation of applying concepts of "lean manufacturing" and "Six Sigma," developing the methods of elimination, conducting an integrated assessment of satisfaction of stakeholders, and others.

Supporting changes in the business model through system management will provide organizations with competitive advantages in the modern economy. Continuous monitoring of stakeholder requirements, which is a necessary element of the proposed management system, will provide the prerequisites for sustainable development.

References

Adams, C.A.: The international integrated reporting council: a call to action. Crit. Perspect. Account. **27**, 23–28 (2015)

Barilenko, V.I.: Business analysis and overcoming the "new normality" of the domestic economy. Manag. Bus. Adm. **4**, 44–50 (2016)

Berdnikov, V.V., Gavel, O.Yu.: Controlling business processes during the creation of competitive business models in agribusiness. Siberian Financ. Sch. **3**, 68–73 (2011)

Davis, A.: Just Enough Requirements Management: Where Software Development Meets Marketing. Addison-Wesley, Boston (2013)

Dawson, A., Hirt, M., Scanlan, J.: The economic essentials of digital strategy. McKinsey Q. **3**, 45–49 (2016)

Freeman, R.E., Harrison, J.S., Zyglidopoulos, S.: Stakeholder Theory: Concepts and Strategies, p. 486. Cambridge University Press, Cambridge (2018)

George, M., Rowlands, D., Price, M., Maxey, J.: Using DMAIC to improve speed, quality, and cost. In: The Lean Six Sigma Pocket Toolbook: A Quick Reference Guide to Nearly 100 Tools for Improving Process Quality, Speed, and Complexity. McGraw-Hill, New York (2005)

Integrated Reporting Committee: The International < IR > Framework (n.d.). http://integratedreportingsa.org/ircsa/wp-content/uploads/2017/05/13-12-08-THE-INTERNATIONAL-IR-FRAMEWORK-2-1.pdf

Jones, T.M., Wicks, A.C., Freeman, R.E.: Stakeholder theory: the state of the art. In: The Blackwell Guide to Business Ethics, pp. 17–37. Blackwell Publishing, Oxford (2017)

Konopatov, S.N., Salienko, N.V.: Analysis of business models based on platforms. Econ. Environ. Manag. **1**, 21–32 (2018)

Kotelnikov, V.Yu.: New Business Models for a New Era of Rapid Change Driven by Innovation. Eksmo, Moscow (2007)

Malcolm, E., Hindle, K., Paul, D., Rollaston, C., Tudor, D.: In: Paul, D., Yeates, D., Cadle, J. (eds.) Business Analysis, p. 273. BCS; The Chartered Institute for IT, UK (2014)

Marinov, M., Angelov, G.: Streamlining the management of corporate accounts receivable. Russ. Technol. J. **6**(2), 67–81 (2018)
Sahlman, W.A., Stevenson, H.H., Roberts, M.J., Bhide, A.: The Entrepreneurial Venture, 2nd edn. Harvard Business School Press, Boston (1999)
Soolyatte, A.Yu.: A business model is a key to business development based on innovation. Innov. Manag. **1**(09), 6–15 (2010)
Vashakmadze, T.T.: Improving the business model as a means of business value management. Econ. Entrepreneurship **5**(28), 384–387 (2012)
Zenkina, I.V.: Methodological foundations and tools of strategic analysis of the organization. Audit Financ. Anal. **1**, 106–115 (2013)

The Features of Innovation Risks Management in Russian Knowledge-Intensive Industries (Enterprises)

Alberto Gabriel Henrique(✉)

RUDN University, Moscow, Russia
alberto.gabriel@yandex.ru

Abstract. The article is devoted to the study of features of innovation risk management in the Russian knowledge-intensive industries (enterprises). The author indicates the considering topic relevance and significance. The theoretical aspects of the innovation risk management methodology in knowledge-intensive industrial organizations (enterprises) are studied as the basis for creating a system and implementing a risk management mechanism. A system for creating and implementing an innovation risk management mechanism in knowledge-intensive industrial organizations, with the goal of a more efficient "Three Defense Lines" model adaptation, has been also developed.

Keywords: Innovation · Risk management · Knowledge · Industries

1 Introduction

It has been more than a decade since the beginning of the 2008 Great Recession. Its consequences and the desire to overcome difficult times forced knowledge-intensive industrial organizations (enterprises) to take a conservative position in the innovation risk management field and transfer responsibility for this process from business units to a second defense line (monitoring function) (Popkova et al. 2019).

At the same time, in the context of general destabilization and changes occurring in the external environment, knowledge-intensive industrial organizations (enterprises) feel the need to revise the current practice (PwC 2017, p. 5). Innovative risk management based on the effective interaction of the first (business function) and second defense lines (with a clear responsibility transfer to the first line) is the key to success on the way to achieving the target state of organizations (enterprises) and ensuring the necessary growth rates.

In turn, the second defense line acts proactively and contributes to making management decisions by focusing on new and emerging innovation risks, providing interaction in maintaining effective practices for such risks assessing, and ensuring timely advice and support. The third defense line (independent function, internal audit) provides independent guarantees, particularly the effective functioning of the innovation risk management system, thereby making an additional contribution to maintaining and increasing the value of a knowledge-intensive industrial organization (enterprise).

A. V. Bogoviz and Y. V. Ragulina (Eds.): ISCI 2019, LNNS 115, pp. 326–335, 2020.
https://doi.org/10.1007/978-3-030-40749-0_39

On the other hand, despite the apparent originality and the implementation simplicity of the *Three Defense Lines* model (Lelekova 2017) being unadapted to Russian realities, its organization and functioning in large business domestic organizations, especially operating in industry, is characterized by a number of the following difficulties (PwC 2015, p. 8): (1) inefficient and (or) unreliable model organization; (2) insufficient interaction in the identifying risks context, developing response measures and monitoring identified risks; (3) insufficient activities coordination, including effort duplication and the lack of the used tools integration; (4) insignificant use of the specialized information systems in terms of internal control automation, risk assessment, and internal audit; (5) insufficient staff competence in identifying and developing response measures and monitoring new emerging risks. Taking the above into account, knowledge-intensive industrial organizations (enterprises) need, first of all, to focus on the issue of creating and implementing a mechanism for managing innovative risks in the context of the stage application of the *Three Defense Lines* model.

2 Materials and Methods

Materials for the study includes scientific papers and publications on the risk management theory and methodology. Such general scientific methods as analysis, synthesis, induction, deduction, grouping, comparative analysis, and processes modeling were used as a methodological basis.

3 Results

In our country, the creation of our own effective management system in knowledge-intensive industrial enterprises in the developing market relations conditions is possible only with the organization of a clear, relevant, and well-thought-out risk management system (Shchiptsov 2013). In recent years, in practical activities of domestic knowledge-intensive organizations, especially those operating in industry, there has been a desire to organize a risk management system and to find the methodological foundation for existing risk management concepts in the ever-increasing need for innovation, which is the integral element of the entire basis for building business processes in an organization interested in investment performance.

The risk management methodological foundation, in terms of knowledge-intensive industrial enterprises innovative design, is based on the established risk management concepts, which form an organic set of methods and tools to influence risks associated with innovative (project) activities. The most well-known innovative risk management concepts are the risk minimization and acceptable risk concepts, as well as the risk as a resource concept.

The classical risk minimization concept is based on the J.S. Mill and N.U. Senior's studies, in which risks are identified with some consequences of the negative person's management decisions (Hislop et al. 2013). From the perspective of the concept under study, the innovation (project) risk is an absolutely negative component of those economic entities being engaged in innovation activities. This concept methodology aims to reduce the innovation risk level to minimal values; however, its key difference is the formal-logical factor in making management decisions that are characterized by the risk absence. This explains the simplicity and versatility of the concept in the innovation context. At the same time, analyzing the behavior of a risk management subject, i.e. a knowledge-based industrial enterprise, it is important to take into account that management decisions (in which risks are absent or minimal) lead to mediocre results for the organization's business processes as a whole. Thus, the possibility of using the risk reduction method is limited by the "loss-risk" paradox, which has a binary nature and can be expressed as the "profitability-risk". It means that the management decision with a high innovation risk adoption can lead to increased losses (damage) from this decision; conversely, the same decision can lead to high incomes.

It is known that choosing a solution with the lowest risk level can be made through the special events organization, or the alternatives selection with certain risk characteristics. The necessary costs evaluation, in the concept's terms, reflects one of the key risk management provisions, the minimizing innovative risks' costs should not exceed the possible losses from the undesirable events occurrence. Due to the fact that there are practically no "random" economic events, the measures developed by a knowledge-intensive industrial enterprise can be effective and reduce risks to any acceptable level. The risk of forcing an innovative product (service) out of the market due to the presence of substitute goods presence may serve as an example. In this case, the risk can be minimized by including in the developed product characteristics the widest useful qualities, important for the consumers.

However, despite the theoretical utility, simplicity, and other advantages of the minimizing innovation risk concept, the risk cannot be reduced to the absolutely minimum value. Based on this, the concept can be applied in a knowledge-intensive industrial organization solely as a part of a risk management strategy in conjunction with other project risk management tools. The second concept is the acceptable risk concept. B. N. Prokofiev is considered to be its founder, because he noted the risk acceptability property in his writings. Despite the fact that it was developed with a focus on the production sphere, it can be applied to almost any economic system. This concept has become a response to the risk minimization methodology ineffectiveness (Blank 2014). The essence of this concept lies in the risk perception, which must be reduced to an acceptable level for a particular project. As a result, the organization's activities will not undergo qualitative changes. This is what distinguishes this concept from the one discussed above. In other words, the difference lies in the direction of managerial interaction, when the level of innovation risk should be reduced not to an absolute minimum, but to a level acceptable for a particular situation.

In addition, the cost factor and its proportionality, the problem of which exists in the risk minimization concept, is solved via the acceptable risk method, by means of rationalizing the comparison of effects and costs to achieve these effects. As in other concepts, risk is identified only with possible damage to the activities of knowledge-intensive industrial organizations. More than that, the concept provisions do not allow estimating the probable income from making management decisions at a given innovation risks level and in its positive implementation event (i.e., such risks existence occurs without negative consequences for business processes and activity results). At the same time, Kuzmin notes that "In the acceptable risk concept, it would be impossible to solve a scientific task of determining the possibility of obtaining additional income as a management decision result under risk and uncertainty conditions" (as cited in Popov 2006, p. 234). Based on this thesis, the concept also cannot be accepted as the only correct and optimal one for reducing innovative risks of knowledge-intensive industrial enterprises.

Thus, despite the concept is apparent universality, it has a number of distinctive features, including: (1) innovation risk management should be based on the separation of starting and final risks; (2) the innovation risk level should not be reduced to the minimum, but to the acceptable level; (3) the risk level of innovation activities of a knowledge-intensive industrial enterprises, traditionally higher than in other commercial activities types. Based on this, both described concepts are aimed at reducing innovation risk. At the same time, a rational approach is implemented in the case of using an acceptable risk system. This means that the costs on anti-risk measures are compared with the size of possible losses and the volume of possible consequences volume.

Kroon and co-authors believe that the acceptable risk concept is not yet fully developed, and further improvement is necessary. However, the disadvantages inherent in the format in which it exists (including the impossibility of using changes in the qualitative risk properties in order to obtain additional benefits) can minimize the "risk as a resource" concept (Blaug 1994). It was designed by M. A. Greenfield and subsequently adapted by A. B. Sekerin (Shchiptsov 2013), who refined the mathematical process of risk management within the framework of the concept under the study. He also scientifically based the idea of combining the risk concept as a resource and the two concepts previously described by us as a single concept, resulting in an integrated risk management concept. The essence of the risk concept as a resource is that risk management in a knowledge-intensive industrial organization must be identical to the resource management system in the classical management theory. In his work, M. A. Greenfield focuses on cost-effectiveness, which is aimed at the implementation of certain measures to reduce the level of risk. The concept "core" is the optimization of risk management costs, which is carried out by comparing them with marginal costs (risk-reducing measures costs) and incomes (its loss amount, in which the risk exposure level is reduced).

The specificity of the considered methodology is that it is the most optimal principle for managing resource-like risks, the key features of which are the following ones: (1) an increase in the level of such risk may lead to additional benefits, i.e. risk is characterized by the presence of a positive factors combination; (2) a resource-like risk can be avoided (for example, avoiding purchasing the equipment for the innovative products manufacturing that has not received a quality certificate, but it is already advertised as being energy-efficient or the most multifunctional one); (3) an increase in this risk level is effective up to a certain point, when a number of positive factors affect the innovation activity results; and after that, new factors become involved, the manifestation of which would be exclusively negative (Filin 2008).

Based on the foregoing, the innovative risk management as a resource within a knowledge-intensive industrial enterprise should consist in supporting its optimal level, which particularly implies the possibility of a conscious increase in its level (Levkovich and Somenkova 2014). However, on the other hand, if the resource-like risk level is higher than optimal, it is appropriate to reduce it. A significant disadvantage is that not all innovation risks are resource-like, i.e. their identification and the assessment possibility for the further adoption of management decision can be carried out only if being used in conjunction with other methods of managing innovative risks.

A comparative analysis allows make a conclusion that, first of all, none of these concepts is the most optimal one, or it is the most complete for use in the implementation system of innovative activities of a knowledge-intensive industrial enterprise. Second, the acceptable risk concept is a qualitative update of the risk minimization concept, eliminating certain shortcomings of the latter, and the risk as a resource concept is focused only on resource-like risks that are not the only ones in the enterprises' innovative activity system. Proceeding from this, a knowledge-intensive industrial enterprise needs to make a decision on the importance and appropriateness of using these methodologies. Talking about the integrated concept, its key element is the risks resource intensity, the assessment analogue of which is the risk coefficient. Its relevance is due to the fact that it conditionally brings together all the concepts together, forming a holistic approach. Consequently, there is a possibility for knowledge intensive industrial organizations to begin with the merit's assessment of this particular concept.

4 Results and Discussion

The *Three Defense Lines* model was developed by the International Institute of Internal Auditors in 2013. This model coordinates the risk management and internal control processes by clearly defining and distinguishing the respective functions and responsibilities. The first defense line (*business function*) is operational management; this is responsible for assessing, managing and minimizing risks, as well as ensuring an effective internal control system. The second defense line (*monitoring function*) focuses on ensuring and monitoring the implementation of effective risk management practices, internal control, as well as compliance with laws and administrative rules/internal regulations and the fraud investigation. The third defense line (*independent function*) relies on a risk-based approach. This line implies the procedure when internal audit provides confirmation on how effectively a particular organization assesses its risks and manages them, including the effectiveness of the first and second defense lines, providing information to the directors board and senior organization management. A well-considered comparison of first and second defense lines, and then the organization of the third line in practice, allows an enterprise to manage various types of risks in the most effective way. However, as it was already mentioned above, this model can be successfully implemented exclusively in a carefully thought out and adapted to the activity type of the enterprise risk management mechanism. This is due to the insufficient conditions for adaptation of this model in Russian companies, particularly in knowledge-intensive industrial organizations, as well as because of certain difficulties in selecting an optimal risk management concept.

We propose our model of a system for creating and implementing a mechanism for managing innovative risks in knowledge-intensive industrial organizations (enterprises). Figure 1 presents an algorithm (system) for implementing innovation activities of the knowledge-intensive industrial, as well as a reference system for implementing innovative activities, which is a necessary prerequisite for the formation of a mechanism for managing innovative risks in knowledge-intensive industrial organizations (enterprises). The proposed scheme makes it possible to identify the driving forces behind the mechanism creation, with the aim of taking a direct part in its implementation. It is developed from the position of a systematic approach to innovative risk management capabilities in knowledge-intensive industrial organizations (enterprises).

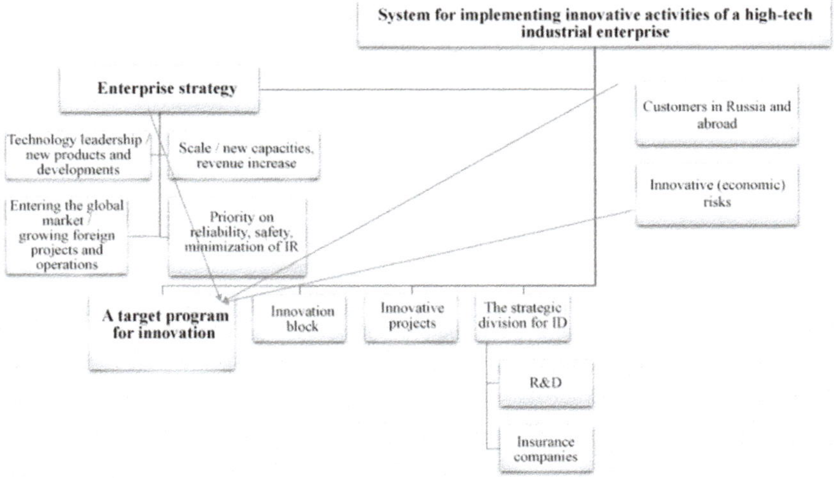

Fig. 1. A system for creating and implementing a mechanism for managing innovative risks in knowledge-intensive industrial organizations (enterprises).

Certainly, a key point in the knowledge-intensive industrial enterprise management is the process of creating a targeted innovation program that would determine the strategic choice of creating and developing innovative projects of an organization. At this point, the directions of allocating own funds or funds from external sources for the strategic goals of a particular enterprise (industry) are determined. This is the crystallization point of many industries' forces, and an innovation mechanism is launched directly from this point. The target innovation program choice, as a starting point for a system innovation mechanism for developing an enterprise (industry), is also conditioned by the need for corporate efforts to implement innovation projects. From our point of view, this is exactly how a system should look like, which could meet two key criteria at the same time: project's "reliability-safety" and the knowledge-intensive industrial enterprise activities, as well as the "investment attractiveness" of a project, company, or industry.

The structural blocks of the innovation risk management model aimed at minimizing them within the systems approach framework are presented in Fig. 2.

Fig. 2. Structural elements of the innovation risk management process in knowledge-intensive industrial organizations (enterprises).

The first block, the block of strategic management is the basis of the mechanism for managing innovative activities of the knowledge-intensive industrial enterprise. One of the most important factors determining the functioning is the enterprise's activity strategic plan and the target program on innovations for the company or industry developing purposes.

Depending on what goals are set, innovative solutions, plans, and projects are developed for achieving the goals. A strategic plan for the innovative development of knowledge-intensive industrial enterprise's (industry) traditionally has spatial and temporal reference points. In this case, it is the breadth of market coverage and the long-term sustainable development within this market. In the innovation management research process, it is determined that only with a strategic focus on reducing various types of innovation risks, as well as safety and reliability, can they be managed from a systematic approach perspective. On this basis, the entire corporate system or a particular industrial system must be configured to solve this problem. As can be seen from the presented scheme, the existing enterprise strategy has three key areas: (1) *technological leadership*, due to the increased R&D efficiency and the introduction of effective management system; (2) *scaling*, i.e. new capacities, increase in the market, revenue, scientific and technical base creation for self-development purposes; (3) *access to international markets*, including an increase in the implementation of innovative projects abroad and trade operations related to the products and technologies developed by the company. However, one more direction is to be put forward with respect to our research. This is the reliability and minimization of innovative risks.

The first block, which is associated with the creation of the so-called "incentive motive," is responsible for the policy of innovative development of a knowledge-intensive industrial enterprise (industry). This means that there is a priority to reduce certain innovation risks, depending on the economic policy task of the enterprise and the target innovation program.

The next block defines the "customer needs," i.e. those structures on which knowledge-intensive industrial enterprise's innovative activity is focused, on those who need products and technologies developed by an organization. It is impossible to create a program for the further innovations implementation without determining the purposes and the target audience. Determining and predicting the clients' needs and specifics in innovative projects is a key task of the mechanism for managing relevant risks, since this determines the method and tools to minimize them; at the same time, it is also necessary to take into account their provision possibilities.

The third block is a structural innovation risk management division. The question is who develops the mechanism for forming a target program to manage such risks within the target innovation program, collects all the necessary information, and monitors the implementation progress. Obviously, such a unit as the interpretation of the *Three Defense Lines* model, which exists in almost all high-tech industrial enterprises, is best suited for this. This structure should be inside of the innovation management unit in such a way that it is closely interlinked with innovation and innovation projects. It is in this division that the creation of an expert group is recommended, such as a strategic innovation risk management department. This subdivision would have meaning only if there is a certain strategic task, the mission of the knowledge-intensive industrial enterprise's long-term sustainable existence. The employees of such a unit can collect and analyze information that is tied to factors influencing the target program implementation for the organization's innovations, as well as cooperate with other expert departments and strategic partnership organizations when it comes to a public company that has several branches in the country and abroad, in particular. This unit allows performing certain functions (technologies and industry forecasts analysis; innovation risks analysis and classification; consolidated cooperation with other knowledge-intensive enterprises and industries; issuing recommendations for management decisions; control over the innovative risk management measures implementation, etc.

The fourth block is the innovation risk management infrastructure. It is well known that the key element involved in the risk management mechanism is the "infrastructure" block. This unit is responsible for the development of a supporting resource subsystem in order to implement innovations. To perform such responsibilities, the institutional environment, material and technical base, as well as a deferred innovative mechanism should be well developed. In the innovation risk management system, the "infrastructure" block is a supporting system for the mechanism operation and includes the following components: (1) *material and technical base*: manufacturing enterprises and the raw material base of a company (industry); (2) *institutional environment*: legislative and legal formalization of the enterprise; (3) *investments*: an ordinary (traditional) element of business support (Fig. 3).

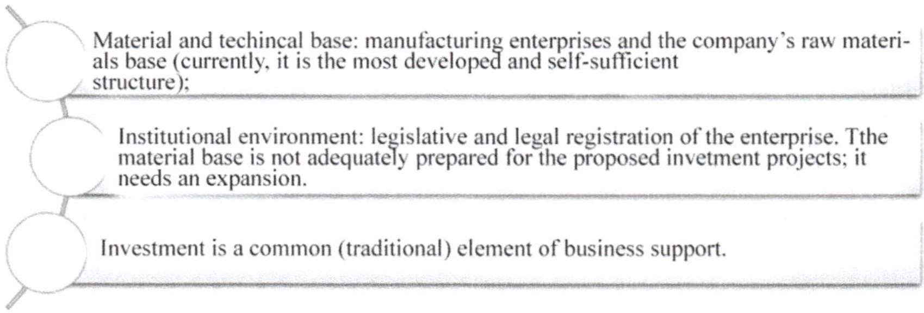

Fig. 3. Elements of the mechanism for managing innovative risks.

The fifth block is the scientific and technical progress (STP). Certainly, such an "engine" as a scientific and technical support of innovation activity of a knowledge-intensive industrial enterprise and STP, which directly participate in the process and affect the rest of the mechanism's elements efficiency, does not remain aside.

Consequently, the mechanism's structural blocks of innovative high-tech industrial enterprise risks provide a process for managing such risks, as well as an algorithm for reducing them. The process formation occurs in the interaction of the associated "mechanism parts."

5 Conclusions

The contemporary enterprises have to introduce innovations, and risk managers must find the optimal risks and benefits balance in order to get profits. As a result, organizations that effectively manage innovation risks more impactfully implement appropriate innovations (PwC 2018, p. 3). However, the question of properly organizing risk management and improving the innovative risk management program arises. The proposed model for managing innovative risks, creating, and implementing a mechanism for managing innovative risks in knowledge-intensive industrial organizations (enterprises) is relevant, because it allows revealing the motivating forces of creating a mechanism with the goal of directly participating in its implementation. The proposed system for creating and implementing an innovation risk management mechanism meets two key criteria at the same time: (1) "reliability-safety" of the project and activities; (2) "investment attractiveness" of a project, company, or industry. Structural blocks of the proposed innovative risk management mechanism provide a process for managing such risks, as well as an algorithm for reducing them. The process formation is carried out in the interaction of the associated "mechanism parts."

References

Blank, I.A.: Financial Risk Management. Nika-Center, Moscow (2014)
Blaug, M.: Economic Thought in Retrospect. Case Ltd, Moscow (1994)
Filin, F.N.: Risk Management. GrossMedia, ROSBUH, Moscow (2008)

Hislop, D., Bosua, R., Helms, R.: Knowledge Management in Organizations: A Critical Introduction. Oxford University Press, Oxford (2013)

Lelekova, M.A.: Practical aspects of implementation. Models of three lines of building the protection of the internal control system. Press Center Interfax (2017). http://old.nuiac.ru/images/122017Lelekova.pdf

Levkovich, A.I., Somenkova, N.S.: Types of innovative risks in industrial enterprises (2014). http://itportal.ru/science/economy/vidy-innovatsionnykh-riskov-na-prom/

Popkova, E.G., Ragulina, Y.V., Bogoviz, A.V. (eds.): Industry 4.0: Industrial Revolution of the 21st Century. Springer, Heidelberg (2019)

Popov, A.I.: Economic Theory: Textbook for Universities, 4th edn. Piter Publishing House, Russia (2006)

PwC: Practical aspects of using the "Three Lines of Defense" model and automation solutions using GRC (2015). https://docplayer.ru/40837524-Prakticheskie-aspekty-primeneniya-modeli-treh-liniy-zashchity-i-resheniya-avtomatizacii-sredstvami-grc-noyabr-2015.html

PwC: Preparing for tomorrow's workforce, today: insights from a global survey of business and HR leaders. The origins of the situation. Collaboration leadership is key to success. A look at the risks: PwC annual survey, 32 (2017). https://www.pwc.com/jg/en/publications/survey-business-hr-leaders.html

PwC: Risk management for sustainable growth in an era of innovation (2018). https://www.pwc.ru/ru/riskassurance/publications/assets/pwc-2018-risk-in-review-russian.pdf

Shchiptsov, A.A.: Features and the need for innovation in industrial enterprises. Sci. Vector Togliatti State Univ. (Ser. Econ. Manag.) **1**, 141–143 (2013)

Minimization of Information Losses in the Management of Innovation Development in the Digital Economy

Aleksandr M. Batkovskiy[1(✉)], Vladimir A. Sudakov[2,3], and Alena V. Fomina[1]

[1] Joint Stock Company "Central Research Institute of Economy, Management and Information Systems "Electronics", Moscow, Russia
{batkovskiy_a, fomina_a}@instel.ru
[2] Keldysh Institute of Applied Mathematics, Russian Academy of Sciences, Moscow, Russia
sudakov@ws-dss.com
[3] Plekhanov Russian University of Economics, Moscow, Russia

Abstract. The paper focuses on the information losses that arise in the case of aggregation of criteria when one solve the problem of selecting samples of innovative products using information and communication technologies. The relevance of using the vector criterion when selecting the samples of innovative products in the information economy is demonstrated. The specificity of solving this problem requires taking into account a large number of criteria that need to be aggregated and distributed into a convolution tree with qualitative, quantitative, and fuzzy rules. A technique for optimal partitioning of criteria scales into generalized gradations for using combined methods of multicriteria analysis of innovative development alternatives is proposed in the paper. The implementation of this methodology in practice allows increasing the validity of management decisions and, accordingly, the innovation effectiveness. A formal approach to building an aggregation tree is presented, and suggestions are made to combine the criteria into groups. It is determined that their aggregation leads to an increase in the dimension of the scales of new generalized criteria. It has been established that many qualitative criteria for selecting innovative product samples do not allow working with high-dimensional scales, so they must be converted to scales with fewer gradations. For this purpose, the derivation of the Bellman recurrence function is shown, which is used to solve the problem of minimizing information losses arising from the conversion of scales. An experiment allowing to assess of the distortions in the judgments of decision-makers when using the aggregation procedures has been carried out. The presented results indicate that a significant correlation between the rankings obtained in different ways indicates the adequacy of the proposed methodology. The practical use of the considered approach to minimizing information losses when choosing equipment for conducting space experiments is noted.

Keywords: Innovative products · Criterion · Aggregation · Management · Entropy · Digital economy · Information losses

A. V. Bogoviz and Y. V. Ragulina (Eds.): ISCI 2019, LNNS 115, pp. 336–343, 2020.
https://doi.org/10.1007/978-3-030-40749-0_40

1 Introduction

According to Schumpeter's classical theory, innovation is a production function that, based on the totality of factors acting on a product, will determine its qualitative changes (Michaelides 2009). In practice, many alternative ways of creating innovative products with different prototypes have to be faced. In this case, it is necessary to evaluate possible alternatives by the vector criterion. The innovative product component leads to the fact that the vector criterion has a rather high dimension (tens or hundreds of components). A decision-maker (DM) is not able to evaluate such a large set of criteria; therefore, it is necessary to build a multi-level tree of criteria (Corrente et al. 2012).

The basis of the tree of criteria aggregation, as a rule, is the integral criterion for the innovative product quality. Thus, the transition from the vector to the scalar allows for solving optimization problems. At the intermediate levels of the aggregation tree, the criteria that characterize certain aspects of the alternative (production costs, consumer properties, competitiveness, innovation, etc.) are presented. Preferably, the primary criteria, which are involved in the calculation of different intermediate criteria, should be mutually independent by preference.

For all of the criteria except the primary criteria, the method of decision support used for their calculation is indicated. Both quantitative and qualitative methods of decision-making support can be used, and the fuzzy modifications of these methods are promising.

When building the tree of criteria aggregation, as a rule, it is impossible to collapse more than 5–10 criteria in one node. The specific maximum number of criteria is determined by the decision support method used. For this reason, and also due to the high dimensionality of the initial vector criterion for evaluating innovative product samples, the aggregation tree has a significant number of levels at which information loss occurs.

2 Materials and Method

Such an approach to the aggregation of criteria is in many ways similar to the analytical hierarchy of criteria proposed by Thomas Saaty (Saaty and Vargas 2000), though Saati used only additive and multiplicative convolution in nodes, which is incorrect for some problems that require the use of high-quality and hybrid approaches. One of the effective methods that can be applied to convolution in nodes is a combined method based on preference areas (Batkovskiy et al. 2017). However, in this method, as in many other methods using qualitative scales, the question of how to determine the gradation of criteria rationally arises.

Let us assume that at the lowest level of the tree of criteria aggregation, N_1 groups are specified. Each i group has q_{1i} criteria. In order to apply qualitative methods, it is necessary to fulfill the condition $q_{1i} < 6$. If n is the power of the initial criteria vector, and the number N_1 is within:

$$\left[\frac{n}{\max_i q_{1i}}\right] \leq N_1 \left[\frac{n}{\min_i q_{1i}}\right] \tag{1}$$

The operation [...] is applied here, rounding to the nearest integer. At the next level of the tree, q_2 is the criteria number in groups, and the role of the input criteria is performed by N_1 groups of the lower level, etc. As a result of applying this procedure, one integral criterion will remain on the basis of the tree.

In decision support tasks, we usually want to form such a system of preferences in the criteria space so that all the alternatives can be distinguished by preference. The number of gradations of an ideal preference scale should be equal to the power of a full set of alternatives. A complete set of alternatives is a theoretical concept that corresponds to all combinations of gradations of criteria in scales. The power of a real set of alternatives is less or equal to the power of the full set. However, when working with innovations, we often cannot unambiguously identify all potentially feasible alternatives. Therefore, at the stage of determining the aggregation tree, we have to work with preferences on a full set. Let us suppose that in one node, we form the convolution of four quality criteria. Each of them has seven gradations, so in this case, the power of the full set is $7^4 = 240$. If the result of such convolution in some aggregation tree is intermediate, then, theoretically, at a higher level, we should use a criterion with a scale containing 2401 gradations, which, admittedly, is difficult to implement in practice. Thus, by lowering the dimension of the criterion, we increase the dimension of the scales, which complicates the use of qualitative methods. It is advisable to apply these methods in the case of dependencies between the criteria according to preferences and in the case of non-linear functions of the values of the decision-maker.

The logical way to solve the problem of high-dimensional scales of criteria is to lower their dimension by combining part of the gradations into new "generalized" gradations. The simplest option that implements such a procedure is a *uniform* splitting of the scale when an equal number of initial gradations is combined into *generalized gradation*. However, will this procedure ensure maximum efficiency?

Any division of the scale into gradations leads to information loss. Effective partitioning should minimize these losses. The amount of information on the original scale and on the generalized scale can be estimated using the concept of entropy (Arndt 2004). The amount of information in the initial scale cannot be influenced. Therefore we must ensure that the initial scale is divided so that the generalized scale retains the maximum amount of information from the initial scale.

3 Results

Let us introduce the following notations:

m – the number of gradations specified for the original scale,
m^* – the number of gradations specified for the generalized scale,
w_r ($r = 1..m^*$) – the positive integer variable, equal to the number of gradations on the original scale, which is included in the rth gradation of the generalized scale,

p_k $(k = 1..m)$ – the probability of the kth gradation in the original scale.

Then the number of the first gradation of the initial scale included in the rth gradation of the generalized scale s_r is calculated as:

$$S_r = \sum_{l=1}^{r-1} w_l + 1, \tag{2}$$

and the number of the last gradation of the initial scale included in the rth gradation of the generalized t_r scale is calculated as:

$$t_r = \sum_{l=1}^{r} w_l \tag{3}$$

The objective function of the problem of maximizing the amount of information in a generalized scale consists of separable components responsible for the amount of information in the corresponding generalized gradations:

$$\max_{w_1, w_2, \ldots, w_{m^*}} \left(-\sum_{r=1}^{m^*} \left(\sum_{k=s_r}^{t_r} p_k \right) \log_2 \left(\sum_{k=s_r}^{t_r} p_k \right) \right). \tag{4}$$

In addition to the restriction on the integer optimization variables, there is only one linear constraint that shows what part of the initial gradation is required to be distributed among the generalized gradations:

$$\sum_{r=1}^{m^*} w_r = m. \tag{5}$$

Since w_r are integers, the optimization problem belongs to the class of discrete integer programming.

Many stochastic models of complex processes and systems, including models for evaluating samples of innovative products, have difficulties associated with the need to know the priori probability distribution. In this case, we need to know the values of p_k. In order to calculate them, it is required to process statistics on the already existing alternatives. If N alternatives were assessed in total, and the number of estimated alternatives for which of the criterion value fell into a gradation with number k equal to M, then the probability p_k can be:

$$p_k = \frac{M}{N}. \tag{6}$$

A reasonable estimate of the probabilities p_k can be calculated using the standard lottery method, which is especially important if statistics are not available.

Since the objective function is separable and monotonous, the discrete optimization problem posed is solved by the dynamic programming method (Eddy 2004). For this, the following Bellman function was derived:

$$\Delta_r(\xi) = \max_{w_1,\ldots,w_r}\left(-\sum_{j=1}^{r}\left(\sum_{k=s_j}^{t_j} p_k\right)\log_2\left(\sum_{k=s_j}^{t_j} p_k\right)\right). \tag{7}$$

$$\sum_{j=1}^{r} u_j \leq \xi, \tag{8}$$

$$r \leq \xi,\ \xi = 1,\ 2,\ldots, m \tag{9}$$

The following recurrence formula allows us to calculate the Bellman function for any number of gradations:

$$\Delta_1(\xi) = \max_{1\leq w_1 \leq \xi}\left\{-\left(\sum_{k=1}^{w_1} p_k\right)\log_2\left(\sum_{k=1}^{w_1} p_k\right)\right\} \tag{10}$$

$$\Delta_2(\xi) = \max_{1\leq w_2 \leq \xi}\left\{-\left(\sum_{k=\xi-w_2+1}^{\xi} p_k\right)\log_2\left(\sum_{k=\xi-w_2+1}^{\xi} p_k\right)+\Delta_1(\xi-w_2)\right\} \tag{11}$$

$$\Delta_r(\xi) = \max_{1\leq w_r \leq \xi}\left\{-\left(\sum_{k=\xi-w_r+1}^{\xi} p_k\right)\log_2\left(\sum_{k=\xi-w_r+1}^{\xi} p_k\right)+\Delta_{r-1}(\xi-w_r)\right\} \tag{12}$$

The Bellman function for $\xi = m$ и $r = m^*$ allows us to find the optimal value of the original objective function:

$$\Delta_{m^*}(m) = \max_{1\leq w_{m^*} \leq m}\left\{-\left(\sum_{k=m-w_{m^*}+1}^{m} p_k\right)\log_2\left(\sum_{k=m-w_{m^*}+1}^{m} p_k\right)+\Delta_{m-1}(m-w_{m^*})\right\} \tag{13}$$

4 Discussion

The aggregation of criteria is heuristic, and depending on how the criteria are combined into groups, we can get different results. Therefore, the issue of the correct use of aggregation for multicriteria analysis of alternatives for the selection of samples of various products is relevant (Batkovskiy et al. 2016). Since the structure is identical for all nodes of the criteria tree, it can be assumed that if the use of the aggregation procedure is adequate to the initial (not aggregated) assessment for one of the generalized criteria, then this adequacy is preserved for the entire criteria tree.

The methods for calculating the integral estimates of alternatives will be considered adequate if, while they are used, the same decision-maker receives a similar ranking of alternatives. To determine the measure of the proximity on the linear order relations, the special measure proposed by Kemeny is used. The formula determines the distance between ranking A and B:

$$d(A,B) = \frac{1}{2}\sum_{i,j=1}^{N} \left|a_{ij} - b_{ij}\right|, \tag{14}$$

where N – the number of the alternatives;

$$aij = \begin{Bmatrix} 1 - \text{if the } i_{th} \text{ alternative is preferred over the } j_{th} \\ 0 - \text{if the } i_{th} \text{ alternative is equivalent to the } j_{th} \\ -1 - \text{if the } j_{th} \text{ alternative is preferred over the } i_{th} \end{Bmatrix}$$

b_{ij} is defined similarly. The maximum distance between rankings can be equal to:

$$\max d(A,B) = N(N-1)/2 \tag{15}$$

As a result of the normalization of the distance between the rankings, the Kendall rank correlation coefficient is obtained:

$$\tau(A,B) = 1 - 2\frac{d(A,B)}{N(N-1)} \tag{16}$$

In addition, the Spearman rank correlation coefficient can be used:

$$\rho(A,B) = 1 - \frac{6}{N^3 - N}\sum_{i=1}^{N} (r_i - s_i)^2, \tag{17}$$

where r_i – the rank of the ith alternative in ranking A, s_i – the rank of the ith alternative in ranking B.

The calculation results for them vary slightly, so it is difficult to know which coefficient should be used. It is necessary to evaluate the correlations between the rankings in both ways. The evaluation was carried out on two typical tasks of ranking the samples of innovative products in four criteria. In order to obtain ranking A, the rule for calculating the ranks of alternatives was applied directly to the primary criteria, and for ranking B, preliminary aggregation of the primary criteria was carried out into two aggregating criteria. Table 1 shows the results of the statistical processing of experiments.

Table 1. The correlation coefficients.

Decision-makers and their rating characteristics	Experiment 1 (rank correlation coefficients)		Experiment 2 (rank correlation coefficients)	
	Kendall (τ)	Spearman (ρ)	Kendall (τ)	Spearman (ρ)
1	0.18	0.25	0.42	0.59
2	0.66	0.82	0.72	0.85
3	0.34	0.45	0.47	0.58
4	0.58	0.74	0.58	0.63
5	0.84	0.96	0.48	0.64
6	0.46	0.63	0.4	0.44
7	0.66	0.83	0.41	0.53
8	0.19	0.2	0.2	0.21
9	0.84	0.96	0.73	0.87
10	0.65	0.8	0.59	0.72
11	0.69	0.87	0.69	0.87
12	0.82	0.95	0.74	0.9
Range	[0.18; 0.86]	[0.2; 0.97]	[0.2; 0.86]	[0.21; 0.97]
Average	0.61	0.75	0.59	0.71
Dispersion	0.05	0.06	0.04	0.04

Source: (Bomas and Sudakov 2011).

Based on the analysis of the experimental results, the following conclusions were made:

- The rankings obtained with and without aggregation are well correlated in all experiments, both according to Kendall (τ) and Spearman (ρ);
- The Kendall correlation coefficient is smaller on average, but even such values indicate a significant correlation;
- The majority of the decision-makers expressed the view that, to a greater extent, their point of view reflects the ranking obtained as a result of aggregation. One may explain this fact by looking into the direct construction of high-quality decision-making with a high dimension of the criterion causing difficulties due to the need to memorize preference levels for a large number of combinations of criteria;
- The low dispersion of these correlation estimates allows us to consider the experiments credible, although there were a small number of observations.

Consequently, the aggregation of criteria does not introduce significant distortions in taking into account the preferences of decision-makers when choosing samples of an innovative product in an innovative economy.

5 Conclusion

The approbation of the proposed methodology for splitting the criteria scales was carried out as part of solving the problem of choosing equipment for conducting space scientific and applied research on the Russian segment of the International Space Station. The calculation of the gain from the optimal partition was performed, and the data from 93 space experiments were used to determine the estimates of the probabilities of falling values in the gradation of scales (Sudakov et al. 2017). The use of the entropy approach in this problem and several other practical problems showed that 15% of the power of the initial preference scale is enough for the amount of information on the generalized scale to be close to the amount of information on the original scale.

The developed original approach can reduce information loss in the evaluation, ranking, and selection of innovative product samples, taking into account the vector criterion of a high dimension. The entropic approach to the partitioning of scales allows the use of high-quality methods of multicriteria analyses of alternatives with high-dimensional criteria. It can be used to create planning and management systems for enterprises in the digital economy.

Acknowledgment. The study was carried out with the financial support of the Russian Foundation for Basic Research in the framework of the research project No. 18-00-00012 (18-00-00011) COMFI.

References

Arndt, C.: Information Measures: Information and Its Description in Science and Engineering. Springer, Heidelberg (2004)

Batkovskiy, A.M., Fomina, A.V., Semenova, E.G., Trofimets, V.Ya., Trofimets, E.N.: Method for adjusting current appropriations under irregular funding conditions. J. Appl. Econ. Sci. **XI** (5), 828–841 (2016)

Batkovskiy, A.M., Semenova, E.G, Trofimets, V.Ya., Trofimets, E.N., Fomina, A.V.: Statistical simulation of the break-even point in the margin analysis of the company. J. Appl. Econ. Sci. **XII**(2), 558–571 (2017). (European Research Centre of Managerial Studies in Business Administration, Romania)

Bomas, V.V., Sudakov, V.A.: Support of subjective solutions in multi-criteria problems. MAI, Moscow (2011)

Corrente, S., Greco, S., Slowinski, R.: Multiple criteria hierarchy process in robust ordinal regression. Decis. Support Syst. **53**(3), 660–674 (2012)

Eddy, S.R.: What is dynamic programming? Nat. Biotechnol. **22**(7), 909–910 (2004)

Michaelides, P.G.: Joseph Schumpeter and the German Historical School. Camb. J. Econ. **33**(3), 495–516 (2009)

Saaty, T.L., Vargas, L.G.: Models, Methods, Concepts, and Applications of the Analytic Hierarchy Process. Kluwer Academic Publishers, Boston (2000)

Sudakov, V.A., Nesterov, V.A., Kurennykh, A.E.: Integration of decision support systems 'kosmos' and WS-DSS with computer models. In: Proceedings of 2017 10th International Conference Management of Large-Scale System Development (MLSD). IEEE, Moscow (2017)

Methodology for the Objective Assessment of the Efficiency of Competitiveness Management

Vasily I. Ponomarev(✉) and Maxim V. Grushkin

State Corporation for Assistance to Development, Production,and Export of Advanced Technology Industrial Product "Rostec", Moscow, Russia
{v.i.ponomarev,m.v.grushkin}@rostec.ru

Abstract. Large, diversified corporations usually combine a significant number of business units, whose goals are set depending on the goals of the corporation, etc. Consequently, there arises the task of comparing the achieved results not only in terms of absolute financial indicators but also relative to each other, at the level of internal benchmarking of the management of business units. The task of an objective assessment of the performance of business units and the competitiveness management of industrial holdings in the military-industrial complex is relevant for the Russian corporations in the face of sanctions and tasks of the diversification, entering new markets, and developing the production and sales of civilian and dual-use products. The paper focuses on the stages of the study and implementation of the rating mechanism according to the results of activities as part of the strategic management and competitiveness system for a diversified industrial corporation. The results of the research can be applied as practical recommendations for updating the strategic management methods and increasing the competitiveness of large corporations and their subsidiaries.

Keywords: Strategic management · Competitiveness · Goal differentiation · Management efficiency · Ratings

1 Introduction

Large, diversified industrial corporations have business units operating in markets of various maturity degrees that significantly differ in terms of competition and growth rates. Accordingly, both the starting conditions and the goals set for business units differ significantly. In these conditions, the task of an objective assessment of the contribution of various business units to the implementation of the corporate strategy as a whole, taking into account both the current situation in the markets of varying maturity degrees and the ambitious goals and objectives, is of particular relevance. To introduce a mechanism for an objective assessment of the performance of business units and manage the competitiveness of industrial holdings, including in the military-industrial complex of the Russian Federation, the authors proposed and implemented a performance rating mechanism at the Rostec State Corporation as part of the system for strategic corporate management.

A. V. Bogoviz and Y. V. Ragulina (Eds.): ISCI 2019, LNNS 115, pp. 344–352, 2020.
https://doi.org/10.1007/978-3-030-40749-0_41

The Rostec State Corporation (hereinafter referred to as the Corporation) was established in accordance with the Federal Law of the Russian Federation of November 23, 2007 No. 270-FZ "On the Rostec State Corporation" (State Duma of the Russian Federation 2007) (hereinafter referred to as the Law) to facilitate the development, production, and export of high-tech industrial products for civil and military purposes.

In accordance with the decrees of the President of the Russian Federation (Presidential Executive Office 2008; Presidential Executive Office 2016a, b; Presidential Executive Office 2018), the Corporation includes more than 800 organizations of the Corporation, which are, for the most part, united in holdings and located on the territory of 60 constituent entities of the Russian Federation; products are delivered to the markets of more than 70 countries. The most famous are AvtoVAZ, KAMAZ, and Helicopters of Russia, Schwabe, Kalashnikov Concern, etc. The consolidated revenue of the Corporation is growing dynamically, and in 2017 approached ₽ 1.6 trillion, net profit—more than ₽ 120 billion. Over the past ten years, the value of the corporation's assets has grown from ₽ 1 trillion to ₽ 3 trillion. The average salary for the corporation in 2017 amounted to ₽ 46.8 thousand, the number of employees—more than 500 thousand people (the second employer in the Russian Federation), and the tax payments of the corporation to the budgets of all levels exceeded ₽ 136 billion. Over the past ten years, the value of the corporation's assets has grown from ₽ 1 trillion to ₽ 3 trillion.

In 2016, a decision was made to transfer Uralvagonzavod JSC to the corporation (Presidential Executive Office 2016b), and in 2018, the United Aircraft Corporation (UAC) became a part of the Rostec (Presidential Executive Office 2018).

The volume and importance of corporations for the Russian industry impose special requirements on the organization of asset development. The system of corporation activity planning is defined by the law and the presidential decree (Presidential Executive Office 2016a). Given the global competition of high-tech corporations in the world markets, economic instability, and sanctions imposed on the enterprises of the Russian defense industry, Rostec has set the task of accelerating development at a pace significantly higher than previously achieved and entering new fast-growing markets in the transition from iron to intellect. A number of classical mechanisms for strategic analysis, such as the BCG (Boston Consulting Group) Matrix and the GE-Mackinsey Matrix, allow us to conclude on the status of business units, etc. at a specific point in time but do not provide the possibility of objective comparison (including with competitive effect) of the quality of business units in time to achieve established and significantly different goals.

2 Materials and Methods

Let us consider options for solving this issue of strategic management in order to increase competitiveness on the example of Rostec.

The Supervisory Board approved the updated development strategy of the corporation (Rostec State Corporation 2018a, b) on December 23, 2015. According to the Development Strategy, the Rostec Group is currently at the stage of active growth. The key imperative of the updated strategy is to reach the scale of leading global competitors.

In devising the development strategy of the Corporation, a comprehensive work was carried out in the part of the markets to determine the target segments and the important areas of action, which allowed us to identify the market directions that could ensure a high rate of revenue growth. It has been established that the Corporation's traditional markets are growing slower than the required growth rates. New markets for smart products are developing more than twice as fast as traditional Rostec markets. The growth of target revenue can be achieved by concentrating the resources on fast-growing global markets for smart civilian products. The Corporation retains the priority of executing contracts under the state defense contracts and military-technical cooperation, while the achievement of goals is planned differentiated by the holding companies, taking into account the fact that their competencies and accessible markets differ significantly depending on the industry. "Diversification issues are on the agenda … and their acuteness will increase" (Chemezov et al. 2017).

In order to implement the development strategy of the Corporation, in 2016–2017, the Corporation organized the devising of the development strategies for holding companies (Rostec State Corporation 2017), specifying the goals and specific markets to the areas of action in the markets and their segments, including civil and export, and aimed at achieving the planned indicators, while the planned rates of average annual growth differ significantly.

For example, some holdings operate in mature markets with few diversification opportunities and have a capital-intensive business with a long investment period. Moreover, the ability to quickly build competencies is limited primarily by the development time and, in some cases, the certification of new products. For these holdings with a large share of existing markets, including those associated with the increased defense capability or traditional exports, the objectively relatively low planned growth rates have been established (Nevertheless, they are several times higher than the forecasted GDP growth rates).

At the same time, the Corporation's strategy determined that a significant share of the Corporation's revenue will be provided by entering the fast-growing segments in which the Corporation has not yet been present, including civil telecommunication equipment, telecommunication networks of a new generation, cybersecurity, and so on. The new markets required setting goals for growth rates, in some cases exceeding the average annual 22% for holdings aimed at the explosive growth in fast-growing markets for smart civilian products, which is associated both with the low-base effect and with the ambitious goals set.

Thus, it is difficult to directly compare the performance of the holding companies of the Corporation operating in markets that are so different in terms of tasks, initial conditions, and required business models. At the same time, "Management decisions should be based on the assessments of the state of the facility" (Artyakov et al. 2014).

To increase the objectivity of the result evaluation, the authors proposed a new mechanism for managing the effectiveness of business units based on the HC [[[AUTHOR: please define 'HC' somewhere in the text before you begin using this abbreviation. Thank you.]]] comparative rating system, with the aim of:

- presenting information to the management of the Corporation about the degree to which their strategic objectives and tasks have been fulfilled
- creating the conditions conducive to the development of positive internal competition between the HC heads for leadership in achieving the Corporation's strategic goals, and
- creating a transparent mechanism for assessing the relative effectiveness of the HC in terms of implementing the Corporation's Strategy.

The comparative ratings of the HC are based on the following principles:

- The position in the rating depends on the fulfillment degree of the HC and the strategic substation control points (SSCP) of their current strategic goals and objectives.
- The management efficiency is assessed by the performance indicators approved in the Corporation Development Strategy: revenue (consolidated), sales of civilian products (PGO) (consolidated), EBITDA (consolidated), and net profit (loss) (consolidated) profitability.[[[AUTHOR: please review and correct 'net profit profitability;' thank you.]]]

The final rating is formed based on multifactor and multilevel calculations (see Fig. 1).

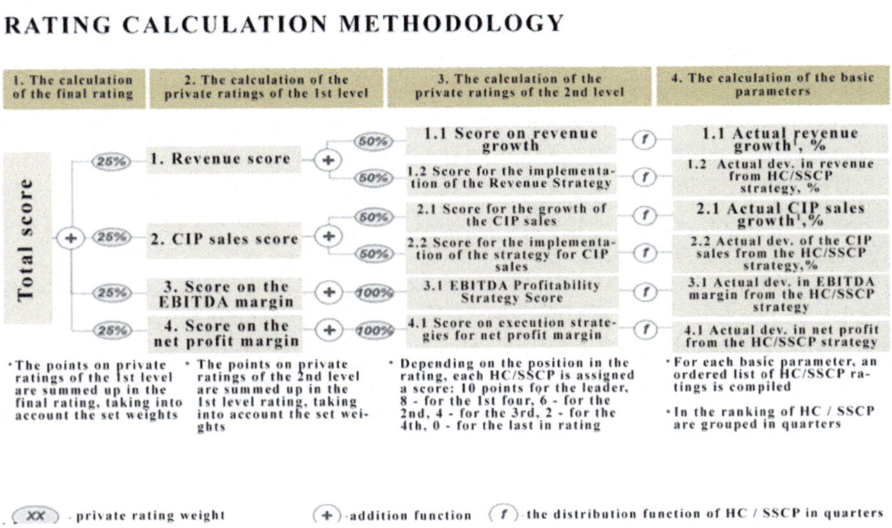

Fig. 1. The illustration of the methodology for calculating the final rating.

The comparative ratings of the HC—an ordered list of HCs, the position of which is determined by the number of points received by the HC as a result of calculation in accordance with this methodology, and which characterizes the effectiveness of the implementation of the Development Strategy of the State Corporation Rostec.

The basic parameters are the parameters calculated based on actual and planned values of financial performance indicators. The basic parameters of growth are actual revenue growth (%) and actual growth in the sales of civilian industry products (CIP) (%).

The basic parameters of the degree of strategy implementation are actual deviation in revenue from HC development strategy (%), actual deviation in sales of CIP from HC development strategy (%), actual deviation in EBITDA margin from HC development strategy (i), and actual deviation by profitability by net profit from the development strategy of HC (i). The reporting period is the final year, relative to which the basic growth parameters are calculated, and the basic parameters of the degree of strategy execution are calculated.

The reporting interval is the time interval between the base period (the year the strategy was launched) and the reporting period (at least one year and not more than six years).

The private ratings of the 2nd level are the comparative ratings of HC and strategic SSCP formed on the basis of the basic parameters. Private ratings of the 2nd level by revenue include rating by the growth of revenue, rating by the execution of the strategy by the revenue, and private ratings of the 2nd level by sales, which are similarly formed CIP.

The private profitability rating of the 2nd level by EBITDA includes rating by the execution of the profitability strategy by EBITDA and, similarly, by net profit.

The HC score for the private rating of the 2nd level includes the number assigned to the HC, which determines the position of the HC in the private rating of the 2nd level, which is determined by such basic parameters as the score for revenue growth determined by the actual growth of revenue (%), the score on the execution of the revenue strategy determined by the actual deviation in revenue from the development strategy of the HC (%), and the score of CIP sales growth, which is determined similarly.

The points for the execution of the strategy for EBITDA margin and net profit are determined by the actual deviation of EBITDA margin and net profit, respectively, from the development strategy of the HC (i).

The 1st-level private ratings are generated based on the 2nd-level private ratings. The HC score of the SSCP for the 1st-level private rating is the number assigned by the HC, which defines the position of the HC in the 1st-level private rating and is determined by the number of the HC's points for the 2nd-level private ratings. The revenue point is determined by the number of points for the 2nd-level private ratings of revenue; the point for revenue growth and the point for the execution of the strategy for revenue similarly determine the point for sales of CIPs for profitability for EBITDA, and for net profits, it's determined by the number of points for the $2^{nd\text{-}level}$ private rating for profitability for EBITDA and net profit, respectively.

The final rating is the comparative rating of the HC and SSCP and is based on the 1st-level private ratings. The total HC/SSCP is the number assigned to the HC/SSCP, defining the position of the HC/SSCP in the final rating, and is determined by the number of points of the HC/SSCP for 1st-level private ratings: revenue score, score of civil products sales, EBITDA margin, and net profit margin.

In the final rating, three high-scoring (with the highest scores) and two low-scoring (with the lowest scores) HC are noted. If the third one in the final HC rating scored an equal number of points with the subsequent HCs, then such HCs are also marked among the highly successful, that is, the list of high scores expands. If the second from the end in the final HC rating scored an equal number of points with the previous HCs, then such HCs are also noted among the low-performing ones, that is, the low scoring list expands.

3 Results

The results of applying the proposed rating system are shown in Fig. 2, regarding the results of 2018, and Fig. 3, in terms of results for four years—from 2015 to 2018. Similar ratings are also formed for the interim years. The decisions on restructuring

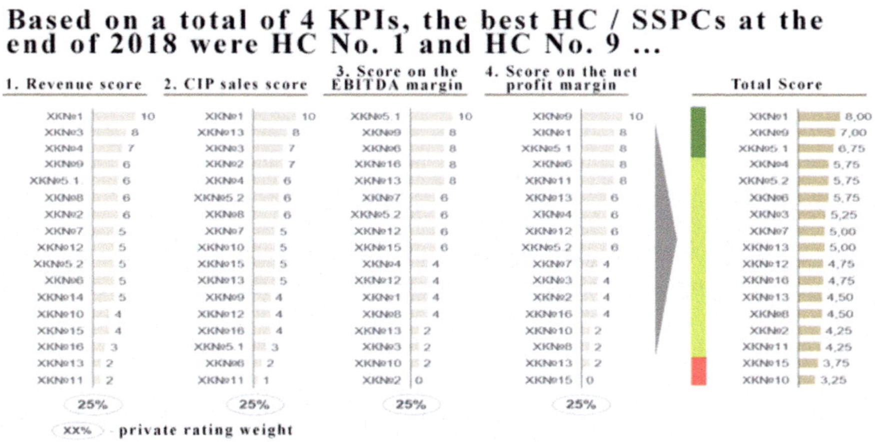

Fig. 2. The rating of the holding companies by the end of 2018.

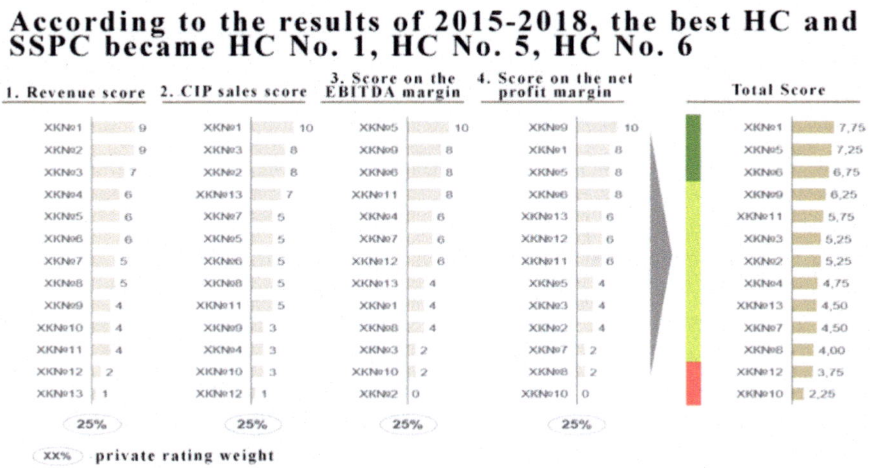

Fig. 3. The rating of holding companies in 2015–2018. * The names are replaced by numbers for privacy reasons.

adopted since the formation of the Corporation's strategy cause the differences in the list of HCs. Subject to the volume requirements, the details of second-level ratings calculated following the methodology are not given in this paper.

4 Discussion

The formation of ratings for a four-factor complex that takes into account both the difference in plans (due to objective factors) and in the results actually achieved (the management achievements) allows us to objectively evaluate the results of the holding companies (business units) against each other. It eliminates the discussion that appears due to significant differences in the conditions of activity and plans of business units.

The construction mechanism, both for the year and for the period, allows us to more accurately determine trends and patterns while eliminating the influence of stochastic factors on the performance.

The construction of Ratings based on the results of reporting intervals completed in 2015, 2016, 2017, and 2018 confirmed the continuity of the results shown by holding companies. It also strengthened objectivity in evaluating the results of various business units (holding companies) and their real contribution to achieving the goals of the Corporation Strategy.

As follows from the results, the dynamics of the HC in the ranking have significant differences. For example, despite the average results of HC No. 1 in operational efficiency following the results of the period, high growth rates amid ambitious goals, including according to the CIP, allow us to recognize this holding company as a leader in the implementation of the Corporation Strategy.

On the example of HC No. 6, it can be noted that relatively low places in terms of growth in both total revenue and sales revenue, which are due to the fact that the holding operates in mature highly competitive oligopoly markets with long investment-intensive new product development cycles (significantly exceeding the terms, those that have passed since the beginning of the implementation of the strategy) were fully compensated by the constant, from year to year, ongoing systematic work to improve operational efficiency and financial management. Thus, a high place in the final rating, according to the results of 4 years, emphasizes the holding's work efficiency and the quality of its management.

On the contrary, a failure to achieve high goals, despite significant growth rates and relatively high profitability (with low EBITDA and profit) allows us to characterize HC management as the least efficient overall.

Also, the example of HC No. 3 demonstrates that high growth rates, calculated in double-digit % in the average annual value, are not sufficient to achieve the goals of the Corporation. One of the key elements of the Corporation's strategy is to increase operational efficiency (EBITDA margin) to global average values in 2020 and the best quarter of global companies—analogs by 2025, required both from the point of view of the Corporation's profitability as a shareholder, and from the point of view of financing opportunities for the investment program. Despite the stage of the rapid growth of the holding, performance indicators are worse than their counterparts provided in the planning documents. For these reasons, the final position of the holding in the aggregate HC rating is relatively low.

5 Conclusion

The ratings are an ordered list of the HCs. They are built for various reporting intervals on the objective data and allow us to evaluate the direct performance of holding companies and take into account the significant difference in the goal setting, the initial conditions (such as the HC at the entrance to a rapidly growing market or an HC from the leaders of the oligopolistic market), the market opportunities, and the quality of the use of these opportunities by the HC management.

The results of the study were implemented at Rostec as one of the mechanisms for monitoring the implementation of the development strategy of the corporation; the possibility of using them in a formal motivation system for holding company managers (business units) is being considered.

The practical results and recommendations were obtained on the development of a system for the objective assessment of the effectiveness of competitive management and the contribution of individual business units operating in markets of varying maturity degrees to the implementation of the strategy of a diversified industrial corporation based on the comparative rating and formed on the basis of a multifactorial approach.

The use of ratings makes it possible to evaluate the effectiveness of the efforts of individual holdings objectively to implement the corporation's strategy, regardless of the difference in the initial conditions of the development and the degree of the maturity of the market, and it shows the relative (in relation to the best practices in the corporation) effectiveness of the general managers of the holding company, which may be an additional element of their motivation.

References

Artyakov, V.V., Chursin, A.A., Rusinov, A.A.: Modeling the sustainability of project management of enterprises in science-intensive industries. Bus. Law **6**, 270–273 (2014)

Chemezov, S.V., Kulikov, S.A.: Priorities and mechanisms for ensuring the unity of state policy in the field of military-technical cooperation and high technologies. Bull. Acad. Mil. Sci. **1**(34), 34–42 (2011)

Chemezov, S.V., Volobuev, N.A., Koptev, Yu.N., Kashirin, A.I.: Diversification, competencies, problems, and tasks. New opportunities. Innovations **4**(222), 3–26 (2017)

Presidential Executive Office: Decree "Issues of the State Corporation for the Promotion of the Development, Production, and Export of High-Tech Industrial Products "Russian Technologies"" (July 10, 2008 No. 1052) (2008)

Presidential Executive Office: The decree "On some issues of the State Corporation for the Promotion of the Development, Production, and Export of High-Tech Industrial Products "Rostec"" (July 22, 2016 No. 356) (2016a)

Presidential Executive Office: Decree "On the property contribution of the Russian Federation to the State Corporation to promote the development, production and export of high-tech industrial products "Rostec" and on amending the list of strategic enterprises and strategic joint-stock companies (December 27, 2016 No. 715) (approved by the Presidential Decree of the Russian Federation of August 4, 2004 No. 1009") (2016b)
Presidential Executive Office: The decree "On the property contribution of the Russian Federation to the State Corporation to promote the development, production and export of high-tech industrial products "Rostec"" (October 24, 2018 No. 596) (2018)
Rostec State Corporation: Order "On the Procedure for Strategic Planning of the State Corporation "Rostec"" (December 28, 2017 No. 166). Rostec, Moscow (2017)
Rostec State Corporation: Order "On Amending Order of State Corporation Rostec of April 19, 2017 No. 47." (October 18, 2018 No. 117). Rostec, Moscow (2018a)
Rostec State Corporation: The updated development strategy of the Rostec State Corporation until 2025 (December 26, 2018). Rostec, Moscow (2018b)
State Duma of the Russian Federation: Federal Law "On the State Corporation Rostec" (November 28, 2018 No. 270-FZ) (2007)

Business Assessment and Goodwill in Ensuring the Competitiveness of Russian Organizations

Victor S. Plotnikov(✉) and Olesya V. Plotnikova

Novosibirsk State University of Economics and Management, Novosibirsk, Russia
vcplotnikov@yandex.ru, o.v.plotnikova@nsuem.ru

Abstract. The competitiveness of an organization is an indicator of the effectiveness of its activity, which correlates with the assessment of the value of a business. In many business organizations in Russia, there is an underestimation of the business value due to the lack of internally created goodwill. Goodwill means a wide range of intangible assets, which reflection in the financial statements of an organization provides a competitive advantage in the future additional income. Therefore, it is necessary to find out what is meant by the internally created goodwill, as well as which accounting opportunities exist so they can be reflected in the financial statements of organizations. The objective of the paper is determined by the need for reflecting the value of internally generated goodwill in the organization's balance sheet. As a result of the study, ithe authors propose to reflect the internally created goodwill as part of intangible assets as a law that has the potential to create economic benefits, with the subsequent transformation of this right into financial assets of the organization.

Keywords: Organization competitiveness · Business valuation · Internally created goodwill · Market capitalization · Business combination

1 Introduction

Modern, effective business management is impossible without the development and implementation of solutions for its improvement, due to both market demands and the increased competitiveness of domestic organizations.

One of the factors that positively affects the assessment of the activities of business structures is manifested in the reflection of internally created goodwill in the financial statements of the organization. Internally created goodwill demonstrates an increase in competitive advantages, an increase in the market value of the organization, and an incentive for the long-term development of the organization. Castedello and Klinbeil analyzed the impact of goodwill and intangible assets on the value of a business. According to the analysis, the goodwill share in the value structure of organizations ranged from 36% (energy production) to 70.4% (Internet and electronic business). However, within only six types of activities, the goodwill share was less than 50% (Klingbeil 2010). All this confirms the need to reflect on the internally created goodwill in the balance structure of Russian organizations.

A. V. Bogoviz and Y. V. Ragulina (Eds.): ISCI 2019, LNNS 115, pp. 353–361, 2020.
https://doi.org/10.1007/978-3-030-40749-0_42

However, the developers of the IAS 38 “Intangible Assets” (Ministry of Finance of the Russian Federation 2011) and the developers of Accounting Rules 14/2007 “Accounting for Intangible Assets” (Ministry of Finance of the Russian Federation 2007) are of the opposite position, which states that “Internally created goodwill is not recognized as an asset.” The developers of the standard explain: “In some cases, costs are incurred to create future economic benefits, but they do not lead to the creation of an intangible asset that meets the recognition criteria established by this standard. Such costs are often called unprofitable, contributing to the formation of the internally created goodwill. The internally created goodwill is not recognized as an asset because it is not an identifiable resource (i.e., it is not separable and does not arise from contractual or other rights that are controlled by the enterprise and can be reliably measured at cost)” (International accounting standards (IAS) 38 2011)[1]. This is not a very convincing explanation, since, according to this criterion, both a trade and a merchandise mark can be excluded from the intangible assets list.

“The treatment of the unidentifiable assets is less obvious because they do not have a semantic interpretation. It means that they do not meet the rule of classifying as assets. However, in accordance with the traditional accounting structure, if funds were spent for generating future income, these costs should be capitalized and carried to the future reporting periods” (Hendriksen 1997). The definition of an unidentifiable asset does not allow assigning internally created goodwill to assets, according to the existing rules. However, the capitalization of the organization is reflected in business valuation, that is, an increase or a decrease (decapitalization) in the value of the organization’s assets. Consequently, the market fluctuations reflected in business valuation should determine the value of a business at a specific point in time. Thus, internally created goodwill can be recognized in the accounting and balance, which “reflects the internal state of the work and its administration” (Eliseeva and Dmitriev 2010).

The developers of Russian regulatory documents and a number of scholars working in this field prefer to use an equivalent to the goodwill definition of business reputation and link its definition with customer relationships, personnel training level, high administrative resources level, knowledge and technology use, etc.

“In the domestic financial science and the accounting theory, the term of the business reputation of a company is recognized as synonymous with the international term ‘goodwill’ (according to N. A. Abdullayev, I. A. Blanka, N. A. Kolayko, Ya. V. Skolova and others)” (Belsky and Smetanin 2010).

In foreign financial analysis, the term goodwill has become widespread since the 15th century. At the end of the 19th century, I. F. Shera and F. Pixley concluded that goodwill is not only the accounting product but also the discounted value of future super-profits. Further, it was estimated from the position of the intangible asset, which increases the company value (E. Schmalenbach, K. D. Larson, P. W. Miller, and others) (Belsky and Smetanin 2010).

However, even this approach is rarely used in the financial statements in Russian organizations. Unfortunately, attempts to capitalize on the above costs as

[1] International Financial Reporting Standard (IAS) 38 “Intangible Assets”. (2011). N 160n. Effective since 11/25/2011. Moscow, Russia: Ministry of Finance of the Russian Federation. (in Russ.).

"non-redeemable" goodwill[2] provide no apparent advantages or changes in the business valuation. The capitalization of these types of costs, taking into account the provisions of the International Standard for Integrated Reporting (International Standard for Integrated Reporting 2013), should form the value of certain types of capital: intellectual, human, client, etc. Internally created goodwill is quite tangible, which should be reflected in the balance of the organization.

Pixley used a slightly different approach in the correlation dependence of the value of shares on the stock exchange from the value of earned income. "The stock price of a particular company on the securities market is a market estimate of the value of the shares as retained earnings already earned and the discounted value of a portion of future excess profits" per share representing a proportionate share of the value of its net adjusted assets. "The difference between the market value of a share and the amount attributable to it of the net assets of the issuing company is part of the discounted value of the expected future super-profits, goodwill per share" (Sokolov and Pyatov 1997; Pixley 1922).

In his definition, Pixley does not mention either the purchase or the sale of shares. A comparative characteristic of internally created goodwill is assumed with the difference obtained between the market value of the shares and the amount attributable to it of net assets. This approach is taken as the basis of the methodological and procedural support of the reflection of internally created goodwill in the financial statements.

2 Materials and Methods

The methodology for researching internally created goodwill is based on the concepts of business valuation, the conceptual framework for the presentation of financial statements (hereinafter "the conceptual framework"), and the concept of "economic profit," allowing us to comprehend the possibility, recognize the necessity, and evaluate this object in the financial statements. The authors believe that the information about internally generated goodwill can be very important for the users of financial statements.

Recognizing internally created goodwill as the difference between the market capitalized value of the organization and its net assets should be based on the fact that: "The field of objective assessments, based on naturally formed market prices, makes it possible to correct the calculated indicators correctly, to take into account the principal shifts in their structure and absolute value, which is achieved by analyzing the interrelation of its elements" (Sokolov 2000).

In this case, the comparison of the internally created goodwill with the "economic resource" given in the Conceptual Framework is of interest (Conceptual Basis for the Presentation of Financial Statements 2018): "An economic resource is a right that has the potential to create economic benefits. For the existence of such a potential, it is not necessary to say that this right will create economic benefits with absolute certainty or

[2] E. S. Hendriksen first used the term "non-redeemable" in the book "Accounting Theory" (Hendriksen 1997). The implication of goodwill, created at an enterprise itself.

even with a high probability. It is only necessary that this right already exists, and that it could create economic benefits for the organization that exceeds the benefits available to all other parties at least in some specific circumstances" (Conceptual Basis of Financial Reporting 2018).

This provision of the conceptual framework is directly related to the need for recognizing internally created goodwill in the financial statements of the organizations since, at the time of the financial statement, this asset reflects only a right that has the potential to create economic benefits. Such a right can be transformed into a real asset under certain circumstances, namely, when acquiring the investee shares. Income from the investee or subsidiary reflects the potential for generating economic benefits. At the same time: "The right can meet the definition of an economic resource and, consequently, be an asset, even if the probability that it will create economic benefits is small. However, this low probability can affect decisions about what information to provide and how to do it, including decisions about whether to recognize an asset and how to evaluate it" (Conceptual framework for the presentation of financial reports 2018).

A similar approach can be used for the internally created goodwill, since it meets the definition of an economic resource and, therefore, can be an asset, even if economic benefits can only be obtained in the future.

The other evidence of the need for a balance sheet to reflect internally generated goodwill can be found in the concept of "economic profit." This concept suggests that income can be decomposed into its constituents, cash flow, and cost, where the cost component reflects changes in respect of expected future cash flows. This concept is universally associated with Hicks's work (Black 2009).

Currently, most economists who are researching profits distinguish between its two main categories: profits reflected in the financial statements and profits as the organization market capitalization. The first category describes the organization activity for a certain period (month, quarter, year); the second category implies the organization's independent stock valuation itself at any given time. This is the basis for financial analysts' conviction that these are entirely different categories.

Financial analysts are partly right, as it should be noted that the accounting profit presented in the financial report and the economic profit do not directly correlate with each other. In these circumstances, the accounting of the actual return on capital markets can be seen as an attempt to offset the difficulties of financial accounting, which increase with the increasing interest in cost analysis for shareholders, whose focus is on the analysis of capital market returns. While the "accounting" and "analysis" of the capital market profitability are closely related (particularly due to the fact that reporting systems must be developed in accordance with an imaginary interpretation scheme), any reporting system must be based on available and reliable data, allowing investors to draw their own conclusions (using additional information and knowledge) on the prospects and profitability of a given company (Black 2009).

However, we should keep in mind that the sum of net profit calculates the first element of the economic profit, namely the profit received, and is confirmed by cash flow.

The second element of the economic profit is cost, which reflects changes in relation to the expectations of future cash flows, which are the additional income that an organization could receive by selling its shares at the financial statement time. This situation is rare. However, the organization should reflect the cost of a possible "additional" income in the form of internally created goodwill, without reflecting which of the business valuations will be underestimated.

Based on the preceding, it can be concluded that internally created goodwill is the difference between the assessment of the organization's market capitalization formed at a certain point in time on the stock exchange valuation and the carrying amount of the organization for the current reporting period. Thus, goodwill is the value that determines the second component of economic profit; that is, the value that can be confirmed by cash flow in the future.

Regarding the determination of the book value of the organization, in this matter we should agree with Bernstein (2002): "The term 'book value' refers to conditional terminology related to the net worth of the assets, i.e., the number of assets reduced by the priority requirements for them."

When abstracting from some details related to the calculation of the value of net assets, it is possible to say that the value of net assets determines the book value of the organization, but the "retained earnings" should be increased to the amount of net profit remaining at the owners' disposal. So, the book value of the organization will be more accurate.

The market evaluates the organization at the value of its securities. So to evaluate the effectiveness and "business reputation" of an organization, it is necessary to explain the results of its functioning in the capital market, that is, an objective market assessment of the organization or market capitalization of the organization, which can be used to calculate the cost "where the cost component reflects changes in future cash flows" (Black 2009).

However, the existing accounting system uses more specific definitions of objects than "economic resource" and "economic profit." Therefore, based on the substantive accounting, which, at a minimum, recognizes double entry and balance, we propose to reflect the internally created goodwill of D-mt 04 "Intangible assets," subaccount "Internally created goodwill" until the sale of the organization's shares on the stock exchange, since the intangible assets mostly reflect the right to receive economic benefits in the future, and not a physical object.

The definition of the corresponding credit account is more complicated, since "unrealized profit" is not reflected according to the existing accounting rules in accounting and reporting, and is absent in the accounts chart for financial and economic activities (Ministry of Finance of the Russian Federation 2000). Therefore, the authors propose the "unrealized profits", in order not to mix it with the consolidated accounting system, to reflect at C-t "non-monetary profit" because it is not yet confirmed by cash flow. Non-monetary profit represents "rights or claims for which neither the date of conversion into money nor the amount into which they will be converted" is known in advance (Hendriksen 1997).

It is not a mistake that only subparagraph (a) is cited, clause 4.16 of the conceptual framework, since it is contracts that presuppose the transformation of a right into real assets or balance sheet liabilities in the accounting system.

The problem of accounting for the internally created goodwill transformation into a real financial asset in a business combination transaction that ends with the investee shares acquisition is also of economists' interest.

Accounting for a business combination using special methods for capitalizing the combined business value involves developing a concept for accounting for these transactions "allowing generalizing information on the capitalized value of an economic association of legally independent enterprises as a single entity in the consolidated financial statements" (Plotnikov et al. 2016).

Based on this, initially, two factors make it possible to comprehend the internally created goodwill concept.

1. Internally created goodwill is an indicator reflecting a change in the market capitalization of the organization: that is, a change in the stock price, but without its selling. At the same time, internally created goodwill, formed when acquiring control over an investee, is an integral part of the price, which generally acts on the stock market. In this interpretation, even IFRS recognize goodwill as an asset.
2. Goodwill is only the result of the valuation of other assets, the control of which the acquirer obtained as a result of financial investments in the investee (subsidiary). Consequently, goodwill represents the difference between the acquisition price and the fair value of all assets (working capital, fixed assets, intangible assets, etc.) and liabilities assumed.

In assessing financial investments, the internally generated goodwill value is not reflected. However, there is a substantial connection between internally created goodwill and the price generated as a result of market relations in a business combination transaction.

To provide a rationalization for this substantive interconnection in accounting for a business combination, let us consider several factors that ensure the exchange business transaction.

"Firstly, in accordance with clause 4.1 of Article 420 of the Civil Code of the Russian Federation, any exchange transaction conducted on the basis of financial instruments is, first of all, an agreement (contract)—an agreement of two or more persons on the establishment, amendment or termination of civil rights and obligations. Therefore, a contract is a legal document confirming the rights and obligations of the participants in an exchange transaction.

Secondly, the contract itself as a financial instrument (forward contract and option) in most cases is bilateral, since mutual rights and obligations arise simultaneously for all the participants of the exchange transaction. In addition, each of the parties acts as an investor in the exchange transaction, since the manufacturer (seller) must invest (or has already invested) its funds in the production of the asset to be exchanged, and the buyer—to invest money (or other assets) to recover the manufacturer's investment.

Moreover, the exchange transaction involves the existence of counter-obligations, [whose] nature depends on the specifics of financial instruments in the exchange transaction, which mainly consist of the seller's obligation to sell the underlying asset and the buyer's obligation to pay cash for this asset.

Thirdly, any exchange transaction formalized by an agreement (financial instrument) in the financial market is a compensable agreement, because the asset seller has the right to demand an appropriate monetary or other remuneration, meeting the conditions of the financial instrument for fulfilling its contractual obligations to present the underlying asset to the buyer.

Fourthly, each exchange transaction in the market of financial instruments is of a consensual nature, since it is considered concluded from the moment of signing the agreement by the participants in the exchange transaction for all the essential terms of the transaction, the procedure, and form of fulfillment of contractual obligations, and, most importantly, it determines the transfer of ownership of the asset" (Plotnikov and Plotnikova 2018).

Goodwill, being the result of the business combination, manifests itself in an exchange broker's board transaction for the purchase and sale of the shares, where the essential point is that options are listed on the stock exchange and that they are the sale objects. Consequently, the option itself has its value and is an exchange commodity that is bought and sold in the stock market.

"There are two main types of options. *Buyer's option and call option; a call option* gives its owner the right to buy the underlying asset on a specific day at a certain price. *Seller option, option to sell, put option* gives its owner the right to sell the underlying asset on a specific day at a certain price. The price fixed in the contract is called the *exercise price* or the *strike price*. The date specified in the contract is called the *expiry date* or the *payment date*" (Hull 2008).

So, let us try to consistently deal with the factor of stock options trading and find what value the above option characteristics have for the internally created goodwill assessment.

The first factor, and perhaps the main one for our study, is that the option that determines the internally created goodwill's market value is an exchange commodity and therefore has its price. What is the basis for determining the option price? It is the difference between the market capitalization of the organization, registered at a certain point in the exchange valuation, and the book value of the organization that forms the valuation of internally created goodwill and is the basis for assessing the intrinsic value of the option. This estimate, reflected in the balance sheet as an intangible asset and subject to some temporary difference, can be recognized as an initial estimate of the option value.

The second factor is that, on the day of the contract signing, the purchase can be determined by the beginning of the financial investments in a business combination transaction. In this situation, the seller's option, a put option, must be recognized in the owner's accounting for the sale act. Consequently, non-monetary profit (unrealized profit) is transformed into sales revenue and can be reflected by d-t acc. "Non-Monetary Profit" and c-t sc. 91, "Other incomes and expenses," with its subsequent reflection in the aggregate income of shareholders.

At the same time, internally created goodwill in the seller's account is transformed from the intangible asset's composition into the financial capital structure, d-t sc. 58 "Financial investments," subaccount "Goodwill," c-t sc. 04 "Intangible assets," and subaccount "Internally created goodwill."

Financial accounting is reflected in the call-option buyer accounting and investments made in the investment object, which should be reflected in the d-t acc. 58 "Financial investments" (subaccount "Goodwill"), and c-t sc. 51 "Current accounts." In the consolidated accounting and reporting, these accounting records are deleted in the future, but this is already a separate issue.

The third factor is that the underlying asset in the option is the seller's stock, which should be reflected in the strike price fixed in the contract. Consequently, during the time the option is exercised (the contract expiration date), the "strike" price on the shares does not change, but the option value changes.

"With an increase in the current stock price, term, volatility, or risk-free interest, the option rate value for shares purchasing increases. With an increase in the strike price or the expected dividends size, the option value for shares purchasing falls. With an increase in the strike price, duration, volatility, or the expected dividends size, the option value for shares selling, as a rule, increases. If the stock price increases or the risk-free interest rate increases, the option value for the stock selling falls.

The acquisition of shares of another organization, of course, is a financial asset that brings benefits, or rather, gives the right to receive income. Therefore, it must be identified with the net assets of another organization from which these shares were purchased" (Plotnikov and Plotnikova 2013).

"It is this fluctuation in the securities price that forms the difference that makes up the true economic, commercial profit. This is an objective market price. It ignores the details. It does not matter how the assets are evaluated or how and by whom the reserves are created. The company cost is a single complex" (Eliseeva and Dmitriev 2010).

The goodwill recognition and measurement are the final accounting step in reflecting a business combination accounted for using the acquisition method. It reflects the result of the completion of the transaction. Two important points should be noted here: "it is the final accounting stage" and "the result of the transaction completion." In essence, goodwill is an accounting method of allocating the financial assets (investments) value to their nominal value, corresponding to the participation share in the net identifiable assets of the investee and financial investments exceeding their nominal value.

3 Results

First of all, existing objective reasons for the accounting procedures' complexity in reflecting the capitalized costs' diversity forming the internally created goodwill value should not be the reason for not reflecting it in the financial statements. At the same time, the information on the factors that influenced the market capitalization of the organization should be provided in the explanatory notes for ensuring its competitiveness.

Second, the difference between the market capitalization of the organization, determined by the market valuations and the organization book value, forms an estimate of the internally generated goodwill and reflects the second part of the "economic profit" component, that is, the cost in relation to the expected future cash flows.

Third, internally created goodwill is the balance indicator, reflecting the changes in the organization's capitalization and ensuring the growth of its competitiveness.

4 Conclusion

In conclusion, the authors would like to point out that the importance of reflecting internally created goodwill for business valuation should gradually decrease as the International Integrated Reporting Council (2013) standard is improved and put into practice, when most of the capitalized costs that determine the internally generated goodwill value will be reflected in such capital types cost as intellectual, client, human, social and so on, that, in fact, determine the organization business reputation demanded in the market.

References

Belsky, I.E., Smetanin, I.M.: Accounting and evaluation of goodwill: international experience and Russian specificity. Int. Acc. **3**(135), 24–29 (2010)

Bernstein, L.A.: Analysis of Financial Statements: Theory, Practice, and Interpretation. Finance and Statistics, Moscow (2002)

Black, A.: Cost Issues. Master the Latest Management, Investment, and Regulation Techniques Based on Company Value. Olimp - Business, Moscow (2009)

Eliseeva, I.I., Dmitriev, A.L.: In Memorial: Yaroslav Vyacheslavovich Sokolov (1938–2010). Nestor-History, St. Petersburg (2010)

Hendriksen, E.S.: Accounting Theory. Finance and Statistics, Moscow (1997)

Hull, J.K.: Options, Futures, and Other Derivative Financial Instruments. Williams Publishing House Ltd., Moscow (2008)

International Accounting Standards (IAS) 38: "Intangible assets" (No. 160n). Ministry of Finance of the Russian Federation, Moscow (2011)

Klingbeil, C.: Intangible Assets and Goodwill in the context of Business Combinations: An Industry Study (2010). https://www.consultancy.nl/media/KPMG%20-%20Intangible%20Assets%20and%20Goodwill-836.pdf

Ministry of Finance of the Russian Federation: A conceptual framework for the financial reports' presentation (2011). https://www.minfin.ru/ru/document/?id_4=125979

Ministry of Finance of the Russian Federation: On approval of the Chart of Accounts for the accounting of financial and economic activities of organizations and instructions for its use (No. 94n). Ministry of Finance of the Russian Federation, Moscow (2000)

Ministry of Finance of the Russian Federation: On the approval of the Accounting Regulations "Accounting for Intangible Assets" (No. 153n). Ministry of Finance of the Russian Federation, Moscow (2007)

Pixley, F.W.: Accountant's Dictionary, vols. 1, 2. Sir Isaac Pitman & Sons, Ltd., London (1922)

Plotnikov, V.S., Plotnikova, O.V.: Accounting for Financial Instruments as a Factor in Ensuring the Reliability of the Information on the Functioning of a Company in the Market. Publishing House FINANCE and CREDIT, Moscow (2013)

Plotnikov, V.S., Plotnikova, O.V.: Business Combinations and Consolidated Financial Statements. INFRA-M, Moscow (2018)

Plotnikov, V.S., Plotnikova, O.V., Shevchuk, A.I.: Estimation of the capitalized value of business processes in the analysis of a business combination transaction. Econ. Anal.: Theory Pract. **12**(459), 139–152 (2016)

Sokolov, Ya.V.: Fundamentals of Accounting Theory. Finance and Statistics, Moscow (2000)

Sokolov, Ya.V., Pyatov, M.L.: Goodwill: a "new" category of accounting. Bukhgalterskiy uchet **2**, 46–50 (1997)

Developing the Toolkit for Evaluating Innovative Activities of Enterprises to Improve Their Competitiveness

Aleksandr M. Batkovskiy[1(✉)], Vladimir A. Sudakov[2,3], and Alena V. Fomina[1]

[1] Joint Stock Company "Central Research Institute of Economy, Management and Information Systems 'Electronics'", Moscow, Russia
{batkovskiy_a, fomina_a}@instel.ru
[2] Keldysh Institute of Applied Mathematics (Russian Academy of Sciences), Moscow, Russia
sudakov@ws-dss.com
[3] Plekhanov Russian University of Economics, Moscow, Russia

Abstract. The paper focuses on the application of new fuzzy methods for decision-making in the multi-criteria evaluation of innovation. It is proposed to determine the areas of preferences that allow obtaining fuzzy integral estimates of alternative development options for innovatively active enterprises in the space of criteria. The relevance of using the vector performance criterion for assessing their performance is determined. The main groups of criteria are selected that take into account the competitiveness of innovatively active enterprises, which include: their value, economic efficiency, and financial viability; the risk level of changes in market conditions; innovation life cycles; social, scientific and technological effects of creating and implementing innovations. When assessing the innovation activity, there is a necessity in taking into account its specificity, which is determined, first of all, by those indicators determining the high technology of products, availability and quality of necessary resources, production characteristics, and scientific and technical potential of enterprises. The authors considered and analyzed the methods and modern approaches to the management of innovative activities of enterprises based on fuzzy logic tools. For this purpose, the use of the approach based on a multi-criteria evaluation of alternatives was proposed. The use of fuzzification and defuzzification procedures has made it easier to find the ranks of alternatives in the management of innovation. The basic relations that allow checking the completeness of preferences description in a fuzzy information environment when solving the problem are listed. An algorithm for using the procedures of a fuzzy analog of weighted summation is presented. The implementation of the proposed web technologies in practice allows us making the procedure for a fuzzy assessment of the innovation activities of enterprises available when solving managerial tasks aimed at improving their competitiveness.

Keywords: Innovation activity · Preference areas · Fuzzy criteria values · Membership function · Fuzzy weighted summation · WS-DSS

A. V. Bogoviz and Y. V. Ragulina (Eds.): ISCI 2019, LNNS 115, pp. 362–369, 2020.
https://doi.org/10.1007/978-3-030-40749-0_43

1 Introduction

Determining the effectiveness of innovation is a complex multi-objective task. This efficiency should be assessed based on the given technological, socioeconomic, and financial statuses, and other criteria. Also, it is necessary to determine not only the final result of the innovation but also to calculate the efficiency at all stages of the process of implementing the innovation for each participant/agent and each resource used (Shelton and Arciszewski 2007). Thus, it is necessary to build an integrated performance assessment based on a vector criterion taking into account the value system of all interested parties and taking into account both objective and subjective expert assessments.

Currently, there are no universal approaches to the assessment of innovative activity (Batkovskiy et al. 2015). Many scholars are developing methods for assessing it, based on the existing theoretical foundations of the analysis of an investment project and the theory of economic efficiency (Veselovsky et al. 2018; Ragulina et al. 2020). However, various sectors of the economy have their specifics, which requires the use of different assessment criteria and a system of priorities between them, as well as project implementation risks, when solving this problem.

The criteria characterizing innovative activity are heterogeneous. From the mathematical formalization of the problem under consideration, it is usually distinguished as numerical (quantitative), lexical (qualitative), and unclear criteria. The evaluation of innovation, as a rule, is carried out according to the following criteria:

- the estimation of the innovative enterprise value at a forecast date;
- the assessment of economic efficiency and financial soundness;
- the risk assessment of the market changes in the innovation field;
- the assessment of the viability of innovation;
- the assessment of the social, scientific, and technological effects of innovation.

When forming a system of criteria for assessing innovation, the specifics of certain enterprises implementing innovations should be considered:

- the industry-specific features of innovative processes, including their high technology intensity;
- the availability and quality of necessary resources;
- the characteristics of the production system, level of labor intensity, capital productivity, etc.;
- the stage of development of the organization (growth, maturity, decline);
- the type of production;
- the innovative potential of the enterprise.

2 Materials and Method

Recently, unclear methods of multi-criteria evaluations of alternatives have been intensively developed in Russia and abroad. Hereinafter, alternative approaches are understood as either alternative ways of developing an innovative enterprise (the so-called task of choosing on a finite set of alternatives) or the state of the enterprise at given moments of the past or forecast moments of the future (the so-called monitoring task).

The development of computer technology and computer networks can significantly simplify the receipt of expert opinions from a wide range of specialists in various aspects of innovation. However, the expert often cannot unambiguously give a clear assessment of the object of examination, or unambiguously determine the relationship of preference between the two objects. As a rule, the interval estimates of certain characteristics of the examination objects are more acceptable. Valuable information from an expert can be an assessment of the degree of his confidence that the assessment assumes a particular value.

The basis of the theory of fuzzy sets proposed by Lotfi Zadeh (Zadeh 1988) is well-suited for such estimates. In the past decade, the application of the theory of fuzzy sets when making managerial decisions has expanded (Deinichenko et al. 1996). Different scholars have conducted research in and continue researching the field of individual functions of fuzzy preferences based on fuzzy areas (Dutov et al. 2018). In several works, fuzzy selection methods are developed based on the Pareto principle and the fuzzy qualitative importance of criteria (Noghin 1994). In the paper (Dutov et al. 2018), it was proposed to introduce fuzzy scales for all criteria consisting of fuzzy gradations (intervals of values with fuzzy boundaries). The intersection of fuzzy gradations within the framework of one criterion is allowed, taking into account restrictions on the minimum and maximum degree of belonging to an arbitrary value. The decision-maker evaluates some combinations of fuzzy criteria values on a fuzzy preference scale. Next, we check the completeness of the preference model: for all points of the criterion space, the level of preference should not be lower than the given.

The definition of a comprehensive assessment of innovation takes place with the help of fuzzy implication, proposed in Terano et al. (1992), and Mamdani (1974) for management tasks based on fuzzy logic. This model allows automatic evaluation of an arbitrary number of alternatives. For all areas of preference, the method allows building membership functions of alternatives.

Clear and/or fuzzy values are input to the multicriteria assessment model. To obtain a final ranking of alternatives, one should either de-fuzz fuzzy ranks of alternatives or determine a fuzzy domination relation for all the alternative pairs.

3 Results

For a multi-criteria evaluation of innovation using fuzzy preference areas, we will offer for the case of n criteria, and m specific evaluated alternatives, we obtained m membership functions corresponding to all alternatives (1):

$$p_t = f_t(y, x_1^t, x_2^t, \ldots, x_n^t) \tag{1}$$

where t – an alternative number (t = 1 … m)

The degree of the distinguishability of alternatives will be high if the membership function is "substantially" different for "significant" differences in the criteria values. In this case, the term "substantially" is subjective and depends on the specifics of the problem being solved and the opinion of the decision-maker (Batkovskiy et al. 2016).

To estimate the proximity of two membership functions f_1 and f_2, the metrics of different function spaces are used. For example, (2):

$$\|f_1 - f_2\| = \sup|f_1(y) - f_2(y)| \tag{2}$$

or (3):

$$\|f_1 - f_2\| = \sqrt[2]{\int (f_1(y) - f_2(y))^2 dy} \tag{3}$$

Since the criteria values, in general, correspond to the membership function, it makes sense to use the following rule: "If the value of at least one criterion is significantly different, then the ranks of the alternatives should also be significantly different." The lexical term "significantly different" can be expressed as the following membership function (4):

$$\mu(\|f_1 - f_2\|) = \begin{cases} 0, \ \|f_1 - f_2\| \le \alpha, \\ \frac{\|f_1 - f_2\| - \alpha}{\beta - \alpha}, \ \alpha < \|f_1 - f_2\| < \beta, \\ 1, \ \|f_1(y) - f_2(y)\| \ge \beta. \end{cases} \tag{4}$$

The constants α and β are selected based on the specifics of the problem under consideration (for many practical problems, a good approximation is $\alpha = 0.3$ and $\beta = 0.7$). In order to check the above rule for the proximity of the two alternatives, t_1 and t_2, it is necessary to calculate the degree of confidence in the premise truth using the formula (5):

$$d(t_1, t_2) = \min_i \mu(\|f_{t_1}(x_i^{t_1}) - f_{t_2}(x_i^{t_2})\|) \tag{5}$$

Then, the membership is calculated by the formula (6):

$$u(t_1, t_2) = \mu(\|f_{t_1}(y^{t_1}) - f_{t_2}(y^{t_2})\|) \tag{6}$$

Furthermore, if there are t_1, t_2 such that d > u, then it is necessary to clarify the function of preferences using quantitative fuzzy methods.

The most common quantitative fuzzy method is fuzzy weighted summation. For all criteria, fuzzy weights w_i with the membership function μ_i(w) are assigned, and for the values of the criteria, the membership function is also given: λ_i(x).

The weighted sum for the case of ordinary (stable) numbers is determined as follows (7):

$$s = S(x_1, x_2, \ldots, x_n) = \sum_{i=1}^{n} w_i x_i \tag{7}$$

It can be upgraded for the case of fuzzy numbers using the generalization principle. For the components of the sum, the membership function is calculated by the formula (8):

$$z_i(g_i) = \sup_{g=w_i x_i} \min(\mu_i(w_i), \lambda_i(x_i)) \tag{8}$$

The membership function of the weighted amount takes the form (9):

$$r(s) = \sup_{s=\sum_{i=1}^{n} g_i} \min_i(z_i(g_i)) \tag{9}$$

The integral evaluation of the alternative in the proposed method consists of two components: the fuzzy preference level y and the fuzzy result of weighted summation s. The integral rank of the alternative $I = y + s$ for the fuzzy case is also found using the generalization principle. Its membership function is (10):

$$\varphi(I) = \sup_{I=y+s} \min(p(y), r(s)) \tag{10}$$

The use of this integral criterion allows for increasing the distinguishability of the ranks of the alternative.

An information model should be developed for representing criteria, fuzzy gradations, preferences, and fuzzy values of criteria for alternatives in order to work with fuzzy areas. Let us consider two alternative approaches to solving this problem: a relational representation and a JSON representation.

In the relational representation, a piecewise linear membership function is defined by separate tuples for each point. There will be quite a lot of such points during the generalization operations by formulas (8) and (9). When processing a large number of entries, the relational scheme is inefficient. In the case of a large number of gradations and criteria, there can also be many preference areas. Also, the relational database management systems (DBMS) do not scale well. The advantages of the relational approach include the ability to control referential integrity at the level, for example, that only valid fuzzy gradations of criteria are chosen for the fuzzy preference area.

An alternative way to represent fuzzy preferences is the JSON format. It contains hashes corresponding to similar relational entities and arrays corresponding to tuples. The entity name or attribute is specified in the hash key. The compactness of data representation in JSON is increased since one object can be nested in another.

Fuzzy sets and numbers are stored in an array accessible by the key *fuzzy_sets*. Entries in this array can be as follows: fuzzy levels of preference for fuzzy preference areas, fuzzy gradations of criteria, or fuzzy estimates of specific innovation activity aspects. To write to the *fuzzy_sets* corresponding to a fuzzy number, the pieces array of piecewise linear approximation, accessible by a key polyline, should be defined. To write in the *fuzzy_sets*, corresponding to a fuzzy set, the elements of this set and the degree of their belonging must be listed.

Criteria are available in the top-level hash by the key *criterions*. The criteria array contains both primary criteria determined from the subject area, and aggregative criteria, allowing to determine the fuzzy rank of the alternative. The criteria aggregation should be structured so that the output has one complex criterion, and other aggregation criteria allow obtaining a hierarchy with the required number of nesting levels. A complex criterion is not required to be explicitly stated. It is defined as a criterion in which value can be calculated, but it does not participate in the calculations of other criteria.

The main attributes of the criterion are the *name* and the *scale*. The criterion *name* is unique and allows referring to it. The criterion *scale* contains an array of gradations, fuzzy numbers, or sets.

The *rules* key defines a set of rules for calculating the criterion. It has two attributes: conditions (conditions for the rule to be fulfilled) and the result (a fuzzy value of the aggregation criterion, which will be achieved when the corresponding conditions are met). The conditions for implementing the rule, in this case, are getting alternatives in the appropriate fuzzy area of preference.

Alternative solutions for managing innovation are listed in the array available on the alternatives key. Each alternative contains a hash, in which the key is the name of the criterion. The value of the hash is the fuzzy value of the criterion. Clear criteria values are set by specifying the membership function at one point with the value $\mu(x) = 1$.

To work with data in JSON format, NoSQL DBMS are usually used, allowing the effective scaling. Though many modern relational DBMSs already contain JSON processing algorithms, they do not scale well. The difficulty of ensuring JSON data consistency is one of the drawbacks of this approach. To control the data consistency, they are encapsulated into instances of classes at the intermediate software layer above the NoSQL DBMS layer. Therefore, the choice of approach to designing a physical data structure for a fuzzy information environment for evaluating innovation depends on the planned load on the system, the programming system used, a centralized or distributed software architecture, and the amount of input data.

4 Discussion

The combination of fuzzy preference areas with the procedure of fuzzy weighted summation on a single criteria space allows us to talk about the originality of the method. The procedure for determining the criteria for distinguishing alternatives on fuzzy areas of preference has a scientific novelty. Based on fuzzy preferences of decision-makers in fuzzy areas and fuzzy criteria weights, the described decision support method allows ranking an arbitrary number of alternatives in an automatic mode. In order to improve the quality of the decisions made, rules and ratios have been developed to evaluate the completeness of the task of fuzzy judgments.

The WS-DSS portal contains open access to the developed software implementation of the fuzzy ranking method (Web Services for Decision Support Systems 2015). This portal allows publishing the console software implementations of mathematical models in C++, R, and Ruby languages on the Internet, and also makes it possible to connect other web services and organize a sequence of calls to a group of mathematical models with the transfer of input/output parameters between them. The queue manager Sidekiq implements model processing and allows model scaling threads to scale on a large number of nodes in a computing cluster. This approach is consistent with a unified decision support methodology based on the concept of extensible programming.

5 Conclusion

The implementation of the presented methodology for assessing the innovation activities of enterprises contributes to the growth of these enterprises' competitiveness since it allows them to solve a whole range of tasks to improve the management of their innovative activities:

- evaluation of the innovation effectiveness following the integral fuzzy estimates obtained based on the vector criterion;
- sensitivity analysis of the criteria in relation to changes in input parameters of the innovative processes' models;
- risk analysis for performance criteria; and
- a search for the most significantly influencing parameters in terms of the innovation performance criteria.

This methodology based on fuzzy preference areas can be used for assessing innovation in terms of achieving its goals, taking into account the resource constraints. Due to the invariance of the considered approach, it can be extended to assessment problems in other economic activity areas.

Acknowledgment. The study was carried out with the financial support of the Russian Foundation for Basic Research in the framework of the research project No. 18-00-00012 (18-00-00011) COMFI.

References

Batkovskiy, A.M., Kalachikhin, P.A., Semenova, E.G., Telnov, Yu.F., Fomina, A.V.: Economic-mathematical model and mathematical methods for substantiating the choice of the company innovation strategy. Indian J. Sci. Technol. **9**(27), 99–111 (2016)

Batkovskiy, A.M., Konovalova, A.V., Semenova, E.G., Trofimets, V.Ya., Fomina, A.V.: Risks of development and implementation of innovative projects. Mediterr. J. Soc. Sci. **6**(4.S 4), 243–253 (2015)

Deinichenko, V., Bikesheva, G., Borisov, A.: The fuzzy approach of economic modeling of economics growth. In: Artificial Intelligence in Economics and Management. Kluwer Academic Publishers, Boston (1996)

Dutov, A.V., Nesterov, V.A., Sudakov, V.A., Sypalo, K.I.: Fuzzy preference domains and their use for selecting an electronic flight bag for flight crews. J. Comput. Syst. Sci. Int. **57**(2), 230–238 (2018)

Mamdani, E.H.: Applications of fuzzy algorithms for control of the simple dynamic plant. Proc. IEEE **121**(12), 1585–1588 (1974)

Noghin, V.D.: Upper estimate for a fuzzy set of nondominated solutions. Fuzzy Sets Syst. **67**, 303–315 (1994)

Ragulina, J.V., Bogoviz, A.V., Lobova, S.V., Alekseev, A.N., Pyatanova, V.I.: Strategy of increasing the global competitiveness of Russia's economy and Russia's becoming a new growth vector of the global economy. Lecture Notes in Networks and Systems, vol. 73, pp. 203–210 (2020)

Shelton, K., Arciszewski, T.: Formal innovation criteria. Int. J. Comput. Appl. Technol. (IJCAT) **30**(1/2), 21–32 (2007)

Terano, T., Asai, K., Sugeno, M.: Fuzzy Systems and its Applications. Academic Press, San Diego (1992)

Veselovsky, M.Y., Izmailova, M.A., Bogoviz, A.V., Lobova, S.V., Ragulina, Y.V.: System approach to achieving new quality of corporate governance in the context of innovation development. Qual. - Access Success **19**(163), 30–36 (2018)

Web Services for Decision Support Systems (2015). https://ws-dss.com

Zadeh, L.A.: Fuzzy logic. Comput. J. **21**(4), 83–93 (1988)

An Assessment of Commercial Banks' Financial Sustainability

Elvira V. Sitnikova[1], Tatiana S. Kolmykova[1(✉)], Elena O. Astapenko[2], and Evgeny A. Grivachev[2]

[1] Southwest State University, Kursk, Russia
0209elvira@mail.ru, t_kolmykova@mail.ru
[2] Yugra State University, Khanty-Mansiysk, Russia
kgtu_fk@list.ru

Abstract. The paper focuses on the results of a study proving that the commercial bank's sustainability is ensured by the quality of its assets and liabilities, liquidity, and profitability. The main indicators were determined, and the aggregated financial sustainability indicator was proposed. The type of financial sustainability of commercial banks is determined based on this indicator. Due to the developed unified methodological approach, measures are being developed to increase the efficiency of a commercial bank.

Keywords: Financial sector · Banking · Commercial bank · Financial sustainability

1 Introduction

The unstable state of the modern Russian economy depends mainly on the reliability and sustainability of the banking system, which, in turn, is subject to the influence of different exogenous and endogenous factors. That is why the primary task of the state regulators at the macro level and credit institutions at the micro level is to ensure sustainability through rational monetary policy. They may analyze and monitor the state of the real and financial economy sectors, reduce all the risks, and generate incomes that exceed total expenses. Recently, however, when the foreign economy is not stable and the policy of the Russian Central Bank (which is the single mega regulator of the country's financial system) is tough, a considerable number of commercial banks fail to cope with the task of ensuring liquidity and solvency (Abramyan 2015).

One result of this is the record number of licenses that were revoked from banks in 2016–2018. In 2016, the Central Bank revoked more than 90 licenses, and in 2015, that number was 88. In 2017, the Central Bank only revoked licenses from 47 banks, which is two times fewer than in 2016 and significantly fewer than in previous periods. For comparison, 29 licenses were revoked in 2013, and 18 were revoked in 2012 (Kolmykova et al. 2018). Thus, the inconsistency in the requirements of financial sustainability, risk, and profitability ratio is the consequence of the elimination of a large number of banks. Financial sustainability is the most important characteristic of

A. V. Bogoviz and Y. V. Ragulina (Eds.): ISCI 2019, LNNS 115, pp. 370–378, 2020.
https://doi.org/10.1007/978-3-030-40749-0_44

financial activity in modern economic conditions, and its maintenance is one of the main problems of the financial management of any commercial bank.

2 Materials and Method

The study examines important parameters of a commercial bank such as profitability, reliability, and financial sustainability. The applied research and calculations are performed on the example of Russian commercial banks. The informational basis consists of the official data of the Russian Central Bank, the Federal State Statistics Service, publications in scientific magazines on the sustainability and reliability of monetary institutions, and materials on the Internet.

3 Results

The most important characteristic of a commercial bank is its financial sustainability. Its provision is becoming one of the most challenging tasks for commercial banks. The result of the assessment depends on the methodology used and the system of indicators chosen. Therefore, to assess the sustainability of a bank, it is important to use a system of indicators that allows a full assessment of the bank's policies related to the allocation of its resources, its profitability, etc., based on reliable and complete information (Kolmykova et al. 2015).

Financial sustainability is influenced by internal and external factors determining the state of the capital, asset quality, resources quality, contingency of assets and liabilities by maturity and amount, the liquidity and profitability level (Tretyakova and Sitnikova 2016) (Fig. 1).

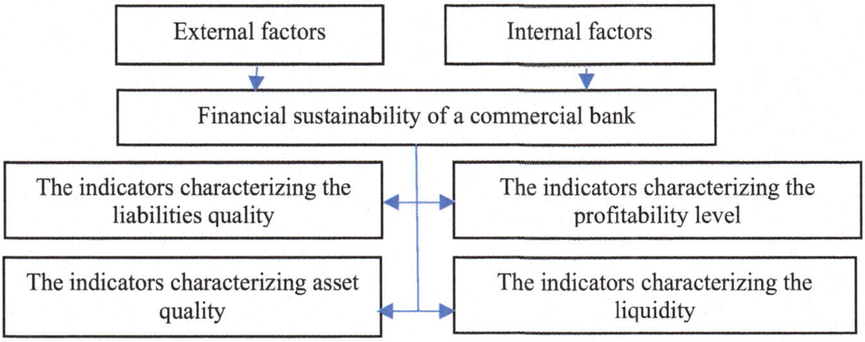

Fig. 1. The components of determining the financial sustainability of a commercial bank.

Thus, the sustainability of a commercial bank is ensured by the presence of such components as asset quality, liabilities quality, liquidity, and profitability.

The indicators characterizing the quality of the assets are the most important for determining the bank's reliability and the effectiveness of its activities. These indicators help define the ability of any commercial bank to achieve its main goal, namely, profitable, stable operation. The indicators characterizing the quality of the assets include the share of income-generating assets and credit activity ratio.

The indicators characterizing the liabilities quality are instantiated by the resource base sustainability, the attraction cost, the sensitivity to changes in the interest rates, and dependence on the external financing sources. The indicators characterizing the quality of assets include the share of income-generating assets and the coefficient of credit activity.

The indicators characterizing liquidity determine the ability of the banks to ensure the timely fulfillment of their obligations to all counterparties. The indicators that characterize liquidity include absolute liquidity ratio, critical liquidity ratio, and total liquidity ratio.

Indicators that characterize profitability are needed to encourage staff members and management to expand and improve operations, reduce costs, improve the quality of provided services, and successfully carry out subsequent emissions and, accordingly, capital growth, allowing the volume expansion and the quality improvement. The indicators that characterize profitability include returns on assets, returns on equity, and general levels of return.

The method of estimating the main indicators for calculating the aggregated indicator of the bank financial sustainability of a commercial is presented in Table 1.

These indicators will help determine the aggregated indicator of each component of the financial sustainability of a commercial bank.

1. Aggregated asset quality indicator:

$$\mathrm{I_A} = \sqrt{DAD \times CKA} \tag{1}$$

2. Aggregated liability quality indicator:

$$\mathrm{I_P} = \sqrt{\mathrm{CISR} \times \mathrm{CFKP}} \tag{2}$$

3. Aggregated liquidity indicator:

$$\mathrm{I_L} = \sqrt[3]{\mathrm{C_1} \times \mathrm{C_2} \times \mathrm{C_3}} \tag{3}$$

4. Aggregated indicator of profitability:

$$\mathrm{I_{Prof}} = \sqrt[3]{R_a \times R_{CK} \times R_{gen}} \tag{4}$$

Based on the above, a cumulative aggregate indicator of the financial sustainability of a commercial bank can be calculated, which is determined as the geometric average of the products of individual indicators:

$$\mathrm{AI_{fs}} = \sqrt[4]{I_a \times I_p \times I_L \times I_{prof}} \tag{5}$$

The value of the aggregate indicator allows for the assessment of the result of the changes in financial sustainability in the activities of a commercial bank. The result of

Table 1. The indicators for calculating the aggregated indicator of the financial sustainability of a commercial bank.

Index name	Calculation method	Contents
Indicators characterizing asset quality		
Share of income-generating assets	The number of funds in credit institutions, net investments in securities measured at fair value through profit; net investments in securities and other financial assets available for sale; net investments in securities held to maturity; net loan receivables to assets	Shows the number of working assets, defining the income
Credit activity ratio	Net loan to assets ratio	Evaluates bank credit operations
Indicators characterizing liabilities quality		
Coefficient on fixed-term resources use	The number of loans, deposits and other funds of the Russian Central Bank, credit institutions' funds; customer funds, non-credit organizations; issued debt to the number of funds in credit institutions; net investments in securities at fair value through profit; net investments in securities and other financial assets; net investments in securities held to maturity; net loan debt	Shows the effect of using the attracted fund
Stock capitalization profit ratio	Common sources to equity	It characterizes the efficiency of the bank's management, their ability to increase its capital at the profit expense, not the additional issue
Indicators characterizing liquidity		
Absolute liquidity ratio	Cash to liabilities	It shows the obligations availability that can be paid from the available funds
Critical liquidity ratio	Cash and net debt to liabilities	Characterizes such a state, when the rapid funds' implementation for the asset covers urgent liabilities
Total liquidity ratio	Cash ratio; funds in credit institutions; net investments in securities at fair value through profit; net investments in securities and other financial assets available for sale; net investments in securities held to maturity; net loan debt obligations	It characterizes the extent to which the liquid part of all bank assets covers urgent liabilities

(continued)

Table 1. *(continued)*

Index name	Calculation method	Contents
Indicators characterizing profitability		
Return on assets	Net profit to an average asset value	Bank investments efficiency
Return on equity	Net profit to equity ratio	Bank capital efficiency
Overall profitability	Net profit to total expenses	Bank capital efficiency

this aggregate indicator makes it possible to determine the type of financial sustainability:

- absolutely financially stable (AI > 1);
- financially stable banks (0.6 < AI < 1);
- financially unstable credit institutions (AI < 0.6).

The obtained assessment of financial sustainability is the first sign of the need to make management decisions aimed at maintaining financial sustainability or exiting from a crisis situation for managers and owners.

While assessing the financial sustainability of a bank, the proposed indicators will help to get an idea about:

- the net effect of a commercial bank;
- the profit sufficiency for the expansion of the activities, based on the specifics of the bank's financial condition;
- the efficiency of the procedures for managing financial flows and probable problems in the activities of bank managers.

The proposed system is universal, i.e., applicable to all commercial banks, regardless of the legal form and banking business policies.

A method for calculating the aggregate indicator of the financial sustainability of commercial banks was developed as a result of the study. This methodology is presented in the form of the subsequent actions to determine and calculate the aggregated measures using the indicators characterizing the quality of the assets and liabilities, the level of profitability, and liquidity. The determination of the type of financial sustainability using this technique will allow management to take preventive measures at the earliest stages of problems, identifying processes and factors that negatively affect the financial sustainability of the bank.

4 Discussion

The main trends in the development of the Russian banking system from 2016–2017 can be characterized as follows:

- the stagnation of the Russian economy by limiting the demand for loans from legal entities;
- the lack of structural liquidity, the tight monetary policy of the Bank of Russia, and the growth of the risk component in the economy negatively affected the cost of resources, which also affected the demand for loans from enterprises;
- the worsening of quality of the loan portfolio;
- continued growth in competition in the banking sector (Kolmykova and Sitnikova 2015).

The activities of most commercial banks are aimed at maintaining the strategic priorities of economic development. Given the changes taking place in the economy, the main objectives of the activities of credit institutions can be formulated as follows:

- ensuring and maintaining the required financial sustainability level;
- ensuring economic efficiency;
- ensuring business growth in line with trends observed in the economy and the monetary sphere;
- optimizing the banking network (Kazarenkova 2016).

The first task in any bank activity will be to ensure and maintain the necessary and proper level of financial sustainability. The authors of the study proposed to assess the financial sustainability of several credit institutions by determining the aggregate indicator of financial sustainability. PJSC Bank Zenit, PJSC Promsvyazbank, and VTB Bank (PJSC) were selected for the study. The results of the calculations of the main indicators in the studied banks are presented in Table 2.

Table 2. The financial sustainability of banks.

Indicator	PJSC Bank Zenit		PJSC Promsvyazbank		VTB Bank (PJSC)	
	2016	2017	2016	2017	2016	2017
Indicators characterizing asset quality						
Income generating assets	0.822	0.868	0.811	0.829	0.899	0.874
Credit activity ratio	0.685	0.589	0.622	0.604	0.680	0.679
Indicators characterizing liabilities quality						
Urgent resources use	1.079	0.975	1.121	1.315	0.927	0.947
Stock capitalization ratio	1.111	0.929	5.496	−8.217	2.085	2.182
Indicators characterizing liquidity						
Absolute liquidity ratio	0.019	0.018	0.016	0.025	0.019	0.033
Critical liquidity ratio	0.764	0.702	0.681	0.563	0.814	0.829
Total liquidity ratio	0.913	1.026	0.884	0.763	1.069	1.058
Indicators characterizing profitability						
Return on assets	–	–	0.005	–	0.007	0.011
Return on equity	–	–	0.071	–	0.051	0.071
Overall profitability	–	–	0.003	–	0.005	0.011

Let us define the aggregated indicator of each component of the financial sustainability for the analyzed banks (Table 3).

Table 3. The aggregated indicators of financial sustainability.

Indicator	PJSC Bank Zenit		PJSC Promsvyazbank		VTB Bank (PJSC)	
	2016	2017	2016	2017	2016	2017
Aggregated asset quality indicator	0.750	0.715	0.710	0.707	0.781	0.770
Aggregated banking passives quality indicator	1.095	0.952	2.482	–	1.390	1.437
Aggregated liquidity indicator	0.237	0.235	0.212	0.221	0.255	0.307
Aggregated return indicator	–	–	1.021	–	1.213	2.048

Table 3 does not include the aggregate indicator values of PJSC Bank Zenit for 2016–2017 or PJSC Promsvyazbank for 2017. These credit institutions have losses for the periods under consideration; therefore, it is not advisable to calculate the components that are part of this indicator since their calculation is based on the presence of net profit, which these organizations do not have. Also, in 2017, PJSC Promsvyazbank did not have the calculated aggregate quality indicator for liabilities since one of the components included in this indicator has a negative value (see table). The ratio of the stock capitalization of profits is negative because PJSC Promsvyazbank had a negative value of the total sources of its own funds in 2017.

Let us move to calculate the aggregated indicator of financial sustainability for the analyzed banks:

PJSC Bank Zenit:

$$\mathrm{AI_{fs}}(2016) = \sqrt[3]{0,750 * 1,095 * 0,237} = 0,579$$
$$\mathrm{AI_{fs}}(2017) = \sqrt[3]{0,715 * 0,952 * 0,235} = 0,542$$

PJSC Promsvyazbank:

$$\mathrm{AI_{fs}}(2016) = \sqrt[4]{0,710 * 2,482 * 0,212 * 1,021} = 0,785$$
$$\mathrm{AI_{fs}}(2017) = \sqrt[2]{0,707 * 0,221} = 0,395$$

PJSC VTB Bank:

$$\mathrm{AI_{fs}}(2016) = \sqrt[4]{0,781 * 1,390 * 0,255 * 1,213} = 0,761$$
$$\mathrm{AI_{fs}}(2017) = \sqrt[4]{0,770 * 1,437 * 0,307 * 2,048} = 0,913$$

The type of financial sustainability of the analyzed banks, determined on the basis of an aggregate indicator, is presented in Table 4.

Thus, no bank has shown a general indicator greater than one. Bank Zenit PJSC demonstrates the annual decline of the aggregate financial sustainability indicator. Zenit PJSC is one of the financially unstable banks. This is evidenced by earlier calculations (the bank has no net profit, which did not allow us to calculate the main profitability indicators).

Promsvyazbank PJSC also shows a decrease in the main indicator that determines the financial sustainability of the bank from 0.785 in 2016 to 0.395 in 2017, which led to a change in the type of financial sustainability. Promsvyazbank PJSC also shows a decrease in the main indicator that determines the financial sustainability of the bank from 0.785 in 2016 to 0.395 in 2017, which led to a change in financial sustainability.

Table 4. The type of financial sustainability.

Company name	2016			2017		
	Aggregated indicator value	Financial sustainability type		Aggregated indicator value	Financial sustainability type	
PJSC Bank Zenit	0,579	$AI_{fs} < 0,6$	Financially unstable	0,542	$AI_{fs} < 0,6$	Financially unstable
PJSC Promsvyazbank	0,785	$0,8 \leq AI_{fs} < 1$	Financially stable	0,395	$AI_{fs} < 0,6$	Financially unstable
PJSC VTB Bank	0,761	$0,8 \leq AI_{fs} < 1$	Financially stable	0,913	$0,8 \leq AI_{\text{ФУ}} < 1$	Financially stable

VTB Bank PJSC has financial sustainability, and the aggregate indicator of financial sustainability is growing.

5 Conclusion

The reduced financial sustainability is associated with an unstable economic environment and increased risks. The reasons for the decline in financial sustainability at the global level include:

- a significant decrease in oil prices. The high oil prices observed in the first half of the year were significantly reduced in the second half, as a result, by the end of the year, the minimum oil quotes for many years were recorded; and
- the slowdown in the growth of the developing countries and the protracted stagnation in the European Union, which caused the outflow of capital from the emerging markets and contributed to the fall in commodity prices (Sitnikova and Tretyakova 2014).

The reasons for the decline of financial sustainability at the state level include:

- sanctions that provoked structural changes in many economic areas (change of trading partners, investment plans, rising financing costs, etc.);
- the devaluation of the ruble, the fall in oil prices, and the transition to a free-floating exchange rate regime;

- financial crises and a tightening of the monetary policy;
- a slowdown in GDP growth, which led to a decrease in the investment activity and an end to the expansion of final household consumption; and
- maintaining a "tough" fiscal policy and a budget balanced by the ruble devaluation in the absence of the substantial growth in expenditures (Kolmykova et al. 2015).

The main types of risks associated with the activities of any credit institution can be divided into general and specific types.

The main general (systemic) risk factors include an increase in overdue debt and the number of defaults on debt obligations of Russian borrowers, the risk of an increase in the market interest rate and a drop in the market value of the debt of Russian borrowers, and the growth of country risks due to a decrease in key macroeconomic indicators.

The specific types of risks include credit, market, liquidity, operating, legal, reputational, and strategic.

The most significant risks for modern credit institutions are credit and market risk (taking into account its varieties). At the same time, the liquidity risk is also central, which management has given priority.

References

Abramyan, G.A.: The imperatives of the financial sustainability of commercial banks in modern conditions. Molodoy Uchenyy **19**, 347–350 (2015)

Kazarenkova, N.P.: Financial intermediation as the essence of banking. Proc. Southwest State Univ. Ser: Econ. Soc. Manag. **1**(18), 151–158 (2016)

Kolmykova, T.S., Sitnikova, E.V.: Long-term bank lending in the Russian economy: the role and development trends. J. Econ. Entrep. **2**(55), 49–54 (2015)

Kolmykova, T.S., Kazarenkova, N.P., Ovcharova, O.N.: Assessment of interaction between banking and real sectors of the economy based on the convergence effect. Int. J. Econ. Financ. Issues **5**(3S), 281–286 (2015a)

Kolmykova, T.S., Lukianykhina, O., Baistriuchenko, N., Lykianykhin, V.: International integration in innovative development of the economy. Probl. Perspect. Manag. **13**(1), 203–207 (2015b)

Kolmykova, T.S., Merzlyakova, E., Bredikhin, V., Tolstykh, T., Ovchinnikova, O.: Problems of formation of perspective growth points of high-tech productions. In: Advances in Intelligent Systems and Computing, pp. 469–475 (2018)

Sitnikova, E.V., Tretyakova, I.N.: Lending and major trends in the industrial sector. Russ. J. Ind. Econ. **1**, 34–36 (2014)

Tretyakova, I., Sitnikova, E.: International financial processes influence on the retail lending market in Russian development. Econ. Ann.-XXI **3-4-1**(157), 100–102 (2016)

Cognitive Modeling in Improving the Organization Competitiveness

Elena A. Alpeeva[1(✉)], Alexander V. Zelenov[1], and Nadezhda A. Serebryakova[2]

[1] Southwest State University, Kursk, Russia
alpeevael@yandex.ru, kgtu_fk@list.ru
[2] Voronezh State University of Engineering Technologies, Voronezh, Russia
nad.serebryakova@mail.ru

Abstract. The automatization of enterprise management based on the economic and mathematical models is a modern technology used for increasing competitiveness. The use of cognitive modeling allows us to make managerial decisions in the conditions of uncertainty in the competitive environment. The paper analyzes various approaches to the use of cognitive modeling and discusses its impact on organization competitiveness.

Keywords: Innovation · Production automatization · Cognitive modeling · Competitiveness

1 Introduction

At present, one of the ways to increase the competitiveness of industrial enterprises is to reach a fundamentally new level of management of financial and economic activity through the development of innovative potential, improvement of the innovative culture (Alpeeva and Kharchenco 2018), and introduction of innovations using a complex of economic and mathematical models and information technologies, including automatization of production accounting.

2 Materials and Method

An enterprise should consider automatization of the production accounting an investment, which should bring returns through improved manageability, increased production efficiency, and cost reduction. The automatization value is comparable to an increase in plant capacity (Alpeeva and Ryabtseva 2012).

The automatization of management based on economic and mathematical models, information technology, and computer technology for almost all enterprises is a critical development stage, bringing business to a whole new level of management (Internet journal "Generalnyj Direktor" 2018).

The newest software tools include, first of all, manufacturing execution systems (MES), which raise the production culture to a higher level (Dilman 2012).

A. V. Bogoviz and Y. V. Ragulina (Eds.): ISCI 2019, LNNS 115, pp. 379–386, 2020.
https://doi.org/10.1007/978-3-030-40749-0_45

The MES is a mandatory organizational component of Western companies' production, regardless of the industry affiliation. In Russia, this system is still rarely used. It is the MES system that should become the technological basis for the introduction of world-class production standards for domestic manufacturers, regardless of the equipment used or the qualifications of workers employed in the production (Zagidullin 2011).

Modern management technologies are based on problematic knowledge and a company's experience making decisions in the face of uncertainty. These technologies take into account not only the accumulated experience of specialists in certain spheres, but also the formation of the basis for computerization of control systems focused on the automatization of the control process.

For structuring information and making decisions in uncertain conditions, a cognitive modeling model is used.

The ideas of cognitive modeling are presented in the works of Makarenko, Maltseva, and Karaev (Marenko and Maltseva 2015).

Cognitive modeling is a visualized construction of causal relationships between the entities describing a system (Savchuk et al. 2015).

The method of symbolic and weight-directed graphs is effectively used to describe cognitive models. The weights of arcs in purely cognitive models are sought either using statistical information processing or expertly. The factor changes are carried out in steps to determine the response of the system; after that, with the help of multi-criteria selection, many favorable scenarios are determined and ranked.

The development of classical cognitive models is fuzzy cognitive models (FCM), which take into account the fact that the interactions between factors caused by the presence of cause-effect relationships can have different intensities, while the intensity of any influence can change over time (Kulinich 2002).

The cognitive map shows only the fact of the influence of factors on each other. It does not reflect either the exact nature of these influences, nor the dynamics of changes in the influences depending on the changes in the situation, nor temporary changes in the factors themselves. Taking all these circumstances into account requires a transition to the next level of information structuring, that is, to a cognitive model. At this level, the relationship between the factors of the cognitive map is revealed by the corresponding dependencies, each of which can contain both quantitative (measurable) variables and qualitative (not measurable) variables. Moreover, quantitative variables are naturally represented in the form of their numerical values. Each qualitative variable is associated with a set of linguistic variables representing various states of this qualitative variable (for example, customer demand can be "weak," "moderate," "rush," etc.), and each linguistic variable corresponds to a particular numerical equivalent in the scale [0, 1]. With the accumulation of knowledge about the processes occurring in the studied situation, it becomes possible to more fully disclose the nature of the relationships between the factors (Maksimov 2004).

This graph can be represented by a matrix, which, in turn, is also called cognitive.

As an example of the use of cognitive modeling technology, a study was conducted on the automation of production accounting of material flows of an industrial enterprise to increase the level of accounting and use of information for the rational functioning of the unit and effective management decision-making. In order to build a rational procedure of intuitive-logical thinking, an expert group of 20 people with the necessary competencies in this field was formed.

At the first stage, the experts identified factors affecting the automation of the production accounting of material flows. Twenty factors affecting the problem under study are systematized in Table 1.

Table 1. The factors affecting the automatization of the production accounting of material flows in an industrial enterprise.

	Factor		Factor
	Production		*Public policy*
1	The production scale	10	Legislative regulation
2	The complexity of the manufacturing process	11	Licensing policy
	The automated Information System (AIS) of the production accounting		*Account Information Effectiveness*
3	The number of automated working places (AWP) – control points (CP)	12	Operational efficiency
4	The manufacturability of the AIS	13	Comprehensiveness
5	The requirements to work with the Internet-Resources	14	Analyticity
	The material and technical equipment	15	Credibility
6	Hardware computer equipment		*Performance management*
7	The additional equipment integrated with the information system	16	Control over resources
	Staff	17	Unit Efficiency
8	The number of administrative staff	18	Unit Cost Level
9	Staff qualifications	19	Quality control
		20	Competitiveness

These factors fall into seven content groups: (1) production, (2) the automated information system (AIS) of production accounting, (3) material and technical equipment, (4) personnel, (5) government policy, (6) the efficiency of information accounting, and (7) the management of the unit performance.

In the second stage, the causal relationships between the concepts (factors) were established, taking into account their influence on the system. The next step was to determine the significance (weightiness) of each factor affecting the automatization of the production accounting of material flows, which was carried out using the expert's method. Two types of causal relationships were identified: positive and negative. With a positive connection, an increase in the value of the factor-cause leads to an increase in the value of the factor-effect, and with a negative connection, an increase in the value of the factor-cause leads to a decrease in the value of the factor-effect.

To establish a cause-effect relationship, a scale has been defined for assessing nature (positive or negative) and the strength of the relationship between the underlying factors (Makarenko 2006). It sets the values of the corresponding variables; each of them is assigned with a number ranging from minus to plus units (Fig. 1).

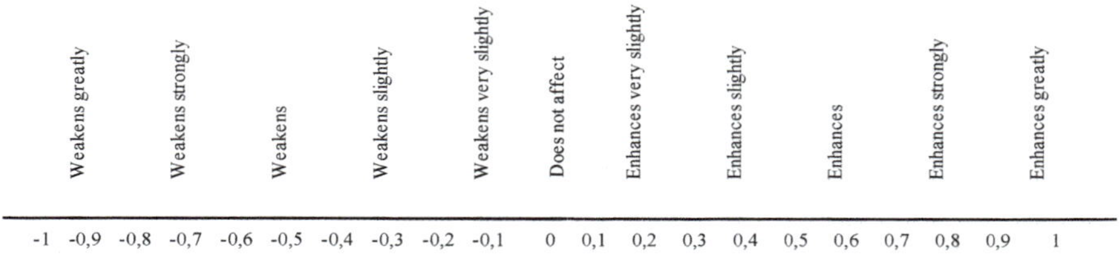

Fig. 1. A scale for assessing the values and strengths of the mutual influence of cognitive map factors.

After structuring the information, the cognitive matrix was made up:

- On the lines of which the concepts-causes are indicated, and on the columns, the concepts-effects;
- The intersection of the row and column contains an average estimate of the influence intensity of one concept on another.

Based on the matrix, a cognitive map is constructed that reflects the causal structure of the system (Fig. 2).

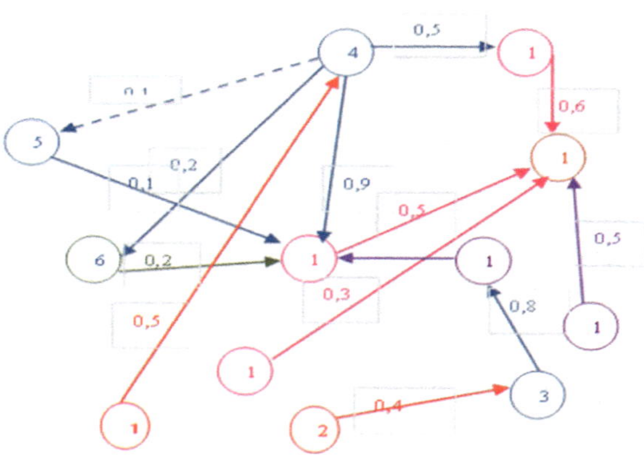

Fig. 2. A fragment of the model of a fuzzy cognitive map of automation of production accounting of material flows of an industrial enterprise unit.

The analysis of the resulting visual model allows us to draw several conclusions. The most significant positive impact on the system, without experiencing a strong influence on its part, is provided by concepts 4 and 9. By influencing these factors, it is possible to achieve a more effective innovative level of production accounting for the material flows.

During the analysis of the constructed map, it turned out that the system itself had the most significant influence on such concepts as 12, 13, 15, and 19. Thus, it can neutralize the strong negative influence of external factors. In order to automate the production accounting of material flows of the slaughterhouse, the strategic and tactical development tools should focus on the factors that have the most powerful impact on the system, namely the manufacturability of the automated information system and the qualifications of personnel.

The concepts 2, 4, 7, 16, and 18 harm the system. On the other hand, the system has a negative impact on concepts 5, 9, 16, 17, and 18. Since, in the case of concepts 16 and 18, the negative influence is bilateral, there arises a negative cycle. This means that with a favorable external impact on this concept, its negative impact on the system will increase over time. If it is negatively affected, it will lead to a positive effect on the system as a whole.

Let us draw the attention to the fact that factor 4 is in the node of positive and negative interactions, indicating that a positive effect on this factor is desirable, but requires great caution and a thorough assessment of the consequences.

In order to study the structure of the system and obtain the forecasts of its behavior under various control influences, analytical processing methods should be applied to the constructed cognitive map.

When analyzing the constructed scheme, two closely related tasks can be formulated: the direct, or how the situation will develop under existing external influences, and the opposite, or what effects should be chosen to get the desired state.

For the analysis, the target factors are singled out, and those factors that change in the right direction must be achieved. Then, the leverage, those factors that can be changed within certain limits, is highlighted. Various scenarios are being analyzed.

The most common is the interpretation of the matrix as the conversion of the percentage changes in causes to percentage changes in consequences.

For example, there is a connection A $\rightarrow$ B on the cognitive map with an overhang of +0.8, which means that if the value of the factor increases by 10%, then the value of factor B increases (the "+" sign) by 8% (=10% * 0, eight).

It allows us to consider the factors on the same model without worrying about units. Obviously, this approach is not very accurate, but the quantitative estimates are not very important for the qualitative conclusions—we need to understand which lever needs to be moved up and which needs to be moved down or evaluate how the result depends on this lever or the other (Makarenko 2006).

In order to increase the level of the production accounting of material flows, management's goal is to find and implement such management (i.e. changes in controlled factors) that would lead to an increase in the factors of "efficiency," "completeness," "reliability," "quality control," and "competitiveness". The possibility of achieving these goals is affected by both the initial state of internal and external factors forming the analyzed fragment of the strategy and management, which is implemented by changing the dynamics of the controlled factors.

Let us conduct the simulation experiments to detect the changes in the controlling and controlled subsystems. The essence of the experiment is that a perturbation is introduced at one or several vertices of the graph at time t. The state of the peaks at time $t + 1$ is determined by the equation (Marenko and Maltseva 2015) (1):

$$X(t+1) = X(t) + P(t+1), \quad (1)$$

where P(t + 1) is the vector of the increments of factor values at the vertices of the graph at time t + 1, and where X(t) and X(t + 1) are the states of the factors at time points t and t + 1

Let us consider a scenario in which the efficiency of the processed information should grow by 50%. As the target factor, we will choose the "operational efficiency" variable, and the control factors will be the variables of "the AIS manufacturability", "the requirements for working with the Internet resources," "hardware computer equipment", "the number of managerial personnel", and "licensing policy".

The result of the experiment shows that to increase the efficiency of the processed information by 50%, it is necessary to increase the adaptability of AIS by 45%, increase the power of hardware for computer equipment by 10%, and increase the requirements for working with Internet resources (protocols that are needed for data transfer) by 5%, expand the number of managerial personnel with the functions of preparing and monitoring reporting by 5%, and accordingly, acquire an additional number of workplace licenses in AIS by 5% more than the existing number. The control factors improve over time.

In order to achieve this goal, it is necessary to increase the activity of all factors, but in different sequences and to different degrees. The factor changes are carried out in steps to determine the response of the system, after which, with the help of multi-criteria selection, many favorable scenarios are determined and ranked. First of all, it is necessary to significantly increase the manufacturability of the used automated information system (AIS). The main efforts should be directed at the outstripping growth of this factor: factor activity should grow faster and more (to the level of 55.2%) by the end of the modeling period – the 18th step.

Other factors, according to the degree of their influence on the result, are distributed as follows. The second important factor is the "hardware computer equipment" factor. The factors "requirements for working with the Internet resources," "the number of managerial personnel," and "licensing policy" have an equal effect.

After the sixth step, the requirements for the activity of these factors can be weakened, since, after this moment, the growth rates of their activity are equalized as they reach a maximum.

Having introduced a 20% indignation to the top of the graph corresponding to the managing factor "Personnel Qualification," we observe an increase in the target factors "analyticity" by 12%, "completeness" by 10%, and "reliability" by 10%. At the same time, the "reliability" factor is growing most actively, reaching 11.4% at the end of the simulation. The requirements for the activity of all factors can be relaxed at the eleventh step of modeling, when the growth rates of their activity are aligned (all indicators reach their maximum).

Having introduced a 20% perturbation to the top of the graph, corresponding to the control factor "The technological ability of the automated information system (AIS)," we observe an increase in the target factors "efficiency" to 21.3%, "reliability" to 20.5%, "completeness" to 11, 3%, "hardware computer equipment" to 4.5%, "requirements for working with the Internet resources" – a decrease to 2.2%.

3 Results

The results of the experiments showed that the factor "Manufacturability of an automated information system (AIS)" is a significant controlling factor for ensuring the growth of the target factors "efficiency," "completeness," and "reliability." In turn, the factors "efficiency," "completeness," and "reliability" have a significant impact on the "quality control" and the "quality control" on the "competitiveness."

Thus, we can conclude that the management of the enterprise should be recommended to pay priority attention to increasing the factor "manufacturability of the automated information system" (AIS). The automatization of the production accounting of material flows of an industrial enterprise and a specific unit is advisable.

In the case of automatization of the production accounting of material flows of a specific unit of an industrial enterprise, the new model of business processes will fully take into account the requirements of the quality management and safety systems based on the principles of HACCP and ISO series standards.

4 Conclusion

The automatization of enterprise management and, as a result, increasing its competitiveness based on the economic and mathematical models, and information technology is one of the main stages of the development for all enterprises. The use of cognitive modeling allows making managerial decisions in the conditions of uncertainty. The paper describes the process of creating the experimental model of automatization of the production accounting of material flows based on the use of fuzzy cognitive maps. A cognitive modeling algorithm is presented. The main advantages of cognitive tools are noted: (1) the possibility to explore the fine structure of managerial decisions (the necessary sequence of inclusion of managerial influences, the necessary degree of activity of these influences, the study of the dynamic stability of strategies, etc.); (2) the possibility to study the dynamics of managerial decisions at a qualitative level, without attracting hard-to-reach and not always reliable quantitative information for this purpose, which is extremely important in a rapidly changing business environment and the growing pace of technological innovation.

It is emphasized that none of the known management support tools have superior capabilities. The dynamic cognitive analysis significantly expands the management tool base, based mainly on the means of static situational analysis and prescription decision-making schemes.

The analysis showed that the proposed model is fully operational and can be used to predict economic activity and increase competitiveness. The results of the work should be considered as a solution to several management tasks.

References

Alpeeva, E.A., Ryabtseva, I.F.: Progress and innovations: analysis of system conditionality. Econ. Anal. Theory Pract. **18**(273), 37–41 (2012)

Alpeeva, E.A., Kharchenco, E.V., Geyderikh, P.V.: Principles of forming effective, innovative company culture. In: Proceedings from the 5th SGEM 2018: International Multidisciplinary Scientific Conference on Social Sciences and Arts. Vienna (2018)

Dilman, A.: Myths and reality of MES-systems implementation (2012). https://www.osp.ru/cio/2012/01/13012614/

Internet journal Generalnyj Direktor. Automatization of management processes: we increase the efficiency of the company (2018). https://www.gd.ru/articles/9757-avtomatizatsiya-protsessov-upravleniya

Kulinich, A.A.: System of cognitive modeling Kanva (2002). https://www.raai.org/about/persons/kulinich/pages/kanva2003.html

Makarenko, D.I.: Models and methods of strategic management of the military-industrial complex. (Dis. of Cand. of Sci. (Economy)). Moscow, Russia (2006)

Maksimov, V.I., Kornoushenko, E.K., Kachaev, S.V.: Cognitive technologies to support management decision-making (2004). https://www.iis.ru/events/19981130/maximov.ru

Marenko, V.A., Maltseva, M.I.: Application of cognitive modeling to analyze the problems of small businesses. News Irkutsk State Econ. Acad. **25**(6), 1014–1024 (2015)

Savchuk, O.V., Ladanum, A.P., Gerasimenko, T.M.: Fuzzy cognitive modeling in the systems of control of the technological process of milk processing plants. Eng. Sci. **1–2**(35–36), 15–19 (2015)

Zagidullin, R.R.: Management of Machine-Building Production with the Help of MES, APS, ERP Systems. TNT, Stary Oskol (2011)

The Management of Sustainability of Innovative Development Using a System of Key Performance Indicators

Elena V. Potekhina[1(✉)], Ludmila I. Antonova[2], and Olga M. Guschina[2]

[1] Russian State Social University, Moscow, Russia
elengapotechina@mail.ru
[2] Moscow City Pedagogical University, Moscow, Russia
antonova.ludmila@mail.ru, ole4ka.gushina@mail.ru

Abstract. The paper focuses on the theoretical and methodological features of the essence, content, and forms manifesting key performance indicators (KPI) of innovation activity. The features of the formation of key indicators (following the unique specifics of innovative activity) and its stages are determined. More than that, methodological provisions for determining such indicators are substantiated. A conceptual model of forming and developing key indicators of the effectiveness of innovative activities based on the proposed organizational measures is also developed and presented in the paper.

Keywords: Key indicators · Effectiveness · Performance · Innovative Activity · KPI principles · Methodology · Organization · Motivational mechanism

1 Introduction

The development of modern society in the informatization and digitalization era is unthinkable without increasing the efficiency of innovative processes in all sectors and areas of the economy. Along with that, the introduction of innovative technologies varies significantly among enterprises, industries, and regions. One of the main reasons for this is the lack of effective mechanisms for managing innovation activity.

In this regard, the economy faces the task of searching and substantiating modern tools for managing innovative activity as a decisive factor, ensuring the implementation of the latest achievements of scientific and technological progress in public production.

Most of the existing tools and methods do not fully contribute to the solution of such problems. That fact causes an objective need to develop and implement adequate tools for assessing the effectiveness of innovative activity in current conditions. Moreover, such tools should allow evaluating the effectiveness of innovative processes, based on the analysis and use of assessing adequate and reliable information (Korniyko et al. 2019).

One of these tools is *the system of key performance indicators (KPI),* which is widely used in Russia and abroad to evaluate and manage performance in innovation activity.

A. V. Bogoviz and Y. V. Ragulina (Eds.): ISCI 2019, LNNS 115, pp. 387–394, 2020.
https://doi.org/10.1007/978-3-030-40749-0_46

This paper focuses on the theoretical and methodological analysis of the current state and the results of scientific research on the problem of using the system of key indicators of innovative activity, substantiating their features and developing proposals for improving the methodology for their application.

2 Materials and Methods

The theoretical and methodological basis for the study of the formation of key indicators of the effectiveness of innovation activity was domestic and foreign publications of scholars on the problems of substantiating and using the KPI system, as well as indicative management and planning of innovative companies.

The methods of functional-logical, system-target, process, and situational analysis, as well as the methods of structuring goals, expert and forecast estimates, and so on, were used in the course of the study.

The analysis was carried out based on a study of domestic and foreign experience in managing the effectiveness of innovative activities using key indicators for assessing the effectiveness of KPI in companies in various sectors of the economy, as well as on the author's theoretical generalizations obtained in the process of researching the problem.

3 Results

Key performance indicators (KPI) are indicators that characterize the final performance of an activity. Such indicators are designed to help in assessing the achievement degree of both the strategic and tactical goals of companies.

In the standard ISO 9000: 2008, the term *Key Performance Indicator* is characterized by two main features: effectiveness and efficiency. Effectiveness is characterized by the achievement degree of the set goals or final results, and efficiency, by the ratio between the results obtained and the resources expended on them.

The main purpose of using KPIs is to link goals, plans, programs, results, as well as the motivation and income of employees in the process of organizing the activities of companies. Such a system of KPIs helps not only to evaluate the results and the effectiveness of personnel and departments but also to form an effective motivational mechanism for company employees (Ministry of Economic Development of the Russian Federation 2015).

The most important advantage of KPIs is their ability to transfer the assessment of the effectiveness of a particular type of activity from the sphere of subjective assessments into the world of real and objective assessments.

Other advantages of KPIs should include their contribution to solving such main tasks as:

- the establishment of causal relationships between the strategic goals of innovation activity and the indicators of their achievement;

- the identification of narrow spaces and weaknesses in the innovation activity of the company;
- the identification of factors and patterns of company performance
- the proof of an inextricable link between the company's development plans, the results of their implementation, and the motivation system

However, studies have shown that these theoretical postulates cannot be entirely attributed to the possibility of evaluating the effectiveness of innovative activities of a company using KPI since such activity has its specific features. These features are predetermined by the essence, content, and technologies of the main stages of innovation activity (i.e., development, implementation, promotion, reproduction, and utilization of innovations). All these stages fundamentally differ from each other.

Innovative activity is primarily characterized by:

- significant uncertainty in innovative goal-setting. In innovative activities, the initially set goal may not always be achieved, but this is not considered a failure, unlike in many other types of activities, including in the commercial sphere. At the same time, the achievement of the original goal is not always adequate to a successful result
- the intensity of innovative activity (the central resource of innovative activity is human capital, creative activity, the ability to generate and translate ideas into real processes, technologies, and products; innovative activities are distinguished from other types of activities by a high level of self-motivation, self-control, the creative nature of work, commitment, and aversion to opportunism while performing professional tasks)
- the contradictory and paradoxical nature of innovation, typical only for its characteristics: creative destruction, as a process of transformation; opportunity (probability) is a characteristic of this process in contrast to other business processes for which clear certainty is an important aspect of their assessment, etc.;
- a high degree of uncertainty and risks (increased uncertainty, high risks, and low predictability of results);
- the ability to initiate structural and other changes in the organization that can not always be formalized;
- the duration of the innovation process in time (the innovation process in any business process is the most time-consuming process because it integrates the cycles of other business processes, which determines the long payback period of innovation), etc.

All these and other features of innovative activity predetermine the specifics of the formation of key indicators of its effectiveness.

According to its socio-economic content and methodological approaches to the formation, the system of KPI in innovation activity will significantly differ from the KPI of enterprises in industries, trade, etc.; that is, those that are directly related to the market economy and which main mechanism is commercialization.

When selecting KPI in a commercial enterprise, it is necessary, first of all, to identify those that fundamentally determine the profit of the organization. The uniqueness of this indicator—its main distinguishing feature—is that with its slight change, the profit of the organization increases significantly. Thus, if the main goal of the organization's activities is to make a profit, then in order to identify KPI, it is necessary to determine first, the key indicators affecting the profit, and second, to identify those responsible for their implementation.

In commercial enterprises, as a rule, operational (leading) KPI is used to evaluate efficiently operating units of workers and organizations. It helps manage the process of production, trading, and other enterprises during a specific reporting period in order to achieve its ultimate goals. The analysis allows us to understand the state of enterprise affairs, the expected financial results in the future, and how much the manufactured products meet the demand and requirements of the market, as well as how satisfied consumers are with it.

At production, trading, and other commercial enterprises, KPI is used to ensure the implementation of final and intermediate goals. All main commercial indicators, as a rule, should be quantifiable and amenable to calculations (level, profitability, turnover, financial flows, etc.). Such criteria should show the dynamics of the company's development in the market (labor productivity, staff turnover, costs for each employee, their motivation, etc.) (Gonin et al. 2014).

Thus, if for enterprises working for profit and commercialization in a market, the main criteria laid down in KPIs are financial (e.g., indicate the potential of the company), then for organizations engaged in innovative activities, the presence of only this criteria is unacceptable.

The system of KPI indicators reflecting the effectiveness of innovative activity should be based on universal indicators characterizing the results of innovative activity.

4 Discussion

The definition of key indicators of the effectiveness of innovation activity should be based on methodological provisions that take into account its specific features.

The formation of a system of key indicators, showing how effective innovations are, should be based on the following targets:

- the innovative result that innovation activity can bring;
- the method that will be used to bring a particular innovation (innovative product) to the market;
- the ways of forming the need for this innovation in the market;
- the degree to which the interests of innovation consumers coincide with their perceptions.

Given the features of innovation as a specific activity, it becomes apparent that we cannot only talk about a financial KPI as the main criterion for identifying the effectiveness of innovation and the potential of the company. Key performance indicators of innovative activities should, first of all, take into account the features of innovative activity as a specific form of manifestation of social activity aimed at accelerating the implementation of scientific and technological progress in social production.

Several universal indicators should be the basis of the KPI, reflecting the effectiveness of innovation. The definition of such indicators should be based on a process approach that indicates what kinds of results were brought about by the innovation, the way a particular innovation activity (innovative product) is brought to the market, the way the need for this innovation is shaped in the market, and the degree to which the interests of innovation consumers coincide with the representations of their needs.

At the same time, the KPI system of innovation should not only measure the effectiveness or efficiency of a particular activity or process but also evaluate the individual results of creative activity of particular employees and teams in general indicators of the effectiveness and efficiency of innovation (Savina et al. 2017).

It should be noted that the KPI of innovation activity cannot (as it happens in the industry, commerce, etc.) give a direct assessment of the work of each employee, the results of that employee's work, or the employee's contribution to the overall result of the effectiveness of innovation. However, each participant in the innovation process must know precisely what criteria are used to form his remuneration.

The organization of the development and implementation of the KPI of innovation requires certain organizational measures.

I. *The formation of the working group and organization of work for substantiating a KPI system of innovation.*

The formation of working groups for developing a KPI system of innovation includes the following steps:

- the determination of the list of manning participants involved in the development and justification of the KPI system of innovation activity, as well as in the distribution of functions and responsibilities between them;
- the selection of specific forms of organizing the work and interaction between all participants engaged in the development of a KPI system of key performance indicators of innovation activity;
- the clarification of the forms and methods of collecting, processing, testing, and presenting the results of the working group to the management of the company to discuss decision making.

The working committees are created in order to competently solve the problems of developing and substantiating a KPI system of the effectiveness of the innovative activity. The functions of these committees are the direct organizational and scientific-methodological management of the development and quality control of the preparation of the KPI of innovation activity. The committee has working groups that provide guidance and control over the preparation of individual sections of the KPI system of innovation activity, which are headed by highly qualified specialists.

Working committees provide direct methodological guidance to the entire process of developing a KPI system of innovation activity. Their functions include:

- the substantiation of the general goals of the KPI system of innovation activity;
- the development of the structure of the KPI system of innovation activity;
- the selection and justification of the most effective KPI of the company's innovation activities;
- the determination of the composition of measures for organizing the development and implementation of the KPI system;

- the conduction of predictive and planned calculations on the effectiveness of implementing a KPI system;
- the determination of the circle of performers and the costs of preparing a KPI system;
- the implementation of financing and control over the progress of the assignment;
- the assessment of the effectiveness and quality of recommendations being developed;
- the coordination, that is, ensuring the target orientation of all developers for solving the main problems, achieving the general goal - creating a KPI system.

The organization of the development of the KPI showing the innovation effectiveness in the company involves several stages:

1. the stage of coordination and approval of KPI concepts by the top management of the region;
2. the stage of formation of teams for the development of management projects and innovative activities;
3. the formation of innovative corporate culture in the organization of project activities;
4. the substantiation of the strategy for developing a system of key indicators of innovation performance;
5. the organization of training seminars on the development and substantiation of the system showing key indicators of innovation activity in the region;
6. the unification of management indicators and development effectiveness with KPI, entering KPI into the database;
7. the selection of KPI of innovation activity in the company;
8. the development of a methodological model of the management process using KPI;
9. the development of a system of regulatory and methodological provisions (regulations) on the use of KPI in management practice;
10. the preparation of guidelines for the development and implementation in the reporting structure of all hierarchical levels of KPI;
11. monitoring with further adjustment of the actualization of the key indicators used to assess the effectiveness of innovation (Ilyas 2015).

II. *The definition of a methodology for substantiating the selection of a model of KPI system*

Different companies use different KPI systems to evaluate innovation, which, of course, requires the development of methodological foundations and methodological recommendations.

The methodology of the development of KPI for assessing the effectiveness of innovation cannot be mechanically transferred from organizations and enterprises directly working in order to maximize profits. For the development of KPI, such criterion as maximizing profit is unacceptable in this setting. Thus, a significant adjustment of the methodology for the development of innovation activity is necessary, taking into account the characteristics of a specific type of innovation process.

The fundamental feature of the formation of a methodology for developing a KPI system of innovative activity is the choice of a methodological approach (what approach should be the basis for developing a KPI system - a process (business process) or project approach).

In this regard, it is obvious that the initial task in the formation of a KPI system for assessing the effectiveness of innovative activity is to determine the approach to the assessment of process and project activities.

A business process is a system of cyclical, focused, sequential, and regulated activities where, through managerial influences, using resources, process inputs are transformed (converted) into outputs. The results of such processes represent real values for consumers.

The project is a unique activity, aimed at achieving fundamentally new results and goals, creating a unique, innovative product with restrictions on time and resources, and requirements for an acceptable level of risk and quality. Based on this, two different approaches to the formation of design and process goals, and, consequently, the KPI system for assessing the effectiveness of innovation activity can be used.

III *Methodological justification of the KPI system for assessing the effectiveness of innovation*

The specific features of developing a KPI system are:

- the lack of the indicator showing a payback period, since such indicators can be applied when making decisions on launching an innovative project;
- the absence of the indicator of growth in labor productivity when innovative activities (innovative projects) are in the investment phase, and, therefore, labor productivity, as a whole for the enterprise, as a result of expenses on research and development, cannot be measured;
- the minimization of the number of indicators included in the KPI system of innovation activity (according to experts, such indicators should be no more than 3-5 at each level). Following the theory of measurements, with an increase in the accuracy of measurements, their reliability decreases;
- the comprehensiveness and consistency in assessing the results of innovation (the ability to evaluate functional, economic, social, environmental, and other results) (Stepanov and Bialoblodskaya 2014).

5 Conclusion

The methodology of selection of KPI for assessing the innovation activities should be based on a decomposition method (a method of constructing the tree of objectives of KPI showing the efficiency of innovative activity). This method can be used to define summarizing KPI, generalize the characteristics of lower hierarchical levels (smaller tasks) and, on the contrary, to disaggregated general indicators (generalized key indexes of efficiency of innovative activity) on several more specifics (Korniyko et al. 2019).

The depth of decomposition of the generalized should not exceed 3–5 levels, reflecting and matching the original methodological principles of selection of KPI.

The development and selection of a KPI system of innovation activity should be carried out using the method of constructing *a tree of goals of KPI.*

After the construction of the tree of KPI, the determination of the composition of the KPI system of innovation activity does not end.

The final selection of indicators of the KPI system of innovative activity is decided based on their comparative assessment, the degree of their relative importance for achieving the general goal of the innovative development of the company.

The degree of the relative importance of indicators included in the preliminary stage in the KPI system of innovation activity is assessed by taking into account their importance for elements of a higher hierarchical level of the goal tree based on the Delphi method (a method for determining the degree of preference, etc.).

In the process of constructing a *goal tree*, key indicators of the performance of each employee are linked to general criteria, indicators, and indicators of the effectiveness of innovative activities of the company as a whole or a unit, based on expert assessments, using the methods of *the tree of KPI constructing*.

Based on general indicators reflecting innovative activity, one can judge the *innovative efficiency* of the team of the innovation project.

References

Gonin, V.N., Kashurnikov, A.N., Khanchuk, N. N.: . Methodological approaches to assessing the effectiveness of innovation in economical systems (2014). https://cyberleninka.ru/article/n/metodicheskie-podhody-otsenki-effektivnosti-innovatsionnoy-deyatelnosti-v-ekonomicheskih-sistemah

Ilyas, A. A. (2015). The formation of a methodological approach to the economic assessment of the effectiveness of corporate strategies of oil and gas companies based on a balanced scorecard. *Petroleum Engineering*, *3*, 543-575. Retrieved from http://ogbus.ru/issues/3_2015/ogbus_3_2015_p543-575_IlyasAA_ru_en.pdf

Korniyko, YuV, Stepanov, I.A., Shcherbakova, M.A.: KPI of regional innovation development. Manag. Econ. Syst. **3**, 121 (2019)

Stepanov, A.A., Savina, M.V., Stepanov, I.A., Zolotareva, A.F., Antonova, L.I.: Concept of creative management in the paradigm of informational society management theory. Int. J. Econ. Res. **14**(7), 35–152 (2017)

Stepanov, A.A., Bialoblodskaya, S.: Methodological aspects of assessing the effectiveness of agricultural systems in an innovative economy. Appl. Econ. Res. **4**, 48–51 (2014)

Ministry of Economic Development of the Russian Federation. Recommendations on a key indicator of the effectiveness of innovation, included in long-term development programs and systems of key performance indicators used to motivate the management of state-owned companies (2015). http://economy.gov.ru/minec/about/structure/depIno/20150703467

Forecasting Financial Results of the Enterprises' Activities Under the Conditions of Fluctuations in Production Volumes

Aleksandr M. Batkovskiy[1(✉)], Valeriy Ya. Trofimets[2], and Nikolai I. Turko[3]

[1] Central Research Institute of Economy, Management and Information Systems "Electronics", Moscow, Russia
batkovskiy_a@instel.ru
[2] Saint-Petersburg Mining University, St. Petersburg, Russia
zemifort@inbox.ru
[3] State Corporation for Assistance to Development, Production, and Export of Advanced Technology, Industrial Product "Rostec", Moscow, Russia
N.I.Turko@rostec.ru

Abstract. In the context of the introduction of Industry 4.0 technologies, the task of increasing the efficiency of Russian enterprises in the transition to fully automated digital production controlled by intelligent systems in real-time is of particular importance. The most important direction of its solution is the optimization of financial planning for the development of enterprises, including forecasting the financial and economic results of their activities. The most important direction for its solution is the optimization of financial planning for the enterprises' development, including the forecasting of the financial and economic results of their activities. The complexity of solving this problem is determined by the fact that the digitalization of economy implies a new level of organization of production and management of the value chain throughout the entire life cycle of products, as well as the development of automation and data exchange in constant interaction with the external environment. In recent years, the uncertainty of the external environment of enterprises has increased dramatically due to changes in the external conditions of their industrial and economic activities. The political situation in the world has changed for the worse, which has a negative impact not only on the foreign economic activity of many Russian enterprises but also on the conditions for their innovative development. The measures are taken (import substitution, state support, etc.) aimed at neutralizing and smoothing the negative consequences of these circumstances do not yet ensure the transition of Russian enterprises to a stable production increase. Many enterprises experience fluctuations in production volumes due to the sanctions, currency fluctuations, lack of financial resources in the required volumes, low profitability, and other reasons. Taking into account these circumstances, the task of forecasting the financial results of enterprises during the implementation of Industry 4.0 technologies, in the context of fluctuations in production volumes caused by the mentioned circumstances, has acquired the most important scientific and practical value. The development of a toolkit for

A. V. Bogoviz and Y. V. Ragulina (Eds.): ISCI 2019, LNNS 115, pp. 395–401, 2020.
https://doi.org/10.1007/978-3-030-40749-0_47

solving this problem is given in this paper. For this purpose, a comparative analysis of applied econometric models is carried out, designed to predict the dynamics of economic processes subjected to influences that affect the volume of production. The advantages and disadvantages of trend-seasonal models are compared with other econometric models that take into account these factors. A modified algorithm for constructing trend-seasonal models is proposed, which increases their accuracy by taking into account the functionality of modern tabular processors. It allows automating complex blocks of this algorithm. The accuracy of models built on the basis of the modified algorithm is compared with models built based on the standard algorithm. Its practical feasibility is proved when introducing the Industry 4.0 technologies.

Keywords: Forecasting · Time series · Econometric models · Seasonal component · Computer modelling

1 Introduction

One of the most critical tasks of planning of the enterprise's activities is to predict their financial results. Its solution is usually associated with the use of specific models and methods of various scholars (Avon 2015; Batkovskiy et al. 2016; Batkovskiy et al. 2017; Gilliland et al. 2016; Hanke and Wichern 2014; Hofmann and Klinkenberg 2013; Runkler 2012). It should be borne in mind that econometric tools for forecasting financial results are not universal. The effectiveness of its use is determined by many factors. Therefore, forecasting financial results of activities at enterprises with fluctuating production volumes differs significantly from doing so at enterprises with approximately equally distributed levels of production and financial flows over time.

The consideration of circumstances in the enterprise activity is essential when planning not only financial flows from operating activities but also paying taxes. In addition, the factor of fluctuations in production volumes can be taken into account when establishing a correcting coefficient of basic profitability K2. Therefore, this factor is important in predicting financial results and, as a consequence, in making adequate management decisions for the development of the enterprise by introducing various types of information and communication technologies, electronics, and industrial robotics into production processes and their automation.

2 Materials and Methods

Among the models intended for predicting the processes under consideration, seasonal varieties of adaptive models, models of autoregression of the integrated moving average (ARIMA model), and trend-seasonal models are widespread.

The adaptive forecasting models are based on an exponential smoothing model (Hamilton 1994). Its numerous modifications led to the emergence of adaptive forecasting models, including those that take into account the seasonal (oscillatory) component in the studied time series. The Holt-Winters model and the Theil-Wage model are best known among such models. (Franses et al. 2014; Box et al. 2015).

The Holt-Winters model is an integration development of the two-parameter Holt linear growth model and the seasonal Winters model. In this model, the time series is represented as a combination of a linear trend with a seasonal component applied multiplicatively.

The Thela-Wage model differs from the Holt-Winters model in that it is additive. The creation of said model is intended to simplify the prediction procedure, since the combination of a multiplicative seasonal model with the linear growth is mathematically cumbersome. When constructing the Holt-Winters and Theil-Wage models, there should be a series of data consisting of at least three periods:

- seasonal factors are calculated according to the first period;
- the forecast model itself is built according to the data of the second period, but the calculation still uses adapted seasonal coefficients obtained from the first period;
- optimal smoothing constants for the seasonal component are selected according to the data of the third period.

The task of constant smoothing is nontrivial, since in the presence of the normalized seasonal coefficients, it is required to solve a nonlinear equation, and if the coefficients are not normalized, it is impossible to determine them at all. It is also necessary to select another s seasonal factor and two coefficients for estimating the trend component. Consequently, the construction of adaptive Holt-Winters and Theil-Wage models is a rather labor-intensive process, and their application in practice involves the use of specialized statistical packages (Yaffee and McGee 2000; Lawrence et al. 2009; Woodward and Gray 2016).

Using the autoregression model of the integrated moving average (ARIMA model), it is possible to build forecasts for a wide range of time series. Certain modifications of these models make it possible to describe the time series with a changing component very accurately. The models of this class have a precise mathematical justification, which makes them one of the most scientifically sound models of the whole set of forecasting models. Until recently, it was believed that the ARIMA models provide the most accurate predictions, as they are more common than the class of other models. However, in several cases, the ARIMA models proved to be no better than the exponential smoothing models, so the opinion about the indisputable advantage of these models changed to the completely logical concept that for each case, a case-specific model should be used.

At least 40 observations are required to build an adequate ARIMA model. Also, building an adequate model requires a lot of time and analyst experience. Thus, the use of ARIMA models in everyday practice, as well as adaptive forecasting models, is difficult if specialized statistical packages are not used.

Other widespread forecasting models that take into account production instability are trend-seasonal. They are based on the assumption that the main factors and trends of the past period will continue for the forecast period, or that the direction and change of trends in the considered perspective can be justified and taken into account, i.e., large inertia of the studied processes is assumed. The practice of using trend-seasonal models shows that, for all their simplicity, they can provide fairly reliable predictive results. One of the significant drawbacks of these models is that they poorly isolate outliers (sharp jumps) in short periods. However, to predict the results of the enterprise activity in the medium and long term, this drawback is insignificant since, in this case, the minimum length of the oscillation period is measured in months or years.

There are various algorithms for building trend-seasonal models. The overwhelming majority of them, regardless of the reasons causing fluctuations in production volumes, include the following main stages (Hamilton 1994; Franses et al. 2014):

- determining the presence of a trend in the time series;
- identifying fluctuations in the time series;
- time-series filtering;
- assessing the quality of the constructed model.

The presence of a trend in the time series can usually be established quite quickly by visual analysis of the time series chart. For a more rigorous proof of the presence of a trend, we can use the analytical methods based on testing the hypothesis of the invariance of the average value of the time series.

To determine whether there are fluctuations in the time series, we can also use a visual analysis of the time series or more rigorous analytical methods based on checking the randomness of the residual component after selecting a trend from the time series. Such testing is usually based on the calculation of seasonal indexes and the construction of a seasonal wave, or on the calculation of autocorrelation coefficients and the construction of a correlogram.

The procedure of the filtering time series involves its division into a trend, a seasonal component, and a random component. For such a separation, the regression, spectral, and iterative methods can be used. As a rule, the iterative methods are used, which, on the one hand, provide acceptable filtering accuracy, and, on the other hand, are quite simple.

The quality assessment of the constructed model is formally determined by accuracy and adequacy. It is advisable to use the average relative error of approximation to assess the adequacy of the model, the properties of the residual components, i.e., discrepancies between levels calculated by the model, and actual observations. The study should include several checks:

- Independence (autocorrelation lack) of the levels of several residues (for example, using the Darwin-Watson dw-criterion);
- Correspondence of some residues to the normal distribution law (for example, using the *RS*-criterion);
- The expectation of the number of residue levels to be zero.

3 Results

The proposed modified algorithm for constructing trend-seasonal models allows for improvements to the accuracy of the typical algorithm; on the other hand, it takes into account the functionality of modern tabular processors commonly used in office activities. The latter circumstance is of particular importance for the practical use of the algorithm since it makes it possible to greatly simplify the use of knowledge-intensive computational procedures for mathematical statistics and non-linear programming.

The main modifications implemented in the algorithm developed for building trend-seasonal models are as follows:

1. For an analytical confirmation of the presence of seasonal fluctuations in the time series, it is proposed to use the F-criterion. This is more effective than either the standard procedure for calculating seasonality indices with the subsequent construction of a seasonal wave graph or the procedure for calculating autocorrelation coefficients with the subsequent construction of a correlogram.
2. In order to smooth the initial time series, the developed algorithm does not use the centered moving average method but instead uses polynomial models that record up to the third degree. The advantage of this method is that all levels are used to smooth the time series. By contrast, during the use of a centered moving average, some levels are lost. This fact has a particularly negative effect on the study of the "short" time series. For example, when using a centered moving average for a 2-year time series with fluctuations at a 12-month frequency, 12 levels out of 24 will be lost (levels 1–6 and 19–24).
3. In order to adjust the seasonal component in the case of the additive model, the optimization problem (1) is solved:

$$\begin{aligned} e_{rel} &\rightarrow \min, \\ \sum_{i=1}^{T_0} s_i &= 0, \end{aligned} \tag{1}$$

where $e_{rel} = \frac{1}{n}\sum_{i=1}^{T_0}\sum_{j=1}^{k}\frac{|\varepsilon_{ij}|}{y_{ij}}$ – the average relative error of approximation;
y_{ij} – the actual value of the original time series in the ith quarter (month) of the jth year;
s_i – values of the seasonal component;
ε_{ij} – values of the random component;
T_0 – a seasonal period in time series;
k – a number of years of observation;
n – a number of observations in the time series ($n = kT_0$).

In order to correct the seasonal component in the case of a multiplicative model, an optimization problem is solved (2):

$$\begin{aligned} e_{rel} &\rightarrow \min, \\ \sum_{i=1}^{T_0} s_i &= T_0. \end{aligned} \tag{2}$$

The proposed algorithm for constructing trend-seasonal models was tested on the array of data on the financial and economic activities of several enterprises that had fluctuations in the production volumes. For assessing the accuracy of the constructed trend-seasonal models, the average relative error of approximation was used. The calculated values of the approximation error did not exceed 3%, which allows us to conclude that the constructed models are sufficiently accurate.

Based on these models, the point and interval estimates of the projected financial and economic indicators of enterprises were calculated.

4 Discussion

The accuracy of the models that are built based on a modified algorithm is compared with the accuracy of the models that are built based on a typical algorithm. It showed that the first models exceed the second models by an average of 2%. It should be noted that the complexity of models that are built based on a modified algorithm is less than the complexity of models that are built based on a typical algorithm.

When constructing a model based on a modified algorithm, an additive time-series model was adopted as a working hypothesis (3):

$$y_t = u_t + s_t + \varepsilon_t \tag{3}$$

where u_t – a trend component; s_t – a seasonal component; ε_t – a random component.

The model (3) is usually used if the seasonal wave has a relatively constant amplitude. It is usually assumed that the error (random component) has a normal distribution with zero mathematical expectation and some constant dispersion:

$$\varepsilon_t N(0, \sigma_\varepsilon^2) \tag{4}$$

If significant seasonal fluctuations are observed, then the multiplicative model is used (5):

$$y_t = u_t \cdot s_t \cdot \varepsilon_t \tag{5}$$

If a multiplicative model (5) is adopted as a working hypothesis, then the accuracy of the model may increase or decrease depending on the initial data. When using model (5), where the error is taken into account multiplicatively rather than additively, it is necessary to present model (5) in an additive form by a logarithm, where, by analogy with (3) and (4), we can assume:

$$\ln \varepsilon_t \sim \ln N(0, \sigma_{\ln \varepsilon}^2). \tag{6}$$

Thus, the error in the multiplicative model will have a log-normal distribution:

$$\varepsilon_t \sim \log N(0, \sigma_{\ln \varepsilon}^2). \tag{7}$$

The normal distribution of errors is one of the formal criteria by which we can verify the correctness of the choice of the model type. If normally distributed errors are obtained after constructing a model from an empirical series of data, then we can conclude on choosing the most suitable type of model.

5 Conclusion

The result of the study is that the main drawback of trend-seasonal models is their reduced sensitivity to sharp jumps in short periods. However, to predict the financial results of the enterprise, when the minimum period of fluctuations is measured in months, this drawback is insignificant, which makes it possible to consider trend-seasonal models as a useful tool for predicting financial results.

Acknowledgment. The study was carried out with the financial support of the Russian Foundation for Basic Research in the framework of the research project No. 18-00-00012 (18-00-00008) COMFI.

References

Avon, J.: The Basics of Financial Modeling. Apress, New York (2015)
Batkovskiy, A.M., Semenova, E.G., Trofimets, V.Ya., Trofimets, E.N., Fomina, A.V.: Method for adjusting current appropriations under changing funding conditions. J. Appl. Econ. Sci. Rom. Eur. Res. Centre Manag. Stud. Bus. Adm. **XI**(5), 828–841 (2016)
Batkovskiy, A.M., Semenova, E.G., Trofimets, VYa., Trofimets, E.N., Fomina, A.V.: Statistical simulation of the break-even point in the margin analysis of the company. J. Appl. Econ. Sci. **XII**(2), 558–571 (2017)
Box, G.E.P., Jenkins, G.M., Reinsel, G.C., Ljung, G.M.: Time Series Analysis: Forecasting and Control. Wiley, Hoboken (2015)
Franses, P.H., Van Dijk, D., Opschoor, A.: Time Series Models for Business and Economic Forecasting. Cambridge University Press, Cambridge (2014)
Gilliland, M., Tashman, L., Sglavo, U.: Business Forecasting: Practical Problems and Solutions. Wiley, Hoboken, NJ (2016)
Hamilton, J.D.: Time Series Analysis. Princeton University Press, Princeton (1994)
Hanke, J.E., Wichern, D.W.: Business Forecasting. Pearson, Harlow (2014)
Hofmann, M., Klinkenberg, R.: RapidMiner: Data Mining Use Cases and Business Analytics Applications. CRC Press, Boca Raton (2013)
Lawrence, K.D., Klimberg, R.K., Lawrence, S.M.: Fundamentals of Forecasting Using Excel. Industrial Press, South Norwalk (2009)
Runkler, T.A.: Data Analytics: Models and Algorithms for Intelligent Data Analysis. Springer Vieweg, Heidelberg (2012)
Woodward, W.A., Gray, H.L.: Applied Time Series Analysis with R, p. 634. CRC Press, Taylor & Francis Group, Boca Raton (2016)
Yaffee, R.A., McGee, M.: Introduction to Time Series Analysis and Forecasting with Applications of SAS and SPSS. Academic Press, Cambridge (2000)

Economic Evaluation of the Corporate Projects' Efficiency

Liliya F. Sukhova[1(✉)], Yuliya A. Glaz[1], and Lyubov V. Agarkova[2]

[1] Stavropol Institute of Cooperation (Branch) of Belgorod University of Cooperation, Economics and Law, Stavropol, Russia
Suhova.lilia@yandex.ru, glazia@gmail.com
[2] Stavropol State Agrarian University, Stavropol, Russia
alv23@mail.ru

Abstract. The modern system of indicators for assessing the economic efficiency of corporate projects does not show the adequacy of the forecasted efficiency level of the project. The purpose of the study is to justify the need for introducing new indicators that can provide a full assessment of project effectiveness. The basis of the study is a systematic approach and its scientific methods, according to which the financial system is considered as a subsystem of the general economic system of the corporation, and its financial results are an integral element of the integral indicator of project efficiency. The results of the study show that for a full assessment of the effectiveness of the project, it is necessary to use new indicators (financial and economic profitability of the project, financial potential of the project), which allow us to evaluate the project effectiveness taking into account the economic efficiency of the designed financial system. The scientific novelty is represented by the following: the system of indicators for evaluating projects has been expanded; the economic content and the purpose of the indicators of financial potential and economic profitability are disclosed; the technique of analyzing and evaluating the project effectiveness and financial stability is provided; new factors and indicators of project performance management are identified. A project can be considered effective at the proper level if its level is not lower than the level of financial and economic profitability.

Keywords: Project performance criteria · Financial and economic profitability · Financial potential · Project efficiency

1 Introduction

The future development of the corporation based on present facts is formalized by various types of projects (corporate foresight projects, investment projects, business plans, innovative projects, etc.). The feasibility is evaluated by a system of economic indicators.

Scientific research in the field of evaluating project effectiveness is carried out in a number of areas, for example, a review and characterization of modern approaches to assessing the effectiveness of investment projects (Krivtsov 2016) and economic

A. V. Bogoviz and Y. V. Ragulina (Eds.): ISCI 2019, LNNS 115, pp. 402–410, 2020.
https://doi.org/10.1007/978-3-030-40749-0_48

evaluation of project effectiveness (Rimmer et al. 2007). A separate area is the development of the integrated systems for evaluating the effectiveness of projects (Chance and Peterson 2002; Damodaran 2012a, b), including direct investment projects (Vilensky et al. 2002), and an integrated system for evaluating the effectiveness of corporate projects taking into account industry specifics, for example, investment and construction projects (Gilekhmanov and Braille 2016; Gorshkov and Silantieva 2015). The development of scientific research in these areas includes a search for criteria indicators (Avrashkov and Grafova 2003; Yanovsky and Goryansky 2009) and better methods for analyzing and evaluating the effectiveness of projects (Beres and Havranek 1995; Masterov 2017; Damodaran 2012a, b; Smolyakova and Sukhov 2015), research on modern principles for evaluating investment projects (Rimmer et al. 2007), and concepts for improving corporate performance (Sidorova and Kovalenko 2014). At the same time, some scholars prefer to evaluate project effectiveness using the indicators of investment analysis based on the assessment of cash flows, while others consider this approach to be limited, justifying the need for the evaluation of economic effectiveness, taking into account the financial stability of the projects.

However, none of the modern approaches to assessing the economic efficiency of projects takes into account the economic effect of the financial policy of a corporation. This effect is expressed, first, by the result of a comparison of the financial costs of mobilizing funds for the implementation of the project with the level of their return and, second, by the level of financial potential that the projected financial system has. The introduction of these indicators into the current system of evaluating the economic efficiency of projects will answer such important questions as:

- Whether the projected level of economic return on the project is sufficient to cover the costs associated with its financial support;
- What is the financial potential of the project, and what is its contribution to ensuring financial stability and development of the financial system of the project?

The financial potential of a corporate project is the internal reserve of financial support for the project, generated by the financial system of the corporation itself, which can be used for self-preservation, self-development, and self-sustainability of the designed financial system of the corporation (Sukhova 2016; Sukhova 2017). The financial potential characterizes the level of price competitiveness of the corporation and its potential access to the capital market in order to financially support the implementation of the project.

Only the existing financial system has financial potential, that is, the one whose financial resources are capitalized and bring returns.

The economically justified size (scale) of the financial system is the volume, composition, and structure of total financial resources, the capitalization of which provides an economically justified level of their return and the formation of financial potential. In accordance with this, the leading indicators reflecting the interconnectedness and interdependence of project efficiency on the designed financial systems can be the financial and economic profitability of the project, financial potential, and the level of financial potential.

2 Materials and Methods

The studies to determine the role of new indicators and justify the need for their inclusion in the current system of indicators of economic efficiency and financial sustainability of projects were carried out based on the systematic approach. In accordance with this approach, the financial system of the corporation was considered a subsystem integrated into the overall economic system of corporate governance. From the toolkit of methods of a systematic approach, the economic and mathematical methods, special financial methods, analysis, and synthesis methods were used, which allowed the essence of such economic categories (indicators) to be revealed as financial resources, financial potential, and as financial and economic profitability of the project. Along with this, as the evidence for the need to introduce these new economic categories, there are requirements for conducting business processes arising from the "golden rule of the economy" and "the rule of financing the activities of the organizations."

The financial system of a corporation materializes in the form of financial resources. Moreover, since the financial system is material, like any material body in nature in general, *it has the potential.* Since the financial system is discussed, this potential is called financial potential.

The financial potential of the corporate financial system is the internal ability of the financial system itself to perform financial work on the accumulation, return, and expenditure of financial resources due to its position (financial, economic, territorial, etc.) relative to other corporate financial systems, as well as due to any other individual properties and features of the corporation's financial system itself (Sukhova 2016; Sukhova 2017).

A rigidly determined economic model is used as a tool for analysis, planning, and management of financial potential. This model includes three elements: the reserve of financial resources generated by the income generation policy (FSM - financial strength margin); the reserve of financial resources generated by the policy of passive capital formation (RPCF); and the reserve generated by the policy on active capital use (RACU). The algorithm for calculating financial potential is as follows (Sukhova 2016):

$$FP = FSM + RPCF + RACU$$

As the corporation's project is being implemented, its financial system itself develops, but only if the financial policy at the previous stages of the project was successful. Under these conditions, all prerequisites are created for increasing the financial potential, that is, the volumes of hidden internal reserves of financial resources are increasing, which can be used either to develop the financial system of a corporate project or to preserve (maintain) it, for example, until the next stage of the project is introduced.

The time aspect of the formation and use of financial potential is essential to consider when implementing the most financially intensive stages of the project and/or those stages of the project, the implementation of which requires the mobilization of funds from expensive sources.

According to the authors, an indicator of financial and economic profitability is significant for the evaluation of projects. The need for this indicator is explained by the fact that, for example, when calculating the internal rate of return of an investment project, the expert is of the opinion that the funds are reinvested at the percentage that resulted from the calculation of this indicator. In this case, the following questions arise:

- Is the internal rate of return (IRR) sufficient to recognize the project as effective;
- What is the necessary level of project profitability in order to recognize it as not break-even, but economically profitable, (that is, overlapping the projected financial costs of mobilizing all the types of funds for its implementation);
- What is the permissible price for mobilizing own, short-term, and long-term borrowed funds? Depending on what funds were mobilized and what assets they were invested in, the range of prices for the corporation's access to the capital market is determined to support the project financially.

The financial and economic profitability of a project is the level of project profitability at which its return level is not lower than the level of market prices for money mobilized for its implementation. Its achievement is possible subject to the positive effect of financial leverage (Sukhova and Semenova 2017).

3 Results

The result of the research is a methodology for analyzing and evaluating the economic efficiency of the project using the proposed indicators.

The financial potential of the project and its level are calculated based on the forecast data. These indicators reflect the target state of the financial system, where the current financial system of the corporation should turn, or which should be created considering the project of the newly created corporation.

During the development and evaluation of the financial policies for the implementation of the corporate project, it is necessary to strive to ensure that the size of the financial potential increases from stage to stage and consists of all the constituent elements. Such an approach will indicate that the project is not just economically feasible but has the proper level of economic efficiency and financial stability.

As the next stage of the project is implemented (designed), the nature of the change in the magnitude of the financial potential should be monitored. If, as a result of the implementation of the next stage of the project, the growth rate of the financial potential is higher than the growth rate of total financial resources, then during the implementation of this stage of the project, financial potential has accumulated and, consequently, increased the level of financial potential. If this is reversed, then there was spending of financial potential for maintaining the financial system. According to the results of the analysis of financial potential, it is possible to find out the changes that occurred in its size, why the financial potential was spent and in what amounts, and what can and should be provided for financial support for the implementation of the next stage of the project due to financial potential (Sukhova et al. 2016).

The methodology for calculating the proposed indicators for the two intermediate stages of the implementation of the conditional corporation's project is presented in Table 1.

According to the data, the project is cost-effective. However, based on the economical nature and results of the calculation of the proposed indicators (Table 1, i. 7.13–18), we can draw the following conclusions.

The scale of the financial system of the conditional corporation at the end of the second year will increase by 14%, both due to an increase in incomes by 8% and due to the influx of cash from the outside by 17.9%.

The project provides the predominant growth of investments in the development of the material and technical base of the corporation and property in general. This conclusion follows from the fact that in the second year, in comparison with the first project, an asset growth rate is planned to be higher than the revenue growth rate (112.8% > 101.4%), i.e., the projected reduction in the return on assets.

Despite the growth in the scale of the financial system, its financial potential by the end of the second year will decrease by 0.5%. This indicates its partial expenditure since the level of the projected economic return of the project in the second year in the amount of 1.5% (i. 12;) is insufficient.

In the second year, compared with the first one, there is a significant excess of market prices for cash in comparison with the level of their return (i. 14 > i. 12). Therefore, the level of the competitiveness of the corporation price in the capital market is reduced, and there is no reserve of financial resources generated by the policy of using the passive capital (−805 mln. ₽; i. 15).

The project level of the return on assets is insufficient. In the first year, the return on assets was lower than the market price of cash by 1.72% [(0.058 − 0.8 × 0.094) × 100 = −1.72%], and in the second year by 6.67% [(0.015 − 0.8 × 0.102) × 100 = −6.67%]. Based on this, the implementation of the project in the first year will require the expenditure of the financial potential in ₽ 179 million, and in the second year, ₽ 805 million, i.e., more than 4.5 times.

The financial and economic profitability of the project in the first year was supposed to be at least 7.52% (0.8 × 0.094 × 100), and, in the second year, to increase by 0.64 points and amount to 8.16% (0.8 × 0.102 × 100). When designing a project at the level of financial and economic profitability, there can be an increase in the financial potential due to an increase in the level of solvent demand of the corporation in the capital market. During the first year, it will increase from 5.8% to 7.52%, and during the second year, from 1.5% to 8.16%.

An increase in the level of the projected profitability to the level of the financial and economic profitability is possible due to the factors of growth in the financial stock, as well as by increasing the effectiveness of the financial policy. The latter is achieved, for example, by reducing financial costs, as a result of the development of a more detailed financial plan. The preparation of a financial plan with a breakdown of cash inflows and outflows not for the whole year or quarterly, but with a breakdown by months, decades, or days (depending on the cash flows volume), provides the clarification of the terms for borrowing, their sizes, and repayment periods. This approach to the preparation of the financial plan, as a rule, leads to significant savings in the financial costs.

Table 1. The proposed indicators for evaluating the cost-effectiveness of a project of the conditional corporation.

Indicators name	Project years		Growth rate, %
	1st	2nd	
1. Revenue, mln. ₽	3 933	3 990	101.4
2. Other income, mln. ₽	3 265	3 805	116.5
3. Depreciation, mln. ₽	419	472	112.6
4. Liabilities (Assets), mln. ₽	10 855	12 249	112.8
5. VAT, mln. ₽	56	70	125.0
6. Reinvested earnings, mln. ₽	488	21	4.3
7. Financial resources, mln. ₽	17 090	19 481	114.0
7.1. Revenues	6 779	7 323	108.0
7.2. Income	10 311	12 158	117.9
8. Cost of sales and administrative expenses, mln. ₽	2 971	3 069	103.3
8.1. Fixed costs	2 065	1 934	93.7
8.2. Variable costs	906	1 135	125.3
9. Financial strength margin, mln. ₽	2 683	2 703	100.7
10. Net profit, mln. ₽	628	188	29.9
11. Equity, mln. ₽	8 369	9 089	108.6
12. Return on assets, in shares	0.058	0.015	26.5
13. The weighted average price of corporation capitals in shares [(i. 13.1 × i. 13.2 + i. 13.3 × i. 13.4 + i. 13.5 × i. 13.6)/100]	0.069	0.027	39.6
13.1. The price of long-term liabilities, %	5.68	5.44	95.8
13.2. The proportion of the long-term obligations in liabilities	0.115	0.139	120.9
13.3. The price of short-term liabilities, %	7.038	6.84	97.2
13.4. The proportion of the current obligations in liabilities	0.0604	0.0617	102.2
13.5. The price of equity, %	7.505	2.079	27.7
13.6. The share of equity in liabilities	0.771	0.742	96.2
14. The market price of cash, in shares[a, *]	0.094	0.102	108.5
15. The financial potential, total, mln. ₽ [i. 9 + (0,8* × i. 14 − i. 13) × i. 7.2 + (i. 12 − 0,8* × i. 14) × i. 7.2]	2 572	2 559	99.5
Including The reserve of financial resources generated by the policy of revenue generation (i. 9)	2 683	2 703	100.7
The reserve of financial resources generated by the policy of passive capital formation [(0,8 × i. 14 − i. 13) × i. 7.2]	68	661	978.6
The reserve of financial resources generated by the policy of using passive capital [(i. 12 − 0,8 × i. 14) × i. 7.2]	−179	−805	450.3
16. The level of financial potential, % (i. 15: i. 7 × 100)	15.5	13.1	87.3
17. Financial and economic profitability, % (0,8 × i. 14 × 100)	7.52	8.16	108.5
18. The proper level of project profitability, % (i. 12 × 100 ≥ i. 17)	≥7.52	≥8.16	108.5

Note to the table.
[*]0,8 – this is the value of the tax corrector (income tax rate of 20%).
[a]Retrieved from http://www.cbr.ru.

The financial potential of the corporation in the amount of ₽ 2.559 million during the second year indicates that the corporation has access to the capital market for the implementation of the next phase of the project. The value of the financial potential shows the extent to which the excess of market prices for cash over the projected level of their return is permissible.

The project provides a decrease in prices for long-term and short-term cash assets by 4.2% and 2.8%, respectively (table i. 13.1 and 13.3) and a decrease in the weighted average price of capital by 60.4%, i.e., by 2.5 times (i. 13). However, this does not indicate an increase in the level of effectiveness of the financial policy pursued by the project developer, since the level of the price access of the corporation to the capital market is significantly reduced:

- at the projected prices for long-term borrowed funds in the amount of 5.68% and 5.44%, they are available for the corporation at the prices of 1.03% and 0.32%;
- at the projected prices for short-term borrowed funds of 7.038% and 6.84%, they are available for the corporation at the prices of 12.92% and 3.7%.

The drawbacks of the proposed corporation financial policy are compensated by the use (expenditure) of the financial potential in the first year in the amount of ₽ 179 million, and in the second year ₽ 805 million (Table, i. 15).

4 Discussion

The economic content, role, and purpose of new indicators of economic evaluation and financial sustainability of projects, as well as algorithms for their calculation, are submitted for discussion.

The position of the authors is as follows. The proposed indicators make it possible to justify the proper level of project effectiveness, and the input indicators in the composition of their calculation formulas are the factors in the management of project efficiency. It is not the statement of the level of project profitability that matters, but whether this level is appropriate (that is, economically justified not only from the standpoint of efficient use of the material but also the financial resources).

The implementation of these indicators allows us to solve and evaluate the significance of the following issues:

- What number of the financial resources can be used to implement the subsequent stages of the project due to the internal reserves that the financial system itself possesses, from the stage of the project from which the project begins to generate revenue for the corporation? The successfully operating corporations have a priori financial potential that can be used to support the upcoming stage of the project financially;
- For what purposes and to what extent can the corporation use the existing financial potential to develop the scale of the financial system, to increase the financial potential, to maintain the size of the financial system, to maintain the stability of the financial system, or make combined decisions on the use of financial potential;

- At what price is it possible to mobilize funds in the capital market for the implementation of a particular stage of the project, as well as the whole project? The economic opportunities for a corporation to access the market for money mobilized for the short and long term are different. The maximum cash prices for a corporation are determined by the profitability levels of current and non-current assets, which are calculated according to new formulas (Sukhova 2017);
- What is the level of economic return of each cash type;
- What is the level of the proper economic profitability of the project?

5 Conclusion

Studies in the field of quantitative and qualitative assessment of the effectiveness and financial sustainability of corporate projects have made it possible to establish the feasibility of including such indicators as the scale of the financial system (total financial resources), financial potential and its level of financial and economic profitability of the project. The application of these indicators significantly expands the information base for making scientifically safe decisions in the field of evaluating and managing economic efficiency and financial stability of corporate projects. Along with these indicators, information is also important on the maximum allowable prices for a corporation for monetary funds (own, borrowed) mobilized to finance the project. The level of these prices is nothing more than the level of critical (zero) profitability of their use.

So far, the financial potential has not been considered as an internal reserve of financial support for the implementation of projects. The results of these studies reveal the possibility of solving this problem.

The approach to the economic content and purpose of the financial potential and financial and economic profitability of the project, which is described in this paper, allows expanding the information base on the economic efficiency and financial sustainability of the project; their calculation algorithms will improve the digital technology base of the business process of the corporation.

The main directions for further research in the field of assessing the effectiveness of projects are the search for algorithms for calculating the time of the development inertia, self-preservation, and self-sustainability of the designed financial system.

The calculation of the proposed indicators can be based on the use of not only linear models, but also on higher-order models (non-linear models). The basis for the development of non-linear models for calculating these indicators is that several indicators included in the formulas (models) can be described by a non-linear relationship. As a rule, nonlinear economic models provide a higher degree of calculation accuracy. However, in practical work, the use of linear economic and mathematical models is more convenient for reading and analysis and, despite some often insignificant losses in the accuracy of calculation results, provides the analyst with adequate information for making economically safe management decisions.

References

Avrashkov, L.Ya., Grafova, G.F.: Criteria and performance indicators of investment projects. Auditor **7**, 38–47 (2003)

Beres, W., Havranek, P.M.: Investment Performance Assessment Guide. Interekspert, Infra-M, Moscow (1995)

Chance, D.V., Peterson, P.P.: Real Options and Investment Valuation. The Research Foundation of the AIMR, Virginia (2002)

Damodaran, A.: Investment Valuation: Tools and Techniques for Determining the Value of Any Asset University Edition. Wiley, New York (2012a)

Damodaran, A.: Investment Valuation: Tools and Techniques for Determining the Value of Any Asset, 3rd edn. Wiley, New York (2012b)

Gilekhmanov, R.A., Braille, N.V.: Methods for assessing the financial and economic efficiency of investment and construction projects. Constr. Unique Build. Struct. **10**(49), 7–16 (2016)

Gorshkov, R.K., Silantyeva, E.E.: Uncertainty factors and their impact on the effectiveness of an investment construction project. J. Econ. Entrepreneurship **6**, 445–448 (2015)

Krivtsov, S.V.: Modern approaches to assessing investment project effectiveness. Molodoy Uchenyy **12**, 1324–1327 (2016). https://moluch.ru/archive/116/31929/

Masterov, A.I.: Economic analysis of investment projects as a tool to improve budget policy. Econ. Taxes Law **5**, 60–67 (2017)

Rimmer, M.I., Kasatova, A.D., Matienko, N.N.: Economic Investments Valuation. Peter, St. Petersburg (2007)

Sidorova, L.V., Kovalenko, V.V.: Modern concepts and tools to improve corporate performance. Sci. J. NRU ITMO (Series: Econ. Environ. Manag.) **3**, 523–530 (2014)

Smolyakova, M.K., Sukhov, V.D.: Features of the analysis of investment project scenarios in the conditions of a significant fluctuation of external parameters. Theor. Econ. **4**, 41–48 (2015)

Sukhova, L.F.: The financial potential of the enterprise: concept, essence, measurement methods. Finan. Anal.: Probl. Solutions **12**, 2–11 (2016)

Sukhova, L.F.: Corporate Finance: Study Guide (Advanced). Fabula LLC, Stavropol (2017)

Sukhova, L.F., Semenova, S.V.: The economic effect of financial leverage. Bull. Belgorod Univ. Cooper. Econ. Law **2**(63), 78–92 (2017)

Sukhova, L.F., Kryuchkova, I.V., Botsyun, I.B.: Analysis and assessment of financial potential: an innovative approach. Bull. Belgorod Univ. Cooper. Econ. Law **2**(58), 106–116 (2016)

Vilensky, P.L., Lifshits, V.N., Smolyak, S.A.: Assessment of Investment Projects Effectiveness. Theory and Practice. Delo, Moscow (2002)

Yanovsky, V., Goryansky, D.: Methods and criteria for evaluating investment projects. RISC: Resour. Inf. Supply Compet. **1**, 122–125 (2009)

The Knowledge Potential in the System of Strategic Management of the Innovative Development of an Enterprise

Natalia N. Pasmurtseva(✉)

Ural State University of Economics, Yekaterinburg, Russia
pnn-kufeu@yandex.ru

Abstract. The paper focuses on the use of the knowledge-based approach in the system of strategic management of the innovative development of an enterprise. The author focuses on the need for the formation and implementation of the innovative knowledge-based strategies in order to increase the competitiveness and ensure the sustainable growth of Russian business. The aim of the study is to develop a model of innovative strategic potential, including a knowledge component, determining its place in the strategic management system of the innovative development of an enterprise. In the process of the research, the author used general scientific methods of cognition: a comparative and logical analysis, a systematic approach, as well as a graphical research method. Various approaches to the interpretation of the essence of such concepts as "knowledge potential," "knowledge assets," and "knowledge resources" are studied. The author's definition of the concept of "knowledge potential" is formulated in the system of strategic management of the development of innovative enterprise. The author offers a model of the innovative strategic potential of the enterprise. The system of strategic management of the innovative development based on the knowledge-based approach is also presented.

Keywords: Knowledge · Knowledge-economy · Knowledge potential · Strategic innovative potential · Strategic management

1 Introduction

In the twenty-first century, key factors in the dynamic development of Russia's economy and the long-term strengthening of the Russian position in the world community are the innovative strategies implemented at all management levels. At the same time, the achievement of the economic efficiency of their implementation becomes possible when innovative strategies are based on knowledge.

At the present stage of society's development, knowledge is "a resource in the economic, political, and social sense, requiring special managerial competencies" (Vasilenok et al. 2017).

So, at the macroeconomic level, long-term socio-economic development of the Russian Federation up to the year 2020 involves "the lack of alternative to the transition of the Russian economy from the export of raw materials to a new innovative one

A. V. Bogoviz and Y. V. Ragulina (Eds.): ISCI 2019, LNNS 115, pp. 411–416, 2020.
https://doi.org/10.1007/978-3-030-40749-0_49

based on knowledge" (Government of the Russian Federation 2008). At the micro-level, the knowledge component is also an integral part of innovative strategies, and a key task of increasing the competitiveness of Russian enterprises is the formation of the knowledge potential of economic entities.

2 Materials and Method

The methodology for determining the economic essence of the concept of "knowledge potential" requires the study of such categories as "knowledge" and "potential." To this end, the author has studied the essence of various approaches used by domestic and foreign scientists.

The term "knowledge" has no universally accepted interpretation and is defined by scholars differently. Ushakov interprets knowledge as "the totality of information, awareness in any field" (Ushakov 2008). According to the Big Encyclopedic Dictionary, knowledge is "a form of existence and systematization of the results of cognitive activity of a person" (Izmailov and Ashmarina 2016). Alfred Marshall noted the importance of knowledge in the economy, considering knowledge as a necessary factor for entrepreneurial activity and "one of the most powerful production engines" (Kalugin and Pogarskaya). According to Drucker, "knowledge is information that has practical value and serves to obtain specific results" (Pudovkina et al. 2018). According to the definition of the famous domestic scholar Kleiner, "knowledge is not just information, not just data, but that data that symbolizes an in-depth understanding of the world that has passed the social or institutional stage of consolidation" (Varavva 2018). Zvonnikov and Chelyshkova argue that "knowledge has not only an independent value but also affects other factors, contributing to the growth of the effectiveness of business behavior and decision-making" (Zvonnikov and Chelyshkova 2015). Despite some differences in views, most researchers note the importance of the knowledge component in the development of economic processes, emphasizing the fact that knowledge is the most important "strategic economic resource for the modern society, the source of the competitive advantage of the state, and the most important factor for sustainable economic growth" (Moutinho 2016).

The concept of "potential", as well as the term "knowledge", does not have exclusively economic meaning (Filatova 2012). The Great Soviet Encyclopedia formulates "potential" as "the funds, stocks, and sources available and able to be mobilized, put into action, used to achieve certain goals, implement a plan, solve a problem; the possibilities of the individual, and society, in a certain area" (Novikova 2017). In economics, potential is considered to be "a combination of various types of resources, including material, financial, intellectual, scientific, technical and other, which help to obtain certain results", for example, intellectual potential, production potential, and entrepreneurial potential (Rumyantseva 2005). The potential can have both a functional orientation, for example, material or human potential, and it can be realized at various levels: country, region, or enterprise. Thus, the analysis of various literary sources allows us to consider the "potential of the enterprise" as a characteristic that includes the ability of an economic entity to use its available resources.

In domestic and foreign literature, there is no single term that characterizes the enterprise knowledge system. Thus, researchers apply such concepts as "knowledge potential", "knowledge assets", "knowledge resources", and "knowledge capital".

According to Dumnaya, knowledgeable assets "can become the basic condition for the emergence of special qualities inherent exclusively to this company, necessary to maintain competitiveness in an ultra-dynamic world" (Dumnaya 2015). Pilipenko, Batalov, and Petrunin consider "knowledge capital" as an economic asset, a product of productive activities of people in the field of spiritual production (Pilipenko et al. 2016). Molodchik claims that the system of knowledge potential includes components of intellectual capital such as human, organizational, and relational capital (Molodchik 2011). According to Pulyaeva, the following should be included in the enterprise knowledge system: intellectual property, internal documentation, various sources of exogenous information and knowledge, information and communication technologies, competencies of employees, and the "knowledge culture" of the enterprise (Pulyaeva 2017).

Modern scholars working on the problems of a knowledge-based economy place emphasis not only on the category of "knowledge" but also on categories such as "information", highlighting the concept of "knowledge-information resource", or the enterprise potential.

In the works of Ashmarina and Filatova, the information and knowledge resources are considered "the total amount of information and knowledge of the organization, which is the source of achieving a given production and organizational goal" (Ashmarina and Filatova 2012). Therefore, information and knowledge resources are the basis of the organization in achieving its strategic, tactical, and operational goals. According to Filatova, information and knowledge resources of an enterprise is a concept that reflects "the level of use of both internal and external information and knowledge, as well as the totality of operations used by the enterprise to process them". In the works of Sorochaykin, information and knowledge resources are considered "the basis for the formation of qualitative characteristics of the information and knowledge potential of an enterprise" (Sorochaykin 2012). Thus, knowledge is the basis for the formation of intellectual, human, and information capital, creating a model of knowledge potential in the structure of the innovative strategic potential of an enterprise.

3 Results

Based on the foregoing and on the results of the researches conducted by modern scholars, the author suggests defining the potential of an enterprise's innovation management system as a system of assets or resources, including human capital, intellectual, and information resources, that may be involved or are already used in the process of making strategic decisions on innovation development of the enterprise. It is important to emphasize that knowledge potential is realized at three levels of management: strategic, tactical, and operational. Each of them has specifics. So, at the strategic level, the knowledge potential, being part of the innovative strategic potential, is used to establish the mission, the strategic goals, the formation of the enterprise's strategic plan by senior management.

In the system of the strategic management of an enterprise's innovative development, knowledge potential should be considered a key element of the model of a modern enterprise's innovative strategic potential according to the author; this is shown in Fig. 1.

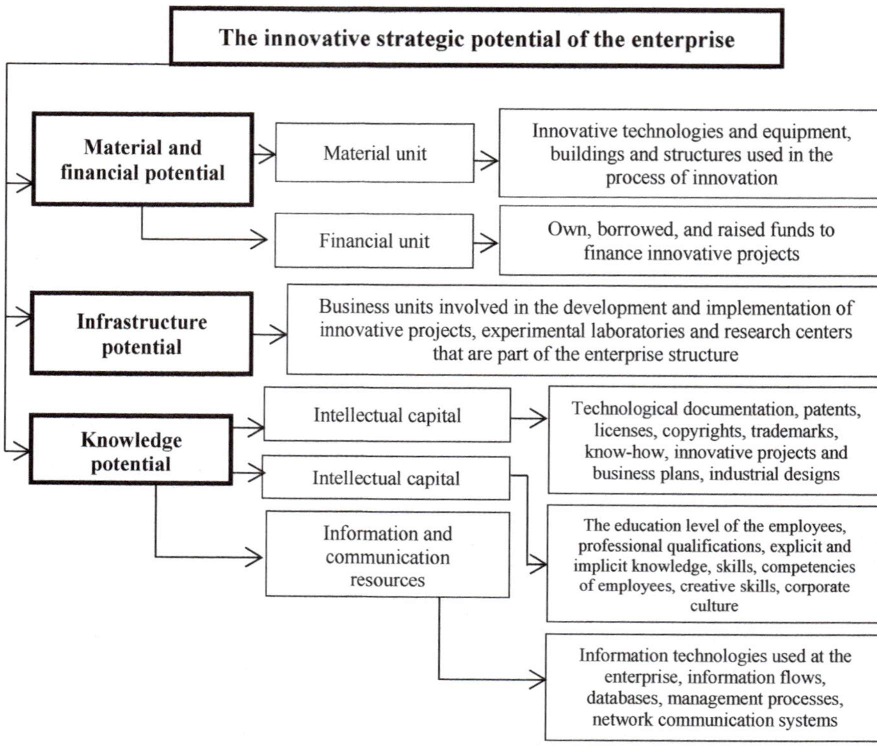

Fig. 1. The structure of the innovative strategic potential of the enterprise.

4 Discussion

Despite the differences between the approaches to the formation of a model of strategic management of an enterprise in modern and domestic scientific literature, given the specifics of innovative strategies (Shaybakova and Pasmurtseva 2009), the mechanism of the strategic management of an enterprise's innovative development should be system-sequential according to the author.

Thus, for modern enterprises implementing innovative projects, the key nature is the systematic nature of the knowledge implementation, which means that knowledge resources are not used occasionally but represent a mechanism for their formation within the framework of the strategy of the innovative development of an enterprise. Thus, according to the author, the system of strategic management of the innovative development of an enterprise, which is formed based on the knowledge approach, can be represented as follows (Fig. 2).

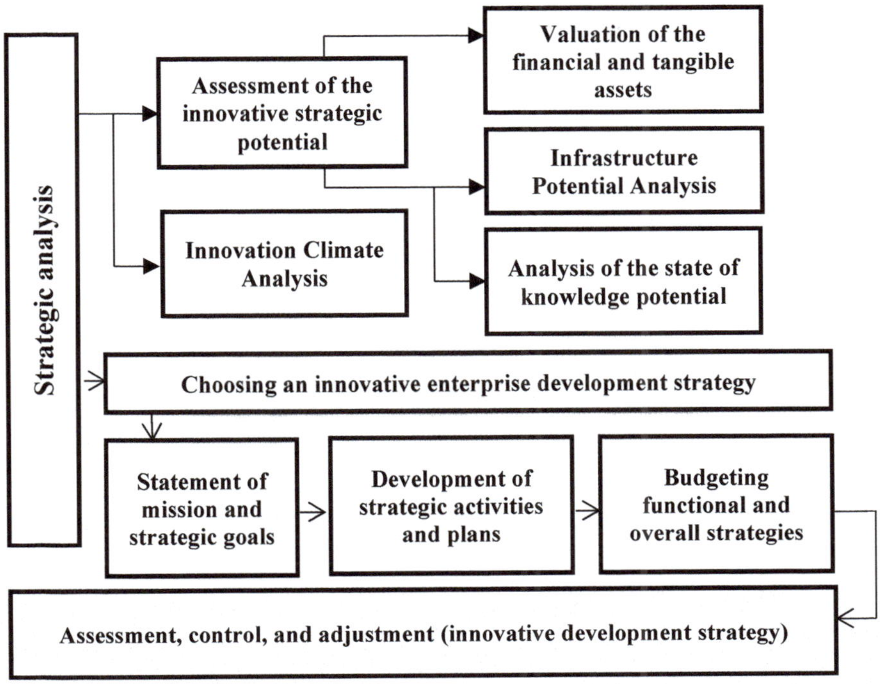

Fig. 2. The model of strategic management of the innovative development of the enterprise and the place of knowledge potential in it

5 Conclusion

The presented models of innovative strategic potential and strategic management of innovative development of an enterprise, based on a knowledge-based approach, allow:

- the accumulation of strategic and knowledge management approaches in order to increase the efficiency of innovative decision-making at the enterprise;
- the focused attention of senior management on the significance of assessing the innovative strategic potential of an economic entity carried out as part of a strategic analysis;
- the details of the structure and content of the innovative strategic potential of the enterprise, including its knowledge component.

References

Ashmarina, S.I., Filatova, A.V.: The genesis of the concept of "information and knowledge resources". Bull. SamSU **4**(95), 11–16 (2012)

Dumnaya, N.N.: Carriers of knowledge assets: the special role of medium-sized businesses. Promoting Dev. Small Medium-Sized Bus.: Foreign Exp. Russ. Pract. **2**, 77–88 (2015)

Moutinho, R.F.F.: Absorptive capacity and business model innovation as rapid development strategies for regional growth. Investigación Económica **75**(295), 157–202 (2016)

Filatova, A.V.: Methods and tools used in the management of information and knowledge resources. Bull. SamSU **4**(95), 67–72 (2012)

Government of the Russian Federation: The concept of long-term socio-economic development of the Russian Federation for the period till 2020 (November 17, 2008 No. 1662-r), Moscow, Russia (2008)

Izmailov, A.M., Ashmarina, S.I.: Mechanism of management of information and knowledge resources. Bull. Voronezh State Univ. Eng. Technol. **1**, 261–266 (2016)

Kalugin, V.S., Pogarskaya, O.S.: The historical aspect of using a knowledge-based approach to the innovative development of the economy (n.d.). http://dspace.bsu.edu.ru/bitstream/123456789/14822/1/Kalugin_Istoricheskiy_16.pdf

Molodchik, M.A.: Methodological aspects of transforming a company's knowledge into its competitive advantages. J. Econ. Theory **3**, 69–78 (2011)

Novikova, N.V.: The potential of new industrial modernization: approaches to the definition and regional economic dimension. Proc. USUE **5**(73), 127–140 (2017)

Pilipenko, E.V., Batalov, Yu.V., Petrunin, Yu.Yu.: Artificial intelligence and methodological issues of knowledge management. Philos. Sci. **8**, 67–74 (2016)

Pudovkina, O.E., Gorokhovitskaya, T.N., Tikhonov, Yu.A.: The formation of information and knowledge potential based on the use of information resources selection methods. Vestnik TSPU **5**, 212–217 (2018)

Pulyaeva, V.N.: Development of knowledge management tools in metallurgy. Econ. Ind. **2**, 121–127 (2017)

Rumyantseva, E.E.: New Economic Encyclopedia. INFRA M, Moscow (2005)

Shaybakova, L.F., Pasmurtseva, N.N.: The formation and implementation of an innovative development strategy for a metallurgical enterprise, Yekaterinburg, Russia (2009)

Sorochaykin, A.N.: Methods for assessing the components of the information and knowledge potential of an enterprise. Basics Econ. Manag. Law **6**(6), 91–93 (2012)

Ushakov, D.N.: Large Explanatory Dictionary of the Modern Russian Language. ROOSSA, Moscow (2008)

Varavva, M.Yu.: Information, knowledge, scientific knowledge as the basic categories of the new economy. Theor. Econ. **2**, 120–127 (2018)

Vasilenok, V.L., Brazevich, D.S., Safronova, G.S.: Cognitive management and innovation strategies. Sci. J. NRU ITMO: Ser. Econ. Environ. Manag. **4**, 119–124 (2017)

Zvonnikov, V.I., Chelyshkova, M.B.: Knowledge management: the fashionable theory and technology of the future. High. Educ. Today **8**, 33–37 (2015)

The Formation of Innovative Leadership on Key Product Features Based on the Unique Technological Competencies

Sergey V. Chemezov[1], Aleksandr I. Kashirin[1,2,3], and Vadim V. Strenalyuk[1,2,3(✉)]

[1] State Corporation "Rostec", Moscow, Russia
chemezov.office@rostec.ru, vstren@mail.ru
[2] People's Friendship University of Russia (RUDN University), Moscow, Russia
kashirin_ai@pfur.ru
[3] Plekhanov Russian University of Economics, Moscow, Russia

Abstract. Market leadership is achieved through the development and launch of globally competitive products and technologies with superior key competitors. The creation of such products is provided based on setting ambitious tasks that meet global challenges. This is accompanied by the formation of the UTC development teams that form the basis for the sustainable excellence of new products and technologies. The authors of the paper critically discuss the formation of innovative leadership on key product features based on the unique technological competencies.

Keywords: Unique technological competencies (UTC) · Competitive strategy · Key features · Innovative leadership · Global technological excellence

1 Introduction

One of the urgent tasks of the country's economic development is to accelerate the innovative development of Russian enterprises. In order to solve this problem, especially in the field of high technology, it is necessary to develop new approaches and mechanisms for innovative development that ensure competitiveness and technological leadership through the creation and launch of globally superior products on the market. It is proved in modern economic scientific literature that one of such innovative development tools is the unique technological (technical) competencies (hereinafter, UTK) of scientific and technical teams, which form potential "growth points" of the entire high-tech industry and the national economy as a whole (Chemezov et al. 2017).

A. V. Bogoviz and Y. V. Ragulina (Eds.): ISCI 2019, LNNS 115, pp. 417–426, 2020.
https://doi.org/10.1007/978-3-030-40749-0_50

2 Materials and Methods

At present, there is a growing interest in the topic of management of the management system for innovative development and increasing competitiveness (Chemezov et al. 2017; Kashirin et al. 2018, 2016; Chursin et al. 2018). The interest, for example, in the topic of increasing competitiveness and the tools for its achievement is also confirmed by the applied research (Ereshko and Kokuytseva 2017; Kiselev et al. 2018).

Given the novelty of the topic, it is necessary to determine the meaning of the terms used here. In the framework of this paper, we introduce the following definitions:

Unique technological (technical) competence (UTC)—a combination of knowledge and skills of a team of scholars and specialists, allowing them to create innovative technologies and products for various fields of application, while the technical characteristics of these products meet the criteria of global superiority (that it exceed the characteristics of the best world analogs) or global competitiveness (that it is comparable with the characteristics of the best world analogs) (Chemezov et al. 2017).

Key feature—technical parameters of the product (system, key element, process, etc.), determining the effectiveness of its application for the intended purpose (functionality), including competitiveness or technical leadership.

Task *(scientific, technological)* —a formulated task to achieve a clearly defined goal, technical result (**key feature**).

If the *Goal* answers the question "*What?*", then the *Task* answers the question "*How or What to do?*". A task is divided into subtasks. The Task Description Syntax: *<action> <object> <criterion>*

Criterion—*quantitatively (or qualitatively) expressed key characteristics.*

Problem—a task where the method for satisfactory solving is not defined. The problem contains a factor that impedes the achievement of the goal and is associated with technological barriers or contradictions.

The problem is a complex issue, a task requiring resolution and research. (Explanatory Dictionary Ozhegova).

It should be noted that competencies reflect the ability of specialists to effectively achieve the desired result at all stages of the product life cycle (development, production, operation, repair, disposal). In the economy of enterprises, as a rule, the main stages of creating a product (or technology) are its development and production. At the stage of product development (including design, construction, and engineering), the technical appearance of the product itself, its key elements are formed, the technical decisions are made, and the production processes are developed. At the same time, the creative stage of product development has the greatest impact on the innovative development of enterprises, when the key characteristics of the product, which determine its global superiority and competitiveness, are laid.

In the framework of this study, we highlight the creative UTC stages of the development of a product or technology of the inventive step, which is creative and has the most significant impact on the creation of innovative products.

The UTC at the stage of developing a product or technology is understood as a combination of knowledge, skills, and abilities of a team of specialists that allows to invent new devices, products, and technologies with characteristics that are unattainable for the competitors, based on the management of specific physical (chemical, biological, social, etc.) processes (Strenalyuk 2018a, b).

Given the transience of innovative processes, an important role is played by the speed of solving set scientific and technical problems in developing new technologies and products. The successful solution of a new, non-standard, previously unsolvable problem forms a new UTC, the use of which in the future will allow the UTC team to solve similar problems in the development of new products and technologies in other fields of the application quickly and successfully. After a certain period, the competence ceases to be unique, as competitors gradually master this competence. The number of UTC is the number of points of innovative growth in the current moment. As a rule, the UTC arises based on existing technological competencies in the process of solving new (non-standard) problems and involves the use of an intelligent, unconventional approach to solving such problems. The UTC can be transmitted through training. The UTC carriers [[[AUTHOR: The meaning of "carrier" here is unclear. You may want to use something like "expert" instead. Thanks.]]] in the organization are scientific and technical teams of specialists or the "UTC teams."

The key characteristics of the products that are the basis of competitive advantage can be achieved through the special skills of the team to implement the following:

(1) the design, product development, including the development as part of the product of the key element(s) or special layouts of elements in the product, and ensuring the inclusion of key technical characteristics of the product that have a global superiority in comparison with existing analogs;
(2) the creation and application of the technologies (specific actions) based on the possession of methods, ways of use, and management of certain physical (chemical, etc.) processes.

Figure 1 shows the decomposition of the description of the UTC, reflecting the causal chain of the formation of the global advantage of the product, including the relationship of the key characteristics (competitive advantages) of the product, and the characteristics of its key elements and processes, which actually ensure the product to achieve leadership characteristics.

Figure 2 shows the decomposition of the UTC of an enterprise of the Ural Optical and Mechanical Plant JSC (UOMP) for the development and production of laser microscopes with the nanometer-resolution unrivaled for this type of microscope. One of the key characteristics of the microscopes is the ability to observe biological objects in the living state of systems, for example, living red blood cells in motion, which is unattainable for the competitors.

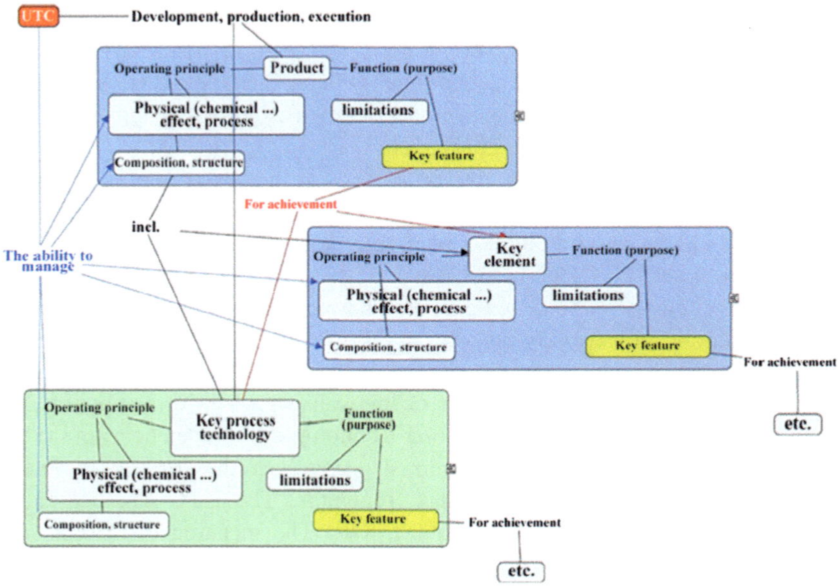

Fig. 1. The structure of the UTC, reflecting the relationship of the key characteristics of the advanced product, its key elements, and technological processes.

Figure 2 schematically shows the UTC of the microscope development team of the UOMP JSC, which is decomposed as follows. The microscope has key technical specifications that match the level of global excellence. In the analysis of the UTC, consistently answering the question of why these characteristics were achieved, we come to the identification of the key elements and processes of the microscope. The achievement of such key characteristics of the microscope became possible, on the one hand, on the basis of the high resolution of the optical system (a key element of the microscope), which, in turn, became possible thanks to special mathematical processing of the optical signal (key process). On the other hand, it is possible due to the accuracy positioning of the coordinate table (a key element of the microscope), ensuring the immobility of the object of study, which, in turn, became possible through the use of aeromagnetic guides (key element) that create the table stillness by achieving the aeromagnetic dynamic equilibrium (key process).

Thus, the UTC of the microscope development team has six levels, which provides the opportunity for the market to create products to meet the requirements of customers, including microscopes, optical systems, coordinate tables, aeromagnetic guides, as well as to develop the systems for mathematical processing of optical signals and to control aeromagnetic systems.

Varabin et al. (2018) supply another example of identifying key competencies.

Let us note that the central feature is a key characteristic, expressed quantitatively or qualitatively, that must be achieved, which is the criterion of technological leadership and product competitiveness.

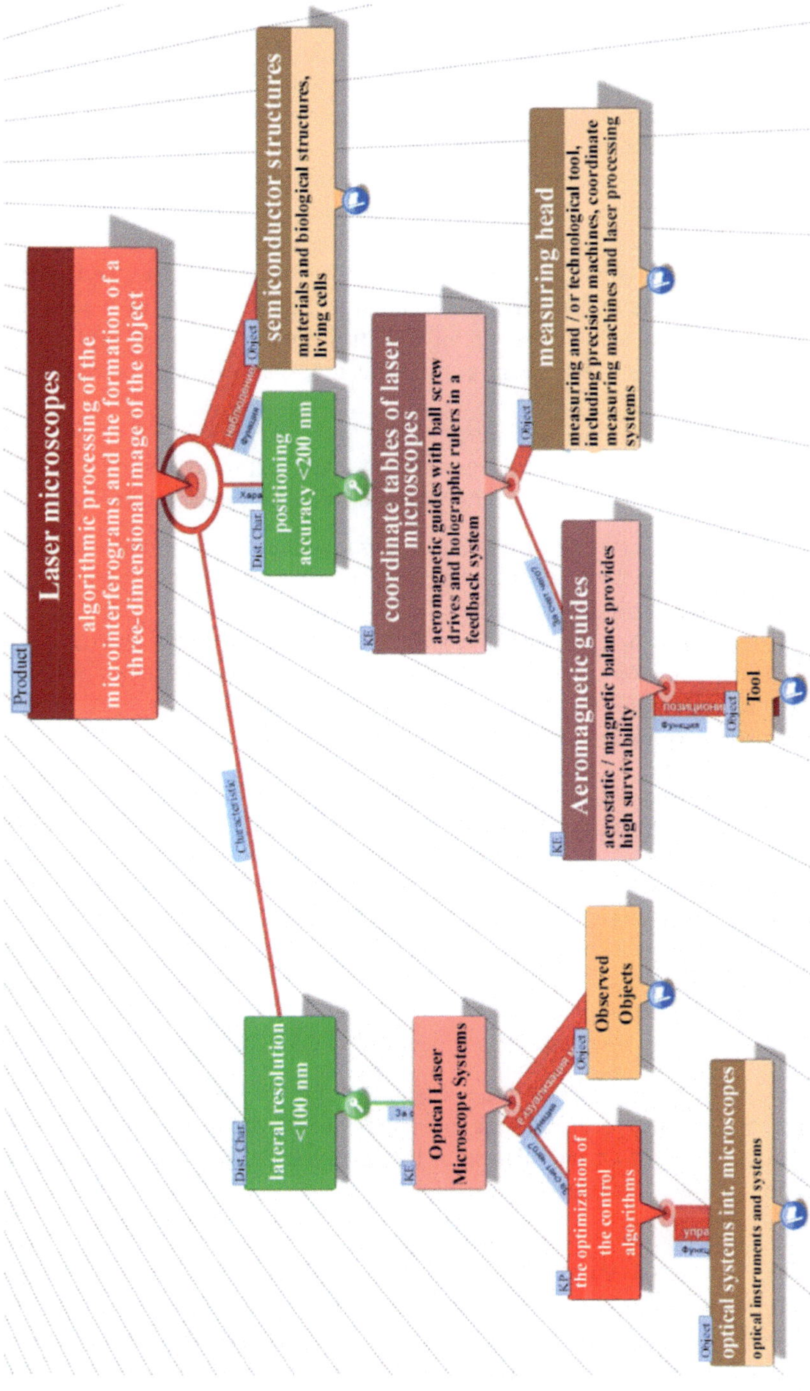

Fig. 2. The example of the structure of the UTC UOMP JSC.

A key characteristic allows us to identify and determine whether a specific technological competence is unique or common. A key characteristic reflects the competitiveness of a product or technology. At the same time, with the help of a key characteristic, we can set the fundamental requirements for the development of the product and its elements.

In a more general case, when setting the task of creating an innovative product, the groups of characteristics that can be conjugate (coordinated) are required when the actions to achieve one characteristic simultaneously lead to the achievement of another (other) key characteristic. Or, on the contrary, these actions lead to the appearance of contradictions. In the second case, the solution is achieved either by a compromise or by setting and then solving the optimal control problem with an explicitly defined optimality criterion and restrictions or by the methods of resolving contradictions using the TIPS tools.

Let's consider the role and place of the UTC in the process of developing an innovative product. The process of developing a new innovative product is a sequence of the following steps that are performed iteratively until reaching the goal:

- The conceptual synthesis (shaping the appearance of the product, including the determinations of the product that is being developed, goals, purpose, and principles of the product/technology);
- Formation of technical requirements (key functions and characteristics);
- Functional, structural synthesis (architecture, basic key elements, processes, and communications);
- Parametric synthesis (the formation of quantitative requirements for the elements and the formation of the component technical specification),
- The optimization (iterative redesign process to meet technical requirements and resolve contradictions).

At the same time, the developer solves design, engineering, and technological problems by meeting and overcoming the corresponding problems.

The solution of the specified problems is possible by means of attraction of necessary own and attracted UTC; at the emergence of inventive tasks, it is expedient to involve the TIPS experts.

This raises two important management tasks:

1. The search and attraction of the necessary external technical training centers or the formation of their technical training centers in the process of solving the problem;
2. The determination of the balance of the involved utilities in the production cooperation and the own used/created utilities.

It is clear that upon the successful solution of a new scientific and technical problem, the corresponding technical and technological complexes will be formed and can be used further as a competitive advantage of the company. At the same time, the formation of their UTCs may require significant time and material resources. Also, if there are available UTCs in the market, it is necessary to assess, taking into account the available time and other restrictions to solving the problem, the feasibility of attracting third-party UTCs.

In any case, an adequate and correct statement and formulation of the task for external or internal performers will be required, which requires some experience. For example, when transferring the competencies from other areas of science and technology, it will be necessary to translate the terms of one subject area into the required area.

Figure 3 schematically shows that the process of developing an innovative product includes a market idea and its technical implementation.

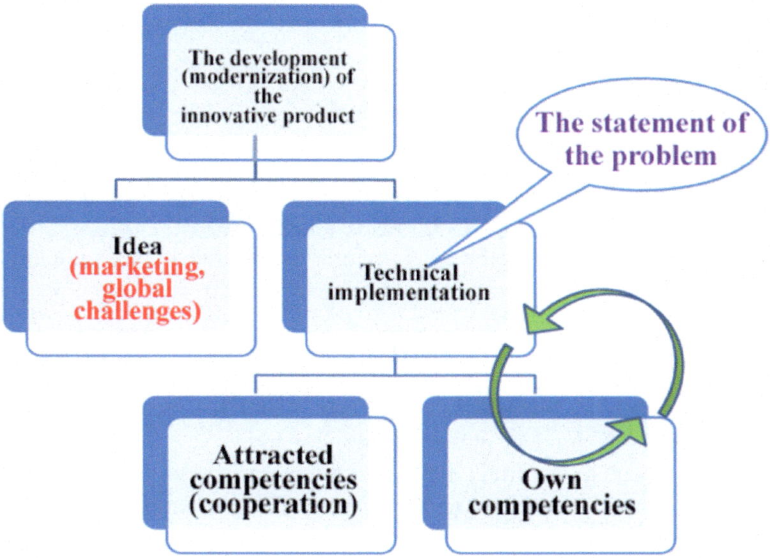

Fig. 3. The development of an innovative product includes the formation of a market idea and its technical implementation.

It should be noted that the root of breakthrough innovations lies in the generation of market ideas. Of the two well-known mechanisms for the formation of technological innovations, *Market Pull* and *Technology Push*, the first one accounts for 85% of successful businesses.

Jason Chen, Executive Director, President, and Chairman of the Board of Directors of Acer Inc., in an interview with Business FM on April 11, 2019, said, "The root of our innovation is that we first determine the market direction, and then we select necessary technology, not vice versa."

This suggests that the commercialization of the UTC is mainly a source of improving innovations, while the breakthrough innovations help in solving market challenges. The in-depth analysis of market needs and the vision of development trends are the basis of breakthrough innovation.

At the same time, the task of choosing a key characteristic, consumer properties of the product, and its quantitative (in the particular case, qualitative) expression as a criterion of breakthrough innovation remains. For example, the Tesla electric car has been changing the car market since 2009 by providing mileage without recharging, acceptable for everyday use (350 km). This key feature has been decomposed into the system requirements and resolved on its own and affected UTC.

3 Discussion

Mathematically, the statement of the problem can be described as follows. The technical quality of the product $\vec{Q}$, which is an integral part of its competitiveness, can be expressed as a vector (tuple)

$\vec{Q} = (\vec{K}, \vec{H})$, including:
$\vec{K}$ – the vector of product key features, and
$\vec{H}$ – the vector of ordinary technical characteristics of the product.

The matrix ***K*** [***N*M***] of competitive offers in the market can be composed by comparing the **N** key characteristics of **M** competing products.

Then the condition for the product superiority is to achieve the extremum of the ith key characteristic:

$$L_i = \max_{j=1,N} k_{ij},$$

where L_i – the superior competitor ith key feature, and k_{ij} – the values of the ith key characteristic of the jth product presented on the market.

The achievement of a certain level of key characteristic, including its superior value, is a function of the technological competencies forming it:

$$k_i = \mathrm{f}(\vec{C}),$$

where $\vec{C}$ – vector of technological competencies of the product development team, including both their unique technological competencies $\vec{U}$, and the general scientific and engineering competencies $\vec{R}$:

$$\vec{C} = (\vec{U} + \vec{R}).$$

Each **t** competency can be described as a function g_t:

$c_t = g_t(a_t, p_t, k_t)$, where
a_t – the action,
p_t – the object,
k_t – the quantified key characteristic

Then the task of the diversification of the UTC is to search for an object $p_r \in P$ for which the condition **T** (technical and technological feasibility) and **E** (economic feasibility) are satisfied.

P – is a set of relevant objects of action application.

The task of scouting (searching for technologies and competencies) mathematically reduces to finding those $a_r \in A$ (a set of suitable actions) for which the required key characteristic is achievable concerning a given object k_0.

In general, a classical optimal control problem can be formulated with a given optimality criterion and constraints.

4 Conclusion

In conclusion, we would like to note that market leadership is achieved by developing and bringing globally competitive products and technologies that have superior key characteristics to the market. The creation of such products is provided based on setting ambitious goals that meet global challenges and is accompanied by the formation of the development teams, which form the basis for the sustainable superiority of new products and technologies.

Also, one of the main mechanisms for accelerating innovative development is the unique technological (technical) competence of scientific and technical teams of the level of global superiority, which form the potential "growth points" of the entire high-tech industry and the national economy as a whole. Depending on the market situation, the company must make a choice and maintain a balance between attracting external innovations and developing its competencies. In critical areas, it is advisable to develop their key competencies.

Acknowledgment. This article is supported by the Program "5-100" of the People's Friendship University of Russia aimed at the creation of centers of the national technological initiative.

References

Chemezov, S.V., Volobuyev, N.A., Koptev, Yu.N., Kashirin, A.I.: Diversification, competences, problems, and tasks. New Oppor. Innov. **4**(222), 3–27 (2017)

Chursin, A.A., Kashirin, A.I., Strenalyuk, V.V., Semenov, A.S., Ostrovskaya, A.A., Kokuytseva, T.V.: The approach to detection and application of the company's technological competences to form a business model. In: IOP Conference Series: Materials Science and Engineering, vol. 312, no. 1 (2018)

Ereshko, A.F., Kokuytseva, T.V.: Computable models of the cooperation of digital economies. In: Proceedings of 2017: 10th International Conference Management of Large-Scale System Development. IEEE, Moscow (2017)

Kashirin, A.I., Semenov, A.S., Strenalyuk, V.V., Filippov, P.G., Ostrovskaya, A.A., Kokuitseva, T.V.: Competence Management in Corporate Practice (Monograph). People's Friendship University, Moscow (2018)

Kashirin, A.I., Semenov, A.S., Strenalyuk, V.V., Ostrovskaya, A.A., Kokuytseva, T.V.: The development and management of core competencies in a corporation: global experience. Int. Res. J. Part 1 **8**(50), 44–49 (2016)

Kiselev, V.N., Degtereva, E.A., Zobov, A.M., Moseykin, Y.N., Kokuytseva, T.V.: Risks and restrictions limiting Russia's participation in international scientific/technical cooperation: a study of characteristics. Int. J. Mech. Eng. Technol. **9**(11), 1629–1636 (2018)

Strenalyuk, V.V.: Creative unique technological competencies of scientific and technical teams of high-tech enterprises – the basis of innovative development of the industry. Drukerovskij Vestnik **6**, 154–160 (2018a)

Strenalyuk, V.V.: The unification of the unique technological competencies of enterprises in Russia and the Arab countries of the Middle East is a new direction in innovative international development. Russia Asia Electron. Sci. J. **3**(4), 55–64 (2018b)

Varabin, D.A., Zaiko, I.N., Kashirin, A.I., Strenalyuk, V.V.: The experience of identifying and commercializing unique technological competence in JSC "VNII Signal". Innovations **12**(242), 11–14 (2018)

Advanced Technologies, Application, Best Practices

Artificial Intelligence in the Training and Decision System in Ophthalmology

Valery V. Bakutkin[1(✉)], Ilya V. Bakutkin[2], and Vladimir A. Zelenov[2]

[1] Saratov SRI SG Rospotrebnadzor, Saratov, Russia
bakutv@bk.ru
[2] MAKAO, Saratov, Russia
Bakutkiniv@mail.ru, zelenov88@yandex.ru

Abstract. The article presents the results of the development of hardware and software systems based on artificial intelligence in the training and decision-making system in ophthalmology. The simulation robot is designed to simulate medical situations, based on which the skills of diagnosis, clinical decision-making, and the implementation of practical techniques of ophthalmoscopy are developed. The main feature of artificial intelligence-based software is the ability to use it in a system of lifelong learning and medical decision-making, primarily in case of the most significant ophthalmological diseases that lead to a decreased vision, disability because of blindness.

Keywords: Artificial intelligence · Ophthalmology · Neural networks · Training · Decision making

1 Introduction

The most priority direction in the development of ophthalmology is the development of digital technologies, primarily with the use of neural networks, artificial intelligence (Gonzalez and Woods 2007). Currently, the Unified State Health Information System of the USHIS is functioning (Decree of the Government of the Russian Federation of May 5, 2018 No. 555 "On the Unified State Information System in the Field of Health" DG of the RF No. 555, Regulation on USHIS). The main areas of focus are telemedicine, smart medical devices, and data statistics. Particular importance is given to the development of artificial intelligence since only with its use becomes possible to analyze a much larger number of parameters and give accurate recommendations. This will make it possible to make diagnoses and give recommendations for treatment based on objective information and not just relying on the subjective opinion of a doctor. The implementation of electronic medical records and devices for remote monitoring of patients is possible only with BigData (Goodfellow et al. 2016).

The use of artificial intelligence in Ophthalmology has excellent prospects, in particular, in the process of training, acquiring practical skills, and making medical decisions (Gribova and Petryaeva 2013; Svistunov 2015). Artificial intelligence is used in high-level hardware and software systems and training robots. There are research and experience using the complex of training simulators with virtual reality, which is in

A. V. Bogoviz and Y. V. Ragulina (Eds.): ISCI 2019, LNNS 115, pp. 429–435, 2020.
https://doi.org/10.1007/978-3-030-40749-0_51

the foundation of the IACPaaS5 platform (Gribova 2011). Information components are the following: virtual environment ontology, a model of a training simulator, multimedia data, and external functions (Fedorishchev 2012). An interactive model with user control is used to create training simulators. The EYESI virtual simulation system is designed for practicing skills in intraocular microsurgery and refining ophthalmoscopy skills (VRmagic, Germany) (Slonimsky and Obrubov 2015). The surgical simulator is a medical training system in the field of eye surgery. The purpose of the simulator is to teach the methodology and sequence of movements and actions of the surgeon in preparation for surgery on the human eye. The virtual system is equipped with a touch screen, two operational interface heads (mask with a model of the eye), a set of tools (tips) for the operation, a stereomicroscope, and a multifunctional foot pedal. Many aspects of the use of artificial intelligence in the educational process and the practical work of an ophthalmologist are actively developing.

2 Materials and Method

The study aimed to create hardware-software systems based on artificial intelligence in the training and decision-making system in ophthalmology.

Ophthalmoscopy is a research method that allows one to examine the inner shells of the eye with the help of specialized optical devices. This technique is the basis for the diagnosis of eye diseases. Fundus examination is performed in all patients, including newborns. This is because retinopathy is possible in premature infants, which can lead to blindness. There are anatomical features of the eye of the newborn, namely, significantly smaller sizes, significant differences in the structures of the retina, macular region, and optic disc. (Bakutkin and Gorelov, 2015). Eye diseases (such as retinal dystrophy, retinal detachment, diabetic retinopathy, glaucoma, optic atrophy, tumors, congenital diseases, and many others) can only be diagnosed by examining the fundus using ophthalmoscopy (Tereshchenkova 2010). The authors have developed a hardware-software complex with an electronic-optical system of its own design, integrated into a learning robot of high realism. It corresponds to the size of the newborn and provides the most approximate conditions for ophthalmoscopy—possible imitation of the child's motor activity, which is created by the software. The possibility of generating real digital images of the fundus obtained during the examination of infants with various pathological conditions is implemented. The database for ophthalmoscopy is available as a cloud storage option on the Internet (Bakutkin and Chichev 2017). The software part for teaching ophthalmoscopy in infants consists of the following components: creating a user's personal account, databases on types and diseases of the eyes in infants, test files related to the educational process, test tasks with the possibility of self-monitoring, an operating system with an interface, the possibility of constantly updating the training database of images, including via the Internet, medical decision-making software according to the standards of diagnosis and treatment (Avetisov 2008), the possibility of current and final control to determine the level of knowledge. It is possible to use both standard programs and individual ones, relying on the students' level of knowledge. The teacher determines the level of difficulty of the tasks.

The program of making medical decisions is based on comparing patient data with similar data in the archive and selecting the closest analogs.

The development of this project was the creation of a hardware-software complex for training in fundus examination and medical decision-making in adult patients. This modification has a different constructive solution, and the conditions for ophthalmoscopy correspond to the real ones in detail, color reproduction, and image sizes. The regularly updated fundus image database complies with international standards and classifications. To create a software part, data from real patients with descriptions of the clinical picture and high-quality fundus images were selected. All digital images of the fundus are obtained on standardized equipment that has certificates of conformity. Experts with international certificates and experience in research with degrees in Medical Sciences were involved in the diagnosis of diseases. Clinical cases were included in the training base only based on an independent opinion of three experts examining clinical information. We have implemented a methodology for identifying pathological zones for subsequent analysis and use in the learning process, monitoring the medical decision-making. This allows one to highlight the areas that are most important for diagnosis, optimize the data processing. Zones can be located in any part of the fundus, but we suggest using a localization connection with the main parts of the fundus, such as macular area, optic disc, papillo-macular bundle, and square distribution on the retinal periphery. Figure 1 shows the allocation of zones with the most significant changes for diagnosis.

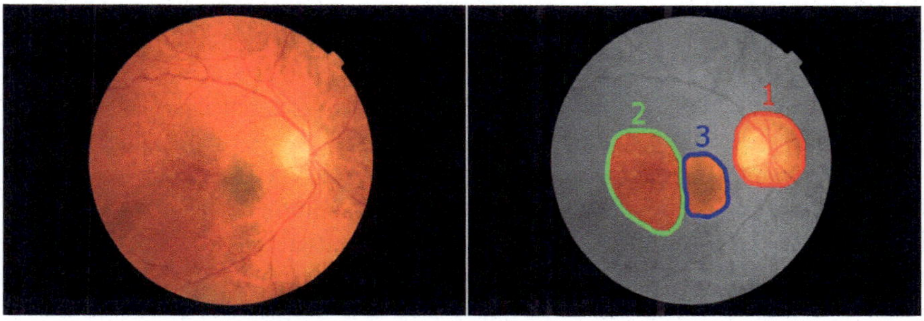

Fig. 1. The allocation of zones with the most significant changes for diagnosis.

Zones are highlighted in borders with various colors. The most significant are highlighted in red, auxiliary ones are highlighted in blue, and related changes are highlighted in green.

In addition to analyzing digital images, a standard description of the condition (which is used in case histories) is introduced. This is necessary to match the graphic and descriptive parts of the program (Fig. 2).

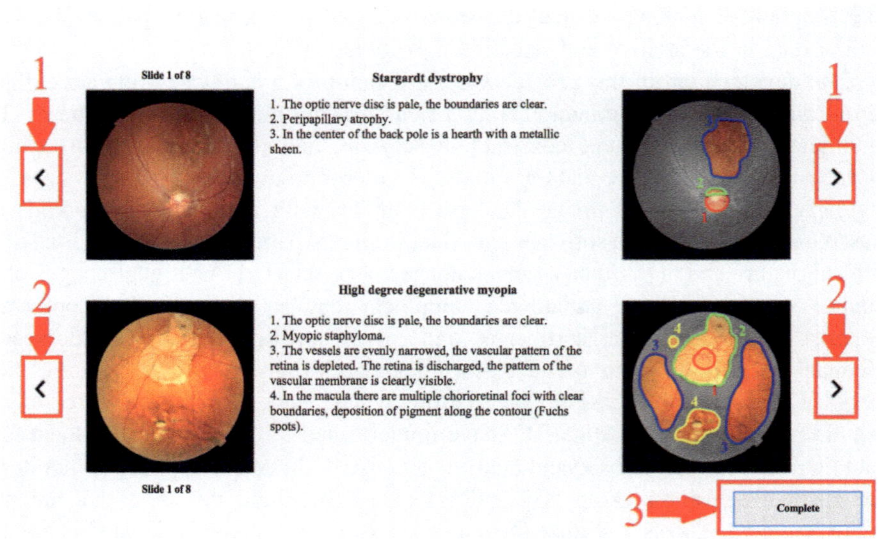

Fig. 2. The description and digital images of the fundus with highlighted diagnostic areas.

3 Results

The simulation robot is designed to simulate medical situations, based on which the development of diagnostic skills, clinical decision-making, the implementation of practical techniques occur by both an individual doctor and a medical team. Management of all actions of the robot, as well as monitoring and analysis of the actions of trained doctors, is carried out using an external computer. A robot dummy is used to objectively evaluate the performance of a doctor's clinical actions since the outcome of treatment entirely depends on the chosen tactics. Monitoring with archiving in a personal account is carried out to control training. The program controls both the time of studying each section and the amount of training. The learner and teacher have a clear picture of what was learned, how much time was spent, which test results were obtained. Figure 3 shows a variant of testing current control in real clinical conditions.

Training and control of acquired knowledge take place in each section, including at the end of the daily and final classes. In each clinical section, there are both standard cases, which are the most common, and in combination with other clinical conditions. In each section, there are options for clinical cases that are selected according to the principle of random generation. The transition to the next section or more complex clinical cases occurs only with 90% of the correct answers in the previous section. An example of testing is presented in Fig. 4.

The main difference of this project for the development of hardware and software systems based on artificial intelligence in the training and decision-making system in ophthalmology is the possibility of a continuous learning process. There is a stage of primary education, which allows one to master practical skills, acquire an initial clinical

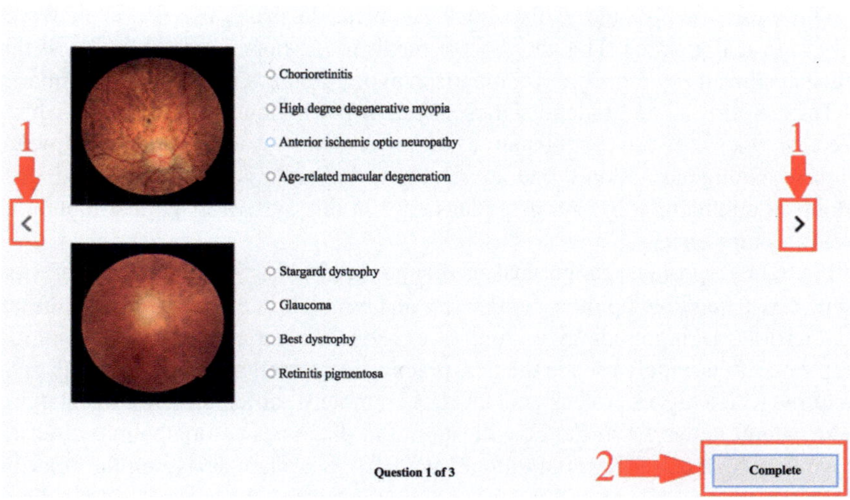

Fig. 3. A variant of testing control in real clinical conditions.

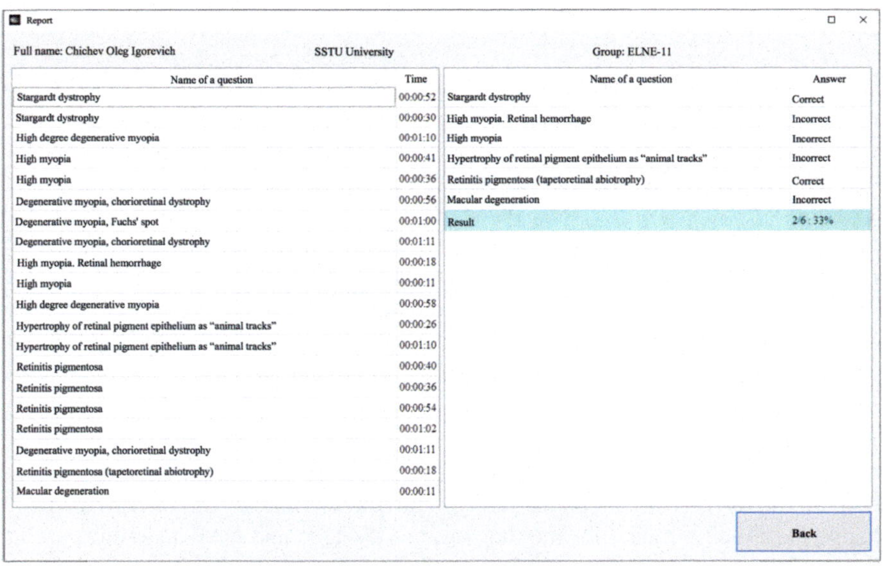
Report

Full name: Chichev Oleg Igorevich SSTU University Group: ELNE-11

Name of a question	Time
Stargardt dystrophy	00:00:52
Stargardt dystrophy	00:00:30
High degree degenerative myopia	00:01:10
High myopia	00:00:41
High myopia	00:00:36
Degenerative myopia, chorioretinal dystrophy	00:00:56
Degenerative myopia, Fuchs' spot	00:01:00
Degenerative myopia, chorioretinal dystrophy	00:01:11
High myopia. Retinal hemorrhage	00:00:18
High myopia	00:00:11
High degree degenerative myopia	00:00:58
Hypertrophy of retinal pigment epithelium as "animal tracks"	00:00:26
Hypertrophy of retinal pigment epithelium as "animal tracks"	00:01:10
Retinitis pigmentosa	00:00:40
Retinitis pigmentosa	00:00:36
Retinitis pigmentosa	00:00:54
Retinitis pigmentosa	00:01:02
Degenerative myopia, chorioretinal dystrophy	00:01:11
Retinitis pigmentosa (tapetoretinal abiotrophy)	00:00:18
Macular degeneration	00:00:11

Name of a question	Answer
Stargardt dystrophy	Correct
High myopia. Retinal hemorrhage	Incorrect
High myopia	Incorrect
Hypertrophy of retinal pigment epithelium as "animal tracks"	Incorrect
Retinitis pigmentosa (tapetoretinal abiotrophy)	Correct
Macular degeneration	Incorrect
Result	2/6: 33%

Back

Fig. 4. Testing in real clinical conditions.

level of knowledge. The next step is to improve qualifications and study more complex clinical cases. Initial access to the training database is subsequently maintained through a personal account.

This makes it possible to expand the clinical database, use neural networks for analysis in real practice. The adoption of medical decisions is possible only if there is sufficient clinical experience and comparison of performance with previous similar cases.

Testing allows one to identify those areas of professional knowledge in which this specialist does not have sufficient experience. The nosological block structure of artificial intelligence allows one to structure the process of training and decision making in ophthalmology. An essential factor in this system is the distribution of the level of competence.

If a trained specialist cannot make a diagnosis, s/he may seek the advice of a teacher and receive a qualified opinion. In practice, this principle is maintained in the discussion and decision-making mode by a group of experts. The variant of consilium solution of complex cases is widely used in medical practice. An expert group in a particular clinical discipline with a higher professional level can remotely participate in a consultation. The use of neural networks in the classification and diagnosis of ophthalmic diseases has shown very high efficiency (Bakutkin et al. 2013). So, when analyzing the colorimetric parameters of the optic nerve head using neural networks, the accuracy of the diagnosis of glaucoma genesis atrophy was 85–90% (Bakutkin 2008).

4 Discussion

The development of distance and virtual medical education based on artificial intelligence is one of the tasks of medicine. The international requirements of the Society for Simulation in Healthcare are reflected in the Order of the Ministry of Health of the Russian Federation No. 334n dated June 2, 2016 *On Approval of the Regulation on the Accreditation of Specialists*. The mandatory use of simulation training methods for programs of secondary, higher, and postgraduate continuing medical education for practicing and objective assessment of practical skills is legislatively fixed. According to existing requirements, technologies for the interaction of several virtual models with each other, with medical equipment, pharmaceuticals, and the external environment, should be used. Complex, interdisciplinary, or non-standard clinical situations, skills in a complex, realistic environment are being worked out. The development of virtual medical education is possible only based on artificial intelligence since traditional diagnostic methods largely depend on the professional experience and knowledge of doctors. The integration of ophthalmology and artificial intelligence can fundamentally change the existing model for the diagnosis of diseases and have a significant clinical impact. The use of hardware-software systems based on artificial intelligence in the system of training and decision-making in ophthalmology in several higher educational institutions indicates the high efficiency of this methodology.

5 Conclusion

The development of programs based on artificial intelligence in the training and decision-making system in ophthalmology provides several advantages: using approved requirements, taking into account domestic and foreign experience, and the ability to quickly modify the course of study, creating individual training courses. Also, the ability to assess the degree of preparedness of a specialist when creating clinical cases of varying complexity arises. This knowledge assessment is independent and objective, which allows corrective training based on the results. The main feature is the possibility of using in the system of continuous training and further use in making medical decisions, primarily in case of the most significant ophthalmological diseases, leading to decreased vision, disability because of blindness.

Acknowledgment. The publication was prepared in accordance with the Agreement with the Russian Foundation for Basic Research № 18-29-02008 "Intelligent Laser System for Eye Surgery." The authors are grateful to the staff of the Center for Continuing Professional Education of I. M. Sechenov First Moscow State Medical University (Sechenov University) and VIRTUMED for help and recommendations.

References

Arzhimatova, G.Sh., Slonimsky, A.Yu., Obrubov, A.S., Samoilenko, A.I.: Possibilities of the virtual simulator EYESI in the system of training and advanced training of ophthalmologists. In: IV Congress of ROSOMED-2015 in the Framework of the International Conference on Innovative Educational Technologies in Medicine. City Clinical Hospital named after S. P. Botkin, Moscow, Russia (2015). https://rosomed.ru/theses/187

Avetisov, S.E. (ed.): Ophthalmology: National Guide. GEOTAR-Media, Moscow (2008)

Bakutkin, V.V., Chichev, O.I.: Virtual technologies for teaching ophthalmoscopy in infants. Virtual Technol. Med. **2**, 12–15 (2017)

Bakutkin, V.V., Bakutkin, I.V., Spirin, V.F., Zayko, Y.: Computer analysis of images of the optic disc in clinical practice. Vestnik Orenburg State Univ. **4**(153), 29–30 (2013)

Bakutkin, V.V., Zayko, Y.N., Lepesko, A.S., Skidanov, A.N.: Development of a system for classification and diagnostics of ophthalmologic diseases based on artificial neuron networks. Proc. Saratov Univ. Ser. Phys. **8**(1), 36–41 (2008)

Bakutkin, V., Gorelov, V., Zayko, Y.: Statistical analysis of the parameters of the optic disk image. J. Biomed. Photonics Eng. **1**(4) (2015). https://doi.org/10.18287/jbpe-2015-1-4-270

Fedorishchev, L.A.: Internet complex for creating learning systems with virtual reality. Distance Virtual Learn. **7**, 4–12 (2012)

Gonzalez, R., Woods, R.: Digital Image Processing. Prentice-Hall, New Jersey (2007)

Goodfellow, I., Bengio, Y., Courville, A.: Deep Learning. MIT Press, Cambridge (2016)

Gribova, V.V., Petryaeva, M.V.: Virtual reality computer training simulator for ophthalmology. Otkrytoye Obrazovaniye **6**, 43–48 (2013)

Gribova, V.V., Kleschev, A.S., Krylov, D.A.: The IACPaaS project is a developed complex for the development, management and use of intelligent systems. Artif. Intell. Decis. Mak. **1**, 34–40 (2011)

Svistunov, A.A.: Certification using simulation. Virtual Technol. Med. **1**(13), 10–12 (2015)

Tereshchenkova, M.S.: Laser treatment tactics for early manifestations of posterior aggressive retinopathy of prematurity using transpupillary contact laser retinal coagulation, Dissertation of Candidate of Medical Sciences. S. N. Fedorov NMRC "MNTK "Eye Microsurgery"", Moscow, Russia (2010)

Intelligent Monitoring as the Basis for Improving the Competitiveness of High-Tech Production

Andrey I. Boginsky[1(✉)], Lidia S. Zelentsova[2], and Alexey I. Tikhonov[3]

[1] Holding "Helicopters of Russia", Moscow, Russia
info@rhc.aero
[2] State University of Management, Moscow, Russia
zelentsovals@yandex.ru
[3] Moscow Aircraft Institute, Moscow, Russia
mai512hr@mail.ru

Abstract. The article considers the problem of increasing the competitiveness of high-tech industries through the active introduction of digital technologies and artificial intelligence in the management cycle and at all stages of the product life cycle. The purpose of the study is to justify the mechanism for integrating the latest online monitoring platforms into existing information and communication systems. New functions of intelligent monitoring (integrative, coordinating, regulatory, information and communication, adaptive and prospective) are defined, and their contents are disclosed. A model with a built-in mechanism for integrating the organization's information infrastructure onto a new platform for intelligent monitoring (IM) is proposed. The main following intelligent monitoring modules are highlighted. The originality of the authors' approach consists in providing the concept of minimally costly intellectualization of monitoring production systems based on the creation of a mechanism for integrating it into the existing information infrastructure, in defining its functions and their content based on the knowledge necessary to generate new business models and increase its competitiveness.

Keywords: Intelligent monitoring · Competitiveness · Integration mechanism · Production system · Value system component · Digital transformation

1 Introduction

The relevance of the study is due to the fact that the most essential task of the development of domestic high-tech industries (HTIs) is to increase their competitive advantages based on the following: (1) taking into account the influence of external and internal factors on the competitiveness of the designed products, which would allow to avoid design errors and thereby reduce unjustified costs; (2) establishing more advanced technical parameters of the designed products, new requirements for technological processes and the level of production excellence; (3) improving quality

A. V. Bogoviz and Y. V. Ragulina (Eds.): ISCI 2019, LNNS 115, pp. 436–443, 2020.
https://doi.org/10.1007/978-3-030-40749-0_52

parameters earlier and more accurately than competitors, providing a certain competitive advantage and faster entry into the market (Baimuratov et al. 2011). So, the practice shows that an increase in investments in improving quality by 2% at the design stage gives a 20% increase in profit from the sale of finished products.

An important role in increasing the competitiveness of high-tech industries belongs to information and communication technologies and monitoring based on digital and intelligent technologies.

Traditionally, monitoring is seen as a means of information and analytical support for management. The development of a knowledge-based economy has expanded the functional field of monitoring already as a means of acquiring new knowledge about managed objects and processes (Isermann 1993). An analysis of relevant research (Narendra 1996; Patton et al. 2000) made it possible to identify such monitoring functions as cognitive, research, methodological, diagnostic, corrective, and motivational. The use of end-to-end intelligent technologies in monitoring expands its capabilities to ensure the competitiveness of high-tech products and the organization producing them already at the design stage but is associated with a number of problems (Zhernakov 2006; Ryzhov 2008). One of the most critical problems is the difficulty of introducing new digital platforms into the existing information infrastructure. In this regard, the article presents a model for transferring an organization's IT infrastructure to a new platform (combination with intelligent monitoring) with an integrated integration mechanism, as well as a transition scheme. The practical application of the development will allow with high economic feasibility to move to a new level of competitiveness of the HTIs.

2 Materials and Method

In the context of digital transformation and the use of artificial intelligence, opportunities for implementing a fundamentally new approach to ensuring the competitiveness of HTIs at various stages of the life cycle of products appear, including at the stage of outline design. The realization of the potential of digital technologies is possible provided that the life-cycle management system of high-tech products is transformed, first of all, by creating effective mechanisms for the development and production of competitive products (Baimuratov et al. 2011).

The concept of virtual development, production, and operation of the product, i.e., preliminary modeling of these processes on a computer and obtaining virtual design, technology, and operation, allows for a quantitative preventive assessment of product competitiveness.

The main purpose of intelligent monitoring is to receive, store, and process information from the external and internal environment; in the regulation and coordination of the entire set of elements of production systems. In the course of the work, the authors identified the following functions of intelligent monitoring, which directly provide sustainable competitiveness of a high-tech organization:

- Integrative is a combination of various components of the PS into a single whole and ensuring their interconnected and coordinated functioning;
- Coordinating is maintaining a constant internal environment, ensuring a coordinated flow of functions of various elements of the PS, successfully adapting them to changing environmental conditions;
- Regulatory is the provision of the ability to restore and maintain an internal stability as a result of reactions that compensate for the impact of external influences;
- Information and communication imply implementing the regulation of the intensity of exchange along highways, every second transmitting a huge amount of information. Information is generated in various sensors and special centers of occurrence and processing, then it enters a control center, where the generation of ideas about the external and internal state of the production system occurs;
- Adaptation is the restructuring of the activities of various elements of the production system in accordance with environmental changes based on the analysis and synthesis of incoming in specialized sensors of various information.
- Prospective is about forecasting the future behavior of objects and entities based on current and retrospective facts, identifying risks and opportunities for predicting future events.

The implementation of functions (IM) is carried out on the basis of the integration of complexes of common information models and the use of specialized software.

3 Results

Intelligent monitoring is a basic element of the information and communication system of high-tech organizations, which includes the following elements: diagnostic complexes, databases (DB) and knowledge bases (KB), expert systems (ES), modeling systems, business analytics and visualization platforms (Fig. 1).

The presence of digital platforms provides not only analytical processing of a huge amount of data but also building forecasts and models of the future behavior of the production system and its individual elements, necessary for making informed management decisions (MD) in real-time.

The main tasks of the module "*Environmental Monitoring*" include the following:

- Assessing the potential demand for high-tech products;
- Requirements of potential customers for HTIs (new competencies and characteristics of HTIs);
- Analysis of competitors and their products;
- Assessment of a company's competitive position in the market.

As indicated in (Baimuratov et al. 2011), evaluating the competitiveness of high-tech products is a prerequisite in the organization's process of bringing it to market. At the exit from this module, information for designers regarding competitive technical and economic characteristics (parameters) of the future model of HTIs is formed.

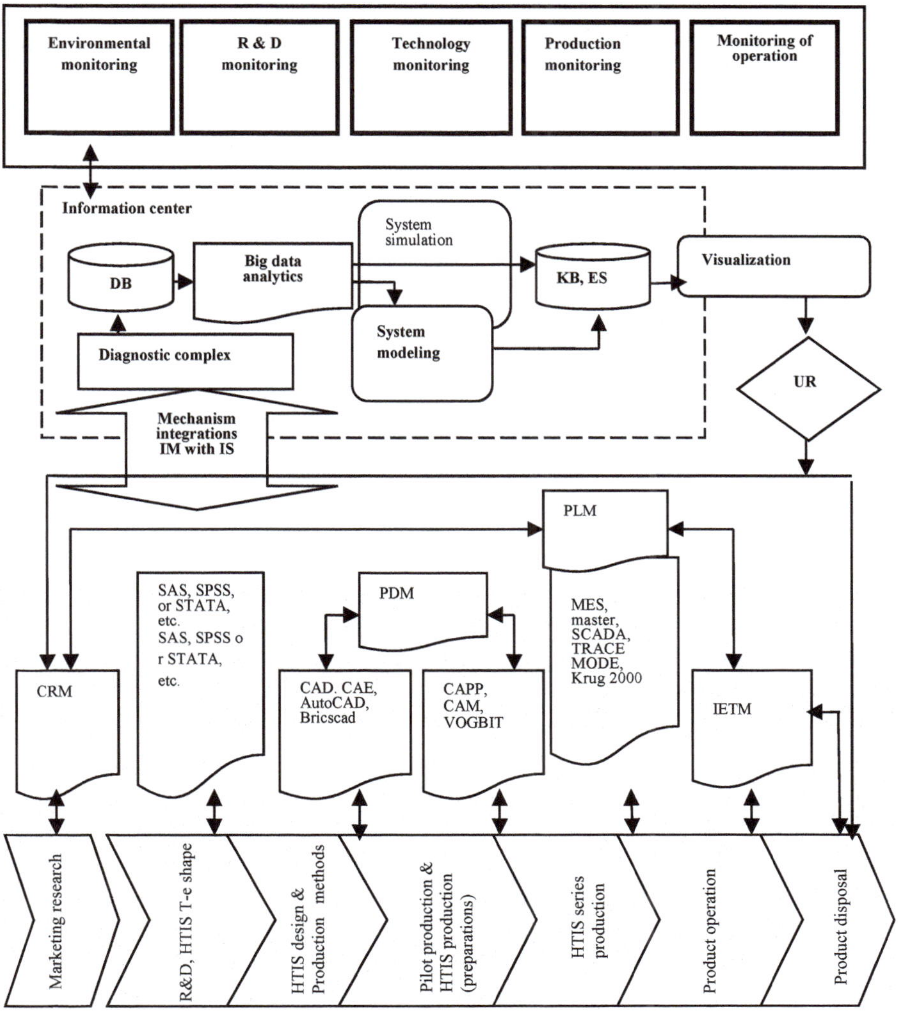

Fig. 1. Model for transferring an organization's IT infrastructure to a new platform (combination with intelligent monitoring). *Source*: Developed by the authors.

The module "*R&D Monitoring,*" from the point of view of determining the competitiveness parameters of the future model of HTIs, should fulfill, first of all, the tasks associated with the following:

- By tracking information flows about the trends in the technological development of HTIs, identifying problems that need urgent solutions, and finding possible solutions to determine the paths of developing processes in various fields of knowledge;
- Accompanied by complex products at all stages of the life cycle, up to disposal;

- Assessment of the current state of the problem based on a comparison of information obtained from the external environment, and assessment of the capabilities of a particular organization to create a competitive HTIs (Ryzhov 2008);
- Predicting and modeling the technical development and competitiveness trends of the HTIs and forming on this basis the technical and economic appearance of the new model;
- R&D planning and monitoring;
- The search for ways to reduce the complexity of R&D and, accordingly, reduce their cost, which would significantly affect the speed of launching new products on the market, in particular, the use of digital and additive technologies, as well as methods of interactive design.

The *"Technological Monitoring"* module is designed to timely identify technological development opportunities and, accordingly, achieve competitive advantages based on the following:

- Comparative analysis of our own technological equipment with equipment from competitors;
- Tracking development trends and opportunities for the emergence of new technologies that can provide a breakthrough in competitiveness;
- Obtaining and accumulating the initial information necessary for developing technological development strategies and modeling their implementation;
- Development of recommendations to reduce external and internal risks when introducing new technologies (Efimushkin 2008; Tsoukas and Shepherd 2004).

The intelligent module *"Production Monitoring"* is possible if all production facilities, namely equipment and workplaces, are combined into a single information network. Thus, the opportunity to use IoT, namely the industrial Internet of things, is provided. It is enough to mount sensors with network access into the equipment, and the conditions for remote monitoring of production sites, timely routine maintenance, prediction of accidents and scheduled preventive repairs, etc. are created for manufacturers and customers. Thus, the entire chain from suppliers to consumers will be carried out without human intervention. A variation of this intelligent module is "Equipment Monitoring," which performs the following functions:

- Equipment load control;
- Fixation and typology of equipment downtime;
- Inspection of labor discipline of staff;
- Warning of emergencies requiring a quick response, messages by e-mail or notifications of managers on a working computer.
- Dynamic analytics for the rapid assessment of the situation, key performance indicators (KPIs) of the organization, its units, individual units of equipment and operators:
- Visualization of analytics results and report generation with a visual display of time losses and causes of equipment downtime, etc.

As a result of the performance of these functions, bottlenecks are identified in a timely manner, time losses and the causes of their occurrence are estimated, equipment loading is optimized, working time reserves are created, etc. The module can be combined or integrated into the production monitoring module.

In terms of increasing the competitiveness of products and their organization, the module *"Monitoring the HTIS operation"* is gaining high importance, because with the help of data mining methods, Data Mining, the organization not only assumes responsibility for operational efficiency but also ensures lower operating costs and operational reliability due to continuous monitoring and diagnostics of the technical condition of the object. Timely detection of defects in the operation of the facility and their exclusion becomes a decisive factor in improving the efficiency of operation and competitiveness of the organization.

The integration of intelligent monitoring with information systems existing in high-tech production does not contain anything critical, in fact. On the contrary, modern monitoring systems of the MDC/MDA class (Machine Data Collection/Machine Data Acquisition) make it possible to improve production without significant investment. In the course of the study, a scheme of transition to a higher level of competitiveness of HTIs based on intelligent monitoring was developed (Fig. 2).

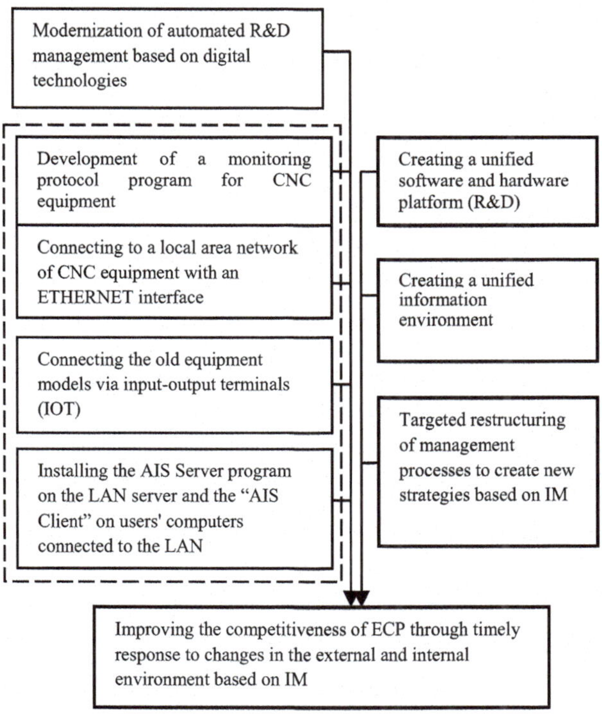

Fig. 2. The scheme of transition to integrated intelligent monitoring in the integrated information environment of the HTIs.

At the same time, the tasks not only to increase production efficiency but also a number of related problems are simultaneously being addressed. Domestic developments are interesting, for example, such as the automated information system AIS "Dispatcher" or "VOGBIT," they contain modules and functions for performing the following operations: managing the production process, warehouse, supply, collecting factual information from real production, building complete statistics on orders, plots, workers, etc., to quickly calculate the planned cost, to speed up the launch of new products, to track the history of each unit released or products with their unique number, and much more. The integration of such systems with ERP, PDM, EAM, etc. implies synchronization of current parameter changes, in order to ensure the relevance of normative and reference and other operational information, eliminating errors and ambiguity of generated reports. An analysis of the compatibility of software products showed the feasibility of the proposed developments and their practical significance.

4 Discussion

The scientific novelty of the authors' research consists in substantiating the concept of minimally costly intellectualization of monitoring based on the creation of a model for translating the information infrastructure of production systems with a built-in mechanism for integrating it with MI, identifying its specific functions and filling them with new content, developing a transition scheme to the intelligent monitoring necessary to generate new business models and increase its competitiveness.

5 Conclusion

The article substantiates the relevance, scientific and practical significance of the problem of increasing the competitiveness of high-tech industries based on the integration of advanced technologies of intelligent monitoring in the existing information infrastructure of a high-tech production system.

The proposed concept of minimally costly intellectualization of monitoring made it possible, on the basis of analysis of the transformation of its functions and their contents, as well as the compatibility of software products that are on the market, to form a mechanism for integrating an information system with intelligent monitoring, built into the model for transferring IS to new platforms, as well as a step-by-step algorithm (scheme) of transition to IM in the integrated information environment of the HTIs.

The blocks of integrated intelligent monitoring presented at the stages of the life cycle, described in detail, provide a timely and preventive receipt of objective, reliable, relevant, predictive, and forming new knowledge information necessary for making managerial decisions that directly affect the competitiveness of high-tech industries.

Acknowledgment. The authors are grateful to the "Helicopters of Russia" Holding, the Peoples' Friendship University of Russia, and the State University of Management for the opportunity to participate in the research project "Methods of Optimizing the Economic Efficiency of a Complex Technical System: Design at a Given Cost and Competitiveness in the Market."

References

Baimuratov, W.B., Makarov, Y.N., Chursin, A.A.: Investment with Innovation: Synergy and Competitiveness of Mechanical Engineering. MACD Publishing House, Moscow (2011)
Efimushkin, I.V.: Expert information system of technological monitoring information society technologies. Spec. Issue T-Comm **7**, 174–177 (2008)
Isermann, R.: Fault diagnosis of machines via parameter estimation and knowledge processing. Automatica **29**, 815–836 (1993)
Narendra, K.S.: Neural networks for control: theory and practice. Proc. IEEE **84**(10), 1385–1405 (1996)
Patton, R.J., Frank, P.M., Clark, R.N.: Issues in Fault Diagnosis for Dynamic Systems. Springer, London (2000)
Ryzhov, A.P.: Information monitoring of complex processes: technological and mathematical foundations. Intell. Syst. **11**(1–4), 101–136 (2008)
Tsoukas, H., Shepherd, J.: Coping with the future: developing organizational foresightfulness. Introd. Futures **36**(2), 137–144 (2004)
Zhernakov, S.V.: Application of neural network technology to diagnose the technical condition of aircraft engines. Intell. Syst. Prod. **2**, 70–93 (2006)

Blockchain Technology in Management of the Energy Sector of the Russian Federation

Natalia G. Shchegoleva[1(✉)], Olga I. Terenteva[1], and Vladimir I. Khabarov[2]

[1] Lomonosov Moscow State University, Moscow, Russia
{shegoleva, terentieva}@spa.msu.ru
[2] Moscow Financial-Industrial University "Synergy", Moscow, Russia
vhabarov@mfpa.ru

Abstract. The research determines general directions of applying the distributed blockchain registry technology in the Russian energy sector management. Blockchain, as a method of modernization of the energy sector of the Russian Federation, is a subject of our research. Methodology: a comparative analysis of foreign practices of using blockchain technology and their generalization. Summary. The system of trading certificates for electricity based on blockchain technology will allow private investors to participate in the expansion of the renewable electricity market in Russia. The development of projects within this system will reduce transaction costs through decentralization, minimizing barriers to entry. The introduction of blockchain technology in the management of the Russian energy sector and the formation of a competitive environment for the emergence of new organizations in the market could make the Russian Federation a world leader not only in the production but also in the effective distribution and accounting of electricity.

Keywords: Energy sector · Blockchain technology · Digital Economy · Competitiveness · Energy certificates

1 Introduction

The fourth industrial revolution leads not only to the creation of new sectors and markets but also to the radical transformation of traditional industries (Chursin and Makarov 2015). So, today, significant changes resulting from the transition to a new technological structure are also observed in the electric power industry.

The Russian Federation, despite the significant advancement of the new technological agenda in the energy sector (specialized startups, approval of the Energy Strategy of Russia for the period until 2035) (Government of the Russian Federation 2014), the Forecast of Scientific and Technological Development of the Sectors of the Fuel and Energy Complex of Russia for the period up to 2035 (Ministry of Energy 2016a), the National Technology Initiative (Ministry of Energy 2016b), is just beginning to resolve issues related to the long-term priorities of sectoral technological policy, as well as optimal timing and mechanisms for the transition to a new energy structure.

A. V. Bogoviz and Y. V. Ragulina (Eds.): ISCI 2019, LNNS 115, pp. 444–449, 2020.
https://doi.org/10.1007/978-3-030-40749-0_53

The relevance of the research topic is because Russia, while having powerful energy resources, spends them inefficiently (Lobova et al. 2019; Popkova et al. 2019; Alekseev et al. 2019). Therefore, developed countries need to improve the efficiency of their energy sector. Additionally, meeting new requirements of consumers without a noticeable increase in prices is another challenge. However, there is a reason to believe that modern technologies of decentralized storage and data transfer are able to answer this challenge. They will help ensure the transition from analog to digital control methods in the electric power industry and support the transformation of consumer behavior patterns, as well as business practices of energy supply and service companies.

2 Foreign Practices in Using Blockchain Technology

Despite attempts to practically use blockchain technology to transform bus processes and actively discuss possible preferences from its implementation, this problem has not been sufficiently studied in the scientific literature. So, among the works of domestic and foreign researchers on the prospects of using blockchain technology in the energy sector, one could note (Tapscott and Tapscott 2016) and (Swan 2015). Also, studies on this topic by the world-famous consulting agencies are of interest (PWC 2016; Booth et al. 2016).

We want to note about the blockchain technology that this is a technological innovation, the development of which is currently being carried out by subjects of various segments of the economy (Shchegoleva and Terenteva 2018). Blockchain is a unique technology, which serves as a basis for conducting operations between equal participants acting without intermediaries, and its decentralized storage of information is used to reflect all data about operations (PWC 2016).

Assessing the objective capabilities of the Russian Federation to strengthen the competitiveness of its position in the global energy market, it is necessary to address the problem of introducing digital technologies into the energy industry of the Russian Federation. The state program "Digital Economy of the Russian Federation" implies the widespread introduction of digitalization both at the level of organizations and sectors of the economy. As for the energy sector of the Russian Federation, this program has laid the potential for the implementation of information technology in existing business processes. One of them is a blockchain technology, in our opinion, the implementation of which is necessary. It should be noted the importance of its application in order to record and control the supply of energy resources. However, in their distribution system, individuals should also be included, not only as consumers of energy but as their suppliers, if they have surpluses of it. The so-called prosumers are the consumers who produce and sell energy themselves. They can get certain advantages. Also, ordinary consumers of electricity and gas will receive specific preferences. In particular, the blockchain technology will reduce the cost of energy.

3 Results

For the first time, the blockchain technology began to be introduced in the energy market in Germany, which is associated with the specifics of this market, including the developed renewable energy production and the preservation of centralized structures in the field of energy trade. Blockchain technology is likely to transform the energy market towards decentralization, allowing consumers to purchase energy directly from producers or prosumers, rather than from central suppliers. In theory, blockchain could create an energy market where supply can flexibly respond to demand.

The Stadtwerke Energie Verbund (SEV), a consortium of eight local public utility companies in the region of North Rhine-Westphalia, announced the implementation of the Grün Strom Jetons ("Green Electric Tokens") project. It is based on the use of smart meters to record energy consumption on an hourly basis, and the blockchain is used to store consumption data. When spending more green energy, the consumer receives "Green Energy Jetons" tokens, which serve as direct evidence of the individual consumption of electricity from environmentally friendly production. In the opposite case, the consumer receives gray ones (Stadtwerken Energie Verbund n.d.). In the future, these tokens can be used as an object of the trade like a cryptocurrency.

Similar projects exist in other countries. One example is the New York Power System LO3, which also includes the German company "Siemens." LO3 uses blockchain to enable local prosumer and consumers to trade energy (LO3 Energy n.d.).

Regarding the pace of implementation of blockchain technology in the Russian energy sector, we can assume that at present, this technology can be used only in private complexes, in which mini-farms produce electricity for themselves and their region (Telegina et al. 2017).

One should note that the modern Russian energy system has a small number of large participants, and strong ties already exist between them. Vertical integration and interaction with the state make the system inactive. Therefore, the energy system of the Russian Federation is the exact opposite of the basic principles of blockchain technology. On the other hand, in recent years, the transition from the continental level to a smaller one (urban, regional, or even micro-level) is beginning to acquire significant economic meaning. The scale of savings has multiplied, and the introduction of blockchain can significantly strengthen this trend. At the same time, the only area in which the authorities see microgeneration as a solution to problems is the energy supply to remote areas (for example, the development of renewable energy sources in the Far North).

It must be emphasized that now, the only company that has expressed interest in using blockchain technology in the energy sector is Qiwi, which is negotiating with "Tavrida Electric" (RNS 2016). So, back in 2017, the World Economic Forum discussed how digitalization is transforming the energy business; has a massive impact on the entire value chain in the energy sector. In general, digitalization in the electricity sector is based on implemented network technologies, which are often generalized under the auspices of smart metering, smart grid, and smart home. From the perspective of energy supply, digitalization and implementation of distributed registry technology will increase labor productivity, asset efficiency, as well as transmission and distribution networks through remote monitoring, diagnostics, management, and automation (PWC 2016).

This type of business processes based on blockchain technology is designed to stimulate and encourage the installation of renewable energy sources. The target audience of this business process is solar energy producers, prosumers, consumers, and miners, on whose capacities the blockchain is deployed. The value proposition for manufacturers and prosumers is that for each generated kWh or MWh of electricity, they receive a unit of *cryptocurrency* or *token* (State Duma of the Russian Federation 2018). Tokens provide evidence of renewable energy generated, i.e., they provide transparent accounting of green energy production.

The process of creating value for issuing coins for the generated energy depends on the tokenization carried out in two ways. The first method involves the use of the Ethereum ecosystem, and tokens are created inside an existing smart contract and have the ERC-20 standard (Buterin 2014). The second way is to use public blockchain and public tokens.

The use of tokens in the production and consumption of "green energy" has the following advantages. First, tokens serve as an incentive for the production of solar energy. In regions where renewable energy subsidies or similar support schemes have not yet been implemented, tokens can shorten the payback time of investments in solar panels, which can lead to large-scale deployment of renewable energy production. Second, the blockchain-based mechanism for issuing coins generates liquidity. Their issue is directly related to the generated solar energy and is easily verifiable due to the nature of the distributed registry technology. Thus, the rather complicated system of issuing certificates for electricity can be simplified by tokenizing the generated energy in real-time.

4 Discussion

In the electricity market, the state purchases or sells significant amounts of electricity. In this case, the state's task is to avoid the formation of excess electricity demand so as not to provoke price increases. Therefore, in our opinion, the possibility of creating an Over-the-Counter (OTC) platform for the implementation of OTC transactions for the sale and purchase of electricity is of particular importance.

The OTC Trading Platform is a distributed market for wholesale OTC trading of electricity; the key actors in this market are the energy producers and utilities.

The value chain is based on a platform that combines the purchase/sale applications of all participants in a given market. Market participants can initiate and physically carry out energy transactions through the platform. Bidding is proposed to be carried out on the condition of anonymity, and only interested parties can identify their counterparties.

Blockchain allows all participants in this network to exchange information with each other, excluding intermediaries. The OTC Trading Platform, designed to reflect aggregate demand and aggregate supply, serves the same purpose. Bidders using the platform can count on reducing their transaction costs and transferring all the benefits of transparency to end-users. Furthermore, due to the low transaction costs, the OTC market could potentially be available to participants with lower trading volumes. The blockchain-based OTC Trading Platform can become an effective decentralized platform that meets security requirements and is not subject to manipulation.

5 Conclusion

In our opinion, a retailer, an incentive scheme for certificates of electricity, and the OTC-trading platform are those system-forming business models that could make positive changes to existing business processes. Since the best practices of process automation are used to increase their efficiency, as well as the technology of a distributed registry for pricing, therefore, administrative and transaction costs will be reduced. Blockchain-based OTC platforms provide new ways to solve problems in the energy sector. In particular, they stimulate the development of small commercial solar photovoltaic systems, and also contribute to the decentralization of the electricity market, reducing barriers to entry into the market.

We also believe that the electricity certificate trading system makes a significant contribution to the development of the renewable electricity market by providing various entities with the right to participate in it. Finally, the use of "green energy" based on blockchain is a new transparency and reporting tool that allows one to identify the actual share of green energy in its consumption, calculate its efficiency, and develop its use as an alternative to traditional energy, following the experience of developed countries.

In sum, we would like to note that blockchain in the management of the energy sector of the Russian Federation can optimize and improve many energy business processes. Also, in our opinion, the emergence of both investors investing in the development of renewable energy sources in the Russian Federation to generate additional income, and operator companies uniting producers and end consumers are radically transforming the market for electricity production and consumption. These trends and processes will lead to its structural and institutional transformation.

References

Alekseev, A.N., Lobova, S.V., Bogoviz, A.V., Ragulina, Y.V.: A critical review of Russia's energy efficiency policies in the construction and housing sector. Int. J. Energy Econ. Policy **9**(4), 166–172 (2019)

Booth, A., Mohr, N., Peters, P.: The digital utility: new opportunities and challenges (2016). https://www.mckinsey.com/industries/electric-power-and-natural-gas/our-insights/the-digital-utility-new-opportunities-and-challenges

Buterin, V.: A next-generation smart contract and decentralized application platform. In: Ethereum White Paper. Blockchain Lab, London, UK (2014). http://blockchainlab.com/pdf/Ethereum_white_paper-a_next_generation_smart_contract_and_decentralized_application_platform-vitalik-buterin.pdf

Chursin, A., Makarov, Y.: Management of Competitiveness: Theory and Practice. Springer, Berlin (2015)

Government of the Russian Federation: Draft federal law “Energy Strategy of Russia for the Period up to 2035”, Moscow, Russia (2014). https://minenergo.gov.ru/system/download-pdf/1920/69055

LO3 Energy: Official Website (n.d.). https://lo3energy.com

Lobova, S.V., Ragulina, J.V., Bogoviz, A.V., Alekseev, A.N.: Energy security performance: a dataset on the member countries of the commonwealth of independent states, 2000–2014. Data Brief **26**, 104450 (2019)

Ministry of Energy of the Russian Federation: The forecast of scientific and technological development of the sectors of the fuel and energy complex of Russia for the period until 2035 is approved (2016a). https://minenergo.gov.ru/node/6365

Ministry of Energy of the Russian Federation: Road Map “Energinet” National Technology Initiative (2016b). https://minenergo.gov.ru/node/8916

Popkova, E.G., Inshakov, O.V., Bogoviz, A.V.: Motivation of energy saving within the corporate market responsibility of economic entities. LNNS, vol. 44, pp. 95–105 (2019)

PWC: Blockchain - new opportunities for electricity producers and consumers? (2016). https://www.pwc.ru/ru/publications/blockchain/blockchain_opportunity-for-energy-producers%20and-consumers_RUS.pdf

RNS, Rambler News Service: Qiwi discusses with “Tavrida Electric” the use of blockchain in the energy sector (2016). https://rns.online/it-and-media/Qiwi-obsuzhdaet-s-Tavridoi-elektrik-ispolzovanie-blockchain-v-energetike-2016-10-07/

Shchegoleva, N.G., Terenteva, O.I.: World monetary system transformation: the future for cryptocurrency? Moscow Univ. Econ. Bull. **2**, 75–93 (2018)

Stadtwerken Energie Verbund: Official Website (n.d.). https://www.kleinerracker.de

State Duma of the Russian Federation: Draft Federal Law “On Digital Financial Assets”, 20 March 2018, No. 419059-7, Moscow, Russia (2018)

Swan, M.: Blockchain: Blueprint for a New Economy. O’Reilly Media Inc, Beijing (2015)

Tapscott, D., Tapscott, A.: Blockchain Revolution: How the Technology Behind Bitcoin is Changing Money, Business, and the World. Portfolio/Penguin, Toronto (2016)

Telegina, E.A., Eremin, S.V., Katjuha, P.B., Bessel, V.V., Salahov, I.I., Kanajama, R.D., Sultanov, E.R.: Transformation of energy markets: a new stage of technological development of world energy. Oil Gas Bus. **10**, 43–52 (2017)

Application of Modern Technologies in Cyberspace and the International States Responsibility

Marianna V. Ilyashevich and Pavel A. Banis(✉)

Peoples' Friendship University of Russia (RUDN University), Moscow, Russia
iliashevich-m@rudn.ru, bpavel24@gmail.com

Abstract. With the development of technologies and their accessibility, the means of committing internationally wrongful acts are radically changing, including in cyberspace. The biggest problem associated with its use is the anonymity of persons acting in cyberspace, as well as the possibility of hiding identification data, which makes it challenging to identify the offender. The implementation of the responsibility of states committing internationally wrongful acts, encroaching on the sovereignty of foreign countries and the interests of their legal and physical persons, is one of the critical aspects of ensuring stability among members of the international community. It is necessary to establish a standard of proof to avoid the erroneous appropriation of responsibility to the state, and, as a result, unjustified, and often, depending on the gravity of the wrongful act, severe consequences.

Keywords: International law · International responsibility · Cyberspace · UN International Court of Justice · Internationally wrongful act · Assignment of responsibility

1 Introduction

The assignment is one of the essential elements of an internationally wrongful act of the state. However, there are still many gaps and differences in scholars' opinions on some significant issues in the law of international responsibility.

Among these, one could note the absence of a universal international treaty in this area. Another problem is the lack of existing norms in the field of global responsibility to meet new challenges and threats, which include the use of cyberspace to commit internationally wrongful acts. At the same time, in the event of a cyber-attack, the essential objects of state infrastructure, national security facilities, databases of commercial and financial structures may suffer, which makes the problem of implementing the responsibility for their commission extremely urgent.

Using the Internet and cyberspace in general, in addition to their obvious benefits to users, poses many problems for law enforcement agencies. In particular, this also applies to the issue of assigning the responsibility of states for internationally wrongful acts in several areas.

A. V. Bogoviz and Y. V. Ragulina (Eds.): ISCI 2019, LNNS 115, pp. 450–456, 2020.
https://doi.org/10.1007/978-3-030-40749-0_54

2 Materials and Methods

In preparation for this study, we use an array of methods such as analysis, comparative-legal, logical, and hermeneutic methods. As a material for the development of the work, the texts of court decisions of various international judicial instances, as well as international agreements and other international legal documents that are not binding, were used to a greater extent.

3 Results

The standard of proof to attribute responsibility for a state committing an internationally wrongful act using cyberspace should vary depending on the consequences that were caused to the "affected" state.

If the consequences of an unlawful act can be equated with the use of kinetic weapons or constitute a violation of a peremptory norm of international law, the standard "beyond a reasonable doubt," and in other cases, the standard "more weighty evidence" should apply.

4 Discussion

International law, as a system of law governing interstate relations, is criticized from time to time. On the one hand, it is sometimes overly progressive. On the other hand, it is not flexible enough to find timely solutions for constantly arising issues.

Confirmation of the above is the lack of a universal international agreement between states in the field of international responsibility. The draft articles on the responsibility of states, prepared by the United Nations Commission of International Law and adopted in 2001, despite the long history of its development, have not received approval from countries as an international treaty. Some of its provisions are codification. They are legally binding as norms of international legal custom, while other rules are nothing more than the progressive development of international law, which does not give rise to legal obligations for states.

According to article 2, an internationally wrongful act takes place when it is assigned to a state under international law and constitutes a violation of its international legal obligations (United Nations 2001).

Together with the development of technologies, as well as their growing accessibility for a wide range of people, the threats of their illegal use are growing, and the Internet is no exception.

The Internet provides an opportunity for individuals not only to act anonymously hiding their data but also to conceal their location, data on the equipment used, as well as virtually change all this data. Currently, means of covering identification data are becoming more accessible, and they are developing as a means of committing unlawful acts faster than effective ways to combat them.

Another problem in the area under consideration is the need to determine what actions or inaction that may be attributed to the state that may be qualified as a violation of its international legal obligations. For example, some scholars consider espionage, including using the Internet, a "tradition" between states. But the use of cyberspace, leading to consequences equivalent to the use of kinetic weapons, can be regarded as the use of force within the meaning of Article 2 of paragraph 4 of the Charter of the United Nations (United Nations 1945a) and entail as a response attack, and the implementation of forms of international legal responsibility, namely sanctions, reprisals, etc.

Consequently, it is worth noting that there is a divergence of views in the international legal doctrine regarding the issue of the applicability of international law to the actions and inactions of states in cyberspace. So, some scholars claim its inapplicability (Schmitt 2013). On the other hand, following interpretation of Article 2 (para. 4) of the UN Charter in the Advisory Opinion of the International Court of Justice of the United Nations (Court) on the Use of Nuclear Weapons, Prohibition of Threats, or Use of Force refers to any use of force and weapons, and the UN Charter "neither directly prohibits nor permits the use of any particular weapon" (International Court of Justice 1996).

In our opinion, international law extends to the behavior of both subjects of international law and individuals and organizations in cyberspace because its misuse can lead to real consequences. Only the ways of committing internationally wrongful acts are changing but not their effects, which are expressed in violation of the rights and interests of states, corporations, and individuals.

The use of cyberspace for unlawful purposes can have consequences for various groups of entities, starting with individuals and ending with critical government structures. The result could be identity theft, disruption of public and private infrastructures (Grassegger and Krogerus 2017), and, in the worst case, failure of critical infrastructures. This could result in many deaths and environmental damage, as could have been the case with the Stuxnet virus attacking Iranian uranium enrichment centrifuges (Kaspersky Lab 2014; Gold 2010). In no case of cyber-attacks, has it been ascertained which state should be held responsible for their implementation, the result was only numerous statements by politicians of the "affected" countries, media reports, and investigations by state and private enterprises.

The cyberattack that Estonia suffered in 2007 gave impetus to the unification of some states in the face of such a new type of commission of illegal acts. As a result, an international group of experts developed the "Tallinn Guide to International Law Applicable to Cyber Conflict" at the request of the North Atlantic Treaty Organization (NATO).

One of the most significant aspects of the subject under consideration is proof that the state has committed an internationally wrongful act using cyberspace. In this case, it is worth noting the difficulties that arise during the collection of evidence because a cyberattack, although initiated from a specific place, can virtually pass through computers located in several different jurisdictions. In this regard, it seems necessary to seek cooperation with relevant states to obtain information.

In accordance with the Tallinn Guidelines for International Law applicable in cyber conflict, even if it is established that a cyber-attack was committed from a public infrastructure facility, this would not be sufficient to assign responsibility to the relevant state (Schmitt 2013). In the same way, if the computers of the state infrastructure were links in the chain through which the cyberattack went to its goal, this would not be enough to assign responsibility to the relevant state (Schmitt 2013).

Consequently, the question of the applicable standard of evidence seems to be one of the most significant in connection with the difficulties arising in the process of collecting information and due to the weight of evidence.

Evidence standards in court proceedings were developed primarily in the common law system. These include measures such as "beyond reasonable doubt" used in the criminal process and the standard "by a preponderance of evidence" (which is softer and applicable in civil litigation).

Direct references to applicable standards of evidence by international judicial authorities, including the United Nations International Court of Justice, in the by-laws and the decisions themselves are rare. An exception, perhaps, can be called regional international courts. Thus, the European Court of Human Rights, in some of its judgments on violations of Article 3 of the European Convention for the Protection of Rights and Fundamental Freedoms 1950 (which enshrines the right to life and freedom from ill-treatment) indicated that the standard should be applied "beyond reasonable doubt" (European Court of Human Rights 1978). The statute of the International Criminal Court also refers to the standard "beyond a reasonable doubt" that it applies (United Nations 1998).

The standard of proof "beyond a reasonable doubt" is enshrined in the Statute of the International Criminal Court in part 3, Article 66 as follows: "In order to convict the accused, the Court must be convinced of the guilt of the accused beyond reasonable doubt" (United Nations 1998). In other words, the accused only needs to provide consistent evidence reasonably justifying the correctness of his version. If there are doubts that the prosecution has proved the version of the case at the end of the process, then the accused must be acquitted (United Nations 2001).

The application of the standard "more weighty evidence" can be found, for example, in cases considered by the World Trade Organization. Its essence can be described as "more likely than not" or "probably more than 50%" (Barceló 2009).

The statute of the International Court of Justice of the United Nations, as well as its rules, does not contain direct references to a specific standard of proof. In the practice of the Court, the absence of a specific standard of evidence in international disputes between states is confirmed: "The court never indicated the applicable standard of evidence" (International Court of Justice 2003). Article 53 of the Statute of the Court states that it must "ascertain not only the jurisdiction of the case, according to Articles 36 and 37, but also whether this claim has sufficient factual and legal justification" (Statute of the International Court of Justice, Art. 53) (United Nations 1945b).

Clarifying the above provision, the Court indicated that it had the freedom to evaluate evidence (International Court of Justice 1986); the requirement to "verify" the legal and factual justification of the dispute should be understood as its validity by the rules of law; and the facts on which it is based, according to the Court, must be "convincing" (International Court of Justice 1986). In another ruling, the Court also indicated the need to ascertain "that all facts are available" (International Court of Justice 1974).

In the case of military and paramilitary actions in and against Nicaragua, the parties to the dispute in which were Nicaragua and the U.S., the court noted that a significant portion of the evidence presented by the Nicaraguan authorities in evidence of US responsibility for the conduct of contras in its territory consisted of a review of media materials, books, as well as statements by government officials, including high-ranking officials. The court pointed out the caution that must be exercised in their assessment, as well as the fact that they cannot be evidence confirming the existence of certain facts (International Court of Justice 1986).

The case considered above seems relevant for the issue under consideration. In the situation of the commission of an internationally wrongful act using cyberspace, a dispute is not excluded between the parties. Such a disagreement is specifically about the presence of specific facts directly related to the use of state cyber infrastructures, the presence or absence of funds for carrying out a cyber-attack, malware.

Thus, in order to implement the responsibility of states, it seems appropriate to use the following approach. If a cyber-attack as a consequence had a result comparable with the effects of applying kinetic weapons, as well as in cases of violations of peremptory norms of international law, it should use the highest standard of proof, namely "beyond reasonable doubt." In cases when it comes to a consequence in the form of malfunctioning of private or public cyber infrastructures, which led to material or non-material damage but was not associated with a violation of fundamental peremptory norms of international law, the standard of proof "the presence of more weighty evidence" can be applied.

It seems that such an approach helps to avoid the misappropriation of responsibility, which is associated with the imperfection or even the absence of provisions that would govern the attribution of responsibility of states for internationally wrongful acts committed in cyberspace.

Such a cautious approach is based, among other things, on the fact that states are considering the possibility of using self-defense in cases where a cyber attack can be equated with the use of kinetic weapons. For example, as outlined in the national cyber security strategy, the UK authorities are considering responding to a cyber-attack in the same way as if they required a response to the use of traditional weapons (Government of the United Kingdom 2016).

In its decisions on cases on the Corfu Strait of April 9, 1949, as well as on the application of the Convention on the Prevention and Punishment of the Crime of Genocide of February 26, 2007, the UN International Court of Justice indicated that evidence must be undeniable in cases of complaints of violations of exceptional severity (International Court of Justice 2007; International Court of Justice 1949).

The application of the standard "the presence of more weighty evidence" seems inappropriate because it may not satisfy the principle of equality of arms in litigation in cases where the disputing states are at different levels of technological development. Thus, it will be more difficult for a less developed nation to present evidence.

However, at present, as practice shows, the possibility of attributing responsibility for the implementation of an unlawful act by the state in cyberspace is absent. The existing documents, such as the Tallinn Guide to International Law Applicable in Cyber Conflicts (as mentioned earlier), trace rhetoric providing a cautious approach to attribution of responsibility.

Perhaps the exception is the "political" appropriation, in which much less evidence is required for the political figures of states than for the international judicial body. For example, political persons blamed Russia for cyber-attacks on Estonia in 2007 based only on the fact that it was possible to track a cyber-attack by IP addresses to the territory of the Russian Federation. At present, however, conclusions based solely on such facts appear erroneous (Schmitt 2013).

Some scholars suggest that the standard of evidence be mitigated by applying the rule of "more weighty evidence." We consider this undesirable because of a high degree of probability of error in the implementation, as well as severe consequences of an erroneous decision.

5 Conclusion

The standard of evidence applicable in the resolution of interstate disputes is only rarely indicated in documents regulating activities of relevant international judicial bodies. This statement is also true in relation to the decisions taken. Despite this, the standards of evidence are used to evaluate evidence and vary depending on the court and the circumstances of the dispute, which sometimes makes the outcome of the case unpredictable for both the plaintiff and the defendant.

The misuse of cyberspace can lead to severe consequences and, in some cases, go unpunished due to the possibility of hiding the identity of the offender.

At the same time, the application of a stricter standard of proof in cases where a peremptory norm of international law has been violated, to establish the fact of an offense due to the consequences of a wrong definition of the offender, seems reasonable. In other cases, the application of the standard "more weighty evidence" seems permissible.

Acknowledgment. The publication has been prepared with the support of the "RUDN University Program 5-100."

References

Alekseev, A.N., Lobova, S.V., Bogoviz, A.V., Ragulina, Y.V.: A critical review of Russia's energy efficiency policies in the construction and housing sector. Int. J. Energy Econ. Policy **9** (4), 166–172 (2019)

Booth, A., Mohr, N., Peters, P.: The digital utility: new opportunities and challenges (2016). https://www.mckinsey.com/industries/electric-power-and-natural-gas/our-insights/the-digital-utility-new-opportunities-and-challenges

Buterin, V.: A next-generation smart contract and decentralized application platform. In: Ethereum White Paper. Blockchain Lab, London, UK (2014). http://blockchainlab.com/pdf/Ethereum_white_paper-a_next_generation_smart_contract_and_decentralized_application_platform-vitalik-buterin.pdf

Chursin, A., Makarov, Y.: Management of Competitiveness: Theory and Practice. Springer, Berlin (2015)

Government of the Russian Federation: Draft federal law "Energy Strategy of Russia for the Period up to 2035", Moscow, Russia (2014). https://minenergo.gov.ru/system/download-pdf/1920/69055

LO3 Energy: Official Website (n.d.). https://lo3energy.com

Lobova, S.V., Ragulina, J.V., Bogoviz, A.V., Alekseev, A.N.: Energy security performance: a dataset on the member countries of the commonwealth of independent states, 2000–2014. Data Brief **26**, 104450 (2019)

Ministry of Energy of the Russian Federation: The forecast of scientific and technological development of the sectors of the fuel and energy complex of Russia for the period until 2035 is approved (2016a). https://minenergo.gov.ru/node/6365

Ministry of Energy of the Russian Federation: Road Map "Energinet" National Technology Initiative (2016b). https://minenergo.gov.ru/node/8916

Popkova, E.G., Inshakov, O.V., Bogoviz, A.V.: Motivation of energy saving within the corporate market responsibility of economic entities. LNNS, vol. 44, pp. 95–105 (2019)

PWC: Blockchain - new opportunities for electricity producers and consumers? (2016). https://www.pwc.ru/ru/publications/blockchain/blockchain_opportunity-for-energy-producers%20and-consumers_RUS.pdf

RNS, Rambler News Service: Qiwi discusses with "Tavrida Electric" the use of blockchain in the energy sector (2016). https://rns.online/it-and-media/Qiwi-obsuzhdaet-s-Tavridoi-elektrik-ispolzovanie-blockchain-v-energetike-2016-10-07/

Shchegoleva, N.G., Terenteva, O.I.: World monetary system transformation: the future for cryptocurrency? Moscow Univ. Econ. Bull. **2**, 75–93 (2018)

Stadtwerken Energie Verbund: Official Website (n.d.). https://www.kleinerracker.de

State Duma of the Russian Federation: Draft Federal Law "On Digital Financial Assets", 20 March 2018, No. 419059-7, Moscow, Russia (2018)

Swan, M.: Blockchain: Blueprint for a New Economy. O'Reilly Media Inc, Beijing (2015)

Tapscott, D., Tapscott, A.: Blockchain Revolution: How the Technology Behind Bitcoin is Changing Money, Business, and the World. Portfolio/Penguin, Toronto (2016)

Telegina, E.A., Eremin, S.V., Katjuha, P.B., Bessel, V.V., Salahov, I.I., Kanajama, R.D., Sultanov, E.R.: Transformation of energy markets: a new stage of technological development of world energy. Oil Gas Bus. **10**, 43–52 (2017)

Methodology for Determining the Prospective Needs of Companies and Society for High-Tech Products

Nikolay A. Okatyev, Anna M. Nasonenko(✉), and Elena V. Butrova

Peoples' Friendship University of Russia (RUDN University), Moscow, Russia
nokatye@mtkgrp.ru, {nasonenko-am, butrova-ev}@rudn.ru

Abstract. The paper presents a methodology for determining the future needs of the company, which is aimed at obtaining a reasonable method for identifying and forecasting the needs of society for high-tech products, as well as determining the potential market share that a manufacturer can take. An algorithm of the market's prospective needs for high-tech products, an assessment of the manufacturer's prospects for obtaining any percentage of this market are reviewed and analyzed. Also, the paper considers those risks associated with the introduction of new high-tech products on the market and the way to maintain the competitiveness of new products.

Keywords: Competitive advantages · Future needs · Methodology · Method · Algorithm · High tech products · Competitiveness · Production · Market · Market segments · Production planning

1 Introduction

In times of fierce competition of products on the market, companies must think about finding competitive advantages, offer additional product options, etc.

It is worth noting that the process of creating a new product always has some risks for the company producing it. After the product is brought to the market, it may turn out that it is not convenient to use, has poor quality, does not interest consumers, etc., which leads to decreasing profits. Today, in the face of constantly changing consumer demands, new technologies, and new competitors, the successful activity of a company directly depends on how successful the new products are developed and introduced by it to the market.

The correct application of appropriate marketing strategies to a high-tech product at all stages of the life cycle, frombirth to being displaced from the market by-products that are more consistent with consumer requirements, is of great importance (Bendikov and Frolov 2001; Golubkov 1999).

However, it would be more rational and efficient to conduct preliminary studies to study the target audience, the strategy of promoting the product at different stages of its life, the course of developing the product on the markets, etc.

To avoid a lack of success, preliminary studies should be carried out to identify future needs. This will allow developing a strategy for promoting new products,

A. V. Bogoviz and Y. V. Ragulina (Eds.): ISCI 2019, LNNS 115, pp. 457–467, 2020.
https://doi.org/10.1007/978-3-030-40749-0_55

avoiding market failures, such as, for example, non-compliance of a new product with the needs of customers, improperly selected target audience, etc. (Guryanov 2014).

The development of a methodology for determining the future needs of a company is an important task, the solution of which is aimed at obtaining a sound method for identifying and forecasting the needs of society in high-tech products. Also, the solution to this problem is aimed at determining the potential market share that a manufacturer may occupy as a result of the sale of products with specific technical and economic characteristics.

By limiting the future needs and their volumes, the manufacturer will be able to determine the most promising market segments and carry out more effective planning for the production of competitive high-tech products.

In parallel with solving the problems of the sustainable functioning of the enterprise, the manufacturer has to build a system of adapting its capabilities to ensure competitiveness in the market niche both domestically and internationally. At the same time, the critical task of enterprise management is to give it the flexibility to respond to new challenges promptly. In matters of innovative development, the development of new mechanisms of interaction with innovative actors (including in the framework of public-private partnerships) from various fields is of great importance. In particular, for large corporations, the implementation of the role of the driver of technological updating of the domestic economy implies a qualitative change in approaches to identifying breakthrough scientific and technical areas and targeted financing of projects, which can give synergistic effects for a particular industry and socio-economic development of the country (Barinov and Okatyev 2019).

2 Materials and Method

The introduction of innovation is always associated with significant risks. A high level of risk is caused by the fact that marketing deals with the external environment of the company (competitors, contractors, and consumers), which is challenging to manage.

Therefore, there is a high probability that, while construction is underway, analog products with highly competitive advantages may appear on the market. Competition is an integral part of market activity. Even if at the moment there is no competition, it would certainly rise when the produced high-tech product is successful. As a result, no matter what line of business and enterprise is engaged in, it would inevitably collide with other firms that have chosen the same field of activity, which carries a higher risk for the manufacturing company.

The study of potential competitors is not only the determination of the leading direct competing enterprises in the market as a whole but also the determination of firms competing indirectly in specific segments and regions, as well as the determination of their strengths and weaknesses. It is necessary to analyze their comparative advantages in the competition (price, quality, the technical level of the product, etc.).

The study of potential competitors allows not only to assess the level of threat but also to study the requests better, also focusing on the wishes and preferences of a prospective buyer for high-tech products.

The main thing to find out is why the buyer would prefer products under your brand. It is also necessary to determine the following: (a) what forms of competitiveness are most important; (b) what features of the product are the most interesting for consumers in this market segment; (c) what changes should be made in the product, in the methods of its promotion, in other components of marketing, in the management and organization of production, engineering and technology; as well as (d) in which sequence and terms.

The ability to successfully sell a high-tech product in a particular market in a certain period determines the competitiveness of a manufactured product.

To timely respond to a potential threat, an enterprise should continuously monitor changes in data on existing competitors, their possible actions in the market, collect and process data on their goods and services, prices, innovations, etc. Also, one needs to monitor new potential competitors, researching information regarding their readiness to enter one's market segments. The following is essential for increasing competitiveness: cooperation with suppliers of components and clarification of the delivery conditions, the possibility of reducing the cost of materials and components, demand trends, the possibility of improving the quality of materials, changing circumstances, etc. After that, one should analyze the needs of the consumer, taking into account the structure of the proposed product, type of competition, and the actions of competitors. One must always keep abreast of the latest market trends to understand which of the innovations may affect the potential market capacity of one's product.

As a result, in the production process, it is necessary to increase the competitive advantages of the manufacturer steadily. Moreover, the strategy for the formation of these advantages should have economic, technological, and socially justified indicators and be based on a variable analysis of the external and internal environment of the company.

Thus, the following factors have a significant impact on the development of competitive advantages: demand dynamics, purchasing power, socio-political constraints, competition intensity, market structure and behavior of its participants, as well as infrastructural and institutional aspects, etc.

Also, in addition to analyzing the firm's capabilities in creating competitive advantages, it is necessary to carefully approach the forecasting of shortcomings that may arise in the enterprise due to not introducing new technologies and not taking into account changing market demand.

Thus, for the company and its high-tech products to be competitive, it is necessary to modernize products by applying innovative technologies and expanding and improving the competencies of the company and its employees.

The process of developing and introducing a new high-tech product on the market is always associated with the risk of failure. At the initial stage of product development, it is crucial to determine what factors can make a new product successful in the future. Also, it is necessary to determine which type of business is more suitable for introducing a new product to the market, namely: small, medium, or large. Each has its characteristics, which are essential to consider when introducing new products to the market.

This technique is used to determine and assess the prospective needs of companies and society for high-tech products, taking into account the dynamics of the development of market segments, focusing on potential consumers of high-tech products.

A prerequisite for determining future needs is the segmentation of consumer markets since specific factors affecting the dynamics of market development and needs in it exist for each market segment.

Also, this technique allows one to determine the long-term needs of companies and society for high-tech products and the potential market share that a manufacturer can occupy as a result of the sale of a manufactured product that has specific technical and economic characteristics.

The methodology for determining the future needs of companies and society in high-tech products is implemented by solving the following tasks:

- Assessing the development of consumer markets (by segments of consumers of high-tech products);
- Predicting the demand for high-tech products in the market segment based on specific indicators and indicators characterizing needs;
- Predicting the available market capacity that a manufacturer can cover, taking into account the characteristics of high-tech products, the development of competitors, barriers to entry to the market, and other factors.

The positive synergistic effect of solving these problems lies in the achievement by a high-tech corporation of a state in which it will be ready for the production of commercially successful and competitively capable high-tech products.

According to the methodology, for the necessary calculations, the initial data are needed, which are the following indicators:

1. Typical indicators for each market segment of high-tech products:
   - Dynamic indicators of market segments (growth in the market; market capacity (available, actual, potential); market saturation level; physical and moral depreciation of high-tech products);
   - Characteristics of consumers of high-tech products in each segment (dynamics of purchase by year, the write-off of products due to wear and tear/emergencies; conformity of equipment samples to consumer expectations, etc.);
   - Product life;
   - Foreign countries' market development indicators.
2. Segment specific features:
   - Indicators of growth in demand for segments (for example, for a sign of the growth of the oil and gas industry, the dynamics of oil prices, the number of offshore platforms and individual vessels with helipads, the growth of oil production on the shelf and inaccessible places, etc. are determining parameters).
3. Environmental indicators:
   - Sanctions restrictions (including a ban on cooperation in promising Arctic, deep-sea or shale projects);
   - Dynamics of prices for related services (repair and maintenance, insurance, etc.);
   - Macroeconomic indicators (exchange rates, inflation, customs tariffs, etc.).

3 Results

The role of determining future needs is huge. Moreover, the timely and accurate determination of the market needs for high-tech products is an essential prerequisite for determining the necessary products and their required quantity provided to the consumer market. If the estimate of the demand for high-tech products is not derived correctly, then either surplus or shortage of products may arise. The occurrence of excess with too high a need estimate will lead to increased costs. With a low assessment of needs, disruptions in the provision of high-tech products, and a shortage of goods on the market arise, from which potential customers can go to direct competitors, leading to a decrease in turnover and loss of market share.

For an uninterrupted sale and guaranteed turnover, the merchant must provide a vast supply of goods in the right quantity.

The purpose of determining the need for products is to provide potential consumers with the necessary high-tech products in the right quantity and at the right time. In turn, this will allow the company to conduct its activities in the consumer market successfully, sell goods, and make a profit through satisfying customer demand.

4 Discussion

The determination of the future market needs for high-tech products and the assessment of the manufacturer's prospects for obtaining any share of this market are carried out following the algorithm presented in Fig. 1.

At the first stage of the formation of the statistical base and marketing analytics, to assess the development of the market, the collection of initial data is carried out according to the list that includes quantitative and qualitative characteristics of the market.

On the one hand, such analytical work demonstrates to the company's management what a buyer wants, what specific properties of the goods are significant for him. On the other hand, in the course of marketing analysis, one can understand how to present new products to potential customers. It may make sense to focus on individual characteristics (when improving a product) and to ensure its promotion on the market. Marketing research provides information on new products or changes in existing products, as well as new prospects for the buyer.

At the stage of segmentation of the consumer market, taking into account the industry sector of the consumer and the choice of potential consumer segments, based on the analysis of the current market situation, a list of segments (consumers distributed by industry) is formed that are currently on the market and can be formed in the future to assess the development of needs in segments.

Market segmentation provides the following opportunities:

- Find out the specifics of the possible buyer of this product; demonstrate the nuances of consumer qualities in different market segments; determine which of the properties of the consumer group are sustainable and therefore more essential for designing the needs and desires of consumers;

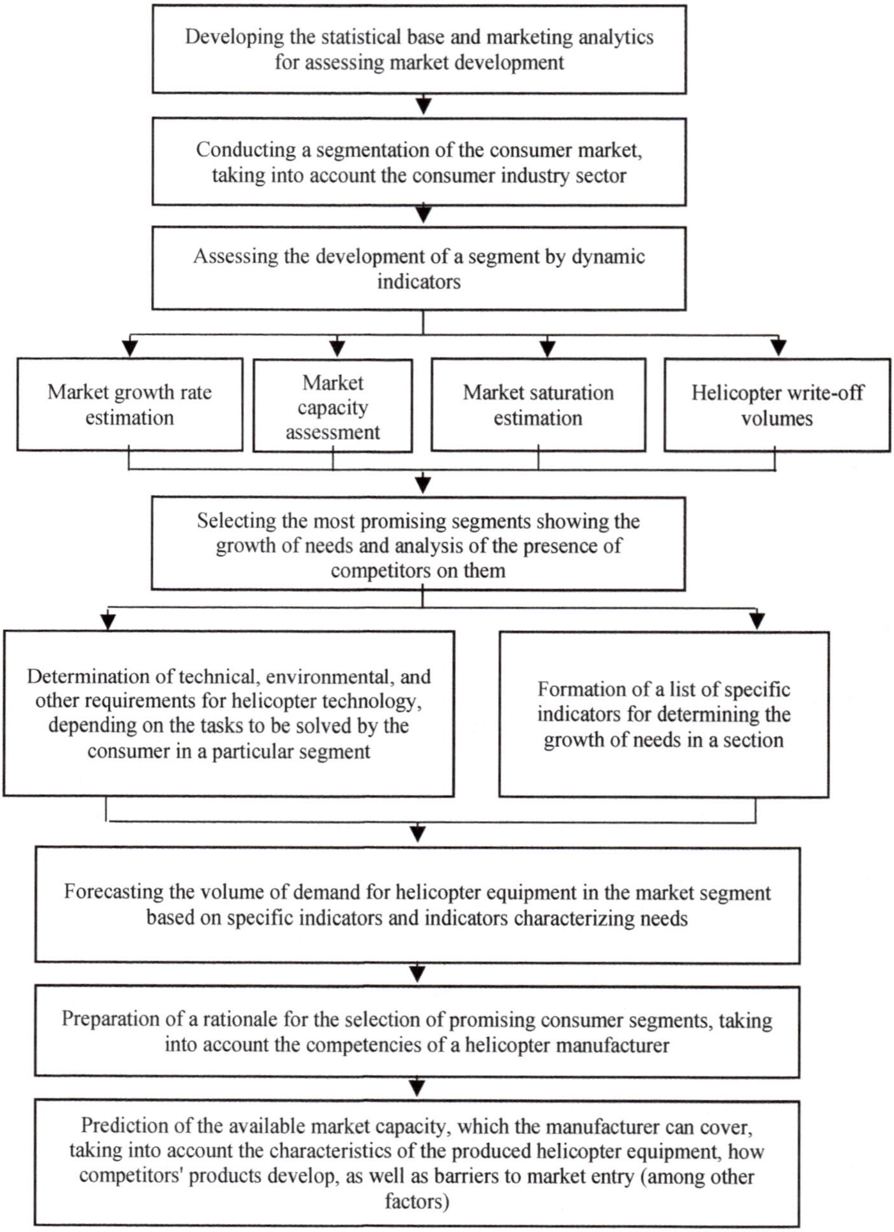

Fig. 1. Algorithm for determining the future needs of high-tech products.

- To clarify (adjust) possible market capacity, simplify sales forecasting;
- Determine how to modify advertising, the organization of points of sale, what changes should be made to the marketing tactics of the company when interacting with different groups of customers;

- Understand how to change the properties of a product (device, cost, delivery, appearance, packaging, etc.) when selling in different market segments.

In terms of higher sales, the company leadership should be aware of the properties of their consumer group, which are in first place at the moment, or will be there shortly. Based on these properties, we can establish a target market segment, i.e., the most significant or promising for the company, one that meets its specifics. The right choice of the target segment (a consumer group in which the most likely buyers of a particular product are collected) is a characteristic difference between a marketing-oriented company.

At stage 3, assessing the development of a segment by dynamic indicators, an assessment of the development of market segments is carried out based on determining the number of indicators, including:

- Market growth;
- Market capacity (available, actual, potential);
- The level of market saturation;
- Write-off volumes (physical and moral depreciation of equipment; natural and other decrease in technology).

Understanding the size of the market in which the company operates, and the degree of decrease or growth is a crucial indicator for assessing the growth potential of future revenue.

The market growth rate is used to assess the two main components:

1. Market growth dynamics for a manufacturer of high-tech products;
2. The dynamics of the economic growth of the consumer sector of high-tech products.

Market volume is measured by the total quantity or total value of goods or services sold over a given period (usually one year). The market growth rate is the ratio of market volume in the current period to market volume in the previous period. If the indicator is less than one, then the market is reduced; if the indicator is higher than one, then the market is growing.

The market growth rate is calculated using the following formula:

$$T = \frac{V_1}{V_0}, \tag{1}$$

where V_1 is the total sales in the market this year;
V_0 is the total sales in the market for the previous year.

At the same time, sales can be expressed both in monetary units and in the number of units sold.

The market growth rate is calculated annually, or quarterly cumulatively within the year (Chief Engineer n.d.).

As a result, market development trends are identified.

In turn, determining the market capacity allows us to assess the economic environment in the dynamics of time periods, which means it helps to make timely strategic decisions. Since the main commandment for success in competition is to identify and respond to the slightest changes immediately.

This indicator characterizes the possible volume of demand in the market. It is determined by the following formula:

$$E = \sum_{i}^{n} (S_i \cdot k \cdot Э_x) - (H - И_ф - И_м) - A - C \quad (2)$$

E is the market capacity; S_i is the number of the i-th group of consumers; k is the level (coefficient) of consumption in the base period or the standard of consumption; $Э_x$ is the coefficient of elasticity of demand from prices and income; H is market saturation; $И_ф$ is the physical depreciation of goods; $И_м$ is the obsolescence of goods; A are alternative forms of meeting needs (substitute products); C is the share of competitors (Kunilova et al. 2011).

If the market capacity indicates the availability of products in the designated geographical area, then saturation is the upper limit of this capacity, above which production or delivery of products to the market can lead to losses. Therefore, to determine the boundaries of market opportunities, it is essential to identify the saturation parameter.

For durable goods, the following formula applies:

$$\text{Ae} = \text{Ab} + \text{P} - \text{D} \quad (3)$$

where Ae is the availability of products at the end of the period; Ab is the availability of goods at the beginning of the period; P is a purchase for a period; and D is the disposal for the period (based on the average duration of service of the goods).

Further, based on these indicators, the ranking of segments according to priority in terms of their potential for acquiring high-tech products is carried out. After identifying the most promising segments, showing the growth of needs and analysis of the presence of competitors in them, the work at the next stages of the algorithm continues (Belyaevsky et al. 2006).

Stage 4. Description of the functional, technical, environmental requirements for high-tech products, depending on the tasks to be solved by the consumer in a particular segment.

In the process of selling high-tech products, the objective needs to develop such a product range that would meet expectations and satisfy the requirements of consumers appears.

In fact, the creation and management of the range of high-tech products of the enterprise consist in the timely development and proposal of a set of goods that meet the profile of its production activities and the requirements of consumers of certain categories.

Requirements for the rationality of the nomenclature are presented depending on the market conditions (values, knowledge, consumer reactions, as well as socio-economic, legal, and other environmental characteristics).

When selling goods and creating a product range, manufacturers of high-tech products must take into account the specifics of the market, since the perception of the proposed products by the consumer is unknown due to the increasing pace of updates (Kostina and Torgashova 2014).

Stage 5. Determining the parameters that a manufacturer must provide to meet the needs of a potential buyer.

An ideal consumer model of high-tech products should fully meet the potential needs of prospective buyers, which is based on information collected in the process of market research. Data from peer-to-peer advertising materials can play a crucial role.

The model includes the following:

- The list of technical and economic indicators of products;
- Consumer assessment of the significance of each indicator;
- Determining the ideal value for a potential consumer of each indicator.

The parameters of high-tech products can be divided into several groups:

- Regulatory parameters are indicators that demonstrate the fundamental possibility of selling products on the target market;
- Technical settings, such as purpose, durability, reliability, aesthetics, and ergonomics. Parameters that cannot be expressed numerically should be evaluated in points;
- Economic parameters, which include the price of consumption and operating costs (Shchetinina 2017).

Stage 6. Forecasting demand for high-tech products in the market segment based on specific indicators and indicators characterizing needs.

Determining the value of market demand and market share indicators for market segments is an essential direction in forecasting markets for high-tech products and choosing target markets for further product sales, which are based on the study of market demand.

Demand for most high-tech products can be estimated based on official statistics, as data on volumes of products sold and services rendered are published for various categories, including for international markets, markets of individual countries and regions, by different industries and enterprises. But for many types of products, reliable statistics are not available.

As a result, marketing research is needed to predict demand and other market characteristics.

The current market demand in monetary terms is defined as

$$Q = n * c * p, \tag{4}$$

where n^* is the number of buyers of this type of product in this market; c^* is the average price of this product; p is the number of purchases of the buyer for the studied period.

Stage 7. Preparation of the rationale for choosing a promising consumer segment, taking into account the competencies of the manufacturer of high-tech products.

After conducting market segmentation research and determining the composition of each of them, the manufacturer needs to decide which market segments to enter (namely, one, several, or all at once). For a more precise understanding, it is necessary to assess the degree of attractiveness of each of the segments by the following factors:

- Segment capacity;
- Opportunities for increasing segment capacity;
- Estimated profit in this segment;
- The interest of competitors in this group of buyers;
- The attitude of customers to competitors' products;
- Availability of appropriate distribution channels.

After analyzing the above indicators, one determines which of the market segments the production will enter. The combination of these segments forms the target market of the company.

5 Conclusion

Thus, to implement the most effective production planning of commercially successful and competitively capable high-tech products, as well as to identify promising market segments for manufactured products, determining the future needs of the company is an important task. For the company to occupy more market segments and meet the requirements of prospective customers, it must also develop its competencies and technologies, since the high-tech product's market is rapidly growing.

The method proposed in work will allow enterprises to substantiate the need for the production of certain high-tech products substantively, determine their need in the market, calculate the payback, etc. These measures will increase the income of companies and avoid losses.

Acknowledgment. The study was carried out with the financial support of the Russian Foundation for Basic Research in the framework of the scientific project №19-010-00619 "Development of a Methodology for Applying the Results of the "Digital Earth" Project to Solve the Problems of the National Economy and a Forecast Model for the Economic Effect of Their Application in the Context of Russia's Digitalization."

References

Barinov, V.A., Okatyev, N.A.: Strategy for restoring the competitiveness of the domestic rocket and space industry. RIOR: INFRA-M, Moscow, Russia (2019). http://www.znanium.com, https://doi.org/10.12737/19037

Belyaevsky, I.K., Danchenok, L.A., Korotkov, A.V.: Statistics of the Market of Goods and Services. Moscow State University of Economics, Statistics and Informatics, Moscow (2006)

Bendikov, M.A., Frolov, I.E.: High-tech product markets: trends and development prospects. Marketing in Russia and Abroad, 2 (2001). https://www.cfin.ru/press/marketing/2001-2/02.shtml

Chief Engineer: Market growth rate (n.d.). http://chiefengineer.ru/organizaciya-proizvodstva/pokazateli-effektivnosti/temp-rosta-rynka/

Golubkov, E.P.: Market research. Marketing in Russia and Abroad, 1 (1999). http://www.mavriz.ru/articles/1999/1/10.html

Guryanov, T.I.: Improving the assessment of the state and development of the consumer market, Dissertation of Candidate of Economic Sciences. Siberian Federal University, Krasnoyarsk, Russia (2014)

Kostina, G.D., Torgashova, A.V.: Features of consumer perception in the sale of goods on the high-tech products market. Econ. Socio-Humanit. Stud. **1**(1), 16–27 (2014)

Kunilova, M.A., Antonenko, O.O., Karaulova, L.V.: Statistics. Vyatka State University, Kirov (2011)

Marketing analysis of the market: types, stages, methods (n.d.). https://www.vvs-info.ru/helpful_information/poleznaya-informatsiya/marketingovyy-analiz-rynka/

Shchetinina, I.V.: Methodology for assessing the competitiveness of high-tech products. Econominfo **1–2**, 68–71 (2017)

Creating a Crewless Ship in the Framework of the Technological Paradigm of the Russian Federation

Vladimir V. Zaslonov(✉), Anastasiya A. Golovina, and Anatolii N. Popov

Admiral Ushakov Maritime State University, Novorossiysk, Russia
djdf3d@yandex.ru, a.a.golovina@mail.ru,
an.popov.mga@gmail.com

Abstract. This article tells about the formation of crewless shipping as a promising area, not only as a branch in the marine transport industry but also as an essential industrial area that affects many specialists in the implementation and operation at all stages. The paper describes one of the many methods that (via conversion) can be adapted into a sophisticated creation system and achieve the most effective result with open data for a quality process. The authors pay special attention to such a stage as the formation of augmented reality. The later is seen as an element of the control system. As part of the study, the authors note that the introduction of crewless ships will require a mandatory review of the essence of the training of swimming personnel, as well as changes (additions) to the provisions of international conventions governing relations in the field in question.

Keywords: Crewless shipping · Unmanned vehicles · Augmented (virtual) reality · Sea transport

1 Introduction

The Russian maritime complex operates in the context of implementing the goals and objectives of the national policy, which are determined by the *Marine Doctrine of the Russian Federation for the Period Until 2020 (Arctic Regional Direction)* (2016). These measures are urgently needed in the context of a systemic crisis, the disintegration of the country's transport complex, and economic reform during the 1990s.

The critical areas in the maritime transport industry of the contemporary period include developing crewless shipping (CS), unmanned vehicles (UV), remote-controlled unmanned underwater vehicles (RUUV), augmented (virtual) reality as a control system (CS) among all types of marine ships (Filimonov 2008; Khodzhaev et al. 2017). For the effective implementation of these areas in conditions of deficit, stagnation, and crises, new approaches to work (design, formation, operation, etc.), as well as improving the legal regulation of relations arising in the area under consideration, are necessary (Fig. 1).

A. V. Bogoviz and Y. V. Ragulina (Eds.): ISCI 2019, LNNS 115, pp. 468–474, 2020.
https://doi.org/10.1007/978-3-030-40749-0_56

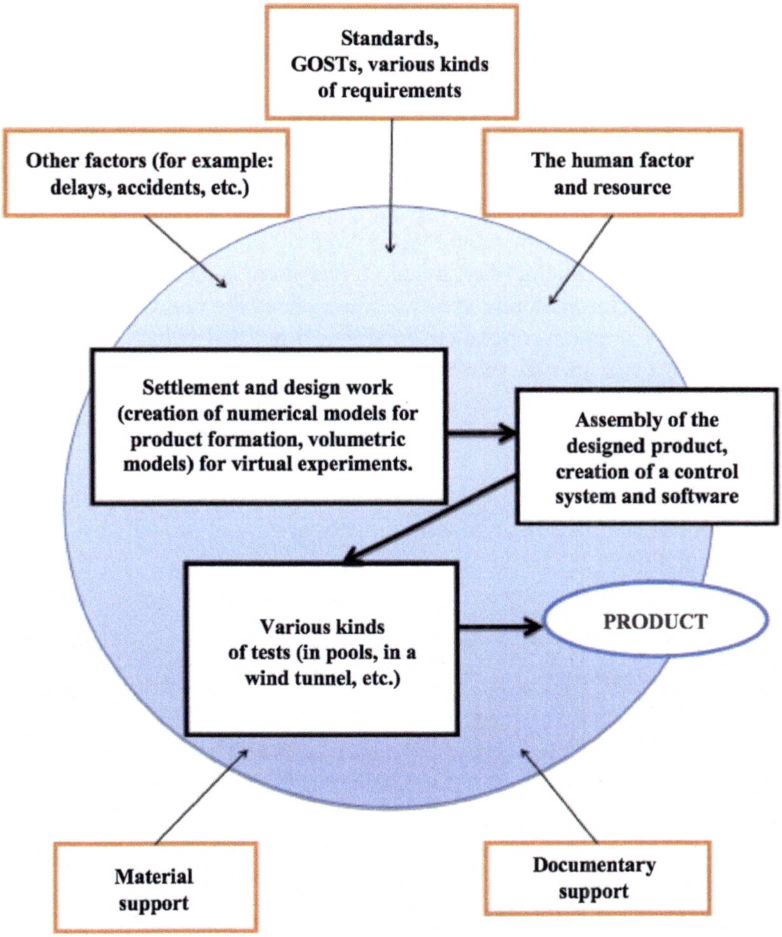

Fig. 1. The general sequence of product sales, taking into account external factors.

2 Materials and Method

The use of computing tools for product development, as one of the important stages, has become a standard approach in engineering. The development of a complex high-tech product, such as RUUV, CS, without automated design systems (ADS), analysis, and modeling of the system, is challenging. In addition to the production process, modern research and science rely heavily on computer modeling (Kravchenko and Zaslonov 2016; Zaslonov et al. 2017). Using simulation and calculations allows engineers to investigate phenomena that are difficult to study experimentally: material properties or physical phenomena occurring in an electric motor. At the same time, it should be noted that such methods and techniques for solving such problems should be governed by the requirements of the standards for the maximum efficiency of their application. In turn, the rules should correspond to the modern technological structure, while the existing standards that apply to the specified field of activity do not meet contemporary needs, the state of science and technology, and they are virtually outdated.

We will consider one of the possible ways to implement the stages of the formation of crewless shipping. We introduce the concept of a multifactor system instead of a multibody. As a rule, a multifactor system is defined as a set of factors interacting with each other and affecting the process of product formation. Modeling a multifactor system, as one of the areas of system modeling, represents a significant improvement in predicting the performance of marine technological complexes (MTC) in comparison with other methods, which are often based on analytical equations or empirical testing. The use of a multifactor modeling system contributes to the rapid and compelling study of variables in the creation, operation, modernization, and behavior of the apparatus compared to the experimental approach. Accordingly, the use of a multi-factor modeling system in the design and creation of products (for example, CS) can reduce the need for physical prototypes, thereby significantly accelerating the overall design process and saving resources.

Figure 2 illustrates the phases of the process of modeling a multi-factor system. The following steps are characteristic of a multifactor system of modeling (on the example of the CS formation):

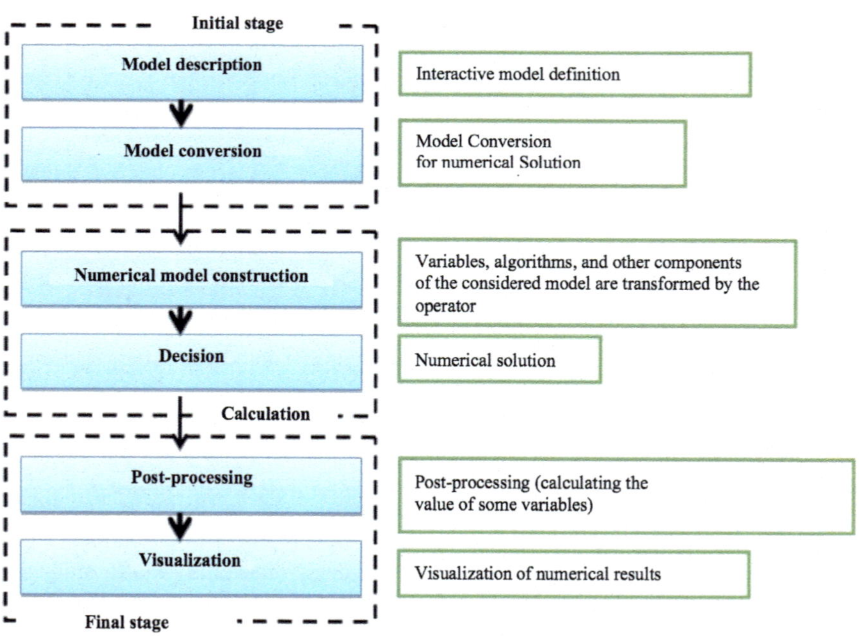

Fig. 2. Phases of a multi-factor system modeling process.

- Pre-processing includes the following: creating the topology of the CS model, determining the components of the model and their parameters, and also often setting the solution parameters for the numerical solver.

- The solution consists in the actual calculation of the numerical results based on the mathematical model of the system (for example, the algorithmization of the navigation process together with augmented reality for CS controlling).
- Post-processing may include the calculation of additional dependent variables, visualization of the simulation results, planning of the components of the results and analysis of the results for the subsequent production of elements and CS as a whole.
- The pre-processing phase of the simulation process describes the production of CS. Besides, as a rule, pre-processing is the most time-consuming phase, and, therefore, it is a natural choice as a starting point for increasing the efficiency of the modeling process.

3 Results

Based on the results of the study, it seems possible to conclude that the use of this method does not solve the main problem, namely the exchange of simulation data and results between a set of software applications in the process of product design. The rapid increase in design data and the need for data exchange in a networked operating environment require new approaches to ensure product development progress. Communication barriers between computing tools, design methods, and groups of people as a whole can be eliminated by revising the standards and register of the Russian Federation, and open exchange formats such as *ACIS2* и *Parasolid3*. If a similar problem can be solved in this way, then what can be said about each stage as a whole? Analysis, mathematical modeling, and design of a crewless ship can be compared with the current shipbuilding complex. Here, similar steps are established, and well-known problems can be solved just by using a multifactor system.

However, there is a stage in design that leaves a lot of gaps both in the legal, educational, and technological fields. The augmented reality control system is such a stage. Crewless navigation implies the complete absence of the ship's crew. Thus, the ship can be controlled from any point with higher quality than the skipper due to the processing and presentation of information to the operator. And also, control of the ship can be performed with less load on the person himself. It should be noted positively that, in this case, the specialist is no longer in a challenging environment to operate the ship. Besides, the load on his/her brain is significantly less since the necessary information not only is in a more convenient form for management, but the "machine" is engaged in its processing along with the presentation in the required form to the operator.

As part of a multifactor system, one of the main tasks for the design stage of the control system is to create a mathematical algorithm for the interaction of devices. So, all the information necessary for the operator for the desired calculations, which must be updated every second, is read. The technology for adapting augmented reality in shipping does not have a clear display. Still, it has only just begun to develop in the Russian Federation, and the final solutions (within which there are working analogs) have not been established. We will present our method using a concrete example. We will use the combinatorial approach in the framework of the concept of "Amount of

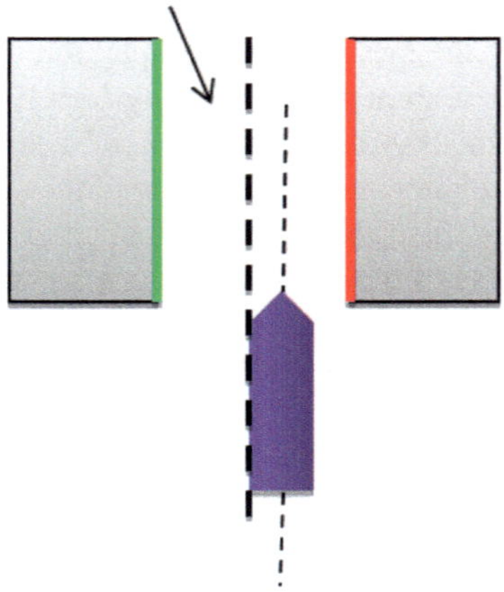

Fig. 3. Passage of the ship through the channel.

information" by A. N. Kolmogorov. We will consider the case of a ship passing through a canal (Fig. 3).

$$\mathrm{H}(\mathrm{X}\_{—ж}) = \frac{\text{logвБ°((sAвИЖR }\frac{\text{n}}{\text{360¬?}}\text{)*0,5)}}{\text{вИЖX}\_{—ж}}, \quad (1)$$

where ΔX_c is the center of the considered section of the channel; ΔR is the radius of the considered section.

$$\mathrm{H}(\mathrm{X}\_{—В}) = \frac{\text{logвБ°(sAвИЖR }\frac{\text{n}}{\text{360¬?}}\text{)}}{\text{вИЖX}\_{—В}}, \quad (2)$$

where ΔX_m is the direction of the current section; n is the degree of the sector in question; ΔR is the radius of the considered circle.

$$\mathrm{H}(\mathrm{v}\_{—В}) = \frac{\text{logвБ°((sAвИЖR }\frac{\text{n}}{\text{360¬?}}\text{)/вИЖ(-Г'}_1\text{--Г'}_2\text{))}}{\text{вИЖv}\_{—В}}, \quad (3)$$

where Δv_m is the flow velocity with a certain direction of the considered section; n is the degree of the sector of the considered section; ΔT_1, ΔT_2 stand for the time interval; ΔR is the radius of the circumference of the considered section.

$$\mathrm{H}(\mathrm{L}\_{—Д}) = \frac{\text{logвБ°(sAвИЖR }\frac{\text{n}}{\text{360¬?}}\text{)}}{\text{вИЖL}\_{—Д}} \quad (4)$$

$\Delta L_{ф}$ is the fairway length of the section in question; ΔR is the radius of the area under consideration.

Consequently, the conclusion is

$$H\left(X_{-\text{Ж}}X_{-\text{В}}v_{-\text{В}}L_{-\text{Д}}\right) = H(X_{-\text{Ж}}) + H(X_{-\text{В}}) + H(v_{-\text{В}}) + H(L_{-\text{Д}}) \quad (5)$$

4 Discussion

By entering constant variables, the values of which we get from devices, we can create a database of values with which the program (algorithm) will calculate the desired values for the operator. The problem is that this work involves the analysis of many options in shipping. However, we get an extensive database that would reduce accidents in shipping thanks to working with information (high accuracy of obtaining sets of variables) and reduce the load on a person. Thanks to the described method of information interchange between stages, the problem in system design will also be solved (Fig. 4).

Fig. 4. The concept of augmented reality in CS.

5 Conclusion

Thanks to the use of a multi-factor production modeling system, it is possible to solve and optimize all stages of production and modern technologies, as well as obtain the necessary base. This base is necessary for updating standards, requirements according to the modern technological structure.

A multi-factor system is one of the possible ways to solve the problem of implementing the stages of the formation of CS in the Russian Federation and other MTC for the development of the industrial component of the country.

Based on the study, we can conclude that the new development trend in the maritime transport industry will give life to new directions, covering areas such as:

- Secondary, higher, additional education;

- Development of current legislation in the field of standardization of CS design processes;
- Formation of new types of specialists.

Thus, by using a multifactor system, we described all the stages of the CS formation, in particular, the control system through augmented reality.

References

Filimonov, G.D.: Development of scientific, technical, and organizational foundations for increasing the innovative level of shipbuilding production to ensure the implementation of the provisions of the marine doctrine of the Russian Federation, Dissertation of Doctor of Technical Sciences, Saint-Petersburg, Russia (2008)

Khodzhaev, S.S., Zaslonov, V.V., Kravchenko, N.A.: Technical requirements for the marine technological complex and its components. Tech. Technol. Eng. **4**(6), 36–45 (2017)

Kortelainen, J.: The semantic data model for multibody system modeling, Doctor of Science (Technology). Lappeenranta University of Technology, Lappeenranta, Finland (2011)

Kravchenko, N.V., Zaslonov, V.V.: The problem of applying mathematical modeling in the creation of marine technological complexes. Molodoy Uchenyi **7**, 88–90 (2016)

Marine Doctrine of the Russian Federation for the Period Until 2020 (Arctic Regional Direction) (2016). https://link.springer.com/referenceworkentry/10.1007%2F978-3-319-24237-8_328

Zaslonov, V.V., Kravchenko, N.A., Khodzhaev, S.S.: Requirements for ensuring the survivability and overall performance of marine technological complexes. In: Akhmetov, I.G. (ed.), Materials of the VII International Scientific Conference: Engineering in Russia and Abroad, pp. 86–89. "Buki-Vedi" Publishing House, Moscow, Russia (2017)

The Apprmulti-Purposet Input Formation for Purchasing Helicopters in Russia

Vyacheslav M. Kraev, Natalya V. Prosvirina, and Alexey I. Tikhonov(✉)

Moscow Aviation Institute (National Research University), Moscow, Russia
vmkraev@mail.ru, nata68.92@mail.ru, mai512hr@mail.ru

Abstract. A helicopter is a specific product, not only because of the unique technical and operational characteristics of the equipment but also because of the relatively high production and operating costs. Until recently, the manufacturers of helicopter equipment in the Russian Federation, when developing a marketing policy, focused mainly on the value of production costs. However, the prospective buyer also takes into account the costs of the entire life cycle of a helicopter. This approach should not only be taken into account by Russian helicopter builders but also actively applied to reduce operating costs by introducing new design and technological solutions.

Keywords: Helicopter · Helicopter market · Helicopter production cost · Helicopter life cycle · Helicopter operating costs

1 Introduction

Helicopters, like any other product of high-tech industry, have gone through various stages of product value formation. During the years of the formation and development of the helicopter industry in Russia, the main buyers of the equipment were the state military and civil aviation structures. In the absence of significant competition in the domestic market, the equipment cost was formed based on the development, testing, and production costs. This approach allowed the buyer to influence the equipment cost only at the state level, during the formation of technical specifications. Since the beginning of the USSR collapse, however, the helicopters of foreign production have lived in the domestic market. Such competition has led not only to the competition of technology models but also to the competition of approaches in forming input costs. Foreign manufacturers have been using the concept of the helicopter's life cycle for a long time. This concept can reduce not only the production cost but also the size of operating costs. It is this totality of costs that is decisive for a potential buyer.

2 Materials and Method

The methods of strategic planning and cost forecasting in the domestic and world markets of aviation equipment mainly served as the authors' research methods. The study was based on the use of predictive analytical materials published by such

A. V. Bogoviz and Y. V. Ragulina (Eds.): ISCI 2019, LNNS 115, pp. 475–485, 2020.
https://doi.org/10.1007/978-3-030-40749-0_57

companies as Russian Helicopters, Airbus Helicopters, Robinson, Bell Helicopter, Boeing, and the Interstate Aviation Committee. The primary method of research was strategic planning, which focused on the analysis of the long-term development of the helicopter market.

3 Results

The civil aviation market is a globally growing market that has no national borders and, at the same time, is characterized by fierce competition of national manufacturers. Difficult technological challenges and high costs are the reason that only a small number of countries and several large companies work in the aircraft industry. Thus, competition in the market of aircraft manufacturers is oligopolistic in nature, or in other words, dominated by several large international companies that have a strong influence on the entire market. The top 10 rankings of global helicopter manufacturers are as follows (MIC News 2019):

1. Airbus Helicopters—20,145% (Eurocopter Tiger H61—modern attack helicopter designed by the Franco-German consortium Eurocopter (one of the world's leading helicopter manufacturers))
2. Robinson Helicopter—18,804% (Robinson R44—light, multi-purpose four-seater commercial helicopter, manufactured in America. It was developed in 1980)
3. Bell Helicopter—15,561% (Bell UH-1D—American multi-purpose helicopter, also known as Huey. One of the most popular and massive machines in the helicopter history. Serially produced since 1960)
4. «Russian Helicopters»—12,497% (Mi-8AMTSH "Terminator"—one of the modifications of the combat Mi-8 family helicopter (one of the most popular in the world). Mi-8 is the second most common military helicopter in the world (after Sikorsky UH-60 Black Hawk)
5. Leonardo—9,181% (AgustaWestland AW139—Anglo-Italian twin-engine, multi-purpose helicopter. Designed by Agusta and Bell Helicopters under the designation Agusta-Bell AB139)
6. Sikorsky—7,658% (Sikorsky UH-60 Black Hawk—American multi-purpose helicopter. The most popular military helicopter in the world)
7. Helicopters Guimbal—2,472% (The Guimbal Cabri G2 is a light commercial double helicopter designed by Bruno Gimbal, a former Eurocopter engineer. First took to the air in 2005)
8. Boeing—2,034% (Boeing CH-47 Chinook—American heavy military transport helicopter of the longitudinal scheme. It has been widely used in the United States since the early 1960s. It is in service with more than 20 countries)
9. Korea Aerospace Industries—1,552% (KAI KUH-1 Surion—a multi-purpose helicopter, developed in 2010 by the joint efforts of aircraft manufacturers Korea Aerospace Industries and Airbus Helicopters)
10. Enstrom Helicopter—1,337% (Enstrom F-28—light helicopter. Received the first flight certificate in the USA in 1965) (Fig. 1).

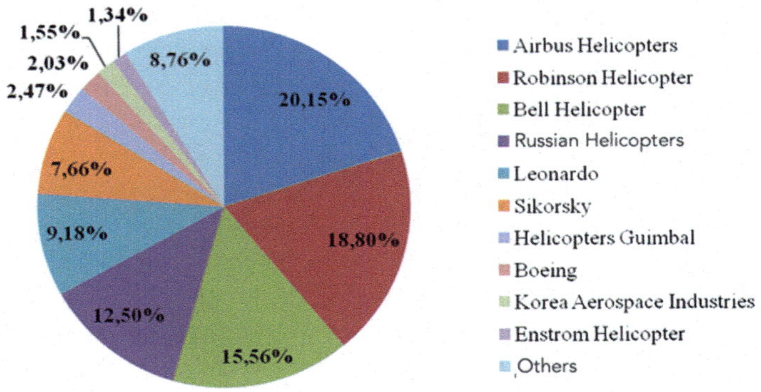

Fig. 1. The market share of the largest helicopter manufacturers in the world.

In the long term (by 2034), the global fleet is to grow by 50%, but in the near future, the weakness of the energy market will put pressure on the production volumes (MIC News 2019).

According to Vorobyov, the number of helicopters delivered in 2017 increased by 17% compared to 2016, despite the slippage of some purchases initiated by customers. In 2016, the holding company Russian Helicopters, the only helicopter developer and manufacturer in the country, sold 189 helicopters, and 221 in 2017.

The holding set the task to increase the share of civilian products in sales to 40% by 2035. In 2016, its share was less than 10%, only 17 civilian vehicles out of 189. In 2017, the sales of civilian helicopters rose sharply to about 70 helicopters.

In recent years, domestic prices for civilian helicopters have risen significantly, and the cost of helicopter operations remains almost unchanged, which reduces the airline's investment opportunities in the fleet. Also, in certain segments, helicopter pilots are experiencing competition from uncrewed aircraft.

In 2017, the holding received ₽228 billion in revenue in accordance with the IFRS (International Financial Reporting Standards) and ₽26.9 billion net profit (the second-best result after 2015). The holding is 87.5% owned by Rostec, 12.5% by the RFPI consortium, and Middle Eastern funds (Vorobyov 2019).

Foreign helicopters are also supplied to the domestic Russian market. These are mainly products of Bell, Robinson, and Eurocopter. The cost of foreign helicopters in Russia is undoubtedly higher than abroad.

The factory price of Robinson R22 and R44 helicopters in the basic configuration (without the equipment required for operation in Russia and without the cost of factory packaging in boxes) starts between $288,000 and $379,000 (carburetor)/$461,000 (injection). However, for objective reasons, it is impossible to get new helicopters in Russia at these prices.

The final price will increase significantly due to the cost of mandatory for Russia (according to the IAC-type certificate data card of the Interstate Aviation Committee) additional equipment (in the manufacturer's sense); accessories and additional options; packaging costs; transportation; insurance; customs broker; terminal handling and other expenses in a temporary storage warehouse at the customs post; final assembly after the export shipment; aircraft overflight; registration of all the necessary documents for commissioning a helicopter in Russia or the CIS, such as the ownership registration certificates, helicopter registration in the State Register of civilian aircraft, airworthiness certificate, etc. (HeliCo Group, 2019a, b).

Helicopters prices intended for individuals sometimes differ tenfold. Such technical parameters determine the difference in price as the type and engine power, load capacity, number of seats, and several other indicators.

The quality of the inner trim also matters. Expensive models offer passengers leather seats with comfortable backs and folding armrests. Finally, several typical aviation parameters can be distinguished, which also affect the cost. They are climb rate, maximum and cruising speed, flight range at one refueling, maximum flight altitude, screw diameter, and other parameters affecting the technical characteristics.

Companies specializing in the sale of civil aviation helicopters are transferring a machine that is already ready for use to the customer. The price of a new helicopter includes all additional costs and fees: shipping by sea or by plane, VAT, the connection of security systems provided for by Russian regulations, control flyby, and registration with government bodies.

Table 1 shows the prices for new vehicles, including VAT and other related payments. The list shows the minimum rates at dealers.

Table 1. The cost of a new helicopter.

Model	Price, thousand dollars
Robinson R22 Beta	421.0
Robinson R44 Raven I	625.0
Robinson R66 Turbine	1.277.0
Bell 505	1.913.3
Airbus Helicopters H130	4.358.0
Agusta Westland AW119ke	5.010.8
Ми-8 AMT VIP	13.200.0

Source: According to the data (Heliport, 2019; HeliCo Group, 2019b).

When purchasing expensive equipment from dealers in a lease, the payment is extended for up to 7 years, and the final amount will be 20–30% higher than the initial cost.

The cost of a military helicopter is quite challenging to estimate (Table 2). Prices for it can vary greatly depending on the supplies volume and the contract terms with a specific state (Sukhov 2019).

Further, we will consider the approaches to the formation of initial cost when buying helicopters in Russia.

There is a tendency to express the aircraft price through the estimated cost of a kilogram of the construction. According to Harris and Scully, this value is $1,000 per kilogram. Often, for the conditional estimation of the cost of helicopters, the price of the engine(s) is taken and multiplied by four (Johnson 1996). Nevertheless, the modern

Table 2. The price of some popular models of helicopters abroad.

Model	Price, thousand $
Robinson R-22	250.000
Robinson R-44 Raven I	340.000
Robinson R-44 Raven II	415.000
Bell B206 JetRanger	700.000
Eurocopter EC120 Colibri Hummingbird	1.700.000

Source: According to the data (ToughNicke, 2019).

electronic equipment of helicopters may be estimated at $20,000 per kilogram, and the programs used by the on-board computers increase this figure even more. However, such a cost estimate is quite rough (Vagefinjad 2000).

There are more accurate ratios for estimating helicopter prices. In addition to the construction weight, they include other important design parameters. The method, developed by Harris, helps determine the estimated cost of the serial production of helicopters (Harris 1997). The applied calculation method is based on the principle of deterministic statistical data processing.

The applied calculation method is based on the principle of deterministic processing of statistical data:

$$C_{est} = C_{\#} C_f M_{emp}^m N_{eng}^n N_b^n, \quad (1)$$

where C_f is the so-called class factor, which is the product of five factors:

$$C_f = C_{td} C_{ua} C_{ins} C_{sh} C_{tsh}; \quad (2)$$

M_{emp} is empty helicopter mass, kg;
M_{eng} is take-off engine power, hp; and
N_b is the number of main rotor blades.

According to the results of statistical analysis, performed by F. D. Harris, coefficient and exponents included in Eq. (1) are: $C_{\#}$ = $ 338.14, m = 0.4638, n = 0.5954, and k = 0.1643.

The accuracy of the price prediction by Eq. (1) for more than 100 helicopters is within $\pm$ 20%. It is known that the more helicopters that are planned to be produced and sold, the lower the labor intensity of their production and, accordingly, the helicopter price. In order to perform the correct calculations, such factors as the number of the helicopters released and the annual inflation rate should be taken into account in Eq. (1):

$$C_{est} = C_{\#} RiC_n CfM_{emp}^m N_{eng}^n N_b^n, \quad (3)$$

where R_i—Inflation rate,

C_n—coefficient taking into account the number of helicopters released.

In the case of a detailed study of the design of the helicopter, the cost calculation should be carried out on the basis of estimates of the complexity of real technological processes, prices for materials, finished products, as well as the overhead costs structure and level that are specific to a particular enterprise and a specific time (Vagefinjad 2000).

According to Imenitov, when comparing helicopters for the main technical and economic indicators, the following expenditure criteria are subjects to analysis:

- Fuel consumption costa (during launch and in-flight);
- Maintenance service costs, based on the cost of 1 m-hour (not a service hour), including the maintenance cost (gaskets, oil seals and oils replacement, etc., except the work on the overhaul of components and assemblies);
- A linear depreciation value, defined as the quotient of dividing the cost of acquiring a new helicopter on its resource in motorcycle hours. The result is the value of the helicopter cost transfer per one-hour resource. It does not take into account the replacement cost of the machine at the time of resource exhaustion; that is, it is assumed that after the resource exhaustion, the machine is subject to disposal. If there is a need to replace engines or units for overhaul during the operation, the value in the numerator is increased by the cost of the corresponding replacement or repair. It is possible to carry out the calculation in a different way. The calculation of the depreciation is also done linearly. It is done by reducing the value in the numerator by the cost of selling the helicopter at the residual or market value after the resource has been exhausted, and increasing it by the cost of refurbishment necessary for the sale. Since there are no stable statistics on such indicators for objects of comparison, this approach does not apply.
- Optional: storage and transportation costs.

The total integrated cost of flight hours (C_i) is defined as the cost of motor-hour, equal to the sum of fuel costs, helicopter maintenance costs, separated by its resource, and linear depreciation value.

$$C_i = F_c + S_c + D_c, \tag{4}$$

where F_c – the cost of the fuel consumed by the helicopter for the conditional hour;

S_c – maintenance costs per operating hour;
D_c – linear depreciation (Imenitov, 2019).

According to Proskurin, the main tasks of the feasibility analysis performed during the helicopter design is to determine the expected costs at the main stages of the machine life cycle, which include (Proskurin, 2014):

- The costs of developing a new helicopter, including research and development, design, development of experimental machines, testing, and drawing up a working draft;

- The costs at the stage of helicopter production, including the costs of manufacturing the main fuselage, main and tail rotors, transmission units, the purchase of components, and assembly and installation work;
- The cost of flight operation of the helicopter, taking into account depreciation costs, staff salaries, maintenance costs, and fuels and lubricants.

The helicopter price, the flight-hour cost, the ton-kilometer cost, and other technical and economic indicators are determined based on the calculation of the cost types listed above. Based on these indicators, the designed helicopter is compared with its counterparts; the quality and availability of the developed project are assessed.

The initial data for the feasibility analysis are the parameters of the helicopter, calculated at the stage of preliminary design: the empty helicopter mass; the mass of the fuselage with the equipment; the mass of the main rotor hub per blade; the rotor fill factor; the rotor diameter; the number of the main rotor blades; main gearbox reduction ratio; torque on the output shaft of the main gearbox; the number of the helicopters produced; and life indicators of the main helicopter units in-flight hours, taken according to the average data.

In the developing process, creating and preparing a new helicopter, the designing cost makes up to 12% of all costs, pilot production—about 50%, and during the preparation for serial production, the amount of work reaches 40% due to the implementation of measures aimed at improving manufacturability, design improvements, the aggregate resource increasing and other works.

The costs of the enterprise producing helicopters in mass production are divided into two parts: (1) the cost of its own production, including the parts manufacturing, components and details producing, fitter and assembly work, final assembly, helicopters testing; (2) the cost of purchasing engines, gearboxes, instrumentation equipment and other components produced by specialized enterprises.

The cost of the helicopter equipment is approximately 15% of the fuselage cost. The cost of the purchased gas turbine engines is approximately 80% of the cost of the fuselage in the kit.

A preliminary price calculation of the helicopter X, ₽, is performed according to the formula:

$$X = \left[P_f + P_{bl} + \left(P_p + P_g + P_{eq} + P_{eng}\right) C^P_{pc}\right] \cdot C^P_{pr}] \tag{5}$$

where P_f, P_{bl}, P_b, P_g, P_{eq}, P_{eng} are the costs, ₽, respectively, of the fuselage in the set, blades, bushings, gearbox, equipment, engines;

$C^{p}_{pc} = 1,04$ – profitability coefficient for purchased components;
$C^{p}_{pr} = 1,27$ – profitability ratio for the company's costs.

The main technical and economic indicators of the helicopter during operation are the costs of flight hour and the cost per ton-kilometer. The effectiveness of various models of helicopter, including the newly designed helicopters with existing prototypes, are compared based on these indicators. The reduced costs are calculated per flight hour for assessing economic efficiency.

Helicopters make up from 40–50% of the global operating aircraft fleet. The growth rate of the civilian helicopter fleet is about 15% per year. In the armed forces, the number of various helicopters has become comparable to aircraft. Further development of the helicopter industry is aimed at improving the flight-technical and technical-economic characteristics and competitiveness of the created machines (Proskurin 2014).

Further, the latest trends in this field from foreign company industry leaders are considered. As stated at the beginning of the article, the concept of the product life cycle is widely used abroad. This approach allows helicopter developers and manufacturers to find key points for the introduction of new designs and technological solutions. Implementation of these solutions can significantly reduce total cost over the entire life cycle of the helicopter, and, consequently, create more competitive equipment samples.

As stated above, the method with the highest potential of reducing client input costs is the introduction of new design and technological solutions with the following goals:

- Reducing equipment cost for the buyer;
- Reducing take-off weight and resulting fuel consumption;
- Reducing the cost aggregates and consumables during routine maintenance;
- Modernization of the design to increase the service interval;
- Simplification of the design and units for the possibility of servicing by a specialist with a lower-level license;
- Improvement of the design to minimize the required equipment during routine maintenance, etc.

The manufacturers of foreign helicopters form their approaches based on the existing scientific, design, and technological potential. As a rule, the ways to reduce the costs at the aircraft operation stage are (Aviation Week Network, 2019) the improvement of components and assemblies based on statistical data on their operation, the application of new designs and technologies to new models, introduction of new technologies into existing models (advanced upgrade kits), updates as the life cycle progresses, and Performance Improvement Packages (PIPs).

4 Discussion

The competitors among foreign models of one segment include Robinson R-44, Bell B206, and Eurocopter EC120 in terms of passenger capacity, range, etc. However, prices for similar models differ significantly, mainly due to different options. Other factors affecting the operating costs (training costs, operating costs, storage costs, etc.), also play an important role.

Airbus Helicopters, as Eurocopter, chose a strategy to simplify maintenance requirements, achieved a reduction in its time, as well as a reduction in the requirements for specialized equipment and training. The purpose of such innovations is the possibility for the operator to carry out maintenance in a place convenient for him with a reduction in costs (AIRBUS 2018). Maintenance and overhaul can be carried out either on its own at a certified Airbus Helicopters service center, or directly at the equipment operator facilities. Airbus Helicopters develops customized solutions to improve the overall performance of the helicopter and reduce its maintenance cost, offers more than 1,000 standard solutions for upgrading existing helicopters, based on service bulletins and additional type certificates.

The manufacturer of helicopters Bell, Bell Helicopter, chose a strategy to reduce operating costs more than ten years ago. For example, in the Bell 407 helicopter, after a series of innovative implementations, part of the maintenance work was excluded from the service bulletins. As a part of ongoing work, several requirements for maintaining airworthiness were revised to reduce operating costs (VERTICAL 2007). According to the company, this innovation will lead to a reduction of more than 12% of the previously published cost estimates for direct maintenance (VERTICAL 2007). In 2015, Bell Helicopter acquired Textron Aviation (provides the components and assemblies repair, their exchange, and the provision of the spare parts). Bell Helicopter chose this path to reduce the customers' operating costs. "Bell Helicopter focuses on expanding after-sales service to reduce operating costs and customer downtime," said Glenn Isbell, executive vice president of customer support and services at Bell Helicopter (Able Aerospace Services Inc., 2012). Chief Executive Officer of Able Aerospace Services (Able Aerospace Services Inc., 2012) Lee Benson said, "The strategy for safely reducing operating costs by providing resource repair components, major repairs and spare parts replacement is now significantly improved when the company became a service center for Bell helicopters." The chosen strategy implies an annual 6% reduction in life cycle costs for new helicopters. Cost reductions should be achieved by reducing domestic costs and reducing the selling price by at least 3% per year. The spare parts cost should also be reduced by 3% or more per year (Bell Helicopter Textron Inc., 2011).

5 Conclusion

In current conditions, the formation of costs for the acquisition and maintenance of the determining trend is the application of the life cycle concept. Initially, this approach was introduced at foreign helicopter companies, and it showed its viability and efficiency. At the enterprises of the Russian industry, the need for a comprehensive implementation of the life cycle concept has been overdue for a long time, since such an approach takes into account the integrated operating costs and motivates helicopter-building enterprises to reduce them in an innovative way. This approach forces helicopter developers to introduce constructive and technological innovations in precisely those elements of helicopter systems that will provide a significant reduction in operating costs.

Acknowledgments. The authors thank the management of the holding company "Helicopters of Russia" and A. Chursin, Doctor of Economics, Prof., for their great help in work and the valuable comments and corrections made.

References

Able Aerospace Services Inc.: A Textron company all rights reserved (2012). http://ableaerospace.com

AIRBUS: Aircraft maintenance, aircraft repair, and on-site assistance (2018). https://www.airbus.com/helicopters/services/helicopter-maintenance.html

Aviation Week Network: Continuous technology upgrades through engine lifecycles (2019). https://aviationweek.com/continuous-technology-upgrades-through-engine-lifecycles

AVIATIONPROS: Bell Helicopter announces able aerospace services as a new service center (2016). https://www.aviationpros.com/aircraft/rotorcraft/press-release/12177488/bell-helicopter-bell-helicopter-announces-able-aerospace-services-as-new-service-center

Bell Helicopter Textron Inc.: Materiel: Cost reduction (2011). https://www.bellflight.com/suppliers/materiel/cost-reduction

Harris, F.D.: Helicopter cost too much (1997). https://vtol.org/store/product/helicopters-cost-too-much-7170.cfm

Harris, F.D., Scully, M.P.: Rotorcraft costs and prices. In: Proceedings of Vertical Lift Aircraft Design Conference, American Helicopter Society, San Francisco (1995)

HeliCo Group: Official website (2019a). https://helico-russia.ru/

HeliCo Group: What is the price of a Robinson helicopter? Why is the factory cost noticeably different from the final one? (2019b). https://helico-russia.ru/blog/iz-chego-skladyvaetsya-tsena-vertoleta-robinson/

Heliport: Official website (2019). https://heliport-moscow.ru/

ToughNicke: How much does a helicopter cost? (2019). https://toughnickel.com/misc/How-much-does-a-Helicopter-cost

Imenitov, E.L.: Comparison of the Mi-34 and analogs in the main economic indicators of operation (2019). https://history.wikireading.ru/399364

Johnson, C.W.: Cost drivers-an an operator's perspective, NASA. In: Proceedings from Operator Rotorcraft Economics Workshop, Sponsored by the NASA Short Haul CivU. Tiltrotor Program, NASA Ames Research Center, Moffett Field, CA (1996)

Kraev, V.M., Siluyanova, M.V., Tikhonov, A.I.: Approaches to the development of models of the life cycle of domestic aviation technology. Moscow Econ. J. **1**, 52 (2019a)

Kraev, V.M., Siluyanova, M.V., Tikhonov, A.I.: Approaches to modeling the life cycle of modern domestic aircraft engines. Financ. Econ. **2**, 177–183 (2019b)

MIC news: Helicopter market forecast published until 2030 (2019). https://vpk.name/news/167100_opublikovan_prognoz_ryinka_vertoletnoi_tehniki_do_2030_goda.html

Proskurin, V.D.: The Helicopter Parameters Calculation at the Preliminary Design Stage. Orenburg State University, Orenburg (2014)

Prosvirina, N.V., Tikhonov, A.I.: Forecasting the demand for the production of Russian aircraft in the world market. Manag. Bus. Adm. **1**, 91–109 (2017a)

Prosvirina, N.V., Tikhonov, A.I.: Improving the efficiency of forms of organization of production in the aircraft engine-building enterprises. RISC: Resour. Inf. Supply Compet. **4**, 166–170 (2017b)

Prosvirina, N.V., Tikhonova, A.I.: Prospects for the development of the Russian aircraft engine building market. Moscow Econ. J. **2**, 43 (2017c)

Prosvirina, N.V., Tikhonova, A.I.: Factors of competitiveness and prospects for the development of Russian civil aircraft. Moscow Econ. J. **3**, 61 (2017d)

Sukhov, V.: What is the price of a Robinson helicopter? Why is the factory cost noticeably different from the final one? (2019). https://yakapitalist.ru/finansy/stoimomt-vertoleta/

Vagefinjad, S.: The choice of light transport helicopters parameters, taking into account the climatic conditions of the operation (Candidate of Science Dissertation) Moscow, Russia (2000)
VERTICAL: Bell helicopter reduces direct operating costs for the proven Bell 407 Helicopter (2007). https://www.verticalmag.com/news/bell-helicopter-reduces-direct-operating-costs-for-the-proven-bell-407-helicopter-html/
Vorobyov, A.: For the first time in five years, Russian Helicopters have increased helicopter deliveries (2019). https://www.vedomosti.ru/business/articles/2018/05/21/770091-narastili-postavki-vertoletov

Foreign Experience in Forming the Cost of Operating Helicopters

Vyacheslav M. Kraev and Alexey I. Tikhonov(✉)

Moscow Aviation Institute (National Research University), Moscow, Russia
vmkraev@mail.ru, mai512hr@mail.ru

Abstract. The helicopter industry, both domestic and foreign, in modern conditions is called upon to provide solvent demand from customers-operators. The authors determine the main trends in changing the characteristics of helicopter technology. Moreover, the paper analyzes how the clients of helicopter companies can influence not only the helicopter technical parameters but also the technological level of the helicopter production. In the foreign market for helicopter technology, this trend is most pronounced, as the research clearly demonstrates. More than that, the paper focuses on the current trends in helicopter pricing. It is shown that leading manufacturers actively apply the life cycle concept in the formation of the cost of helicopters.

Keywords: Helicopters · Helicopters market · Helicopter production cost · Helicopter life cycle · Helicopter operating costs

1 Introduction

Helicopters, like any other high-tech industry product, have gone through various stages of product value formation. At the dawn of the industry, the customer was willing to pay and paid the required price for the unique characteristics of this transport: the helicopter can take off almost anywhere where there is a flat platform slightly more than the propeller diameter, the helicopters are able to reduce the horizontal and vertical speeds to zero, and the helicopters can carry trucking on the external load, which allows carrying bulky goods, as well as carry out installation works. Subsequently, with the development of a full-fledged helicopter market and the competition emergence, the manufacturers started not only meeting the requirements of the customers in terms of the equipment characteristics but also minimize the cost of the helicopters. In order to successfully compete, it was necessary not only to reduce the helicopter selling price but also to ensure acceptable operating costs, which are significantly important.

2 Materials and Method

The research methods are the methods of strategic planning and value forecasting in the global aviation market. The main research method is strategic planning, which is focused on the analysis of the long-term development of the market for modern helicopters. The study is based on the use of analytical and forecast materials published by

A. V. Bogoviz and Y. V. Ragulina (Eds.): ISCI 2019, LNNS 115, pp. 486–498, 2020.
https://doi.org/10.1007/978-3-030-40749-0_58

such companies as the British Helicopter Association (n.d.), Airbus Helicopters (Eurocopter), Robinson, Bell Helicopter, Lufthansa, and the US Federal Aviation Administration.

3 Results

Since the first serial helicopter appeared in 1946, this transport type has been developing rapidly. Modern business aviation helicopters can carry up to fourteen passengers at speeds of over 300 km/h. Each type of helicopter is a compromise in payload, capacity, comfort, speed, and price (The British Helicopter Association, n. d.). The formation of the helicopter cost is an important step in the entire business process of helicopter production. Given the planned nature of the domestic economy in the USSR period and the absence of the market mechanisms at that time, let us turn to the foreign experience, which has been developing for decades.

Studies related to the analysis of the cost of the operating helicopters for industrial purposes were conducted from the end of the 20th century (Born).

According to experts from the British Helicopter Association, the acquisition and annual operating costs for a helicopter will be greater than any other transport type of similar capacity and power (The British Helicopter Association, n.d.).

When purchasing a helicopter, one must analyze its operating cost and selling price, which can hardly be significantly lower due to the helicopter's high cost. Let us consider their power structure (The British Helicopter Association, n.d.).

Fixed costs are calculated as annual costs and do not depend on flying hours. These include such spending patterns as depreciation, insurance, taxes, crew costs, overhead costs, equipment costs, etc.

Hourly expenses are expenses that directly depend on the number of flight hours and compensate for the costs of:

- Fuel and oil;
- Scheduled maintenance;
- Unscheduled maintenance;
- Engine overhaul;
- Overhaul of the airframe.

Only after accounting all costs is it possible to formulate the cost of a helicopter for a client (The British Helicopter Association, n.d.).

From the above, it follows that:

- General operating costs decrease as flight hours increase, and
- Fixed costs for a new machine make up a large part of the total costs (but can be reduced by using a lower reported value of a used helicopter), and direct fuel costs play an important role.

In Table 1, a comparison of the cost and basic characteristics of helicopters is provided.

Meyer and Schrage (2011) have conducted the cost analysis associated with the development, production, and operation of the main aerospace systems. In the modern

Table 1. A comparison of the cost and basic characteristics of the helicopter.

Helicopter type	The new helicopter cost, th. $.	Direct operating expenses, $ per hour	Max. take-off weight, lb.	Weight (dry), lb.	Disposable load, lb.	Max. weight on the external sling, lb.	Max. distance, mile	Cruising speed
R-66	800	–	2700	1280	1420	–	325	120
Schweizer 333	900	260 (2010)	2550	1250	1300	–	310	105
Enstrom 480B	800	–	3000	1695	1305	1000	350	110
Bell 206 B3	1100	–	3200	1713	1487	1337	374	110
EC120	1700	–	3780	2127	1653	1540	393	122
MD 500E	1150	320 (2008)	3000	1481	1519	1769	239	135
MD 520 N	1650	345 (2008)	3350	1585	1765	1965	200	125
MD 530F	1400	357 (2008)	3100	1591	1509	1809	206	135
AS350 B2	1900	–	4960	2690	2270	2522	360	133
AS350 B3	2150	–	4960	2716	2244	3086	359	140
EC130B4	2200	–	5351	3020	2331	2557	329	130
Bell 206 L4	2000	–	4450	2327	2123	1941	324	110
Bell 407	2400	–	5000	2653	2347	2646	330	133
MD 600 N	1700	500 (2008)	4100	2100	2000	2100	342	134
Agusta 119	3300	–	5997	3152	2845		360	144
Bell 210	3000	–	–	1-	–	–	–	–

Source: According to the data (HeliNews, n.d.).

helicopter industry, the concept of life cycle cost is widely used. The helicopter life cycle cost can be divided into three phased parts: research and development, initial investment or production, and operating costs.

The most common method for cost estimation is the use of the cost estimating relationship (CER). To determine the costs of operating a rotary-wing machine, one must proceed from the model of the tasks facing the operator and sum up all the costs associated with personnel, training, and spare parts throughout the life of the system (Meyer and Schrage 2011).

At the state level of the United States, the Federal Aviation Administration has established a method for calculating estimates of variable and fixed operating costs for aircraft (The Federal Aviation Administration 2016). The authors of the regulatory document assumed that when evaluating investments in aviation equipment, it is necessary to take into account not only its acquisition costs but also operating costs.

Considering the results of economic reports analyzing the recession of 2007–2009, the authors talk about a significant decrease in the demand for air transportation and a simultaneous rise in the price of jet fuel (The Federal Aviation Administration 2016). So, it focuses on the profitability of air travel by standardizing the operating costs.

The US Federal Aviation Administration divides the costs into variable and fixed costs. Variable costs vary in proportion to aircraft use and include fuel and oil, maintenance, crew costs, etc. The fixed costs do not practically change in proportion to aircraft use.

The ratio of the main operating costs is given in Table 2.

Table 2. The ratio of the main operating costs.

Expense group	Expense type	Share in expenses	
		Civil aviation	Cargo aviation
Direct	Air freight operation costs	52%	43%
	Subtotal: air traffic operating costs	**52%**	**43%**
Indirect	Advertisement	1%	0%
	Aircraft maintenance	5%	4%
	Aircraft depreciation costs	0%	0%
	Technical equipment depreciation	0%	0%
	General and administrative expenses	6%	9%
	Maintenance and depreciation of the ground equipment	1%	2%
	Passengers service costs	7%	0%
	Costs for booking and selling services	5%	1%
	Dispatching costs	8%	9%
	Subtotal: maintenance, sales, and total running costs	**34%**	**25%**
	Transportation costs	14%	32%
	Subtotal: air transportation indirect costs	**48%**	**57%**
General operating expenses		**100%**	**100%**

Source: According to the data from (The Federal Aviation Administration 2016).

Categories of variable and fixed costs are given in Table 3.

Table 3. Categories of variable and fixed costs for aircraft operation.

Variable	Categories
Fuel and oil	Aviation fuel
	Aviation oil
	Taxes and others
Maintenance	Labor costs for aircraft design
	Aircraft structure repair
	Propulsion systems repair
	Aircraft replacement costs
	Consumables for aircraft structure
	Engine installation supplies
	Airworthiness of the aircraft structure
	Postponed overhaul of the aircraft structure
	Ensuring airworthiness for propulsion systems
	Postponed overhaul of propulsion systems
	On-board maintenance equipment
	Obsolescence and equipment wear
Crew	Salary of the pilot and co-pilot
	Salary of other flight personnel
	Staff costs
	Training
	Expenses for bonuses and pensions
	Taxes
Fixed	**Categories**
Depreciation	Aircraft structure
	Propulsion systems
	Spare parts for aircraft structure
	Spare parts for propulsion
	Other flight equipment
Rent	When replacing the aircraft
	Aircraft rental
	Flight equipment
Insurance	Insurance payment
Other	Injuries, damages, losses
	Other expenses
	Other purchases
	Other payments to specialists

Source: According to the data from (The Federal Aviation Administration 2016).

Moreover, the US Federal Aviation Administration recommends several electronic services for comparing more than 500 aircraft types (Conklin and de Decker 2014).

In one of the last works, Scott (2016) created a new approach based on the helicopter life cycle. An integral hypothesis of many integrated design methodologies is that the fulfillment of the requirements of the technical specifications for aircraft design involves various designs and technological ways of solving problems. Moreover, the helicopter's efficiency and its cost, respectively, depend on these paths. Any innovation invariably leads to growth in structural complexity and production costs.

The authors (Scott 2016) propose a methodology for assessing the technologies used, the effects obtained, and the cost of these technologies, both in the production and for the buyer. Moreover, the model considers the costs of the buyer, not only at the stage of buying a helicopter but also during the operation of the equipment.

The analysis of the components of the cost of helicopter technology is usually carried out on the stages of the life cycle of the equipment. The first group includes the costs associated with the new helicopter development, i.e., the cost of the research, design and experimental works, related equipment, and building prototypes for conducting strength and flight tests. The second group includes expenses for the preparation for mass production, i.e., the cost of creating new technology, designing, and manufacturing equipment, developing technological documentation, and upgrading production equipment. The third group covers the costs associated with logistics (the cost of materials, raw materials, and purchased products, engines, equipment, and weapons). This group also includes the costs of labor and of researching new materials. The fourth group takes into account the operating costs.

The first group of costs does not depend on the production scale; i.e., it does not matter if only one prototype or several thousand aircraft are built. Naturally, after the launch of mass production, these costs are distributed evenly among all the machines built so that the helicopter price is always lower.

Let us consider the helicopter cost for the client. The acquisition of a particular type of helicopter must be the result of careful cost analysis by the client (Chase 2016). Helicopter operational costs can be divided into fixed and variable (Chase 2016). Initial expenses are the purchase with the one-time payment or the installment. The following expenses are the insurance and the pilot services payment. Further such operating expenses are the rental of premises for the hangar and, possibly, an office follows. The basis of variable costs in aviation is fuel and helicopter maintenance.

When determining the potential models for purchase, the client initially selects those models that are suitable for such main parameters as (Chase 2016) maximum take-off weight, fuel tank capacity, fuel consumption, maximum payload with fuel, maximum payload without fuel, maximum flight range, overall cabin dimensions, etc. At the next stage, the buyer considers the magnitude of the variable costs in the helicopter operation per flight hour. The next step is determining the equipment depreciation.

At the final stage, the productivity index is determined, which includes the range, speed, cabin size, etc., which is compared with the helicopter cost (Chase 2016).

Experts (Chase 2016) draw attention to the fact that the estimated operating costs, compiled by the manufacturer, are often underestimated, so a preliminary cost analysis should be carried out independently.

Langer (2010) agrees with this point of view. In addition to the manufacturer using the lowest prices for fuel and oil for preliminary calculations, the operator has to fulfill the requirements for the inspections/replacements of components and assembly, which are sent by the manufacturer in separate technical bulletins.

To compare helicopters at the cost of both purchase and operation, Langer (2010) proposes a helicopter comparison index based on a 4-h operation. Even though the technique was originally intended to analyze the cost of helicopters for fighting forest fires, it may be applicable in other areas.

Most helicopter manufacturers offer their potential customers calculations that illustrate the operating costs for each of their models, but it is hard to say how accurate such estimates are. The main goal of the manufacturer at this stage is to sell the helicopter. Also, the numbers they provide usually refer to a specific level of use. In the case of Robinson, this level is 500 hours per year.

To obtain more accurate and individual data, the author (Morgan) developed a cost model for operating helicopters, which gives the figures specific to work. The model is based on a combination of official information from Robinson, real statistics collected from the owners, and real indicators of maintenance services from the maintenance companies.

The input data for the calculations are:

- An estimated number of flying hours per year; the helicopter age (year);
- Flying hours;
- Fuel costs per gallon;
- Hour-work costs of the mechanic;
- Insurance costs;
- Costs (per month) for helicopter storage.

The calculation results are:

- Main operating costs (fuel and oil) per flight hour;
- Maintenance costs per flight hour;
- Overhaul costs per hour;
- Insurance and hangar costs per flight hour;
- Total operation costs per flight hour;
- Total operation costs per month;
- Total operation costs per year.

The cost of the aviation equipment maintenance, repair, and overhaul accounts for a significant fraction of the total helicopter life cycle cost (Seemann, Langhans, Schilling and Gollnick 2011). Therefore, to assess the economic efficiency of commercial aircraft, it is extremely important to estimate their maintenance and repair costs. The largest share of them is the maintenance costs of aircraft engines.

For the owners, services exist that minimize maintenance and repair costs and optimize cash flows and operational flexibility. The Lufthansa Technik AG offers cost reduction by optimizing the forms of maintenance and repair, managing maintenance and repairs, minimizing operational risks, etc.

Such services for cost estimation have already found their niche among buyers (*Compare helicopters, piston, turbine…*, 2018). The Aircraft Cost Calculator

(ACC) helps users accurately determine the actual helicopter costs in minutes, not in hours or days, using a database of more than 500 types of aircraft and helicopters. The user also can use the ACC Show Wizard tool to create professional reports containing graphics, including the cost of repairing equipment failures in the time interval from a day to a year, and more.

In the work *"A Life Cycle Cost Approach for Military Aviation,"* the authors propose to decide on the choice of one or another type of aircraft based on a comprehensive analysis of the cost of the entire vehicle life cycle, taking into account the risk component. The life cycle concept provides data for analysis, comparison, and decision making. Understanding the operating costs of servicing equipment is important not only to increase profitability but also to systematize costs and make informed decisions in the future.

At the stage of the life cycle of a helicopter, expenses include the provision of spare parts (new or restored), thorough incoming inspection, the work of a technical specialist, etc. According to the work *"A Life Cycle Cost Approach for Military Aviation,"* these costs usually account for 70% to 80% of the cost of the entire life cycle.

Over the past 15–20 years, the situation with the formation of the value of helicopter equipment has changed. In the past, cost estimates were based on production and marketing costs. Today, client choice is affected by the cost of the entire life cycle (Eigner and Stelzer 2009). Customers are not ready to pay higher prices to service the equipment (Bierer and Götze 2011). This approach has significantly changed the business model of helicopter manufacturers, which vary from standard sales models to models with stronger service integration (Zhao and Harrison 2011). The same opinion is shared by Seemann, who analyzed the aircraft engine's life cycle, the most expensive unit to maintain as a part of the aviation equipment (Seemann 2010).

(Zhao, Harrison, Roy and Mehnen 2011) agree that the decision to purchase aviation equipment is always made after a cost analysis at the stage of product operation.

Several criteria are given in the works by Bagul, Zeid and Kamarthi (2008); Dragomir, Gouriveau and Dragomir (2009); Gorjian, Ma, Mittinty, Yarlagadda and Sun (2009); Luo, Namburu, Pattipati, Qiao, Kawamoto and Chigusa (2003); and Schwabacher (2005). Fernandes, Rajkumar and Andrew (2011) have grouped the methods of the assessment life-cycle.

- Model methods based on physical representation;
- Expert methods;
- A combination of the above methods;
- Statistical methods based on system operating data.

The helicopter cost formation, or rather, its production and operation cost, takes place in several stages (Boer, Stevens, Vankan and Lamme 2007). At the first stage, the appearance of the future helicopter is formed with a set of characteristics and properties. The next step takes into account the costs of maintenance during the operation. Often these two processes have different directions; i.e., between construction, technological solutions, and minimal operating costs, there is a need for a trade-off. The pre-design phase requires an interdisciplinary optimization approach. In other words, even at the development stage, it is necessary to represent the market segment for which the equipment is designed, the flights' specifics, etc. As a result, developers are analyzing the structure's "sensitivity" and its options for a future increase in operating costs.

Boer, Stevens, Vankan and Lamme (2007) propose a new software product called "Value Improvement through a Virtual Aeronautical Collaborative Enterprise" (VIVACE), which represents the whole design environment of the entire life cycle of the aviation equipment. This software environment will support the full cycle design of the equipment and its engines, providing virtual results for each life cycle phase. The software product has been adapted for use with helicopters such as the Vivace MDO Eurocopter (EC). The life cycle model reflects the influence of both the basic technical parameters and the basic requirements of the customers and their goals. Preliminary results led to several conclusions about the need to reduce the helicopter weight (i.e., the need to widely use composite materials).

(Loukopoulos, Katsiropoulos and Pantelakis 2018) set forth the impact on the cost of the helicopter of its life cycle but rather the formation of the value of the helicopter through the calculation of its life cycle. The models have been developed to quantify the costs of various structural and technological solutions (Table 4).

Table 4. The price of some popular helicopter models abroad.

Model	Min. price, th. $.
Robinson R-22	250.000
Robinson R-44 Raven I	340.000
Robinson R-44 Raven II	415.000
Bell B206 JetRanger	700.000
Eurocopter EC120 Colibri Hummingbird	1.700.000

Source: According to the data from (ToughNicke 2018).

Helicopter manufacturers struggling for the buyer are forced to reduce the costs of both the equipment itself and its operation costs. The only way to reduce the cost and the solution with the highest potential is the innovative introduction of new design and technologies solutions that allow the following:

- Reducing the equipment cost for the buyer;
- Reducing the equipment take-off weight, and as a result, reducing fuel consumption;
- Reducing the aggregates and consumables cost during routine maintenance;
- Improving the structure to increase the interval between service forms;
- Simplifying the structure to potentially carry out the forms (or parts) of service by a specialist without a compulsory license;
- Improving the structure to minimize the mandatory equipment when carrying out regulatory maintenance forms, etc.

Each foreign helicopter manufacturer chooses its own strategy. Innovative ways to reduce costs during the operation phase of aircraft are as follows (*Continuous technology upgrades…*, 2014):

- The improvement of components and assemblies based on statistical data about their operation;
- The application of new designs and technologies to new models;
- The introduction of advanced upgrade kits;
- Updates as the life cycle progresses;
- The Performance Improvement Packages – PIPs.

4 Discussion

Robinson R-44, Bell B206, and Eurocopter EC120 can be attributed to one group of competing products, at least by their basic characteristics, passenger capacity, range, etc. However, the producers' prices differ substantially. There is no doubt that the presence of various options justifies the cost of Bell B206 and Eurocopter EC120. There are many other factors (training costs, operating costs, storage costs, etc.) that can help the manufacturer compete within the helicopter market.

Airbus Helicopters (Eurocopter) took the path of simplifying maintenance requirements and achieved a reduction in its time, a decrease in the specialized equipment share, and technical personnel training. So, simple maintenance forms can be performed either at the flight transit point or in the equipment operator premises (AIRBUS 2018). More complex forms of maintenance and overhaul for the Airbus helicopter can be carried out either at a certified Airbus Helicopters service center or directly at the operator's facilities.

Airbus Helicopters offers customized solutions to increase the overall performance of the helicopter, reduce the cost of its maintenance, and ensure better preparation for its tasks. Airbus Helicopters offers over 1,000 standard retrofit solutions available through its worldwide network, based on service bulletins (SB) and optional service type certificate (STC). The standardization of such services (including kit preparation and dual certification) can reduce delivery times and increase competitiveness.

Bell Helicopter, more than ten years ago, took the path of reducing operating costs. For example, in the Bell 407 helicopter, after a series of innovative implementations, part of the maintenance work was excluded from the service bulletins. As part of an ongoing effort to reduce the Bell 407 direct operating costs, engineers revised a number of requirements in order to reduce the operator's maintaining airworthiness cost (VERTICAL 2007). According to the company, this innovation will lead to a reduction of more than 12% of the previously published cost estimates for direct maintenance.

In 2015, Bell Helicopter acquired Textron Aviation, which provides the components and assemblies repairs, exchanges, and the provision of spare parts. This is the leading direction of the corporation, which is aimed at reducing customers' operating costs.

"Bell Helicopter focuses on expanding the after-sales service to reduce operating costs and customer downtime," said Glenn Isbell, executive vice president of customer support and services at Bell Helicopter (Able Aerospace Services Inc. 2016).

The chief executive officer of Able Aerospace Services (Able Aerospace Services Inc. 2016), Lee Benson, said, "The strategy for safely reducing operating costs by providing resource repair components, major repairs and spare parts replacement is now significantly improved when the company became a service center for Bell helicopters".

The chosen strategy implies an annual 6% reduction in life cycle costs for new helicopters. Cost reductions should be achieved by reducing domestic costs and reducing the selling price by at least 3% per year. The spare parts cost should also be reduced by 3% or more per year (Bell Helicopter Textron Inc. 2011).

5 Conclusion

Foreign helicopter manufacturers in the production cost formation actively apply the product life cycle concept. This concept has found its application among helicopter operators. Its application allows us to identify the main ways to reduce operating costs, see what buyers are interested in, and introduce a new structure and new technological solutions for creating more attraction from the point of view of total costs throughout the helicopter technology life cycle. Such an approach has already been reflected in applied software solutions, which make it possible to make the best decision for the manufacturers and operators of the helicopters.

Acknowledgment. The authors thank the management of the holding company "Helicopters of Russia," as well as A. Chursin, Doctor of Economics, Professor, for their great help in this work and the valuable comments and corrections made.

References

Able Aerospace Services Inc.: Bell Helicopter announces able aerospace services as new service center (2016). https://www.ableengineering.com/press-releases/bell-helicopter-announces-able-aerospace-services-as-new-service-center/

AIRBUS: Aircraft maintenance, aircraft repair and on-site assistance (2018). https://www.airbus.com/helicopters/services/helicopter-maintenance.html

Bagul, Y., Zeid, I., Kamarthi, S.: Overview of remaining useful life methodologies. In: 2008 ASME International Design Engineering Technical Conferences and Computers and Information in Engineering Conference, DETC 2008, vol. 3, New York (2008). http://proceedings.asmedigitalcollection.asme.org/proceeding.aspx?articleid=1628295

Bell Helicopter Textron Inc.: Materiel: cost reduction (2011). https://www.bellflight.com/suppliers/materiel/cost-reduction

Bierer, A., Götze, U.: A target costing-based approach for design to energy efficiency. In: Hesselbach, J., Herrmann, C. (eds.) Glocalized Solutions for Sustainability in Manufacturing. Braunschweig, Germany. Proceedings of the 18th CIRP International Conference on Life Cycle Engineering (LCE 2011) (2011)

Boer, J., Stevens, J., Vankan, W., Lamme, W.: Multi-role helicopter life cycle cost (LCC) optimization: the pre-design strategy (2007). http://toc.proceedings.com/08121webtoc.pdf

Chase, M.: Helicopter comparison: airbus AS350-B3 versus Bell 206L-4 (2016). https://www.avbuyer.com/articles/helicopter-comparison/helicopter-comparison-airbus-as350-b3-versus-bell-206l-4-69360

Conklin, A., de Decker, W.: The aircraft cost evaluator, Orleans (2014). https://www.conklindd.com/t-conklindedeckerfreeaviationcosttools.aspx

Continuous Technology Upgrades Through Engine Lifecycles (2014). http://aviationweek.com/continuous-technology-upgrades-through-engine-lifecycles

Compare Helicopters, Piston, Turbine & Other Aircraft with Aircraft Cost Calculator (2018). https://www.controller.com/blog/how-to-and-tips/2018/03/compare-helicopters-piston-turbine-and-other-aircraft-with-aircraft-cost-calculator

Dragomir, O., Gouriveau, R., Dragomir, F., Minca, E., Zerhouni, N.: Review of the prognostic problem in condition-based maintenance (2009). https://hal.archives-ouvertes.fr/hal-00418761/document

Eigner, M., Stelzer, R.: Product lifecycle management. In: Ein leitfaden für product development und life cycle management. Springer, Heidelberg (2009)

Fernandes, P., Rajkumar, R., Andrew, H.: An overview of degradation modeling for service cost prediction (2011). https://link.springer.com/content/pdf/10.1007/978-3-642-19689-8_54.pdf

Gorjian, N., Ma, L., Mittinty, M., Yarlagadda, P., Sun, Y.: A review on degradation models in reliability analysis (2009). https://link.springer.com/chapter/10.1007/978-0-85729-320-6_42

HeliNews (n.d.): Single engine turbine helicopter comparisons. http://www.helinews.com/turbinecomparison.php

Langer, M.: The real cost of helicopter ownership (2010). https://www.aneclecticmind.com/2010/12/28/the-real-cost-of-helicopter-ownership/

Loukopoulos, A., Katsiropoulos, Ch., Pantelakis, S.: Life cycle assessment and cost analysis evaluation of a helicopter's canopy production using different manufacturing processes (2018). https://www.matec-conferences.org/articles/matecconf/abs/2018/47/matecconf_iceaf-v2018_01020/matecconf_iceaf-v2018_01020.html

Luo, J., Namburu, M., Pattipati, K., Qiao, L., Kawamoto, M., Chigusa, S.: Model-based prognostic techniques (2003). https://www.researchgate.net/publication/4042785_Model-based_prognostic_techniques_maintenance_applications

Meyer, S.A., Schrage, D.P.: Parametric cost analysis of helicopters and advanced rotorcraft designs (2011). https://www.tandfonline.com/doi/pdf/10.1080/10157891.1987.10472808?needAccess=true/

Schwabacher, M.: A survey of data-driven prognostics (2005). https://www.researchgate.net/publication/245094037_A_Survey_of_Data_Driven_Prognostics

Scott, R.: A new rotocraft design framework based on reliability and cost. (Dissertation of Doctor in Economics) (2016). https://www.aircraftcostcalculator.com/helicopter-operating-costs

Seemann, R.: Modeling the life cycle cost of jet engine maintenance (2010). https://ru.scribd.com/doc/42773201/Modeling-the-Life-Cycle-Cost-of-Jet-Engine-Maintenance

Seemann, R., Langhans, S., Schilling, T., Gollnick, V.: Modeling the life cycle cost of jet engine maintenance (2011). https://elib.dlr.de/75837/

The Federal Aviation Administration: Aircraft operating costs (2016). https://www.faa.gov/regulations_policies/policy_guidance/benefit_cost/media/econ-value-section-4-op-costs.pdf

ToughNicke: How much does a helicopter cost? (2018). https://toughnickel.com/misc/How-much-does-a-Helicopter-cost

VERTICAL: Bell helicopter reduces direct operating costs for the proven Bell 407 helicopter (2007). https://www.verticalmag.com/news/bell-helicopter-reduces-direct-operating-costs-for-the-proven-bell-407-helicopter-html/

Zhao, Y., Harrison, A.: Aircraft engine component deterioration and life cycle cost estimation. In: Hesselbach, J., Herrmann, C. (eds.) Glocalized Solutions for Sustainability in Manufacturing. Springer, Heidelberg (2011)

Centers of Global Technological Excellence – Drivers of Innovative and Sustainable Development in Corporations

A Competitiveness Assessment of the Russian Companies: Global Landmarks

Darina A. Sizova(✉), Elena K. Voronkova, and Tatiana V. Sizova

Plekhanov Russian University of Economics, Moscow, Russia
Darina3@yandex.ru, Voronkova.EK@rea.ru, Stv-60@mail.ru

Abstract. In the contemporary world, in the era of production globalization and internationalization, one of the most strict criteria for the effectiveness of a company and the relevance of its products is competitiveness. The concept of competitiveness is very relative since a product being competitive in some markets can completely lose in competition in other markets. And this fact forces companies and states to more carefully develop their competitive strategy and improve the quality characteristics of their products. The goal of the study set by the authors was to study the issues of assessing the Russian companies' competitiveness concerning their management practices. In the process of studying the issue of the Russian companies' competitiveness, the authors of this article used such methods of scientific knowledge as the method of generalizations and comparisons, statistical and factor analysis, etc. The article develops ideas for systematizing approaches to assessing the competitiveness level of the Russian companies. The authors analyze the reasons and factors affecting the competitive environment of the Russian market. Also, they discuss the competitiveness indicators of the Russian Federation at the macro and micro levels. The need to develop a system for assessing the competitiveness of Russian companies based on global methodologies is justified.

Keywords: Competitiveness · Competitive environment · National economy · Financial competitiveness indicators · Global Competitiveness Index

1 Introduction

The strategy of strengthening the Russian economy as a goal determines the formation of conditions for creating high-quality growth in domestic production (Chursin and Makarov 2015; Chursin and Tyulin 2018; Ragulina et al. 2020). The growing competitiveness of Russian companies will make their full inclusion in the global division of labor possible, strengthen their international position by improving the technical base, improving the quality of products. Active assistance to the development of competition in the Russian Federation has been declared as a priority area of activity of the President of the Russian Federation, the Federal Assembly of the Russian Federation, the Government of the Russian Federation, the Central Bank of the Russian Federation, federal executive bodies, legislative (representative) and administrative bodies of state power of the constituent entities of the Russian Federation, as well as local governments.

A. V. Bogoviz and Y. V. Ragulina (Eds.): ISCI 2019, LNNS 115, pp. 501–508, 2020.
https://doi.org/10.1007/978-3-030-40749-0_59

The National Plan for the Development of Competition in the Russian Federation for 2018–2020 (Presidential Executive Office 2017) is valid in our country.

Competitiveness is a multi-purpose indicator that characterizes the stability, innovativeness, success of the subject (Sizova and Sizova 2017a). State competitiveness is primarily determined by corporate competitiveness, and it is directly related to the objectivity of its assessment. Ensuring a decent level of the country's competitiveness implies choosing the right direction of movement. Consequently, this is also about supporting those companies whose activities are most likely to lead to positive results (Slepov 2017). This condition is achievable if the state has a universal system for assessing the competitiveness of companies that combines the methodologies and business practices of global and national markets.

2 Materials and Method

In the process of studying the competitiveness of Russian companies, the authors used such methods of scientific knowledge as the method of generalizations and comparisons, statistical and factor analysis. Methodological approaches to assessing competitiveness, specific factors affecting competitiveness in the Russian Federation and abroad, are established. The dynamic series of the global competitiveness index, financial competitiveness indicators, such as solvency of Russian companies, fixed capital investments, depreciation of fixed assets in the Russian Federation, as well as the indicators of the Forbes international rating, reflecting the level of corporate competitiveness of the largest companies in the world, have been built.

3 Results

Geopolitics led to a tightening of operating conditions in the Russian market. In comparison to the previous year, only in 2018, the outflow of capital from Russia almost tripled, amounting to an unprecedented $ 67.5 billion. It broke the trend of moderate indicators of capital export from the country over the past two years. In total, over five years, from 2014 to 2019, the outflow of capital from the Russian Federation amounted to 316.1 billion US dollars, which is comparable with the annual size of the international reserves of the Russian Federation (Fig. 1).

The number of non-resident companies operating in Russia has declined. So, over the past five years, a total of 1,300 German companies have left the Russian market. Major companies included the MediaMarkt retailer, AirBerlin airlines, EasyJet low-coster, as well as the C&A department store chain (Buhelle 2019). Companies from other countries also left Russia, such as the Finnish retailer Stockmann, automobile manufacturers Opel and Chevrolet, the British home goods store "LauraAshley," the Swiss insurance company "Zurich," etc.

Until 2018, the dynamics of foreign direct investment was negative. In 2018, the rate of foreign investment in the Russian economy was the lowest in the last 11 years. In comparison to 2017, it decreased by 14 times and 30 times if considering the "pre-crisis" 2013 (Central Bank of Russia n.d.).

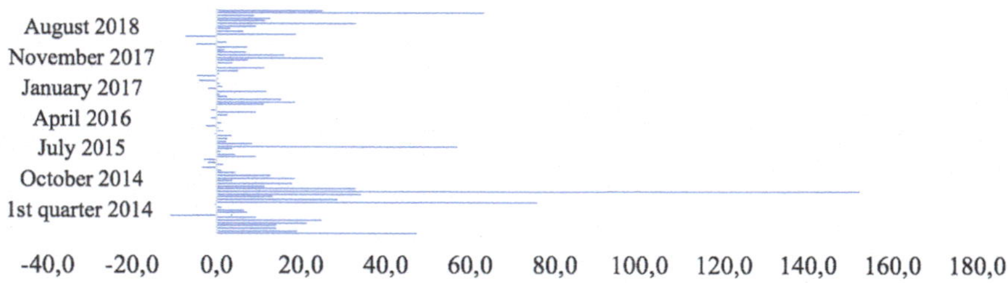

Fig. 1. The balance of financial operations of the private sector (net lending (+)/net borrowing (−), according to the balance of payments of the Russian Federation in USD bln. *Source:* (Central Bank of the Russian Federation (Bank of Russia n.d.a).

Consequently, it is highly necessary to seek new drivers for the development of the domestic economy, areas of foreign economic activity, and build new partnerships. In recent years, the trade turnover of the Russian Federation with the countries of the CIS, BRICS, EAEU, SCO, APEC has been increasing. So, according to the results of 2018, the most significant increase in the Russian Federation's trade turnover was noted with the BRICS countries (amounted to 22.4%). With the SCO countries, it amounted to 20.8%. Trade with the APEC countries increased by 19.8%, and by 10.8% if considering the CIS countries (Ministry of Economic Development of the Russian Federation n.d.). However, Russian companies were not able to fully restore their international trade scale, primarily due to their low competitiveness.

We consider that the aggravation of the competitiveness problem for Russian companies is the result of the inconsistency of the current situation. On the one hand, in the absence of proper competition, Russian companies gained extensive opportunities in the national market. On the other hand, the lack of an adequate competitive environment is holding back innovation in all areas, namely technical, financial, and management. To solve their problems, companies often use extensive methods. For example, rising costs are not offset by increased labor productivity through financing the development of production, but by a banal increase in prices. The result is the extreme backwardness of the technical base of Russian companies. Depreciation of fixed assets is about 50%, and it does not tend to decrease (Fig. 2).

At the same time, there are no existing investments in fixed assets (Table 1). It is noteworthy that a similar picture is typical for the economy as a whole and our leading exporters. More than that, it is typical for research and development (Voronkova and Dolgova 2019).

Following the Global Competitiveness Index (GCI), which is determined by the World Economic Forum, the Russian Federation ranks 38th in the world rating (in 2017–2018). It is well known that the Global Competitiveness Index points to the most vulnerable points of the national economy. Conclusions of the World Economic Forum for Russia exactly determine the cause-effect relationships of the competitiveness of companies and their investment activities. In particular, they indicate that unique problems in the Russian Federation are related to the level of development of the sphere of innovation, the ability to conduct business (Fig. 3). Although one should consider that the WEF also notes some progress on the Russian market, raising the rating of the country as a whole.

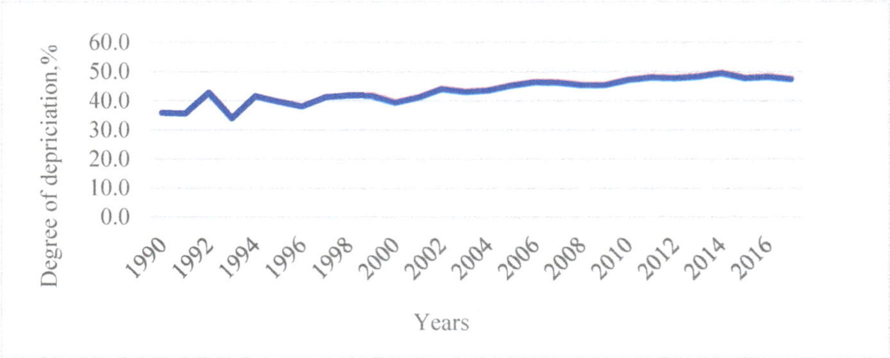

Fig. 2. The degree of depreciation of fixed assets of the Russian Federation at the end of the reporting year. *Source:* (Federal State Statistics Service (Rosstat) n.d.a).

Table 1. Investments in fixed assets in total, and for some types of economic activity, bln. rub.

Investments in fixed assets	2014	2015	2016	2017	2018
Total, billion rubles	**13902.6**	**13897.2**	**14748.9**	**16027.3**	**17595.0**
Including by type of activity:					
Mining	2144.8	2385.2	2710.5	3023.2	3199.6
Manufacturing industries	2080.9	2167.6	2103.3	2296.5	2584.9
Research and development	137.3	184.5	174.7	165.3	191.8

Source: (Federal State Statistics Service (Rosstat) n.d.c).

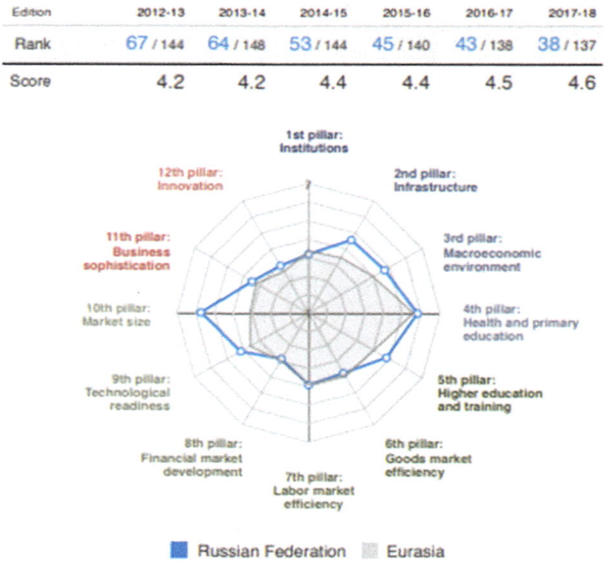

Fig. 3. Assessment of the national development of the Russian Federation following the GCI. *Source:* (Schwab et al. 2017).

The traditional financial competitiveness assessment adopted in our country also gives alarming signals. The autonomy coefficient by Russian standards fluctuates around optimal values. But the current liquidity indicator is below the optimal level. In some periods, it approaches a problem line. With the ratio of own working capital, the situation is even worse (Fig. 4).

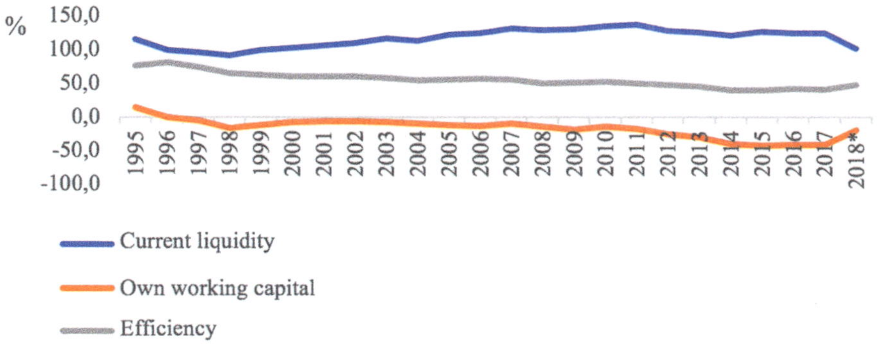

Fig. 4. Dynamics of organizations' solvency ratios (excluding small businesses) in the Russian Federation (according to financial statements, in %). *Source:* (Finance. Official Statistics n.d.).

At the same time, the profitability indices of the domestic manufacturing sector look quite calm (Fig. 5). In our opinion, this discords with those trends of competitiveness indicators mentioned earlier. However, the analysis reveals those reasons used for the "carelessness" of Russian companies in the innovation sphere.

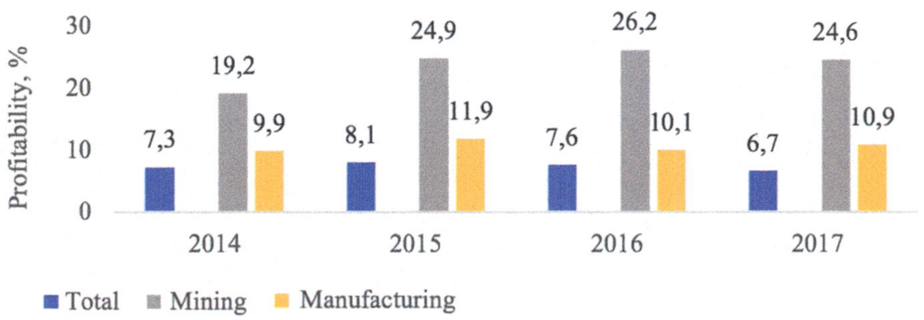

Fig. 5. The dynamics of the profitability of goods sold, products (works, services) for the period 2014–2017. *Source:* (Federal State Statistics Service (Rosstat) n.d.b).

4 Discussion

The analysis allows us to raise the question of the need to develop a system of indicators to assess the competitiveness of Russian companies. Although for the first time since 2015, business representatives noted an increased competition both directly and through indirect indicators in 2018, the share of those entrepreneurs who indicate a decrease in the number of competitors is increasing. It's not bad that companies are working to reduce costs if the question of competition arises. Such companies make up the majority. But we can recognize the following trend as unfavorable. The trend is because only in specific sectors (namely food, metallurgy, agriculture, and some others), such methods of increasing competitiveness as technical re-equipment, R&D, the introduction of patents, know-how, licenses, etc. get benefits (Kovaleva et al. 2019).

In our opinion, among the indicators for assessing competitiveness, those that characterize the modernization of production and innovative activities of the company must necessarily appear (Sizova et al. 2017). Particular attention should be paid to the indicators by which our companies are definitely behind the best companies in the world. First of all, this is market capitalization, which reveals not only the company's competitiveness but also the indicator of investment attractiveness associated with it. The well-being of this indicator is of interest to investors, i.e., the company solves financing issues and to increase competitiveness. Not without reason, the ranking of the largest companies in the world Global 2000 (which is Forbes) is based on this indicator and also includes indicators of revenue, profit, assets. Most of the 25 Russian manufacturing companies (27 companies in 2017) included in this list in 2018 are located behind the line of the third hundred. Only the Gazprom PJSC (43rd place; 40th place in 2017), Rosneft PJSC (73rd place), and Lukoil PJSC (98th place) (Table 2) made an exception.

Table 2. Place of Russian companies in the Global 2000 ranking of Forbes in 2018, billion US dollars.

Place	Company name	Revenue	Profit	Assets	Market price
43	Gazprom	112.2	12.2	316.8	57.8
73	Rosneft	94.8	3.9	214.2	69
98	Lukoil	99.9	7.2	92	60.4
335	Surgutneftegaz	19.8	3.3	74.5	17.2
551	Novatek	10.5	2.2	18.9	37.9
577	Tatneft	11.9	2.1	19.2	25.6
660	Norilsk Nickel	9.2	2.1	16.6	27.5
667	Transneft	15.2	3.3	50.5	4.4
739	ROSSETI	16.3	1.8	42.2	2.7
861	NLMK	10.7	1.6	11.2	15.3
982	Severstal	8.3	1.5	7.1	13.4
1173	Inter RAO	15.7	0.923	10.9	6.7
1214	Magnitogorsk Iron and Steel Works	7.9	1.2	7.9	8.3
1366	Alrosa	4.7	1.3	7.4	10.5
1552	Polus Gold	2.7	1.2	6.5	8.3
1777	RusHydro	6	0.412	18	5.1
1822	United Aircraft Company	7.7	−0.23	17.2	4.8
1993	Uralkali	2.8	0.874	9	5

Source: (Forbes Media 2019).

The proposed set of indicators for assessing competitiveness should find practical application in the domestic competition support system. It will help orient Russian companies to the most critical growth points and ensure their systematic development.

5 Conclusion

A competitive economy is a productive economy that can provide economic growth and a decent level of income. At present, in the Russian Federation, a steady request to increase corporate competitiveness by strengthening motivation for the effective use of progress, optimizing costs and revenues, and distributing profits have been formed. The national project "International Cooperation and Export" is aimed at the development of competitive industries (Presidium of the Presidential Council for Strategic Development and National Projects 2018). At the same time, the state support for companies must be based on a wide range of assessment indicators measuring potential competitiveness and reflecting both global trends and the specifics of the Russian reality. A healthy competitive position of Russian companies will create conditions for increasing the stability of the national market to crises, sanction pressure, internal and external market fluctuations (Sizova and Sizova 2017b).

Acknowledgments. The authors thank their colleagues for their help.

References

Buhelle, V.: New US sanctions may prevent partnerships between the Russian Federation and the Federal Republic of Germany (interview) (2019). https://www.interfax.ru/interview/648256

Central Bank of Russia: External sector statistics (n.d.a). https://www.cbr.ru/vfs/statistics/credit_statistics/bop/outflow.xlsx

Central Bank of Russia: External sector statistics (n.d.b). http://www.cbr.ru/statistics/?ch=PAR_17218&prtid=svs#CheckedItem

Chursin, A., Makarov, Y.: Management of Competitiveness: Theory and Practice. Springer, Heidelberg (2015)

Chursin, A., Tyulin, A.: Competence Management and Competitive Product Development: Concept and Implications for Practice. Springer, Heidelberg (2018)

Federal State Statistics Service (Rosstat): Depreciation of fixed assets at the end of the year (n.d.a). http://www.gks.ru/wps/wcm/connect/rosstat_main/rosstat/ru/statistics/enterprise/fund/

Federal State Statistics Service (Rosstat): Finance: Official statistics (n.d.b). http://www.gks.ru/wps/wcm/connect/rosstat_main/rosstat/ru/statistics/finance/

Federal State Statistics Service (Rosstat): Investments in fixed assets in the Russian Federation by type of economic activity (n.d.c). http://www.gks.ru/free_doc/new_site/business/invest/tab_inv-OKVED.htm

Forbes Media: The world's largest public companies: Ranking 2018 (2019). https://www.forbes.com/global2000/list/41/#tab:overall

Kovaleva, E., Volkov, A., Malih, S., Gerasimova, E. (eds.): Assessment of the competitive environment in Russia (2019). http://ac.gov.ru/files/publication/a/21420.pdf

Ministry of Economic Development of the Russian Federation: The results of foreign trade in 2018 (n.d.). http://www.ved.gov.ru/files/images/slaids%202018.pdf

Presidential Executive Office: Decree of the President of the Russian Federation "On the main directions of state policy on the development of competition" (2017). https://fas.gov.ru/documents/614337

Presidium of the Presidential Council for Strategic Development and National Projects: Passport of the national project (program) "International Cooperation and Export", No. 16, 24 December 2018 (2018). http://static.government.ru/media/files/FL01MAEp8YVuAkvbZotaYtVKNEKaALYA.pdf

Ragulina, J.V., Bogoviz, A.V., Lobova, S.V., Alekseev, A.N., Pyatanova, V.I.: Strategy of increasing the global competitiveness of Russia's economy and Russia's becoming a new growth vector of the global economy. In: Lecture Notes in Networks and Systems, vol. 73, pp. 203–210 (2020)

Schwab, K., Sala-i-Martin, X., Samans, R. (eds.): The Global Competitiveness Report 2017–2018 (2017). http://www3.weforum.org/docs/GCR2017-2018/05FullReport/TheGlobalCompetitivenessReport2017%E2%80%932018.pdf

Sizova, T.V., Sizova, D.A.: Methodological approaches to the diagnosis of the financial crisis of the company. J. Leg. Econ. Stud. **2**, 254–260 (2017a)

Sizova, T.V., Sizova, D.A.: Making managerial decisions based on the results of the diagnosis of the crisis financial condition of the company. J. Leg. Econ. Stud. **3**, 135–141 (2017b)

Sizova, T.V., Sizova, D.A., Volobuev, N.A.: Diagnostics of the level of economic security of organizations in contemporary conditions of economic development. In: Public and Private Finance in an Innovative Economy, pp. 205–214. Plekhanov Russian University of Economics, Moscow (2017)

Slepov, V.A. (ed.): Financial Sources of Growth of the Russian Economy. Molodaya Gvardiya, Moscow (2017)

Voronkova, E.K., Dolgova, M.V.: Foreign exchange priorities of the financial policy of Russia: expectations and results. Financ. Money Invest. **1**, 3–12 (2019)

Improving the Functioning of High-Tech Sectors in the Russian National Economy on the Basis of Cluster Policy and in the Conditions of Digitization: Perspectives from the Defense Industrial Complex

Arsen S. Abdulkadyrov[1(✉)], Natalya A. Rykhtikova[2], and Aleksandr I. Tarasov[3]

[1] Institute of Socio-Political Research of the Russian Academy of Sciences, Moscow, Russia
ars.rggu@mail.ru

[2] Russian Presidential Academy of National Economy and Public Administration, Moscow, Russia
rykhtikova-na@ranepa.ru

[3] Military University of the Ministry of Defense of the Russian Federation, Moscow, Russia
tarasov.a.i@mail.ru

Abstract. The evolutionary development of the leading world states leads to the post-industrial, and subsequently the newest economy creation, which is the knowledge, innovation, intellectual, scientific, and digital economy. The paper covers the functioning and improvement of the Russian national economy's high-tech sectors based on their clustering, taking into account the digital development aspect declared and implemented by the government. The authors identified the following problem points: (1) the research insufficiency in the scientific analysis of cluster policy features in the economy's high-tech sectors; (2) the need to study foreign experience in conducting similar studies, evaluating innovative processes and potentials of high-tech industries, and introducing digital technologies. The scientific novelty of this study is represented by two key provisions. First, the author proposes a typology of key factors for improving the activities of the high-tech sector of the national economy (military-industrial complex). The typology is based on the industry clustering approach, in accordance with the strategic planning documents adopted in Russia in 2017 with the purpose of advancing the introduction of digital technologies. Second, we provide recommendations for improving the functioning of the national defense industry, using its competitive advantages, such as the main directions of developing and implementing the cluster strategy to improve the functioning of this state system-forming industry. The main takeaway is that the effective implementation of the cluster strategy in the defense production field will provide an opportunity to form and develop competitive high-tech clusters with a strong market position, digital technological superiority in the defense industry, and the ability to innovate.

A. V. Bogoviz and Y. V. Ragulina (Eds.): ISCI 2019, LNNS 115, pp. 509–515, 2020.
https://doi.org/10.1007/978-3-030-40749-0_60

Keywords: High-tech cluster · National security · Military-industrial complex · Digital economy · Innovations · Competitiveness

1 Introduction

Since the beginning of the 21st century, a point of view has been steadily forming among the Russian and foreign scholars that one of the ways to increase the competitiveness level of the national economic system and gain leadership in world markets is the cluster approach, which implies the creation of clusters in various industries and areas, suggesting innovative approaches in implementing the strategies of economic development (Bogoviz et al. 2017). The relevance of the chosen research topic is related to the need to improve the functioning efficiency and enhance the competitive advantages of high-tech sectors, which are the locomotives of the national economy, in the context of the ongoing implementation of the State Program "Digital Economy of the Russia Federation" (Government of the Russian Federation 2017). It is worth noting that a number of legal and normative acts have been already adopted in Russia to stimulate the development of the digital economy in the country (Federal Law of the Russian Federation 2015; Federal Law of the Russian Federation 2014; Federal Law of the Russian Federation 2012).

The research object is the military-industrial complex of Russia, which is the high-tech sector of the national economy. Management and economic relations arising in the high-tech enterprises operation in the military-industrial complex during the implementation of cluster development policy are considered to be the subject of our research. The paper's object is to study the foundations and promising opportunities for the clusters formation in the national economy's high-tech sectors as one of the key digital economy areas, which are based on cluster theory and its implementation in the Russian defense industry, in order to study possible ways of increasing its competitiveness.

2 Materials and Methods

In the contemporary scholarship, the authors used such traditional cognition methods for solving the tasks as structural analysis methods, dialectical, logical, scientific generalizations, methods of tabular and graphical interpretation, which made it possible to ensure the theoretical conclusions and practical recommendations validity. It should be noted that according to the 2017 Global Innovation Index (GII), Russia occupies only the 48th place (Dutta et al. 2017). At the same time, the index has a total of 79 indicators reflecting both the state's innovative capabilities and the available indicators, out of which the "weak links" and "cluster development" are the weakest indicators in Russia's performance in the "infrastructure" block. It is worth noting that Russia actively participates in international integration processes and implements a number of federal targeted development programs. The problem of creating various cluster structures is particularly acute, the necessity of which makes the presented research highly relevant.

Among many foreign and domestic authors, we should highlight the contribution of scientists involved in studying these issues: (Alekseev et al. 2018), (Khachaturyan et al. 2016), (Chen and Kamall 2016), (Enright 2003), (Etzkowitz and Leydesdorff 2000), (Martin and Sunley 2003),

It is also important that the authors of this very paper are actively involved in the clustering problems of the high-tech industries based on the innovative products manufacturing (Khachaturyan et al. 2017; Khachaturyan et al. 2018; Khachaturyan et al. 2018). Another important research field is the intensification of development based on digital technologies (Abdulkadyrov 2017).

3 Results and Discussion

The authors analyzed and reviewed the country's high-tech military-industrial complex as a potential platform for the development of clusters. We note that those economic resources that are not so productively used in other industries can be transferred to the sector. Since this is a mutual process, the resources transfer can be considered mutually beneficial. Such cluster structures fit in and function effectively within the digital economy framework. A feature of the cluster mechanism is the spread of competitiveness factors up the technological chain. Therefore, the creation of clusters in the complex would allow defense enterprises to make demands on improving the quality of components, which are supplied to them, thereby forcing their suppliers to increase their competitiveness. The authors proposed a typology of key factors for improving the high-tech sectors of the national economy (focusing on the defense industry) activities based on clustering. The authors also determined the system properties of the proposed high-tech cluster, which are graphically presented in Fig. 1. The typology is based on the following characteristics:

- Specialization;
- Geographic concentration;
- A large number of economic agents;
- Achieving the necessary "critical mass" in the cluster size, which makes it possible to obtain a positive effect (externalities) in the defense and other fields from its formation and development;
- Competition and mutually beneficial cooperation;
- Involvement of the cluster companies in the innovation process by including them into the technological, product, market, innovation, and digital technologies processes;
- A long-term viability of clusters, which characterizes the presence in the "research baggage" of breakthrough digital technologies in promising development areas.

The scientific review and analysis of domestic and foreign monographic literature and publications made by the authors clearly shows the significance and timeliness of our research, which essence is to update the data, update the scientific theoretical and methodological provisions. The proposed approach to building a high-tech sector in the military-industrial complex of Russia having a cluster structure is based on a systematic view of the main cluster links, including the management impact on the organizational

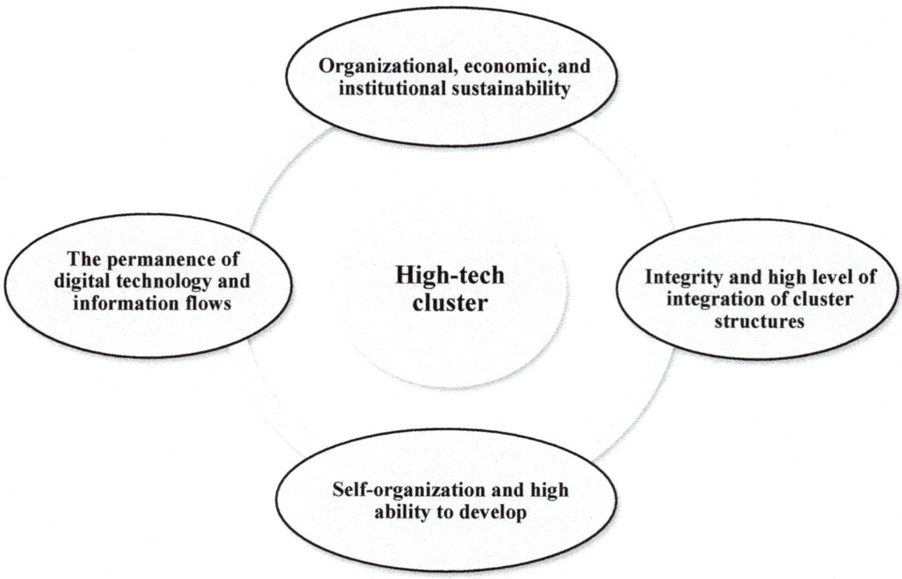

Fig. 1. The structure of the high-tech cluster system properties.

and economic production, as well as scientific, technical, technological, digital aspects of its creation and development. We strongly believe that a created cluster should be integrated into a regional or national innovation system, because such system integration can generate a synergistic effect in the conditions of digitalization. In connection with the adoption and implementation of the Digital Economy Program by the Russian Government, serious adjustments are necessary in the conditions of conducting innovative and investment activities in the domestic defense industry, which should be used as the basis for the so-called "cluster strategies." The emerging cluster strategy in the defense industry should have such key innovative priorities as nanotechnology, digital technology, latest aerospace technology, new defense-technical systems and security systems, biotechnology and biomedicine, etc. (Fig. 2).

An assessment of the cluster structure effectiveness being created in the high-tech industry should be carried out using quantitative and qualitative methods.

Quantitative methods for evaluating the cluster structures effectiveness are very diverse. For example, one could use the following: (1) indicators of the individual cluster members' financial performance (based on the cost-benefit ratio traditional approach); (2) summary indicators characterizing individual areas of cluster activity (production, research, innovation, etc.); (3) direct indicators of the cluster participants' integration effectiveness (total cluster participants costs, their assets market value, gross profit, etc.). The proposals made by the authors to substantiate promising areas, which ensure the construction of high-tech clusters within the digital economy, as well as their development through implementing the cluster strategy effectively, do not obviate the need for further detailed elaboration of this issue.

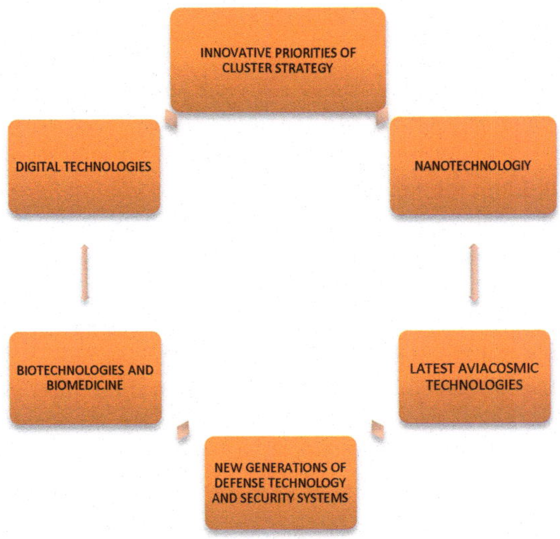

Fig. 2. The main innovation priorities of a cluster strategy.

4 Discussion

The weakness in introducing innovation and the scientific and technological backwardness of the Russian national economy are particularly clear in the framework of further digitalization. And the high-tech sectors, including the leading defense industry, are not an exception (unfortunately). Cluster policy, which is based on open intense competition between companies and industries, largely determines the increase in competitiveness. The innovation development strategy in the scale of the entire state economy with cluster elements looks somewhat limited. As a rule, they are narrowed down to general theoretical discourses on the measures content taken in this area at the level of the Russian Government, which nevertheless does not lead to quick and concrete results, and such results are so necessary for the most innovative structures. The program on Digital Economy of the Russian Government provides promising opportunities for the formation of competitive high-tech clusters, including those in the defense industry, which combine innovative and defense-industrial potentials in order to obtain a synergistic effect of innovative-technological and digital development. This is especially important due to the fact that the defense industry is a branch of the national economy that remains competitive in foreign trade markets. In this regard, today there is a need to develop a conceptual framework for the formation of cluster policy, highlighting the fundamental principles on which it should be implemented, determining its most promising areas and tools.

5 Conclusion

The Russian and foreign literature analyzed by the authors on the chosen topic makes it possible to determine the range of further studies related to the formation of clusters in high-tech industries, namely, in the system-forming and vital military-industrial complex of Russia within the framework of transformations in the field of digital economy. The scientific research novelty is represented by two key provisions: (1) a typologization is proposed, describing the key factors of improving activities of the high-tech sector of the national economy (military-industrial complex) based on clustering the industry in the framework of ongoing digitalization; (2) recommendations are formulated for improving the functioning of the national defense industry complex, based on using its competitive advantages, including the main directions in the formation and implementation of the cluster strategy for improving the state's system-forming industry efficiency.

References

Abdulkadyrov, A.S., Ryzhov, I.V., Strokov, A.I., Kamzolov, Y.V.: Actual aspects of improving the organization of high-tech products. Proc. High. Educ. Inst. Technol. Text. Ind. **5**(371), 18–27 (2017)

Alekseev, A.N., Evdokimov, S.Y., Khachaturyan, K.S., Tarasova, A.Y., Khachaturyan, A.A.: Financial strategy of development of industry 4.0 in the countries with developing economy. Espacios **39**(12), 1 (2018)

Bogoviz, A.V., Ioda, E.V., Ioda, Y.V., Kuranova, V.B., Bobrova, V.V.: Cluster development of innovational entrepreneurship: new possibilities and priorities in the conditions of the innovational economy creation. Eur. Res. Stud. J. **20**(3), 530–538 (2017)

Chen, W., Kamal, F.: The decision of the law and the international law of the world. J. Int. Bus. Stud. **47**(5), 563–576 (2016)

Dutta, S., Lanvin, B., Wunsch-Vincent, S.: The global innovation index 2017: innovation feeding the world (2017). https://www.wipo.int/edocs/pubdocs/en/wipo_pub_gii_2017.pdf

Enright, M.J.: Regional clusters: what we know and what we should know. In: Bröcker, J., Dohse, D., Soltwedel, R. (eds.) Innovation Clusters and Interregional Competition, pp. 99–129. Springer, Berlin (2003)

Etzkowitz, H., Leydesdorff, L.: The dynamics of innovation are from the national systems and the international standardization. Res. Pol. **29**, 109–123 (2000)

Federal Law of the Russian Federation: The Federal Law on the Advanced Research Fund (October 16, 2012 No. 174-FZ). Rossiyskaya Gazeta, Moscow, Russia (2012)

Federal Law of the Russian Federation: The Federal Law on Industrial Policy in the Russian Federation (December 31, 2014 No. 488-FZ). Rossiyskaya Gazeta, Moscow, Russia (2014)

Federal Law of the Russian Federation: The Federal Law on public-private partnership, municipal-private partnership in the Russian Federation and amendments to certain legislative acts of the Russian Federation (July 13, 2015 No. 224-FZ). Rossiyskaya Gazeta, Moscow, Russia (2015)

Government of the Russian Federation: Order on Approving the Program "Digital Economy of the Russian Federation" (July 28, 2017 No. 1632-p) (2017). http://base.garant.ru/71734878/

Khachaturyan, K.S., Ogurtsov, E.S., Rylov, A.A., Durdyeva, D.A., Lebedev, N.A., Safyan, A. M., Rochev, K.V., Tsareva, G.R.: The role of business incubators in supporting economic growth and advancement of small business of the Russian Federation. Indian J. Sci. Technol. **9**(29), 89339 (2016)

Khachaturyan, A.A., Abdulkadyrov, A.S., Zhigulina, E.P., Sirotkina, N.V.: Issues of improving the investment climate and investment attractiveness of Russian industries in the medium term. Proc. High. Educ. Inst. Technol. Text. Ind. **5**(371), 40–43 (2017)

Khachaturyan, A.A., Khachaturyan, K.S., Abdulkadyrov, A.S.: Model of innovational development of modern Russian industry. Adv. Intell. Syst. Comput. **622**, 44–50 (2018)

Khachaturyan, A.A., Ponomareva, S.V., Melnikova, A.S.: The impact of digitalization and industrialization on asset planning and scientific and technological development of the production and economic system of the Russian Federation. St. Petersburg State Polytech. Univ. J. Econ. **11**(4), 60–64 (2018)

Martin, R., Sunley, P.: Deconstructing clusters: chaotic concept or policy panacea? J. Econ. Geogr. **3**, 5–35 (2003)

Securing the Border Territories of the Russian Federation: Features of the Implementation of Industrial Policy in the Arctic

Irina V. Makarova[1(✉)], Grigory V. Lepesh[2], and Olga D. Ugolnikova[2]

[1] Ural State University of Economics, Yekaterinburg, Russia
k5ll@mail.ru

[2] St. Petersburg State University of Economics, St. Petersburg, Russia
ctoubt@mail.ru, olga_ugolnikova@mail.ru

Abstract. The study aims to identify the vector of improving the security of the border areas of Russia by forming a new industrial policy. The territories of the Arctic are selected as the object of study. For the study, we relied on comparative, informational, and statistical analysis, logical modeling, the Foresight method. We identify the objects of interaction between industrially developed regions and border regions and focus on the primary forms of inter-regional economic relations. Besides, the model of optimal long-term industrial development of the Arctic using the economic potential of neighboring industrially developed territories. For the first time, in the course of the study, in order to clarify certain elements of the industrial policy of the Arctic territories, an integrated approach is applied. This approach takes into account the features, directions, and potentials of global, national, regional, corporate, interregional, and intercountry development. The results can be used to analyze the effectiveness of implementation and adjust the industrial policy of border areas.

Keywords: National security · Industrial policy · Border areas · The arctic

1 Introduction

The topic of border security is trendy in the world research community. Seven border zones are identified in the Russian Federation: Norwegian-Finnish, Baltic, Belarusian, Ukrainian, North Caucasian, Kazakhstan, East Siberian, and Far Eastern border areas. Each zone has its characteristics, and the territories included in it have unique systems of relations with the Russian Federation and its subjects, as well as foreign participants in cross-border cooperation.

Currently, cross-border cooperation in the Arctic is a strategic priority of the national policy of the Russian Federation, since most countries associate medium and long-term development prospects with the development of this particular territory. As previously noted by (Zhuravel 2018; Kuznetsov 2015; Konovalov 2012; Pavlov, Selin and Selin 2012; Pashchenko 2019), the complexity of studying the Arctic space is determined by the duality of its development. One should consider this region as a

A. V. Bogoviz and Y. V. Ragulina (Eds.): ISCI 2019, LNNS 115, pp. 516–523, 2020.
https://doi.org/10.1007/978-3-030-40749-0_61

territory with abundant geo-economic resources, as well as an element of the national socio-economic system. This asymmetry "leads to a chronic deepening of regional imbalances and serves as a constant reason for political, social, and fiscal overstrain" (Kuznetsov 2015). This determines the urgency of solving the problem of the Arctic territories' security. Notably, this region is considered as key one in the formation of a new system of global, national, and regional development, as indicated in the Strategy for the Development of the Arctic zone of the Russian Federation and Ensuring National Security for the period until 2020, approved by the President of the Russian Federation on February 20, 2013.

The national security strategy provides for the creation of conditions for the protection of the economy of the Arctic territories from internal and external conditions, processes, and factors, primarily through the industrial development of the Arctic. (Granberg and Lazhentsev 2004; Dodin 2005; Mazur 2010), and others dealt with conceptual issues of developing such a fundamentally new Arctic export-raw material orientation policy. The problems of the strategic development of the Arctic sectors in the United States, Canada, Denmark, and Norway are reflected in the strategic documents of these countries and examined by many experts. Such researchers as (Anderson 2010; Byers 2010; Howard 2009; Williams 2010; Dmitrieva and Buryy 2019; Lazhentsev 2018), and many others actively dealt with issues of spatial development and distribution of productive forces in the Arctic. The methodology of their research includes methods of regional statistics, geoinformatics, spatial planning, geoplanning, and geographic expertise. At the same time, the issues of modeling the development of industry in the Arctic, taking into account the optimal use of the potential of neighboring territories, remain open.

In this context, the aim of the study is a comprehensive analysis of the potential for industrial cooperation between the Arctic and other Russian territories bordering this region to form the vectors of their mutually beneficial partnership.

2 Materials and Method

The territories of the Russian Federation, directly attributable to the Arctic zone in whole or in part, will be considered in the study. Namely, we focus on the Republic of Sakha (Yakutia), the Murmansk, Arkhangelsk, and Krasnoyarsk regions, as well as Nenets, Yamalo-Nenets, and Chukotka Autonomous Districts (Fundamentals of State Policy of the Russian Federation in the Arctic for the period until 2020 and beyond, 2008). They border the Kingdom of Norway (Finnmark, Troms, and Nordland), the Republic of Finland (Lapland, Oulu, Cain), Sweden (Norrbotten and Westerbotten), Canada, and the USA.

In methodological terms, the basis of the research consists of provisions of theories of regional and sustainable development, economic security, the network economy, and it also uses comparative, informational, and statistical analysis.

3 Results

Vast natural resources (oil, gas, precious metals, diamonds) are concentrated in the Arctic zone. There is also the potential for organizing and developing global transport corridors. Today, the industry of the Arctic territories produces 10% of Russia's GDP, and the population living there is only 1.6% of this indicator for the country. The basis of the industrial complex is hydrocarbon production. More than 90% of Russian gas and about 10% of oil comes from Arctic enterprises (Pashchenko 2019).

The need for the industrial development of the Arctic is due to the problem of the non-renewability of many natural resources and the increasing demand for fuel and raw materials. Extraction and use of raw materials are often difficult to access due to lack of mining and production technologies, underdeveloped infrastructure (transport, engineering, business, social, etc.), extreme environmental conditions, etc. Therefore, the main tasks of industrial development include the following: creating and comprehensively developing the industrial sector; focusing on long-term investment projects of a production nature that do not violate the environmental balance, carried out using innovative technologies. As part of solving these problems, the following features of the Arctic must be taken into account:

- Lack of sustainable motivation to attract highly qualified specialists from industrially developed regions of the country with a high level of material and spiritual needs, value orientations;
- The specificity of the technical and technological policy of industrial development of territories, which consists of the need to use low-populated technologies, robots, automated machinery, machines, and equipment in the northern version. Besides, due to the high transport costs, the need to create mini-factories, mobile stations or equipment for the production of consumables, primary processing, and transportation of raw materials, creating favorable living conditions arises in such territories;
- Large investment capacity of industrial development projects, which is associated not only with high production costs, but also the lack of infrastructure facilities, underdeveloped social sector of the economy, etc.;
- Fragile ecological balance, the violation of which could hurt industrial development;
- Particular forms of economic, cultural, and everyday life of the indigenous population, which, if possible, must be preserved as alternative economic development options for the Arctic.

Systemic industrial interaction of industrially developed regions with the Arctic territories occurs on the conditions of obtaining bilateral positive effects. On the one hand, the development of new areas contributes to the expansion of sales markets, the development of the existing research base, the increase in cargo flow, which affects the increase in economic activity of economic entities of the industrial region. The analysis of the production needs of the northern, polar, and Arctic territories, carried out during an expert survey of specialists, showed the following. The most relevant projects for them are those focused on the extraction, primary processing, and transportation of natural resources, solving environmental and nature management problems, developing the Northern Sea Route, developing infrastructure, including social. On the other hand, industrial regions can not only meet the needs of the Arctic territories in technology, equipment, and materials, but they also carry out the construction and modernization of infrastructure. They are creating an impetus for the development of the following industrial territories: (a) mining and metallurgical, mining and chemical enterprises of the Kola Peninsula; (b) the Yamal gas complex; (c) the Norilsk industrial region; (d) the mining industry of the Republic of Sakha (Yakutia), Chukotka Autonomous District; (e) timber exporting enterprises of the Arkhangelsk region, Yakutia; (f) individual industrial enterprises, construction sites, airports, expeditions, research stations, small settlements and other things (Smirnova 2011). From an economic point of view, the idea of balancing a resource-oriented economy with the development of new industries seems useful. Therefore, we can consider the following objects as objects of interaction between the industrialized regions and the newly developing regions:

- In the mining industry, we can discuss the industrial development of solid mineral deposits, the creation of powerful integrated mining, metallurgical, and other corporations;
- In petrochemical chemistry, we can consider the implementation of projects to create contemporary petrochemical complexes;
- In mechanical engineering, we can find service and maintenance enterprises. One could also count those enterprises producing the aviation service system of the northern regions based on the modernization of the airport network and the development of small aircraft. Based on enterprises in industrialized territories, specialized engineering industries for the fuel and energy complex, transport construction, transport, mining and forestry complexes can be created;
- In the energy sector, we can consider the expansion of energy supplies and the development of energy-intensive industries in neighboring regions;
- In the forest industry, we can find modern timber processing complexes that allow us to increase the depth of timber processing to produce paper and cardboard and furniture (office, upholstered, and built-in furniture) products. The development of the timber industry complex may give an impetus to the development of the chemical industry in the northern, polar and Arctic territories, including due to increased exports to Central Asia and western China;
- In the construction industry, we can consider new production of energy-efficient and environmentally friendly building materials, products and constructions;
- In socially-oriented sectors, we can discuss the production of consumer goods, which would give impetus to the development of small businesses, including handicrafts among the indigenous peoples of the Far North, as well as the medical industry;

Developing interregional economic and transport-infrastructure integration ensures the complementarity of neighboring regions' economies, rationalization of spatial relationships in the new market schemes of industrial and technological cooperation. It becomes a decisive condition for the further economic growth of enterprises and regions.

The primary forms of interregional economic ties may include the following:

- Material exchange of labor results. At present, intersectoral relations in the following areas are predominant for adjacent industrialized territories: metallurgy, mechanical engineering, shipbuilding, chemistry and petrochemistry, building materials industry. However, the analysis of existing inter-farm relations in the Ural and the Leningrad region showed that the potential of inter-regional cooperation is not used enough. For the Arctic territories, increasing the competitiveness of products not only of the dominant raw materials industries but also of products with high added value is becoming increasingly important, which increases the relevance of using new forms of resource-commodity exchange. The sale of raw materials and semi-finished products from the Arctic territories to the industrially developed regions of Russia for their further processing is not enough, which requires the expansion of existing and the creation of new production bases in the Arctic. The need to introduce new forms of industrial interaction of these territories follows from this. Commodity exchange as the primary form of industrial communication of the regions should be supplemented by scientific, technical, technological, and industrial cooperation, the creation of transport, information and business infrastructures, as well as joint development institutions;
- Participation in associations (corporations) of economic interaction. Researchers distinguish between "soft" and "hard" forms of financial interactions. The first one involves the unification of regions in the framework of the formation of various associative organizations. The second one implies creating the state (with state participation) structures and inter-regional executive and legislative authorities in a macro-region. Interregional clusters belong to one of the most promising forms of regional industrial interaction within the framework of creating associative organizations. From the perspective of creating the state (with state participation) structures and inter-regional executive and legislative bodies in macro-regions, corporations with state participation are a valid form of applying an integrated approach to the development of territories (industrial, social, infrastructural, etc.), attracting various large investors, ensuring state guarantees, participating in federal and regional development programs and, therefore, implementing mega-projects;
- Implementation of socio-economic development programs in the framework of national and local target programs requires coordination of interregional target programs, activities for their application, and regional subprograms. A feature of investment projects implemented in the Arctic territories is their high capital intensity. Because of the specifics of climatic conditions, there are great transport and heat and energy costs, burdening social and infrastructure construction costs, etc. One of the most important mechanisms for the practical support of the industrial interaction of industrialized regions and territories of new development is the implementation of large complex investment projects, including projects in the field of infrastructure arrangement of the Arctic territories;
- Joint performance of work of the production and non-production nature. The problem areas of industrial development of the newly developing regions are technical, technological, and scientific, as well as the support of projects implemented in the territories of new development. In particular, today, more than 90% of all oil refining processes under construction are imported both at the level of technical ideas and equipment (Medovnikov and Rozmirovich 2015). Specialized

scientific and design organizations are absent, a low degree of geological, cartographic, and geodetic exploration of the territory, the lack of necessary information for the development of deposits are preserved. This leads to the fact that most orders for these works are placed abroad.

Interregional integration should not be limited to solving the problems of developing the mineral resource complex of the Arctic territories. Regions of new development, together with industrially developed territories, can carry out projects on the development and bringing to the production of various products in the Arctic.

Based on the results of the study using foresight tools, a complex of models for the optimal long-term industrial development of the Arctic using the economic potential of neighboring industrially developed territories has been developed. The technology of such a simulation may include the following sequence of steps:

1. Analysis of the conjugation of the potential needs of the Arctic territories in products, technologies, personnel and the capabilities of individual industrial territories;
2. Development of the concept of industrial interaction of individual territories of the Arctic with industrialized territories and the construction of a comprehensive model for optimizing such interaction with the display of the following elements: production and technological relations, taking into account the stages of the technological process and the specifics of their modeling, the potential of the territories of interaction; scientific-technological and scientific-educational ties;
3. Development of a complex of models (based on the proposed model and based on the principle of phased modeling) ensuring the interconnection of the proposed technological scheme of industrial interaction with schemes for the balanced development of internal and external markets, development strategies of industry and its sectors, interstate economic integration, business and financial integration, strategic guidelines for the development of interested territories, etc.;
4. Construction of strategic maps to take into account the leading and lagging effects of the industrial interaction of enterprises of the Arctic zone of the Russian Federation and adjacent territories.

4 Discussion

The integration of global, national, regional, and corporate interests is considered as the primary scientific approach to solving issues of industrial development of the Arctic. It is proposed to solve the problem of restructuring the single-industry economy of the Arctic zone, modernizing the infrastructure of the Russian Arctic, as well as preserving, developing, reproducing human capital in the framework of a coordinated national program (program method). These measures will unite the potentials of territories of different levels and private businesses in the form of public-private partnerships. However, such a solution to the problem cannot replace the introduction of specific approaches to each individual industry, production, enterprise. Therefore, the optimization of the development and deployment of the Arctic industry will require the development of new methodological techniques and unique economical and mathematical models, taking into account their characteristic features.

5 Conclusion

Thus, the study showed that the model of new industrial development is formed on the basis of the following principles: (a) complexity is ensured by a combination of industrial development with social, economic, socio-cultural, and other types of development; (b) consistency is ensured by the harmonious integration of the developed regions into the economic, political, social, cultural, and other life of the base territory.

The condition for the effective industrial development of the Arctic is their interaction with the industrially developed regions of Russia. The main areas of such interaction are the following changes: (a) changes in the technological basis for the development of industry in the Arctic due to the rational use of resources of adjacent industrialized regions (focus on high-tech processing of raw materials); (b) changes in the quality of human resources (advanced training and labor productivity of employees, increase in the quality of life of the population); (c) changes in technologies for the reproduction of resources (growth of greening industry). These trends will positively affect the security of the border areas of the Russian Federation.

References

Anderson, A.: After the Ice: Life, Death, and Geopolitics in the New Arctic. Harper Paperbacks, New York (2010)

Byers, M.: Who Owns the Arctic? Understanding Sovereignty Disputes in the North. Douglas & McIntyre, Vancouver (2010)

Dmitrieva, T.E., Buryy, O.V.: Support zones for the development of the Russian Arctic: content, ratings, and projects. ECO **49**(1), 41–59 (2019)

Dodin, D.A.: Sustainable Development of the Arctic: Problems and Prospects. Nauka, St. Petersburg (2005)

Granberg, A.G., Lazhentsev, V.N.: Principles and directions of state policy in the North. In: Granberg, A.G. (ed.) North of Russia: Actual Problems of Development and the State Approach to Their Solution, pp. 15–19. Publishing House of the Komi Scientific Center, Ural Branch of the RAS, Moscow; Syktyvkar (2004)

Howard, R.: Arctic Gold Rush: The New Race for Tomorrow's Natural Resources. Continuum, London, New York (2009)

Konovalov, A.M.: Arctic: national interests in the context of globalization. Digest Financ. **1** (205), 32–40 (2012)

Kuznetsov, S.V.: The role of single-industry economies in the formation of the geo-economic space of the Russian Arctic zone. Econ. North-West: Probl. Dev. Prospect. **3**(48), 30–39 (2015)

Lazhentsev, V.N.: Socio-economic space and territorial development of the North and the Arctic of Russia. Region's Econ. **14**(2), 353–365 (2018)

Mazur, I.I.: The Arctic is a bifurcation point in the development of global peace. Cent. Glob. **2**, 93–104 (2010)

Medovnikov, D., Rozmirovich, S: Entrepreneurial challenge (2015). https://www.hse.ru/news/143357672.html

Pashchenko, L.V.: National interests of Russia in the Arctic and problems of their implementation. Polit. Sci. Issues **2**(42), 298–305 (2019)

Pavlov, K., Selin, V., Selin, I.: Russian Arctic: current status and development prospects. Soc. Econ. **3–4**, 180–192 (2012)

Smirnova, O.: The Arctic is the territory of the partnership of the future. Econ. Strateg. **10**, 2–7 (2011)

Williams, G.: Arctic Labyrinth: The Quest for the Northwest Passage. University of California Press, Berkeley, Los Angeles (2010)

Zhuravel, V.P.: The Arctic as a constantly evolving multidimensional space. Arct. North **31**, 62–79 (2018)

Image-Making of the Republic of Tatarstan Through Investment Activity in Industry

Taliya I. Minullina(✉)

Tatarstan Investment Development Agency, Kazan, Russia
taliya.minullina@tatar.ru

Abstract. The article presents the relationship of investment activity in the industry and the growth of recognition of the Republic of Tatarstan as a brand. An analysis of investment activity in the industry is based on statistics from the Bank of Russia, the Federal State Statistics Service, and its Territorial Authority for the Republic of Tatarstan, the United Nations Trade and Development Commission, and departmental data. The presence of a developed industry in Tatarstan provides the republic with the opportunity to use the results and potential of the industry to promote the region. At the same time, processing industries prevail over other areas of economic activity in terms of investment in fixed assets. The republic uses high investment activity in the industry for promotion. And such investment activity gives a synergistic effect in the form of a constant influx of new business partners.

Keywords: Industry · Processing industry · Investments · Investments in fixed assets · Investment activities · Promotion · Brand · Image · Image-making

1 Introduction

Industry in the Republic of Tatarstan is a leading sector of the economy, and the most significant amount of investment in fixed assets is allocated there. In Tatarstan, the most significant number of technology transfers of leading world companies occurs precisely in the industry. Over the past 15 years, such world-famous brands have placed production as the automobile giants "Ford" and its suppliers "Kiekert", "Coskunoz", "Magna", "Clariant", "Mefro Wheels", "Sisecam", as well as "Daimler", "Federal-Mogul", "Cummins", "ZF", "Koluman" in Tatarstan. Leaders of the chemical, paper, woodworking, and other industries (such as "3M", "Hayat", "Kastamonu", "Armstrong", "Rockwool", which opened their plants in the republic) are reinvesting in increasing capacities.

In the republic, own funds are actively invested in development. In 2018, this source accounted for 65.5% of the total. The largest brands in Russia's industry, widely known both in our country and abroad, are present in Tatarstan, including "KamAZ", "Nizhnekamskneftekhim", "Kazan Helicopters", "Kazan Aviation Plant", "Zelenodolsk Plant named after A. M. Gork", and many others.

Creating a positive image of the republic is necessary for marketing the products of its enterprises, for attracting external funds, technologies, and ensuring knowledge transfer into its economy. The flagship for promoting the image of the Republic of

A. V. Bogoviz and Y. V. Ragulina (Eds.): ISCI 2019, LNNS 115, pp. 524–533, 2020.
https://doi.org/10.1007/978-3-030-40749-0_62

Tatarstan in the international and interregional arena is the state executive authority of exclusive competence, namely the Tatarstan Investment Development Agency. Creating a positive brand in our region, including, is possible thanks to the current investment activity in the industry.

2 Materials and Methods

An analysis of investment activity in the industry is based on statistics from the Bank of Russia, the Federal State Statistics Service, and its Territorial Authority for the Republic of Tatarstan, the United Nations Trade and Development Commission, and departmental data. The period for the last two years (2017–2018) is under study, relying on the retrospective of the previous ten years (from 2007 to 2017), which is confirmed by data at the global, national, and regional levels.

An array of empirical and quantitative methods are used in the research. In addition, a theoretical analysis of the scientific literature is applied in work.

The relevant foreign and domestic literature on branding and promotion of the territories is also reviewed and analyzed.

3 Research Results

The competitiveness of the territory in the international and national arena directly depends on the positive image of the region.

Today, the brand of the territory is an integral part of its socio-economic development. The leadership of any area aimed at its strategic development actively promotes its brand to attract investment, increase trade and tourism flows, develop innovation, industry, culture, education.

The key definitions of "brand" and "image" are not identical concepts. The image concerning the brand is perceived as a generalized portrait of the branding object. Image demonstration shapes public opinion.

Territory branding is a relatively new direction in marketing. In 2002, Simon Anholt introduced the concept of place branding (Anholt 2009). Marketing classics such as Philip Kotler (Kotler et al. 2002) and guru Seppo Rainisto also worked on place marketing (Rainisto 2003). Simon Anholt introduces such a concept as a "competitive identity cycle," which shows the process of becoming a territory as a brand, a method of continuous, gradual, systematic, coordinated, and targeted implementation of a territory recognition strategy. This implies the work of the state on the effective use of external communication channels and maintaining a constant high level of socio-economic indicators of the territory to achieve correspondence between its image and reality. Eric Brown, who considers branding of regions at the city level, defines territorial branding of the area as traditional branding plus the activities of national governing bodies, which include all interested stakeholders, government agencies, and businesses.

In many countries of the world, the formation and promotion of the image of territories begin at the state level as part of strategic planning. Based on regulatory documents of national significance, a regional development strategy is adopted in which the role of the subject within the country is determined, and identification of its unique qualities occurs.

A global study of industrial policy by the United Nations Trade and Development Commission shows that over the past 10 years, at least 101 economies in developed and developing countries (which account for more than 90% of global gross domestic product) have adopted formal industrial development strategies (United Nations Committee on Trade and Development 2018).

The study shows that current industrial policies are becoming more diverse and complex. It addresses new topics and includes many goals in addition to conventional industrial development and structural transformations, such as the integration and modernization of global supply chains, development of a knowledge economy, building those sectors related to the sustainable development of goals, and competitive positioning for a new industrial revolution (United Nations Committee on Trade and Development 2018).

About 90% of the contemporary industrial policy documents involve the use of specific investment policy instruments, mainly incentives for productivity, special economic zones, simplification of procedures, and new investment selection mechanisms.

Contemporary industrial policy is a critical factor in investment policy trends. More than 80% of investment policy measures registered since 2010 are directed to the industrial sector (production, industrial infrastructure), and about half of them serve the purpose of industrial policy. Most of them are cross-industry.

Incentives remain the tool most commonly used for industrial policy. Significant progress has been made in creating more efficient means for industrial development. About two-thirds of incentive schemes applicable to the production goals of several or individual industries are typically focused on R&D.

SEZs continue to spread and diversify. In most countries, the transition from clean export processing zones to value-added zones continues, and new types of zones are still emerging. Targeted strategies for engaging in specific industries and linking multiple zones have contributed to industrial development and the integration of global production chains in some countries that have taken incentive measures. High-tech zones or industrial parks are also becoming a vital tool for industrial policy oriented towards a new industrial strategy.

Contemporary industrial policy has increased the contribution to investment, although it played a secondary role in the investment policy until recently. Many developing countries have made to simplify the entry of investors of one of the critical tasks in the industrial development strategies. Targeted investment promotion (other than incentives and SEZs) is also essential. Two-thirds of investment promotion agencies are guided by the industrial policy in identifying priority sectors for attracting investment, and three-quarters of agencies have special promotion schemes for attracting technology to the region's industry.

Thus, the investment policy (in particular, the policy on foreign direct investment) is a crucial tool for the region's industrial policy. Different models of industrial policy imply different mechanisms of investment policy. At the same time, the investment policy toolkit is being developed using industrial policy models.

We will consider this relationship by the example of the Republic of Tatarstan. Today, the region is a widely recognized brand. The Republic has 170 foreign trade partners and 66 investor countries from 195 countries of the world (this is data for departments of the Territorial Authority of the Federal State Statistics Service for the Republic of Tatarstan). In Russia, Tatarstan is traditionally perceived as an oil region, although its share of oil production in the country's volume is about 6–7% (Federal State Statistic Service 2018). At the same time, according to the most important macroeconomic indicators, the Republic of Tatarstan is traditionally among the leading regions of the Russian Federation. The high level of development of the mining and processing industries plays a significant role in this success (Fig. 1).

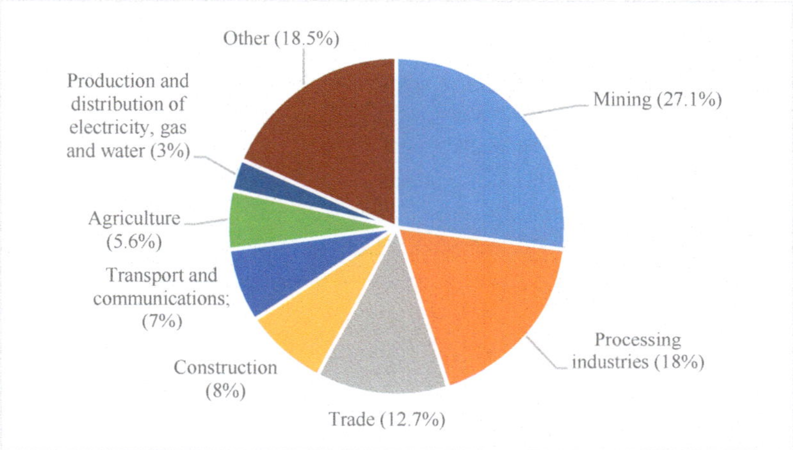

Fig. 1. Structure of the internal regional product for 2018, in %. *Source*: (Federal State Statistic Service 2018).

In the structure of the economy, the industry traditionally occupies the largest share, namely, 48.1% (including, mining - 27.1%; manufacturing industries - 18.0%; electric energy, gas, steam, air conditioning - 2.3%; water supply, water disposal, waste collection and disposal, pollution elimination activities - 0.7%), wholesale and retail trade constitute 12.7%; construction occupies 8.1%; and agriculture is 5.6%.

Analysis of statistical data shows that the primary industry in which the investment in fixed assets is placed in the Republic of Tatarstan is the processing industry. In January-December 2018, a significant amount of investment in fixed assets was directed to the development of processing industries (it amounted to 36.0%). Including, 11.0% of investments were in the production of coke and petroleum products, 10.7% was in the production of chemicals and chemical products, 5.6% was in the manufacture of motor vehicles, trailers, and semi-trailers. For comparison, in 2017, from processing industries, 22.0% of investments were in the production of coke and petroleum products, 7.2% was in the production of chemicals and chemical products, 3.1% was in the manufacture of motor vehicles, trailers, and semi-trailers.

The flow of foreign investment in the industry in the Republic of Tatarstan is growing every year. Their dominant part is directed to the development of processing industries. In 2018, according to Tatarstanstat, it amounted to 84.9% of the total foreign investment received, in 2017, it amounted to 41.7%, in 2016, it amounted to 39.5%. In 2017, mining was in 4th place from industries by revenue, and it was insignificant in 2018 (Territorial branch of Federal State Statistic Service in the Republic of Tatarstan 2017, 2018) (Table 1).

Table 1. Foreign investment by type of activity, %.

Type of activity	2017	2018
Total,	100.0%	100.0%
Agriculture, forestry, hunting, fishing, and fish farming	0.1%	*
Mining	12.1%	0.1%
Processing industries	41.7%	84.9%
Food production	0.3%	0.2%
Manufacture of wearing apparel	*	0.0%
Wood processing and the manufacture of wood products and cork, except furniture; manufacture of articles of straw and plaiting materials	7.9%	3.9%
Production of paper and paper products	13.2%	9.3%
Production of chemicals and chemical products	0.5%	28.6%
Manufacture of rubber and plastic products	4.3%	0.2%
Manufacture of other non-metallic mineral products	0.4%	0.2%
Manufacture of fabricated metal products, except machinery and equipment	*	1.0%
Production of computers, electronic and optical products	0.2%	*
Electrical equipment production	0.2%	5.1%
Manufacture of machinery and hardware not included in other categories	0.0%	0.5%
Manufacture of motor vehicles, trailers, and semi-trailers	14.8%	36.0%
Furniture manufacture	0.0%	*
Production of other finished products	0.0%	*
Repair and installation of machinery and equipment	0.0%	*
Providing electric energy, gas, and steam; air conditioning	0.6%	0.0%
Water supply; water disposal, organization of waste collection and disposal, pollution elimination activities	0.0%	0.1%
Construction	9.1%	1.2%
Wholesale and retail trade; repair of motor vehicles and motorcycles	14.0%	0.3%
Transportation and storage	*	*
Activities of hotels and catering establishments	*	*
Information and communications activities	*	*
Financial and insurance activities	16.0%	9.0%
Real estate activities	2.3%	0.1%
Professional, scientific and technical activities	4.1%	4.3%
Administrative operations and additional related services	0.0%	0.0%
Health and social work activities	0.0%	0.1%

*Note**: Investments amounted to less than 0.0%.
Source: Territorial authority of the Federal State Statistics Service for the Republic of Tatarstan (data for departments).

As the Table 2 shows, the main types of manufactured products were motor vehicles, chemicals, and paper products in 2018. In 2017, the main types of manufactured products were motor vehicles, wood products, and paper products.

Table 2. Processing industries by type of activity, %.

	2017	2018
Processing industries	100%	100%
Food production	0,7%	0,2%
Manufacture of wearing apparel	*	0,0%
Wood processing and manufacture of wood products and cork, except furniture; manufacture of articles of straw and plaiting materials	18,9%	4,5%
Production of paper and paper products	31,6%	11,0%
Production of chemicals and chemical products	1,1%	33,7%
Production of rubber and plastic products	10,3%	0,2%
Production of other non-metallic mineral products	0,9%	0,3%
Production of fabricated metal products, except machinery and equipment	0,0%	1,1%
Production of computers, electronic and optical products	0,4%	*
Production of electrical equipment	0,4%	6,0%
Manufacture of machinery and equipment not included in other categories	0,0%	0,6%
Manufacture of motor vehicles, trailers, and semi-trailers	35,6%	42,4%
Furniture manufacture	0,0%	*
Manufacture of other finished products	0,0%	*
Repair and installation of machinery and equipment	0,0%	*

Note *: Investments amounted to less than 0.0%

Source: Territorial authority of the Federal State Statistics Service for the Republic of Tatarstan (data for departments).

Indeed, in recent years, the most significant foreign investments have been made in the production of Ford cars, Hayat paper products, 3 M chemical coatings, Kastamonu flooring, and plastics at the "Nizhnekamskneftekhim" plants.

In the Republic of Tatarstan, the accumulated foreign direct investment in processing industries is the most significant among the types of economic activity of direct investment enterprises. So, according to the Bank of Russia, over the past four years (from 2014 to 2017), the share of processing industries ranged from 45.5% to 63.5% of the total accumulated funds (Bank of Russia, n.d.). Mining (from 11.7% to 15.3%) and financial activities (from 5.4% to 15.7%) are the next most significant liabilities of foreign investment. And for nine months of 2018, the share of manufacturing industries amounted to 70.5% of the total amount of foreign direct investment accumulated in the republic. Among other things, this means that in the industry, investments are delayed in larger volumes than in other activities. Namely, investors in industrial sectors save financial resources in the invested company and use them to develop production.

In my perspective, the most significant factor in the attractiveness of the region is the willingness of its leadership to solve problems, respond to feedback quickly.

A vital consequence of the successful implementation of investment projects in the industry is the synergy effect. Businesses react positively to the presence in a specific territory of the national brands that are familiar to them. For example, in due time, the domino effect made itself felt concerning Turkish investors. After successfully placing the production of Kastamonu Entegre, other Turkish investors went to Tatarstan for this company. The recommendations of industry leaders and business partners are a characteristic trend for promoting the territory. It is important to note that Turkey always expresses a keen interest in investing in Tatarstan. Turkish companies make up one-third of current residents in the "Alabuga" SEZ.

In the country, in almost all cases, foreign direct investment occurs in the supply chain relationship than in relation with local competitors. Carefully monitoring market conditions and attracting investors to free niches or links in the supply chain makes the territory more attractive for investment activity. This works particularly well with a cluster implementation. So, several suppliers pulled up behind the "Ford Sollers" company, which opened a car manufacturing plant in the "Alabuga" SEZ. And now, paints, body parts, and auto components are also produced in Tatarstan.

The "Alabuga" SEZ is an excellent example of creating a brand of a zone of industrial-production type. Real performance indicators of the site confirm its leadership among all SEZs of this type in Russia. At the same time, this territory has a prestigious rating of the Fdi Intelligence magazine (Financial Times group). It was three times marked as the best special economic zone in Europe for hosting large enterprises, and it was noted in the nomination "New Investments" in 2018 (fDi Intelligence 2014, 2016, 2017, 2018). The drafters of the rating highly praised the project of the current resident launched in the SEZ to expand production under a new brand. Thus, a combination of real achievements and awards of prestigious ratings is another tool for promoting a brand of a region's territory due to its allocated area.

Along with the SEZ, industrial parks and sites provide investors with conditions for the implementation of investment projects, mainly in industry. In Tatarstan, about 100 sections of various levels have been created (Government of the Republic of Tatarstan 2019). And today, placing a project on a site means not only receiving multiple preferences in the form of state support measures but also the high quality of the management company's services for the operational organization of production. The technopolis "Khimgrad" and "Master" Kama industrial park were selected as leaders among industrial parks and special economic zones in terms of investment attractiveness by the "Expert RA" Rating Agency. Consequently, they are evidence of this and recognizable brands.

On average, annually, about 100 companies with foreign participation are registered in the Republic of Tatarstan, the same amount transfers funds to their Tatarstan subsidiaries, and nearly half of them are the funds of new investors. The other half are the funds of investors on an ongoing basis. In 2018, 14 foreign companies reinvested in the republic. These are large international industrial companies such as "Ford", "Hayat", "Haier", "Daimler", "Kastamonu", "Drylock", "3M", "Saria", and others. Thus, the image of the republic as a territory of continuous business development has been created.

The direct participant in the investment activity of the republic is the Tatarstan Investment Development Agency, which is responsible for attracting Russian and international investments in the economy and supporting large investment projects (Government of the Republic of Tatarstan 2011). To attract investment in industry, the Agency is organizing and conducting advertising campaigns to promote the investment potential (brand) of the republic.

FutureBrand conducts corporate brand research annually (FutureBrand, 2019). Company experts note the importance of the country of origin indicator in understanding the country's brand strength. In particular, they talk about the role of this indicator in the choice of consumers who historically prefer products and services that are authentic to their country of origin. And the lack of a "made in" strategy in a particular state makes it less competitive (FutureBrand 2014).

Today, in Tatarstan, the following three brands are actively promoting: "Visit Tatarstan" (tourism), "Made in Tatarstan" (industry), and "Invest in Tatarstan" (investing). The Agency has identified many tools and channels for promoting these brands.

One of the fundamental aspects of promoting the image of the Republic both at the federal and international levels is participation in major exhibition events and forums. So, the Agency draws up a stand, stand form and advertising materials in the same style, with the mandatory use of the "Invest Tatarstan" logo, as well as the "Made in Tatarstan" logo, if the size of the stand and the representation of enterprises of the republic are sufficient. About 15 expositions are built up per year, geography expands and varies from year to year.

Another tool is the organization of events on its territory. The most important aspect is the definition of the original theme, authentic for the country and other territorial divisions, as macro-regions. So, every year, the Agency holds the International Economic Summit "Russia - Islamic World: Kazan Summit". The Russia's largest halal industry exhibition "Russia Halal Expo" is traditionally organized within its framework. The scale and popularity of the exhibition is growing every year. In 2018, the exposition occupied an area of 1.4 thousand square meters, in 2019, it increased by more than 3.5 times and reached 5 thousand square meters. The Republic of Tatarstan is one of the leaders in the production of halal products in Russia, and it presented 32 local companies at the last exhibition, among which were Tatarstan producers of food products (meat, flour products, spices), carpets, cosmetic oils and perfumes.

The creation of centers for demonstrating the history and achievements of the territory in the format of museums is an established practice. The Agency has opened in its own building an exhibition hall "The United Presentation Center of the Republic of Tatarstan". The center is intended to become a tool for promoting Tatarstan and a place for visiting guests and partners of the republic. In the main hall of the Center, equipped with video panels and video screens, the video content about the Republic of Tatarstan is shown in various areas of its life, such as economy, industry, construction, trade, tourism, culture, natural and human resources and many others. There are also "tricks" in the presentation center, namely: the last word in technology is a 5D cinema and a built-in Tatar hut that conveys national flavor and allows one to warmly welcome guests with tea with Tatar treats. To demonstrate the products manufactured by enterprises of Tatarstan, a wall under the brand name "Made in Tatarstan" was created. This is not only a showroom for inspecting the achievements of the industry but also a way to influence the audience through their direct study, all exhibits are available to visitors, they can be picked up.

4 Discussion Results

Currently, the brand promotion provides a synergistic effect in conjunction with other activities of socio-economic development and foreign economic activity. Evaluation of the effectiveness of activities to promote the investment brand of the Republic of Tatarstan can be carried out using modern methods of financial analysis, such as planning and analyzing the results of returning investments on funds spent on promotion.

So, the republic's participation in congress and exhibition events is included among the measures to promote the investment brand of Tatarstan. Therefore, in the framework of the Comprehensive Plan of Action of the Government of the Republic of Tatarstan on the Implementation of the Message of the President of the Republic of Tatarstan to the State Council of the Republic of Tatarstan for 2019, an event to develop a regulation of interaction between executive bodies of state power, local authorities of municipalities of the Republic of Tatarstan, and the Agency has been introduced. It is applied when organizing participation in international investment-oriented events and at the stands of the Republic of Tatarstan in terms of improving the efficiency and quality of the presented investment projects.

Thus, we can state that the brand of the Republic of Tatarstan has a high level of identification at the international and regional levels. At the same time, capitalization instruments are not applied to a local brand, unlike commercial ones. The last time, the Republic of Tatarstan issued external loan bonds was in 1998 (Realnoe Vremya 2017). At the same time, securities of Tatarstan producer brands are actively quoted on international exchanges. In the Moscow Stock Exchange, shares and corporate bonds of "Tatneft", "Nizhnekamskneftekhim", "Nizhnekamskshina", "KamAZ", and others are widely represented. Capitalization of the regional brand of the republic may become one of the financial instruments for promoting the territory.

5 Conclusion

The presence of a developed industry in Tatarstan provides the republic with the opportunity to use the results and potential of the industry to promote the region. At the same time, processing industries prevail over other areas of economic activity in terms of investment in fixed assets. The republic uses high investment activity in the industry for promotion. And such investment activity gives a synergistic effect in the form of a constant influx of new business partners. The relationship between promotion and growth of development results is evident both in industry and investment, individually and in synergy. Thus, the promotion of investment results in the industry has a positive effect on the image of the territory.

References

Analytical center "Expert": Investment attractiveness rating of industrial parks and SEZ (2018). https://expert.ru/ratings/rejting-investitsionnoj-privlekatelnosti-industrialnyih-parkov-i-oez/

Anholt, S.: Places: Identity, Image and Reputation. Palgrave MacMillan, New York (2009)

Bank of Russia: External sector statistics (n.d.). http://cbr.ru/statistics/?PrtId=svs

fDi Intelligence. "SEZ «Alabuga»: largest in Russia and best in Europe, 2018" (2018). https://www.fdiintelligence.com/News/SEZAlabuga-largest-in-Russia-and-best-in-Europe

fDi Intelligence. "fDi Global Free Zones of the Year 2017 – the winners" (2017). https://www.fdiintelligence.com/Rankings/fDi-Global-Free-Zones-of-the-Year-2017-the-winners

fDi Intelligence. "Global Free Zones of the Year 2016" (2016). https://www.fdiintelligence.com/Rankings/Global-Free-Zones-ofthe-Year-2016

fDi Intelligence. "fDi Global Free Zones of the Year 2014 – Winners" (2014). https://www.fdiintelligence.com/Locations/fDi-Global-Free-Zones-of-the-Year-2014-Winners

Federal State Statistic Service: Socio-economic situation in Russia (2018). http://www.gks.ru/free_doc/doc_2018/social/osn-12-2018.pdf

FutureBrand: Made In: The value of country origin for future brands (n.d.). https://carlobattisti.files.wordpress.com/2015/06/made_in_final_hr.pdf

Germany Trade & Invest (GTAI): 2008 Anholt-GfK Roper Nation Brands IndexSM Report Highlights (2008). https://www.gtai.de/GTAI/Content/CN/Invest/_SharedDocs/Downloads/Studies/diw-wochenbericht-9-2010.pdf

Global market research and consulting "Ipsos": Press Release. Germany Retains Top "Nation Brand" Ranking, U.S. Out of top five again (2018). https://www.ipsos.com/sites/default/files/ct/news/documents/2018-10/nation-brands-index-2018-pr-2018-10-24-v2.pdf

Government of the Republic of Tatarstan: Report on the activities of the executive authorities of the Republic of Tatarstan for 2018 (2018). http://prav.tatarstan.ru/rus/pdf-otchet-o-deyatelnosti-organov-ispolnitelnoy-3522697.htm

Government of the Republic of Tatarstan: "Order of the Cabinet of Ministers of the Republic of Tatarstan of June 6, 2011 No.460 Issuesof Tatarstan Investment Development Agency" (2011). http://invest.tatarstan.ru/upload/iblock/c6d/postanovlenie_km_rt_ot_6_iyunya_2011_g._n_460_voprosy_agentstva_investitsionnogo_razvitiya_respubliki_tatarstan.pdf

Government of the Republic of Tatarstan: "Rustam Minnikhanov addressed the State Council of the Republic of Tatarstan with theannual message" (2019). http://president.tatarstan.ru/poslanie-prezidenta.htm

Kotler, P., Hamlin, M.A., Rein, I., Haider, D.H.: Marketing Asian Places: Attracting Investment, Industry, and Tourism to Cities. States and Nations. Wiley, New York (2002)

Market research institute "GfK": Press Release. GfK Research: Anholt-GfK Nation Brand Index 2017 (2017). https://www.gfk.com/fileadmin/user_upload/dyna_content/RU/Documents/Press_Releases/2017/GfK_Rus_Press_Release_NationBrandsIndex_Dec2017.pdf

Ministry of Economy of the Republic of Tatarstan: On the results of the activities of the Ministry of Economy of the Republic of Tatarstan for 2018 and tasks for 2019 and for the period until 2024 (2019). http://mpt.tatarstan.ru/file/File/Об%20итогах%20деятельности%20Министерства%20экономики%20РТ%20за%202018%20год%20и%20задачах%20на%202019%20год%20и%20на%20период%20до%202024%20года.pdf

Moscow Exchange: Official website (n.d.). https://www.moex.com/

Rainisto, S.: Success factors of place marketing: a study of place marketing practices in Northern Europe and the United States (Dissertation of Doctor of Science in Technology). Helsinki University of Technology, Institute of Strategy and International Business, Helsinki, Finland (2003)

Realnoe Vremya: "Borrow from the center: Tatarstan refused to acquire debt financing" (2017). https://realnoevremya.ru/articles/69525-tatarstan-otkazalsya-vypuskat-obligacii

TatCenter: Talia Minullina, AIR RT: "I have married Tatarstan and so far have not been unfaithful to him" (2017). http://tatcenter.ru/rubrics/lyudi-nashej-respubliki/taliya-minullina-air-rt-quot-ya-vyshla-zamuzh-za-tatarstan-i-poka-emu-ne-izmenyayu-quot/

Territorial branch of Federal State Statistic Service in the Republic of Tatarstan: Analytical review "Basic indicators of investment and construction activities in the republic of Tatarstan in January-December 2018" (2018). http://invest.tatarstan.ru/upload/iblock/f75/analiticheskiy-obzor-investitsionnoy-i-stroitelnoy-deyatelnosti-v-respublike-tatarstan-za-2018-god.pdf

Territorial branch of Federal State Statistic Service in the Republic of Tatarstan: Analytical review "Basic indicators of investment and construction activities in the republic of Tatarstan in January-December 2017" (2017). http://invest.tatarstan.ru/upload/iblock/5bd/analiticheskiy-obzor-investitsionnoy-i-stroitelnoy-deyatelnosti-v-respublike-tatarstan-za-2017-god.pdf

United Nations Committee on Trade and Development: World Investment Report 2018 (2018). https://unctad.org/en/PublicationsLibrary/wir2018_en.pdf

Cote d'Ivoire in Regional Integration of the Economic Community of West African Countries

N'djah Amoin Jocelyne-Sandra(✉)

Peoples' Friendship University of Russia (RUDN University), Moscow, Russia
ndjah@mail.ru

Abstract. According to B. Balassa's theory of economic integration, we live a rather high, effective, and promising stage in the development of the world economy, a qualitatively new and more complex stage in the internationalization of economic relations. Regional economic integration involves the cooperation of individual countries in order to more efficiently use resources by creating favorable conditions for the implementation of productive business activities simultaneously in the markets of several countries (regional markets). The purpose of the article is to determine the role of Cote d'Ivoire in the regional integration of the Economic Community of West African Countries (ECOWAS). The article analyzes the contribution of Cote d'Ivoire to the development of regional integration of the Economic Community of West African Countries (ECOWAS), its position based on the African Regional Integration Index. The article assesses the country's effectiveness in the regional market, allocated particular qualities of customers of Cote d'Ivoire in the domestic market.

Keywords: Economy · Regional economic integration · ECOWAS · Cote d'Ivoire

1 Introduction

The African Development Bank and the World Bank identify regional economic integration as an essential driver for Africa's competitiveness. The successes achieved in ensuring economic growth in the economies of African countries must be accompanied by measures to stimulate competitiveness in the long term, that is, Africa must ensure sustainable development to improve living standards.

Regional integration is a process of development of stable economic relations and division of labor of national economies, which, covering foreign economic exchange and the sphere of production, leads to a close intertwining of national economies and the creation of a single economic complex on a regional scale (Aksyuk 1994; Carrere 2004; Fituni 2010). In Africa, regional integration is a pivotal tool to help increase competitiveness, diversify the economic base, and create enough jobs for young people (Morozenskaya 2000; Morozov 2002). Integration in West Africa is very important because: gives the opportunity to expand the range of products that Africans can buy in the local markets, facilitate travel between countries in order to visit tourist routes or

A. V. Bogoviz and Y. V. Ragulina (Eds.): ISCI 2019, LNNS 115, pp. 534–541, 2020.
https://doi.org/10.1007/978-3-030-40749-0_63

look for work, accelerate the transfer of funds to family members between states and expand opportunities for obtaining seed capital for business. Economic Community of West African States ECOWAS is a regional integration grouping in West Africa. It was enshrined in the Lagos Treaty, which was signed on 28 may 1975, and included fifteen countries in West Africa: Benin, Burkina Faso, Cote d'Ivoire, Gambia, Ghana, Guinea, Guinea-Bissau, Liberia, Mali, Niger, Nigeria, Senegal, Sierra Leone, Togo. In 1976, Cape Verde joined ECOWAS, and Mauritania withdrew from it in 2000.

2 Materials and Methods

According to the index of regional economic integration, which was published in 2016, among the ECOWAS member countries, Nigeria ranks first in the region (75% of total GDP), but the country is not included in the countries with the highest rating of regional integration. Cote d'Ivoire provides 6% of the GDP of the regional community (Journal Le Point-Économie 2016).

Figure 1 presents a rating of ECOWAS-member countries according to the index of regional integration in 2016. The figure shows that Cote d'Ivoire ranks first with a rating of 0.675, Togo with a rating of 0.66 is in second place, and Senegal is in third place with a rating of 0.65.

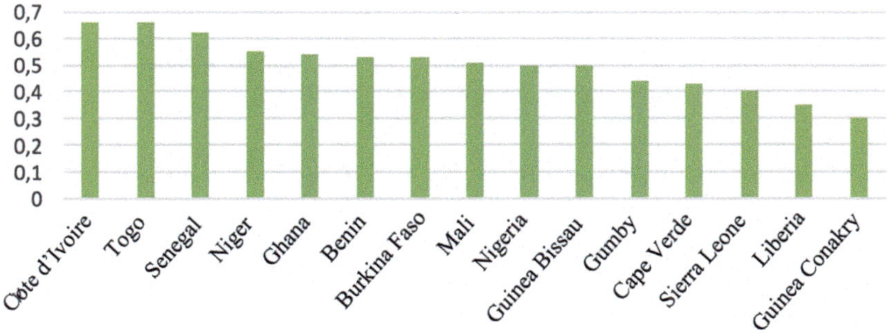

Fig. 1. Rating of ECOWAS member countries according to the regional integration index in 2016. *Source*: Compiled by the author based on www.integrate-africa.org.

In Fig. 2, one may see the status of all groups of indicators in Cote d'Ivoire. As we see, the best indicators are in the field of commercial as well as financial integration and convergence of macroeconomic policies. COnsequently, Cote d'ivoire occupies a significant place in the economy of ECOWAS after Ghana and Nigeria. An analysis of the regional integration index shows that Cote d'Ivoire is a prosperous country in terms of the free movement of people, with a value of 0.8. About 26 African countries can enter Cote d'Ivoire without obtaining a visa or with a visa upon arrival, which allows Cote d'Ivoire to be in ninth place in Africa and first place in the ECOWAS zone. Also, Cote d'Ivoire has ratified the protocols on the free movement of people from the regional economic communities [Community of Sahel-Sahara States (CEN-SAD) and ECOWAS] of which it is a member.

As regards trade integration, it should be noted that Cote d'Ivoire, within the framework of the regional economic community, has made significant progress towards the elimination of tariff barriers to imports. In fact, the average tariff applied to imports from ECOWAS member countries is 0.7% (third largest among ECOWAS members).

Trade between Cote d'Ivoire and ECOWAS countries is developing very intensively, as evidenced by the dynamics of imports and exports. In terms of exports, the country of Cote d'Ivoire has a high degree of integration with the rest of its subregion and ranks second in the ECOWAS region with an index value of 0.97.

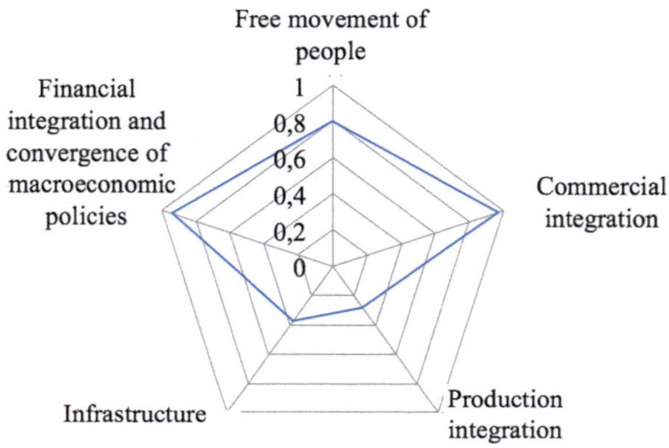

Fig. 2. Performance of Cote d'Ivoire according to the regional integration index. *Source*: Compiled by the author based on www.integrate-africa.org.

Industrial integration Cote d'Ivoire is characterized by the index of productive integration and the index of commodity complementarity of products.

The index of productive integration is based on the shares of subregional exports of intermediate goods (% of the total volume of subregional exported goods) and the shares of subregional imports of intermediate goods (% of total subregional imported goods). Cote d'Ivoire ranks 7th in the Product Commodity Complementarity Index, reflecting the total absolute value of the difference between the share of imports and the share of exports of a member state in the regional economic community.

In terms of infrastructure, Cote d'Ivoire is in ninth place with an index value of 0.37; in terms of per capita electricity production, it ranks eighth in Africa and fourth among all ECOWAS member countries.

Based on financial integration and the convergence of macroeconomic policies, the regional convertibility of national currencies and the difference in inflation rates (based on an agreed consumer price index) are estimated. Cote d'Ivoire by this index takes the 7th place with value 0.94.

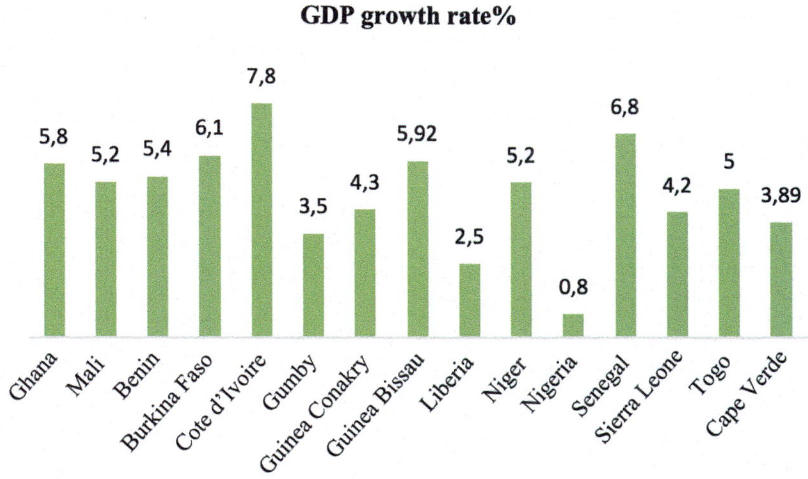

Fig. 3. The GDP growth rate of ECOWAS countries in 2017, in %. *Source*: Compiled by the author based on www.bceao.int.

In 2017, Cote d'Ivoire was the leader among ECOWAS countries in economic growth (+7.8%), then Senegal and Burkina Faso follow with values of 6.8% and 6.1% respectively (Fig. 3). This growth is driven by government action to implement the National Development Plan (NDP), which aims to make the private sector the engine of economic growth, the creator of economic well-being, and the job provider. The National Development Plan also meets the country's strategic industrial policy goals.

Economic integration affects the competitiveness of economies of the countries, first, due to dynamism and strengthening. Second, this is due to the elimination of contradictions and the decrease in barriers to competitiveness. It all depends on the right economic policy, the correct assessment of the economic and political situation in the country, the presence or absence of economic resources (Kurbanov 2016; Skubko 2012; Voinov 2011). The World Economic Forum (WEF) considers competitiveness as a set of institutions, policies, and factors that ultimately determine the level of productivity and, consequently, the standard of living of a country. Under this methodology, for each of the countries surveyed, composite competitive indices are calculated, measured based on macro- and microeconomic indicators, which are most important for assessing a country's economic growth and characterize its dynamics in the medium and long term. WEF data is essentially a benchmarking tool that can be applied by politicians, economists, large businesses in the analysis of the state of the economy of countries or their comparison. Such a comparison can be used to identify the competitive advantages of some countries with respect to others, in the international expansion of business, in finding ways out of the crisis, and the effective development of the economic system at both micro and macro levels.

The World Economic Forum (WEF) uses a new methodology. On a scale from 0 to 100, they calculate 98 indicators, which are combined into 12 essential criteria, of which four groups are composed: a favorable environment, human capital, markets, and an innovative ecosystem.

The efforts of Côte d'Ivoire are quite successful, which is reflected in its position in the competitiveness rating of the World Economic Forum (Table 1).

Table 1. ECOWAS countries' competitiveness rating 2018–2019.

Country	World ranking/138 countries	ECOWAS rating/10 countries
Ghana	106	1
Mali	125	9
Guinea	126	10
Cote d'Ivoire	**114**	**4**
Burkina Faso	124	8
Gumbi	119	6
Benin	123	7
Nigeria	115	5
Cape Verde	111	2
Senegal	113	3
Liberia	132	11

Source: Compiled by the author based on Report "The Global Competitiveness 2016–2017". Note: Guinea Bissau, тоґо, Niger, and Sierra Leone are missing.

As we can see, in the ratings of 2018–2019, Cote d'Ivoire takes fourth place among ECOWAS member countries, and 114th[1] in the world.

In 2016, the share of ECOWAS trade was 1.1% of total world trade. Moreover, the share of exports was 0.5%, imports amounted to 0.6%. The ECOWAS acts as an area focused on trade outside the European Union since only 10.6% of exports and 7.8% of world imports are between member countries of this union (Fig. 4).

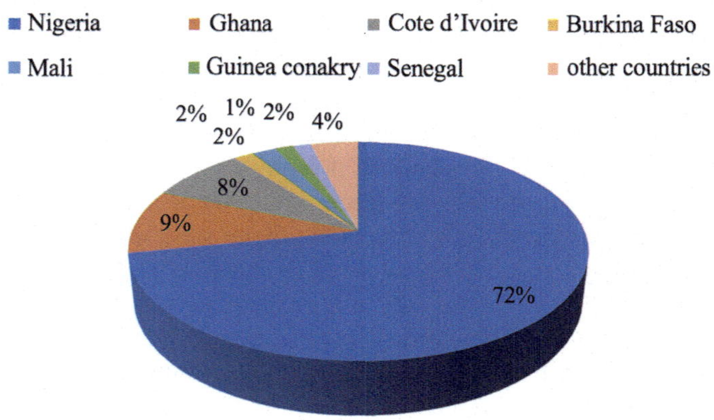

Fig. 4. Export structure of ECOWAS countries in 2010–2016. *Source*: Compiled by the author based on UN Statistics.

[1] Report "The Global Competitiveness 2016–2017".

On average, Nigeria accounted for 72% of the exports of the ECOWAS region in 2010–2016, Ghana accounted for 9.3% and Cote d'Ivoire had 8%. Mali and Burkina Faso are in 4th and 5th place, with a share of 2% and 1.7%, respectively. Guinea and Senegal account for 1.5% of total exports. The rest of the member countries is less than 4% of total exports. The commercial imports of the ECOWAS countries range from $ 100 billion to $ 150 billion per year (Fig. 5).

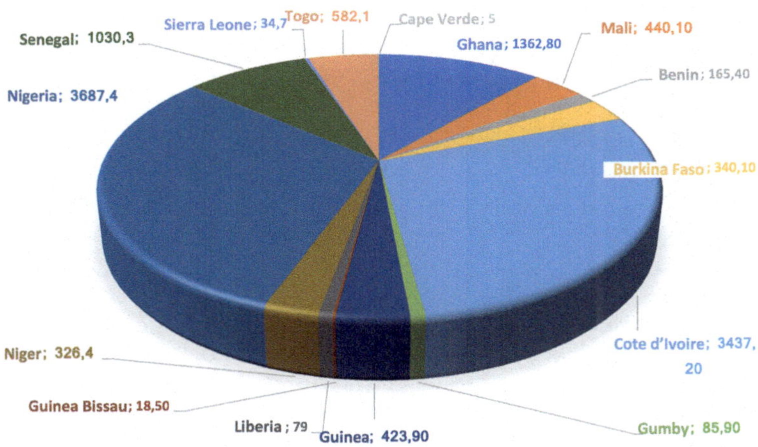

Fig. 5. Exports among ECOWAS member countries in 2016. *Source*: Compiled by the author based on <www.ecowas.int?>.

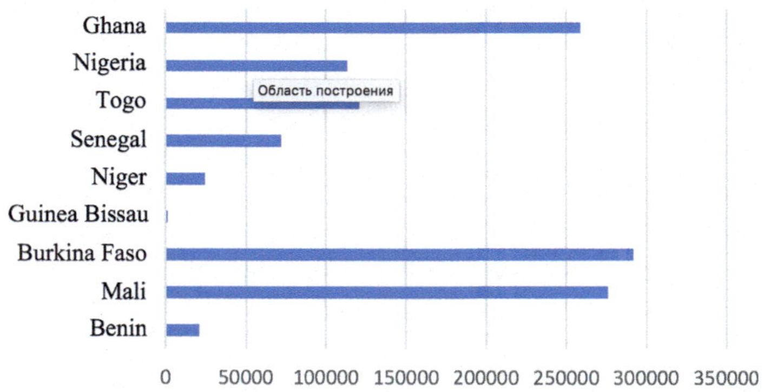

Fig. 6. Cote d'Ivoire exports to ECOWAS countries in 2016 (mln. CFA franc). *Source*: Compiled by the author based on UEMOA data (Central Bank of West African States).

Figures 5 and 6 show the high export activity of Cote d'Ivoire. Cote d'Ivoire supplies its products to most ECOWAS member countries, and on the level of export is in second place after Nigeria.

In 2016, Burkina Faso, Ghana, and Mali were the most active partners of Cote d'Ivoire in the ECOWAS zone. Burkina Faso and Mali import palm oil and petroleum products into Cote d'Ivoire. As for Ghana, it supplies finished products such as petroleum products and plastics. At the same time, import between states is approximately 10%, or from 10 to 20 billion US dollars per year (Fig. 7).

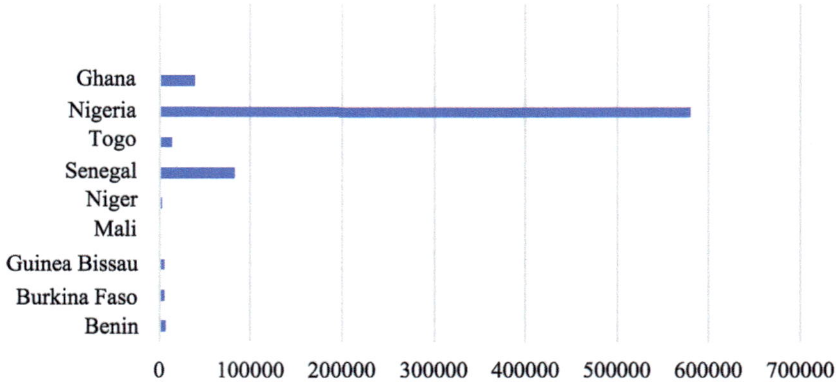

Fig. 7. Cote d'Ivoire imports to ECOWAS countries in 2016 (mln. CFA franc). *Source*: Compiled by the author based on BCEAO (Central Bank of West African States).

As we can see, Nigeria leads in the ECOWAS zone, accounting for 9.8% of Ivorian imports. Ivory coast buys crude oil from Nigeria for use as raw materials for the domestic Ivorian refining company (Societe Ivoirienne de Raffinage).

3 Conclusion

Thus, economic data demonstrate that Cote d'Ivoire plays a critical role in the ECOWAS zone in comparison with the other member states. Among the West African countries, after Ghana and Nigeria, Cote d'Ivoire ranks 3rd, although its contribution is only 6% of the region's GDP. Expanding Cote d'Ivoire's economic performance and reaching the maximum use of the high potential of regional integration of this country requires the following: (a) developing and improving the state's economic policy; (b) strengthening partnerships with all ECOWAS member countries; and (c) creating favorable conditions for doing business and attracting investments.

Acknowledgments. This study is supported by the RUDN 5-100 Program.

References

Africa Regional Integration Index: Official website (n.d.). www.integrate-africa.org
Aksyuk, L.N.: Raw Policy of the Developing Countries of Africa. Nauka, Moscow (1994)

Carrere, C.: African regional agreements: impact on trade with or without currency unions. J. Afr. Econ. **13**, 2 (2004)
Fituni, L.L.: African economy: challenges of post-crisis development. Asia Afr. Today **8**, 54 (2010)
Journal Financial Afrik: Regional trade and non-tariff barriers in ECOWAS (2016). https://www.financialafrik.com/2016/06/13/commerce-regional-et-barrieres-non-tarifaires-au-sein-de-la-cedeao/
Journal Le Point-Économie: Economy customs union ECOWAS can do better (2016). https://www.lepoint.fr/economie/union-douaniere-la-cedeao-peut-mieux-faire-17-08-2016-2061961_28.php#xtmc=cedeao&xtnp=2&xtcr=17
Kurbanov, R.A.: African integration: theoretical and legal aspects. Law State **2**(71), 77–81 (2016)
Morozenskaya, E.V., et al. (eds.): Africa: new trends in economic policy. Institute of Africa RAS, Moscow, Russia (2000)
Morozov, V.P.: Collection of articles. Regional Aspects of African Economic Development (90s). Institute of Africa RAS, Moscow, Russia (2002)
Skubko, Y.S., Shubin, G.V.: Some aspects of social policy in South Africa in 1994–2006. Sci. J. "SPERO" **6**, 23 (2012)
United Nations Economic Commission for Africa: Official website (n.d.). www.uneca.org
Voinov, Y.A.: Economy and foreign economic relations of the Republic of South Africa, Moscow, Russia (2011)
ECOWAS: Official website (n.d.). www.ecowas.int
World Economic Forum: Opinions and analysis of key sessions (n.d.). https://www.weforum.org/events/world-economic-forum-on-africa-2013/programme

Investment Process Mechanisms Based on Improving Regional and Municipal Infrastructure in the Context of Digital Development

Nikita A. Lebedev[1(✉)], Ruslan A. Mammaev[2], and Stanislav Yu. Evdokimov[3]

[1] Institute of Economics of the Russian Academy of Sciences, Moscow, Russia
swonson@bk.ru

[2] Research Institute of Management, Economics, Politics, and Sociology of the Dagestan State University of National Economy, Makhachkala, Russia
marshalnew@mail.ru

[3] Moscow Regional Division of the Russian Presidential Academy of National Economy and Public Administration, Krasnogorsk, Russia
stanislavyurevich@list.ru

Abstract. The entire set of economic, geographical, and social factors creates a significant imbalance in the development of the territories of Russia, which has been repeatedly noted by Russian scholars and economists as a significant obstacle to the sustainable growth of the national economy. This paper provides an answer to the urgent issues of enhancing the investment activity at the regional and municipal levels, which should be based on the territorial infrastructure projects in the context of the processes of universal digitalization of the country's national economy. In the course of the study, the authors identified such problem points as (1) the lack of a balanced and progressive strategy for the development of regional and local infrastructure; as well as (2) the urgent need to fast-track a process of the investment intensification in the priority infrastructure projects. Two key provisions represent the scientific novelty of this research. First, the authors develop an organizational and economic model of the project activation of the launch and implementation of the investment mechanism for the development of priority regional and local infrastructure. Second, authors' conclusions and suggestions, made as a result of the study, can be further used both in developing the financial policy of the state administration of territorial development at the federal and regional levels.

Keywords: Investments · Territories · National economy · Infrastructure projects · Digitalization · Spatial development · Municipal districts

1 Introduction

The analysis carried out by the authors in the course of the study makes it possible to conclude that a key feature that has historically been critical for Russia is the significant heterogeneity in the development of the territories and regions. The research topic is

A. V. Bogoviz and Y. V. Ragulina (Eds.): ISCI 2019, LNNS 115, pp. 542–547, 2020.
https://doi.org/10.1007/978-3-030-40749-0_64

relevant due to the need for intensification of the financing of infrastructure investment projects, such as the renovation of railway networks, the construction and laying of a new roads and bridges, the modernization of the airport network, the formation of fundamentally new, innovative industries that meet the requirements of the digital economy, etc.

The purpose of this study is to propose, based on an analysis of domestic and foreign specialized regulatory literature, a model for the sustainable development of the activation of the process of attracting investments in the infrastructure at the regional and local levels. One of the priority ways is issuing municipal infrastructure bonds. According to the authors, the development of the infrastructure system can and should become the starting point in the evolution of the country's economic system. An increase in tax revenues, an increase in turnover between regions and territories, an improvement in the dynamics and turnover of fixed and working capital, an increase in the interests of potential investors in regions and territories, and, of course, and the formation in this space of a fundamentally new qualitative standard of living are all key goals of the social function of the state.

2 Materials and Methods

In this paper, the authors used the general scientific methods of scientific knowledge of research materials, and patterns of economic science, using such traditional economic methods as: statistical and economic analysis (systemic, causal, and functional), synthesis method, economic and information processes modeling, mathematical methods, and a systematic approach, as well as using the hypothetical-deductive method of scientific knowledge and modeling, and a model for activating the investment process in infrastructure projects at the local level through the use of municipal infrastructure bonds. The theoretical and methodological basis of the research consists of the scientific works of Russian and foreign scholars and economists on the problems of state regulation of investment processes at the regional and municipal levels. The subject of this scientific research has recently been quite popular among scholars. Among foreign and domestic authors, the contribution of the following scholars should be highlighted (Davidson and Mariev 2015; Katunina 2018; Kokovikhin et al. 2017; Leksin and Porfiryev 2016; Berezinskaya 2017).

3 Results and Discussion

The study of the modern foreign and domestic experience makes it possible to conclude that the already established model of a simple redistribution of revenue between the unevenly developing territories through the budget mechanism proved its economic failure a long time ago. The acute relevance of the country's competitiveness in world markets clearly indicates the need for drastic measures in the field of collapsing territorial and regional cohesion. Within the framework of such a redistribution between the territorial units, the prerequisites for intensive growth of all economic systems involved in the process do not form, and the heterogeneity and large extent of the country's territory exacerbate this fact once again. The key goal in this situation should not be a simple redistribution mechanism, but a mechanism for intensifying economic and production processes that serves as the foundation for the conditions favorable for long-term development. The key method is the qualitative and quantitative growth of the investments in the regional infrastructure projects. In the current conditions of budgetary constraints and the global importance of the infrastructure component in the development of the state and its territories, one of the ways to attract investment in infrastructure projects of various levels is a public-private partnership (Davidson and Mariev 2015).

At the state level—except the additional possibility of reducing budget expenditures and the implementation of the infrastructure projects—it is also the additional profitability of tax revenues to the budgets of various levels. Therefore, expanding the possibilities of alternative financing on a long-term debt basis is one of the priorities of the state policy regarding regional socio-economic development. Financing of the infrastructure projects in the framework of the public-private partnerships is possible based on a bank lending mechanism, a budget co-financing mechanism, and issuing of bond loans. The implementation and effective realization of the projects on the basis of public-private partnerships can help solve one of the main tasks of the Russian national economy system – the transition from the redistribution of monetary resources from donor regions to subsidized regions to the redistribution of opportunities and the averaging of the investment attractiveness of territories (De Groot et al. 2009). The municipal infrastructure bonds can be an effective element in attracting private investment capital and, at the same time, real development. The authors of the study consider program municipal infrastructure bonds as well as the scheme for the accumulation and redistribution of income by the Bank for Infrastructure Development (proposed to create a financial and credit structure), starting from the issue of bonds of the municipal infrastructure project. The system mechanism for issuing program bonds is presented in Fig. 1.

The program bonds' functional is financing the development of programs and plans for the development of infrastructure of municipalities. This type is primarily for the release by those municipalities that do not have the well-developed plans of the infrastructure development that define growth points at the regional and municipal levels, taking into account the institutional features of the region and the requirements of the investment structures. The type of bonds presented to the attention is possible to be redeemed both from general tax revenues at the municipal level and from the revenue side of the effective investments of ongoing projects (Katunina 2018). The model of the sustainable infrastructural regional development of the Russian Federation is created on the basis of four elements: economic, and social, to which information and digital security are added in modern conditions, which, according to the authors, gives an additional impetus to expanding the investment opportunities of the regions, simplifying the access and accelerating the speed of operations carried out in the context of digitalization.

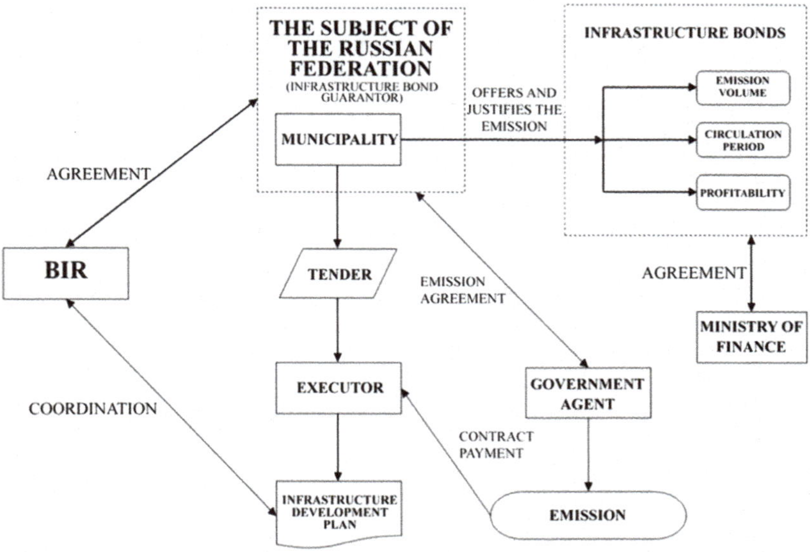

Fig. 1. A system for generating financing for planning the development of infrastructure of municipalities based on the program bonds.

4 Discussion

The above-described advantages of attracting the municipal infrastructure bonds, which were described above, in order to intensify investments in the field of financing infrastructure projects, in our opinion, leave no doubt about the mutual benefit for the state and private business representatives. However, several factors hindering this process should be noted. The main one is the uncertain status of municipal infrastructure securities in the institutional field of the state. In the Budget Code of the Russian Federation, there is no direct prohibition on issuing municipal infrastructure bonds; in fact, project (infrastructure) bonds are not municipal, and municipal bonds are not infrastructural. It should also be borne in mind that the investment projects to finance the development of the infrastructure facilities are long-term. Accordingly, one of the required characteristics of the issued bonds is a significant duration of the bond circulation period. At the same time, the current circulation periods of municipal bonds rarely exceed five years.

It should be noted that at present, best practices and investment tools used in the renovation of the infrastructure of Russian municipalities are constrained by the readiness of the municipalities themselves to organize the activities to attract investments to their territories. The Russian government regularly raises the issue of strengthening the effectiveness of public finance management, which resulted in the Budget Code of the Russian Federation of July 31, 1998, No. 145-FZ (as amended on April 15, 2019); unfortunately, the section regarding municipal bonds repeated almost the same content as that from the previous edition. If significant amendments are made to the Budget Code of the Russian Federation, detailing the rules for financing loans for the municipalities' infrastructure, as well as the approval of the manager of the newly created credit and monetary structure, on the basis of a large bank - a regional development bank or an infrastructure development bank, establishing an institutional framework and acquiring the legislative status of the established institution, in order to regulate and coordinate the investment projects of the infrastructure development. These actions will enable the municipalities to use their own power and, at their discretion, to attract funds directly from various markets for the development, which is important without state subsidies.

5 Conclusion

The decision made by the country's leadership on the transition to the latest model of development of the Russian national economy is undoubtedly associated with the understanding of fundamental changes in the paradigm of the world economic thought, strengthening the relationship between the business and the social component of society. The implemented program of the digital economy is directly focused on a qualitatively different institutional approach of both state and social development (Kokovikhin et al. 2017). It also manifests itself in a regional context, taking into account the mandatory participation in the management process of the municipalities and local communities. This process involves stimulating the activities of the business community and enhancing the investment activities, which is directly related both to an increase in the investment image and to a significant increase in the responsibility of the regions and municipalities in meeting the established criteria and priorities of the state policy of spatial development.

An important aspect is the creation of conditions, both institutional and a real, competitive environment, to maximize the favor of business processes, industrial production, trade, and tourism operations, while taking into account local characteristics (Leksin and Porfiryev 2016; Wu et al. 2014). We also consider it important to note the regular monitoring and control of the situation on the ground—in the regions and territories, especially relevant in the context of a large extent and significant differences in the development of Russian regions. The key priority in the creation and effective development of the appropriate institutional conditions in the national economy of the country in the context of universal digitalization may be the formation and implementation of mechanisms for investing funds in financing projects of regional and municipal infrastructure.

References

Berezinskaya, O.: Investment drought in the Russian economy: structural characteristics and turnaround perspectives. Russ. J. Econ. **3**(1), 71–82 (2017)

Davidson, N.B., Mariev, O.S.: The impact of spatial concentration on enterprise performance. Econ. Region **4**, 95–105 (2015). https://doi.org/10.17059/2015-4-8

De Groot, H., Poot, J., Smit, M.J.: Agglomeration externalities, innovation, and regional growth: theoretical perspectives and meta-analysis. In: Capello, R., Nijkamp, P. (eds.) Handbook of Regional Growth and Development Theories, pp. 256–281. Edward Elgar, Cheltenham (2009)

Katunina, I.V.: Organizational project management in Omsk region companies: current state and development constraints. Econ. Region **14**(1), 69–78 (2018)

Kokovikhin, A.Y., Ogorodnikova, E.S., Williams, D., Plakhin, A.E.: Institutional factors in the evaluation by the entrepreneur of municipality investment climate. Econ. Region **13**(1), 80–92 (2017). https://doi.org/10.17059/2017-1-8

Leksin, V.N., Porfiryev, B.N.: Evaluation of the effectiveness of government programs of the socio-economic of the regions of Russia. Stud. Russ. Econ. Dev. **27**(4), 418–428 (2016)

Wu, J., Wu, G., Zhou, Q., Li, M.: Spatial variation of regional sustainable development and its relationship to the allocation of science and technology resources. Sustainability **6**, 6400–6417 (2014)

Modeling Strategic Decisions to Improve Enterprises Competitiveness of in the Territorial Production Cluster

Nikita Y. Fomin(✉), Alena N. Dyrdonova, and Galina N. Larionova

Nizhnekamsk Institute of Chemical Technology "Kazan National Research Technological University", Nizhnekamsk, Russia
ya.juventino@yandex.ru, dan-home@yandex.ru, galalar@yandex.ru

Abstract. This paper focuses on the cooperative interaction between business entities in a territorial-production cluster. The essence of the cluster as an integration and competitiveness tool is discussed by the authors. The methodological section outlines the approach proposed by the authors to the extension of strategic directions for cluster development. The indicated approach relies on the methods of regression analysis and optimization modeling. The authors propose a resulting indicator measuring the effectiveness of a territorial-production cluster. This indicator reflects the amount of profit generated by the cluster-forming enterprises. According to the developed algorithm, the key factors of the total economic effect are identified, and an optimization model for its maximization is developed. The paper presents the main results of the experimental testing of the developed algorithm.

Keywords: Territorial production cluster · Economic effect · Optimization · Production and economic condition · Production resources

1 Introduction

The current conditions for the function of business entities are characterized by steadily increasing competition and growing public needs. In this case, the determining factor in the competitiveness of an enterprise is intensive development, which is realized through the introduction of advanced technologies in production activities and the intensification of innovative activities. It is worth noting that innovative development requires many resources, including human, material, and financial. Independent business entities do not always have sufficient resources. Integration is one of the main ways to cover such a deficit. It must be emphasized that enterprises can gain access to a wide resource base only in conditions of integration with the social and innovative infrastructure of the territory.

The integration of enterprises, based on the combination of share capital, leads to the monopolization of production and, consequently, to the loss of the main development engine: competition. That is why the modern economy requires the

A. V. Bogoviz and Y. V. Ragulina (Eds.): ISCI 2019, LNNS 115, pp. 548–555, 2020.
https://doi.org/10.1007/978-3-030-40749-0_65

organization of a different cooperation form based on the implementation of large-scale, joint projects for joint development. Nowadays, this interaction form is increasingly taking the form of territorial production clusters.

Harvard Business School professor Michael Porter introduced the concept of a cluster in economics. According to his definition, "a cluster is a group of interconnected companies, specialized suppliers, service providers, firms in relevant industries, as well as organizations related to their activities in certain areas that are competing, but at the same time, working together, geographically concentrated" (Porter 2005). The most crucial distinguishing feature of the development of the economic clusters, from Porter's point of view, is the combination of cooperation and competition: firms simultaneously cooperate and compete with each other. This paradoxical effect of coexistence within the cluster is called co-competition.

The cluster approach to the development of territories has recently become widespread around the world. In particular, the prospects for the clustering of Russian territories can be traced within the framework of the priority project of the Ministry of Economic Development of the Russian Federation: "the development of innovative clusters – the leaders of world-class investment attractiveness" (Ministry of Economic Development of the Russian Federation 2016).

Domestic literature contains many theoretical studies related to the nature and principles of how clusters function. The works devoted to the search for optimal management decisions for the development of cluster-forming enterprises based on the mathematical modeling and analysis methods are less common. The authors of this paper focused on solving this problem, i.e., expanding the methodological base for the development of managerial solutions for optimizing the cluster operation using economic and mathematical methods.

2 Materials and Methods

Turning to the question of the used methodology, we should note that, due to the heterogeneity of cluster members, scholars often practice modeling and analysis of the indicators of aggregate blocks or target groups of cluster members, for example, its production sector or its innovation or financial infrastructure (Markov 2015).

It is reasonable to assume that the development trends of most manufacturing enterprises and the cluster infrastructure are determined by the trends in the production activities of its largest business entities. The above has become the main prerequisite in developing an approach to the modeling of the production processes cluster. The modeling object is the core of the cluster, i.e., the set of cluster-forming anchor cluster enterprises. The main goal of modeling is the identification of patterns in the formation of the resulting production indicators of the cluster core. In the process of analyzing the methodology for modeling territorial production clusters, it was revealed that classical methods of mathematical modeling of economic systems are applicable for solving this problem, in particular, the correlation and regression analysis and linear programming. First of all, there was a need to determine the most significant technical and economic indicators whose values can characterize the main results and efficiency of the production activities of cluster enterprises. The classic absolute indicator of production activity is the level of profit received by the enterprise. Many scholars aggregate the profits of the organizations participating in the cluster to one of the most common indicators used in the analysis and modeling of clusters (Fomin 2018).

Thus, the expediency of using the total profit of enterprises as an indicator of the economic effect created by the cluster-forming enterprises was revealed. The economic effect is calculated according to Formula 1:

$$EE = \sum\nolimits_{i=1}^{n} Pbt_i, \tag{1}$$

where EE – total economic effect of cluster-forming enterprises; Pbt_i – annual profit before tax of the ith cluster-forming enterprise.

In the framework of the proposed algorithm, the next step is a comprehensive study of the factors of the resulting indicator (x_1, x_2, …, x_n). It is proposed to implement it by identifying the correlation of the economic effect with a wide range of indicators of the supply of industrial resources of the cluster enterprises.

The next task is to develop an optimization model of the economic effect. It has been proposed to build it as a linear programming problem (LPP) for maximization. The target function of the LPP is the regression model of the economic effect. The LPP limitations (b_1, b_2 …, b_n) are the functions of the dependence of the relative indicators of cluster performance on the factors of the target indicator.

The cluster is a dynamic system whose effectiveness depends on the economic condition of the territory and cluster-forming enterprises. In this regard, it is advisable to search for optimal values of the economic effect of the cluster for its various states. The type of economic condition affects the values of the LPP's constraints, as presented in Table 1.

Table 1. The types of cluster economic status.

Type of economic condition	Values of LPP constraints	Brief description of the condition
Favorable	Maximum parameter values demonstrated by cluster-forming enterprises	The favorable economic situation reflects the maximum degree of intensification of the use of production resources by cluster-forming enterprises. The economic effect generated by the clusters is marginal in the current economic environment. Further growth is possible only due to the extensive expansion of the cluster resource base
Standard	Average values of parameters demonstrated by cluster-forming enterprises	It reflects the average level of the production efficiency of cluster-forming enterprises. The cluster is in this state most of the time. If the cluster indicators show a standard state, then this is the indicator showing those reserves used for the economic growth due to more efficient use of human, material, and financial resources of the cluster

(*continued*)

Table 1. (*continued*)

Type of economic condition	Values of LPP constraints	Brief description of the condition
Crisis	Minimum parameter values demonstrated by cluster-forming enterprises	The crisis state is characterized by the minimum values of the indicators of production efficiency, which were typical for the enterprises based in the analyzed cluster. This state indicates the non-optimal management of cluster resources and requires the development of operational and strategic measures to improve cluster efficiency

The developed approach was tested in the process of modeling the Nizhnekamsk Petrochemical Cluster, which is described in the research results.

3 Results

The Nizhnekamsk Petrochemical Cluster (NPChC) was chosen as an object for testing the developed models. The development trends of manufacturing enterprises and cluster infrastructure are determined by the trends in the production activities of cluster-forming enterprises. In the structure of the NPChC, such enterprises are PJSC Nizhnekamskneftetekhim (PJSC Nizhnekamskneftetekhim, n.d.), JSC TAIF-NK (JSC TAIF-NK 2019), JSC TANECO (JSC TANECO, n.d.), and PJSC Nizhnekamskshina (PJSC Nizhnekamskshina, n.d.).

A statistical study of these enterprises for the period 2011–2018 made it possible to develop regression models of the target indicator and constraints for the linear programming problem. The function of the economic effect of the Nizhnekamsk petrochemical cluster is presented as Formula 2:

$$EE = 457 * x_1 + 416 * x_2 - 52, \tag{2}$$

where x_1 is the average share of inventories on the balance sheet of a cluster-forming enterprise; x_2 is the average share of cash on the balance sheet of a cluster-forming enterprise.

The task of the linear programming on maximizing the economic effect of the Nizhnekamsk petrochemical cluster is presented as Formula 3:

$$\begin{gathered} EE = 457 * x_1 + 416 * x_2 - 52, \\ \begin{cases} 26 * x_1 + 13 * x_2 - 0,39 \leq b_1, \\ 1,95 + 23 * x_1 - 9,21 * x_2 \leq b_2, \\ 8348 - 42281 * x_1 + 166027 * x_2 \leq b_3 \\ 0,35 * x_1 - 0,73 * x_2 - 0,008 \leq b_4, \\ 0 \leq x_1 \leq 1, \\ 0 \leq x_2 \leq 1, \end{cases} \end{gathered} \tag{3}$$

where b_1 is the average return on assets of a cluster-forming enterprise; b_2 is the average value of the turnover ratio of current assets of a cluster-forming enterprise; b_3 is the average output per employee of a cluster-forming enterprise; b_4 is the average value of return on assets ratio of a cluster-forming enterprise.

The linear programming problem was solved in three variations following the three types of the economic status of the cluster described in Table 1. In other words, a search was done for three optimal solutions for the LPP with the alternate substitution of the minimum, average, and maximum values of the indicators of production efficiency of the NPChC. A multivariate approach to the solution of LPP will allow us to fix and analyze the dynamics of optimal solutions due to changes in the efficiency of production activities of cluster-forming enterprises.

The optimal LPP solutions are presented in Table 2.

Table 2. The optimal LPP solutions for maximizing the economic effect of the NPChC.

Indicator		Optimal value	Allowable value
CRISIS ECONOMIC STATE			
LPP target	Economic Effect (EE) – the total profit of the cluster enterprises, ₽ million.	6 359	EE → max
Arguments of the objective function of the LPP	x_1 – the average share of inventories in the property structure of a cluster-forming enterprise	0,08	$0 \leq x_1 \leq 1$
	x_2 – the average share of cash in the property structure of a cluster-forming enterprise	0,05	$0 \leq x_2 \leq 1$
Indicators-limitations of the LPP	b_1 – the average return on assets of a cluster-forming enterprise	2,47	$b_1 \leq 2{,}47$
	b_2 – the average value of the turnover ratio of current assets of a cluster-forming enterprise	3,63	$b_2 \leq 3{,}65$
	b_3 – the average value of the output per employee of a cluster-forming enterprise, ₽ thousand/person	14 335	$b_3 \leq 14\,406$
	b_4 – the average value of return on assets ratio of a cluster-forming enterprise	0,056	$b_4 \leq 0{,}056$
STANDARD ECONOMIC STATE			
LPP target	EE	33 371	EE → max
Arguments of the objective function of the LPP	x_1	0,12	$0 \leq x_1 \leq 1$
	x_2	0,07	$0 \leq x_2 \leq 1$
Indicators-limitations of the LPP	b_1	3,7	$b_1 \leq 3{,}79$
	b_2	4	$b_2 \leq 4$
	b_3	15 623	$b_3 \leq 15\,623$
	b_4	0,09	$b_4 \leq 0{,}09$
FAVORABLE ECONOMIC STATE			
LPP target	EE	53 898	EE → max
Arguments of the objective function of the LPP	x_1	0,15	$0 \leq x_1 \leq 1$
	x_2	0,09	$0 \leq x_2 \leq 1$
Indicators-limitations of the LPP	b_1	5,12	$b_1 \leq 5{,}14$
	b_2	4,67	$b_2 \leq 4{,}67$
	b_3	16 222	$b_3 \leq 16\,222$
	b_4	0,13	$b_4 \leq 0{,}13$

For clarity, it is appropriate to present the dynamics of the optimal values of the economic effect in Fig. 1.

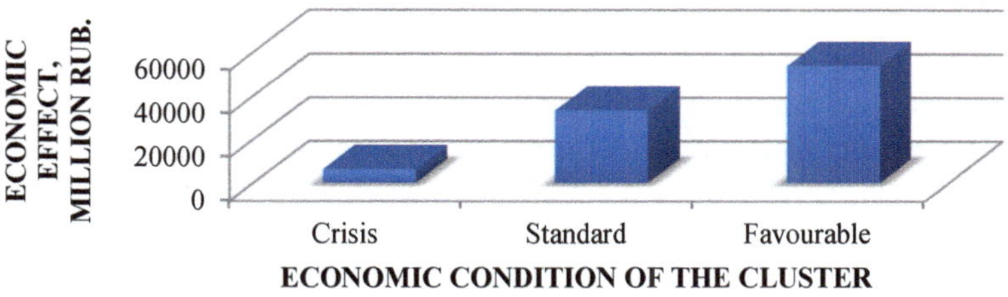

Fig. 1. The dynamics of the economic effect of the NPChC.

In Fig. 2, it is advisable to reflect the dynamics of the factors of economic effect.

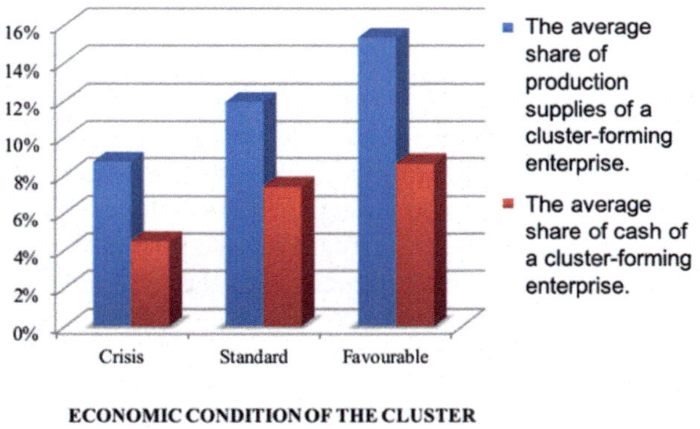

Fig. 2. The dynamics of factors of the NPChC economic effect.

The analysis of the dynamics of the results allowed us to conclude that, as the economic condition of the Nizhnekamsk petrochemical cluster improves, it is recommended to increase the volume of circulating production resources, thereby expanding the current production activity. We can also interpret the results from the perspective of determining the optimal strategy for managing current assets. Since, in the context of economic growth, as the dynamics show, enterprises need to create additional inventories of stocks and cash, a conservative working capital management strategy characterizes them. It is due to the risks arising from the extensive expansion of production activities.

The average shares of inventories and cash presented in Table 2 can serve as standards for the respective economic conditions of the cluster.

4 Discussion

The results obtained during the testing allowed us to identify several characteristics of the developed approach:

- The proposed modeling algorithm is versatile. With its implementation, various performance indicators of integrated production systems can be investigated, and a wide range of production factors can also be covered;
- The approach was based on a linear approximation of economic functions. Most production processes are non-linear, which may adversely affect the accuracy of the results. However, it should be noted that all developed models were evaluated in terms of statistical significance. In particular, in the process of obtaining the results, regression models with a coefficient of multiple determination below 0.9 were not used;
- From the standpoint of natural economic trends, the methodology gave adequate results. In particular, an increase in the economic effect was revealed as relative indicators of production efficiency improved;
- In the process of testing, many factors of the economic effect were investigated, which included the shares of various categories of assets in the total property of enterprises, the absolute values of various production resources, various categories of expenses of the enterprise, and others. The volume of the analyzed factors gives rise to some randomness of the results. In other words, according to the results of the testing of the model under conditions of different clusters, various dependencies can be revealed. However, the experiment revealed the possibility of an adequate interpretation of the results, which characterizes the method of significance.

Thus, the approach proposed by the authors for optimization modeling can be characterized as requiring an adequate tool for analysis, forecasting, and optimization of indicators of territorial production clusters.

5 Conclusion

In the course of the study, we once again revealed that the territorial-production clusters are complex systems of interaction between business entities and territorial infrastructure. The results of this interaction depend on the quality of managerial decisions.

The quality of management decisions largely depends on their preliminary study. In this section, the use of mathematical modeling and analysis tools is relevant. The authors tried to expand the methodology for mathematically modeling the performance indicators of territorial production clusters.

The testing allowed us to identify the key factors of the economic effect generated by the Nizhnekamsk petrochemical cluster, as well as to describe the trends of the economic effect in various economic conditions of the cluster. Based on the results of the study, a strategic solution for managing the working capital of cluster-forming enterprises was proposed, and the standards for certain categories of current assets were obtained.

Thus, we can conclude that the goal set in the framework of the presented study was achieved.

Acknowledgment. We express our gratitude to the administration of the Kazan National Research Technological University and the Nizhnekamsk Institute of Chemical Technology for financial and organizational support.

References

PJSC Nizhnekamseneftekhim: Corporate website (n.d.). https://www.nknh.ru/

JSC TAIF-NK: Corporate website (n.d.). https://www.taifnk.ru/

JSC TANECO: Corporate website (n.d.). http://www.taneco.ru/

PJSC Nizhnekamskshina: Corporate website (n.d.). http://shinakama.tatneft.ru/

Fomin, N.Y.: Modeling and evaluation of the effectiveness of the cluster form of the organization of petrochemical production (Dissertation of Cand. of Technical Sci.). Kazan National Research Technological University, Kazan, Russia (2018)

Markov, L.S.: Theoretical and methodological foundations of the cluster approach. Institute of Economics and Industrial Engineering of the Siberian Branch of the Russian Academy of Sciences, Novosibirsk, Russia (2015)

Ministry of Economic Development of the Russian Federation: Order "On the priority project of the Ministry of Economic Development of Russia 'Development of innovation clusters – investment attractiveness leaders world-class'" (27 June 2016 No. 400). Moscow, Russia (2016)

Porter, M.: Competition. Williams, Moscow (2005)

Natural Resources and Energy: Building Competitive and Resilient Systems

Searching for an Innovative Model of the Future Energy Development with Respect to Digitalization

Vladimir F. Ukolov(✉) and Sergey E. Grishin

Peoples' Friendship University of Russia (RUDN University), Moscow, Russia
ukolovdom@mail.ru

Abstract. The research rationale is a significant increase in the need for energy resources and their irrational use. The world should follow the path of implementing the innovative energy concept. The basis of its long-term development is not only mining, production, and export of traditional energy carriers, but also their substantial rationalization based on the latest scientific achievements, which would ensure the development of a new energy system and provide energy consumption quality while maintaining the indigenous environment. The global future development vision of the world energy buildout should be based on the predicted changes in the economy and the consumer society, which form the demand for energy products and services, as well as on its defining requirements, setting its social and political development priorities. The possibilities of changing the global energy system (financial, organizational, and technological), which is forced to adapt to qualitatively new demands, cannot be overlooked as well. Even the most developed states' efforts are not enough to solve this task in case if they are single forces. The whole world community should be concerned about this and mobilize resources for building the innovative energy future. An urgent task is the institutional regulation of energy conversion processes on an interstate scale, attracting the world scientific potential to solve its strategic development tasks, searching for new breakthrough energy sources and nonstandard technological solutions, developing the energy-saving culture among energy producers and consumers. This would allow to increase the various energy types production, reduce losses, work steadily, move to such a work with fair competition conditions in the free energy services market. The research purpose is to present a systemic, global vision of ways (a) to create a new model for the sustainable energy development, which would be based on a qualitative change in the energy potential; (b) to ensure sustainable development of territories, energy security, and energy digitalization; (c) to change the structure of energy sources; (d) to preserve the indigenous human environment and industry intellectual potential; (e) to ensure the competent use of alternative and small power energetics; (f) to form the technological and organizational base of innovative energy. The article materials can be practically used by power engineers and economists, politicians, state and municipal authorities, educational institutions, and all those interested in the issues of innovative energy development.

Keywords: Energy research · Sustainable energy · Digitalization · Innovative energy

A. V. Bogoviz and Y. V. Ragulina (Eds.): ISCI 2019, LNNS 115, pp. 559–564, 2020.
https://doi.org/10.1007/978-3-030-40749-0_66

1 Introduction

Many world community countries have substantial energy resources and dominate the free market, providing energy services to corporate and individual consumers. The overall energy consumption results, which are advisable to be more effectively used in the world energy consumption system, are made up of such companies' individual performances (Bushueva and Kalamanova 2011). Russia is one of such countries. It has a huge energy potential and takes the leading place in providing global markets with various energy types. Energy plays an important role in achieving the sustainable development of states, as well as their energy security (Bogoviz et al. 2017; Bogoviz et al. 2018). It is necessary to have a systemic image on the future energy development, taking into account its economic and technological formation trends, based on the predicted transformation trends in Russia and the entire world community, which shape energy demand, energy requirements, and set appropriate priorities, ensuring the political balance of competing states.

A major development concern is the search for mechanisms to ensure the dynamic balancing state between the growing demand for energy and the power economy impact on the natural environment, which has an appropriate resource potential. The dynamic balancing state is essential for a smooth achievement of the sustainable development goals, for the energy sector to adapt to external changes without destroying the natural environment (Vera and Langlois 2007; Kaygusuz 2012).

Such things as creating, introducing, and exporting new innovative technologies, using adequate principles for organizing the energy systems functioning, which are based on artificial intelligence, machine learning, energy and related industries digitalization, training a new generation of people being sensitive to natural resources and their conservation traditions, are also very important to acknowledge (Nijhuis et al. 2015; Zhao et al. 2019; Nagy and Hajrizi 2018).

The creation and testing of new digital systems for managing business processes in the electric power industry, the scheduling of all energy flows, and the organization of their work in an optimal energy-saving mode are ahead (Alekseev et al. 2019).

The global energy industry is in badly need of new energy sources, its transformation principles, and "smart" energy supply systems, which are based on the achievements of science and modern practices centered around building a digital relations system between the subjects of energy management system.

An important global task is to change the sources' energy structure, underlying its production. Traditionally, these sources have always been oil and coal; however, they change over time. Nowadays, natural gas is more acceptable and demanded as the most environmentally friendly and widely used energy resource. Foreign market competition for providing this resource has increased significantly and continues to grow. Even a lower price per gas cubic meter is not always the main competitive advantage. Global politics is involved in the struggle for sales markets, creating serious obstacles to the implementation of profitable gas supply projects (for example, to European consumers and in other world regions) (Paltsev 2014; Visenescu 2018).

The future lies in alternative renewable energy sources, creating electricity and heat by using unconventional energy sources, such as wind, water, tides, geothermal energy, ocean wave power, solar energy, biomass, ocean thermal energy conversion, salinity power, and a shape-memory alloy (Nižetić et al. 2017; Nižetić 2018).

The implementation of these energy development, along with other innovative developments in this area, should contribute to significant energy costs reduction in Russia and other countries, even despite the fact that the situation is much better there. So, on average, the power consumption per $1 GDP in economically developed countries is equal to 0.46 kW/h, while it is 4.7 kW/h in Russia (i.e. ten times higher). Also technological losses are even higher (almost 15 times), leading to the fact that the electricity share in the Russian GDP cost structure reaches 50%, while it is only 5% in the economically developed countries (ten times lower) (Rossteplo, n.d.). Reducing electricity costs is not only an important factor in improving the competitiveness of companies and states, it also saves the natural resources and improves both the environment and population health.

2 Materials and Methods

The following methods were used for studying the problem: (a) an expert assessment of the global energy future vision; (b) a comparative analysis of the dynamics in the energy costs volume and structure changes, as well as their share in various countries' GDP; (c) theoretical generalization of the published materials, which reveals the global energy future and the mechanisms for translating it into reality; (d) assessment of the energy impact on increasing the sustainability of state development and competitiveness of companies.

3 Results and Discussion

All the developed world states are concerned about the energy future. Its achievement is one of their primary development tasks. The study shows that Russia is strategically focused on the traditional energy system reproduction and modernization, paying secondary attention to the small and non-traditional energy development, but considering them to be a promising energy saving component. It invests the vast resource majority in the hydrocarbon and nuclear energy sectors, carrying out their innovative development.

The energy research conducted on the global scale is of the great interest for scientists and specialists regarding its future. The possible development trajectory and the ways of its realization by the world energy leaders (Russia, China, and the USA) are of particular interest. Obviously, each of them has an energy future that is unequal and already begins to pass in different ways. But there is a lot in common, which should be adopted by each leading country.

In our opinion, the world community' energy problems should be addressed in a comprehensive manner, in close connection with the security problems solution. This approach should be reflected in the long-term energy policy pursued by the world community in each states' common and private interests.

The following opinion of experts, who provide certain reasons (Kobets 2017) contributing to the accelerated and balanced development of alternative energy sources, is worth noting. They believe that it is necessary to make alternative energy sources cheaper and not too expensive (for example, through the government subsidies), otherwise they cannot compete with the modern nuclear and hydrocarbon energy technological development. An important factor in reducing prices are new technologies and inventions in the alternative energy field, concerning the new solar cell generation. Unlike silicon-based batteries, their production is much cheaper, allows renewable energy developing everywhere, and their use scale expansion significantly reduces the cost per energy unit. Currently, Russia can produce up to 15% of alternative energy, while it is 30% in the world. Its safety and efficiency are the most important competitive advantages (Proskuryakova and Ermolenko 2019).

The alternative energy promotion creates the image of the "traditional" energy as wasteful, irrationally using natural resources, demanding their well-considered use and high-tech processing of generated waste. Many experts believe that this would serve as additional incentives for the development of new energy production areas, which are based on bioenergy and thermonuclear fusion. Thus, due to the comprehensive, systemic renewable energy development, supported by state development institutions, it is possible to reduce energy costs and ensure its wider use (Gupta 2017; Baretto 2018).

Under the influence of certain factors, there should be a turning point in the civilization development, when the hydrocarbons share in the energy sector is less than the alternative energy share. In the meantime, we are witnessing a gradual nuclear energy dominance, supported, albeit by emerging growth of alternative energy sources. In order for this combination to develop, and the nuclear power industry with unlimited fuel reserves to remain competitive in the future, it also requires organizational, managerial, and technological modernization using the digitalization capabilities (Alekseev et al. 2019).

Modern world energy is characterized not only by the tendency of increasing the role of renewable-alternative energy sources, accompanied by a certain energy systems decentralization, but also by a new trend emergence, which is known as digital coding. The digitalization of energy is becoming a new, key innovation area in the infrastructure and management of personnel development, transferring it from the digital area. According to Romanova (2018), "Digital technologies are becoming a solid foundation for all new technological solutions" (p. 423). As noted by a number of scientists, digitalization has become a significant driving force and one of the modern business drivers (Bower et al. 2018).

New digital technologies will make it possible to efficiently solve specific problems of trouble-free, rational use of equipment and increase energy efficiency. However, their implementation will require global changes in the entire energy system, starting with the production and logistics of enterprises and up to the final product transfer to the consumer, along with the need to solve those security problems solution that are accompanied with financial costs.

Now the ICT development statistics (Seafarers 2017), which is the base for digitalization, is such that the share of enterprises, involved in the information networks in Russia introduction, is 1.7% of GDP, while Japan, one of the leaders in this area, has 7%. However, the cumulative transition potential to digital energy in our country is estimated as the largest one, and the funds for its implementation volume increase annually by 21.3%. This percentage growth depends on the energy sector personnel potential, on the staff's ability to make non-standard decisions based on big data, artificial intelligence, identifying possible threats and quickly adapting to external changes (Gitelman and Kozhevnikov 2018).

4 Conclusions

Russia occupies one of the leading places in the global energy market. Given the changes taking place, emerging trends, and new trends in the energy sector, the states' position on the energy market may change. Expert forecasts regarding its future are very good, taking into account the state of energy resources and assessments of current changes in the industry. They show that Russia holds the first place (19%) in about 50 years, with China being second (18%) and the USA on the third place (17%) (Kobets 2017). Such success is impossible without the state's innovative industrial policy in the energy sector, as well as the international economic relations reorientation towards mutually beneficial cooperation (Warwick 2013).

Acknowledgments. The study was carried out on the basis of work with the RFBR grant No. 19-010-00040 "Interaction Between the Real and Virtual Sectors of the Economy in Order to Increase Their Efficiency and Ensure a Balance of Interests Between Business, Society, and the State". The authors are grateful to the Peoples' Friendship University of Russia for providing very confortable conditions for conducting this study. The publication has been prepared with the support of the "RUDN University Program 5-100".

References

Alekseev, A.N., Lobova, S.V., Bogoviz, A.V., Ragulina, Y.V.: Digitalization of the Russian energy sector: State-of-the-art and potential for future research. Int. J. Energy Econ. Policy **9**(5), 274–280 (2019)

Barreto, R.A.: Fossil fuels, alternative energy and economic growth. Econ. Model. **75**, 196–220 (2018)

Bogoviz, A.V., Lobova, S.V., Ragulina, Y.V., Alekseev, A.N.: Russia's energy security doctrine: addressing emerging challenges and opportunities. Int. J. Energy Econ. Policy **8**(5), 1–6 (2018)

Bogoviz, A.V., Lobova, S.V., Ragulina, Y.V., Alekseev, A.N.: A comprehensive analysis of energy security in the member states of the Eurasian Economic Union, 2000-2014. Int. J. Energy Econ. Policy **7**(5), 93–101 (2017)

Bower, V.P., Podvoisky, G.L., Kotova, N.E.: Adaptation strategies for US companies to digitize production areas. World New Econ. **12**(2), 78–89 (2018)

Bushueva, V.V., Kalamanova, V.A. (ed.): World eEnergy - 2050 (White Book). Energy Publishing House "Energy", Moscow, Russia (2011)

Gitelman, L.D., Kozhevnikov, M.V.: Management education paradigm for technological breakthroughs in economics. Econ. Region **14**(2), 435 (2018)

Gupta, K.: Do economic and societal factors influence the financial performance of alternative energy firms? Energy Econ. **65**, 172–182 (2017)

Kaygusuz, K.: Energy for sustainable development: a case of developing countries. Renew. Sustain. Energy Rev. **16**(2), 1116–1126 (2012)

Kobets, N.: Energeytics of the future: a hundred years ahead (2017). http://expert.ru/siberia/2017/37/energetika-buduschego-na-sto-let-vpered/

Nagy, K., Hajrizi, E.: Beyond the age of oil and gas – how artificial intelligence is transforming the energy portfolio of the societies. IFAC-PapersOnLine **51**(30), 308–310 (2018)

Nijhuis, M., Gibescu, M., Cobben, J.F.G.: Assessment of the impacts of the renewable energy and ICT driven energy transition on distribution networks. Renew. Sustain. Energy Rev. **52**, 1003–1014 (2015)

Nižetić, S.: Carbon free electricity production from the alternative energy concepts based on the utilization of the convective vortex systems as a heat engines: review of the current status and perspective. J. Clean. Prod. **170**, 85–95 (2018)

Nižetić, S., Penga, Ž., Arıcı, M.: Contribution to the research of an alternative energy concept for carbon free electricity production: concept of solar power plant with short diffuser. Energy Convers. Manag. **148**, 533–553 (2017)

Paltsev, S.: Scenarios for Russia's natural gas exports to 2050. Energy Econ. **42**, 262–270 (2014)

Proskuryakova, L.N., Ermolenko, G.V.: The future of Russia's renewable energy sector: trends, scenarios and policies. Renew. Energy **143**, 1670–1686 (2019)

Romanova, O.A.: Industrial policy priorities of Russia in the context of challenges of the fourth industrial revolution. Part 1. Econ. Region **14**(2), 420–432 (2018)

Informational part of "Rossteplo": Problematic article of the Energy Saving Group: "Innovative energy as a resource for development (n.d.). http://www.rosteplo.ru/Tech_stat/stat_shablon.php?id=208

Seafarers, P.V.: Expert opinion: development of digitalization of the energy industry in Russia (2017). https://www.elec.ru/articles/mnenie-eksperta-razvitie-cifrovizacii-energetiches/

Vera, I., Langlois, L.: Energy indicators for sustainable development. Energy **32**(6), 875–882 (2007)

Visenescu, R.S.: Russian-ASEAN cooperation in the natural gas sector: lessons from the Russian-Vietnamese relation. Energy Policy **119**, 515–517 (2018)

Warwick, K.: Beyond industrial policy: emerging issues and new trends. OECD Science, Technology and Industry Policy Papers, 2 (2013). https://doi.org/10.1787/5k4869clw0xp-en/

Zhao, Y., Li, T., Zhang, X., Zhang, C.: Artificial intelligence-based fault detection and diagnosis methods for building energy systems: advantages, challenges and the future. Renew. Sustain. Energy Rev. **109**, 85–101 (2019)

Investment Projects Based on Geothermal Energy Sources: Priority Directions for the Russian Fuel and Energy Complex

Andrey D. Astremsky[1(✉)], Elena I. Danilina[2], and Irina A. Terentieva[2]

[1] Law Institute of Russian University of Transport, Moscow, Russia
astr_11@mail.ru

[2] Educational Autonomous Non-Commercial Organization of Higher Education "Institute of Educational Technologies and Humanities", Moscow, Russia
danilinav01@list.ru, ir.terent99@yandex.ru

Abstract. The paper focuses on the key aspects of the concept of geothermal resources and provides their detailed description. The Russian potential in geothermal resources is considered, and the global geothermal energy use is also discussed. The particularly acute relevance of effective environmental protection from the negative impact of the existing industrial fuel and energy complex facilities is also emphasized. It is concluded that the projects on using geothermal energy are quite promising since Russia has a significant energy potential in geothermal resources (which energy potential is 10–12 times higher than the entire potential of the fossil fuels). Their effective use is capable of compensating a significant part of the hydrocarbon fuel consumed nowadays. A high ecological component of geothermal energy is also emphasized. In general, geothermal resources in Russia are characterized by almost universal distribution and cover 95% of the country's area. Based on the potential for using geothermal resources, the authors propose a highly effective investment project to create an energy-biological complex. They also systematize theoretical, methodological, organizational, economic, and institutional mechanisms that contribute to the intensification of geothermal energy reproduction processes.

Keywords: Geothermal resources · Geothermal energy · Energy sources · Investment project · Energy potential · Ecology · Economic efficiency

1 Introduction

Nowadays, there is a rapid rise in energy consumption, along with the limitations of non-renewable natural resources, while it is necessary to note the limited hydrocarbons resources. Moreover, is aggravated by the constant deterioration in the state of reproduction of the mineral resource base (Alekseev et al. 2019).

The energy policy adopted in the world dictates the priority to low-carbon energy sources. They become increasingly attractive in terms of environmental safety due to lower levels of harmful emissions (Popkova et al. 2019; Lobova et al. 2019). In this

A. V. Bogoviz and Y. V. Ragulina (Eds.): ISCI 2019, LNNS 115, pp. 565–570, 2020.
https://doi.org/10.1007/978-3-030-40749-0_67

regard, as part of the development of the global low-carbon energy policy, close attention is paid to the production of energy resources on the basis of geothermal energy sources.

Considering the energy potential of geothermal resources available in the Russian Federation, such sources can compensate for a significant part of the currently consumed hydrocarbon fuel. Also, greener energy sources and technologies will increase enterprises' competitiveness (Chursin and Vlasov 2016).

The relevance of the chosen topic is related to the need for reorientation from using obsolete hydrocarbon resources as a key energy source to the use of alternative energy, geothermal in our case. The research objective is to study practical applications of geothermal energy sources in the industry and to evaluate the effectiveness of using geothermal resources on the example of a specific investment project. The research object is the intensification of using geothermal energy sources (on the example of a specific investment project). The subject of the study is the construction of a modern high-tech energy and biological complex based on the proposed methods and mechanisms that contribute to the activation and increase of investment activity.

2 Materials and Methods

The authors used a commonality of methods and generalizing principles of scientific knowledge, a sequence of the economic science conclusions, and such economic methods as monographic, method of expert and analytical estimates, statistics, experimental, abstract-logical, comparative analysis, etc. The degree of development of the scientific problem, as well as the study and research in the field of geothermal energy sources, were published in the works of such scholars as (Khachaturyan et al. 2018; Ghaebi et al. 2018).

From the perspective of implementing innovative energy projects in current conditions, as well as introducing innovative digital technologies, special attention was paid to this issue in the research conducted by (Alkhasov et al. 2000; Wang et al. 2019). In addition, a retrospective analysis provides the opportunity to conclude that the research on economic projects in the field of geothermal energy sources was published in the works of such scientists as (Kulcar et al. 2008; Mahbaz et al. 2020; Compernolle et al. 2019). Based on these works, the authors actively engaged in the study and development of this scientific issue.

3 Results and Discussion

What is understood by the term "geothermal energy"? This is the heat energy that has been released from the inner zones of the Earth. Geothermal resources are the Earth's internal heat reserves that can be used to generate thermal energy or electricity. They are usually divided into hydro geothermal and petro geothermal. Geothermal energy resources, in contrast to organic energy resources, are cheaper, more environmentally friendly, and cleaner.

Geothermally rich areas are found in almost all parts of the world. Russia has significant potential in the geothermal energy field.

There are over 3000 wells for the use of geothermal resources in Russia. The authors of the paper suggest considering the potential of geothermal resources in various regions of Russia in comparison with the potential oil reserves (Table 1).

Table 1. The comparative analysis of geothermal and oil reserves by Russian regions as of 2018.

Regions	Geothermal resources, trillion t us.ton	Oil reserves, mln. t.
Nothern	132,2	38,2
Northwest	18,1	31,1
Central	35,4	20,4
Central Black Earth	19,8	47,2
Volgo-Vyatka	12,5	69,7
Volga	59,1	70,4
North Caucasus	45,0	41,4
Ural	64,9	282,2
Siberian	258,1	47,8
Far Eastern	696,3	23,7

Source: Compiled by the authors based on the statistical data coming from (Federal State Statistics Service, n.d.).

Based on the data above, the vast majority of Russian geothermal potential is concentrated in the Far Eastern region; in general, these resources reserves are impressive. Thus, Russia has sufficient potential and opportunities to switch to the use of geothermal resources as the main energy source.

The authors propose a project to create an energy-biological complex (EBC) based on the geothermal energy sources, that should be considered not only from its financial (commercial) efficiency for potential investors but also from its economic and budgetary efficiency.

The introduction of the proposed energy-biological complex through the use of geothermal energy resources, based on the proven deposits in the region that stand idle in modern conditions, will ensure the production and supply of electric energy in the total amount of more than 20 million kWh per year, and more than 400 thousand Gcal per year for thermal energy.

Based on the study and analysis of the initial data, the authors propose a planned structural aggregate of investments in the creation of a EBC project (Table 2).

The valuation of fixed assets (equipment) is made at the current prices of domestic and foreign suppliers. Assessing the cost of installation and construction work, as well as the other expenses, was made by domestic and foreign analogs.

Table 2. The investment structure in the proposed investment project creation.

Name	The estimated investment level, mln. RUB
The required investment amount, taking into account the working capital and the inflation of previously invested funds	5,526.21
The amount of used equity	1,159.90
The amount of used credit	4,366.31

Source: Compiled by the authors based on the project data.

The values of the final performance indicators of the proposed investment project are presented in Table 3 below.

Table 3. The summary indicators of the financial efficiency of the investments in a project for the creation of energy-biological complex based on the geothermal sources.

Indicators name	The performance indicators value
NPV of the investment project, mln. RUB	2074,87
Internal return rate, %	28%
Simple payback period, years	5,38
Discounted payback period, years	7,00
Yield index	3,47

Source: Calculated by the authors according to the investment project data.

As a result of the analytical evaluations performed to assess the proposed effectiveness of the project, and taking into account the above-mentioned values of the initial indicators, the basic criteria values for the investments effectiveness were obtained. More than that, the authors use these data and conclusions about economic efficiency and feasibility of the implementation and realization of the project in modern conditions.

4 Discussion

The development and construction of the energy-biological complex based on the proven geothermal deposits in one of the country's subjects, that is rich in geothermal energy resources, according to the authors, fully complies with the main directions of the state policy in the energy efficiency of the electric power industry based on the renewable energy sources use. The main tasks planned to be solved while implementing the proposed investment project are the formation of additional jobs; the involvement in the regional economy of the previously unused explored isolated, remote and reserve geothermal energy resources; receiving the electrical and thermal energy resources at their account to meet the needs of the complex with the possibility of sale of the energy surpluses subsequent; the organization of the fast-payback industries in the immediate vicinity of the developed free geothermal resources; an increase in the revenue of the regional and federal consolidated budgets; the long-term increased entrepreneurial demand and the social component of the geothermal energy attractiveness.

5 Conclusion

The introduction and implementation of an investment project for constructing a EBC based on the geothermal energy sources will also solve some local problems, e.g., provide locals with jobs, supply the markets with products manufactured by the complex units, provide additional tax revenues to the budgets of various levels, extrabudgetary contributions, etc. The project's discounted income, despite critical conditions adopted for consideration, with the high necessary equipment cost, high capital costs, and operating expenses, with the existing product sales risks, is positive, it amounts to more than 2 billion rubles.

The project implementation for creating a EBC creation based on the geothermal fields (as the renewable energy sources components) is extremely attractive not only for the high values of financial efficiency indicators of the investment project but also from the point of view of increasing the general economic and budgetary benefits from the implementation of this project. Also, the global issue of developing the energy sector for the future periods is of interest, and the implementation of this project corresponds to the main priorities for the development of low-carbon energy.

We would like to highlight the key points of the scientific novelty obtained during the study. First, on the basis of the explored potential of geothermal energy sources in Russia, a fundamentally new investment project for the creation of modern energy-biological complex is economically justified. This project fully corresponds to the main directions of the state policy in the field of the increasing the energy efficiency of the power industry. Second, theoretical, methodological, organizational, economic and institutional tools, contributing to the intensification of the reproduction processes in the geothermal energy field, are systematized.

References

Alekseev, A.N., Lobova, S.V., Bogoviz, A.V., Ragulina, Y.V.: A critical review of Russia's energy efficiency policies in the construction and housing sector. Int. J. Energy Econ. Policy **9** (4), 166–172 (2019)

Alkhasov, A.B., Aliev, R.M., Ramazanov, M.M., Abasov, G.M.: Study of complex heat exchange with account for phase transitions in the secondary contour of the geothermal power plant. Renew. Energy **19**(1–2), 155–161 (2000)

Chursin, A., Vlasov, Y.: Innovation as a Basis for Competitiveness: Theory and Practice. Springer, Heidelberg (2016)

Compernolle, T., Welkenhuysen, K., Petitclerc, E., Maes, D., Piessens, K.: The impact of policy measures on profitability and risk in geothermal energy investments. Energy Econ. **84**, 104524 (2019)

Federal State Statistics Service: Official website (n.d.). www.gks.ru

Ghaebi, H., Namin, A.S., Rostamzadeh, H.: Exergoeconomic optimization of a novel cascade Kalina/Kalina cycle using geothermal heat source and LNG cold energy recovery. J. Clean. Prod. **189**, 279–296 (2018)

Khachaturyan, A.A., Khachaturyan, K.S., Abdulkadyrov, A.S.: Model of innovational development of the modern Russian industry. Adv. Intell. Syst. Comput. **622**, 44–50 (2018)

Kulcar, B., Goricanec, D., Krope, J.: Economy of exploiting heat from low-temperature geothermal sources using a heat pump. Energy Build. **40**(3), 323–329 (2008)

Lobova, S.V., Ragulina, J.V., Bogoviz, A.V., Alekseev, A.N.: Energy security performance: a dataset on the member countries of the commonwealth of independent states, 2000−2014. Data Brief **26**, 104450 (2019)

Mahbaz, S.B., Dehghani-Sanij, A.R., Dusseault, M.B., Nathwani, J.S.: Enhanced and integrated geothermal systems for sustainable development of Canada's northern communities. Sustain. Energy Technol. Assess. **37**, 100565 (2020)

Popkova, E.G., Inshakov, O.V., Bogoviz, A.V.: Motivation of energy saving within the corporate market responsibility of economic entities. Lect. Notes Netw. Syst. **44**, 95–105 (2019)

Wang, W., Hong, T., Li, N., Qi Wang, R., Chen, J.: Linking energy-cyber-physical systems with occupancy prediction and interpretation through *WiFi* probe-based ensemble classification. Appl. Energy **236**, 55–69 (2019)

Application of Financial Technologies to Assess the Competitiveness of Enterprises of the Mineral Resource Complex of Russia

Zinaida M. Nazarova(✉), Victor M. Zaernyuk, and Nurali H. Kurbanov

Russian State Geological Prospecting University, Moscow, Russia
nazarovazm@inbox.ru, zvm4651@mail.ru, nurali.k@mail.ru

Abstract. Widely used in foreign companies, financial technology is an essential competitiveness management tool. But they have not yet received proper development in Russian practice. Therefore, the study aims to develop diagnostic and managerial tools of a financial nature, focused on improving the competitiveness of enterprises of the mineral resource complex (MRC). System analysis, statistical, financial and economic, retrospective, and trend analysis, along with the principles of consistency and development, formed the methodological basis of the study. The authors examine the mechanisms for implementing financial instruments to ensure competitiveness on the example of Russian enterprises of the mineral resource complex. The empirical research methodology is justified, and financial factors determining the level of competitiveness are determined. The resulting integral indicator of the competitiveness of MRC enterprises is proposed. The indicator makes it possible to assess its level both in the industry as a whole and for an individual MRC enterprise.

Keywords: Competitiveness · Financial instrumentation · Mineral resource complex enterprises

1 Introduction

Competitiveness is the ability of particular objects or entities to surpass a competitor in given conditions. Unlike "competition," this concept is more complex and synthetic, since it involves the consideration of a chain of objects of research, although being interconnected at different levels (Rubin 2017).

The relevance of the problem and its great applied significance are caused by the fact that competitiveness both, in the broad sense and in terms of research, has become an area of increased interest for scholars and practitioners. So, in different periods, such scholars as (Tuyakova and Cheremushnikova 2017; Mityai 2015), and many other Russian scientists investigated the competitiveness of enterprises.

A study of economic literature has shown that competitiveness is one of the deeply investigated problems in Russian economic science. Along with this, the financial accents in the study of the category of competitiveness and competitive advantages in

A. V. Bogoviz and Y. V. Ragulina (Eds.): ISCI 2019, LNNS 115, pp. 571–579, 2020.
https://doi.org/10.1007/978-3-030-40749-0_68

monetary terms are practically not represented in the assets of domestic economists. It seems that the marketing nature of competitiveness can largely explain this.

2 Materials and Method

In the economic literature, many external and internal factors of the company's competitiveness, along with their other classifications, are presented. One should note that internal factors, including material, intangible, financial resources, and the efficiency of their use, form the competitive advantages of enterprises and organizations.

Competitive advantage is understood as an exclusive value, the possession of which gives the company superiority over competitors in the economic, technical, and organizational fields of activity and allows more efficient management of available resources. At the turn of the 1970s and 1980s, Michael E. Porter, a recognized specialist in strategies and competition, scientifically substantiated this concept (Porter 1988). The foundations of the theory of competitive advantages were laid down by him, supplemented and subsequently developed by many domestic authors, made it possible to classify competitive advantages according to the following criteria:

- Relation to the system (company, country, etc.);
- Sphere of origin;
- Content of benefit;
- Method and means of obtaining a benefit;
- Benefit location;
- Benefit lead time;
- Type of effect gained.

It should be noted that in this and other classifications, the lack of financial emphasis is apparent. One could note the availability of financial infrastructure and sources of financing and the financial stability of the company, which are both significant in the long term for potential partners and customers, liquidity, and solvency. More than that, settlement and pricing policies of interest to established partnerships, lenders, and customers, capitalization of the company are the parameters of great importance for the company's owners, as well as investors and potential strategic partners.

The above is extremely important for modern organizations, and foreign companies are successfully working with these competitive advantages in the commodity and financial markets. However, attention is essentially not focused on these competitive advantages in Russian theory and practice. In our opinion, this is one of the reasons that the vast majority of Russian companies lose to foreign companies not only global but also national markets.

The reorientation of Russian business to financial competitiveness factors and competitive financial advantages necessitates the development of financial instruments and appropriate management technologies that would improve the market position of Russian companies, and, consequently, sectors of the national economy and the country market as a whole.

The research by Kameneva and Khotinskaya (2014) identifies the factors of companies' competitiveness at the intersection of sciences (finance and marketing), along with financial instruments to increase competitiveness applicable to any of the sectors of the economy.

In our opinion, for the "Mining" industry, financial instruments to increase competitiveness can be represented by the following determinants:

- Ability to generate income, value-added profit;
- Asset generating capabilities;
- Growth efficiency;
- Risks and financial stability characterizing the company's ability to survive;
- The market value of the company as a characteristic of its market position;
- Intellectual capital as a sustainable competitive advantage, significant in the knowledge economy.

3 Results

We will consider each of the noted determinants as applied to enterprises of the mineral resource complex (MRC).

First of all, it is the *ability to generate income*. As the macro analysis showed, the studied sector of the national economy is very competitive in generating income (Table 1).

In general, from 2007 to 2017, revenues, gross value added, and profits of the MRC enterprises increased 2.7–3 times in the analyzed period. At the same time, revenues and total value added grew exponentially, and profit rose linearly (Zaernyuk 2019).

It seems that the indicator characterizing the ability to generate income is the turnover of assets, determined by the ratio of revenue to the average book value of assets. This coefficient characterizes the effectiveness of the company using all the resources at its disposal, regardless of the sources of their attraction.

It should be noted that the rather high dynamics of absolute growth indicators as a factor in the competitiveness of enterprises in the industry under study characterizes not only the potential for generating sales, gross value added, but also the strategic priorities of modern enterprises of the MRC. It allows one to organically fit into the financial model of measurement and management, in contrast to the cost-effective profit-oriented analysis and management model that has existed for many years.

Second, *asset generation opportunities*. Assessing the Russian business' potential for generating revenue as a whole and its industries, including the mining industry, it is essential to compare these absolute indicators with total capital (assets). In economic sources and analytical practice, the revenue generation coefficient BEP (Basic Earning Power) is used for these purposes (Kovalev 2015). It is an indicator of asset generating capabilities. The paper (Kameneva and Khotinskaya 2014) suggested that one should rely not only on operating profit, while calculating this indicator, but also on all components of revenue generation, namely sales, value-added, and net profit. In particular, the researchers propose an income generation coefficient determined by the ratio of income and value-added assets; a coefficient of value-added generation, which

Table 1. Dynamics of absolute indicators of enterprises of the mineral resource complex of Russia.

Indicator	Year											2017 in % to 2007
	2007	2008	2009	2010	2011	2012	2013	2014	2015	2016	2017	
Turnover of organizations, billion rubles	4603	5285	5061	6249	8267	8906	9176	9777	11339	11771	13708	298
Output, billion rubles	4383	4973	4748	5950	7361	8356	8892	9434	10856	11119	12173	278
Gross value-added, billion rubles	2866	3285	2885	3837	4 918	5581	5792	6162	7066	7285	7891	275
Balanced financial result, billion rubles	947	850	907	1298	1811	1691	1558	2124	2452	2186	2596	274
Number of operating organizations, units	8438	9136	9804	10116	10539	11584	12023	12623	12598	18200	17600	209
Share of unprofitable organizations, %	31.9	37.6	43.7	38.3	37.4	36.8	43.5	46.6	44.9	29.2	27.5	−4.4 points
Fixed assets	4977	6365	7861	9085	10572	12242	14110	15733	18518	20707	21830	439
Degree of depreciation of fixed assets, %	53.4	50.9	49.6	51.1	52.2	51.2	53.2	55.8	55.4	56.1	56.4	−3 points

Source: Rosstat data. Retrieved from www.gks.ru.

is calculated by the ratio of value-added assets; and a coefficient of profit generation, determined by the ratio of profit before/after tax and assets.

We will evaluate the generating capabilities of the analyzed industry using the data in Table 1, as well as Rosstat data on asset dynamics (Rosstat 2016, 2018). The analysis showed that the generating capabilities of the industry for mining almost do not change. The increment in sales is based on the increment of assets in a predominant way, which is evidence of the extensive nature of the development of Russian enterprises of MRCs and may indicate a decrease in the efficiency of production, financial, and economic activities of the industry.

In our opinion, the indicator of profitability by the economic added value used in the financial analysis could be used as an indicator characterizing the generating capabilities of assets, that is, the efficiency of resource use. This indicator is defined as the quotient of net operating profit after taxes (NOPAT) minus capital costs by average asset value.

Third, growth efficiency. The study showed that the growth indicators of the vast majority of MRC enterprises in Russia do not correspond to an adequate increase in the ability of production, as well as financial and economic activities. From 2007 to 2017, the unstable profitability and overall decreasing trend were observed both in terms of cost-effectiveness of production and assets for MRC enterprises on average. It should be noted that the exponential growth of the absolute indicators of the enterprises of the mineral resource complex (turnover, output, gross value added) in the context of declining profitability may mean that the most MRC enterprises in Russia have inefficient use of resource potential due to undeveloped financial technologies. The industry under study continues to be at the stage of large-scale growth with ineffective management.

In our opinion, the development/growth efficiency can most accurately be estimated using the indicator of return on assets for a profit before interest and taxes:

Fourth, risks and financial sustainability. Chances are one of the critical factors of competitiveness since they determine the very fact of business survival in the existing economic environment and the foreseeable future. Therefore, in the framework of assessing the competitiveness of MRC enterprises, it is essential to interpret the risks in terms of terms that determine financial stability.

The economic literature differently considers indicators (indicators) of financial stability. Within the framework of a narrow interpretation, these are the following coefficients: autonomy, dependencies, financial leverage, and, finally, the coefficient of long-term structure. An expansive analysis considers short-term financial stability within one year, characterized by current, quick, and absolute liquidity, which can be calculated according to the articles of an accounting organization.

From the above indicators, Rosstat annually publishes the current liquidity ratio and the autonomy coefficient, which we use to assess the risk of financial stability of Russian business. At the same time, one should note that economic literature is replete with different interpretations of evaluation criteria.

We consider the coefficient of autonomy, characterizing long-term financial stability. The foreign practice is considered normal when borrowed capital is twice its sources. The practice widespread in Russia, where quite expensive borrowed resources are present until recently, allowed having an equal proportion of own and borrowed sources of financing. Although, in recent years, when the Bank of Russia managed to repay inflation (and to reduce the key rate in this context), there were prerequisites for cheaper loans and borrowings. This circumstance allows us to establish ourselves in the opinion that a formal excess of equity over-borrowed capital may become a fulfillment of the condition determining financial stability. Foreign accounting and analytical practice focus on the value of 1.5 as the lower critical value of the current liquidity indicator, characterizing short-term financial stability. As for private practice, a coefficient of 2 is taken as the average level of current liquidity. This can be explained by the fact that the forced sale of property due to financial difficulties or bankruptcy of the company, and the real amount received from the sale of funds may be significantly lower than the price put up for auction. It is generally accepted that double coverage of current assets with short-term payables may be considered acceptable to creditors.

The data in the analysis table show that, on average, over the entire mineral resource complex industry, over the whole analyzed period, equity capital exceeded borrowed sources of financing, which suggests long-term financial stability as a system macroeconomic characteristic of the industry. At the same time, a decreasing trend is observed, which allows diagnosing a certain economic instability.

The liquidity problem is typical for most Russian companies since the current ratio did not exceed 1.73 for the MRC enterprises, on average, in the analyzed period. Although this indicator is slightly higher than the average for Russian enterprises and organizations (historical maximum is 1.36); however, it is continuously at a level below the standard (Zaernyuk 2019).

Thus, the data presented allow us to conclude that in the context of large-scale growth accompanied by insufficiently effective management. The risk of financial instability has accumulated, which negatively affects the competitiveness of Russian enterprises and organizations of the MRC sector, their presence markets.

To assess their risk and financial stability, characterizing the ability of a business to survive, we can use the indicator of financial leverage, which is widely used in financial analysis.

This indicator characterizes, on the one hand, the degree of financial risk and stability of the company. On the other hand, it focuses on the possibility of increasing the return on equity.

Fifth, market capitalization. Herbert Simon, an American economist, first proposed the concept of the priority of the owners' economic interests (Simon 1978). He formulated the target concept of economic behavior, the content of which is the need for priority satisfaction of the owners' interests. The applied value of this concept is formulated by maximizing the market value of the company.

Introduced in the mid-80s of the 20th century, the new corporate ideology of VBM (Value-Based Management) has changed the transformation of priorities. VBM is a concept that differs from the previous goal setting, focused on maximizing profits, i.e., the value of a business.

According to experts, to date, in foreign countries, the VBM concept is applied in the order of 30% to 50% of companies (Internet portal for students "Lektsii.org", n.d.). In domestic practice, this concept has not yet become widespread because not all of the above indicators can be measured. As a result of this, we can count no more than three dozen companies participating in international comparisons at market value in Russia.

In 2018, 25 companies from Russia were included in the Global 2000 Ranking (there were 27 in 2017) (Forbes 2018).

The leader among Russian companies in the Forbes Global 2000 ranking remained "Gazprom" for the third year in a row. Its market value was $57.8 billion, and assets are valued at $316.8 billion; profit reaches $12.2 billion. "Rosneft" is one of the three largest Russian public companies. In the world ranking, the company took 73rd place (capitalization is $69 billion, $214.2 billion in asset value, $ 3.9 billion in profit). However, the most significant companies representing the Russian business in the Global 2000 ranking account for only 1.7% of the total market capitalization.

In order to calculate the indicator of market capitalization (company value), the authors propose using a dynamic characteristic determined by the growth rate of the market value of share capital.

Sixth, intellectual capital. As interpreted by renowned foreign scientist Annie Brooking (Brooking 2001), the term "intellectual capital" refers to intangible assets. Without the existence of these assets, companies cannot exist, because this term greatly enhances the competitive advantages of companies. In his opinion, intellectual capital is determined by the presence of the following:

- Human assets (knowledge, experience, skills, willingness and ability to innovate, general culture, firm values);
- Intellectual property (rights and privileges);
- Infrastructure assets (patents, licenses, trademarks, electronic networks, databases);
- Market assets characterized by a system of reliable and long-term mutually beneficial relations with partners and customers.

One should note that with the enormous intellectual potential of Russia, the results from its use in the economy are incredibly unsatisfactory. In Russia, on average, only 1 out of 265 scientific results obtained is subject to legal protection. And the share of the contribution of value added generated from the turnover of intellectual property in the GDP of the Russian Federation is less than 1%, while this share is 12% in countries such as the United States, 7–8% in Germany, 20% in Finland.

Current financial practice in working with intangible assets (IA) allows one to use it in many aspects. In financial practice, intangible assets can reflect the assessment of resource potential, generating value, a way to increase investment attractiveness can finally be considered as an argument for sustainable competitive advantage in the financial market.

As an analysis in Russia showed, a change in dominants in terms of competition has not yet occurred. The Russian companies, including companies belonging to the MRC sector, still underestimate intangible assets as an instrument of competitive advantage. There is not any significant effect of intangible assets on production, financial, and economic activities.

Even though in the analyzed 10-year period, the dynamics of intangible assets of Russian enterprises of MRC is almost two times higher than the growth rate of total assets, their share in the property of enterprises and organizations of the studied industry remains extremely low and amounts to less than 0.5%. At the same time, at the beginning of 2000, the largest foreign companies had a ratio of tangible and intangible assets in the proportion of 30/70% (Doyle 2001).

We can use the share of intangible assets in assets as an indicator characterizing intellectual capital as evidence of a sustainable competitive advantage that is significant in the knowledge economy.

4 Discussion

We propose to determine a general level of competitiveness (based on financial instruments) by the formula of geometric mean weighted from partial coefficients (asset turnover, return on economic value added, return on assets before interest and taxes, financial leverage, growth rate of the market value of share capital, and the share of intangible assets).

Further work with this model involves its quantitative testing on empirical data of the MRC enterprises in Russia, which would make it possible to assess their competitiveness.

5 Conclusion

For the period 2007−2017, a macro analysis of the absolute indicators of the enterprises of the mineral resource complex of Russia led to the conclusion that a large-scale growth in key economic indicators (due to insufficient management) drives to the accumulation of risks of financial instability. This cannot but adversely affect the competitiveness of Russian mining organizations and their presence markets.

The empirical research methodology is justified, and financial factors analyzing the level of competitiveness are determined. The resulting integrated indicator of competitiveness, based on financial indicators both for the industry as a whole and for an individual enterprise, is proposed.

References

Brooking, A.: Intellectual capital. In: Kovalik, L.N. (ed.). Piter, St. Petersburg (2001)

Doyle, P.: Cost Oriented Marketing. Piter, St. Petersburg, Russia (2001)

Forbes: 25 Russian companies ranked Forbes Global 2000 (2018). https://www.forbes.ru/biznes/362743-25-rossiyskih-kompaniy-popali-v-reyting-forbes-global-2000

Internet portal for students “Lektsii.org”: The concept of company value management in the practice of modern corporate finance (n.d.). https://lektsii.org/3-84747.html

Kameneva, E.A., Khotinskaya, G.I.: Financial instruments for improving the competitiveness of Russian business: problem statement. Financ. Life **2**, 78–84 (2014)

Mityai, O.V.: Marketing tools to increase the competitiveness of agricultural enterprises. Mod. Res. Innov. **8–1**, 113–120 (2015)

Porter, M.E.: The Competitive Advantage of Nations, 2nd edn. Free Press, New York (1988)

Rosstat: Russian Statistical Yearbook 2016. Rosstat, Moscow (2016)

Rosstat: Russia in numbers 2018. Rosstat, Moscow (2018)

Rubin, Y.B.: On the constructive theory of competition in entrepreneurship. Mod. Compet. **5**(65), 114–129 (2017)

Simon, H.A.: Rationality as process and as a product of thought Richard T. Ely lecture. Am. Econ. Rev. **68**(2), 1–16 (1978)
Tuyakova, Z.S., Cheremushnikova, T.V.: The system of key performance indicators and its use in rating the competitiveness of telecommunication companies. Econ. Anal.: Theory Pract. **15** (1), 54–68 (2017)
Zaernyuk, V.M.: The use of financial instruments in assessing the competitiveness of Russian mining organizations. Econ. Anal.: Theory Pract. **18**(3), 462–478 (2019)
Kovalev, A. V.: Industrial parks establishment and functioning as a factor of accelerated regional economic growth. Newslett. Adygey State Univ. - Ser. 5: Econ. **3**(165), 154–158 (2015)

Comparative Competitiveness Assessment of Enterprises in the Gold Mining Industry

Victor M. Zaernyuk[1(✉)], Valentina P. Leonova[2], and Elena N. Egorova[2]

[1] Russian State Geological Prospecting University named after Sergo Ordzhonikidze, Moscow, Russia
zvm4651@mail.ru
[2] Russian State Social University, Moscow, Russia
leonova.vp@mail.ru, egelni@yandex.ru

Abstract. Most modern approaches to assessing the competitiveness of economic entities cannot adequately assess the competitiveness of a gold mining enterprise since they do not take into account the influence of the positive effect of integrating some indicators that have a significant impact on the results of production activities. The paper substantiates the need to develop a methodological approach to assessing the competitiveness of a gold mining enterprise, which takes into account all essential aspects of its activities. Such an approach has to solve the task of assessing the current level of competitiveness of the production complex, the dynamics of its change, highlighting the "application points" of control actions, and assessing their effectiveness. A resulting integrated indicator of the competitiveness of gold mining enterprises is proposed, which provides an opportunity to assess its level both in the industry as a whole and for an individual gold mining enterprise.

Keywords: Competitiveness · Gold mining enterprises · Dynamic approach

1 Introduction

The competitiveness of a gold mining enterprise determines its viability in the conditions of competition not only in domestic but also in the world markets. Consequently, each enterprise must study and elaborate measures to improve and maintain its competitiveness. This problem is of particular importance for the enterprises of the gold mining industry of the Russian Federation in the current political and economic situation.

The Russian gold mining industry is one of the leaders in the world of gold mining; at the same time, it has its unique characteristics. According to the State balance of mineral reserves, the volume of gold reserves in Russia exceeds 13.8 thousand tons, which allows it to occupy second place in the world in this indicator. Gold mining is one of the most perspective sectors of the Russian industry because it has significant growth potential. 14% of the world's proven gold reserves are concentrated in Russia. Forecast gold resources are more than three times higher than proven reserves, which, along with the presence of large undiscovered areas, indicates a high potential for growth in the volume of confirmed reserves and gold production (Belyaev 2010).

A. V. Bogoviz and Y. V. Ragulina (Eds.): ISCI 2019, LNNS 115, pp. 580–588, 2020.
https://doi.org/10.1007/978-3-030-40749-0_69

However, the economy of the gold mining industry in Russia faces significant difficulties due to unprecedented changes in the balance of power in the global commodity market. Only those players who will most competently build their work on the background of the complexity of the risks will be able to overcome the existing difficulties, be competitive to prepare for the next phase of growth (Zaernyuk, Chernikova and Zabaykin 2017).

2 Materials and Method

The main task of studying the problem of assessing enterprise competitiveness is to search for and identify the criteria of competitiveness, its sources, and factors. Analysis of the economic literature on the subject allows us to identify several approaches to the solution of the formulated task.

Speaking about the classification of existing methods, first of all, we note that many economists have proposed a wide variety of different methods for assessing the competitiveness of enterprises. According to this variety of methods, there are many classifications based on relevant theoretical content, a form of displaying the evaluation results, a form of mathematically connecting indicators, and many others (Voronov 2014a and b).

The study of economic literature allowed us to identify the following main methods of assessing the competitiveness of enterprises.

First, these are product methods. The only important indicator used to assess the competitiveness of the enterprise is the competitiveness of its products. This indicator is complex; its calculation requires a large number of parameters of the object, including technical and economic characteristics. The critical point in determining the competitiveness of the product will be finding the price-quality ratio of the goods under study.

Often, the competitiveness indicators of different products of one enterprise are reduced to an integral indicator of the competitiveness of the company's products with the help of weight coefficients, where the weights are the volumes of sales of the corresponding product.

Secondly, there are matrix methods. They were developed in the 1960s as a tool of strategic management, which investigated the methods of development and implementation of actions that contribute to the long-term improvement of the level of performance of the enterprise. Besides, the task of assessing the degree of competitiveness of a company, taking into account the full range of its functions and long-term goals, was first set precisely within the framework of strategic management.

One of the most popular matrix models is the development of the Boston Consulting Group's "Market growth rates." This method relies on the relative market share of the company in each product area as an indicator of competitiveness. This model is characterized by duality. Since being a full-fledged strategic planning tool, on the one hand, the Boston consulting group matrix has retained some principles of the product approach, assessing the degree of competitiveness of a company through a single competitiveness factor of its products.

The third group of methods for assessing the competitiveness of an enterprise includes operational methods. Their selection as an independent tool occurred with the development of the tools of the matrix models of strategic planning. In accordance with this method, the degree of competitiveness depends on how well the work of all departments and services is organized at the enterprise. In some sources, this group of methods may be known as "methods based on the theory of effective competition."

Dyatlov (2011) and Chekmarev (2014) also pay attention to the process of the emergence of a new type of competition – hypercompetition, or innovative competition. It implies integrating companies into a single vertical-horizontal network of global structures, which includes the latest methods of influencing demand, motives, the economic behavior of people, as well as the goals of firms and competitors. As a result, new global institutions of management, control of national and world economic processes appear. Concepts, models, mechanisms for regulating the economy and competition are changing; they are becoming more flexible.

Based on these processes, there are severe changes in methodological approaches to the analysis of enterprise competitiveness. Before, the basis for determining competitiveness was material and financial resources, while modern analysis also uses information resources, IT-technologies, and intellectual capital. That is, competitiveness is considered multilevel and multidimensional.

Also, not only static methods but also the assessment of economic dynamics become a tool for analyzing the competitiveness of structural changes in the economy. The effectiveness of each of the company's services depends on the influence of many factors, including the resources of the enterprise. Evaluation of the effectiveness of each of the units involves evaluating the effectiveness of its use of these resources. Moreover, the enterprise's resources include not only capital in financial and material form but also their concept in a wider spectrum (for example, a company's personnel, management status, the quality of contacts with contact audiences, and its marketing organization). Each resource of an enterprise (defined in this way) can be evaluated by an appropriate quantitative or qualitative indicator. Thanks to this approach, the company's competitiveness appears as a set of private indicators of the effectiveness of its performance of certain aspects of economic activity – operations.

Fourth, there are combined methods that combine product and operational approaches, which is why they are based on the same indicators. They differ from operational methods only in form, and the separation of methods into operational and combined is very conditional, valid only from the theoretical justification of the mathematical model used.

Finally, the fifth model is the widely used so-called dynamic method, which is based on the consideration of the leading economic indicators of the company in dynamics. This method is based on two fundamental principles, which are the definition of enterprises' key indicators (for example, profitability, dynamics of market share) and their application concerning the dynamic analysis.

One should note that any information about the dynamics of the economic indicator displays a "picture" of the production and economic activity of the enterprise many times more fully. This method allows one to abandon the collection and processing of dozens of parameters "in statics," significantly increasing the reliability of the assessment of the competitiveness of the enterprise.

Porter (2016) also dealt with the issue of assessing the competitiveness of an enterprise. He developed a model for analyzing the "five forces" of the competitive environment. Porter's "five forces" model has the potential for further development.

Based on the analysis of the objectivity of the applied methods, we can conclude that the dynamic method is the most indicative, as it allows not only to visually display the competitive state of the enterprise using mathematical models but also makes it possible to trace the development path and evaluate the possibilities of increasing competitive advantages in the short-term and long term.

As the main indicators, the authors of this approach consider the following: (1) operational efficiency (profitability of economic activity); (2) strategic positioning (market share); and (3) financial stability, which consists in determining the liquidity indicator (Voronov 2014a and b).

In our opinion, the liquidity indicator, determined by the ratio of current assets to current liabilities, should be used to assess the company's competitiveness in the short-term, while strategic planning is carried out several years in advance. Besides, the assets are not differentiated, which means that converting them into money is difficult.

The authors propose to replace the liquidity indicator by the indicator of financial independence. For its calculation is taken into account the amount of equity capital, which reflects the autonomy of the enterprise from external companies and other lenders. The coefficient of financial independence takes into account the share of own funds in the total cost of all assets of the enterprise, which reflects its dependence on industry specifics.

Also, it is proposed to supplement the model taken as a basis with a fourth indicator reflecting intellectual capital. In the interpretation of Bruking (2001), the term "intellectual capital" refers to intangible assets, without the existence of these assets, companies cannot exist, and this term greatly enhances the competitive advantages of companies.

Note that with the enormous intellectual potential of Russia, the results of its use in the economy are incredibly unsatisfactory. Thus, the share of value-added generated by the turnover of intellectual property in the GDP of the Russian Federation is less than 1%, while in countries such as the US it is 12%, 7–8% in Germany, and 20% in Finland.[1]

The contemporary practice in working with intangible assets allows using it in many aspects. In financial practice, intangible assets can reflect the assessment of resource potential, generating value, a way to increase investment attractiveness, can finally be considered as an argument for sustainable competitive advantage in the financial market (Zaernyuk 2019).

Next, we consider in more detail the calculation algorithm for each of the above indicators of the competitiveness model proposed for enterprises of the gold mining industry in Russia.

1. Operational efficiency implies the performance of similar activities better than competitors, ensuring a profit in the process of selling surplus value (Voronov 2014a and b). Thus, it is the profit of the enterprise that is the most significant criterion for

[1] Message from the President of the Russian Federation V. V. Putin (December 12, 2013).

operating activities. At the same time, specific difficulties arise in comparing large and smaller enterprises, since the former often have hefty profits, which ensures their competitiveness. Based on the scale of activity, there is a need to compare the "profitability" of economic activity and not the amount of profit.

The operational efficiency of the economic activity of gold mining enterprises (*Oe*) is calculated as the ratio of revenue from the sale of products (works, services) of the enterprise over the reporting period to the costs of production and sales of products (works, services), including production costs, commercial, administrative expenses, payments to budgets of all levels for the evaluated company for the reporting period.

The calculation of operational efficiency for the sample is carried out similarly. The sample implies a set of competing enterprises, necessary and sufficient for comparison with the studied economic entity. A sample can consist of several competing enterprises in the industry. As a result of such an analysis, an industry-specific indicator of competitiveness is calculated for the enterprise under evaluation. The sample can also consist of one economic entity. In the latter case, an individual indicator of the competitiveness of the enterprise in question is determined in comparison with the selected competitor.

2. The concept of "strategic positioning" is that the analyzed company seeks to create and maintain a unique position in the market, which consists of the implementation of activities that are not carried out by competing enterprises. Strategic positioning provides the possibility of the process of realizing surplus value as a whole, creating, supporting, and expanding sales markets. The critical indicator and criterion of this ratio is the change in revenue from sales of products (goods or services) compared with the previous period.

At the same time, the market, in this case, does not mean a specific industry or "niche" occupied by an enterprise, but the entire set of economic segments in which the enterprise and the sample operate. Thus, there is a situation when the enterprise with a more significant share of the market is estimated as more competitive, thereby reducing the adequacy of an assessment of strategic positioning. In this regard, it seems appropriate to consider market shares not in absolute terms but a change in market share compared with the previous period.

Concerning the gold mining enterprises, it is advisable to use the indicator of the volume of gold production.

The coefficient of strategic positioning (*Sp*) is defined as the quotient of dividing the gold production of the enterprise in the reporting period to the gold production of the company in the previous period.

3. Among the indicators characterizing the financial stability of the enterprise, the coefficient of financial independence of the concentration of equity takes a vital place, which shows which part of the assets is formed from its own sources. Financial independence of the studied enterprise can be calculated by the ratio of equity (capital) to the balance sheet currency.

The calculation of the financial independence coefficient (*Fi*) is carried out similarly to the coefficients of operational efficiency and strategic positioning, that is, by comparing the corresponding indicators of the analyzed enterprise and the enterprises included in the sample.

4. As an indicator characterizing intellectual capital as evidence of a sustainable competitive advantage that is significant in the knowledge economy, one can use the proportion of intangible assets in assets, determined by the ratio of book value to intangible assets to the book value of assets (Zaernyuk 2019).

If one assessed the possibility of applying this approach to gold mining enterprises, s/he would take into account the fact that the share of intangible assets in their balance sheets is only 0.2–0.5%. Therefore, the influence of this component on the overall competitiveness indicator will give a more objective assessment if we use the so-called index approach when the economic indicator is considered in dynamics.

The coefficient of intellectual capital (*Ik*) will be calculated as the ratio of the intangible asset of the reporting year to the corresponding indicator of the previous year.

Before dwelling on the calculation of the final competitiveness indicator as applied to gold mining enterprises, we note the following.

The basic methodology for determining the indicator of overall competitiveness (Voronov 2014a and b), which the authors of this work are guided by, is based on the multiplicative model when particular indicators are multiplied by each other. Preliminary testing for several gold mining enterprises, within the framework of the methodology noted above, showed a fairly wide scatter of the generalized (final) indicator of competitiveness even for the years of the analyzed period.

In this regard, it seems appropriate to take the well-established method for calculating the aggregate index of the provision of banking services in the Russian regions as a basis for calculating the final competitiveness indicator, using the geometric mean weighted from private coefficients formula employed by the Bank of Russia for more than 15 years. For this, after calculating the coefficients *Sp, Oe, Fi*, and *Ik,* it will be necessary to calculate the indices *Isp, Ioe, Ifi,* and *Iik,* given to the total for the sample.

So, the cumulative index of competitiveness (CIS) will be calculated using the formula of the weighted average of the partial coefficients:

$$CIC = \sqrt[4]{Ioe \times Isp \times Ifi \times Iik},$$

3 Results

Based on the proposed methodology, it is supposed to assess the level of competitiveness of JSC “Gold of Kamchatka” and determine the competitive advantages. The source of the analysis data will be the enterprise documentation: annual reports and consolidated financial statements of JSC “Gold of Kamchatka” according to IFRS for the period from 2015 to 2017, as well as the relevant documents of the main competitors of the analyzed enterprise in the Russian gold market. The calculation data for the coefficients of strategic positioning, operational efficiency, financial independence, and intellectual capital are presented in Table 1.

Based on the data in Table 1, we calculate indicators for assessing the level of competition for Russian gold mining enterprises (Table 2).

Table 3 shows the cumulative competitiveness indices of Russian gold mining enterprises, calculated according to the above formula.

Table 1. Dynamics of criteria for assessing the level of competition for gold mining enterprises in Russia for 2015–2017.

Name of the gold mining company	Coefficient of strategic positioning **(Sp)**			Coefficient of operating efficiency **(Oe)**			Coefficient of financial independence **(Fi)**			Coefficient of intellectual capital **(Ik)**		
	2015	2016	2017	2015	2016	2017	2015	2016	2017	2015	2016	2017
PJSC Polyus	0,720	0,768	0,762	1,699	1,595	1,582	0,387	0,090	0,087	1,050	1,130	1,115
Petropavlovsk LLC	0,206	0,167	0,160	1,141	0,938	2,573	0,397	0,396	0,361	1,000	0,946	0,866
PJSC Vysochayshiy (GV Gold)	0,074	0,065	0,078	2,443	2,378	2,050	0,456	0,504	0,518	0,900	1,086	0,806
Total on Sample	x	x	x	**1,645**	**1,543**	**1,663**	**0,392**	**0,171**	**0,171**	**1,023**	**1,067**	**1,041**
JSC Gold of Kamchatka	0,007	0,009	0,021	1,400	4,070	3,571	0,169	0,249	0,221	1,000	1,056	1,168

Source: Calculated by the authors according to the official financial statements of the following companies: PJSC Polyus (2017), Petropavlovsk LLC (2017), PJSC Vysochayshiy (2018) (GV Gold), JSC Gold of Kamchatka (2017).

Table 2. Dynamics of indicators for assessing the level of competition for gold mining enterprises in Russia for 2015–2017.

Name of the gold mining company	Index of strategic positioning **(Isp)**			Index of operational efficiency **(Ioe)**			Index of financial independence **(Ifi)**			Index of intellectual capital **(Iik)**		
	2015	2016	2017	2015	2016	2017	2015	2016	2017	2015	2016	2017
PJSC Polyus	1.014	1.067	0.992	1.03	1.03	0.95	0.99	0.53	0.51	1.03	1.06	1.07
Petropavlovsk LLC	1.144	0.811	0.958	0.69	0.61	1.55	1.01	2.32	2.11	0.98	0.89	0.83
PJSC Vysochayshiy (GV Gold)	1.057	0.878	1.200	1.49	1.54	1.23	1.16	2.95	3.02	0.88	1.02	0.77
Total	**1.000**	**1.000**	**1.000**	**1.00**	**1.00**	**1.00**	**1.00**	**1.00**	**1.00**	**1.00**	**1.00**	**1.00**
JSC Gold of Kamchatka	1.000	1.286	2.333	0.85	2.64	2.15	0.43	1.46	1.29	0.98	0.99	1.12

Table 3. A cumulative index of competitiveness of gold mining enterprises in Russia for 2015–2017.

Name of the gold mining company	A cumulative index of competitiveness **(CIC)**		
	2015	2016	2017
PJSC Polyus	1,03	1,06	1,07
Petropavlovsk LLC	0,98	0,89	0,83
PJSC Vysochayshiy (GV Gold)	0,88	1,02	0,77
Total on sample	**1,00**	**1,00**	**1,00**
JSC Gold of Kamchatka	0,98	0,99	1,12

4 Discussion

The data obtained in the analysis of the competitiveness of gold mining companies showed that the annual growth of the competitiveness index of the valued enterprise JSC "Gold of Kamchatka" amounted to 0.98 in 2015, 0.99 in 2016, and 1.12 in 2017. Its performance was thanks to an increase in market share and continuous improvement in operating activities.

Stable indicators of PJSC "Polyus" show a small annual growth. The PJSC "Polyus" occupies the highest level in the rating of the Top-20 gold mining industry, at which in 2017, the share of the market of gold made 22.8%. Decline total score of competitiveness was recorded at the "Petropavlovsk" LLC (from 0.98 to 0.83) and PJSC "Vysochayshiy" (from 0.88 to 0.77).

Further work with the proposed methodology involves its quantitative testing on empirical data of Russian gold mining enterprises, which will not only assess their competitiveness but also propose measures to strengthen it.

5 Conclusion

As a result of the analysis of approaches and methods, assessing the competitiveness of industrial enterprises, their weaknesses and strengths were highlighted. The advantages of a dynamic approach, consisting of the availability of practical use, were formulated for a third-party assessment of competitiveness in the conditions of having access only to information officially published in open sources.

The study and analysis of the current state of domestic and foreign non-ferrous metallurgy, the situation on the world market of non-ferrous metals made it possible to identify macro factors that influence the results of the activities of participants in this market: strategic positioning, operational efficiency, financial stability, and intellectual capital.

Practical testing of the methodological approach to assessing the competitiveness of the production complex was carried out using the case of the gold mining enterprise JSC "Gold of Kamchatka." The results of the study allow us to determine the current level of their competitiveness, as well as develop a strategy for its increase in the near future.

The results of these evaluations can be used by the company itself and similar enterprises in solving strategic development problems associated with maintaining and strengthening the competitive positions of the enterprises of the gold mining industry.

References

Belyaev, E.V.: Apatite-bearing greenstone belts in Russia. Russ. Geol. Geophys. **51**(8), 857–862 (2010)

Bruking, E.: Intellectual capital (trans. from Eng.). In: Kovalik, L.N. (ed.). Piter, St. Petersburg (2001)

Chekmarev, V.V.: Competition as a process of ensuring the economic security of business entities. Theor. Econ. **1**, 25–32 (2014)

Dyatlov, S.A.: Global innovative hypercompetition as a factor of leadership in the global economy. Eurasian integration: politics, economics, law. Int. Sci. Anal. J. Inter-Parliament. Assem. EurAsEC **9**, 50–56 (2011)

JSC Gold of Kamchatka: The financial statements of JSC Gold of Kamchatka for 2017 (2017). https://e-ecolog.ru/buh/2017/7710659873

Petropavlovsk LLC: The consolidated financial statements of Petropavlovsk LLC for 2017 (2017). https://ru.investing.com/equities/petropavlovsk-balance-sheet

PJSC Polyus: The consolidated financial statements for 2017 PJSC Polyus (2017). http://polyus.com/ru/investors/disclosure/ifrs-financials/

PJSC Vysochayshiy (GV Gold): The financial statements of PJSC Vysochayshiy (GV Gold) for 2017 (2018). http://www.gvgold.ru/upload/iblock/4ae/GV-GOLD_FS_2018-RUS.pdf

Porter, M.Yu.: International competition. Competitive advantages of countries. In: Komissarov, S.D., Kamennova, N.E. (eds.)Trans. from Eng. Alpina Publisher, Moscow (2016)

Voronov, D.S.: A dynamic approach to assessing the competitiveness of enterprises. Mark. Russia Abroad **5**, 92–102 (2014a)

Voronov, D.S.: A dynamic approach to the assessment of the competitiveness of enterprises. In: Tatarkina, A.I., Krivorotova, V.V. (eds.) Competitiveness of Socio-Economic Systems. Ekonomika, Moscow (2014b)

Zaernyuk, V.M.: The use of financial instruments in assessing the competitiveness of Russian mineral resource mining enterprises. Econ. Anal.: Theory Pract. **18**(3), 462–478 (2019)

Zaernyuk, V.M., Chernikova, L.I., Zabaykin, V.Yu.: Trends, problems, and prospects for the development of the gold mining industry. Financ. Anal.: Probl. Solut. **10**(9), 972–986 (2017)

A Model for Achieving Comprehensive Efficiency of the Energy Projects Implementation in the Sustainable Development Framework

Arsen S. Abdulkadyrov[1(✉)], Ekaterina P. Zhigulina[2], and Mikhail A. Kovnerev[3]

[1] Institute of Socio-Political Research Under the Russian Academy of Sciences, Moscow, Russia
ars.rggu@mail.ru

[2] Gubkin Russian State University of Oil and Gas, Moscow, Russia
zhigulina.ep@mail.ru

[3] Military University of the Ministry of Defense of the Russian Federation, Moscow, Russia
kovnerevm@mail.ru

Abstract. The paper deals with the main implementation components of investment projects in the fuel and energy complex, including innovative ones. The problem highlighted by the authors is as follows: the interests imbalance arising during project implementation predetermines the need to form such a model of business processes that would comprehensively take into account the parties' interests in the sustainable development framework. The formation of a basic indicator system, reflecting the compliance of its implementation process with the established requirements, norms, standards, and characterizing the project effectiveness for the parties is a critical issue. The scientific novelty consists of the following points important from the sustainable economic development: (1) on the basis of the process approach to management, a universal basic model with the indicators system, ensuring a comprehensive assessment of the project effectiveness in the fuel and energy complex implementation, has been formed; (2) a criteria composition to be monitored and analysed during the projects implementation; (3) a scheme for the reverse impact of the high-tech projects implementation on workers, population living standards and socio-economic development was built. Our system model includes indicators to ensure sustainable development, taking into account the existing restrictions, needs to ensure an adequate socio-economic development level, while meeting the need for high growth rates of production indicators. The model will also take into account economic efficiency for investors.

Keywords: Fuel and energy complex · Process approach · Sustainable development · Investment projects efficiency · Innovations

A. V. Bogoviz and Y. V. Ragulina (Eds.): ISCI 2019, LNNS 115, pp. 589–595, 2020.
https://doi.org/10.1007/978-3-030-40749-0_70

1 Introduction

The issue of increasing the economic development and innovation pace in Russia and most developed countries has recently received close attention. At the same time, this period was not characterized by an increased control over maintaining a balance of interests between the breakthrough science and technology achievements, country's economic growth, and degrees of such tendency's security for the society. For ensuring maximum development rates (priority development), implementation of innovative projects mainly in those sectors that can provide maximum margins becomes a priority (given the profitability correlation and projects riskiness) is relatively stable. These criteria are met by the fuel and energy complex (FEC). The large volume of natural resources reserves and the consistently high demand for products is the reason for implementing the thriving energy projects, even despite the presence of several internal and external disincentives. The most important task is to coordinate the goals set by the projects with their influence on the population living standard, the social sphere, and socio-economic development indicators, while coordinating the impact of priority development on ecological aspects and the environment in general. The research object is the economic development process and the conditions for implementing investment scientific and technological projects in the fuel and energy sector. The research subject is the factors ensuring an overall maximum efficiency of project implementation to balance the parties' interests. The purpose of the study is to form a model of a balanced energy project performance based on an integrated factors system that is business processes elements, taking into account existing requirements and constraints within the sustainable development framework.

2 Materials and Methods

Within the study, the authors used a set of methods and general scientific research principles, economics laws, traditional economic methods: statistical and economic analysis (systemic, causal, functional), synthesis, economic and information processes modeling, mathematical methods, systemic and integrated approaches, current methodical approaches and modern software products were used in the field of evaluating energy projects and analyzing their effectiveness. The model building was performed with the ICT capabilities and analysis software tools, modeling methodologies, and process-oriented management tools.

We would like to note a number of prominent scholars working on this subject. They are (Alekseev et al. 2018) and (Bresnahan and Yin 2017). From the standpoint of the innovative energy projects implementation in modern conditions through new digital technologies, we also would like to acknowledge contributions made by (Alkhasov et al. 2000; Abdulagatov et al. 2016). The authors are also actively engaged in the study and development of this issue. The research results are reflected in scientific publications (Abdulagatov et al. 2016; Abdulkadyrov et al. 2017).

3 Results and Discussion

Introducing the results of scientific and technological progress and advanced developments in the main industrial sector activities (energy complex, engineering, oil and gas, transport, food industry, housing construction, and other large-scale industries) will contribute to the development of related economic sectors. Of course, the results of the resource-innovative development implementation and the introduction of innovative development will affect the overall socio-economic indicators. At the same time, they will depend not only on the participants' actions and the private companies' policy but also on state policy and different external factors.

STP is a fundamental factor in the accelerated economic development of most states. Concerning the most critical energy area and the investment projects implementation in the fuel and energy sector (while taking into account the problems that have arisen in the industry today), the introduced innovations and advanced developments into activity become an objective necessity. It happens due to the production processes complexity in the fuel and energy complex and the presence of a number of internal features (raw materials deposits remoteness, complex geography of their location, deposits aging, growth in the share of reserves challenging to extract), as well as due to the high competition level in the world market and the need for maintaining the level of raw materials and products prices.

It is precisely the fuel and energy complex that is one of the bases for the breakthrough country's economic development. The energy projects implemented in the past decades have been given maximum attention to ensure overall priority development, including the high-performance achievement possible due to the development's introduction (both domestic achievements and borrowed, acquired innovations) into production and control processes. Within most innovative projects, the main criterion was the maximum commercial project efficiency, their profitability for investors. The most crucial factor was the innovative development, information, and digital technologies introduction in order to implement the provisions of the country's innovative development strategy (Chen and Kamal 2016).

Thus, overall economic efficiency of the implementation (benefit for the state, society) was not the primary factor. When considering innovations not only as a progress engine but also in the context of sustainable development, special attention should be paid to the issue of their introduction consequences on social development, population living standards, human capital development, and environmental aspects. An indirect impact occurs when implementing and introducing any innovations. Therefore, it is interesting to develop such an indicator system, in which a "complex achievement" could be an indicator of the maximum (optimal) efficiency of the investment energy project implementation in the sustainable development framework, taking into account the existing restrictions. A detailed model of balanced performance indicators for implementing investment projects in the fuel and energy sector in compliance with the main provisions of the sustainable development concept, developed as part of the process approach to management, is presented in Fig. 1.

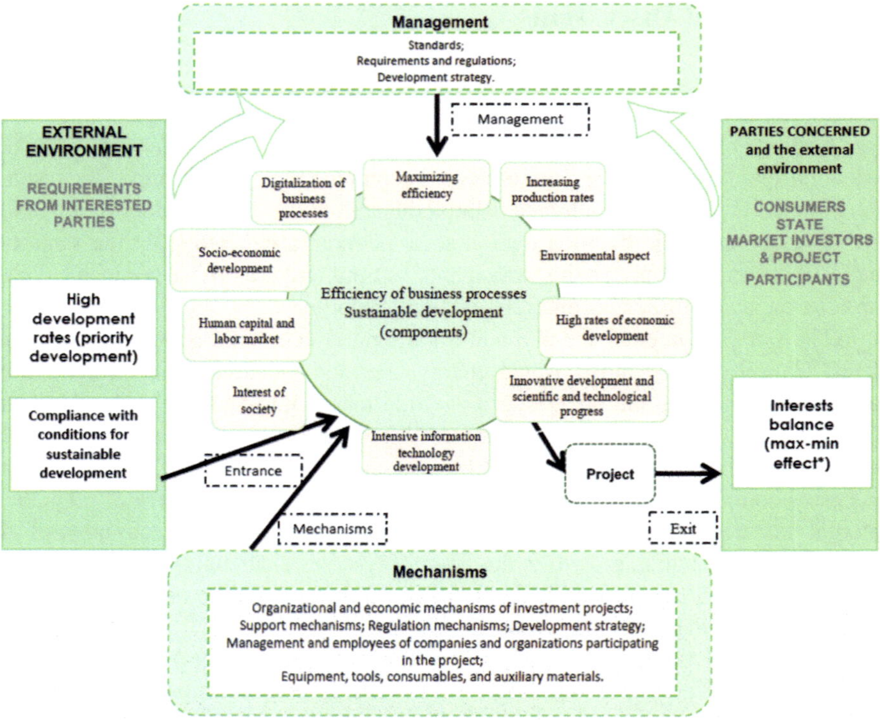

Fig. 1. The model of balanced factors in the energy projects implementation based on the process approach in the sustainable development context (priority development). * The max-min effect in the context of the model implies achieving the maximum financial and economic efficiency (maximum profitability with minimum costs) from project implementation and minimization of negative consequences for the environment. *Source*: Developed by the authors as a result of the study.

The ultimate goal is to maintain a balance of interests of the parties involved (as well as indirect participants), which can be affected by the project implementation effects. In order to ensure a balance of interests, the most important issue is "ecology" since some introduction is faced with modern legislation tightening, for example, with restrictions on harmful substances emissions and increased emissions charges. Some STP achievements have a destructive impact on the environment. The implementation of some projects is possible only with an environment detriment (deforestation for the fields and infrastructure development, wastewater and materials discharge into water bodies, chemical emissions) and the interests' disservice of the people living in the surrounding areas. However, some modern technologies allow taking these requirements into account and reducing the negative impact on the environment.

From the human resources and socio-economic development perspective, the introduction of innovative solutions in industrial sectors, in particular, in the energy projects implementation, is a factor contributing to the creation of new jobs both in the fuel and energy complex and in many related industries. At the same time, when introducing new technologies in the advanced development framework, a multidirectional process can be observed. The scheme of the main elements in the implementation of innovations in production and the reversibility of their impact on human capital and labor resources is shown in Fig. 2.

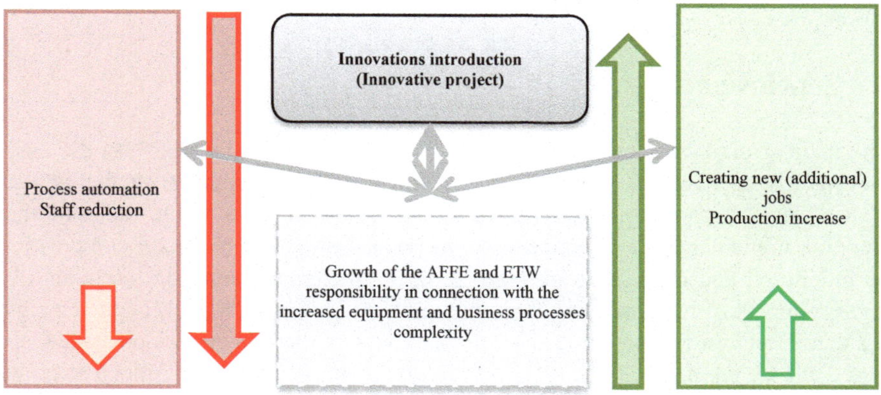

Fig. 2. The reverse impact of introducing new technologies on human resources and the labor market. *Source*: Developed by the authors as a result of the study.

When new innovative technologies are introduced, the situation with labor resources is characterized by multidirectional processes that can occur both separately and simultaneously as part of single innovative project implementation. (1) Creation of additional jobs due to increased production volumes. (2) Partial reduction of the production staff due to the specific production processes automation through implementing new technologies (Arntz et al. 2016). The role and the level of responsibility of the administrative, managerial, and engineering and technical personnel is increasing because of the new equipment increasing complexity. The large-scale project's introduction in the basic sectors affects both the state socio-economic indicators and the general living standard in the country (Government of the Russian Federation 2008, 2009, 2011, 2017). The multidirectional project's impact on company employees will entail significant changes in the labor market. That will lead to changes in the level of population well-being, changes in state spending on social security.

4 Discussion

STP is an absolute advantage and allows reducing production costs, contributes to increased production volumes, margins, efficiency, as well as increased project efficiency for investors. However, the overall project effectiveness is much broader than commercial one; and for sustainable development, the interests of all parties must be taken into account. A basic model with the indicators system is proposed to ensure a comprehensive assessment of the effectiveness of the project in the fuel and energy complex. The indicators highlighted by the authors in the model are universal and applicable to determine the complex effectiveness of the project. It is possible to diversify their composition for individual projects, taking into account their implementation specifics. Each indicator can be refined to a quantitative indicator, with the establishment of a values range for each industry. The most important factor is human capital and consideration of the multidirectional consequences of the innovative project implementation on workers, the labor market, and the social sphere.

5 Conclusion

The main scientific novelty provisions in the study are as follows: (1) on the basis of the process approach to management, a universal basic model with the indicators system, ensuring a comprehensive assessment of the projects in the fuel and energy complex implementation effectiveness, has been formed; (2) a criteria composition to be monitored and analysed during the projects implementation; (3) a scheme of the reverse impact of implementing high-tech projects on the personnel, population living standards, and socio-economic development was built. The model presented in the study allows taking into account the multidirectional interests of the parties when implementing projects by determining the selected factors balance for specific projects in order to ensure sustainable development.

References

Abdulagatov, I.M., Badavov, G.B., Akhmedova-Azizova, L.A., Aliev, R.M.: Measurements of the density, speed of sound, viscosity, and derived thermodynamic properties of geothermal fluid. J. Chem. Eng. Data **61**(1), 234–246 (2016)

Abdulkadyrov, A.S., Ryzhov, I.V., Strokov, A.I., Kamzolov, Y.V.: Actual aspects of improving the organization of high-tech products. Technol. Text. Ind. **5**(371), 18–27 (2017). Proceedings of Higher Educational Institutions

Alekseev, A.N., Evdokimov, S.Y., Khachaturyan, K.S., Tarasova, A.Y., Khachaturyan, A.A.: The financial strategy of development of industry 4.0 in countries with developing economies. Espacios **39**(12), 1 (2018)

Alkhasov, A.B., Aliev, R.M., Ramazanov, M.M., Abasov, G.M.: Study of complex heat exchange with account for phase transitions in the secondary contour of the geothermal power plant. Renew. Energy **19**(1–2), 155–161 (2000)

Arntz, M., Gregory T., Zierahn, U.: The risk of automation for jobs in OECD countries: a comparative analysis. In: OECD Social, Employment, and Migration Working Papers, vol. 189 (2016)

Bresnahan, T., Yin, P.-L.: Adoption of new information and communications technologies in the workplace today. Innov. Policy Econ. **17**(1), 95–124 (2017)

Chen, W., Kamal, F.: The impact of information and communication technology adoption on multinational firm boundary decisions. J. Int. Bus. Stud. **47**(5), 563–576 (2016)

Dmitrievsky, A.N., Komkov, N.I., Mastepanov, A.M., Krotova, M.V.: Resource-innovative development of the Russian economy. In: Mastepanov, A.M., Komkova, N.I. (eds.). 2 edn. Institute for Computer Studies, Moscow (2014)

Government of the Russian Federation: Order "The concept of long-term socio-economic development of the Russian Federation for the period up to 2020" (17 November 2008 No. 1662-r) (2008). http://government.ru/info/6217/

Government of the Russian Federation: Decree "Russia's energy strategy for the period up to 2030" (13 November 2009 No. 1715-p), Moscow, Russia (2009)

Government of the Russian Federation: Order "About strategy of innovative development of the Russian Federation for the period till 2020" (8 December 2011 No. 2227-r) (2011)

Government of the Russian Federation: Order "On Approval of the "Digital Economy" of the Russian Federation Program" (28 July 2017 No. 1632-p) (2017). http://base.garant.ru/71734878/

Ministry of Energy of the Russian Federation: Official website (n.d.). https://minenergo.gov.ru/

Prospects for Biofuel Production in the Selected Regions of Russia

Ekaterina S. Titova(✉)

The Federal Research Centre "Fundamentals of Biotechnology" of the Russian Academy of Sciences, Moscow, Russia
es_titova@inbox.ru

Abstract. The study focuses on analyzing the prospects for biofuel production in the Russian Federation, taking into account the existing opportunities for innovation and world practice. The objects of the research are the ways to enhance innovative processes the processing of biomass for the production of biofuels in Russia. The analysis revealed the existence of a corresponding extensive raw material base in various regions of Russia and the ability to attract innovative technologies for organizing efficient energy production with parallel processing of accumulated biomass. The authors also present original data on the potential volume of biomass suitable for processing on the territory of Russia. Additionally, the groups of innovation achievements in the production of biofuels are identified, and the main prospects for biomass processing are described.

Keywords: Innovative activity · Biofuel · Biomass · Alternative energy sources · Organic waste · Biotechnology · Bioethanol · Biodiesel

1 Introduction

The statement of Marx and Engels that "demand determines supply and, conversely, supply determines demand" is widely known (Marx and Engels 1961). Accordingly, the consumption of various energy carriers as an essential condition for the development of the economy is mostly determined by their cost characteristics, accessibility to consumers, and the existence of alternatives, which form the aggregate demand for this type of resource. At present, even in the countries rich in traditional hydrocarbon energy resources, there is a need for alternative and renewable energy sources (hereinafter AES, RER), including the production of biofuels. The limited and exhaustible nature of traditional fuels, the need to reduce budgetary dependence on the sales of traditional fuels, the development of innovative technologies, the availability of resources for the production of AES, the need to protect the environment, etc. are the objective prerequisites for the use of AES. Even though Russia has significant hydrocarbon reserves, the production of AES, including biofuels, has already begun in various regions of the country. For organizing such production, it is considered important to achieve economic and environmental effects from processing biofuels of steadily accumulating biomass, which includes industrial, agricultural, and household waste. Therefore, it seemed appropriate to analyze the prospects for innovation in the

A. V. Bogoviz and Y. V. Ragulina (Eds.): ISCI 2019, LNNS 115, pp. 596–604, 2020.
https://doi.org/10.1007/978-3-030-40749-0_71

use of biomass as a raw material in the production of biofuels in Russian regions, taking into account differences in geographical, climatic, economic, and other conditions.

2 Material and Methods

The study was carried out using the methods of system analysis, comparative analysis, and synthesis, as well as the methods of expert assessment. The research materials were the works of domestic and foreign scholars, published data from the Federal State Statistics Service of the Russian Federation (gks.ru), the database of the International Agency for Renewable Energy (irena.org), the National Biofuel Association (biotoplivo.ru), and regulatory acts of the Russian Federation.

3 Results

The scientific community determines the organizational and economic prerequisites for the development of AES (including biofuel), which together form a qualitatively new economic development trend based on sustainable development and determine the development of the sixth innovative wave (Silva and Di Serio 2016; Kurki and Wilenius 2015; Chen et al. 2018). They include:

- ensuring competitiveness by *reducing the cost of finished products*, for example, through supply chain management (Ahmed and Sarkar 2018);
- *increasing production efficiency* through the use of innovative waste management technologies (Nidumolu et al. 2009);
- *a transition to the environmentally efficient and safe production* through the reduction of emissions into the environment (Ma et al. 2018; Ratner and Ratner 2017);
- *a depletion of available resources*.

Breakthrough technologies, in this case, include energy storage devices, electric vehicles, unmanned aerial vehicles, solar, nuclear, alternative and renewable energy, nano and bioengineering technologies, neuro and bioinformatics, information and telecommunication systems and networks (Maslennikov 2017), which is reflected in the works of a number of authors (Golova et al. 2017; Silva and Di Serio 2016). It is believed that the effective development of the economy and sustainable development of the state in modern conditions is impossible without paying due attention to the rational use of resources and environmental management (Pinto et al. 2018; Lindow et al. 2018; Rafaj et al. 2018; Das 2017). Accordingly, there is a need to determine the balance of economic, environmental, and social factors in order to create a sustainable environment for the economic development of entities (Silva and Di Serio 2016). It is also assumed that, for example, "the introduction of the AES will allow demonopolizing and decentralizing the electricity market, ensuring energy security through effective competition and the support of independent energy producers" (Alferov 2010). Following the Russian electric energy balance for 2017, the annual increase in the use of renewable energy sources is 13.1% (Ministry of Energy 2018). The power dynamics of the respective facilities in the Russian Federation are presented in Fig. 1.

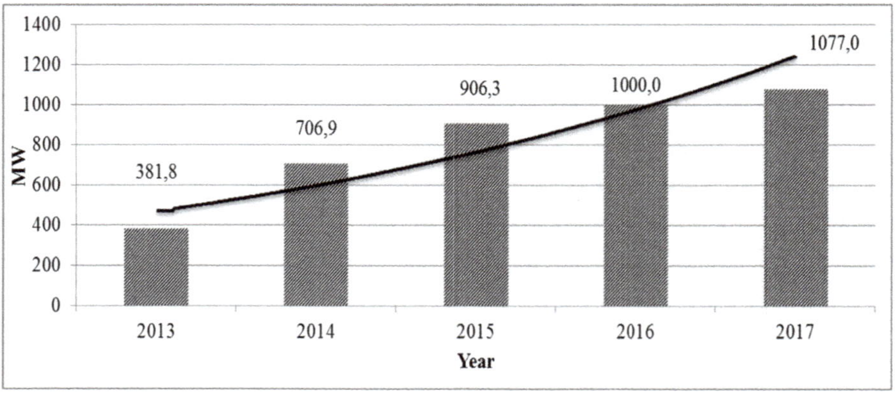

Fig. 1. The capacity of generating facilities operating based on the use of renewable energy in the Russian Federation (excluding hydroelectric power plants with an installed capacity of more than 25 MW), MW, 2013–2017 (Rosstat n.d.).

It is gratifying to note that, according to the available data (EU Biofuels Annual 2018), the Russian Federation has been one of the three leading countries in pellets export to the EU countries for more than five years (Fig. 2).

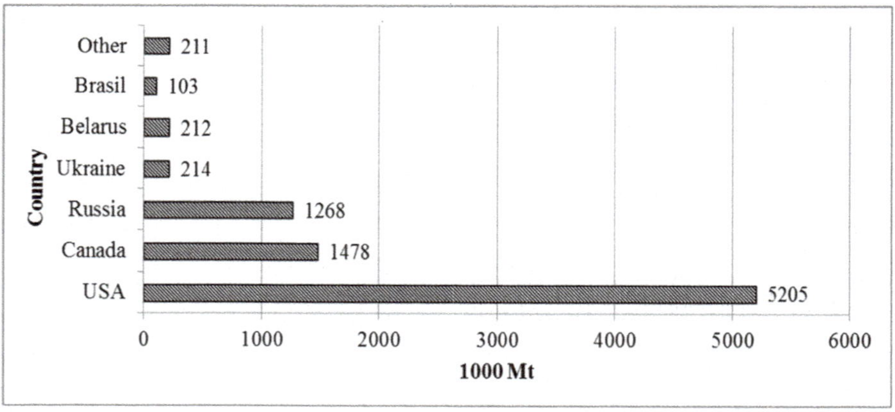

Fig. 2. The leading countries for the export of pellets to the EU countries, Mt, 2017 (Flach et al. 2018).

Obviously, there are tendencies towards the intensification of biofuel production and the increase in its socio-economic role: both social, as a source of created jobs, and economical, as a new economically viable type of fuel (Titova and Bondarchuk 2017).

As many scholars have noted (Kang et al. 2015; Correa et al. 2019), in order to create a stable biofuel production market, it is necessary to conduct an appropriately focused active policy, which, of course, is also necessary for the Russian Federation. One of the recent steps to implement this policy has been the introduction of the amendment to Federal Law of November 22, 1995, No. 171-FZ "On State Regulation of the Production and Turnover of Ethyl Alcohol, Alcohol and Alcohol-Containing Products, and on the Limitation of Consumption (Drinking) of Alcoholic Products" (Federal Law No. 171-FZ 1995). Following this amendment, several restrictions on bioethanol production are removed. The law also establishes a procedure for licensing the production of bioethanol.

The existing positive trends in the intensification of the innovative activities in the production of AES in the regions of Russia are noteworthy. It was shown (Bondarchuk and Titova 2017a, b) that in four large regions of the South of Russia (Astrakhan, and Rostov regions, the Republic of Crimea and the Stavropol territory), within the framework of state and regional programs, as well as through private initiatives, economic activity began to develop renewable energy sources taking into account regional features. An excellent example of the development of renewable energy is the "Luchki" biogas station, which operates in the Belgorod region and processes livestock waste. It annually produces about 29.3 million kWh of electricity as well as 27.3 thousand Gcal of thermal energy, utilizing 95 thousand tons of biomass (AltEnergo 2019). In general, the analysis showed that the start of biofuel production in some areas of Russia is organized, taking into account the existing geographical, climatic, economic, and other conditions.

Current qualitative global transition to new levels in biofuels production (the second and third generations of technologies and products) is due to at least the following four groups of achievements that were directly related to the innovation: the expansion of the range of processed biological resources, the optimization of the known and introduction of new technologies, the development of biotechnological solutions, the processing of the production by-products (Bondarchuk and Titova 2017a, b). Among the main types of biological resources that can serve as raw materials for biofuel production are starch-containing agricultural plants: oil crops, palm oil, cellulose-containing biomass, wild-growing oil plants, micro- and macroalgae, cellulose- and starch-containing wastes from processing industries, wastewater, and municipal solid and organic waste (Ballesteros and Manzanares 2019). Many of these types of raw materials have already found applications in biofuel production in certain Russian regions. The structure of the capacity of generating facilities operating on the RER basis in the Russian federal districts is presented in Fig. 3.

The European Bioeconomic Strategy (European Commision, 2018) notes on this issue that a significant problem is the lack of a forward-looking analysis of the total amount of biomass that can be processed and the socio-economic and environmental compromises. From a management point of view, it is essential to determine the potential volume suitable for biomass processing in relation to economically safe innovative technologies, adequate geographic location of production, potential supply, and marketing channels, as well as the possibility of scientific support of production and the prospects for obtaining and subsequent use of by-products.

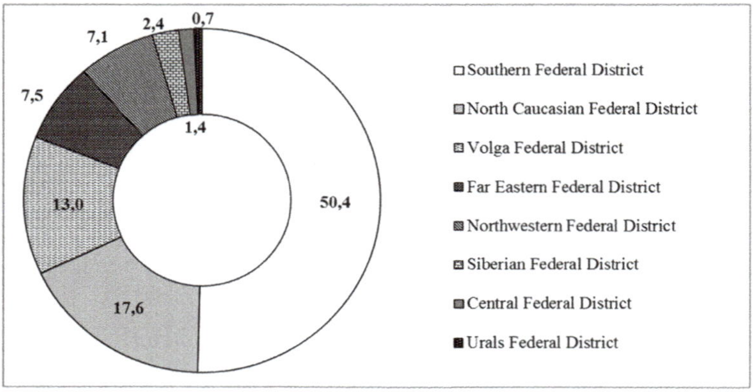

Fig. 3. The capacity of generating facilities operating based on renewable energy in the Russian federal districts (excluding hydroelectric power plants with an installed capacity of more than 25 MW), %, 2013–2017 (Rosstat n.d.).

In Russia, practically no second-generation and third-generation technologies have been used for the production of biofuels (Titova et al. 2017). This is mainly due to the high level of capital costs. While improving the technology for the production of biofuels from biomass, the researchers are faced with the excess of the amount of the received energy over production costs (Rodionova et al. 2017). Many scholars point to the excellent prospects for the use of microalgae in fuel production, which is associated with a short growing season (1–10 days depending on the type of technological process) (Ballesteros and Manzanares 2019), which allows high yield. In the production of biofuels from microalgae, it was possible to come closer to solving this problem and to obtaining an economically feasible technology (Tseng et al. 2019).

There was a message (Correa et al. 2019) that in order to develop the production of this type of energy and to introduce innovative technologies, a certain state stimulation was needed in the form of large-scale investments and the definition of an appropriate regulatory framework. However, this issue can be approached from the other side. The analysis of the experience of individual Russian regions showed that the most effective starting step for determining the path of AES production could be the processing of agricultural waste, municipal solid waste, logging and other biomass, which is continuously formed, as well as the use of the overproduced agricultural products.

According to the optimistic estimates by the International Renewable Energy Agency (IRENA 2017), the potential amount of biomass possible to be processed in Russia for energy production could be 14084 PJ per year by 2030. The average costs reached 4.6–7.3 $USD/GJ, with the potential power of electricity generated from biomass reaching 8.4% of the total volume by 2030, and the production of thermal energy from biomass 9.52% (IRENA 2017). Below, the results of the analysis of the potential for the production of biofuel from two types of overproduced agricultural products, sugar beets and sunflowers, are presented as an example (Table 1).

Table 1. The use of overproduced agricultural products in order to obtain biofuels in the Russian Federation.

Specifications	Vesture	
	Sugar beet	Sunflower
Gross harvest in farms of all categories, thousand tons, 2017	51913	10481
Sales, thousand tons, 2017	46494	9385
Overproduction, thousand tons, 2017 (marketability, %)	5419 (89.5)	1096 (89.5)
The biofuel output rate, l/ton	104	32
Potential biofuels production (bioethanol, biodiesel), l	563.6	35.1

Source: developed by the authors based on the materials (Rosstat n.d.; Hammond and Mansell 2018; NBA n.d.).

The presented data indicates that the potential volume of bioethanol production from the overproduced agricultural products may reach 563.6 thousand tons (using sugar beets), which is 1.5% of the gasoline produced in 2017 (38151.1 thousand tons) (Rosstat n.d.). The potential volume of biodiesel can reach 35.1 thousand tons (when using sunflowers), which is 0.05% of the diesel fuel produced in 2017 (75457.4 thousand tons) (Rosstat n.d.). Thus, in certain Russian regions with developed agricultural production, there are obvious prerequisites for the development of biofuel production, which can be used as a part of the fuel mixtures that are widely used in world practice, E10, E20, E25, E85, etc. (Kassem 2019). As a result, we can assume not only a reduction in the use of hydrocarbon fuel due to its replacement by biofuel but also the economic effect of reducing the total consumption of the fuel mixture with biofuel components (Kassem 2019). Considering the use of waste for biofuel production, it is worth noting that in 2017, the waste from agriculture, forestry, hunting, fishing, and fish farming amounted to 41,499.2 thousand tons, from food production—26,264.4 thousand tons, paper and paper products—5545.6 thousand tons (Rosstat n.d.). A forecast of the potential volume of biomass that can be processed in Russia for energy production is presented in Fig. 4.

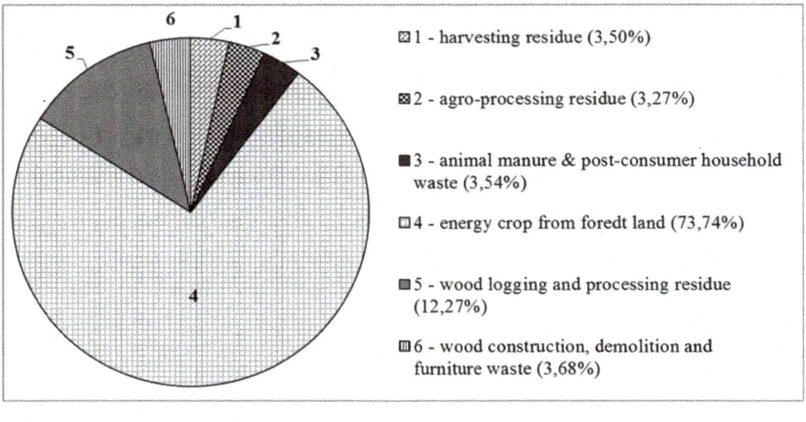

Fig. 4. A forecast of the potential volume of biomass possible for processing in the Russian Federation for energy production by 2030 (IRENA 2017).

There is no doubt that the production of biofuels at the present stage will not be able to replace traditional energy sources completely. However, the existing volumes of biomass that can be processed using well-known technological solutions can, for example, contribute to the energy self-sufficiency of agricultural enterprises. Also, the processing of excess biomass can contribute to the implementation of measures to protect the environment and support the thesis that "energy efficiency and fuel economy remain a key element of Russian environmental policy" (Romanova 2017).

4 Discussion

The study of the prospects for the production of biofuel in various regions of Russia indicates the presence of an extensive base of raw materials and the possibility of attracting innovative technologies for organizing an efficient production of energy with the parallel processing of the accumulated biomass. The creation of a system for the rational use of industrial, agricultural, and household wastes of biological origin will create a solid foundation for the development of innovative activities in the production of biofuels and obtainment of the corresponding synergistic effect. Moreover, unclaimed volumes of agricultural products (e.g., sugar beets) open the way to launch bioethanol production without the formation of additional sown areas, which is important at the current stage of the development of biofuel production in Russia. As a result, such innovative activities can ensure competitiveness when using, for example, a mixture of traditional fuels and bioethanol, reducing the cost of fuel. Many additional opportunities are opening up, such as reducing anthropogenic environmental impact and rational use of resources. At the same time, there remains a number of tasks requiring solutions, such as comparing the costs of energy production from biomass, applying and adapting the obtained results depending on the economic and geographical location of the production, the availability of certain types of biological resources, and the analysis of generating equipment, including cost and quality characteristics. The constraining factors include the absence of a fundamental managerial impact (e.g., tariff and tax regulation), an actual assessment and classification of available biomass, a regulatory framework (including the introduction of standards), government subsidies, and a low rate of managerial decision-making. These factors reduce the opportunities for realizing the existing potential, including the creation of export products.

5 Conclusion

In many countries, there has been an increase in the capacity of the industrial facilities operating on the basis of AES. However, domestic and foreign scholars show that the prospects for the development of AES are much more extensive than the existing volumes. Many resources that can be processed into biofuel remain unclaimed. Russia possesses an extensive base of raw materials, which can become a reference point in the formation of the market of biofuel products.

Moreover, the development should be carried out, taking into account regional economic and social prerequisites (the targeted use of raw materials, the introduction of new technologies, the protection of the environment, the development of human resources, etc.). Also, a significant economic effect can be obtained through the production of biofuel by-products, for example, fertilizers or feed additives for farm animals, as well as preparations for the cosmetic industry. Thus, from developing innovative technologies that are designed to process biomass into biofuel, we can expect a gradual transition to the environmentally friendly production, an increase in the competitiveness of the biofuel products and the spread of applied knowledge. The introduction of the innovative technologies for the production of biofuels will confirm the classic rule "production determines the market and the market—production" (Marx and Engels 1961), while maintaining waste and unclaimed biological resources as "new elements of production" and leaving them as the "carriers of the exchange value" (Marx and Engels 1961).

References

Ahmed, W., Sarkar, B.: Impact of carbon emissions in a sustainable supply chain management for a second-generation biofuel. J. Clean. Prod. **186**, 807–820 (2018)

Alferov, J.I.: To the release of the Russian edition of the book "The Long Road to IRENA". Ecology and Life, vol. 3 (2010)

Ballesteros, M., Manzanares, P.: Liquid biofuels (chap. 3). In: Lago, C., Caldés, N., Lechón, Y. (eds.) The Role of Bioenergy in the Bioeconomy, Resources, Technologies, Sustainability, and Policy, pp. 113–144. Academic Press, Cambridge (2019)

Bondarchuk, N.V., Titova, E.S.: Prospects for renewable energy as one of the areas of sustainable development in some regions of the South of Russia. South Russ.: Ecol. Dev. **12**(4), 12–31 (2017a)

Bondarchuk, N.V., Titova, E.S.: Biofuel production as an innovative way to protect the environment: economic risks and some prospects. Ecol. Ind. Russ. **21**(6), 48–53 (2017b)

Chen, J., Yin, X., Mei, L.: Holistic innovation: an emerging innovation paradigm. Int. J. Innov. Stud. **2**(1), 1–13 (2018)

Correa, D.F., Beyer, H.L., Fargione, J.E., Hill, J.D., Possingham, H.P., Thomas-Hall, S.R., Schenk, P.M.: Towards the implementation of sustainable biofuel production systems. Renew. Sustain. Energy Rev. **107**, 250–263 (2019)

Das, G.G.: Food–feed–biofuel trilemma: biotechnological innovation policy for sustainable development. J. Policy Model. **39**(3), 410–442 (2017)

Flach, B., Lieberz, S., Lappin, J., Bolla S.: EU Biofuels Annual (2018). http://biotoplivo.ru/upload/private/information_items_property_205.pdf

Federal State Statistics Service of the Russian Federation (Rosstat) (n.d.). http://www.gks.ru/

Golova, I.M., Sukhovey, A.F., Nikulina, N.L.: Problems of increasing the innovative sustainability of regional development. Econ. Reg. **13**(1), 308–318 (2017)

Hammond, G.P., Mansell, R.V.M.: A comparative thermodynamic evaluation of bioethanol processing from wheat straw. Appl. Energy **224**, 136–146 (2018)

International Renewable Energy Agency (2017). https://www.irena.org/

Kang, J.-S., Kholod, T., Downing, S.: Analysis of Russia's biofuel knowledge base: a comparison with Germany and China. Energy Policy **85**, 182–193 (2015)

Kassem, M.G.A., Ahmed, A.-M.M., Abdel-Rahman, H.H., Moustafa, A.H.E.: Use of Span 80 and Tween 80 for blending gasoline and alcohol in spark-ignition engines. Energy Rep. **5**, 221–230 (2019)

Kurki, S., Wilenius, M.: Organizations and the sixth wave: are ethics transforming our economies in the coming decades? Futures **71**, 146–158 (2015)

Lindow, K., Kaluza, A., Stark, R.: Study on sustainability developments in industrial practice. Proc. Manuf. **21**, 345–352 (2018)

Ma, Y., Hou, G., Yin, Q., Xin, B., Pan, Y.: The sources of green management innovation: does internal efficiency demand-pull or external knowledge supply push? J. Clean. Prod. **202**, 582–590 (2018)

Marx, C., Engels, F.: Writings, vol. 25. Political Literature Publishing House, Moscow (1961)

Maslennikov, M.I.: Technological innovations and their impact on the economy. Econ. Reg. **13**(4), 1221–1235 (2017)

Ministry of Energy of the Russian Federation: Basic characteristics of the Russian power electricity (2018). https://minenergo.gov.ru/node/532

National Biofuel Association (NBA) (n.d.). http://biotoplivo.ru/

Nidumolu, R., Prahalad, C.K., Rangaswami, M.R.: Why sustainability is now the key driver of innovation. Harvard Bus. Rev. **87**(9), 56–64 (2009)

Official site of AltEnergo LLC (2019). http://altenergo.su/

Pinto, G.M.C., Pedroso, B., Moraes, J., Pilatti, L.A., Picinin, C.T.: Environmental management practices in industries of Brazil, Russia, India, China, and South Africa (BRICS) from 2011 to 2015. J. Clean. Prod. **198**, 1251–1261 (2018)

Ratner, S., Ratner, M.: Developing a strategy of environmental management for electric generating companies using DEA-methodology. Adv. Syst. Sci. Appl. **4**(17), 78–92 (2017)

Rafaj, P., Kiesewetter, G., Gül, T., Schöpp, W., Cofala, J., Klimont, Z., Purohit, P., Heyes, C., Amann, M., Borken-Kleefeld, J., Cozzi, L.: Outlook for clean air in the context of sustainable development goals. Global Environ. Change **53**, 1–11 (2018)

Rodionova, M.V., Poudyal, R.S., Tiwari, I., Voloshin, R.A., Zharmukhamedov, S.K., Nam, H. G., Zayadan, B.K., Bruce, B.D., Hou, H.J.M. Allakhverdiev, S.I.: Biofuel production: challenges and opportunities. Int. J. Hydrogen Energy **42**(12), 8450–8461 (2017)

Romanova, O.A.: The innovative paradigm of new industrialization in the context of the formation of an integrated world economy. Econ. Reg. **13**(1), 276–289 (2017)

Silva, G., Di Serio, L.C.: The sixth wave of innovation: are we ready? RAI Revista de Administraçãoe Inovação **13**, 128–134 (2016)

State Duma of the Russian Federation: Federal law "On State Regulation of the Production and Turnover of Ethyl Alcohol, Alcoholic and Alcohol-Containing Products, and on Limiting the Consumption (Drinking) of Alcoholic Products" (No. 171-FZ), Moscow, Russian Federation (1995)

Titova, E.S., Bondarchuk, N.V.: Analysis of innovative approaches to the creation and use of alternative energy sources on the example of biofuels. Russian Economic Online Magazine 1 (2017). http://www.e-rej.ru/publications/168

Titova, E.S., Bondarchuk, N.V., Romanova, E.V.: Economic aspects of the cultivation of some plants used as raw materials in the production of biofuels. Int. Agric. J. **1**, 54–61 (2017)

Tseng, Y.-H., Mohanty, S.K., McLennan, J.D., Pease, L.F.: Algal lipid extraction using confined impinging jet mixers. Chem. Eng. Sci. **X**, 1 (2019). https://www.sciencedirect.com/science/article/pii/S2590140018300029

European Comission: A sustainable bioeconomy for Europe: strengthening the connection between economy, society and the environment (2018).https://ec.europa.eu/research/bioeconomy/pdf/ec_bioeconomy_strategy_2018.pdf

Rosstat: Official data (n.d.). https://www.gks.ru/compendium

The System of Key Performance Indicators of Innovative Activity as Management Innovation in Oil and Gas Companies

Mikhail V. Rybin[1(✉)], Alexander A. Stepanov[1], and Nadezhda V. Morozova[2]

[1] Department of Innovation Management, Moscow State Institute of International Relations of the Ministry of Foreign Affairs of Russia, Moscow, Russia
rybinml958@yandex.ru, step-916@yandex.ru
[2] Department of Industrial Management, National University of Oil and Gas "Gubkin University", Moscow, Russia
dyubiks@mail.ru

Abstract. The paper focuses on the urgent problem of assessing the effectiveness of innovation in oil and gas companies based on the use of key performance indicators (KPI). The methodological and methodical problems of their application are revealed. Considerable attention is paid to improving methodological approaches to justifying the choice of a KPI, taking into account the characteristics of this activity as being applied to oil and gas companies. The efficiency indicators of innovative activity are presented and discussed by the authors. They are adequate to the features of innovative processes in information and digital economy, as well as reflect the total useful effect from creation and operation of innovations.

Keywords: Innovation · Oil and gas companies · Key performance indicators · Innovative and creative potential · Innovation effectiveness

1 Introduction

The effective development of the fuel and energy complex in the twenty-first century, ensuring its sustainable development through the introduction of scientific and technological progress, requires a transition to fundamentally new models for managing the innovative activities of energy companies under the conditions of information and the digital economy. One of these models is a management system based on the use of key performance indicators (KPI) of innovation.

The problem of evaluating the effectiveness of innovation in companies in the fuel and energy complex is currently one of the most critical tasks.

Great attention is paid to solving this problem in the Russian Federation, both at the federal and economic levels.

The issues of evaluating the effectiveness of innovative activities are reflected in government programs for the innovative development of Russian corporations, as well

A. V. Bogoviz and Y. V. Ragulina (Eds.): ISCI 2019, LNNS 115, pp. 605–612, 2020.
https://doi.org/10.1007/978-3-030-40749-0_72

as in the fundamental documents that determine the directions of innovative development of the fuel and energy complex (Government of the Russian Federation 2009, 2011, 2014; The Federal Agency for the Administration of State Property 2014).

Considerable attention to the development of indicators of innovation effectiveness is given in the innovative development programs of joint-stock companies with state participation, state corporations, and federal state unitary enterprises. In particular, the Ministry of Economic Development of the Russian Federation has developed recommendations on a crucial integrated indicator of the effectiveness of innovative activities, which is included in long-term development programs (The Federal Agency for the Administration of State Property 2014).

At the corporate level, oil and gas companies pay serious attention to the search for and implementation of KPI of innovation.

At the same time, the applied KPI have a number of serious drawbacks: a significant overload of secondary, additional indicators; the lack of clear logical schemes for their construction, which would take into account causal relationships between strategic innovative goals and intermediate results of innovative activity between the efficiency (effectiveness) of innovative activity and the employee motivation system. Also, many key indicators do not always correspond to the principle of relevance.

In this regard, science and practice have an essential task of finding reliable tools for substantiating the selection and application of critical indicators of the efficiency of innovation activity for energy companies.

2 Materials and Methods

The methodology and methods for assessing the effectiveness of innovative activities using KPI constitute a separate area of research that goes beyond the traditional approaches and methods for assessing the effectiveness of innovative investment projects.

The research results presented in the article were based on the methodology of studying the innovation process as a special specific type of activity whose main features are uncertainty, human intensity, inconsistency and paradox, a high degree of uncertainty and risks, the duration of the innovation process in time, etc. (Andryushchenko and Stepanov 2015).

To analyze the practice of applying key indicators to assess the effectiveness of innovation, the methods of system-creative analysis were widely used, in particular, the *tree of goals* method, the method of expert evaluations; brainstorming, Delphi method; Eisenhower priority analysis; method of morphological analysis, the method of collective generation of ideas; and scripting method, as well as methods of strategic analysis of innovative ideas, methods for identifying, analyzing and solving nonstandard problems, etc. (Stepanov and Bialoblodskaya 2014).

3 Results

The study of using KPIs to evaluate innovation activity in Russian oil and gas companies allowed us to determine the main indicators (Table 1).

Table 1. The list of main applied KPI of innovative activity in Russian oil and gas companies.

The share of R&D expenses in revenue	Specific operating costs per one ton of gross production of enterprises in the exploration and production segment	Reducing losses and processing needs
Patents of the Russian Federation on the balance	Specific operating costs per one ton of gross production of enterprises of the *segment* of the *small mining company*	Labor productivity
International patents on the balance	Specific energy costs per unit volume of oil produced by enterprises from the *segment of exploration and production*	Revenues, the *segment of exploration and production*
The share of highly skilled workers (candidates and doctors of sciences) in the total number	Unit costs of energy resources per unit volume of oil, produced by enterprises of the *small mining company segment*	Revenues, the *small mining company segment*
The number of publications in the journals of the HAC list, per year	Oil spills, number	Revenues, enterprises from the *segment of Services and STC*
The increase in recoverable hydrocarbon reserves in categories B + C1 of *enterprises of the exploration and production segments, small mining enterprises*	Work injuries, number	Net profit, enterprises in the *segment of Services and STC*
The production volume of enterprises in the *exploration and production segments*	Accident rate	Net profit margin, enterprises in *the segment of Services and STC*
The production volume of the *small mining enterprises*	An increase in the processing depth	
The use of the Associated Petroleum Gas	An increase in the proportion of light oil products	

The analysis of the use of key indicators for assessing the effectiveness of innovation in energy companies showed the existing imperfection of the current methodological approaches to their formation.

The main problems of the currently used KPI system are as follows:

(1) the used scorecard is significantly overloaded. The number of such indicators totals to a dozen, while there should be no more than three or four indicators. At the same time, KPI should reflect, first of all, the degree to which strategic criteria and development goals of companies are achieved. However, in practice, the indicators included in KPI in the vast majority of cases characterize individual technical and technological aspects, which significantly overloads their assessment system and does not focus the attention of performers on achieving the main strategic priority goals of innovative development. Thus, the applied set of indicators, numbering to several dozen items, is in conflict with the two main principles of KPI: (a) to be a tool to simplify the procedure of analysis and improvement of the efficiency of managerial decisions; (b) to meet the requirements of simplicity in measurement and accessibility in perception of KPI by performers;
(2) in most cases, the used KPI does not allow us to establish causal relationships between specific strategic goals, and the results of innovative activity, the effectiveness of this activity, and the system of motivation of performers. As a result, there is no targeting and communication with direct performers responsible for the result of activities;
(3) the used KPI of innovation activity does not always meet the requirements of relevance. So, for example, the indicator *the share of R&D expenses in revenue* primarily reflects the volume of funds invested in R&D but does not indicate their effectiveness and efficiency (KPI of the innovation activities of the state corporation at the investment stage).
(4) Another example is *the growth of labor productivity,* which is not quite correct due to the complexity of measuring the growth in labor productivity in the innovation process as a result of the costs of the incurred R&D, along with other costs. The use of the growth rate of labor productivity in innovation is also doubtful because innovation projects are only in the investment phase, and, therefore, labor productivity cannot be determined in the whole company as a result of research and development, and engineering work. On the contrary, in the long run, this indicator can be considered as a strategic KPI for the top management of the company, as a substantial and stable increase in labor productivity can be achieved only through ongoing innovation.

All companies resort to the use of patent statistics. At the same time, several companies evaluate the number of used patents and licenses, while others track the increase in the number of patents, which is more indicative.

Some companies use indicators such as cooperation with scientific organizations, the number of publications in journals of the HAC list, the share of highly qualified employees (candidates and doctors of sciences) in the total number. It should be noted that these indicators do not show the effectiveness of the innovative activity but only reflect the characteristics of innovative potential.

Significant congestion and other shortcomings noted above in the use of the KPI regarding the effectiveness of the innovative activities of oil and gas companies are evidenced by a comparative analysis of the use of these indicators in Russian and leading foreign oil and gas companies.

It turned out that only about five KPIs are used in leading Western companies, while in Russian companies, there are more than two dozen.

Thus, the analysis of the application of the evaluation system for measuring the effectiveness of innovative activities in the fuel and energy companies reveals the lack of relevance, system orderliness, and validity in their integrated use.

KPI of innovative activity is, first of all, a set of basic (decisive) performance indicators that affect the final results of innovative and, ultimately, the activities of companies as a whole (profit or other most important indicators of its activity and development). The purpose of applying the KPI system of innovation is to simplify the analysis and increase the efficiency of managerial decision-making in order to achieve the main tasks of the innovative development of companies, to set up a favorable environment and creative atmosphere that promotes the activation of all participants in the innovation process (Stepanov et al. 2019).

According to their socio-economic nature, the KPI of the effectiveness of innovative activities should ensure the transfer of those innovative activities, which effects have been assessed, from the sphere of subjective assessments into the world of real and objective assessments.

Such indicators will allow:

- the establishment of causal relationships between strategic goals and indicators of their achievement;
- the identification of bottlenecks and weaknesses in the company's innovation activity;
- the identification of factors and patterns that determine the effectiveness of innovative activities of the company;
- the provision of an inextricable link between innovation plans, the results of their implementation, and the motivation system.

The KPI system of the effectiveness of innovative activities should provide an opportunity not only to evaluate the results and the effectiveness of personnel and departments but also form an effective motivational mechanism for the personnel of the teams of the innovative project.

The main purpose of the application of the KPI system of innovation should be reduced to linking goals, plans, programs, and results, as well as mechanisms of motivation and income of employees in the process of organizing innovation. As domestic and foreign experience shows, it is precisely such a KPI system that can solve the tasks.

4 Discussions

The KPI system for assessing the effectiveness of innovative activities should include both financial and non-financial indicators. At the same time, diagnostic, prognostic, and dynamic generalizing indicators of innovative activity should find a place.

In order to assess the effectiveness of innovation, it is necessary to use integrated indicators that reflect both the state and effectiveness of using the innovative potential of the company. The analytical assessment of the state and effectiveness of using the innovative potential of the enterprise, the expected innovative results, as well as the economic (financial, economic), social, and environmental consequences should be carried out taking into account the interests of all actors involved in each stage of the innovation process: the development of innovative ideas; the mechanism of technology for turning ideas into an innovative product; the assessment of the effectiveness of business processes; the assessment of the stakeholders' needs, etc.

Currently, more attention should be paid to solving the issues of improving methodological approaches to the formation of a relevant KPI system of the effectiveness of innovative activities. It is evident, nowadays, that traditional methodological approaches, including those based on the method of discounting cash flows, do not fully meet the requirements.

Further improvement KPI of innovation activity should be based on its specificity and uniqueness, which are predetermined by:

- the time lag and the specifics of the formation of the results and effects of innovation, which are manifested in the mismatch (delay) of their manifestation to the period of investment in research and development;
- the uncertainty and unpredictability of the innovation results;
- the complexity of determining the impact of innovation on the performance of companies in general;
- the difficulty of identifying the results of a synergistic effect in the company;
- the ambiguity of evaluations of the effectiveness of innovative activities by various parties of the innovation process (differences of interests);
- the specifics of the formation and use of information, analytical bases, and so on.

The study of the used KPI of innovative activities of oil and gas companies of the fuel and energy complex of Russia and foreign countries showed that such indicators make it possible to assess the effectiveness of the implementation of innovations, rather than the effectiveness of innovation activity.

Traditional approaches, which are based on target standards for the implementation of specified KPI, meticulous budget execution, and a short-term focus reduce the possibility of using a creative approach in the process of innovative activity and negatively affect its effectiveness, including long-term (Stepanov et al. 2016).

Under the conditions of the organizational dualism of innovative activity, an assessment of its effectiveness through evaluating the effectiveness of using innovative potential (evaluating the effectiveness of innovative activity) is becoming increasingly important.

The innovative activity is, first of all, creative activity. The main factor in innovative activity, in general, and in increasing its effectiveness is the innovative activity of people, their ability to creatively (non-standard, original) develop a new, highly effective innovative solution to the problem.

In this regard, it is evident that the creative (creative-innovative) potential of companies is the main factor in increasing the efficiency of innovative activity. The indicators for assessing the effectiveness of innovative (creative-innovative) potential should become the main criteria characterizing and reflecting the features of evaluating the effectiveness of innovative activity from efficiency production activities (Savina et al. 2017).

KPI is a tool to improve staff performance. Therefore, they should orient and motivate management personnel to achieve innovative results.

The indicators of evaluating the effectiveness of using the creative potential of the company should be a priority in the KPI system of innovation activity.

5 Conclusion

In the future, the development of new approaches to assessing the effectiveness of innovation requires the improvement of methodological tools and, firstly, based on the evaluating the effectiveness of the creative potential of the company as a key strategic resource, which determines the development efficiency of all other components of the innovative potential of the company.

In order to assess the effectiveness of innovation, along with the abovementioned, several other indicators should be used. These will reflect the total beneficial effect of the creation and operation of innovations and will be widely used for evaluating the performance of company managers.

It is proposed to introduce the *efficiency of financing innovative activity* as a key indicator of the effectiveness of innovative activity, where the efficiency of financing innovative activity equals a real financing/project financing plan. This indicator makes it possible to assess the degree of real financing of innovation.

The indicator should also use *the intensity of innovation*. This indicator will allow us to estimate the number of expenses for creating the results of research and technological production improvement.

References

Andryushchenko, G.I., Stepanov, I.A.: Risk management problems of microfinance institutions. Int. J. Econ. Financ. Issues **5**, 151–158 (2015)

Federal Agency for the Administration of State Property: The guidelines for the development of long-term development programs for Strategic Open Joint-Stock Companies and Federal State Unitary Enterprises, as well as open joint-stock companies, the share of the Russian Federation in the authorized capital of which in total exceeds fifty percent (2014). http://www.rosim.ru/documents/190521

Government of the Russian Federation: Order "On the Energy Strategy of Russia for the Period until 2030" (November 13, 2009 No. 1715-r) (2009). http://energoeducation.ru/wp-content/uploads/2015/11/LAW94054_0_20151002_142857_54007.pdf

Government of the Russian Federation: Order "On the Strategy for Innovative Development of the Russian Federation for the Period Until 2020" (December 8, 2011 No. 2227-r) (2011). https://www.garant.ru/products/ipo/prime/doc/70006124/

Government of the Russian Federation: Order of the Government of the Russian Federation (July 3, 2014 No. 1217-r) (2014). https://www.garant.ru/products/ipo/prime/doc/70589496/

Mitskevich, A.: KPI of the innovation activities of the state corporation at the investment stage (2016). http://upr.ru/article/rossiyskie-praktiki-upravleniya/KPI_INNOVACIONNOJ_DEYATEL_NOSTI_GOSKORPORACII_NA_INVESTICIONNOJ_STADII__CHAST__1__SITUACIYA_%C2%ABKAK_EST_%C2%BB.html

Stepanov A.A., Savina M.V., Stepanov I.A., Zolotareva A.F., Antonova L.I.: Concept of creative management in the paradigm of informational society management theory. Int. J. Econ. Res. **14**, 135–152 (2017)

Stepanov, A.A., Bialoblodskaya, S.: Methodological aspects of assessing the effectiveness of agricultural systems in an innovative economy. Appl. Econ. Res. **4**, 48–51 (2014)

Stepanov, I.A., Korniyko, Yu.V., Scherbakova, M.A.: KPI of regional innovation development. Manag. Econ. Syst. **3** (2019). http://www.uecs.ru/regionalnaya-ekonomika/item/5426-2019-03-16-08-39-45

Stepanov, I.A., Stepanov, A.A., Morozova, N.V.: Actual problems of developing tools and methods of creative management. Manag. Econ. Syst. **9**(91) (2016). http://uecs.ru/index.php?option=com_flexicontent&view=items&id=4073

The Directions for Forming an Effective Strategy System in the Mineral and Raw Materials Complex

Alexander R. Kalinin, Lyudmila P. Ryzhova(✉), and Alexander M. Kurchik

Sergo Ordzhonikidze Russian State University for Geological Prospecting, Moscow, Russia
kalinal@yandex.ru, kafedra520@mail.ru, nich.rggru@mail.ru

Abstract. The necessity of developing an effective strategic system for a mining and exploration company is considered by the authors in the paper, taking into account the modern requirements of a highly competitive and politicized global market for mineral raw materials. The studies of the scholars of the School named after Academician M. I. Agoshkov allow to change conceptual approaches to the development of mineral resources in assessing the strategic advantages of Mineral and Raw Materials Complexes. The authors also consider rare and rare-earth metals as strategic ones for knowledge-intensive technologies and equipment of the 5th and 6th generations of modern industries (military-industrial complex (MIC), aerospace, telecommunication systems, and other high technologies).

Keywords: Strategy · Economic incentive system · Digitalization · Development system · Global market for mineral raw materials · Technogenic deposit · Innovation

1 Introduction

Recently, an increasing number of fundamental requirements for a highly competitive and politicized global market for mineral raw materials have been intensively formed for the modern strategic system of domestic mining and exploration companies. The key changes are determining, fundamental, and expected to be carried out in the natural, material, and technological spheres of enterprise activity in the mineral resource industry. Even at the stages of preliminary assessment and future concept formation for the planning of the mineral resource complex, it is necessary to revise and adapt the basic principles, mechanisms, and planned production results to the requirements of the modern market. Only timely and comprehensive interaction between the state and private business within the framework of the public-private partnership will allow obtaining significant positive results adapted to modern market requirements at a reasonably close planning horizon.

Systemic changes in the planning of the mineral resource complex and its individual enterprises must begin with a qualitative adaptation of the regulatory framework

A. V. Bogoviz and Y. V. Ragulina (Eds.): ISCI 2019, LNNS 115, pp. 613–618, 2020.
https://doi.org/10.1007/978-3-030-40749-0_73

for subsoil use at the federal, sectoral, and local levels. Participants in the mineral market have many questions about the legitimacy of the current tax system, mechanisms for regulating the ownership of the resources used, the effectiveness of the current system of state (corporate) regulation and management of mineral companies, the monopolization of the domestic raw materials market, etc. In the Strategy for the Development of the Mineral Resources Base of the Russian Federation until 2035, approved by the Government of the Russian Federation, dated December 22, 2018 No. 2914-r, the sustainable development of mineral raw materials of a high quality for modern domestic consumers, existing and planned export obligations, is designated as a priority for the development of the mineral resources sector. Such results can only be achieved in a dynamic mode of adaptation of all domestic participants in the process of rational use of natural resources to the stringent requirements of the market.

A key feature of the current market situation has been the transition to a fundamentally new level of strategic planning for the mineral raw material complex; this is the digitalization of subsoil use. The same Strategy for the Development of the Mineral Resources Base has already formed the basic principles and requirements for the formation of the "digital geoeconomics". It includes robotic technologies for the collection, processing, accumulation, and provision of geological and mining-technological information, the introduction of digital technologies in the technological processes of extraction and processing of mineral raw materials, etc. However, these activities are not sufficient. Systemic digital changes of a fundamentally new level are needed in the entire mineral raw material complex, as well as in related industries. These changes should be based on the use of artificial intelligence (R-technology), technologies for managing large amounts of data (Big Data, etc.), digital marketing and management (CRM), etc.

According to the McKinsey Global Institute, by 2025, digitalization of the Russian manufacturing sector will be able to increase the GDP of the country by ₽ 1.3 trillion to 4.1 trillion annually, labor productivity can double, and the time it takes for commercial output to enter the world market can be reduced by half as well (Kalinin et al. 2019).

It is necessary to implement other elements of the digitalization of raw materials production. These are the concepts of innovative technological marketing, the introduction of digital junior companies in the domestic exploration market, fundamentally new human (intellectual) capital management systems, digitalization of R&D, operational and current planning systems, digitalization of equipment, and all technological processes of exploration, production, and processing of mineral raw materials, breakthrough qualitative changes in the manufactured production, etc. This approach will solve the problems that have long been accumulating and arising in modern conditions of a highly competitive and politicized global market for mineral raw materials.

2 Materials and Method

The current stage of scientific and technological progress (STP) is associated with the development of mineral and raw material bases of rare and rare-earth metals, which are strategic raw materials for high-tech technologies and equipment of 5–6 generations of modern industries: military-industrial complex (MIC), aerospace and telecommunication systems, communications, and other high technologies. Therefore, the struggle for the ore-metal mines and carriers today is associated with the designation of the importance of a particular moment with the most resulting indicators of the effects of their development. In order to move in the wake of the world economies, state monitoring is necessary by advanced countries of the world community, with the analysis and evaluation of long-term trends and goals for the use of mineral resources of rare, including rare-earth, metals and ore-metal deposits (Kaplunov 2018; Kremenetsky and Kalish 2011).

The use of the achievements of previous generations of geologists opens up new horizons for technologies and equipment that are relevant today. So, studies of the school named after academician M. I. Agoshkov (Agoshkov et al. 1983) allow us to change the conceptual approaches to the development of mineral resources at the present stage. They allow us to make a breakthrough with the most significant effect on the systematic choice of strategic advantages of the development of modern industries affecting the development stages of high technology and technology designed to manage the state subsoil fund, which is exceptional for the consistent development of industries and which will maintain priorities in the high level of scientific and technical progress. All this will allow them to develop and carry out production with reasonable sufficiency, as indicated by Academician K. N. Trubetskoy, and to optimize both current and predicted worldwide demands of the individual minerals and the raw potential of grades and types of metals. In other words, the study of the value of the assortment of ores and metal types of ore-metal deposits for modern industries should begin by taking into account the geological and industrial components in the development of mineral wealth. The study should also consider how the military, economic, and scientific-technical components of the spheres of influence of the raw material complex affect scientific and technological progress. It is important to know who uses the mineral and raw material potential of the Atlas of the subsoil of ore-metal deposits, and how they use it. Production schemes are characterized by the presence of associated rare components, including rare-earth metals.

In recent years, Russia has accumulated about 50 billion tons of industrial waste, in which the content of metals often exceeds its content in ores extracted from the bowels and received for enrichment. This mainly applies to old waste and tailings, which were formed in the 1940s and 50s, a time when there were other technologies, and the conditions of production and processing were much higher than the modern ones.

A technogenic deposit is an accumulation of mineral substances on the surface of the earth or in mine workings and is formed as a result of these substances' separation from the massif and storage as the waste of various industries (metallurgical, mining, and enrichment), which, in terms of quantity and quality, is suitable for industrial use (for extracting the necessary useful components and obtaining building materials and fuel).

Technogenic deposits include

- mining dumps,
- concentrator tailings of the enriching factories,
- ash dumps and slag dumps of central heating and power plants, and
- stored wastes from the metallurgical and other industries.

There are examples of the successful use of technogenic deposits. In the 1970–1980s, the Solnechnyy and Khrustalnensky plants began to audit the dumps of past years and the extraction and use of substandard ores for additional products (tin, lead, zinc, etc.). Still, to date, human-made deposits are not processed sufficiently. The main reason for this is that for their widespread involvement in processing, the construction of practically new industries is required, and these must implement new technological principles and solutions that are developed, as a rule, at the level of scientific discoveries or laboratory or semi-industrial research and are rarely brought to industrial production.

Hence, the high capital intensity of new construction and reconstruction with the consistent replacement of existing production aligns with new production. Despite these difficulties, the prospect of the strategic use of the human-made field is obvious since its use will simultaneously solve many environmental, social, and economic problems (Balasyan and Ryzhova 2018).

In order to manage the state subsoil fund with the intended purpose of achieving international cooperation, a flexible system of the variability of subsoil development models is needed. All this will reveal the essence of the latent period in the globalization of world markets for minerals, where the main factor determining the trends in the innovative development of demand for rare and rare-earth metals is the price level. Demand is developing stepwise, as a rule, which is usually associated with the development of new application areas that significantly change the structure of consumption and its volumes; these also need to be paid attention to. It is essential to evaluate the logical constructs of operational grade management, the completeness of judgments about when and why the vector of industrial production capacity is turned on by 30 to 50 to 70%, which provide a mechanism for the success of business and commerce.

It is important in order to achieve management continuity and planning organization for the production goals, a strategic assessment of the options (Kaputin 2002) for the geometrization of ore-metal deposits by geological and technological mapping. This allows us to switch to cluster communications, to effectively use the development of natural resources of the subsoil with trend technologies for the use of self-propelled equipment, and to direct the mined minerals in ore passes (shops), the decisive role of which is to harmonize as an industry industries, and business, and commerce.

3 Results

The strategic regional development, together with the development and implementation of a new management system in MSCs and the improvement of the institutional environment for the development of integrated deposits, should become an integral element of economic incentives made to increase the efficiency of development of ore deposits, taking into account the export-oriented orientation and investment attractiveness of the regions of the Russian Federation.

The need to use an integrated and systematic approach to building models of economic incentives to increase the efficiency of developing complex deposits, taking into account the export orientation and investment attractiveness of the regions of the

Russian Federation based on the use of advanced and promising technologies for geological exploration, is due to the high efficiency of using innovations. For economic incentives, there is a need for a broader toolkit for forecasting and substantiating such key parameters as the discount rate, the price of marketable products, and indicators of not only budget and economic efficiency but also production and socio-economic efficiency of developing complex deposits as mineral-resource centers of socio-economic development.

The development of the ideas of the school named after Academician M. I. Agoshkov, which are of an international character, lies in the new possibilities of compiling grade maps and types of metals, allowing to create new types of minerals in mining technological schemes based on self-propelled equipment with their value and purpose in information technologies that will allow efficient use of natural options subsoil resources capital (Kuzmin et al. 2016).

4 Conclusion

Thus, for the formation of a strategically-quality modern system of economic incentives to increase the efficiency of developing complex deposits, the following main tasks must be implemented (Kalinin and Desyatkin 2019; Kalinin and Desyatkin 2018):

- The formation of state policy, including the creation of the equal access conditions to the subsoil for all domestic subsoil users;
- The improvement of the state management system for the development of complex deposits, ensuring their sustainable development;
- The information and personnel human resourcing for the development of complex deposits;
- the formation of a diversified system in the development of deposits;
- the diversification of mining enterprises in the development of integrated deposits;
- the additional exploration and revaluation of promising deposits;
- the detailed digital geological study of regions;
- the allocation of ore-bearing areas that promise large-scale mineralization based on a metallogenic analysis of the territories of the regions;
- the geological study of promising areas and manifestations in order to identify significant and unique reserves of complex deposits;
- the formation of an investment environment in the regions;
- the provision of investment-active enterprises with licenses for the study and development of complex deposits;
- the assistance to mining enterprises in energy capacities and developing heat-generating sources;
- the promotion of construction and production of building materials; and
- the provision of regional benefits to enterprises.

In the modern world, more and more enterprises that deal with the issues of geological exploration use specialized software and information systems in their work. The use of such systems can significantly accelerate the processing and analysis of information and also automate the processing and interpretation of geological exploration data; additionally, they can be used to model deposits and perform all kinds of calculations and estimates (Groshong 1999).

There are several systems (including those with global brands) for modeling the geological structure of deposits and calculating reserves. Unfortunately, most of the geological models that are created in these systems are designed to perform estimates, calculate reserves of deposits, and analyze the prospects for developing deposits. After an assessment, these models become practically unsuitable for further use; that is, they cannot be used in the operation of the field and development of its reserves. This is connected, first of all, with a different accuracy shown to geological models. If, when calculating reserves for a field, an error is allowed in up to 20–30% of the calculations, then with the operational accounting, it should not exceed 1–3%. The constructed geological models cannot provide such accuracy.

The K-MINE complex supports several different systems for calculating reserves: domestic and several international ones, in particular, the JORC system, the UN framework classification, and others. Moreover, this makes it possible within the same system to carry out calculations in various systems for the classification of reserves.

References

Agoshkov, M.I., Jofin, S.L., Budko, A.V.: Generalization of the experience and assessment of the prospects for wide application of the systems development mechanical production of ore. Gornyi Zhurnal **6**, 34–38 (1983)

Balasyan, S.A., Ryzhova, L.P.: On the analysis and development of technogenic placer deposits of platinum in the north-eastern part of Kamchatka Krai. Financ. Econ. **7**, 1655–1658 (2018)

Groshong Jr., R.H.: 3D Structural Geology: A Practical Guide to Surface and Subsurface Map Interpretation. Springer, Heidelberg (1999)

Kalinin, A.R., Desyatkin, A.S.: Economic mechanisms of complex development of porphyry copper deposits in the Russian Far East. Innov. Invest. **12**, 271–275 (2018)

Kalinin, A.R., Desyatkin, A.S.: Improving the system of mechanisms economic incentives for the development of complex copper porphyry deposits of the Far East. Innov. Invest. **1**, 231–236 (2019)

Kalinin, A.R., Desyatkin, A.S., Ryzhova, A.P., Kurchik, A.M., Borovkov, Y. (eds.): Proceedings from NIinES 2019: XIV International Scientific and Practical Conference New Ideas in Earth Sciences. MGRI, Moscow (2019)

Kaplunov, D.R.: Combined geotechnology in sustainable ecologically balanced subsoil management. Gornyi Zhurnal **1**, 14–17 (2018)

Kaputin, Yu.: Mining Computer Technology and Geostatistics, 1st edn. Nedra, Saint-Petersburg (2002)

Kremenetsky, A.A., Kalish, E.A.: Russia on the world market of rare metals. Mineral recourses of Russia. Econ. Manag. **5**, 63–67 (2011)

Kuzmin, M.B., Krasavin, A.G., Ryzhova, L.P., Usova, T.Yu. (eds.): Proceedings from PPCDCEI 2016: Problems and Prospects of Complex Development and Conservation of the Earth's Interior. Institute of Comprehensive Exploitation of Mineral Resources Russian Academy of Sciences, Moscow (2016)

Regulation, Compliance, and Anti-corruption Mechanisms for Improving Governance and Competitiveness in Industry

Officer Cost Control: Russian Experience in the Context of Global Corruption Response Trends

Marina V. Chudinovskih(✉) and Oleg A. Kozhevnikov

Ural State University of Economics, Ekaterinburg, Russia
Chud-marina@mail.ru, jktul976@yandex.ru

Abstract. The work systematizes the results of a study of the Russian regulatory framework at the federal, regional, and local levels. The authors describe the procedure for organizing the anti-corruption control of expenses and summarize the existing judicial practice. They argue that the quantity and quality of normative acts allow controlling the expenses of officials, while law enforcement practice shows that if there are violations, the number of cases brought to court is minimal. The conclusion about the point nature of cost control and its low efficiency is justified. The authors argued the feasibility of a phased transition from selective to complete control over the expenses of officials based on the use of modern digital technologies.

Keywords: Income · Costs · Control · Officials · Corruption · Civil confiscation

1 Introduction

Corruption is one of the threats to national security, and it undermines the foundations of the state system, resulting in public distrust of the authorities. In 2003, the United Nations General Assembly adopted the *United Nations Convention against Corruption*. The Russian Federation ratified this Convention in 2006. Since that time, the active creation of a regulatory framework in the field of anti-corruption has begun. Despite a large number of efforts undertaken, the severity of the problem of corruption for Russia has not been reduced. According to Transparency International, Russia ranked 138th in the world in the Corruption Perception Index in 2018, being next to the level of countries such as Papua New Guinea, Honduras, and Nigeria (ANO "Transparency International" 2018). Russia lags significantly behind the BRICS countries in the effectiveness of the fight against corruption. For comparison, South Africa and China were able to achieve significantly better results (Kurakin and Sukharenko 2018).

In foreign practice, more than thirty years of experience in combating corruption has been accumulated (Lyrio and Lunkes 2018). In the countries of the post-Soviet space, Eastern Europe, and Southeast Asia, experience in combating corruption is much less. The historical lag, the weakness of civil society institutions, and a slow justice system lead to the fact that the fight against corruption in these countries is not yet compelling enough (Popova and Post 2018). Besides, high levels of corruption in

A. V. Bogoviz and Y. V. Ragulina (Eds.): ISCI 2019, LNNS 115, pp. 621–627, 2020.
https://doi.org/10.1007/978-3-030-40749-0_74

countries with economies in transition can be attributed to low incomes and living standards (Uberti 2018). The data presented prove the need to find the most effective anti-corruption methods and the relevance of the topic chosen for the study.

The purpose of the article is to analyze the law enforcement practice in the field of control over the expenses of officials and to assess its compliance with international trends, UN Conventions, the practice of the European Court of Human Rights.

Control over the expenses of officials is one of the methods of combating corruption. In the Russian Federation, this issue began to be most actively discussed in the period 2012–2014, which was associated with the development of the draft law and the adoption of the law "On Control Over the Conformity of Expenses of Persons Holding Public Offices and Other Persons to Their Income" (Zaikov 2014). Already at this stage, fair remarks were made that transactions on the acquisition of cultural property, antiques, precious metals, and jewelry were removed from state control (Shuraleva and Vankov 2013). Smirnova (2018) assessed the property liability of officials, analyzed the judicial practice of the Constitutional Court. In a separate direction, one can single out a study of the regional (Kabanov 2018) and municipal regulatory frameworks (Flavier et al. 2017), which is of particular importance for the Russian Federation due to its federal structure. The question of the possibility of seizing property from officials as a result of monitoring the conformity of income and expenses remains among the least studied.

2 Materials and Method

The study included the following steps. In the first stage, the systematization of the Russian legislative base in the field of organization of control over the expenses of officials was carried out. In the second stage, the array of regulations at the regional and local levels is studied. In the third stage, the critical positions of the judicial practice of the Constitutional Court are summarized. Finally, the fourth stage analyzes the judicial practice regarding the seizure of property from officials if declared income was less than expenses.

The main regulatory act at the federal level is the law of December 3, 2012 No. 230-FZ "*On Control Over the Conformity of Expenses of Persons Holding Public Offices and Other Persons with Their Income.*" The provisions of the Law No. 230-FZ (2012) are based on the Federal Law of December 25, 2008 N 273-FZ "*On Combating Corruption.*" The Federal Law No. 230-FZ of December 3, 2012 defines the list of persons covered by cost control, the grounds for making a decision on exercising control, the general procedure for audit and its consequences. In accordance with the provisions of subparagraph 8 of paragraph 2 of Article 235 of the Civil Code of the Russian Federation, in accordance with the legislation of the Russian Federation on combating corruption, property for which evidence of its acquisition is not presented for legitimate income is subject to appeal to the income of the Russian Federation by a court decision.

The provisions of this article of the Civil Code of the Russian Federation were the subject of judicial review in the framework of federal constitutional proceedings. The Constitutional Court of the Russian Federation, in the Resolution of November 29,

2016 No. 26-P, indicates that the conversion of property belonging to a public official and members of his family listed in the law into Russian Federation income by a court decision is aimed at protecting constitutionally significant values. The procedure for applying this measure of state coercion allows for a balance of public interests in the fight against corruption. It should be noted that we are talking about the so-called "civil confiscation," in this case, the experience of which is available in foreign countries (Simser 2008). The European Court of Human Rights takes a similar position (ECHR judgment of 12 May 2015 in the case of "Gogitidze and Others v. Georgia").

Due to the fact that the legislation on control over expenditures applies not only to officials at the federal level of government but also to a number of people who fill positions at the level of constituent entities of the Russian Federation and municipalities, law-making bodies of the regional and municipal level are actively forming their respective legal framework. Regional and municipal acts are heterogeneous in their structure and content. There are several approaches to their formation. In the first approach, individual laws of the subjects are adopted. This approach is implemented, for example, in the Tula region, the Republic of Khakassia, the Trans-Baikal region. Amending existing anti-corruption legislation and legislation on a state or municipal service is the most common. This approach was used in Moscow, Novgorod, Omsk region, the Republic of Bashkortostan, and other entities. In the framework of the third approach, a separate regional law was not adopted, legal regulation is carried out, for example, on the basis of a decree of the governor. In order to develop Law No. 230-FZ (2012), Order of the General Prosecutor of Russia No. 179 (2015), Order of the Ministry of Labor of Russia No. 206n (2015) were also adopted. In addition, the Ministry of Labor repeatedly issued letters and methodological recommendations to that law. Thus, the regulatory framework is represented by a wide range of federal and regional regulations and local regulatory acts.

The consideration of disputes on anti-corruption control of expenditures by the Constitutional Court is also essential. As the Constitutional Court of the Russian Federation repeatedly noted in its decisions, one of the main measures for the prevention of corruption in the framework of state control over the property status of civil servants in Federal Law of December 25, 2008 No. 273-FZ *"On Combating Corruption"* is the provisions of paragraph 4 of Article 6 of the said Law, which provides for the application of measures of legal responsibility to an official for failure to provide information about his/her income, expenses, property. Such a legal regulation aimed at preventing and overcoming corruption is determined by the specifics of public service. Entering the civil service, a citizen exercises the right to freely dispose his/her abilities to work (Article 37 of the Constitution of the Russian Federation) and voluntarily chooses a professional activity, which implies the existence of certain prohibitions and obligations related to the implementation of public law powers (Determination of the Constitutional Court of the Russian Federation of January 29, 2019 No. 95-O). Summarizing the analysis of the regulatory framework, we can conclude that a sufficiently large number of sources (more than 1,000) takes into account regional and local acts. Further, an assessment of the practice of applying the norms of the law of December 3, 2012 No. 230-FZ regarding the seizure of property to state revenue was carried out to assess the effectiveness of the implementation of control over the expenses of officials.

3 Results

From January 1, 2013 to January 1, 2017, the courts completed proceedings in 19 cases, in 12 cases (63%) of which, the prosecutor's claims were fully or partially satisfied, in 7 cases (37%) the claims were refused. In the Consultant Plus database, as of January 1, 2019, no more than 100 cases are listed by the law of December 3, 2012 No. 230-FZ. The data presented show that the prosecutors of the Russian Federation use the created regulatory framework very poorly; the number of cases brought to court is ridiculous.

Significant difficulties in law enforcement practice are caused by the question of the procedure for comparing incomes, expenses, as well as the number of exemptions to state revenue. The original version of article 17 of the law of December 3, 2012 No. 230-FZ contained only the rule on the prosecutor's appeal to the court with a statement on the application to the income of property for which no information was provided about the acquisition from legal sources. The practice of applying this article showed that the courts gave a different interpretation of the procedure for the seizure of property.

For example, when monitoring the expenses of a municipal employee, the fact of acquiring a car for 2 million 800 thousand rubles was established. The total family income amounted to 2 million 702 thousand rubles. The trial court decided to seize the vehicle. However, the appellate court "recognized this excess of expenses over incomes (by 3.5%) as insignificant and came to the conclusion that it was necessary to recover in the state's income not the entire cost of the car, but only the amount of 98 thousand rubles (based on Subparagraph 8 of Paragraph 2 of article 235 of the Civil Code of the Russian Federation). This amount is the monetary equivalent of the value of a part of the property, the legality of the acquisition of which has not been proven." Subsequently, this norm was reflected in the new edition of paragraph 3 of Article 17 of the Law No. 230-FZ (2012). This norm is consistent with the principles of the UN Convention Against Corruption.

An analysis of judicial practice shows that the courts must evaluate evidence of the lawful origin of funds spent on the acquisition of disputed property. If a defendant refers to the receipt of funds for civil transactions, the court should examine this fact as a circumstance that is important for the proper resolution of the case, the question of the reality of receiving funds for such transactions, and whether there were any funds are used to purchase the disputed property. For example, in the case of obtaining money under a loan agreement, the courts assess the validity of such an agreement relying on witnesses' testimonies. Income under a gift agreement is recognized as reasonable and legal. However, some courts seek a more in-depth assessment of evidence and analyze the real financial ability of the donor or lender to provide funds. So, when assessing the legality of funds for the purchase of an apartment, the court took into account the defendant's explanations and the testimonies of witnesses that the money was donated to acquire housing for himself and two minor children by his former mother-in-law relying on a verbal agreement. Certificates of the donor in the form of 2-NDFL (Personal Income Tax), a sale agreement, information on cash flows in a bank account can be presented as evidence.

Often, in judicial practice, there are cases the circumstances of which indicate an attempt by individual officials to evade responsibility stipulated by the Law No. 230-FZ (2012). So, in the Primorsky Territory, an official bought a Land Cruiser Prado car. In the documents, the cost was underestimated to 250 thousand rubles, although the actual acquisition costs amounted to about 1 million 850 thousand rubles. This amount exceeded the official's income for three years. The respondent took various actions to conceal expenses. Because the legality of the origin of the funds was not proved, the cost of the car was recovered in state revenue (Appeal Ruling of the Primorsky Regional Court of July 4, 2017 in the case N 33-6620/2017).

The practice also reflects cases when the prosecution authorities and the courts do not assess the real value of the property. For example, in the Saratov region, the following claim was considered. For 2012, a deputy of the municipal district declared annual income of 60 thousand rubles (i.e., 5 thousand per month), for 2013, he declared 36 thousand rubles (3 thousand rubles per month), for 2012 he declared 52 thousand rubles. For the indicated period of time, he acquired eight land plots with a total area of more than 400 acres. Each plot of land was estimated at between 10 and 30 thousand rubles; the total value was estimated at 180 thousand rubles. According to the results of the case, part of the plots was taken into state revenue (Appeal Ruling of the Primorsky Regional Court of July 4, 2017 in Case N 33-6620/2017). In this case, the plot of the case itself indicates flagrant violations of the law. From the perspective of common sense, it is evident that the actual income of the deputy could be significantly higher than declared. Those incomes from which taxes were paid were less than the cost of living. At the same time, the court and prosecution authorities did not consider the issue of the cadastral value of the land. It can be assumed that the value of land in contracts of sale was repeatedly underestimated. It seems that when assessing the value of the real estate and land, it is necessary to focus on the cadastral value. The tax legislation already contains similar rules, establishing the need for real estate valuation at the cadastral value when paying personal income tax.

During the investigation of corruption crimes, the following questions arise the subject of the limitation period and the possibility of verifying the income for which property was acquired before the entry into force of Law No. 230-FZ (2012). In 2018, in St. Petersburg, the trial was completed, in which the income and expenses of a public civil servant were compared for the period since 2002 (Appeal ruling of the St. Petersburg City Court of June 20, 2018 N 33-12888/2018). Despite the sufficient evidence base, the very question of the possibility of comparing income and expenses for the period earlier than 2012 seems controversial. Moreover, in the Determination of the Constitutional Court of the Russian Federation of June 6, 2017 No. 1163-O, the federal constitutional control body has already given its assessment of the moment the provisions of the law of December 3, 2012 No. 230-FZ came into force.

4 Discussion

According to the results of the study, we concluded that a sufficient regulatory framework for the organization of control over the expenditures of state and municipal employees and the seizure of illegally acquired property in state revenue was created in the Russian Federation. To unify the cost of the control procedure in Law No. 230-FZ, it is necessary to reduce the number of reference norms, to exclude the adoption of unnecessary legislative acts at the level of the subject of the Russian Federation and at the municipal level.

Despite the fact that Law No. 230-FZ has been in force for six years, the practice of its application is still very insignificant. The number of cases brought to court is significantly less than the number of adopted federal, regional, and local acts. It is necessary to apply the developed legislation more actively. The current control procedure is selective, initiated by the decision of the head of the state authority or local government. To combat corruption, it is advisable to consider the possibility of a phased transition from selective to continuous control. The implementation of constant control can be carried out based on the use of modern digital technologies, compilation, and the use of databases (Karishina 2016). Computer analysis of information will allow for comparative calculations over several years, to compare data from declarations, information from federal registers. The use of information technology will allow the transition to the monthly implementation of measures to control the reliability of the information on the income and expenses of officials.

Current legislation does not imply control over the acquisition of precious metals, jewelry, antiques, watches. However, the practice of investigating criminal cases in the field of anti-corruption shows that illicit proceeds are used to purchase such luxury goods (Konov and Yakovlev 2013). In some foreign countries, officials even declare the costs of treatment and education (Flavier et al. 2017). When assessing the value of a property, it is essential to evaluate the real value of the objects and not the amount specified in the contract. Cadastral valuation should be used for real estate, land. The option of defining the concept of "large expenses" or "significant discrepancy between income and expenses" can be considered as a possible way to further develop legislation on control over the officials' expenses.

The procedural procedure of civil confiscation also needs to be clarified. Forfeiture of property in criminal proceedings is based on the collection and assessment of evidence by state authorities. In a public process, on the contrary, a person himself must prove the legality of the property origin (Hamin et al. 2015). Besides, the issue of protecting the rights of third parties, for example, family members, when seizing property from a corrupt official requires further study (Kennedy 2008). Thus, the issues of controlling the expenditures of officials in the context of global anti-corruption trends need to be further researched.

5 Conclusion

Based on the study, the authors concluded that an effective mechanism for controlling the expenses of officials in the Russian Federation has not yet been formed. In 2018, there was a transition from the point-based application of the law to the consideration of such cases when significant damage to the state, society, and the entire institution of public service was inflicted. For the real application of the already created regulatory framework, political will and a true desire to eradicate the causes and conditions for the manifestation of corruption are needed.

Acknowledgment. The authors are grateful to the Ural State University of Economics for their assistance in preparing the work.

References

ANO "Transparency International": The Corruption Perceptions Index 2018 (2018). https://transparency.org.ru/research/indeks-vospriyatiya-korruptsii/rossiya-v-indekse-vospriyatiya-korruptsii-2018-28-ballov-iz-100-i-138-mesto.html

Flavier, H., Chikireva, I., Ivanova, K.: Disciplinary responsibility's application to the local government employees violating the anti-corruption legislation in the Russian Federation, Europe and the United States of America. BRICS Law J. **4**(4), 116–144 (2017)

Hamin, Z., Omar, N., Hakim, M.: When property is the criminal: confiscating proceeds of money laundering and terrorist financing in Malaysia. Proc. Econ. Finance **31**, 789–796 (2015)

Kabanov, P.A.: Anti-corruption procedural legislation of the Russian Federation subjects: formation and content. Actual Probl. Econ. Law **2**, 300–333 (2018)

Karishina, I.E.: Increasing the quality of verifying the information about expenditures as a means of corruption prevention. State and municipal administration in the 21st century. Theory Methodol. Pract. **22**, 134–138 (2016)

Kennedy, A.: Designing a civil forfeiture system: an issues list for policymakers and legislators. J. Financ. Crime **13**(2), 132–163 (2008)

Konov, A., Yakovlev, A.: Anti-corruption declarations In Russia: current problems and possible approaches to reforming. Public Adm. **1**, 5–30 (2013)

Kurakin, A., Sukharenko, A.: Anti-corruption in the BRICS countries. BRICS Law J. **5**(1), 56–77 (2018)

Lyrio, M.V.L., Lunkes, R.J.: Thirty years of studies on transparency, accountability, and corruption in the public sector: the state of the art and opportunities for future research. Public Integr. **20**(5), 512–533 (2018)

Popova, M., Post, V.: Prosecuting high-level corruption in Eastern Europe. Communist Post-Communist Stud. **5**(3), 231–244 (2018)

Shuraleva, S.V., Vankov, A.V.: Government and other officials' expenditures control: new in legislation. Perm Univ. Herald. Jurid. Sci. **1**, 72–79 (2013)

Simser, J.: Perspectives on civil forfeiture (2008). http://siteresources.worldbank.org/INTTHAILAND/Resources/333200-1089943634036/475256-1201245199159/2008Mar-asset_recovery-civil-forfeiture.pdf

Smirnova, L.A.: Assessment of property liability of officials in explanations of the Constitutional court of the Russian Federation and its meaning for exercising of the prosecutor's expense control authorities. State Power Local Gov. **3**, 30–34 (2018)

Uberti, L.: Corruption in transition economies: Socialist, Ottoman or structural? Econ. Syst. **42**(4), 533–555 (2018)

Zajkov, D.E.: Legal regulation of control over compliance between the employees' expenditures and incomes for corruption counteraction. Legislation **8**, 60–67 (2014)

Environmental Disclosure as a Factor of Sustainable Development of the Russian Business

Irina N. Tkachenko(✉), Irina V. Pervukhina, and Oksana E. Sokolovskaya

Ural State University of Economics, Ekaterinburg, Russia
{Tkachenko, ivp}@usue.ru,
oksana-sokolovskaya2013@yandex.ru

Abstract. The purpose of the paper is to study the social disclosure of information related to their environmental performance in 30 Russian industrial companies. It also attempts to investigate to which extent social disclosure can have an impact on the firm value. The article surveys both Russian and international literature on environmental disclosure. The authors also fully describe the research methodology and present the empirical data that served as the basis for calculations and conclusions. The findings of the study demonstrate that there is no significant relationship between a firm's ecological disclosure and its stock market performance. The article also stresses the importance of the university and in-house eco-training in industrial companies.

Keywords: Sustainable development · Corporate social responsibility · Environmental responsibility · Non-financial reporting · Sustainability reporting · Environmental and social disclosure · Industrial companies · Global Reporting Initiative (GRI) · Market capitalization Russia

1 Introduction

For the last few decades, there has been a rethinking of corporate development goals: the company owners start realizing the need for renewing trust and confidence in business. Businesses need to create value for all their stakeholders: customers, employees, and suppliers. The interests of stakeholders need to be aligned so that decisions benefit multiple stakeholders, rather than one group at the expense of the other. This brings us to new thinking - "conscious capitalism" (Mackey and Sisodia 2014). Introducing the principles of conscious capitalism has not yet become a common practice. However, it is believed that the significant impact of conscious capitalism on business is forcing the shareholders to re-evaluate the market opportunities, reviewing traditional capitalist values, as well as using added benefits to increasing sustainability development of the society (Tkachenko 2018).

In the realities of the 21st century, corporate social responsibility (CSR) is treated as an integral part of the concept of corporate sustainability (Daizy and Das 2014). This is considered as one of the main tools to increase investment attractiveness, business

A. V. Bogoviz and Y. V. Ragulina (Eds.): ISCI 2019, LNNS 115, pp. 628–635, 2020.
https://doi.org/10.1007/978-3-030-40749-0_75

value, and market capitalization of companies and competitiveness, as well as to safeguard from non-commercial risks and achieve sustainable development in three main areas – environmental, economic, and social.

Owners of companies try to convey information about the various CSR aspects to all stakeholders. Social disclosure allows firms to make any information public in the form of reports. According to the KPMG report, for the period from 1993 to 2017, the level of CSR reporting among N100 and G250 companies increased from 12% to 75% and from 35% to 93%, respectively (KPMG 2017).

In the process of socio-economic development, along with the growth of population and industrial output, there is an increase in the negative impact on the environment. In 2017, the Organization for Economic Cooperation and Development (OECD) issued a "Green Growth Indicators" report that provides information on the results achieved by OECD and G20 countries since 1990 in the field of "green economy" (OECD 2017). Unlike the countries that have succeeded in the area of environment, Russia cannot yet boast of great achievements: socially responsible companies are still rare in the Russian business. However, it is worth noting that the Russian business community is beginning to understand the perspective of economic benefits resulting from implementing social and environmental projects. The 2018 collection of corporate practices ("Russian Business and Sustainable Development Goals") presents best ones among the 25 Russian companies engaged in various sectors of the economy, which reflects the goals and objectives of firms' sustainable development in relation to the sustainable development goals set by the UN until 2030 (Feoktistova et al. 2018).

What is sustainable development? The Global Reporting Initiative (GRI) defines this term as a model of the progressive development of society, which is achieved by meeting the current generation needs without depriving future generations of these opportunities (Global Reporting Initiative n.d.). This paper focuses on the environmental component of sustainable development and social disclosure in industrial companies. The authors discuss the effect of a firm's exposure to environmental policies on the performance of its stock and, consequently, its investment attractiveness. The paper analyzes the content of ecological reports and offers some guidelines on what can be done to engage more companies in providing environmental information. One of the proposed ways out is the environmental training model.

With advancing social responsibility (SR), a transition to the principles of "green economy" has generated considerable recent research interest in Russian academic studies (Uskova and Kopytova 2018). They believe that using non-financial reporting as an information base for environmental analysis will enable to measure an impact of the company's activities on the environment (Ilysheva and Baldescu 2014). However, in Russia, the transition is accompanied by numerous obstacles (Khutorova 2015) as the practice of describing the company impacts on society, in comparison with the traditional reporting on the critical financial and operational indicators, is rather new (Preuss and Barkemeyer 2011).

Criticizing the existing Russian practice of SR following the GRI standard, some researchers (Morozova 2013) highlight limitations of the environmental disclosure: the environmental expenditures are reported without a breakdown by types of activity or ecological components. Examining the relationship between environmental disclosure and CSR (Sorokina 2018) suggests linking reported environmental data with indicators of statistical reporting in the field of environmental protection. Other authors (Medvedeva et al. 2017) attempt to address the issue of environmental damage estimation. They argue whether there is a link between disclosure and company market value (Brooks and Oikonomou 2018). In their recent research, Kuzubov and Evdokimova (2017) analyze the sample drawn from the body of companies located in BRICS nations. They prove that the SR based on the GRI standard resulted in an increase in the company's market value within the year of the report release and a year after being released.

The primary type of report used by Russian organizations is a report on the triple bottom line, which includes information on the company's economic, environmental, and social performance (Glebova et al. 2013). Slightly less than 50% of companies have chosen only one reporting model – the GRI. About 20% of companies use two reporting models: as a rule, these might be the GRI combined with the industry-specific corporate reporting standards or with the Social Charter of the Russian Business or RSPP Reporting indices (Feoktistova et al. 2017). However, while presenting social and environmental information, firms try to show mainly the areas where the positive results were achieved. In turn, the problematic issues of the company are not disclosed. As a result, reports reflect an ideal situation which is presented in a form that meets stakeholder expectations.

2 Materials and Method

Using the "All-Russia Classifier of National Economy Sectors," we identified six industrial sectors, namely: oil and gas, energy, woodworking and pulp and paper, chemical and petrochemical, metallurgical and mining, as well as machine building. We analyzed the companies whose non-financial reports are available in the "National

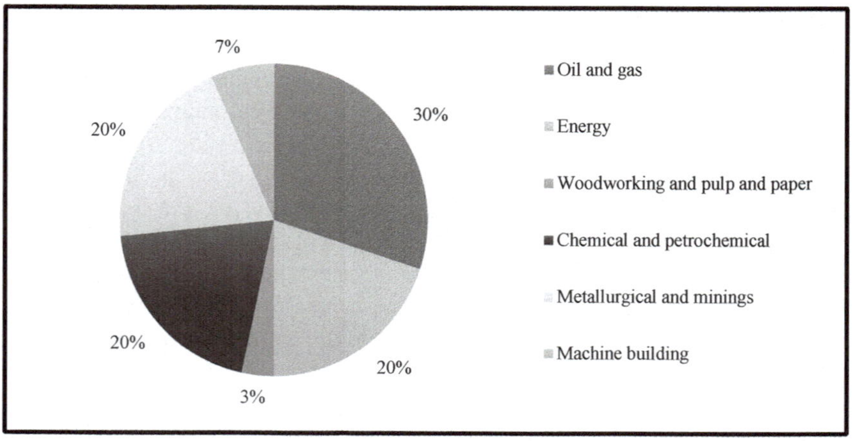

Fig. 1. Sample companies analyzed by the economic sector.

Register and Library of Corporate Non-Financial Reports" (Russian Union of Industrialists and Entrepreneurs 2019). As for possible sectorial differences, mining and processing enterprises are likely to report environmental information (Fifka and Pobizhan 2014). We selected 30 corporations. Figure 1 presents a breakdown of the analyzed companies by economic sector.

Of the analyzed companies, the oil and gas sector was represented by nine public companies: Gazprom, Gazprom Neft, Lukoil, Sakhalin Energy Investment Company, Zarubezhneft, Surgutneftegas, Transneft, Novatek, Rosneft. Six companies were analyzed in the energy sector: RAO UES, RusHydro, FGC UES, Concern Rosenergoatom, IDGC of Centre, and Rosatom. In the woodworking and pulp and paper industry, only one company entered the sample: Segezha Group. The chemical and petrochemical industry sample included six companies: EuroChem, Sibur Holding, Siberian Chemical Combine (SHK), Uralkali, Angarsk Electrochemical Combine, Ural Electrochemical Combine. The metallurgical and mining sector was represented by six companies: Rusal, Nornickel, Siberian Coal Energy Company, Severstal, Polymetal, and Metalloinvest. Two machinery building companies were Kovrov Mechanical Plant and SNIP.

3 Results

The sample companies' non-financial reporting was presented by the availability of environmental reports (10%), sustainability reports (60%); and integrated reports (30%). Of 30 analyzed companies, 27 complied with the GRI G4 Sustainability Reporting Guidelines.

In 12 out of 30 sample companies' non-financial reports mentioned the existence of eco-training/eco-management. Thus, the authors' hypothesis states that there is a relationship between a firm's environmental disclosure and environmental investment, including eco-education and a firm's stock market performance and capitalization.

All non-financial reports of the companies for the period from 2013 to 2017 were analyzed for the availability of the following categories:

1. Air protection
2. Water use and water-resources conservation
3. Waste management
4. Land and soil conservation
5. Energy saving
6. Renewable energy use
7. Biodiversity conservation.

Table 1. The most and least reported categories by industry.

Sector	Most reported category	Least reported category
Oil and gas	Air protection Biodiversity conservation	Renewable energy use
Energy	Waste management Land and soil conservation	Renewable energy use
Woodworking and pulp and paper	Air protection Waste management Energy saving	Water use and water-resources conservation Land and soil conservation Biodiversity conservation
Chemical and petrochemical	Air protection Water use and water-resources conservation Waste management	Renewable energy use Biodiversity conservation
Metallurgical and mining	Air protection Water use and water-resources conservation Waste management	Renewable energy use
Machine building	Air protection Energy saving	Renewable energy use Land and soil conservation Biodiversity conservation

The analysis of the data obtained allowed us to conclude that the most disclosed categories were air protection (in 28 out of 30 companies) and waste management (in 25 out of 30 companies). The least disclosed indicators were renewable energy use (in 10 out of 30 companies) and biodiversity conservation (in 16 out of 30 companies). At the same time, a complete disclosure was reported by oil and gas companies. Table 1 summarizes the most and least frequently reported indicators for individual companies by sector.

The authors attempted to analyze the impact of a firm's environmental policies on its capitalization. Due to the format of the article, we limited ourselves with the data related to five companies only: Gazprom Neft, Lukoil (in the oil and gas industry); FGC UES (in the energy sector); Uralkali (in the chemical and petrochemical industry); and Nornickel (in the metallurgical sector). These relationships are illustrated in Fig. 2.

Though four of five studied companies (Uralkali was an exception) did show a relatively positive relationship between firms' environmental policies and their stock market performance, this evidence was not significant.

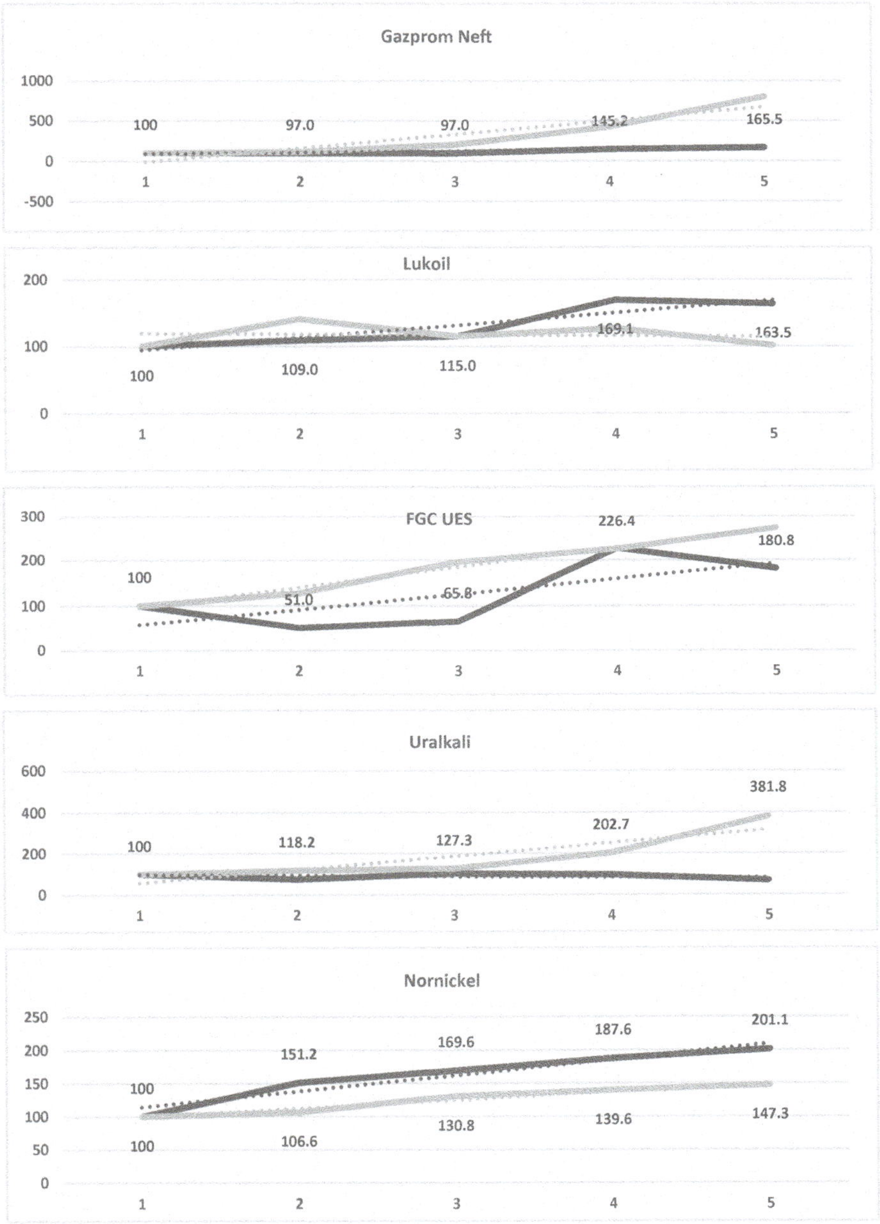

Fig. 2. Relationship between a firm's environmental policies on its capitalization.

4 Discussion

A limiting format of the article does not allow us to fully present the findings of the study and answer the question about a relationship between a firm's capitalization and the volume of its environmental investments. Along with a direct dependence, there is also an inverse relationship: investment in environmental-related activities depends on a firm's available financing, its stock market performance and investment attractiveness. This aspect can be identified as a potential venue for further research. In order to provide a real and considerable benefit, environment-related activities should be promoted and coordinated by professionals. In this regard, an issue of university eco-education and in-house corporate training is becoming of special importance. The difficulty in identifying the impact of individual types of environmental investment on a company's market value can be explained, among other reasons, by the fact that these environmental costs are not broken down by categories as it is not required by the GRI. This conclusion is in line with other research findings (Morozova 2013). The data for a detailed analysis may be obtained only through a field study. Clarification of the level and significance of individual investment items in environmental reporting offers new research opportunities.

5 Conclusion

The present study showed that there is no significant relationship between a firm's ecological disclosure and its stock market performance. At the moment, Russian industrial companies' level of investment in environmental protection activities is linked to their financial capabilities. The choice of a reporting standard is tailored to sector-specific expectations. In fact, effective implementation of measures that lead companies to environmentally responsible behaviour is more important for a "green economy" than a reporting format.

Acknowledgment. The research was carried out within the framework of the RFBR Project No. 18-310-00352 "Stakeholder Model of Environmental Public and Private Partnership as a Tool for Regional Sustainable Social Development".

References

Brooks, C., Oikonomou, I.: The effects of environmental, social, and governance disclosures and performance on firm value: a review of the literature in accounting and finance. Br. Account. Rev. **50**(1), 1–15 (2018)

Daizy, Das, N.: Sustainability reporting framework: comparative analysis of global reporting initiatives and Dow Jones sustainability index. Int. J. Sci. Environ. Technol. **3**(1), 55–66 (2014)

Feoktistova, E.N., Alenicheva, L.V., Dolgih, E.I., Kopylova, G.A., Ozeryanskaya, M.N., Honyakova, N.V.: Analytical review of corporate non-financial reports: 2015–2016 years of release. RSPP, Moscow, Russia (2017)

Feoktistova, E.N., Kopylova, G.A., Ozeryanskaya, M.N., Moskvina, M.V., Hofmann, N.I., Purtova, D.R.: Russian Business and the Goals of Sustainable Development. Compendium of Corporate Practices. RSPP, Moscow, Russia (2018)

Fifka, M.S., Pobizhan, M.: An institutional approach to corporate social responsibility in Russia. J. Clean. Prod. **82**, 192–201 (2014)

Glebova, I., Rodnyansky, D., Sadyrtdinov, R., Khabibrakhmanova, R., Yasnitskaya, Ya.: Evaluation of corporate social responsibility of Russian companies based on nonfinancial reporting. Middle-East J. Sci. Res. **13**, 143–148 (2013)

Global Reporting Initiative: Sustainability reporting (n.d.). https://www.globalreporting.org

Ilysheva, N.N., Baldescu, E.V.: Non-financial reporting as an information base for environmental analysis. Russ. Agrifood Policy **10**(34), 78–80 (2014)

Khutorova, N.A.: Green growth as a new policy direction for Russian economic development. Lesnoi Vestnik **1**, 190–198 (2015)

KPMG: Survey of corporate responsibility reporting (2017). https://home.kpmg.com/xx/en/home/campaigns/2017/10/survey-of-corporate-responsibility-reporting-2017.html

Kuzubov, S.A., Evdokimova, M.S.: Does the company value increase through the publication of non-financial reports under GRI guidelines? (On the example of brics countries). Account. Anal. Audit **2**, 28–36 (2017)

Mackey, J., Sisodia, R.: Conscious Capitalism, with a New Preface by the Authors: Liberating the Heroic Spirit of Business. Harvard Business Review Press, Boston (2014)

Medvedeva, O.E., Mikerin, G.I., Medvedev, P.V.: Integrated non-financial reporting in Russia and the issue of including the environmental component. Prop. Relat. RF **11**(194), 6–14 (2017)

Morozova, E.V.: Non-financial reporting: characteristics, social and environmental data disclosure. Econ. Sociol. Law **12**, 50–53 (2013)

OECD: Highlights: green growth indicators (2017). https://www.oecd.org/environment/indicators-modelling-outlooks/Highlights_Green_Growth_Indicators_2017.pdf

Preuss, L., Barkemeyer, R.: CSR priorities of emerging economy firms: is Russia a different shape of BRIC? Corp. Gov. **11**(4), 371–385 (2011)

Russian Union of Industrialists and Entrepreneurs: Russian business and sustainable development goals, corporate practices collection. RSPP, Moscow, Russia (2019)

Sorokina, Ye.M.: Non-financial reports of Russian companies: topicality and terms of compilation. Baikal Res. J. **9**(2), 2 (2018). http://brj-bguep.ru/reader/article.aspx?id=22101

Tkachenko, I.N.: On the way to “conscious” capitalism: from Marx to modernity. Econ. Reg. **14**(3), 725–739 (2018)

Uskova, T.V., Kopytova, E.D.: Ecologization of the modern Russian economy: problems and ways to solve them. Vestnik Permskogo Universiteta (Seria Ekonomika) **13**(1), 37–57 (2018)

Tax Regulation of Customs Payments in the State Policy of Russia

Elena Yu. Sidorova(✉) and Lubov I. Goncharenko

Financial University under the Government of the Russian Federation, Moscow, Russia
ejsidorova@yandex.ru, goncharenko@yandex.ru

Abstract. The purpose of the study is to determine the levels of state regulation of the economy in order to determine the place of its tax regulation in terms of customs payments. When writing the article, methods such as generalization, comparison, expert assessment and system analysis were used. The following three main developments can be distinguished as the main results: levels of state regulation of the economy with emphasis on tax regulation in terms of customs payments, determination of the main methods and tools in this field of knowledge. Thus, the author identifies four levels of state regulation, namely: Mega level "State regulation of the economy", macro level "State regulation of international trade", meso level "Tax regulation of the economy", micro level "Tax regulation of the economy in terms of customs payments". For the effective implementation of the fourth level, the author proposes to classify the methods of tax regulation in terms of customs payments into the following: administrative, integrated and economic; and tools, he suggests dividing into general, specific and other tools.

Keywords: Customs payments · State regulation of the economy · Tax regulation · VAT · Excise taxes · Customs duties

1 Introduction

Tax regulation of the economy has become an integral part of the general state policy of regulating the country's economy. Instruments, methods, forms of tax regulation of the economy are well studied and effectively applied in the Russian Federation. However, now, in the economic literature, the question is raised about the allocation of tax regulation in terms of customs payments in a separate branch of knowledge and independent study of these issues (Bogoviz et al. 2019). This is due to several circumstances, for example, to what extent the instruments of state regulation of international trade are applicable in tax regulation, to what extent all elements of customs payments in full can be used for tax regulation of the economy. And these are just a few methodological issues and questions.

A. V. Bogoviz and Y. V. Ragulina (Eds.): ISCI 2019, LNNS 115, pp. 636–642, 2020.
https://doi.org/10.1007/978-3-030-40749-0_76

2 Materials and Methods

Analysis and synthesis of specialized literature, publications in periodicals devoted to the conceptual provisions of tax regulation in the field of foreign economic activity, system analysis, comparison, are used in the article.

3 Results

In order to determine what tools can be used to implement tax regulation in the field of customs payments, we need to determine its place in the general system of state regulation of the economy. In our opinion, a tiered approach to tax regulation in terms of customs payments can be presented in the form of the following figure.

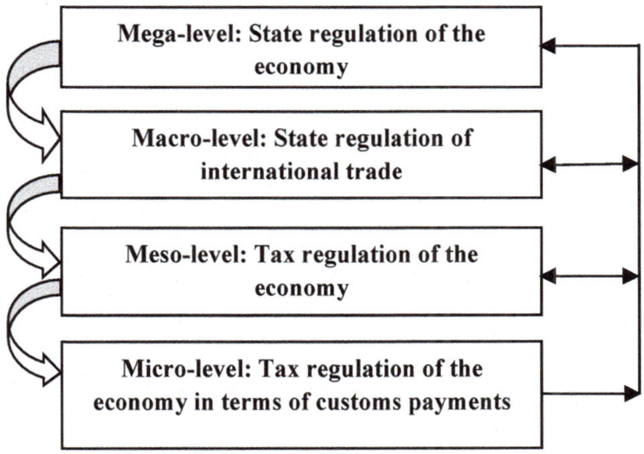

Fig. 1. The tiered approach to state regulation of the economy (developed by the author).

With this approach, the principle of consistency and coherence of the goals of state policy at all levels of its implementation is implemented. At the same time, systematically, the methods of tax regulation of customs payments can be represented in the form of three main groups: administrative, integrated and economic (Fig. 2). The difficulty in describing the issue of tax regulation tools in terms of customs payments lies in the choice of their universal classification. Methodologically, this issue has not yet been resolved, although many types of classifications that indirectly affect it have already been developed (Sidorova et al. 2018; Pozdnyakova et al. 2019). To begin with, we will consider some of them. At the first stage of the study, it is proposed to classify all instruments by their types into the following: tax rates, tax benefits, tax sanctions, formation of the tax base. And the author offers a second classification, which divides customs payments into the following types: general, specific and other tools.

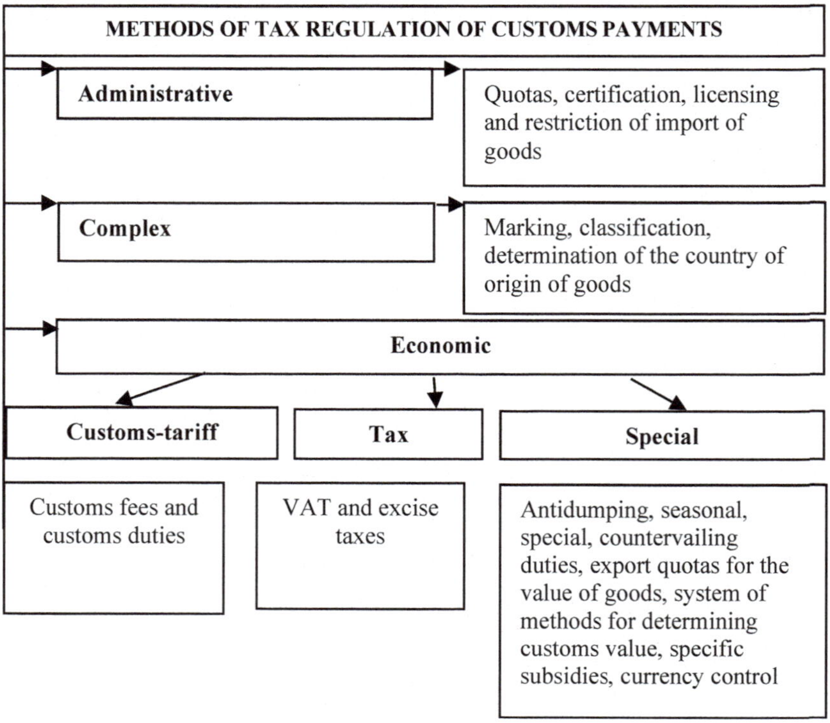

Fig. 2. Methods used in tax regulation of the economy of the part of customs payments.

4 Discussion

Now, we can discuss the results in more detail. Within the framework of the Mega-level "State regulation of the economy" (Fig. 1), the following are mainly distinguished: customs regulation (setting quotas, installment plans, the choice of methods and procedure for determining customs value), tax regulation (tax preferences, structure of the tax system, etc.), currency regulation (types of taxes and fees, principles and forms of their establishment, tax procedures, rights, duties and functions of tax authorities, distribution of taxes at the levels of the budget system).

In the framework of this study, the following two levels will be considered: "State regulation of international trade" (macro), "Tax regulation of the economy in terms of customs payments" (micro). We will consider these levels from the perspective of the tools that are used on them to achieve the goals of this level.

For the level "State regulation of international trade", we can offer the following scheme in terms of tools (Fig. 3).

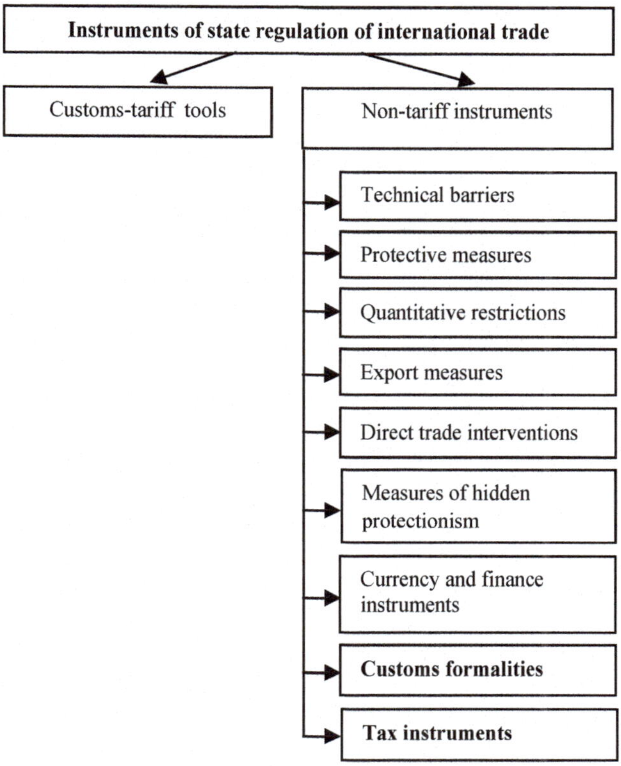

Fig. 3. Instruments of state regulation of international trade.

Moreover, in the framework of tax regulation regarding customs payments, only the following two non-tariff types of instruments are applied:

1. Customs formalities, which include the rules for determining the country of origin of goods;
2. Tax instruments, that is, indirect taxes, payments, and other para-tariff measures that increase the value of imports.

The definition of a system of tools at the micro-level "Tax regulation of the economy in terms of customs payments" is complicated by the fact that it is rather difficult to formulate the goals of the level. The objectives of tax regulation in terms of export and import operations are most correctly formulated. At the same time, the purpose of tax regulation of import operations can be defined as protecting the domestic market from foreign goods, which can compete with local producers by the price factor. They can also provide the domestic market with the necessary quantity of products that are not produced or produced in insufficient quantities within the country.

The goals of tax regulation of export operations include ensuring a rational ratio of import and export of goods, to ensure conditions for the integration of the national economy into the world economy (Sidorova 2015).

The next essential question relates to which of the customs payments are included in the tax regulation system. On this issue, there are two principal positions. Some economists believe that all types of customs payments participate in tax regulation, while others understand that only VAT, excises, and customs duties are involved in tax regulation. Customs duties do not participate in tax regulation (Sidorova and Tishonova 2017). The authors attribute this to the fact that customs duties can actually be considered as fees that compensate for the costs of the FCS of Russia, as well as traditional tax regulators are not used in their calculation. Thus, if we assume that customs payments perform a regulatory function, then this statement is true. Although there are many discussion questions, they are not the subject of research in this work. Therefore, we will not dwell on this.

In more detail, we will consider the second classification concerning the types of customs payments. The second classification divides customs payments into general, specific, and other instruments. Under the general tools of tax regulation in terms of customs payments, the author understands the mechanisms related to all types of customs payments and used in determining them. For example, this is the definition of the country of origin of goods, the formation of the customs value of goods, the declaration of products, the most favored nation treatment.

Under the specific tools of tax regulation in terms of customs payments, the author understands the traditional tax tools applicable to VAT, excises, customs duties, and taxes and having features.

By other instruments of tax regulation in terms of customs payments, the author understands other instruments used for regulatory purposes, but they are not presented in the above groups. For example, one may consider anti-dumping, seasonal, special, countervailing duties, as well as specific subsidies, export quotas for the value of goods.

And in conclusion, we would like to examine how the customs procedure can become an instrument of tax regulation of the economy in terms of indirect taxes paid as part of customs payments (Table 1).

Other tax regulation tools are used in the legislation, for example, the "Deferred" tax regulation tool. In this case, it is understood as granting a delay in the payment of customs payments when the goods are under a specific customs procedure. The most striking example of this tool is the customs procedure "Customs Warehouse." In the Kyoto Convention, "storage procedure at a customs warehouse" means a customs procedure according to which the imported goods are stored under customs control in a designated place (at a customs warehouse) without paying import duties and taxes (EUR-Lex n.d.). Thus, we can conclude that the two above formulations of the definition of the customs procedure and customs warehouse are identical from an economic point of view. Following Article 155 of the Customs Code of the EEU, goods must be stored in a customs warehouse that has such a status. Requirements for owners of a temporary storage warehouse are described in detail in Chapter 57 of the Customs Code of the EEU.

Table 1. Taxes paid in the implementation of individual customs procedures.

Size of the excise tax	Customs procedure	VAT amount
Excise tax not paid	Customs transit	VAT not paid
	Processing in the customs territory	
	Re-export	
	Free trade	
	Destruction	
	Denial in favor of the state	
	Free warehouse	
	Special customs procedure	
Excise tax not paid, or amounts paid are refunded	Export	
	Customs warehouse	
	Free customs zone	
Excise tax is not paid in full or in part	Temporary export	
Excise tax paid in full	Output for domestic consumption	VAT paid in full
	Processing for domestic consumption	
	Processing outside the customs territory	Full or partial exemption from VAT applies
The excise tax (from which it was released or returned) is paid	Reimport	The VAT is paid in the amount of tax from which it was exempted, or the amount that was returned to it in connection with the export of goods.

The economic sense for the importer of placing imported goods under the customs procedure of the customs warehouse is also to be closely considered. When the goods are actually imported into the customs territory of the union, the importer has legal grounds for receiving payments deferment from customs until he determines the final sale or use of the goods on the union territory. In this case, the importer is not to be forced to return customs payments if they are paid when these goods are placed under the customs procedure for domestic consumption (Interstate Council of the Eurasian Economic Community 2017).

The mechanism for the implementation of such an economic scheme is as follows. The importer imports the goods into the territory of the union and places it under the customs procedure of a customs warehouse, where it can be stored for a total of no more than three years (article 157 of the Customs Code of the EEU) (Interstate Council of the Eurasian Economic Community 2017).

An essential feature of the implementation of this customs procedure is the fact that when goods are in stock, they retain the status of foreign goods, and this leads to certain restrictions on their circulation in the domestic Union market. Further, as buyers are found and purchase and sale agreements are concluded with them, the importer sells the goods and accordingly changes the customs procedure for the goods being sold from the customs warehouse to release for domestic consumption. This entails the payment of customs payments and indirect taxes in their composition provided for this product.

5 Conclusion

Thus, in the article, the four levels of state regulation are highlighted in Fig. 1: mega-, macro-, meso-, micro levels, which interact among themselves to create a system of tax regulation of the economy in terms of customs payments. The author proposes the following classification of methods and tools for the implementation of the Micro-level "Tax regulation of the economy in terms of customs payments." The methods of tax regulation of customs payments can be represented in the form of three main groups: administrative, integrated, and economic (Fig. 2). At the first stage of the study, the author offers two classifications of tax regulation tools in terms of customs payments:

- By types: tax rates, tax benefits, tax sanctions and the formation of the tax base;
- In relation to the types of customs payments: general, specific and other instruments.

References

Bogoviz, A.V., Zhuravleva, T.A., Popkova, E.G., Zarudneva, A.I., Alpidovskaya, M.L.: Effective tax policy of the state: conceptual foundations and methodology of evaluation. Stud. Syst. Decis. Control **182**, 91–98 (2019)

EUR-Lex (n.d.): International Convention on the Simplification and Harmonization of Customs Procedures (1973). https://eur-lex.europa.eu/legal-content/EN/TXT/?uri=LEGISSUM%3Al06025

Interstate Council of the Eurasian Economic Community: Customs Code of the Eurasian Economic Union (April 11, 2017, Appendix No. 1 to the Treaty on the Customs Code of the Eurasian Economic Union) (2017). http://www.consultant.ru/document/cons_doc_LAW_94890/

Pozdnyakova, U.A., Bogoviz, A.V., Lobova, S.V., Ragulina, J.V., Popova, E.V.: The model of well-balanced taxation for overcoming the shadow economy in modern Russia. Stud. Syst. Decis. Control **182**, 207–215 (2019)

Sidorova, E.Y.: Improvement the methods of tax regulation mechanism effectiveness assessment in the Russian Federation. Econ. J. XXI **11–12**, 47–53 (2015)

Sidorova, E.Y., Tishonova, A.A.: Assessment of the fiscal effect of the tax reform options until 2019. Econ. J. XXI **3–4**, 25–32 (2017)

Sidorova, E.Y., Goncharenko, L.I., Artemev, A.A., Nazarova, N.A.: Dividend-based and interest-based schemes of minimization of customs value of goods in deals between related parties: Russian practice. Espacios **39**(14), 12–19 (2018)

Creating the Web Service of the Laboratory Information and Management System for Analytical Quality Control

Andrey V. Yaroslavtsev(✉) and Olesya V. Matukhina

Nizhnekamsk Institute of Chemical Technology, Nizhnekamsk, Russia
andryplayvl@gmail.com, ovmatukhina@gmail.com

Abstract. Automated systems and services are the foundation of the digital economy, which are gradually introduced in all areas of industry. One such system is a laboratory of information and management systems allowing to monitor the quality of products manufactured by enterprises. Working with such a laboratory information system requires to supplement and update the information stored in the database periodically. Due to a large amount of data, the process of updating the database can take a long time and become accompanied by input errors. The paper addresses this problem by reducing the transfer of information from the initiator of changes to the direct introduction of data into the system. Also, it develops a service that partially automates the administration of the information and management system. The developed web service is easy to implement due to an intuitive interface. A separately designed database excludes the possibility of user exposure to the work of the laboratory information system, which ensures information safety. This web service has high prospects for implementation in the enterprise due to the simplification of quality control and the lack of analogs in the software market.

Keywords: Software · Database · Automation · Processing · Service · Information-management system

1 Introduction

In our perspective, achieving a higher level of digitalization in the industrial sector of the economy requires introducing the automated systems at all levels of enterprise management. Today, there are various specialized solutions in the field of information collection and automation of processes related to quality management (Chursin et al. 2018; Tyulin et al. 2018). Quality management implies an impact on the production process to achieve specific characteristics of the products created by quality control, accompanied by laboratory-research and further decision-making on quality. Quality control is carried out by testing and analytical laboratories with modern equipment and highly qualified specialists. Laboratory information and management systems are the specialized software systems aimed at automating the work of the analytical laboratory (Tereshchenko and Yanin 2006). The LIMS (Laboratory Information Management System) is a foreign analog of the term (Nazarenko and Osipova 2005). Many enterprises use such solutions from the LabWare company as LIMS, allowing to obtain

A. V. Bogoviz and Y. V. Ragulina (Eds.): ISCI 2019, LNNS 115, pp. 643–649, 2020.
https://doi.org/10.1007/978-3-030-40749-0_77

reliable information on the test results and optimize the resulting information for its use for decision-making (Laboratory information system (LIMS) n.d.; *Flagship product LabWare's LIMS* n.d.)

The laboratory assistant daily enters into LIMS a lot of data on samples, each of which has its initial parameters specific to this sample only. One may distinguish the following parameters: an object of analysis, a place of selection, a point of the technological chain, studied parameters, sampling time, and special parameters stipulated by the research methodology or necessary laboratories (Tereshchenko and Yanin 2006). Therefore, to work in LabWare's LIMS, first of all, it is required to draw up templates for convenient and quick work when filling out the results, and reference books of test results for each controlled product. All these and other data are necessary for quality control of products and intermediates.

Quality control implies that the laboratories carry out various tests, the results of which are recorded in the laboratory system, where they are automatically checked against the standards set by the control plans. At the target enterprise, there are plans for analytical control (ACP) of intermediate products, commodity production, and finished products, drawn up by engineering technologist. ACP is a table containing a list of tests and their norms for each controlled product, product sampling points, component determination procedures, and a sampling schedule. All control plans are edited and filled in manually in a Microsoft Word document. After the management approves a document, it is sent to the LIMS administrators, who also manually enter the information into the system for further work on current data.

It becomes obvious that LIMS operates with a huge amount of static and dynamic data, which makes the work of laboratory workers easier when filling out test results. But with an increase in the amount of data, in the case of changes in any norms, standards (or when starting up new sections), the processing time of the information and the number of potential errors when entering changes manually increases. Taking into account the human factor, many errors can occur during the introduction of new data, made by both the technological staff (when making the adjusted norms in a document) and the LIMS administrators (in the process of making changes to the system). The longer the chain of entering new data, the higher the probability of distortion of information (for example, entering incorrect values, dimensions, skipping the necessary changes due to oversight, due to a large amount of data processed and so on). The LIMS administrator needs to compare the new edition of the analytical control plan with the previous one, find, delete, or add the necessary product and sampling point to the table, change, delete or add new tests in the product, make changes or add new controlled components and their norms. The tree-like display of data in the LabWare's LIMS complicates the work, forcing each parameter to be searched separately. Due to the lack of automation and a lot of manual labor required, making ACP changes to LIMS involves a lot of time and attention. Otherwise, the results of analyses can be incorrectly interpreted by a system.

From all of the above, we conclude on the need to develop software that would automate most of the work made by staff, upgrade current opportunities for maintaining the LIMS database, and facilitate the execution of tasks to make changes to the database.

2 Materials and Method

Imagine a system that allows users to interact with an existing database. The system should have a user-friendly, intuitive interface, preferably in the same format as the plans of control adopted on production. It is planned to create an additional database containing data entered by users and data uploaded from the LIMS database. Creating an intermediate database will protect software users from direct interaction with the existing database, thereby ensuring the safety of information and security in using the software. After editing, all changes should be sent to the administrator for further verification. If the administrator has questions to the user who has made the changes, he would be able to contact the user for additional clarifications before making changes to the central database. Then, after having approval by the administrator, all changes from the secondary database should be automatically made to the main LIMS database. This model of user interaction with the system allows minimizing the time of data updating by reducing the path of information from the initiator to the endpoint.

It was decided to create this system in the form of a web service using the ASP.NET MVC Framework which realizes the model-view-controller template – MVC (*ASP.NET MVC* n. d.).

The controller accepts requests, processes a user input, interacts with the model and view, and returns the result of processing the request to the user. The model is a layer that describes the logic of data organization in the application. The view receives data from the controller and generates elements of the user interface for display of information.

The model and controllers are implemented in the object-oriented C# programming language; the representation is implemented in the html markup language using JavaScript code. In particular, we used jQuery and Vue.js. to expand the possibilities of user interaction with the web page.

Vue.js is an open-source JavaScript framework for building web user interfaces. (*Vue.js* n.d.). Thanks to this framework, each element of the form has reactive properties, which allows the page to respond quickly to user actions, as well as dynamically create and update page data. Also, Vue.js makes it possible to split the entire project into separate components, with the ability to call each part separately to perform the operations laid down in it and transfer the resulting result back to the parent component.

jQuery is a JavaScript library that focuses on the interaction of JavaScript and HTML. The jQuery library helps to get easy access to any element of the object model of the document (Document Object Model, DOM), to address the attributes and contents of the DOM elements and to manipulate them. Also, the jQuery library provides a convenient API for working with AJAX (technology of accessing the server without reloading the page) (jQuery n.d.).

To communicate with the main and auxiliary database, the entity Framework is used. This is the object-oriented data access technology from Microsoft, representing the ability to interact with objects using Entity SQL (*ASP.NET Entity Framework* n.d.).

This set of frameworks and libraries allowed us to create a full-fledged web application that interacts with both databases. It allows having a possibility of maintaining registration these users of a system and exposure of the necessary rights to them.

3 Results

Using the Entity Framework, the Code First approach created an intermediate database (Fig. 1) relying on a pre-designed application data model.

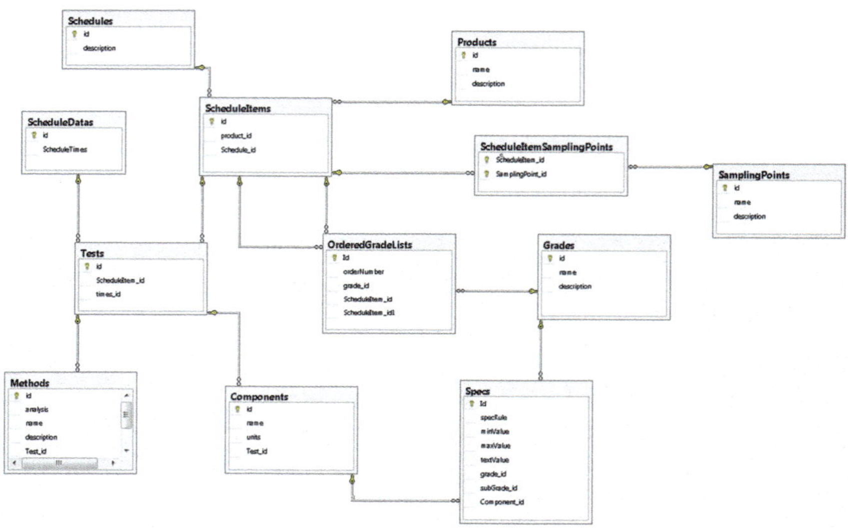

Fig. 1. A diagram of the intermediate database.

The view is implemented by the VUE.JS framework in the form of a table similar to ACP (Fig. 2), with the possibility of interactive data interaction. The entire project of the client part is divided into components with its own functionality, interacting with each other in accordance with their hierarchy. For example, a schedule cell component (a schedule editor component) is one of the scheduled type components (daily, weekly, monthly, yearly) (Fig. 3).

Наименование контролируемого продукта	Точка отбора	Наименование контролируемых показателей	Единица измерения	Норма контроля		Частота контроля	Часы отбора	Метод выполнения измерений	Лаборатория	Примечание
Тяжелая дизельная фракция	SC0001	Фракционный состав		**I вида**	**II вида**	1 число (янв.)	04:00	ASTM D 86		
		Начало кипения	C	н/м 5	н/м 80					
		Конец кипения	C		-2 — 4					
Бензин газовый стабильный	NA0002			**Бензин АИ-95-К5**	**Бензин АИ-92-К5**	1 раз в сутки	00:00	ГОСТ Р 51069 97		
		Плотность при 15 С	КГ/М3	н/б 54	не норм					

Fig. 2. A web interface.

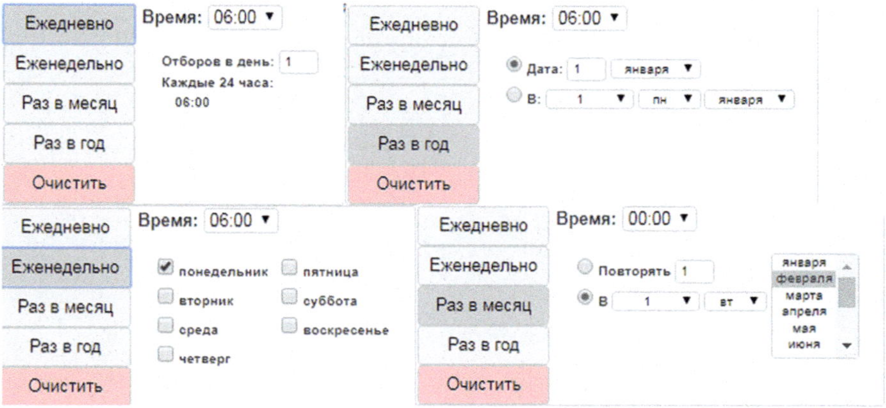

Fig. 3. A schedule editor.

Each individual parameter (product, analysis, component, specification, etc.) corresponds to a separate controller for communication of the server part of the project with the database. The division of the server part into different controllers was made to simplify the development and editing of the project.

When loading the main page, the user has a choice of a list of control plans. When choosing the necessary ACP on the page, the program creates a ready-to-use table with the data corresponding to the selected plan of control. The user can interact with the page, edit all tables, or add new data, depending on the rights s/he has. Those users without rights can only view this page.

A user who has the right to edit tables can interact with each cell using context or dialog boxes, allowing her to enter new data (Fig. 4) and select one or more items from the drop-down list of existing data in the LIMS database (Fig. 5). When loading a page or user request, a controller processes the received data, makes a request (if any) to the database and sends a response to the client station in the form of a JSON string, where it is processed and displayed in the form of a page.

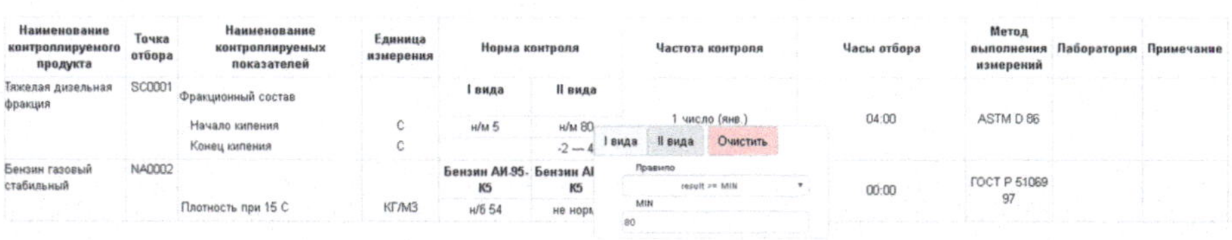

Fig. 4. The dropping-out window of specifications of a product.

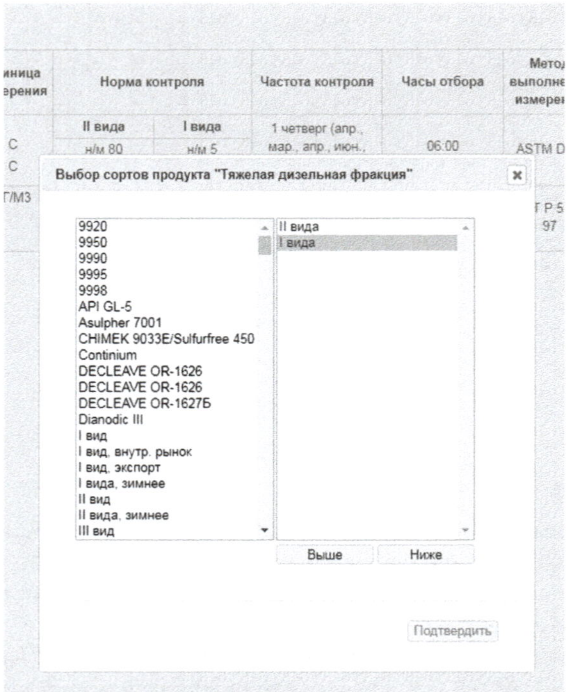

Fig. 5. Dialog box varieties of the product.

The software product is a web page, so there is no need to install additional software separately on each client station. The service does not require high power from the user station and is suitable for any computer with Windows XP and older. The Web service and intermediate database were introduced into the work on the existing server equipment, which did not require additional material investments during the implementation of the project.

4 Discussion

The developed web service is a convenient tool for working with the LIMS database, and already at the current stage of development, it shows its performance and copes with the task. The software product allows us to fill in the table ACP and to enter necessary changes to the intermediate database. At the moment, uploading data from the intermediate database to the target one directly from the form is not available. However, this feature will be available to the LIMS administrator from the web interface. In the future, it is planned to implement additional functionality, such as notifying the administrator about changes by corporate mail, generating a document for printing, and others.

5 Conclusion

The introduction of digital manufacturing management systems allows improving the quality of products. At the moment, many different enterprises, beginning from oil refineries to pharmaceuticals, use LabWare's solutions as a laboratory information system, with all its advantages and disadvantages. This web service does not have any ready-made analogues in the software market. Therefore this project can be considered a new solution that would dramatically reduce the time for updating production control plans, as well as the number of potential errors when entering information. If the project is successfully implemented, then this web service can be used in any enterprise that uses LabWare's LIMS due to the same database structure. This software product will be in demand because of ease integration and use, as well as due to the solution of many problems in the LIMS administration.

Acknowledgment. This work was financially supported by the Russian Foundation for Basic Research, Project No. 19-08-00261.

References

ASP.NET Entity Framework (n.d.). https://ru.wikipedia.org/wiki/Vue.js
ASP.NET MVC Framework (n.d.). https://ru.wikipedia.org/wiki/ASP.NET_MVC_Framework
Chursin, A.A., Kashirin, A.I., Strenalyuk, V.V., Ostrovskaya, A.A., Kokuytseva, T.V.: The approach to detection and application of the company's technological competences to form a business-model. In: IOP Conference Series: Materials Science and Engineering, vol. 312, p. 012003 (2018)
Flagship product LabWare's LIMS (n.d.). https://www.labware.com/ru/p/0401
jQuery (n.d.). https://ru.wikipedia.org/wiki/JQuery
Laboratory information system (LIMS) (n.d.). http://www.technolink.spb.ru/resheniya-po-avtomatizatsii/resheniya/148/
Nazarenko, G.I., Osipova, G.S.: Medical information systems. Theory and practice. FIZMATLIT, Moscow, Russia (2005)
Tereshchenko, A.G., Yanin, A.M.: Laboratory information systems in the domestic market. Ind. ACS Control. Mon. Sci. Technol. Prod. J. **7**, 12–16 (2006)
Tyulin, A., Chursin, A., Yudin, A.: Production capacity optimization in cases of a new business line launching in a company. Espacios **38**(62), 20 (2018)
Vue.js (n.d.). https://ru.wikipedia.org/wiki/Vue.js

Planning, Development, and Quality Systems of Helicopters Production in Russia

Alexey I. Tikhonov[1(✉)], Andrey A. Sazonov[1], and Andrey I. Boginsky[2]

[1] Moscow Aviation Institute (National Research University), Moscow, Russia
mai512hr@mail.ru, sazonovamati@yandex.ru
[2] Russian Helicopters, Moscow, Russia
1062182004@rudn.ru

Abstract. The paper focuses on the planning, development, and quality system of helicopter equipment in Russia. The introduction and mastering of the digital design and development technology are possible only with the participation of senior management, because the digital-based design affects basic business processes and enterprise structures. The authors note that in the process of designing aviation technology, the most important place is given to the layout existing as an independent section of the documentation at all technological stages of product development. The ERP-system, necessary for organizing the efficient production process, is considered in detail. The ERP-system is an enterprise management information system focused on balanced resource planning using specialized software. The key feature of the ERP system is the approach to organizing the interaction between the business processes and operations, regardless of their territorial or functional separation. The main design and preparation stages of the helicopter production, including the preliminary and conceptual design, and the production of the experimental helicopter for the ground and flight tests are considered. The development strategy of the corporate quality management system in the aviation industry is given, which is aimed at creating conditions for improving the quality of helicopters, its operational characteristics, reliability, and safety. In the conclusion, the authors consider the main areas to improve product quality and offer ways to solve the issues of quality assurance in the aviation industry.

Keywords: Aircraft engineering · ERP-system · Industry 4.0 · Corporate quality management system · Helicopter industry

1 Introduction

The Russian economy is currently undergoing a process of transition from an export-raw model to a fundamentally new innovative model of economic development, which is characterized not only by the formation of a balanced research sector but also by a sharp increase in business innovation activity and the dynamics of the emergence of innovative companies. Russian innovative and economic development is dependent upon the use of unique solutions and technologies (including advanced digital technologies available under the Industry 4.0 concept), as well as the application of

A. V. Bogoviz and Y. V. Ragulina (Eds.): ISCI 2019, LNNS 115, pp. 650–662, 2020.
https://doi.org/10.1007/978-3-030-40749-0_78

particular approaches that can significantly increase the growth of various researches and developments that will be interesting not only for the state but also for various business structures. The transition of the domestic enterprises to the Industry 4.0 concept will increase the energy efficiency and the economic competitiveness, blurring the boundaries between industries, and reduce the anthropogenic impact on the environment. The most promising areas for development should be digital design and modeling, new materials, additive technologies, industrial Internet, and robotics (Shu et al. 2012; Phylonin 2018).

2 Materials and Method

The methods of strategic planning and value forecasting in the global aviation products market are used as the research methods. The main research method is strategic planning focused on the analysis of the long-term development of the modern helicopter market. The study is based on the use of analytical and forecast materials published by such companies as British Helicopter Association, Airbus Helicopters (Eurocopter), Robinson, Bell Helicopter, Lufthansa, as well as the US Federal Aviation Administration.

3 Results

Helicopter engineering in Russia is one of the largest industries included in the structural composition of the aviation industry, which is characterized by rather stable dynamics and growth of economic indicators, the key of which is profit. According to estimates by leading Russian experts, profit in this sector in 2017 amounted to ₽ 1.24 trillion. In terms of production and export of helicopters, Russia is currently the third-highest in the world; supplies are made to several dozen countries. Approximately 40% of products are supplied to the domestic market. The helicopter industry in Russia, with a range of medium and heavy helicopters, occupies one of the leading places in the world market. In this class, the market competition is rather limited, and Russia has well-established and well-known products with favorable price characteristics. At the end of 2018, the total planned volume of investments for supporting the industry under the State Program of the Russian Federation "Development of the aviation industry for 2013–2025" amounted to more than ₽ 59,105,000,000. In 2018, 135 aircraft and 234 military and civil helicopters were delivered to customers. According to the management of the Ministry of Industry and Trade, the growth of industrial production in 2018 stopped at the level of 0.3% (Ministry of Industry and Trade of the Russian Federation 2016; Ministry of Industry and Trade of the Russian Federation 2013).

In the process of designing aviation equipment, the most important place is given to the layout, which exists as an independent section of the documentation at all stages of product development. Even when the helicopter is already mass-produced, the layout is still necessary for further product modifications. In the same way, it is important in computer design technology. The layout, the structural-power circuit, and the theoretical bypass are the "basic control structure," the conceptual image of the designed aviation product. Also, there is complete documentation on the machine in the form of a detailed electronic layout. In a perfect case, an associative link should be created between the aircraft's basic control structure and the detailed documentation in the electronic layout. Such technology is conventionally called "Electronic Layout Technology." If this can be achieved, then it turns out to be a very powerful tool for designing technology in the enterprise, as well as for its modernization and state tracking during the operation (Russoft 2017; Timofeev and Yagopolskij 2016).

In the implementation of digital design technology and the CCI in a single environment, there is another major factor. Such technology is effective only if it is paperless. This does not exclude drawings, sketches, technological maps, etc., which may appear as reports or working papers if necessary, but the main principle is that the original documentation is an electronic layout of the designed product at all development and mass production stages. The "working tool" is not the product specification but its structure. The specification, as well as the drawing, is reporting documentation only on a specific configuration. The product structure contains all the possible configurations, performances, and product modifications. Both the designer and the technologist must work with the product structure at all the creation stages. Thus, the changes are made only on an electronic layout, and the drawings, if needed, are printed again. It is impossible to make changes in the drawing, as it is not original (Russoft 2017; Pavlevich et al. 2018; Timofeev and Yagopolskij 2016).

To implement the approach of the design in a single environment, in the context of the product, domestic enterprises developed a rather complex organizational technology. It is essential not only to get an electronic mock-up of the product but also to make it an original document, which is very difficult to achieve. Otherwise, designers and then technologists will have to do lots of extra work. At the same time, there are no guarantees that in the engineering development and production processes, errors that occurred in the pre-computer era will not start emerging. After all, there can be only one original document, either an electronic layout or a paper carrier. Practice shows that electronic drawings are not an option, because, at some point, there is a temptation to make a change only in the drawing (Pavlevich et al. 2018).

The original design documentation in the electronic layout form must have specific properties, and, firstly, it must be of high quality. This means that all the components must be modeled appropriately and linked in the assembly context; in the PDM system, the product structure must be formed, and configuration rules should be written to describe all logically related versions of the product. It is important to understand that with such an approach, the labor intensity at the design stage in the general cycle of product creation increases. However, if achieving a state when the original is an electronic model is possible, then at the CCI and production stages, the timing and cost will pay off in full (Timofeev and Yagopolskij 2016).

Russian helicopter-building enterprises mastered the product design stage in the CAD (Computer-aided Design) system for the development of completely new and unique products. The CAD system is a computer-aided design system intended to perform design work using computer hardware, as well as allowing the creation of design and technological documentation for individual products, buildings, and structures. The middle- and top-level CAD systems are special programs for the 3D-modeling of products, calculations, and automation of the design of electrical, hydraulic, and other auxiliary systems. Data in such systems can be stored both in a conventional filing system and in a single environment of electronic document and data management (PDM and PLM systems) (Pavlevich et al. 2018; Sobolev 2016a, b; Timofeev and Yagopolskij 2016). Often in the middle-class systems, there are programs for the preparation of control programs for CNC machines (CAM-systems) and other programs for technological design. Figure 1 shows an example of a special program that allows creating complex models that contain a large number of components with complex geometry.

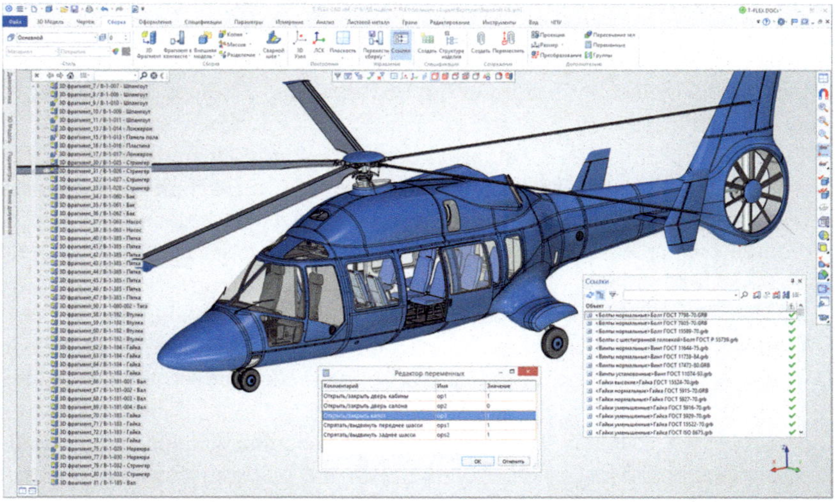

Fig. 1. An example of working with large assemblies.

The ERP system is used to build an effective system for planning and monitoring the production process from accepting an order for a helicopter to issuing a shift/daily task to a workshop worker. The ERP (Enterprise Resource Planning) is an information system that stores and processes most of the critical data for a company. Critical data is a list of data that makes the company's work possible. This data contains information about the work of the sales department and production (if the company is a manufacturer). Some enterprises use ERP primarily for production management since there are no better ways of producing the best solutions. Other enterprises that are not manufacturers (e.g., distributors) can also successfully apply the ERP. Distribution, personnel management, goods, and services sales are of critical importance for them. All ERP systems, regardless of their developer, combine a typical architecture, which can be described as follows (Biankina 2016; Dorokhov 2015; Sobolev 2016a, b).

1. *Platform*. Basic features and environment for modules and components. Only the developer can make changes to the platform code. Users and specialists do not have access to this software code. The platform includes (Borovkov et al. 2012):
   1.1. *Core*. The software environment in which the work is done, for which the users can write some add-ons and components.

1.2. *Basic functionality*. The list of the directories and functions, which makes the operating of the company possible. This is a user directory with access rights, a customer directory, a product/service directory, etc. This functionality is built into the platform; unlike modules, it cannot be disabled.

2. *Data management*. Database, including data storage and processing methods (interpretation). This category includes data storage on the server, software for working with databases (SQL or any alternative), tools for interpreting and processing data, and the process of sending data to software modules.
3. *Modules*. Components that connect to the platform as needed. They work with a single database and use basic functionality (as needed). The modules work independently of each other; they can "seamlessly" connect and disconnect without problems if the need for them has disappeared. Such a modular structure is an important distinguishing feature of the ERP systems. Modules are divided, in turn, into several types:
   3.1. *Internal use modules*. This level contains plug-ins that are used by the company employees. This includes warehouse management, manufacturing, accounting, CRM, etc. Modules can be connected, disconnected, and configured by implementation specialists. The standard set usually includes MRP, HR, CRM, Supply, and Procurement Management.
   3.2. *Modules of work with external users*. This layer contains the modules necessary for organizing direct interaction with various external users, potential and actual customers of the enterprise, partners, product users, suppliers, and customers.
   3.3. *Connectors*. Ready-made solutions for communication with third-party applications. Most often use API from the platform core. Allow telephony integration and set up the data exchange with the website or any software products and systems. Connectors are intended only for data exchange and are usually used with EDI, CMS, CAD, BI, OLAP, etc., i.e., with systems that are not part of EPR but are used in the enterprise.

In the aviation industry, the corporate quality management system's development strategy is aimed at creating conditions to improve the quality of the helicopter, its performance, reliability, and safety, and it includes the following steps (Dorokhov 2015; Borovkov et al. 2012; Tikhonov et al. 2018):

The First Step. The production is carried out by leading domestic enterprises (KVZ, U-UAZ, SMPP, AAK Progress, KumAPP, Rostvertol, Reductor-PM) and includes:

1. AS/EN 9100C certification. AS/EN 9100C standard includes requirements for ISO 9001: 2008 quality management systems and establishes additional requirements, definitions, and notes for the aviation, aerospace, and defense industries. The requirements outlined in this standard are additional (not alternative) to contractual and regulatory requirements. In case of a conflict between these standard requirements and the applicable legal or regulatory requirements, the latter should take precedence.
2. The introduction of automated control systems for product quality, test automation. All the test benches are based on measurement and control systems and include: stands based on the unbalancing mechanism for testing main and tail rotor blades

together with anti-icing system tests, which are similar in the operation principle; stands for testing helicopter stabilizers, and dust protection devices; stands for testing rotating and non-rotating parts of the swashplate; stands for testing leads and rotating stands for testing the swashplate assembly; and test stands for rotor sleeves that create loads simultaneously for all six degrees of freedom, test stands for hydraulic systems and helicopter control systems. The control system simulates the pilot work in all the helicopter operating modes (small gas, transitional, cruising, maneuverable) and the helicopter control from the autopilot.

3. Guaranteeing the production quality assurance. The general methodological principles for building a quality system for manufacturing and servicing airplanes are formulated in the requirements of ISO 9001: 2001, the Aviation Rules of the Interstate Aviation Committee, and the European "Procedures for the certification of aircraft and other aeronautical products and parts." The relevant sections are devoted to the practical aspects of the process approach implementation to the AT quality management. The management system directs and monitors the organization's activities in the quality field (DSTU ISO 9000: 2001). The quality management system (QMS) must be documented and organized to ensure that the requirements of the approved AT sample type design and the AT instance state meet the safe operation.

The Second Step. The design and development are carried out together by enterprises (Mil Moscow Helicopter Plant, Kamov) and includes: (Borovkov et al. 2012; Tikhonov et al. 2018)

1. Certification of the QMS according to the international standard AS/EN 9100. AS/EN 9100 certification is irrefutable proof of compliance with quality standards. Successful certification can help get new contracts in a competitive environment.
2. Comprehensive quality assurance at all stages of the helicopter's onboard equipment development. The AT quality management system regulatory framework includes legislative acts, documents entered into force by the CAA (aviation regulations, airworthiness directives, and other directive and regulatory documents), government regulations (DSTU, GOSTs, OSTs), industry-specific production instructions, supervising process materials, technical conditions, etc.

The Third Step. Procurement and supply chain management is carried out by the enterprise (Procurement and Logistics Center for the Helicopter Industry) and includes (Ministry of Industry and Trade of the Russian Federation 2013):

1. Development of various corporate standards for working with suppliers in the quality field. In the corporate AT standards development and implementation, the requirements of specific standards are taken into account. They are:
   - International standards, in the field of product quality management ISO 9000 series;
   - Unified system for design documentation (USDD);
   - Unified system for technological documentation (USTD);
   - Intergovernmental standards, systems for organizing the delivery of products to the manufacturer;

- Unified system for technological preparation of production (USTPP); and
- The state system for ensuring uniform measurement.

2. Providing quality assurance of primary and secondary information and materials throughout the supply chain. The quality management activities are regulated by the quality manual, methods, standards, organizational, and technical documents of the enterprise.

The Fourth Step. The after-sales service is carried out by a single complex of enterprises (VSK, NARZ, service centers) and includes:

1. The development organization of the unified regulatory quality service system according to the international requirements and the implementation of the international standards ISO series 9000, along with the Aviation Regulations. The implementation of these standards contributes to the improvement of the internal organization of production, as well as the processes of manufacture, maintenance, and restoration of AT. This has a positive impact on the security of the Armed Forces and the financial and economic activity of the enterprise.
2. The use of a single standard of quality service and repair throughout the network of service centers.

The product quality is estimated by a complex quality level indicator (QLI), which is the difference between the basic quality indicator (100) and the sum of complex and individual indicators calculated using the following formula (Russoft 2017; Borovkov et al. 2012):

$$QLI = 100 - SI = 100 - (PCI + \mathrm{IDPQ}\ ) \tag{1}$$

where QLI is the quality level indicator;

SI is the sum of complex and single quality indicators;

PCI is a comprehensive measure of the product quality by the operator, which characterizes customer satisfaction (determined based on individual quality indicators (number of accidents and incidents, claims, calls));

IDPQ is a comprehensive indicator of the decline in the quality of the products, which also characterizes the shortcomings and inconsistencies identified in the manufacturing of the products (determined based on individual indicators characterizing the finished product state, technological discipline, production culture, action plans implementation).

The comprehensive measure of the product quality level is calculated by the following formula (Russoft 2017; Borovkov et al. 2012):

$$PCI = \frac{(a_1 \times Nai_1 + a_2 \times Nai_2 + b_1 \times Nc_1 + b_2 \times Nc_2 + c_1 \times Ncl_1 + c_2 \times Ncl_2)}{0,01} \times N_{ip} \tag{2}$$

where the individual quality indicators characterizing customer satisfaction are:
Nai_1, Nai_2—the number of aviation incidents (including the enterprise);
Nc1, Nc2—the number of calls made by the operator to the enterprise due to product malfunctions (telegrams, letters) without a complaint (including the enterprise fault);
Ncl1, Ncl2 – the number of claims filed by the operator (including the enterprise fault);
a_1, a_2, b_1, b_2, c_1, c_2 – weight coefficients of product quality indicators ($a_1 = 7$, $a_2 = 50$, $b_1 = 0{,}5$, $b_2 = 10$, $c_1 = 1$, $c_2 = 20$ (points));
N_{ip} – the industrial personnel number (all company employees engaged in the production of the goods whose work is directly related to the main enterprise activity).

Measures taken for product development and quality improvement of the domestic helicopter manufacturing enterprises consist of introducing new technological equipment and processes and include the following areas (Russian Helicopters 2016; Sobolev 2016a, b; Phylonin 2018):

1. Technical re-equipment, mechanical processing, and aggregate-assembly production.
2. Reconstruction and technical re-equipment of blade production.
3. The creation of specialized production of parts for automatic helicopter skewing.
4. The mechanical processing modernization and the chemical-thermal production creation.

Let us consider the main problematic issues in the helicopter industry, related to the quality and delivery time of purchased integrated parts (Russian Helicopters 2016; Ministry of Industry and Trade of the Russian Federation 2016; Ministry of Industry and Trade of the Russian Federation 2013; Ministry of Transport of the Russian Federation 2014):

- Most suppliers are uncontested for manufacturers, as they are determined by the lead developer;
- Insufficiently effective work with the supplier companies in analyzing the suppliers' activities (compliance assessment with the technological process requirements, compliance with the state and military standards requirements for product quality), including joint analysis of unsatisfactory activities and the establishment of mutually beneficial cooperation conditions;
- The absence in contracts of the requirements for liability for the defective products supply, the supplier's liability for the deadline for recovering a defective product (no more than 20 days in accordance with GOST RV 15.703) and the supplier's penalty amount for the contract breaches;
- The outdated design of some applicable purchased integrated parts; lack of measures for their modernization and insufficient activity of the head developers for their replacement in the helicopter design;
- The lack of the necessary finances from the final manufacturers and suppliers for forming (at the operator's request) the exchange fund or technical first-aid kits, the purchased integrated parts, that often break down, the insufficient state role in

regulating this issue (there existed a system to order the integrated parts in the Ministry of Defense of Russia);

- Long terms and not always objective research of the noncurrent and advertised purchased integrated parts. GOST RV 15.703 has several shortcomings that it allows to PIP manufacturers:

- Not visiting the operating organization to study the PIP defects causes;
- To not send conditional PIP to the operating organization instead of failed parts (until their guilt is established);
- Researching the reasons for the PIP refusal without the consumer (operator, final producer) participation;
- Establishing the final cause of the defect and make claims to the alleged (in their opinion) perpetrators (including operators), without having the necessary evidentiary materials for this.

As a result, the reasons for the appearance of PIP defects are either not established or established after a long period. There is prolonged helicopter downtime in the disrepair. Measures connected with the defects of PIP are worked on extremely slowly.

The system of risk management of the domestic helicopter manufacturing enterprises is aimed at understanding and managing all risks through the system of the identification process, assessment, and control. At the corporate level, a risk base is being maintained that contains a description of the key strategic risks of enterprises, their descriptions, and plans and procedures necessary to manage these risks. At the level of each business process, risk management is carried out through the internal control system, enshrined in the regulatory documents governing the activities of the enterprise. The risk manager of the enterprise monitors the effectiveness of risk management. Let us consider the main risk types (Russian Helicopters 2016; Borovkov et al. 2012):

Risks Associated with the Changes in the Political and Economic Situation in Russia and the World. Adverse changes associated with the systemic risks in the global financial system (e.g., the risk of lower oil and gas prices) can lead to slower growth or recession in the Russian economy and adversely affect the capital availability and its cost for enterprises, as well as the enterprises' financial results, financial stability, and development prospects. In addition, it is necessary to take into account the sectoral sanctions of the United States and European Union imposed in 2014 against certain sectors of the Russian economy. The imposed sanctions, including the one toward Russian banks, may indirectly have a negative impact on the aviation enterprises' activities. In particular, the following consequences may occur:

- Restrictions on the supply of imported equipment and components;
- An increase in the borrowed resources cost due to the limited access to the foreign capital markets for state-owned banks, which will cause the Russian debt market volume to decrease;
- Difficulties with the passage of the payments in foreign currency (dollars, euros) through the US and EU banks;

- A potential ban on the use of SWIFT temporarily paralyzes all settlements using foreign currency with external counterparties, which may adversely affect the revenues received by enterprises, which, in turn, may entail the inability to cover current debt obligations (primarily contract financing).

In order to minimize the impact of restrictions on the supply of imported components and equipment on the activities of domestic aviation enterprises, measures are being taken to organize early deliveries of the required products in the required quantity (formation of reserves), as well as measures under the import substitution program.

Risks of the Helicopter Industry. The demand for helicopters largely depends on consumer countries' long-term budget priorities. Russian enterprises do not have the opportunity to influence them. Moreover, the helicopter market, especially in the civilian sector, is at risk of a significant technological breakthrough, including the related areas, which can lead to a drop in the demand for existing helicopter models, which will require substantial financial investments to create competitive technology. In this regard, the market position of helicopter-manufacturing enterprises in Russia is directly dependent on the following factors (Russian Helicopters 2016; Dorokhov 2015):

- The economic and political stability of the main consumer countries, including the long-term stability of their national internal security policies;
- The possibility of maintaining a leading position in the domestic market, including the timeliness and quality of the fulfillment of the commitments to ensure the state defense order of the Ministry of Defense as a key customer;
- The success of the holding's projects on the new helicopter models' development and the modernization of the existing ones for ensuring their competitiveness in the domestic and foreign markets.

Risks of Interaction with Suppliers. A significant proportion of materials and components for the helicopter's production is purchased from third-party organizations. Also, the key components are procured from suppliers who, in turn, are monopolists in their industry. In this regard, the aviation enterprises are directly dependent on the performance of key suppliers, including such factors as (Russian Helicopters 2016; Tikhonov et al. 2018):

- The financial stability of the suppliers, which directly affects the timeliness and quality of the fulfillment of contractual obligations for providing enterprises with the necessary components for the production of the helicopters;
- The ability to maintain a competitive price level through effective control over the prices of the industrial inputs.

Due to the current unstable geopolitical and macroeconomic situation, there is a risk of the Ukrainian PIP and material short supply, as well as of the PIP and materials produced by the US, the EU, and the states that have joined the sanctions.

Operational Risks. The customer focus is the main priority of Russian aviation enterprises. In this regard, for providing the client with the highest quality product at a reasonable price, enterprises are always working on analyzing and improving the main operational processes and increasing their efficiency. The key risk factors in this area are (Ministry of Industry and Trade of the Russian Federation 2013):

- The management of personnel effectiveness, including recruitment, training, retention and the development of unique competencies, as well as maintaining the optimal size and creation of a balanced organizational structure that allows and motivates staff to achieve the goals;
- The success of the processes of promotion and sales of helicopter technology, pricing, and contract management, as well as mechanisms for managing production costs to ensure a sustainable supply of helicopters and profit growth;
- The effectiveness of the existing production platform and the reliable management systems and IT support that support it, which ensure the most efficient allocation of resources throughout the entire production cycle;
- The effectiveness of the after-sales service system to provide customers with a reliable warranty and post-warranty service, as well as providing the necessary spare parts.

Financial Risks. The main financial risks include the availability of funds to meet business needs (liquidity, capital, and credit risks), as well as fluctuations in interest rates and foreign exchange rates.

Regulatory Framework and Compliance with Applicable Law. The activities of aviation enterprises are governed by various laws and regulations relating to the antimonopoly legislation, intellectual property protection, currency, customs legislation, and requirements for foreign trade operations, as well as the international accounting and reporting standards. Since the legal regulation of business activity in Russia is rapidly changing still and can be interpreted ambiguously, there is a risk of negative impacts of these changes on the activities of the enterprise, in particular, changes in the state defense order and state subsidies regulation, etc.

4 Discussion

Here are the main issues of the product quality assurance inherent in the Russian aviation industry (Dorokhov 2015).

1. Uncoordinated management systems, the quality of the AT management throughout the life cycle, and the lack of a single quality management system with the authority responsibility of all the participants' establishment.
2. The lack of industry and common industrial requirements for quality services, including quality assurance, organization of technical control with the participation of production, and control personnel.
3. The complexity of control operations is not taken into account when calculating the number of employees of the control apparatus. A lack of uniform standards for calculating the number of control apparatuses, including the introduction of high-performance equipment.

4. The insufficient provision of qualified personnel, including the employees of the Quality Control Department.
5. Insufficiently high motivation of the employees, including the Quality Control Department, for defect-free production.
6. The insufficiently effective control system of the primary and secondary input materials, leading to a decrease in consumer satisfaction for the domestic aviation equipment due to the frequent PIP failures and long demurrage. The PIP suppliers are not sufficiently interested in ensuring the helicopters' operability during the operation.
7. The absence of requirements for the presence of industry standards as a type in the federal law "On Technical Regulation" and in the draft of the federal law "On Standardization," which establish the basic requirements for the quality of products and their development, production, and operation, which is the basis for building systems' industry quality assurance and management.
8. The presence at the enterprises of fundamentally different methodologies for accounting and analysis (classification) of costs due to the defective goods and the absence of typical industry-specific methods for assessing the quality cost recommended by GOST RV 0015-002-2012 "SRPVT. QMS. General requirements."

5 Conclusion

According to the results of the analysis and identified problems in the manufactured products' quality, the following actions are proposed:

1. The optimization of terms (frequency) and quality reporting forms with the participation of specialized research institutes and the development of uniform quality assessment methods based on the experience of industry enterprises. The development of a system of industry leadership and quality management (or QMS) with the establishment of powers and responsibilities of all participants.
2. To update industry-standard requirements throughout the AT life cycle with the participation of relevant research institutes. Decide the level of corporations and holdings on the obligatory requirements of industry standards for the entire AT life cycle.
3. The development of a unified industry-wide methodology for recording and analyzing costs due to defects in the production process and at the guarantee operation stage, and recommendations for assessing quality costs based on the enterprises' experiences.
4. The development of an industry standard on the procedure for filing, reviewing claims, and determining the defects' causes in the guarantee equipment, and the development of effective measures. Reconsider GHOST RV 15.703-2005, taking into account the final product manufacturers' proposals for enhancing their role in ensuring customer and operator satisfaction.
5. The development of industrial control systems for accreditation and certification in the industry by QM integrated structure and large final product manufacturers (ICOP IAQG scheme). The usage of the International Aerospace Quality Group

experience on supplier management: the creation of an oversight system for certification procedures by a group of end-use manufacturers of aerial vehicles (aircraft, helicopter) and a database of approved suppliers.

Acknowledgment. The authors thank the management of the holding company "Russian Helicopters" for their great help and the valuable comments and corrections made.

References

JSC "Russian Helicopters": Analytical materials presented by the JSC "Russian Helicopters" (2016). http://russianhelicopters.su/upload/iblock/699/годовой%20отчет%202016%20BP.pdf

Biankina, A.O.: Digital technologies and their role in the modern economy. Econ. Soc.: Modern Models Dev. **16**, 15–25 (2016)

Borovkov, A.I., Burdakov, S.F., Klyavin, O.I.: Modern Engineering Education. Publishing House of the Polytechnic University, St. Petersburg (2012)

Dorokhov, S.B.: The quality state of the helicopters production and the problematic issues of ensuring the supplies quality and warranty implementation (2015). http://www.aviationunion.ru/Files/16_Dorohov.pdf

Ministry of Industry and Trade of the Russian Federation: State program "Development of the aviation industry for 2013–2025" (2013). http://www.minpom.gov.ru

Ministry of Industry and Trade of the Russian Federation: State program "Development of industry and increase of its competitiveness for the period up to 2020" (2016). http://www.minpom.gov.ru

Ministry of Transport of the Russian Federation: On amendments to the Air Code of the Russian Federation regarding the implementation of international standards in the field of aviation security: Explanatory note (2014). http://www.intrans.ru/documents/detail.php?ELEMENT_ID=19962

Pavlevich, A.L., Staroverov, N.N., Khitrykh, D.P.: An effective platform for applied research and comprehensive numerical modeling based on ANSYS solutions. Observer **4**, 70–75 (2018)

Phylonin, I.E.: Key problems of innovative development in Russia. Bus. Soc.: Electron. J. **4**(20) (2018). http://busines-society.ru/2018/num-4-20/4_filonin.pdf

Russoft: Prospects for Russian IT developments in the global market (2017). http://www.russoft.ru/files/ITexport.pdf

Shu, G., Anderl, R., Gauzemayer, Yu., Hompel, M., Valsetre, V.: Industry Maturity Index 4.0, the Acatech study: Management of digital companies' transformation (2012). http://www.i40mc.de

Sobolev, L.B.: Avionics in the modern aircraft industry. Economic analysis: theory and practice. Finance and Credit, Moscow, Russia (2016a)

Sobolev, L.B.: Restructuring of the military-industrial complex. Finance Credit **10**, 67–79 (2016b)

Tikhonov, A.I., Novikov, S.V., Sazonov, A.A.: Import substitution in Russia, as a backbone factor in the development of the aviation industry. Dvigatel' **3**, 6–8 (2018)

Timofeev, P.G., Yagopolskij, A.G.: The role and importance of PDM-systems in the development of process equipment. Proc. High. Educ. Inst.: Mach. Build. **10**, 73–81 (2016)

The Analysis of Foreign Planning, Development, and Quality Systems for the Production of Helicopter Technology in the World Market

Alexey I. Tikhonov[1(✉)], Andrey A. Sazonov[1], and Alexander A. Chursin[2]

[1] Moscow Aviation Institute (National Research University), Moscow, Russia
mai512hr@mail.ru, sazonovamati@yandex.ru
[2] Russian Space Systems, Moscow, Russia
achursin2008@yandex.ru

Abstract. The paper analyzes the planning system, development, and quality of the production of helicopters on the world market. The process of planning and developing helicopters is carried out in a PLM system (Product Lifecycle Management) by American and European enterprises. The authors of the paper note that the development of products and projects by foreign aviation enterprises, as a rule, is carried out using advanced electronic layout technology, with the original design documentation in the form of an electronic model. The structure of the PLM system, which is a part of the "Teamcenter" complex system, is considered in detail. It solves the most important tasks related to the organization of the enterprises management (including aviation), increasing the efficiency of applying product knowledge and stimulating the application of this knowledge to implement innovative solutions at all stages of the product life cycle. In order to assess the level of the availability of technology in developing various aviation projects, the TRL (Technology Readiness Level) technique is used. The authors note that the process of the development, production, and release of the advanced helicopters is directly related to the innovative technological triad, which includes technology readiness levels, production readiness levels, and technology integration levels in the current production process. In conclusion, the authors present an example of building an aviation project using the technology readiness level.

Keywords: Technological triad · Product life cycle system · World helicopter industry · Electronic modeling

1 Introduction

The world fleet of the civilian helicopters as of 2016 consisted of approximately 38,360 units, of which the largest share is occupied by the markets of North America, Europe, the Asia-Pacific region, and Latin America. The Russian civil helicopter park ranks 5th among the world's regions. In 2016, civilian helicopters were delivered to 70 countries,

A. V. Bogoviz and Y. V. Ragulina (Eds.): ISCI 2019, LNNS 115, pp. 663–674, 2020.
https://doi.org/10.1007/978-3-030-40749-0_79

in most significant numbers to the United States and China. Less than 50 machines were delivered in the rest of the country. In 2016, the civilian helicopter market continued to decline in supply to just around 810 new helicopters. A downward trend has been noted in Africa, the Middle East, Latin America and the Caribbean, North America, Russia, and other CIS countries. Growth was detected only in the European market. The markets of the Asia-Pacific region, China, and India remain stable. In terms of the take-off weight classes, moderate growth was noted in the light helicopter class (2–5 tons). The middle class (5–7 tons) is more or less stable. In other classes, a decrease in deliveries was recorded. In the super-heavy class of helicopters (weighing more than 16 tons), only one helicopter was delivered in 2016, whereas in the previous two years, no deliveries were registered. Following the prediction of the Forecast International (FI), the capacity of the world market for civil helicopters in 2017–2026 was estimated at $67.5 billion. The FI forecast was made only for new helicopters without modified and improved machines. In recent years, the emphasis of most military procurement programs has shifted in favor of the new helicopters, and part of the modernization programs has been transformed into new developments (MH-60R, UH-60M, CH-53K). The development and implementation of production technologies have dramatically reduced the cost difference between new and modernized models (JSC "Russian Helicopters" 2016; *A National Advanced Manufacturing Portal* 2017).

2 Materials and Methods

The methods of strategic planning and value forecasting in the global aviation products market were used as the research methods. The main research method was strategic planning, focused on the analysis of the long-term development prospects of the market for modern helicopter technology. The study was based on the use of the analytical and forecast materials published by companies such as British Helicopter Association, Airbus Helicopters (Eurocopter), Robinson, Bell Helicopter, and Lufthansa as well as the United States Federal Aviation Administration.

3 Results

Currently, the world military fleet (excluding Russia) has approximately 22,117 helicopters, where more than 74% are military transport helicopters. More than 50% of the military helicopter fleet is in the armed forces of the United States and Europe. Following Forecast International's prediction, the world market's capacity for military helicopters in 2017–2026 is approximately $138.9 billion. In 2014–2016, the helicopter fleet decreased slightly. The most significant decline in a park's number of helicopters was observed in North America (a 3% decline) and the CIS (2%). At the same time, industry experts are very cautious about forecasts for the military helicopter segment, and they rate them less optimistically than a few years ago. The retrofitting cycles for helicopter parks in North America and Europe have come to an end. The situation is exacerbated by many governments' efforts to cut military spending, especially in key markets. Some countries have faced budgetary difficulties. In the US, the trend is further aggravated by a drawdown caused by the Iraqi and Afghan conflicts. The defense money flow is depleting, and, as a result, a number of procurement programs have stretched out in time and significantly reduced in volume. The military orders volume will continue to decline until the end of existing programs' forecast period, except for a small number of newly emerging ones that will be able to keep the industry afloat (JSC "Russian Helicopters" 2016).

A return to sustainable growth in the medium term can only occur after 2030. A possible driver is the US program "FVL", which provides the development of a new generation of helicopter equipment. In accordance with the current plans, platforms of various sizes will be developed that can potentially generate contractual capabilities for many manufacturers. While the market is shrinking against the backdrop of tough competition, forcing the major producers to cooperate, new players from China and Korea have appeared. The global high-tech company Leonardo, one of the key players in the aerospace and defense sector, is in the difficult process of restructuring. Airbus Aircraft Company has begun this process two years ago and is in a better position relative to Leonardo now. Both companies have their lines of medium/heavy helicopters and are members of the NH Industries consortium that manufactures the multipurpose NH90 (*A National Advanced Manufacturing Portal* 2017; Curns 2005).

The development of products and various projects by foreign aviation enterprises, as a rule, is carried out using the advanced technology of electronic layout, with the original design documentation in the form of an electronic model. In the process of creating aviation products, the engineers and technologists actively use the product life cycle management system (PLM), which is a unified information environment that allows companies to create global information networks necessary for successful development and product manufacturing in a highly competitive market. Built to industry standards and having the application programming interfaces (APIs), the PLM systems minimize the costs of data conversion and increase the ability of various employees to participate in the enterprise. The PLM systems also help to ensure full transparency of decision-making procedures and work processes throughout all phases of the aviation product life cycle. The PLM opens up several unique opportunities for aviation enterprises (Curns 2005):

1. To maximize the use of innovative developments at all stages of the product life cycle, which helps to increase the company's turnover, increase its market share, accelerate the release of the product that is in demand on the market.
2. To transform enterprise processes to meet market needs.
3. To increase the product knowledge value through working with the information as an intellectual asset of an enterprise within various programs, projects, and initiatives aimed at generating income.
4. To minimize the design and production processes costs by replacing labor-intensive manual processes with accelerated, automated processes.

The leading aerospace and defense industry manufacturers use Siemens PLM Software for product life cycle management (PLM) solutions to increase productivity, reduce time to market, production costs and risks at every stage of the product life cycle, including concept development, part design, production, and logistics/support. The Siemens PLM Software offers Teamcenter, a system of supporting lifecycle that is already used by aircraft manufacturers in the United States and Europe. Teamcenter promotes innovations and increases productivity by providing the professionals of the enterprise with access to knowledge about products and processes to effectively complete assigned tasks in a distributed environment. It should be noted that the Mil Moscow Helicopter Plant uses Siemens PLM Software digital technologies to design and manufacture helicopter technology in Russia. The capabilities of the Teamcenter system in terms of working with the intellectual resources of an enterprise allow creating a single knowledgebase founded on the data about products, processes, production, and service. Since the accumulation, integration, and management of the information are carried out within a single PLM environment, the Teamcenter system allows using this knowledge as part of workflows that automatically synchronize aspects of the activities of teams of various profiles. Teamcenter allows developing helicopters within the concept of a digital project, as well as correcting errors in the early design stages and, therefore, there are no breakdowns during the testing (Curns 2005; US Department of Defense 2015a, b).

Teamcenter also allows implementing a unified and scalable visualization environment in a PLM enterprise environment. The working groups get the opportunity to visualize product data in both 2D and 3D formats, even in cases where this data is obtained from different development systems. The digital model of the helicopter dramatically facilitates the preparation of production, gives a full visual representation at all stages of the production process, provides the ability to track it, and eliminates the stages of transferring paper documents. 3D models also greatly facilitate the development of assembly equipment for an assembly plant. Machined and partially stamped parts, as well as tooling, are measured using digital control and measuring machines (Curns 2005; US Department of Defense 2015a, b).

The visualization functions that Teamcenter has are based on open JT technology, the standard for collaborative work in a distributed environment on 3D projects. It is also possible to realize several additional possibilities in the PLM environment, which can help the specialists working at different stages of the product life cycle (US Department of Energy 2004). These possibilities are as follows:

1. Full-featured visualization in 2D and basic 3D functionality to get an overview of the products at various life cycle stages
2. An extended selection of viewing and layout functions in 2D and 3D, providing additional options for viewing the product data
3. Full-featured analysis tools in 2D and 3D for joint analysis of the assemblies created using various CAD systems
4. Special features for creating digital mockups for assembling full-fledged digital prototypes and carrying out versatile analysis to analyze the shape, suitability, and functionality of the products.

Leading aerospace and defense companies are developing global networks to create innovative solutions that open up opportunities for continuous product and process improvement, as well as to optimize the value chain, ensuring innovation effectiveness. Global innovation networks enable partners, suppliers, and customers to connect through the creation of a common knowledge base of the products and processes. As a result, geographically distributed participants in the value chain have the opportunity to create innovations as a single enterprise. Such networks provide the performance of the following business tasks (Curns 2005; US Department of Defense 2015a, b):

Intensifying the Creation of Innovative Development. Global networks for creating innovative solutions accelerate the innovation process and expand innovation capacity, providing the ability to review new products and analyze their compliance with the consumer and design and technological requirements.

The Reduction of the Time for Entering the Market. Global innovation networks enable enterprises to analyze their existing processes and projects, and suppliers to evaluate the effectiveness of their product development and marketing.

Compliance with Management and Regulatory Requirements. By introducing compliance as an integral part of the product life cycle, global innovation networks reduce the nonconformity risk.

Resource Optimization. Global innovation networks provide the opportunity to reuse designs, processes, resources, best practices, experiences, and relationships with suppliers with maximum efficiency. This enables the value chain participants to respond to new industry initiatives quickly and cost-effectively.

Maximizing the Globalization Benefits. Combining geographically distributed participants in the value chain at all levels, global innovation networks enable enterprises to disperse design, production, and support geographically.

To assess the technology readiness level when developing various aviation projects by aviation enterprises, the Technology Readiness Level (TRL) technique is used. The assessment methodology was adopted and successfully tested by such major technology leaders as Airbus SAS and The Boeing Company. The strengths of the TRL methodology are not only an understanding of the overall project level but also the realization of the possibilities of supporting (at different stages) project financing and making decisions on the technology transfer. The classic TRL availability levels include (National Science and Technology Council 2016):

TRL 1. A detailed description of the technology: needs, requirements, properties, behavior.
TRL 2. Analysis of the selected architectural solution: used technology parameter, interface, critical elements.
TRL 3. Concept validation: feasibility and benefits. GO/NOGO solution 1.
TRL 4. The verification of the individual component, pre-integration, and detailed specification.
TRL 5. Basic integration demonstration.
TRL 6. Production of a fully functional prototype on the pilot line and the development of scaling effects during the production.

Technology readiness levels also help discuss what the company has, describe it according to the criteria, and show how far the technology has been able to move. The key benefits of using the TRL include:

- the ability to control risks;
- the technology assessment in terms of its development;
- the ability to monitor the amount of technology development at each stage.

With the help of the TRL, it is possible to describe in detail the technology not only in terms of its universality but also in terms of its demand, and the presence of a uniform methodology allows specialists of different levels to use unified approaches. Figure 1 shows technology readiness levels.

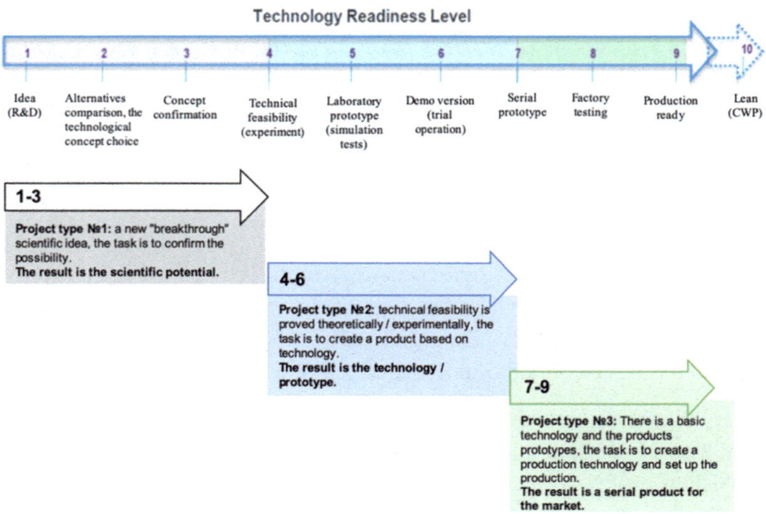

Fig. 1. Technology readiness level.

With the high complexity of the developed projects in the aviation industry in addition to the scale of the TRL level, a level of the manufacturing readiness is added. (Manufacturing Readiness Level, MRL). The description of the levels of the technological and manufacturing readiness (TRL/MRL) are given in Table 1 (The US Department of Energy 2004; National Science and Technology Council 2016; Systems Engineering Guide for Systems of Systems 2008).

For such technologies that are not developed by the companies such that the companies are not involved in the development process but consume the technologies in the form (as part) of finished products (services), a simplified scale is used for expert assessment of the stage (level) of the development (application), based on the fact that acquiring products and services gives access to these technologies:

0 – the technology is not applied in the company (the corresponding products/services are not purchased);
1 – the technology is at the stage of pilot implementation and testing (lots of trial products are purchased and tested; services are used for the first time at a pilot facility);
2 – the technology is applied in the company (the corresponding products/services are purchased systematically).

Table 1. Levels of technological and manufacturing readiness (TRL/MRL).

Level	Technology readiness levels, TRL	Level	Manufacturing readiness levels, MRL
1	2	3	4
TRL 1	**Basic concepts study** Basic research. Transition to the applied research and development has started	MRL 1	**Basic production introduction formation** Basic production concepts are defined at the theoretical level. Capacity assessment as per product requirements
TRL 2	**Technology concept formation** The technological concept and possibilities of its practical application are formulated	MRL 2	**Production concept determination** The production concept is defined according to the application scope. Production Line Design
TRL 3	**Analytical approbation of the concept** Development, including analytical studies and laboratory research, aimed at physical verification of the analytical hypotheses on individual elements of the technology	MRL 3	**Production concept verification** Experimental manufacturing processes are developed. The laboratory studies for design studies verification are produced
TRL 4	**Approbation of the model in the laboratory** The main technological components are integrated to verify that individual components can work together	MRL 4	**Production process in the laboratory** The possibility of manufacturing technical equipment (demonstration samples) in the laboratory has been achieved. Supply chain requirements have been identified
TRL 5	**Approbation of the technology components in conditions close to real ones** The main technological components are integrated with other suitable ("supporting") elements, and the technology can be tested in the simulated conditions	MRL 5	**Production process elements in vevo** Identification of the critical components and technologies is complete. Materials, tools, testing equipment, and staff competencies were verified. The value model was identified according to the value stream
TRL 6	**Demonstration of the prototype in conditions close to real ones** A representative model or prototype of the system, more advanced than the layouts	MRL 6	**Systems and subsystems prototypes in the presence of ready-made elements of the main production** Ability to manufacture a prototype system in the presence of the ready-made elements (primary production) is achieved

(*continued*)

Table 1. (*continued*)

Level	Technology readiness levels, TRL	Level	Manufacturing readiness levels, MRL
1	2	3	4
TRL 7	**Demonstration of the prototype in the operational conditions** The prototype reflects the planned staffing system or is close to it	MRL 7	**Production of systems, subsystems or their components in conditions close to real ones** The possibility of manufacturing systems, subsystems, or their components in conditions close to real. The vendor chain is evaluated
TRL 8	**Technology verification** The technology has been tested for operability in its final form and the expected operating conditions	MRL 8	**Testing a pilot production line** The necessary quality of production processes is proven. The supply chain has been established and is stable. The readiness to start full-scale production is reached
TRL 9	**Putting technology in production** The actual use of the technology in its final form and terms of the task	MRL 9	**Small batch production** The possibility of small-scale production has been successfully demonstrated t, the basis for the full-scale production is prepared. The cost model of full-scale production is justified
		MRL 10	**Full production** Full-scale production with subcontractors. Using lean manufacturing and quality management systems (QMS)

Source: According to the data provided by (US Department of Energy 2004; National Science and Technology Council 2016; *Systems engineering guide for systems of systems* 2008).

In the process of developing a system of designed technologies, the correct solution to the issue, which is related to the possibility of connecting the designed technologies within the boundaries of one system, is of particular importance. As part of the technologies' creation, it is necessary to determine the degree of their joint integration into a single system. For these purposes, aviation enterprises use integration readiness levels (IRL). Table 2 provides unified characteristics and additional descriptions of each of the nine IRL levels (*Systems engineering guide for systems of systems* 2008).

Table 2. A detailed description of the integration readiness levels.

IRL	IRL level value	A detailed description of the IRL level
1	2	3
IRL 1	The interface between technologies is defined with detail sufficient for the further interaction design	It is the lowest level of the readiness for integration at which the integration environment is selected
IRL 2	The specification characterizing interactions (ability to exert influence) between technologies through the interface has been determined	After defining the integration environment, such a method should be chosen that the two integrable technologies can influence each other through the selected environment. The integration concept is established at this stage
IRL 3	The compatibility (common language) technology, allowing to ensure an orderly and efficient integration and interaction, is achieved	A minimum level required for ensuring successful integration. Two technologies are able not only to influence one another but also to transmit the interpreted data. This is the first real level of maturity in the integration process
IRL 4	The interaction quality is achieved, and the integration between technologies is guaranteed	Many technology integration processes have failed at the IRL 3 level due to the assumption that if the two technologies can successfully exchange information, then they are fully integrated. IRL 4 goes beyond simple data exchange and requires that the received data match the sent data, and there is a mechanism to verify this
IRL 5	An adequate level of technology control is achieved to establish, maintain, and terminate the interaction	The ability of one or more integrable technologies to independently manage the integration (to establish, maintain, and terminate interaction) is indicated
IRL 6	Integrated technologies can accept, transform, and structure information according to their intended purpose	The highest technical level possible to achieve. It includes the ability not only to manage the integration but also to determine what information to share, tags that define what kind of information, the ability to translate data from external to internal
IRL 7	Technology integration has been verified and validated with enough detail to use	It represents a significant step compared to IRL 6, integration works not only from a technical point of view but also from the point of view of the requirements. IIRL 7 confirms that the integration meets the requirements for the performance, throughput, and reliability

(*continued*)

Table 2. (*continued*)

IRL	IRL level value	A detailed description of the IRL level
1	2	3
IRL 8	The real integration is completed and checked by testing and demonstration in the system	It represents not only the compliance of the integration requirements but also a demonstration as a part of the system in the appropriate environment. It allows identifying any unknown errors/defects that cannot be detected until the interaction of the two integrable technologies is checked as a part of the system
IRL 9	The integration possibility has been tested in the application	Technologies are used in the system. In order for the technology to reach IRL 9, it must be integrated into the system and then tested in the appropriate environment. The transition to the IRL 9 also affects the achievement of the technology maturity level of IRL 9

Source: According to (Systems engineering guide for systems of systems 2008).

Integrated systems may contain components with different levels of availability, affecting the assembly state as a whole. In such cases, when using readiness levels, we can highlight some areas and set their priorities for achieving the greatest success. The strengths of the TRL methodology are not only the understanding of the general level of the project but also the realization of the possibility of supporting (at various stages) the financing of the project and the possibility of making decisions on technology transfer.

Let us say an aviation enterprise develops a project that has an associated TRL value in the range from 1 to 9. The TRL column is calculated using Formula 1 and consists of one column containing the TRL values for each technology in the system; i.e., TRL_i is the TRL value for technology *i.* The TRL column is as follows (Curns 2005, Systems Engineering Guide for Systems of Systems 2008):

$$[TRL]_{nx1} = \begin{bmatrix} TRL_1 \\ TRL_2 \\ \cdots \\ \cdots \\ TRL_n \end{bmatrix} \qquad (1)$$

In order to calculate the IRL, it is necessary to build a technology integration matrix to determine the interaction degree between various technologies. The IRL matrix shows how technologies are integrated with each other from the perspective of a common system. For a system of N technologies, IRL_{ij} is the IRL between technologies I and j. The technology integration matrix is as follows (Curns 2005; National Science and Technology Council 2016):

$$[IRL]_{nxn} = \begin{vmatrix} IRL_{11} & IRL_{12} & \ldots & IRL_{1n} \\ IRL_{21} & IRL_{22} & \ldots & IRL_{2n} \\ \ldots & \ldots & \ldots & \ldots \\ IRL_{n1} & IRL_{n2} & \ldots & IRL_{nn} \end{vmatrix} \quad (2)$$

The assessment of the project technological readiness degree is carried out according to the following formulas in the TRL method (The US Department of Defense 2015a, b):

$$TRL_i = i - 1 + \prod_{k=1}^{K_i} A_{ik} \quad (3)$$

provided that:

$$A_{ik} = \begin{cases} 0, & \text{if not done} \\ 1, & \text{if done} \end{cases}$$

where, i – TRL number, $i = \overline{1, 9}$;
TRL_i – the numerical value of the ith level TRL;
A_{ik} – the numerical value of the kth indicator included in the ith level TRL;
k – TRL index number included in the ith level of TRL, $k = \overline{1, K_i}$;
K_i – the number of indicators in the ith level of TRL.

4 Discussion

Given the time frame, readiness levels help to reflect the direction of development or the timing of the next generation of technologies. Availability levels also offer the ability to evaluate complex systems at a high level, such as the electrification of helicopters, and focus on key components, while IRLs allows evaluating infrastructure integration.

5 Conclusion

The techniques discussed in the paper are innovative and are actively used by foreign enterprises operating in the aviation industry. There is a necessity in constant updates for the used software and active development of the additive technologies for effective use of the presented techniques by the aviation enterprises. In general, a detailed study of the US experience can essentially help European countries with the technical expertise at various stages of the life cycle of development, production, complex, and operation of high-tech systems, such as aircraft systems. The technical examinations, such as the assessment of the technology readiness level and the assessment of integration levels, are common international practice. This expertise is systems engineering tools that provide an opportunity to assess the progress and maturity of a technology, as well as how this technology moves through various phases and processes of the life cycle.

Acknowledgment. The authors thank the management of the holding company "Russian Helicopters" for their great help and the valuable comments and corrections made.

References

Analytical materials presented by the JSC "Russian Helicopters" (2016). http://russianhelicopters.su/upload/iblock/699/годовой%20отчет%202016%20BP.pdf
A National Advanced Manufacturing Portal (2017). http://www.manufacturing.gov
Curns, T.: Improving ISV applications to advance HPC (2005). http://www.hpcwire.com
National Science and Technology Council (2016). http://www.mgi.gov
Systems engineering guide for systems of systems (2008). http://www.acq.osd.mil/se/docs/SE-Guide-for-SoS.pdf
US Department of Defense: Department of Defense directive 5000.01 (2015a). http://www.dtic.mil/whs/directives/corres/pdf/500001p.pdf
US Department of Defense: Department of Defense instruction 5000.02 (2015b). http://www.navysbir.com/docs/500002p.pdf
US Department of Energy: Department of Energy High-End Computing Revitalization Act (2004). http://www.nsf.gov

Digitalization in Small Business Tax Administration

Diana Y. Boboshko(✉)

National University of Science and Technology "MISiS", Moscow, Russia
dboboshko@gmail.com

Abstract. Digitalization of the tax administration is an integral part of the implementation of the national program "Digital Economy of the Russian Federation." The purpose of the study is to analyze the digitalization of the tax administration of small businesses, the level of involvement and coverage of electronic services by small and medium-sized enterprises (SMEs) in interaction with the tax authorities, as well as information about the security of this process. The analysis showed that the identified areas of digitalization of the tax administration of small businesses are in the process of development and require additional efforts for their distribution. The prospects for the development of a new experimental taxation regime based on the "occupational income tax" implemented, based on the *My Tax* mobile application, are considered separately. Suggestions are made for improving the security of using electronic documents and obtaining electronic digital signatures (EDS) by amending the law "on electronic digital signature."

Keywords: Digitalization · Tax administration · Small business · Electronic document management · Electronic digital signature · Occupational income tax

1 Introduction

Russia is currently implementing a national program, "Digital Economy of the Russian Federation," which has the main task of developing the digital economy from all sources based on the creation of stable and secure information and telecommunication infrastructure for high-speed transmission, processing, and large amounts of data storage. The implementation of the national project includes activities in such areas as the normative regulation of the digital environment, personnel of the digital economy, digital technologies, IT infrastructure, information security, and digital public administration (Ministry of Finance of Russia 2018). In this context, there is a large-scale digitalization of the tax administration and the introduction of digital technologies into the interactions of taxpayers and tax authorities.

In recent years, the Federal Tax Service has been actively creating an information system for digitalizing the tax administration and integrating all sources of information and data flows into it, a system which includes analysis of this data that is based on the modern technologies used for processing large amounts of information. However, it is still too early to talk about a single information space where the processes of data transmission, processing, and storage can be implemented and made available to the

A. V. Bogoviz and Y. V. Ragulina (Eds.): ISCI 2019, LNNS 115, pp. 675–683, 2020.
https://doi.org/10.1007/978-3-030-40749-0_80

appropriate extent for all organizations and households. Neither can it be fully involved in the process of fulfilling the functions of taxpayers, tax agents, and participants in foreign economic activities.

At the meeting of the Community Advisory Group on Digital Transformation of Tax Authorities in September 2018, as part of the transformation of the Organization for Economic Co-operation and Development (OECD) Forum on Tax Administration, a new work plan for 2019–2020 was presented, where the digital transformation has the essential role. Experts at the forum noted that tax administrations are becoming more and more like leading digital companies working with data arrays, paying attention to the interaction with clients, structure flexibility, as well as to the acquisition and development of new competencies (Federal Tax Service of Russia 2018). A project was adopted to create a global strategy for the digital development of tax authorities until 2030 and to form a vision of the future for the tax administration. "In modern conditions, the tax administrations have no choice but to create their digital platforms which will be the basis for the formation of the ecosystem for taxpayers and participants in foreign economic activity" (Mishustin 2018). In the formation of such an ecosystem by the government bodies, a particularly important element is the feedback from all the participants of the interaction. It is important to understand how implemented technologies are effective not only in terms of the tax collection but also in terms of business, especially by small enterprises. However, the researches in this area are primarily focused on the analysis of the impact of digitalization on the efficiency of the tax administrations (ICAEW 2019; Manturova 2018; Migacheva 2018; Grudina 2018; Smirnova 2017).

This paper focuses on the challenges that small businesses face in realizing their duties as taxpayers in the digital economy. This category of tax relations subjects does not always have sufficient resources for effective inclusion in the digital ecosystem. Often, the limitation here is the level of financial and digital literacy, which remains a limiting factor for the effective interaction of tax authorities and SMEs.

2 Materials and Methods

In the course of the study, the authors analyzed legislation as well as analytical materials, published studies, and statistical reports in the field of digitalization of tax administration. The experience of working in small business and as an individual entrepreneur was summarized. The surveys and the consultations of individual entrepreneurs on tax issues were conducted.

3 Results

The following areas were identified as the main directions for analyzing the digitalization of tax administration of small businesses:

(1) the submission of reports to the tax authorities in electronic form;
(2) the use of the cash register equipment (CRE);

(3) the use of the mobile app *My Tax* for the self-employed.

1. *The submission of reports to the tax authorities in electronic form.*
   In 2019, those categories of taxpayers that meet the following criteria should submit declarations in electronic form via telecommunication channels (TCS) to the tax inspectorate:

- the organization classified as the largest taxpayer;
- organizations with the average number of employees (for the previous year) exceeding 100 people;
- VAT payers; all VAT payers are required to submit a VAT return in electronic form (the obligation was introduced on January 1, 2015, when the new version of paragraph 5 of Article 174 of the Tax Code of the Russian Federation entered into force);
- organizations where 25 or more people received income for the year; Reporting on personal income tax in electronic form is mandatory for these organizations. In this case, there are two reports: 2-personal income tax (annual) and 6-personal income tax (quarterly);
- all employers with 25 employees or more; Reporting to the Pension Fund of the Russian Federation and the Social Insurance Fund of the Russian Federation are required to be submitted in electronic form only for all employers with 25 employees or more (State Duma of the Russian Federation 1998).

Thus, the obligation for small businesses to submit reports electronically has not been established. However, the exchange of information with the tax authorities in electronic format has several advantages. That is why small and micro-enterprises use it voluntarily. The advantages of electronic document management for taxpayers include the following:

- There is no need to come to the tax office, as the reports can be sent from the taxpayer's office at any time of the day (saving time costs);
- The reduction in the number of technical errors (the reports are formed in the approved format with the use of the output control means, through which the correctness of the reporting forms is checked);
- The efficiency of updating reporting formats (when changing tax and/or accounting reports or introducing new reporting forms before the reporting deadline, the taxpayer will be able to update electronic versions of the forms);
- Confirmation of documents delivery (the ability to receive the confirmation of obligations fulfillment by the taxpayer within 24 h, as well as to see the personal card online so that there is no need to wait for the reconciliation acts);
- No duplication of paper documents is required;
- It is possible to obtain the following papers in the electronic form: a statement of the state of settlements with the budget; statements of transactions with the budget; a list of tax and accounting reports submitted to the tax office; an act of reconciliation of calculations for taxes, fees, penalties, and fines; current clarifications of the Federal Tax Service of Russia on tax legislation, as well as to send an informational request to the tax authorities (Sidorova and Boboshko 2019).

The reporting in electronic form is carried out either through electronic document management operators or through a personal taxpayer account on the website of the Federal Tax Service. In both cases, it is necessary to issue an electronic digital signature (EDS), the procedure is governed by the Federal Law "On Electronic Signature" (State Duma of the Russian Federation 2011). An analysis of the legislation showed that for obtaining an EDS, the original or notarized documents are not required; the copies of documents certified by the organization's stamp are enough. The degree of the authenticity of copies of documents, guaranteed by the organization's stamp, can be understood by analyzing the market proposals for the manufacture of stamps. A typical announcement about the manufacture of a stamp looks like: "The most accurate and urgent production of seals and stamps of any complexity without documents. The specimen of the stamp (an exact copy of the stamp on paper) can be made in just an hour. Free delivery to any subway station!" (Production of stamps without documents n.d.). Thus, the procedure for obtaining EDS contradicts the stated in the National Project "Digital Economy of the Russian Federation" goal of increasing information security.

Recently, cases of fraud with electronic signatures related to tax reporting have become more frequent. An electronic digital signature is issued on fake copies of documents for the existing organization, and the updated declarations, which indicate the amount of taxes payable, are submitted for this enterprise. As a rule, this applies to VAT and income tax declarations, which allows dishonest taxpayers to create a bona fide counterparty through which they can falsify documents, imposing the obligation to pay taxes on them and thereby regulating their tax burden.

An analysis of a specific situation showed that it is challenging to correct system errors within digitalization. It is impossible to exclude from the database a properly filed declaration via telecommunication channels. According to the procedure, another revised declaration must be submitted, but it should be done within three years, and fake reporting was submitted a few days before this period expired. The investigation of this situation is not complete, and the practice has not been worked out. In the process of solving the problem, however, it turned out that this is not an isolated case.

It should be noted that small businesses do not have sufficient resources to conduct tax disputes with the IFTS, including through the court; therefore, to involve such organizations and individual entrepreneurs in electronic document management, it is necessary to increase the security of work with EDS.

2. *The use of the cash register equipment (CRE)*

   An important step in the digitalization of the tax administration of small businesses is a new procedure for the use of cash registers introduced for more effective control of cash turnover. As of February 1, 2017, the cash registers must send the electronic versions of receipts to the fiscal data operator as well as to the customers. In this case, the protected electronic control tape (PECT) in the CRE has been replaced by a fiscal drive. The cash register should now be connected to the Internet. A service contract should be concluded with the fiscal data operator. Only after this can the CRE be registered with the Federal Tax Service and send receipts electronically. If the cash register does not have a data transfer function, it cannot be used in most cases. The list of such cases has expanded significantly since July 1, 2018.

Firstly, amendments to Law No. 54-FZ (State Duma of the Russian Federation 2003) made it mandatory for the organizations and individual entrepreneurs to issue all non-cash payments with any payment method by a fiscal receipt. Non-cash payments include payments with

- cash electronic money,
- a payment order,
- a credit card,
- an online bank transfer, and
- a receipt through the bank (which requires the business to incur additional costs for updating the CRE).

Secondly, since July 1, 2018, the list of small businesses obligated to apply the CRE has expanded. It has grown to include organizations and individual entrepreneurs who were previously not required to use cash registers and who applied special taxation regimes in the form of a single tax on imputed income (UTII) and the patent taxation system (PTS). Such organizations and individual entrepreneurs, who do not have wage workers, were allowed to delay switching to CRE until July 1, 2019.

There remains a small list of types of activities that may not require the use of the CREs (State Duma of the Russian Federation 2003).

It is important to note that for some categories of small businesses, namely the enterprises and entrepreneurs applying UTII and PTS, the state executes the direction of digitalization at its own expense, providing a tax deduction for the number of acquired cash registers.

The domestic experience in the implementation of the CRE was positively assessed by the Organization for Economic Cooperation and Development (OECD). Currently, the Federal Tax Service of Russia oversees the OECD initiative to study the global practice of using online cash registers and is responsible for the digitalization and transformation of tax administrations in general (Manturova 2018).

3. *The introduction of the occupational income tax*
   On January 1, 2019, Federal Law No. 422-FZ of November 27, 2018, "On conducting an experiment to establish a special tax regime "occupational income tax" entered into force in the federal city of Moscow, the Moscow and Kaluga regions, as well as in the Republic of Tatarstan. As an experiment, this law introduces a new special tax regime called "occupational income tax" in these four regions of Russia. This taxation regime can be applied by the individuals, including entrepreneurs, receiving incomes from the activities in which they do not have an employer or employees, the so-called self-employed. The law establishes restrictions on the types of activities, as well as on the amount of annual income, which should not exceed ₽2.4 million per year; this amount cannot be reduced during the experiment, which will last through December 31, 2028.

The new special tax regime is a fully digital project; it is based on the use of *My Tax* mobile application that works on any computer device (mobile phones, smartphones, or computers, including tablets) (Consultant Plus 2018). With its help, self-employed individuals register with the Federal Tax Service Inspectorate and conduct settlements with customers, automatically generating electronic receipts. It also interacts with the tax authorities, which calculate the amount of tax payable based on information received through this application. The tax rate on occupational income in settlements with individuals is 4% of income, and that with legal entities is 6%; the self-employed are exempt from paying insurance premiums to extra-budgetary funds and from the use of CREs, which functions the *My Tax* mobile application performs.

The introduction of a special regime for the self-employed was preceded by a two-year grace period from 01/01/2017, when the self-employed could register without paying any taxes. Let us turn to statistics on the registered self-employed in the regions (Table 1).

Table 1. Registered self-employed in Moscow, in the Moscow and Kaluga regions, and in the Republic of Tatarstan.

Region	The number of registered self-employed, people		
	On 01/01/2018	On 01/01/2019	On 03/01/2019
Moscow	79	303	347
Moscow region	61	225	271
Republic of Tatarstan	20	50	58
Kaluga region	4	10	13
The total amount of self-employed participating in the experiment	164	588	689

Source: (Federal Tax Service of Russia 2019).

The statistics show that this experiment would not be called large-scale today. In total, on January 3, 2019, 3,416 self-employed people were registered in the Russian Federation. Figure 1 shows the structure of the registered self-employed as of 03/01/2019 by type of activity.

About 70% of registered self-employed people are tutors, educated people, able to teach, and, therefore, able to learn independently. This tax system implies a high level of taxpayer self-organization in the study of new digital technologies and the basic principles of tax payment, which is not typical for people engaged in those types of activities.

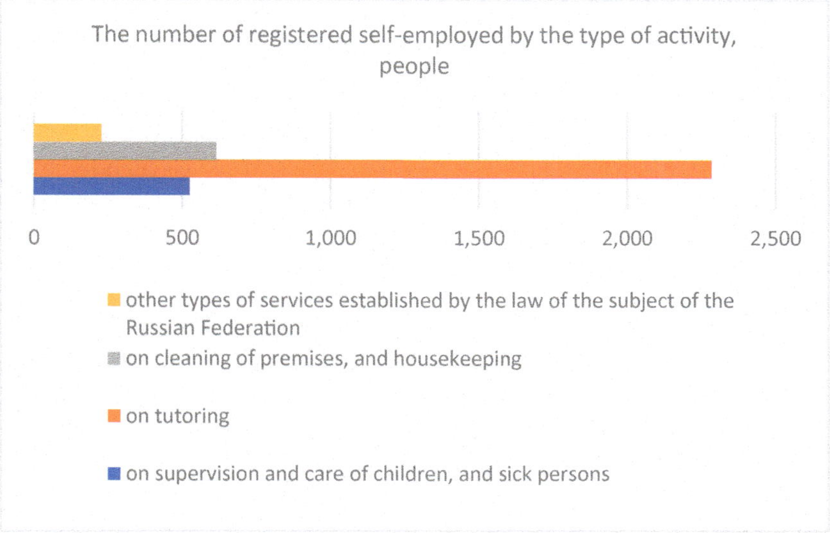

Fig. 1. Distribution of self-employed by type of activity as of 03/01/2019. *Source*: (Federal Tax Service of Russia 2019).

4 Discussion

For the further development of tax administration digitalization, the tax authorities need to involve small businesses in the digital ecosystem through electronic document management, introduce electronic cash registers transmitting real-time data on taxpayer income to the tax authorities, increase the level of digital skills and, most importantly, the level of tax-payment culture. On the other hand, small businesses need support in this process, both in terms of ensuring information security and in minimizing the costs of this process.

As a measure for increasing information security, it is necessary to clarify the provisions of the Federal Law "On Electronic Signatures," namely, clause 2, article 18, which says, "When contacting an accredited certification center, the applicant … submits the following documents or their *duly certified* copies." The power of attorney, which allows the organization's representative to receive an electronic digital signature, must be notarized. The introduction of this requirement will reduce the risk of electronic-signature fraud. Here, an opinion may arise that the falsification of paper statements is also possible, as its delivery does not require a notarized power of attorney, and it is possible to fake the signature of the head and stamp of the organization at the same level. However, to identify the falsification of paper documents submitted to the state bodies, there are well-established procedures that are not yet available in the digital field of tax administration.

The second important direction in the development of the digitalization of the tax administration should be explanatory work with the unregistered self-employed. At this stage in the development of a culture of tax payment, the category of registered self-employed will be replenished by individual entrepreneurs who, having abandoned the status of individual entrepreneurs, will go into the category of self-employed. It applies to entrepreneurs applying the patent taxation system and receiving income from individuals in a case in which the real income of the entrepreneur is equal to or lower than the imputed calculated by the tax authorities. The cost of a patent is 6% of imputed income plus the insurance premiums to the extra-budgetary funds that do not reduce the cost of a patent, as well as a new requirement for the CRE in use from July 1, 2019: For the self-employed, 4% of professional income tax for income received from individuals, the absence of the obligation to pay the insurance premiums to the extrabudgetary funds, and the CREs are replaced by the *My Tax* mobile application. The tax benefit is obvious, and the individual entrepreneurs already have experience in calculating and paying taxes. Attracting the people who have never paid taxes on income from professional activities on their own to the category of self-employed will require the development of additional motivational incentives.

5 Conclusion

The digitalization of the tax administration of small businesses is the area of activity that will require the tax authorities to develop digital technologies, convenient services, personnel training, and additional efforts in order to motivate the transition to the use of developed services.

It seems impractical to attempt to make the interaction between SMEs and tax authorities digital without fail. As the analysis showed, the microbusiness is not ready for this. At the moment, there are no statistics on how many entrepreneurs will cease to work when the use of CREs becomes mandatory officially; perhaps this measure will not have a negative impact. However, electronic tax reporting should remain voluntary for small businesses. An increase in the level of digitalization in this direction will occur with the further development of the service provided by the Federal Tax Service through the resource "Personal Account of the Taxpayer."

The digitalization of the tax administration in the Russian Federation is at an advanced level, and the involvement of small businesses in this process should be phased and accompanied by additional security measures that should, in particular, ensure that electronic signatures are protected by law. Increasing the confidence of small businesses in electronic services of tax authorities will ensure the further development of the digitalization of the tax administration and increase the efficiency of tax authorities.

References

Consultant Plus: Starting January 1, 2019, self-employed persons are subject to professional income tax (2018). http://www.consultant.ru/law/hotdocs/55771

Federal Tax Service of Russia: OECD (2018). https://www.nalog.ru/rn77/about_fts/inttax/oecd/

Federal Tax Service of Russia: Statistics on the self-employed citizens (2019). https://www.nalog.ru/rn77/related_activities/statistics_and_analytics/selfemployed./

Grunina, D.K.: Information digital technologies in the activities of tax authorities. Taxes and Finance, 1 (2018). https://cyberleninka.ru/article/v/informatsionnye-tsifrovye-tehnologii-v-deyatelnosti-nalogovyh-organov

ICAEW: Digitalization of tax: international perspectives (2019 edition) (2019). https://www.icaew.com/-/media/corporate/files/technical/digital-tax.ashx?la=en

Manturova, T.: Digitalization of tax administration: Russia at the forefront. Method, 2 (2018). http://gosmetod.ru/article/356082/

Migacheva, E.V.: Tax control in a digital economy. Financial Right, 8 (2018). https://urfac.ru/?p=507

Ministry of Finance of Russia: The main directions of the budget, tax, and customs tariff policy for 2019 and the planning period 2020 and 2021 (2018). http://www.consultant.ru/document/cons_doc_LAW_308390/

Mishustin, M.: The head of the Federal Tax Service told how to improve the work of tax administrations (2018). https://www.gazeta.ru/business/news/2018/09/07/n_12006235.shtml

Production of stamps without documents (n.d.). https://pechatimetro.ru/

Sidorova, E.Yu., Boboshko, D.Yu.: Taxes and Taxation: A Textbook. INFRA-M, Moscow, Russia (2019)

Smirnova, E.E.: Tax control in the digital economy. Finance **11**, 32–34 (2017)

State Duma of the Russian Federation: Federal Law "Tax Code of the Russian Federation. Part 1" (July 31, 1998 No. 117-FZ) (1998). http://www.garant.ru

State Duma of the Russian Federation: Federal Law "On the use of cash registers when making settlements in the Russian Federation" (May 22, 2003 No. 54-FZ, as amended on December 25, 2018) (2003)

State Duma of the Russian Federation: Federal Law "On Electronic Signature" (April 6, 2011 No. 63-FZ, as amended on June 28, 2014) (2011)

The "University – Industry – Science – Market" Nexus: Creating Competence Networks and Enhancing Integration

The Role of Russian Flagship Universities of Russia in the Formation of the Scientific and Human Potential of the National Centers of Technological Excellence

Sergey P. Sazonov, Irina A. Ezangina(✉), and Ekaterina E. Kharlamova

Volgograd State Technical University, Volgograd, Russia
sazonovsp@mail.ru, ezangirina@rambler.ru, sikaterina@mail.ru

Abstract. The increasing role of the flagship universities in regional development is driven by the consistent policy of the state in forming the centers of technological excellence. To become a regional center of excellence, a university must have a set of proven specific competencies in a particular field, both in terms of educational and research activities. This paper summarizes the advanced experience of the flagship universities in Russia in creating the centers of technological excellence on their basis. More than that, the authors identify the key directions for the development of regional flagship universities as the most important structural units, considering them as a factor in the formation of a system of technological excellence centers.

Keywords: Flagship university · Performance indicators · Center for technological excellence · Scientific and human resource · Strategic projects

1 Introduction

The state's awareness about the importance of ensuring national priorities, including national security by positioning the country on the world markets for high-tech products and services, actualizes the scientific and technological strategy and policy to support the development of the best institutions in certain areas of science and technology.

These are the "centers of excellence", the subjects (associations) possessing the unique material, technical, intellectual and personnel resources, as well as leading fundamental and applied research and development in the breakthrough areas, allowing permanent reproduction of specific competitive advantages in the medium and long term.

National centers of technological excellence (NCTE) are both the state structures and the subjects of commercial interests that are the benchmark for other units of a similar profile (national laboratories, research institutes, research centers, universities, industrial companies, public-private partnership institutions, and also territorial conglomerates, or clusters).

A. V. Bogoviz and Y. V. Ragulina (Eds.): ISCI 2019, LNNS 115, pp. 687–695, 2020.
https://doi.org/10.1007/978-3-030-40749-0_81

The distinctive properties of the NCTE are as follows: global landmarks; high-quality research and development, integration activities based on knowledge and technology transfer, development of interdisciplinary communications, a cluster network aimed at promoting results of project activities (Sazonov 2017). As a part of the study, our attention is focused on their fundamental system-forming role, ensuring the formation and distribution of financial, labor, and material resources between various scientific and technological areas of activity with the effect of synergy and reducing reduction of transaction costs (Zaichenko 2008).

A special place in this regard is given to the innovative ecosystem of regional flagship universities with the uniqueness of their scientific schools and experimental-technological base.

The problems of creating regional support universities in Russia and the features of their activities were discussed in more detail in the works of S. P. Sazonov and others (Sazonov 2016, 2017).

2 Materials and Methods

In justifying and developing the key provisions of the study, the methods of systemic, comparative, and situational analysis are used. Also, we relied on the formalized and heuristic methods for analyzing the effectiveness of the activities of educational organizations using the flagship universities as the example.

3 Results

The main effectiveness indicators of the research and educational activities of flagship universities in Russia are shown in Table 1.

Table 1. The effectiveness indicators of the research and educational activities of the regional flagship universities in Russia in 2018.

The name of the flagship university	Efficiency indicator	
	Educational activities	Research activities
Altai State University	63.71	316.77
Belgorod State Technological University named after V.G. Shukhov	64.87	295.23
Vladimir State University named after Alexander and Nikolay Stoletov	61.55	180.17
Volgograd State Technical University	56.99	514.05
Voronezh State Technical University	61.4	233.15
Vyatka State University	65.96	108.25
Don State Technical University	57.9	117.17
Kalmyk State University named after B. B. Gorodovikov	60.8	77.77
Kemerovo State University	65.28	195

(*continued*)

Table 1. (*continued*)

The name of the flagship university	Efficiency indicator	
	Educational activities	Research activities
Kostroma State University	61	166.41
Nosov Magnitogorsk State Technical University	60.04	311.69
Mari State University	62.83	103.26
Murmansk Arctic State University	63.91	76.83
Nizhny Novgorod State Technical University named after R.E. Alekseev	65.44	854.84
Yaroslav-the-Wise Novgorod State University	61.46	301.47
Novosibirsk State Technical University	64.67	320.74
Omsk State Technical University	61.58	264.84
Orel State University named after I. S. Turgenev	63.57	103.87
Petrozavodsk State University	65.78	374.47
Pskov State University	60.44	121.94
Samara State Technical University	63.44	421.43
Yuri Gagarin State Technical University of Saratov	61.43	295.33
Siberian State Medical University	68.46	203.02
Reshetnev Siberian State University of Science and Technology	56.61	221.92
Sochi State University	64.03	95.49
Pitirim Sorokin Syktyvkar State University	63.05	190.07
Togliatti State University	61.07	230.82
Tula State University	62.15	160.61
Industrial University of Tyumen	64.54	120.66
Ulyanovsk State University	61.18	250.46
Ufa State Petroleum Technological University	66.64	368.31
Cherepovets State University	62.16	162
P.G. Demidov Yaroslavl State University	68.47	252.34

Source: Information and analytical materials on the results of monitoring the effectiveness of educational institutions of higher education (Main Information and Computing Center of the Ministry of Education and Science of Russia, n.d.)

As Table 1 shows, the Siberian State Medical University and the P.G. Demidov Yaroslavl State University are the leaders among the flagship universities in the field of educational activities, and the Volgograd State Technical University and the Nizhny Novgorod State Technical University named after R.E. Alekseev are the leaders in the research activities.

In total, 24 out of the 33 regional flagship universities (73%) have the status of a university center for innovative, technological, and social development of the region.

Indeed, ensuring the sustainable global competitiveness for at least 5 in 2018 and at least 10 leading Russian universities in 2025; creating in the Russian subjects no less than 55 (currently 51 universities) and no less than 100 in 2025 university centers for innovative, technological, and social development of the regions are the declared goals of the priority project "Universities as the Centers for Creating Innovation." The planned federal budget funds for the implementation of the project in the period

2017–2019 are 39.5 billion rubles, and extra-budgetary funds are 5 billion rubles (Presidium of the Presidential Council for the Strategic Development and Priority Projects 2016; Ministry of Education and Science of the Russian Federation 2017a, b).

The distinctive characteristics of the university center for innovative, technological, and social development of the region are presented in Fig. 1.

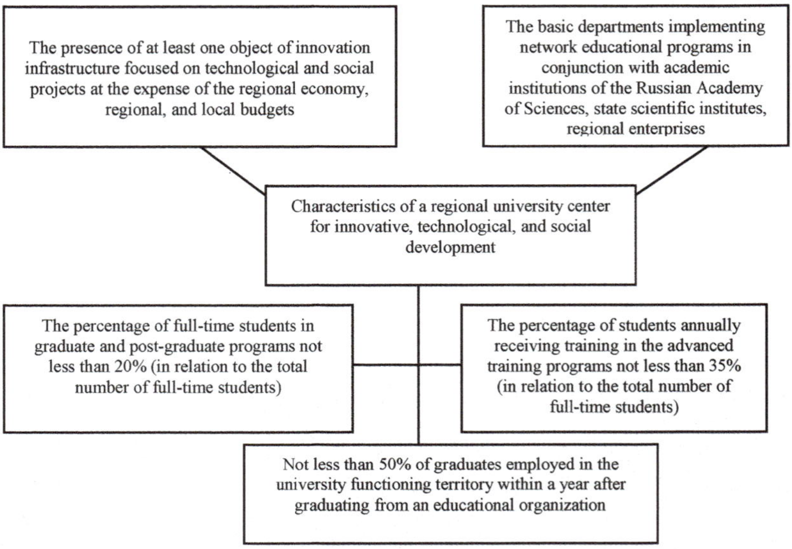

Fig. 1. Distinctive characteristics of the regional university center for innovative, technological, and social development.

In this case, the priority is given to the:

- Project-oriented educational programs involving the team-based implementation of the full life cycle projects;
- Project-oriented joint master's programs in technology entrepreneurship and technology project management with subjects of the real sector of the economy and development institutes.

Thus, since 2017, the Ufa State Petroleum Technological University (particularly, the Department of Economics and Management in the Oil and Gas Industry and the Center for Technological Entrepreneurship) has implemented the project-oriented Master program in "Technological Entrepreneurship in the Fuel and Energy Complex (Ufa State Oil Technical University 2017)." The students work on the end-to-end integrated projects of their start-ups, and each academic semester includes an additional acceleration program. The solutions made during the work process and at the end are supported by the unique Petrochemical Territorial Cluster established in the Republic of Bashkortostan (RB), which includes the following:

- Building and installing a RB cluster;
- A radio-electronic RB cluster;

- An oil and gas machine building RB cluster;
- An oil and gas engineering RB cluster;
- A low-tonnage petrochemistry RB cluster;
- A geophysical -"Quant" cluster.

The cluster infrastructure (Engineering Center, Hardware and Software Complex, Center for Collective Access to High-Tech Equipment, Prototyping Center in the Petrochemistry Field) provides its participants with flexibility and competitiveness, allows product manufacturing with high added value, quickly expanding and changing the range of products in response to changes in the market conditions.

- The engineering centers for innovative and scientific development that promote import substitution in industry. Thus, on the basis of the flagship university of the Mary-El Republic, the Mari State University established an engineering center in the production of the onboard radar systems for the remote Earth sensing. At the same time, this engineering center is also aimed at training personnel and raising the competence level of those specialists working in high-performance programming platforms of the VPX-standard based on FPGA and DSP modules;
- Involving students, graduate students, and scientific and pedagogical workers in innovative and business activities as a tool for ensuring the effective reproduction of the international reputation of Russian higher education and its entry into the global market for education and research;
- An increase in the number of publications of the joint research results in high-rating journals that are part of the Scopus and Web of Science scientometric databases.

4 Discussions

The Voronezh State Technical University (a regional flagship university) has a productive experience in creating the above-mentioned basic departments together with the research and educational centers headed by the leaders of the real sector (Voronezh State Technical University 2017) (Fig. 2).

Currently, the key personnel initiatives for transforming regional core universities into university centers for innovative, technological, and social development of regions include:

- Establishing project offices, integrating the youth technology park with technological entrepreneurship centers, and conducting events with relevant partner organizations;
- Creating conditions for self-determination of individual learning paths for professional and personal competencies (practice-oriented training groups, project-oriented training, network programs, optional modules with the possibility of "further education" for additional advanced training programs, "cross-cutting" educational trajectories) (National Training Foundation 2017).
- Establishing individual grants to support talented students;
- Developing academic mobility programs and internships for students in leading educational and research centers, innovative companies".

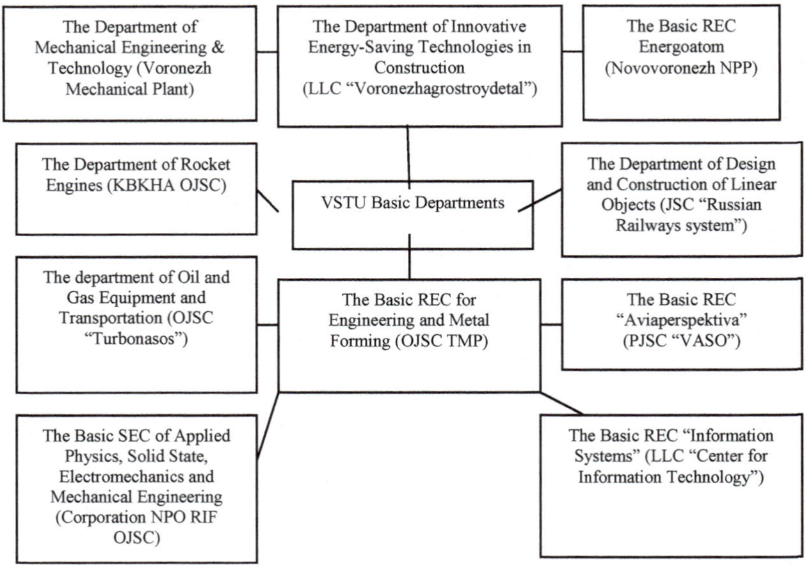

Fig. 2. The existing basic departments of the Voronezh State Technical University.

Thus, the declared strategic guidelines of the regional university center for innovative, technological, and social development and the regional flagship university do not contradict but complement each other, ensuring territorial leadership in the education system (therefore, contributing to the development of the market for social services and social entrepreneurship in 32 Russian regions on the territory of six federal districts) and extraterritorial industry scientific and technological leadership in the world and all-Russia scales. On this basis, a weighty competitive advantage of the region as a whole and a regional university in particular is formed due to the strengthening cooperation ties in the "knowledge triangle" (education – science – innovation/business).

In turn, the implementation of the project "Universities as the Centers for Creating Innovation" was considered effective, if there would be a coordinated renewal of the structure of the regional economy, its filling with effective competitive competencies, the creation of an attractive social environment (including educational), the development of industries in the region knowledge economy capable of commercial technology transfer to real production sectors.

Data on the existing centers of excellence in some Russian regional flagship universities are presented in Table 2.

Table 2. Centers of excellence in the Russian regional flagship universities.

Flagship university	Centers of excellence
Orel State University named after I. S. Turgenev	Center on Automated Control Systems and Industrial Safety
Vyatka State University	Pharmaceutical Biotechnology
Mari State University	High-Performance Computing and Big Data Processing
Don State Technical University	Prospective Robotic Machines and Complexes, Perspective Materials for High-Tech Applications
Voronezh State Technical University	Center of Excellence in the Field of Thermoelectricity
Kemerovo State University	Center for the Introduction of Uncrewed Systems in the Operational Activities of Industrial and Agro-Industrial Enterprises of the Siberian Federal District
Ufa State Petroleum Technological University	Regional Research and Education Center for Excellence in Oil and Gas Reserves Center of Excellence in Chemical Network UGNTU
Orel State University named after I. S. Turgenev	Center of Excellence at the National Level in the Direction of Automated Control Systems and Industrial Safety
Volgograd State Technical University	Center on Organic Compounds and Polymers New Materials Center Control Systems and Robotics
Altai State University	Engineering Center "Prombiotech"

Thus, in the Volgograd State Technical University (VolgSTU), the concentration of efforts of the highly qualified scientific personnel, financial resources, and modern research and technological equipment will provide a breakthrough in the following areas:

- Organic Compounds, Polymers, and Hybrid Materials;
- New Promising Materials;
- Innovative Food;
- Civil and Defense Transport and Cybernetic Systems;
- Building Technologies and Design (Navrotsky 2016).

The formation and development of the centers of excellence are one of the directions for implementing strategic projects of flagship universities, forming a brand of university specialization (Ministry of Education and Science of the Russian Federation, 2017a, b).

The strategic projects also include solutions for regional industry, urban space, in the field of science, technological entrepreneurship, education, youth policy, social policy, tourism and university transformation (Strategic Projects of Supporting Universities).

One of essential features of the systemic-forming function of regional flagship universities is the clustering practice in cooperation with the subjects of the real sector of the regional economy. In particular, the Cherepovets State University is a member of the International Cluster of Wooden Housing and Woodworking, the Cluster of Information Technologies & Biotechnologies.

The Siberian State Medical University, together with six other region medical institutions, initiated the creation of the Scientific and Educational Medical Cluster of the Siberian Federal District ("Siberian"). This cluster operates in the following areas: education, health, science, infrastructure, and personnel.

The Ufa State Petroleum Technological University, on the basis of the existing Center of Excellence "The Chemical Network of USPTU," plans to further integrate through the formation of an intelligent network system that includes 5 interrelated levels of management and digital modeling (synthesis of chemicals, products of the synthesis of chemicals, technologies and methods for the synthesis of chemicals, technological processes and methods for their optimization, organizational processes). The clustering process within the framework of the research involves: creating the network inter-university research laboratories of the geochemistry and ecology of oil and gas, resource-saving technologies for the deep processing of oil and gas, biotechnologies with a focus on providing new substances and materials on the STI market "FoodNet" and "HealthNet," systems management of enterprise architecture and business processes, a network of basic departments at the institutes of the Russian Academy of Sciences, the Academy of Sciences of the Republic of Belarus.

The Pitirim Sorokin Syktyvkar State University was one of the founders of the National Arctic Research and Education Consortium, focused on creating a common research and education space that provides high-quality information and analytical support for the development projects in the Russian Arctic Zone. It should be noted that, along with the associations, research centers and public organizations, 23 Russian universities participate in the project, 3 of which also have the status of a regional flagship university (Petrozavodsk State University, Novosibirsk State Technical University, Murmansk Arctic State University) (Members of the National Arctic Scientific and Educational Consortium).

5 Conclusion

The study allowed us to identify the following development areas of the regional flagship university as the most important structural unit and factor in the NCTE system formation:

- Improving a set of tools used the program-targeted approach. In particular, it is essential to further development of the practice of national and regional programs and projects aimed at strengthening the network of federal universities, the network of flagship universities and scientific and educational clusters;
- Selecting advanced innovation institutions (teams) with the aim of their targeted support (in particular, through the adaptation of the grant system in the Russian educational environment);
- Supporting cooperative research projects of regional support universities for the development of the innovation ecosystem of the state as a whole;
- Ensuring a continuous monitoring of university centers of innovative, technological and social development of regions (taking into account their socio-economic development strategies);

- Using innovative investment mechanism for financing regional support universities, including public-private partnerships;
- Providing institutional and legal support for new self-regulatory mechanisms of universities as the NCTE (in particular, expanding the university autonomy).

References

Main Information and Computing Center of the Ministry of Education and Science of Russia. (n. d.). Information and analytical materials on the results of monitoring the effectiveness of educational institutions of higher education. http://indicators.miccedu.ru/monitoring/?m=vpo

Ministry of Education and Science of the Russian Federation: Order “About the approval of the list of the educational organizations of the higher education recognized by the University centers of innovative, technological, and social development of regions”, Moscow, Russia, December 17, 2017 (2017a)

Ministry of Education and Science of the Russian Federation: Order: Model and parameters of monitoring in the University centers of innovative, technological, and social development of regions, September 1, 2017 (2017b). http://fgosvo.ru/uploadfiles/letters/LO-1754_05.pdf

National Training Foundation: Strategic projects of supporting Universities. Materials of the interuniversity forum “Supporting universities-drivers of regional development” (2017). http://flagshipuniversity.ntf.ru/sites/default/files/Приложение%2041.pdf

Navrotsky, A.V.: Features of the program of the development of the VSTU – the support of the University of the Volgograd region. High. Educ. Russia **7**(203), 107–116 (2016)

Northern (Arctic) Federal University: Members of the national Arctic scientific and educational consortium (n.d.). http://arctic-union.ru/contacts/

Presidium of the presidential Council for the strategic development and priority projects: Universities as centers of the space for creating innovation. Russia, Moscow (2016)

Sazonov, S.P., Ezangina, I.A., Kharlamova. E.E., Pridachuk, M.P.: Small innovation businesses attached to higher schools as new funding sources of Russian higher education. In: Proceedings from TTIESS 2017: International Conference on Trends of Technologies and Innovations in Economic and Social Studies 2017, Tomsk, Russia, pp. 168–174. Atlantis Press (2017)

Sazonov, S.P., Kharlamova, E.E., Gorshkova, N.V., Polyanskaya, E.A.: Competitive advantages of the regional support university and its role in the regional development strategy. Sci. Soc. **3**(1), 180–189 (2016)

Sazonov, S.P., Kharlamova, E.E., Ezangina, I.A., Gorshkova, N.V., Kovazhenkov, M.A., Polyanskaya, E.A.: Theory and methodology of the financial management of the regional supporting university. J. Adv. Res. Law Econ. **8**(1), 211–219 (2017)

Ufa State Oil Technical University: Program of the transformation of the Ufa State Oil Technical University into the University center of innovation, technological, and social development of the Republic of Bashkortostan (2017). http://www.rusoil.net/files/2018-04/Programma-transform.pdf

Voronezh State Technical University: The development program of the VSTU (2017). http://cchgeu.ru/upload/programma-razvitiya/Программа%20развития%20опорного%20университета%20Воронежской%20области.pdf

Zaichenko, S.A.: Centers of excellence in the system of scientific policy. Forsythe **1**(5), 42–50 (2008)

A Didactic Shift in the Organization of Training for the Field of Education

Olga A. Fiofanova(✉)

Russian Presidential Academy of National Economy and Public Administration (RANEPA), Moscow, Russia
fiofanova@mail.ru

Abstract. The article analyzes the changes in educational training programs for education. Based on the analysis of educational programs of the master's program using the method of content analysis of university sites, particularly important changes in the areas of programs, their content, and implementation results are identified. Also, the paper analyzes a didactic shift in the implementation of higher education programs in connection with changes in anthropological ideas about a person, changes in labor market requirements, a changing structure of scientific knowledge about educational activities as a system of human capital reproduction.

Keywords: Modernization of education · Educational training programs · Labor division · Labor deepening · Scientific and technological progress · Human capital

1 Introduction

An analysis by the university community of the changing anthropological ideas about an educated person, more complex types of activities, changing types of reproduction of human potential, as well as the organization of new areas of master's programs in the logic of developing new competencies for new types of labor or deepening traditional types of labor cause the system of division of labor to be updated.

The system of division and deepening of labor determines the scientific, technological, and socio-economic development as a social division of labor (division of professions in society), production division of labor (within the industry), the division of labor between organizations in the production chain, the division of activity between sectors/chains of production (Shchedrovitsky and Kuznetsov 2014; Bogoviz et al. 2019).

The changes taking place in the logical chain "what, whom, why, what, how, and when to teach" can be called a didactic shift, and the changes themselves can be considered in the context of the reproduction of "what and how is reproduced."

A. V. Bogoviz and Y. V. Ragulina (Eds.): ISCI 2019, LNNS 115, pp. 696–702, 2020.
https://doi.org/10.1007/978-3-030-40749-0_82

2 Materials and Method

The paper analyzes the performance results of the countries' educational systems according to international studies (PISA, PIRLS, PIAC, TALIS, a ranking of those countries with the best higher education system (*QS World University Rankings* 2019). We also review the results of the social, technological, and economic progress of these countries, relying on several indexes (*The Social Progress Index* 2018; *International Innovation Scorecard* 2019; *The World's Biggest Economies* 2019; *Global Competitiveness Index 4.0.* 2018). Our analysis allows us to conclude that education and technology development in the new technological order promote country leadership.

Russian universities reflect changes and strive to take their place in the new system of division of labor, including through the design of new educational programs and training areas, rebuild capitalization systems of human resources, and change the tools for organizing educational, research, design, and entrepreneurial activities.

The methodology of the cultural-historical concept of development (Vygotsky 2003) is the basis for the reflection of changes in collective instrumental and sign systems. Instrumental systems are the means of human transformative activity. In turn, sign systems are the means of changing the thinking of a person himself. In connection with a change in the knowledge structure (ideas about the future, about a person, ways of his education, ways of organizing activities, ways of developing society, technology, economics, management), instrumental and sign systems are changing, both in general and in the education system in particular.

Education (in particular) and culture (in general) affect the modernization of society, its socio-economic and scientific-technological development. They are the leading institutions for the formation of values among generations of people and for the creation of human capital (Fig. 1) (Inglehart 1997).

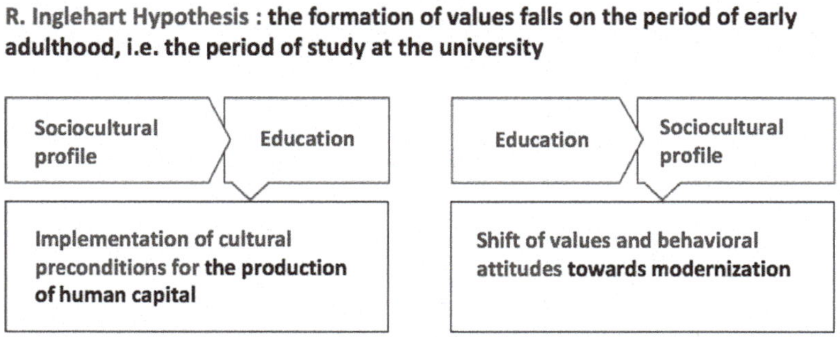

Fig. 1. Educational, cultural prerequisites for the formation of human capital.

We would like to analyze the following: how the approaches at universities to design and organize a portfolio of master's programs in the training direction "Pedagogical Education" and related areas of training managerial, legal, methodological personnel for the education sector have changed (in the following areas of training:

"State and Municipal Administration," "Innovation," "Management," "Law," "Economics") have changed.

The growing demand for new competencies in the labor market in the field of education in its traditional institutional formats (school, pre-school educational organizations, additional education organizations, universities, educational authorities), as well as in new institutional arrangements (towns of professions, interactors, private schools, consulting and expert centers for the development of education, as well as specialized centers for assessing education qualifications), leads to a change in approaches to the organization of training programs for education.

In order to study the parameters of the above changes, a content analysis of educational programs and information materials for applicants posted by universities on their websites is carried out. In particular, the current state of development of the market for educational services for master's programs in higher education (in the direction of "Pedagogical Education," "Management," "Innovation," "State and Municipal Administration," "Law," "Economics" for training in the field of education) is analyzed.

3 Results

It is widely believed that new professional competencies have become in demand in educational practice both in general educational organizations and in universities, including in extra-systemic (extracurricular) educational practices. This is happening in connection with the change in organizational approaches in the field of education. One could note a difference in the technological structure, the need for the formation of new types of functional literacy (competencies of the 21st century), changes in the educational methodology in the logic of human development. These trends are also caused by the increasing role of individualization and personalization in education. At present, such competencies are rather narrowly represented in the labor market of pedagogical workers due to the lack of specialized development programs.

Fixing the "new" teacher's competencies in the teacher's professional standards raises the stakes on the targeted development of such competencies among those persons interested in planning to receive higher education, vocational training, or advanced training.

Research results on the changing functions of a teacher (Fiofanova 2014; Fiofanova 2017) and available analytics of the demand on the labor market for this kind of competence demonstrate a clear need for the following educational programs of higher and additional professional education:

(1) For teachers, deputy heads, and heads of educational organizations (in the direction of "Pedagogical Education"), the following additional programs are in demand: "Organizational Design of the Educational Environment" (including the digital educational environment), "Pedagogical Design of Educational Programs Based on the Assessment of Educational Results";
(2) For education managers (in the direction of "Management"), these programs are "Project Management for the Development of Educational Systems," "Change

Management in an Educational Organization Based on Data," "Educational Qualimetry and Education Quality Management," "Software Project Management in Education: Designing Electronic Educational Resources and Services";

(3) Deputy resource managers, teachers, organizers of extra-budgetary educational services, accountants, contract managers in educational organizations, specialists of the departments for the formation of public tasks and public works in the regional Departments (Ministries) of Education and Science (in the direction of "Economics") might be very interested in the following programs: "The Economics of Education and the Marketing of Educational Services," "Formation of Qualified Orders for Government Work in the Field of Education";

(4) For specialists of the state (municipal) service in Education Departments, heads of regional Departments, Ministries of Education and Science (in the direction of "State and Municipal Administration", "Jurisprudence" (Educational Law)," the following programs could be in high demand: "Analysis of Educational Reforms and the Design of Education Development Policies," "Educational Policy and Educational Law."

In higher education, because of changes in the functions of universities and the transition to the 3:0 and 4:0 university models (research and business universities), there is a shortage of personnel with research competencies (*Monitoring the performance…*, n.d.). In the context of implementing the national projects "Science" and "Education," the Strategy for Scientific and Technological Development of the Russian Federation, the National Technological Initiative, as well as the formation of new markets for the academic profession, the demand for personnel with the competencies of "research, development, innovation management" (managers of R&D, small innovative enterprises, and university development projects) is growing. In the market of educational services, a share of higher education programs of this orientation is tiny, and the existing supply does not cover the demand (Noskova 2018). Therefore, for any potential university teachers, applicants planning an academic career, a career as a researcher, or a technological entrepreneur, the universities offer master's programs in the areas of "Innovation" and "Management," – "Management of University Research, Development, and Innovation."

In the direction 44.04.01 "Pedagogical Education," the following new programs are implemented at Moscow universities. First, "Educational Environment Design," "Expert Activity in the Field of Education," "Design and Evaluation of Educational Programs and Processes," are implemented at the Moscow State Pedagogical University (*Official website of the Moscow State Pedagogical University,* n.d). Second, "Pedagogical Design of Additional Education of Natural Science and Technology," "Management of a System for Assessing the Quality of General Education," "Assessment and Monitoring of Educational Results in the General Education System, "Educational Policy" are at the Moscow City Pedagogical University (*Official website of the Moscow City Pedagogical University,* n.d.).

In the direction of 27.04.05 "Innovation" for the field of education, several programs are implemented in Moscow. First, the programs are taught at the Higher School of Economics - "Management of Research, Development, Innovation in the Organization" and "Project Management: Project Analysis, Investments, Implementation

Technologies" (*Official website of the Higher School of Economics*, n.d.). Second, there is a program "Psychological and Pedagogical Innovation" at the Moscow State Pedagogical University (*Official website of the Russian State Social University*, n.d.).

In the direction 38.04.05 "Management", the following programs are run: "Management in Higher Education" at the Higher School of Economics (*Official website of the Higher School of Economics*, n.d.); "Management in the Education System" at the Moscow State Pedagogical University (*Official website of the Moscow State Pedagogical University*, n.d); "Management of Socio-Pedagogical Activity" at the Russian State Social University (*Official website of the Russian State Social University,* n. d); "Management in the Field of Education" at the Moscow Higher School of Social and Economic Sciences (*Official website of the Moscow School of Social and Economic Sciences,* n.d.).

In the direction of 38.04.04 "State and Municipal Administration," the following programs are implemented in the leading Moscow universities. There is a program entitled "Evidence-Based Education Policy, Education Management" at the Higher School of Economics (*Official website of the Higher School of Economics*, n.d.). Second, such programs as "Management in the Field of Culture, Education and Science, Project and Program Management, Education Management in the Region," and "Management of Social Changes" are mastered at the Russian Academy of National Economy and Public Administration (*Official website of the Russian Presidential Academy of National Economy and Public Administration (RANEPA),* n.d.).

In the direction 38.04.01 "Economics," there are several programs in the leading universities in Moscow. The programs "Public Sector Finance Management," "Investment and Innovation Management in Economics," "Economics and Law," "Contract Manager" are run at the Russian Academy of National Economy and Public Administration (*Official website of the Russian Presidential Academy of National Economy and Public Administration (RANEPA),* n.d.).

In the direction 44.04.01 ("Jurisprudence"), one could mention the programs "Jurisprudence" (with a specialization in Educational Law and Comparative Law) at the Peoples' Friendship University of Russia (*Official website of the Peoples' Friendship University of Russia,* n.d.) and "Educational Law" at the Moscow City Pedagogical University (*Official website of the Moscow City Pedagogical University*, n.d.).

The presented list of programs is concentrated in a thin layer of universities (2–3 universities). These universities practice network forms of educational programs. The leading (top) universities are designing exclusive educational programs based on the definition of promising markets of competencies/qualifications and employment of graduate students. Thus, the universities build a trajectory of continuing education for students at the next stage of their careers (including academic careers), using the capabilities of training programs for scientific and pedagogical personnel (44.06.01, "Education and Pedagogical Sciences"). The new content of master's programs under the new competency profile of professional tasks in the field of education is developing.

In the context of the development of the digital economy and the urgent task for developing the mobile electronic education, some universities are designing narrow programs, such as "Digital Curator" and "Management of Software Projects: Designing Electronic Educational Resources and Services."

4 Discussion

What factors did influence changes in educational programs for training personnel in the field of education? On the one hand, the organizational and didactic shifts in the implementation of higher education and teacher training programs are associated with changes in anthropological ideas about a person, changes in labor market requirements, a changing structure of scientific knowledge about the educational activity as a system of human capital reproduction. On the other hand, this is a consequence of targeted reforms in the system of teacher education.

In 2014, as part of the Federal Targeted Program for the Development of Education (Government of the Russian Federation 2014), the reform of teacher education was launched on the basis of the Concept for the Modernization of Teacher Education (*The project of modernization of teacher education* 2015). Within the framework of the project, work on the design of modules for new educational programs, their examination and testing were carried out.

In 2018, as part of the modernization of educational standards and the development of educational programs (taking into account professional standards), approximate basic educational programs in the direction of preparation "44.04.01 Pedagogical Education" based on GEF or (FGOS) 3++ were developed (*Official website of the Federal Educational and Methodological Association "Education and Pedagogical Sciences,"* n.d.). Universities, as participants in the project of modernization of teacher education, have designed changes in the content, technologies, and organization of educational programs for training the personnel in the field of education.

But the content analysis of the educational programs of the pedagogical direction of the training posted on the websites of Russian universities showed changes in the organization and content of university programs that were not officially participants in the project of modernization of teacher education. This observation may indicate that the following organizational and legal norms are: GEF, professional standards, AQ regulations (distribution of admission quotas). And the cultural norms, initiated by the professional community of participants in the transformations, have a longer period of institutionalization. However, they might have a more sustainable result of modernization.

5 Conclusion

Thus, the growth of demand for new competencies in the labor market, the competition of universities for applicants, the growth of requirements for the quality of education, the changing policy in the distribution of admission control figures, and the development of a national qualifications framework lead to the initiation by universities of new master programs for training personnel for the education sector.

Competencies turn into labor if they are included in different systems of organization and cooperation of activities with different depths of the division of labor, as well as its different productivity. In this regard, the universities make education practice-oriented, being integrated with the activities of organizations and enterprises of the real sector of the economy. They also use tools for an independent assessment of the formation of professional competencies.

Those factors stimulating/limiting the capitalization of a human resource are interconnected. On the one hand, education is a sphere of development of human potential. On the other hand, a socio-economic system is the realization of human capital in the system of division of labor, overcoming those barriers which reduce the return on human capital.

References

Bogoviz, A.V., Alekseev, A., Akopova, E., Przhedetskaya, N., Ragulina, J.: Formation of remote education as a means of restoration of Russian recessive regions' economy. Int. J. Educ. Manag. **33**(3), 438–445 (2019)

Fiofanova, O.A.: New approaches to teacher training. High. Educ. Today **10**, 21–29 (2014)

Fiofanova, O.A.: Training of research teachers at cognitive-type universities (organizational principles of the master's educational program "Management of research, development, innovation at the university"). High. Educ. Today **10**, 17–24 (2017)

Global Competitiveness Index 4.0. (2018). https://basetop.ru/10-samyih-konkurentosposobnyih-stran-mira-2018/

Government of the Russian Federation. Order of the Government of the Russian Federation "On approval of the Federal target program for the Development of Education for 2016–2020" (December 29, 2014 No. 2765-p) (2014). http://static.government.ru/media/files/mlorxfXbbCk.pdf

Inglehart, R.: Postmodern: changing values and changing societies. Polis **4**, 18–28 (1997)

International Innovation Scorecard (2019). https://www.cta.tech/Policy/Innovation-Scorecard/International/Results.aspx

Monitoring the performance of educational institutions of higher education (n.d.). http://indicators.miccedu.ru/monitoring/_vpo/material.php?type=2&id=10301

Noskova, A.: The most popular universities and educational programs in 2017 (2018). https://postupi.online/journal/reyting-vuzov/samie-vostrebovannie-vuzi-i-programmi-2017/.go

Official website of the Federal Educational and Methodological Association "Education and Pedagogical Sciences" (n.d.). https://www.fumoped.ru/kopiya-fgos-vo

Official website the Moscow School of Social and Economic Sciences (n.d.). https://msses.ru/en/

Official website the Peoples' Friendship University of Russia (n.d.). http://www.rudn.ru/

Official website the Russian Presidential Academy of National Economy and Public Administration (RANEPA) (n.d.). https://www.ranepa.ru/eng/

Official website of the Higher School of Economics (n.d.). Retrieved from https://www.hse.ru

Official website of the Moscow City Pedagogical University (n.d.). https://www.mgpu.ru

Official website of the Moscow State Pedagogical University (n.d.). http://мпгу.рф

Official website of the Russian State Social University (n.d.). https://www.rgsu.ru

QS World University Rankings (2019). https://www.qs.com/rankings/

Shchedrovitsky, P.G., Kuznetsov, Yu.V.: From division of labor to division of activity. Philos. Sci. Philos. Econ. **6**, 49–64 (2014)

The project of modernization of teacher education (2015). http://педагогическоеобразование.рф/events/show/31

The Social Progress Index 2018 (2018). https://www.socialprogress.org

The World's Biggest Economies (2019). https://basetop.ru/rejting-ekonomik-mira-2019-tablitsa-vvp-stran-mira/

Vygotsky, L.S.: The history of the development of higher mental functions, vol. 3. Pedagogy, Moscow, Russia (2003)

Effective Functioning of the Labor Market Based on the Interaction of Market Subjects and the Formation of Competitiveness of an Educational Institution

Elena S. Joy(✉), Alexey A. Ilyukhin, and Svetlana V. Ilyukhina

Ural State University of Economics, Yekaterinburg, Russia
kafedra_ekonomic@mail.ru

Abstract. The paper presents an analysis of the labor market of the Sverdlovsk region (Russia), focusing on the problems of its functioning. The most significant factors affecting the functioning of the labor market are identified. The role of educational institutions is considered, and the approach to the formation of directions of a strategy to increase competitiveness is defined, taking into account the specifics of the main goals of consumers of the educational services market, employers and the state. A model of the effective functioning of the labor market is proposed, taking into account market needs and the quality of the workforce.

Keywords: Workforce · Competitiveness · Labor market · Educational institution · Employer · Workforce quality

1 Introduction

This article analyzes the current problems of the labor market, identifies their causes, and suggests solutions. The relevance of this article is enhanced by the lack of a single approach to the analysis of the functioning of this market. Such scholars as K. Marx, A. Smith, A. Marshall, J. M. Keynes, L. Dearden, S. Harkness, S. Mackin, A. Chevalier, I. Walker, S. Mackintosh, and others dealt with these issues for a long time. In European countries, attention to the quality and structure of the workforce was updated thanks to a study by E. Denison, devoted to studying the sources of further economic growth (Denison, 1962). The intensification of research on assessing the expected structure of the workforce (depending on its professional affiliation) was caused, first of all, by the increased understanding of the importance of the quality of the workforce for economic growth (Young 1989; Porter 2002; Vishnevskaya and Zudina 2017). Furthermore, this was established among researchers of various directions and government officials. Another reason was the fear of a shortage of different categories of workers, especially skilled ones, in the conditions of the economic recovery of the 1960s. In Russia, such scientists as L. Abalkin, R. Kapelyushnikov, M. Nikitin, V. Belkin, and many others dealt with these issues.

A. V. Bogoviz and Y. V. Ragulina (Eds.): ISCI 2019, LNNS 115, pp. 703–709, 2020.
https://doi.org/10.1007/978-3-030-40749-0_83

Scientific novelty:

- Based on the methodological approach, the labor market of the Sverdlovsk region was analyzed, which allowed identifying the problems of its functioning;
- A model of the effective functioning of the labor market, taking into account the competitiveness of educational institutions, the needs of consumers, employers, and the state, has been developed.

We also would like to highlight the theoretical and practical significance of the results:

- The use of the results obtained (on the example of the Sverdlovsk region) can contribute to the formation of a better approach to the formation of competitive advantages of an educational institution;
- The proposed model of interaction between employers, educational institutions, and consumers will solve the problems with the competitiveness of educational institutions, allowing them to obtain complete information about the educational institution itself. It could also address the problem of job search for persons under 18 years of age, providing security guarantees, as well as develop an effective strategy to increase the competitiveness of educational institutions.
- This model will allow, through increasing the employment of young people under the age of 18, to influence the social environment in terms of reducing crime and increasing the economic component of the labor market;
- The submitted proposals can be used by state and public organizations that control the activities of educational institutions in the process of developing regional policies in the field of regulation of the educational services market.

2 Relevance

Confirmation of the relevance of this study is a comprehensive analysis of factors affecting both the functioning of the labor market and the social sphere. This analysis will allow us to develop a model of effective market functioning.

3 Results and Discussion

Analysis of the problems of the functioning of the labor market on the example of the Sverdlovsk region.

The development of new technologies, the emergence of new forms of ownership, and challenging institutional changes raise the question of the effectiveness of the functioning of the labor market before many scientists. Modern realities are such that it is not enough for the state to occupy a vast territory; the struggle is on the information level. Furthermore, this requires more and more attention to the quality of the workforce every year. Despite the fact that according to the results of a study conducted by T. A. Klyachko, E. A. Semionova, which reveals the contribution of each level of education to the socio-economic development of the constituent entities of the Russian

Federation, the higher education system is leading, the substantial contribution of the remaining levels cannot be denied (Klyachko and Semionova 2018).

Of course, general education institutions do not produce ready-made labor. Nevertheless, analyzing the situation, the first serious problem of the efficiency of the labor market functioning can be called the quality of training by institutions of general education of future applicants or students of primary, secondary vocational education.

The signing of the Bologna Convention did not bring the education system to the results that were initially mentioned. As a result, in recent years, we can note a severe deterioration in the quality of education (Friedman 1968). Analyzing the indicators of several European countries, this becomes not surprising. Since back in 2010, the data showed that about 20% of schoolchildren in European countries could not read (Belousova 2005). The question arises whether we wanted to get this result in the end. Even according to the students themselves, universal testing does not lead to a desire to learn, but to a desire to memorize more quickly. As a result, such a certification system and a paid education system allowed students with fragile knowledge of the educational program in the field of both the Russian language and mathematics to study at the university, which is the main subject for most specialties. As a result, the parents of the future student had a problem in finding such an educational institution that is not just training for tests but gives knowledge, after which he can choose what he wants. Also, the comprehensive development of the child is not unimportant. Moreover, because most parents cannot drive their children in circles and sections, the search for a school in which this problem will be solved further actualizes this situation.

The next problem is the problem of employment of adolescents under the age of majority. So, according to statistics from the Sverdlovsk Department of Labor and Employment, 1,033 agreements on the creation of 16,147 jobs for children aged 14 to 18 were concluded with enterprises in the region in 2016. According to the press secretary of the department Olga Ruleva, this figure is a quarter less than for the same period of the previous year (Robinson 1933). In 2017, we can observe another decrease in employment. Taking into account that in the period 2013–2015, the operational situation in the Sverdlovsk region in the field of teenage crime was quite tense, even with its improvement in the period 2016–2017, the issue of employment of adolescents is relevant (Ministry of Economics and Territorial Development of the Sverdlovsk Region 2017).

The smooth transition of the youth from school to work is one of the most pressing social problems. When discussing this topic, it is customary to rely on indicators of unemployment and economic inactivity, the values of which among youth groups in different countries far exceed the average values for the labor market as a whole. For example, the proportion of unemployed people aged 15–24 in France was almost 25% in 2015, while the average unemployment rate in the country was 10%. In Spain and Greece, youth unemployment rates reached 50%, with an average unemployment rate in these countries of 20–24%. In Russia, in 2015–2016, the corresponding rates for unemployed people aged 15–19 years old exceeded 30%, for unemployed people aged 20–24 years old made up about 15%, and the average unemployment rate in Russia did not exceed 5–6% in these years (UNESCO, n.d.; Federal State Statistics Service, n.d.).

In accordance with the methodology of the International Labor Organization (ILO), all these values are calculated in relation to the number of labor, and it is usually small at young ages. For example, in 2015, in Russia, the level of economic activity is only 9% among those aged 15–19 years. According to sample studies conducted by Rosstat, the workforce amounted to 76.1 million people in January 2017. This makes up 52% of the total population of the country. More than that, 71.8 million were engaged in economic activities, and 4.3 million people had no work, but they were actively looking for it. The employment rate of people aged 15 and over is 59.1%. In 2016, the employment rate also from 15 to 19 years old was 4.7%. The current unemployment rate is 5.6% (Office of the Federal State Statistics Service for the Sverdlovsk Region and Kurgan Region, n.d.).

Thus, unemployment rates poorly reflect the real situation on the youth labor market, which leads to a shift in perceptions and, accordingly, solutions. It is not a big secret that for graduates of professional educational institutions of any level, it is residually challenging to get a profession having no professional experience. So, according to the Federal State Statistics Service for 2017, the average number of employed graduates falls in the range of 20–30% (according to ITAR-TASS) (Joy 2018a; b). The number of graduates working in another specialty is 55%, and in some specialties, the number of unemployed in general reaches 70–90%. In July 2017, employees of the independent public organization "Association of Managers and Specialists in Human Resource Management," together with the Department of Economics of USUE, interviewed graduates of Yekaterinburg universities. By this time, 60% of respondents managed to get a job. Only 18% of them found a profession in their specialty (Joy 2018b). The rest postponed their diplomas "until better times" and went to work where this document was optional. So, G. E. Zborovsky and P. A. Ambarova offer integration "University - to the City." This will solve several problems, partly the problem of student employment, with real help to the city (Zborovsky and Ambarova 2018; Kurbakov and Demidova 2015).

According to T. V. Gladkova, Deputy Minister of Economics of the Sverdlovsk Region, demographic processes also exacerbate the situation in the employment sector of the Sverdlovsk Region. Taking into account the reducing working-age population, there is a decrease in the number of labor resources per 100 thousand people forecasted by 2023, in comparison with the fact of 2015 (Ministry of Economics and Territorial Development of the Sverdlovsk Region 2018). This situation exacerbates the problem of training quality specialists at various levels. The training programs of higher and secondary educational institutions do not always correspond to the requirements of existing industries, many of which have modernized production facilities, and the transition to a qualitatively new level of technology has occurred, which still causes personnel problems. Of course, programs like the WorldSkills International, a non-profit movement, partly solves these problems, but it is not a panacea for all labor market problems. Modernization of training programs for new specialists is still lagging behind the pace of technology improvement, which leads to the need for "further education" and long periods of adaptation of young personnel (Joy 2006; Grigoriev and Bobylev 2014; Dudnik 2002).

A model of the effective functioning of the labor market based on the interaction of market entities and the formation of the competitiveness of an educational institution also allows solving the problem of youth employment under the age of 18.

To start the study, we tried to derive the concept of competitiveness of an educational institution and the factors influencing it. According to the author, the issue of the quality of education affects several elements. These elements include, firstly, the competitiveness of the educational institution, which directly affects the quality of educational services provided by educational institutions of primary, secondary, and higher education. Secondly, one should consider the availability of education for different segments of the population. Also, this is the quality of the labor force entering the labor market (the correspondence of the quality of the labor force to the demands of employers). Finally, this is the participation of the state in this process (Baumgarten 2005; Joy 2018a).

There are many approaches to determining the quality of education, which is manifested in demand for a graduate in the market, the quality of educational services themselves reflected in the satisfaction of the student. Finally, the institution's competitiveness itself reflects the conformity of this institution (in particular, its specialties) with requests the market (Abramkina and Kovalev 2011; Baumgarten 2005; Joy 2006). Improving the quality of competitiveness of an educational institution is not only relevant but also a socially significant problem. At the present stage, the methodology for studying the competitiveness of an educational institution by structuring factors (parameters) has become widespread. These factors make it possible to assess the level of competitiveness from a quantitative perspective. This approach allows one to create a list of indicators that optimally generate a quantitative list of requirements to assess the competitiveness of the institution. The main differences in the methods are the composition and specificity of the characterizing indicators in the field of expanding the number, content, and refinement of the indicators themselves. According to the author, many methods have several disadvantages (Joy 2018c).

In our opinion, considering market entities, a clear picture of the interaction that allows us to solve the problems that have arisen is built.

First, the creation of a unified information system with data on educational institutions, which allows obtaining complete information about what resources these institutions have, about staff, about learning outcomes, about circles and sections, will allow parents to choose an educational institution not only guided by their location but also the future of his child.

Second, for an educational institution, this platform will serve as an opportunity to stimulate improving the quality of services provided, which may affect the level of teachers' salaries (the more the teacher is in demand, the higher the salary).

Third, for educational institutions of primary, secondary, and higher professional education, this platform will serve as a clear indicator of the quality of student learning. Moreover, the presence of a (possibly electronic) portfolio of students will provide complete information that does not give a single state exam.

Fourth, this platform will allow employers to post information on vacancies for students and people under 18 years of age. By paying for the placement of this information, employers will be able to obtain information about the future employee. The proceeds for posting information can serve as the financing of this platform. At the same time, employers, providing jobs for minors, could apply for preferential taxation, as well as the opportunity to reduce the cost of finding inexpensive labor. In turn, workers would receive guarantees that wages would be paid in full and on time. In turn, the state receives several advantages. First, working young people earn income, which increases the demand for goods and services. Second, due to the employment of this part of the population, the reduction of criminal incidents occurs, which favorably affects the socio-economic sphere. Third, those young people already started working strive for professional growth, which leads them to the desire to improve their skills and, accordingly, education.

4 Conclusions

Consideration of the contradictions and development trends made it possible to identify fundamental approaches to solving pressing socio-economic problems and suggest solutions that can be used to solve similar problems not only in the Russian Federation but also in the countries of near and far abroad. As a result of the proposed model, the author proposes an approach to solving labor market problems by increasing the competitiveness of an educational institution and attracting an increasing number of non-disabled people to the market in the form of persons under the age of majority. Accordingly, this will affect the quality of the workforce and will reduce unemployment in Russia.

References

Abramkina, A.A., Kovalev, A.I.: The system of assessing the competitiveness of educational services of universities. Vestnik Omsk Univ. **2**, 232–238 (2011)

Baumgarten, L.V.: Analysis of methods for determining the competitiveness of organizations and products. Mark. Russia Abroad **4**, 72–85 (2005)

Belousova, E.V.: Assessment of the competitiveness of a higher educational institution in the regional market of educational services (on an example of Primorsky Krai) (Dissertation of the Candidate of Economic Sciences). Vladivostok State University of Economics and Service, Vladivostok, Russia (2005)

Dudnik, S.I.: Educational policy in modern Russia: problems and prospects. Educ. Civ. Soc. **1**, 20 (2002)

Federal State Statistics Service (n.d.). http://www.gks.ru/

Friedman, M.J.: The methodology of positive economics. In: Breit, W., Hochman, H. (eds.) Readings in Microeconomics, pp. 44–45. Holt, Rinehart & Winston, New York (1968)

Grigoriev, L.M., Bobylev, S.N. (eds.): Report on Human Development in the Russian Federation "Human Development in Conditions of Economic Instability". Analytical Center under the Government of the Russian Federation, Moscow (2014)

Joy, E.S.: The formation and development of labor centers as a condition for the balance of the labor market in a transitive economy (Dissertation of the Candidate of Economic Sciences). Chelyabinsk State University, Chelyabinsk, Russia (2006)

Joy, E.S.: Solving the problems of the education system through the information environment of the educational services market. Economy: Yesterday, Today, Tomorrow, 8(5A), 27–32 (2018a)

Joy, E.S.: The impact of the education system on the functioning of the labor market. Econ. Entrepreneurship **6**(95), 275–279 (2018b)

Joy, E.S.: Development of a model of effective functioning of the labor market based on the formation of educational institution competitiveness. In: Proceedings from III International Research-to-Practice Conference MITEC – 2018, Taganrog, Russia (2018c)

Klyachko, T.A., Semionova, E.A.: The contribution of education to the socio-economic development of Russian regions. Region's Econ. **14**(3), 791–805 (2018)

Kurbakov, S.I., Demidova, A.V.: Cooperation between universities and large specialized companies as a condition for the effective functioning of the educational system. Mod. Res. Innov. **5**(4) (2015). http://web.snauka.ru/issues/2015/05/54450

Ministry of Economics and Territorial Development of the Sverdlovsk Region. Materials of the Meeting of the Coordinating Council of the Government of the Sverdlovsk Region on Staffing the Economy of the Sverdlovsk Region (2017). http://economy.midural.ru/2017

Ministry of Economics and Territorial Development of the Sverdlovsk Region. Meeting of the Committee of the Sverdlovsk Regional Union of Industrialists and Entrepreneurs for the Development of Vocational Education and Human Resources (2018). http://economy.midural.ru/content/zasedanie-komiteta-sverdlovskogo-oblastnogo-soyuza-promyshlennikov-i-predprinimateley

Office of the Federal State Statistics Service for the Sverdlovsk Region and Kurgan Region (n.d.). http://sverdl.gks.ru/

Porter, M.E.: Competition. Williams, Moscow (2002)

Robinson, J.: The Economics of Imperfect Competition. Macmillan and Company, London (1933)

The Financial University under the Government of the Russian Federation (n.d.). http://www.fa.ru

The Ministry of Education and Science of the Russian Federation (n.d.). http://www.mon.gov.ru

UNESCO (n.d.). http://en.unesco.org/education

Vishnevskaya, N.T., Zudina, A.A.: The professional structure of the workforce in Europe: what are the forecasts? Vestnik Int. Organ. **12**(4), 4 (2017)

Young, M.: The Technical Writer's Handbook. University Science, Mill Valley, CA (1989)

Zborovsky, G.E., Ambarova, P.A.: Higher education as a factor in the conservation of cities in the Urals macroregion. Region's Econ. **14**(3), 914–926 (2018)

The Formation of a Labor Payment System in Production Holdings with a Regional Structure

Ekaterina V. Korneva[1(✉)], Irina Yu. Eremina[1], and Evgenii N. Mazhara[2]

[1] Gubkin Russian State University of Oil and Gas, Moscow, Russia
korneva-ev@yandex.ru, irinargung@mail.ru
[2] Financial University Under the Government of the Russian Federation, Moscow, Russia
mascharal02@mail.ru

Abstract. The paper focuses on the problem of establishing fair and reasonable payment for the staff of regional divisions (branches, business units, subsidiaries) of production holdings. The aim of the study is to develop approaches for building an effective and unified pay system at the scale of holdings (large enterprises), taking into account the regional aspect. In the course of the study, the authors identified the following problems: (1) the labor market of certain regions is unattractive for workers, which should be taken into account when building a wage system; (2) salaries in the regional divisions should be based on the principles common to the company; (3) the payment level in the regions should be competitive; (4) the payment level should be economically justified. The paper considers different approaches to the formation of a wage system for regional divisions, taking into account both external factors (labor markets in the region, average wages, consumer spending) and internal factors (job evaluation, wage rates). Also, the authors focus on the issue of regional discrimination in payment. The result of the study is a model for constructing the payment system for the regional divisions of holdings.

Keywords: Wages · Regional wage scale · Wage review · Pay discrimination · Key performance indicators

1 Introduction

The issue of establishing the optimal level of wages for employees is relevant for most companies. On the one hand, the level of wages must be sufficient to attract and retain workers from the labor market and, therefore, be competitive in the region. On the other hand, the labor costs should be economically justified and provide the necessary level of profitability to the company. Employers are forced to seek a compromise in order to carry out their current activities with maximum efficiency. For the companies with a regional structure, as well as the holdings with subsidiary companies and branches throughout Russia, the task is becoming more complicated: It is necessary to

A. V. Bogoviz and Y. V. Ragulina (Eds.): ISCI 2019, LNNS 115, pp. 710–715, 2020.
https://doi.org/10.1007/978-3-030-40749-0_84

establish a single pay structure and, at the same time, take into account the regional aspect. To answer the question of how much to pay an employee, the employer has to take into account a large number of external factors, including:

- The availability of necessary staff in the region;
- The living standards and consumer spendings in the region;
- The number of similar vacancies in other companies in the region;
- The payment level in the competing companies in this region;
- The attractiveness of the region for workers (urban infrastructure, culture, education, climate), etc. (Zabirova, Daminova and Syreyshikov 2017).

At the same time, the unity of the approaches to the payment should be maintained: a branch or subsidiary should not become a "state in a state." The policies and practices applied in branch or subsidiary companies in all functional areas of the holding, including in the field of personnel management, must be subject to uniform laws and contain unified approaches that allow the holding to function as a single system.

The purpose of this paper is to analyze and evaluate the model for constructing a regional tariff grid proposed by the authors.

2 Materials and Method

The authors of the paper repeatedly addressed the problem of establishing a fair level of payment for employees of regional divisions, analyzed current policies and practices on the construction of a payment system, including payment of regional divisions. In the course of the research, the authors used the integrity of the methods, the general principles of scientific knowledge, the laws of economic science as well as the traditional economic methods. The study identified several problematic issues that need to be addressed as part of the development of a staff motivation system for the regional units. The result of the study is the developed phase model for constructing a wage tariff system for a company with regional structural divisions.

3 Results

The development of wage scales when building a pay system for companies with regional divisions includes several steps:

- Step 1. The development of a basic wage scale.
- Step 2. The development of a regional wage scale.
- Step 3. The assessment of the competitiveness of the regional wage scale.
- Step 4. The conformity assessment of the regional wage scale to the financial and economic indicators of the regional unit (Eremina and Korneva 2018).

The basis for building wage scales is two principles: the principle of internal justice and the principle of external competitiveness.

For each position level, previously determined through the evaluation and ranking, the sizes of wages (tariff rates) are determined. At the same time, the target payment structure, inter-level (inter-bit) coefficients, as well as the current level of payment (including the size of salaries) are taken into account.

Since the target payment structure for certain personnel categories can differ significantly, in some cases, the separate wage scales for various personnel categories can be developed. The alternative is the development of a single salary scale for all personnel categories, using various types of bonuses.

Inter-level (inter-bit) coefficients represent the difference between the adjacent job levels. As a guideline, we can use 15%—a standard inter-level coefficient, according to the Korn Ferry Hay Group methodology.

The current level of wages is one of the significant factors determining tariffs. The introduction of a unified payment system always leads to an increase in labor costs. In most cases, the loss of key employees, social tensions, legal and reputational risks, reduced labor productivity, and other aspects and risks prevent employers from reducing wages. According to the information from Korn Ferry Hay Group, the salaries of the employees who are paid more than expected for this position are not reduced; they are just not increased for some time. The payments to the employees in the undervalued positions, on the contrary, are gradually increased, within the financial possibilities to increase the payroll (Goncharova 2006).

Since in the process of unifying wages for most workers, the level of wages either does not change or increases, the current level of wages becomes a key deterrent in the modeling of tariff schedules, first of all, in the development of the basic wage scale.

The basic wage scale is formed without taking into account regional coefficients and allowances for the work in the Far North and equivalent locations. The basic wage scale can be formed for the region in which the management company is located (for example, the city of Moscow) or the main assets are held.

As part of the next step, a regional wage scale is formed by applying a coefficient to the base wage scale based on the location of the regional unit. The base region (e.g., Moscow) is applied per unit, after which a regional coefficient is applied to the wage scale of the base region.

The next step assesses the competitiveness of the simulated tariff network.

If the comparison was carried out earlier within the organization, then in the case of external competitiveness, the second strategic line with respect to payment, a comparison is made with other employers outside of the organization.

For the positions related to the main production, as well as for managers of functional areas, it is advisable to compare the data of industry wage reviews with the data of specially prepared (custom-made) wage reviews for the corresponding reference group of comparable companies. For the administrative staff and several auxiliary production professions wherein experience in a particular industry is not critical, industry-wide reviews, in which companies from various sectors of the economy take part, can be used. The number of participants depends on the specifics of the review and varies from one hundred to five hundred companies.

Based on the results of comparison with market data, the analysis of the causes of the deviations is carried out and, if necessary, wages (tariff rates) are adjusted.

Further, for each regional unit, an analysis of the economic feasibility of applying the wage scale is carried out, taking into account the projected financial and economic indicators. Based on the budget model of the regional unit, the following are analyzed:

the fulfillment of the planned production volumes and the forecasted number, personnel costs and the projected level of net profit, labor productivity indicator (in physical or in money terms), and personnel profitability (net profit per employee) (Krasilnikov 2017).

After the development of tariff schedules for all regional divisions, the final calculation of the number of additional costs is made, and a schedule for the transition of divisions to unified wages is drawn up.

In some cases, when establishing different salaries, organizations are faced with issues of regional discrimination in pay. According to some experts, wages for the same positions in different regions of Russia should be the same. Following Art. 37 of the Constitution of the Russian Federation, the discrimination on pay is prohibited. It should be kept in mind that the establishment of different salaries in different legal entities (for example, in subsidiary companies of the holding) cannot be regarded as discrimination by the employer. If we are talking about one employer—that is, one legal entity—then there is a risk of discriminating against the establishment of different salaries in the same positions, even in different regions (Mironov 2014).

The legal precedents regarding wage discrimination are mixed. In many cases, employers in court succeed in upholding the lawfulness of establishing different salaries for the same positions, citing the facts as evidence (Moskaleva 2014).

In order to reduce the discriminatory risks and to establish balanced approaches to the payment by employers, a system of incentive co-payments and allowances is widely used. Regarding the establishment of salaries in different regional units, employers need to apply rational regional coefficients.

Based on the foregoing, it can be concluded that the presence of clear and reasonable criteria for calculating wages in the framework of the development of regional tariff networks will minimize the risk of allegations of discrimination and establish fair approaches to the payment for the holding.

4 Discussion

In recent years, employers have faced an unfavorable trend: workers and job seekers are becoming less mobile. According to a survey conducted by *Antal*, among 8528 respondents working in Russia, two years ago, 40% of respondents were ready to move to another city in Russia for work, and today only 31% (see Fig. 1) (Antal Russia 2018).

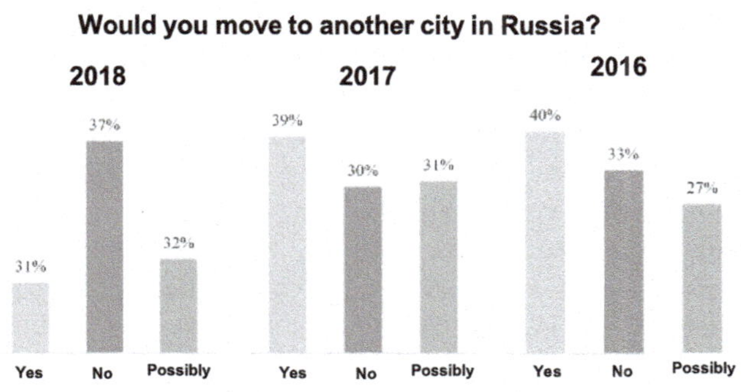

Fig. 1. The survey results, according to Antal's review.

The reasons for the decrease in staff mobility are related to the fact that the Russian labor market is a system of regional markets that are rather weakly interconnected and vary greatly in their "success." Some labor markets are characterized by high employment, high vacancies, and low unemployment and offer a high level of pay, while others cannot provide jobs for a large part of the population and do not allow a decent level of wages (Gimpelson, Kapelyushnikov and Roshin 2017).

The presence of "strong" and "weak" labor markets must be taken into account when implementing any policy, including the personnel policy. Centralized systems, policies, and procedures (including in payment) simplify the procedure of relocation between the regional divisions of the holding. At the same time, the issue of attracting personnel to a region with a "weak" labor market and low wages remains unresolved (Nizamutdinov and Oreshnikov 2017).

This problem is topical not only for Russia but also for other countries, including the USA. According to the annual Corporate Relocation Survey, in 2010, 87% of companies with more than five thousand people faced the problem of workers' refusing relocation (Williams 2011). In order to solve the problem of attracting personnel to the regions, additional benefits are used for relocated workers. Also, to compensate for the inconvenience associated with moving and living in another region, the employee may be set a relocation allowance.

The use of the relocation allowance permits, on the one hand, attracting the necessary personnel to a specific region, and, on the other hand, maintaining the integrity of the unified system of payment.

5 Conclusion

The paper focuses on a phased model for the construction of payment in holdings with a regional structure. The approaches to the development of the regional wage scale are identified, including the development of the basic wage scale (for the region where the management company or the main assets of the holding are located) and the procedure for calculating the regional coefficient for the holding units. The necessity for analyzing the competitiveness of the simulated remuneration level and assessing the conformity of the tariff grid to the financial and economic capabilities of the regional unit are substantiated.

The issue of regional discrimination in pay is considered. According to the results of the study, it was found that to minimize the risk of accusations of discrimination, as well as to comply with the principle of fairness in payment, the employer needs to use clear, reasonable, and objective criteria when calculating wages for regional units.

The problems of regional relocation are also considered. Based on the analysis, it was found that in recent years, the staff has become less mobile. For employers, the issue of attracting personnel to a region with a "weak" labor market and low wages remains an acute issue. The measures to solve the problems of regional relocation, including the provision of additional benefits for displaced workers and the establishment of a relocation allowance, are considered.

The model considered in this article can be applied when introducing a unified system of payment in regional divisions (branches and subsidiaries) of large industrial holdings and companies in other sectors of the economy, with adaptation to the specifics of the organization.

References

Antal Russia: Labor market research and salary survey. Russia (2018). https://antalrussia.ru/upload/medialibrary/d54/antal_issledovanie-rynka-truda-i-obzor-zarplat-2018_rus_2.pdf

Eremina, I.Y., Korneva, E.V.: Unification of salary system in production holdings. J. Econ. Entrepreneurship **12**(6), 1084–1089 (2018)

Gimpelson, V., Kapelyushnikov, R., Roshin, S. (eds.) Report. Russian Labor Market: Trends, Institutions, Structural Changes. Center for Strategic Research (2017)

Goncharova, O.: How job evaluation allows to give each employee what he deserves (2006). https://www.vedomosti.ru/management/articles/2016/03/24/634849-kak-otsenka-dolzhnostei-pozvolyaet-kazhdomu-sotrudniku-vozdat-po-trudu

Krasilnikov, S.: Problems of correlation of growth and financial stability of the holding's subsidiaries. Probl. Manag. Theory Pract. **12**, 73–81 (2017)

Mironov, V.I.: Discrimination in salary. Labor Law Russia Abroad **1**, 22–27 (2014)

Moskaleva, O.: Unequal salary: is it always discrimination? J. Labor law **11** (2014). http://www.top-personal.ru/lawissue.html?2451

Nizamutdinov, M.M., Oreshnikov, V.V.: Modeling of the influence of socio-economic factors of the development of Russian regions on the formation of migration processes. Audit Financ. Anal. **2**, 445–450 (2017)

Williams, J.A.: More workers going places, but many do not "dig" ideas of new digs. Workforce Manag. **90**(7), 12–14 (2011)

Zabirova, L.M., Daminova, L.I., Syreyshikov, A.I.: The compensation policy of companies and related risks. Bull. Econ. Law Sociol. **3**, 22–26 (2017)

Labor Motivation as a Pledge of Successful Development of the Organization

Elena M. Klimova[1,2(✉)] and Vladimir M. Litvishkov[1]

[1] Research Institute of the Federal Penitentiary Service of Russia, Moscow, Russia
klimova_em@mail.ru, mavlad67@mail.ru
[2] Moscow Region State University, Mytishchi, Russia

Abstract. The article presents the results of an empirical study of the interconnection between labor motivation and successful socio-psychological adaptation of the employees of the penal system. The motivation of most employees comes from the content of their work, i.e., the work process is essential for them (working conditions, possibilities for their career growth and professional development, opportunities for taking personal responsibility and ensuring permanent employment). Workers showed the highest level of socio-psychological adaptation with the resulting orientation of the communicative type of labor motivation. It is caused by the fact that communication is of particular importance in the professional activities of the employees of the penal correction system, that is why people with a focus on interacting with others most quickly adapt to such work. The least adapted are employees with the resulting orientation of cooperative motivation. Employees with this type of motivation are most likely do not experience job satisfaction. The obtained results can be applied when developing a professional selection program for the position of a penal system officer, developing a SPA program, taking into account the identified factors and criteria. In this case, it is necessary to diagnose the level of SPA before assuming the position to identify the reserves of the employee and to conduct diagnostics after passing the SPA program to identify the results.

Keywords: Labor motivation · Socio-Psychological adaptation · Success of the staff adaptation · Criminal executive system

1 Introduction

The problem of labor motivation was acute at all times and in all organizations. The need to resolve this issue is especially acute for the employees exposed to adverse performance factors (Batishchev 2017; Kamyshanov 2001; Kamyshanov 1998; Kishikov 2001; Putivtsev and Noss 2011). The management of organizations of the penitentiary system needs to stimulate the staff to perform the activities aimed at achieving the goals of the organization.

In recent years, a significant number of studies have been conducted on the issues of increasing motivation and job satisfaction and stimulating law enforcement officials (Gerchikov 2005; Dmitrieva and Druzhilov 2000; Ilyin 2000), including the employees

A. V. Bogoviz and Y. V. Ragulina (Eds.): ISCI 2019, LNNS 115, pp. 716–727, 2020.
https://doi.org/10.1007/978-3-030-40749-0_85

of the penal system; however, the influence of socio-psychological adaptation on the motivation of employees attracted little attention from scholars. The management style (Kishikov 2004) and the chosen means of professional training (Poruchkina 2017), including training methods (Kamyshanov 2016), have significant impacts on the success of the adaptation of employees.

In a broad sense, the socio-psychological adaptation of an employee means the formation of interpersonal and business relations with colleagues, the development of values, and group norms of behavior. The consequence of the successful socio-psychological adaptation of an employee is the productive labor activity in this team. It is in the interests of the organization that the adaptation process of the employee to take place as soon as possible.

The theoretical bases of the study were the works of domestic and foreign authors on the problem of labor motivation (F. Herzberg, J. Atkinson, J. Adams, L. Porter, E. Lower, V. I. Gerchikov, V. D. Shadrikov, R. A. Piloyan, et al.) and socio-psychological adaptation (A. A. Rean, L. Philips, A. G. Dikaya, A. A. Nalchadzhyan, S. A. Larionova, et al.). The scholars do not have an unequivocal opinion regarding the nature of socio-psychological adaptation in professional activities. Each organization is a special organism with its own mechanisms of existence. The driving force of this organism is people engaged in labor activities. Each organization represents a unique social environment formed by those individuals who are included in it. A person, falling into this already formed environment that is not familiar to him, has to adapt to it. However, in the same way, a person has to adapt to any other social environment, not only the professional one.

2 Materials and Methods

In the study, we used the following methods:

1. **Technique "Dictionary" developed by I. G. Kokurina.** It is designed for the diagnosis of labor motivation. Labor motivation is defined as the process of filling the meanings of the most significant incentives for work, which results in the formation of stable semantic ideas that can have an opposite effect on the labor behavior of the worker.
2. **Methodology for the diagnosis of labor motivation developed by V. I. Gerchikova.** The test is intended to establish the type of motivation in the activity. The test determines the prevailing type of motivation, that is, only one component of the respondent's personality profile.
3. **Questionnaire of socio-psychological adaptation developed by C. Rogers and R. Diamond.** It is designed to study the characteristics of socio-psychological adaptation and related personality traits.

3 Results

The study of labor motivation, according to the "Dictionary" technique developed by I. G. Kokurina.

Let us consider the severity of the resulting orientation of different types of basic motifs presented in Fig. 1. The resulting motivation is expressed as a goal-setting, i.e., associated with obtaining a specific result.

The results of the study show that in the resulting orientation, the transformative motive prevails in 27% of the subjects. A utilitarian-pragmatic motive prevails among 22% of the respondents. A communicative motive prevails in 16% of the respondents. A cooperative motive prevails in 14% of the respondents. Eleven percent of the respondents choose a competitive motive. In 9% of the respondents, an achievement motive prevails.

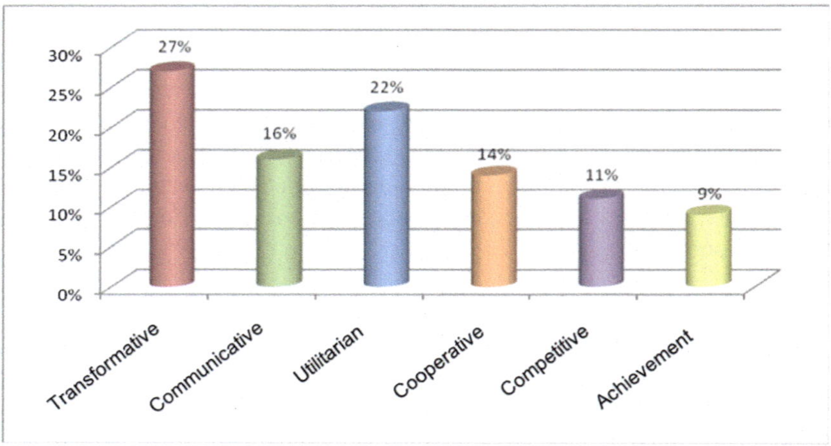

Fig. 1. The severity level of the resulting orientation of different types of basic motives

Next, let us consider the severity of the process orientation of different types of basic motives, as presented in Fig. 2. The procedural motivation is expressed in the need for activity.

Figure 2 shows that in the process orientation, the respondents are dominated by a transformative motivation (32% of the subjects). Seventeen percent of the respondents are dominated by a competitive motivation. Thirteen percent of the respondents are dominated by communicative, utilitarian-pragmatic, and achievement motivations. The cooperative motivation prevails in 12% of respondents.

The general level of the severity of various types of motivation is clearly shown in Fig. 3.

According to the results of the study, the transformative motive prevails among the respondents both in the resultant and in the process orientation (27% and 32% of the respondents, respectively). Most of the respondents (32%) have a pronounced orientation toward the activity for the sake of the process itself; in other words, the process

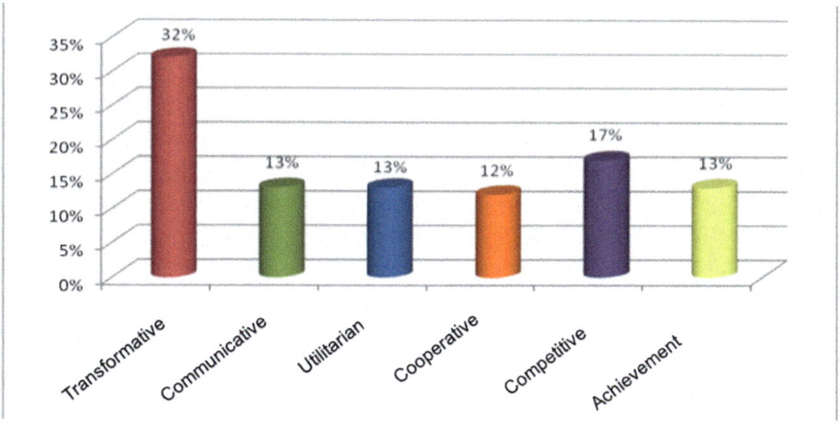

Fig. 2. The severity of the procedural orientation of different types of basic motives

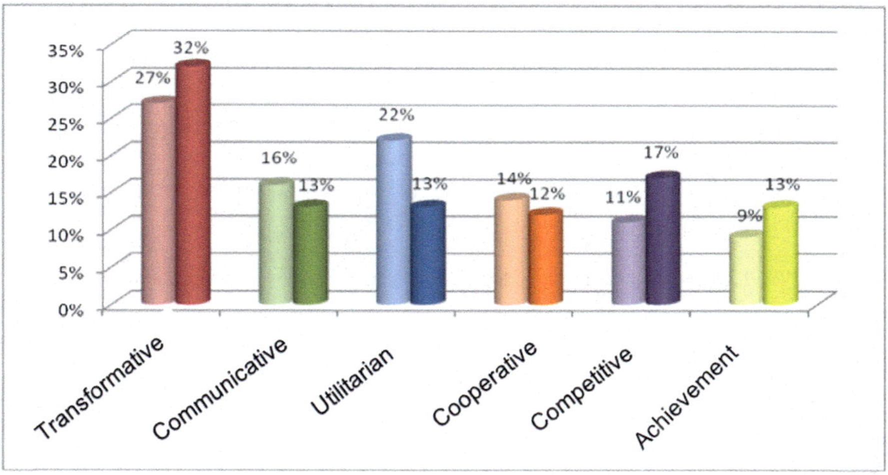

Fig. 3. The general level of severity of different types of basic motives according to the Kokurina technique

of work is important to them. 27% of the respondents are focused on achieving results in professional activities.

22% of the respondents are dominated by the resulting orientation of the utilitarian-pragmatic motivation, i.e., for them, labor is a means of satisfying their non-work needs.

17% of the respondents are dominated by the procedural orientation of competitive motivation. They need to be like everyone else, not stand out and not be worse than others.

In 16% of the respondents, the resulting orientation of communicative motivation prevails. They are focused on active interaction with their colleagues in the process of work.

14% of the respondents are dominated by the procedural orientation of cooperative motivation. This group of respondents considers labor in terms of its social utility.

13% of respondents are dominated by the procedural orientation of the communicative, utilitarian-pragmatic and achievement motivation. The respondents with a pronounced procedural orientation of a communicative motive have an orientation toward maintaining positive relations with their colleagues. The respondents with a pronounced procedural orientation of utilitarian-pragmatic motivation consider labor as a waste of energy and strive to maintain health at work. The respondents with a pronounced procedural orientation of achievement motivation are characterized by a focus on self-improvement and development in the professional activity.

In 12%, there prevails the procedural orientation of cooperative motivation. It is typical for them to consider their work in terms of usefulness for their friends and relatives.

The least pronounced is the resulting orientation of achievement motivation (9%). It is common for them to set super tasks for themselves and to focus on overcoming obstacles.

Thus, according to the results of the *Dictionary questionnaire* developed by I. G. Kokurina, it can be concluded that the majority of respondents in their professional activities are oriented to the work process, as well as to the achievement of results. The minority is characterized by orientation to overcoming difficulties in professional activity.

The study of labor motivation according to the method of V. I. Gerchikov.
The results of the study of labor motivation translated into ranks, according to the method of V. I. Gerchikov, are shown in Fig. 4.

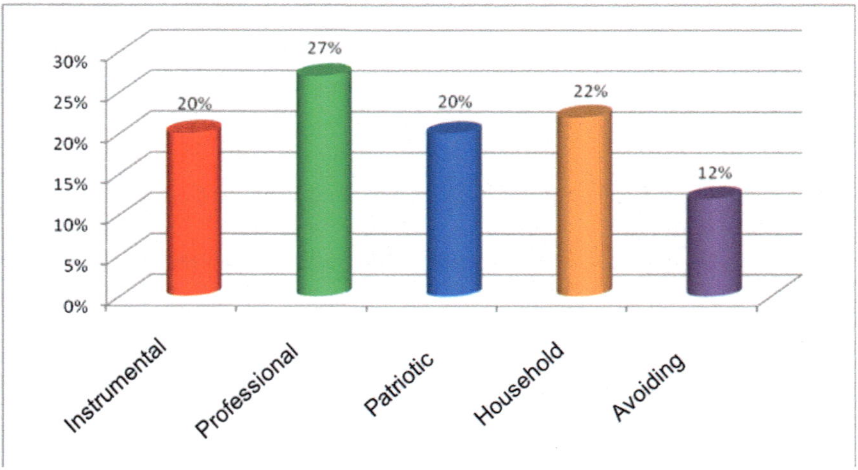

Fig. 4. The severity of types of labor motivation according to the Gerchikov's method.

So, the majority of respondents (27%) are dominated by the professional type of motivation. It is characterized by the value of the content of labor for the employee, the ability to prove himself or herself in his or her professional activity, and independence in his or her work.

22% of the respondents are dominated by the master type of motivation, i.e., they are characterized by the voluntary assumption of responsibility and the performance of work with the maximum return without requiring control of themselves.

An instrumental type of motivation prevails among 20% of the respondents, which is expressed in relation to work as a means of earning and in the most effective work with decent earnings.

The patriotic type of motivation also prevails in 20% of respondents, i.e., public recognition and participation in common affairs are important for them.

An evasive type of motivation is expressed in only 12% of respondents. They are characterized by the minimization of labor efforts, avoiding responsibility, and passivity. However, at the same time, it is essential for them that their level of wages does not differ significantly from other workers' wages.

Thus, according to the results of the study of labor motivation conducted in the method of V. I. Gerchikova, the majority of employees are dominated by motives related to the content of the work, the possibility of career growth and professional development, independence in work, and the possibility of taking initiative. A minority of workers are encouraged by such negative incentives as punishment, the threat of losing their jobs, and paternalism (caring for the employee).

Findings:
Based on the results of both methods to study labor motivation, the motivation of most employees comes from the content of the work, i.e., the work process is important for them (working conditions, the possibility of career growth and professional development, the possibility of taking the initiative, permanent employment). The manager should pay attention to the employees who are characterized by avoiding difficulties in the work process: passivity, avoiding responsibility, and minimizing labor efforts. Such employees should be motivated by negative incentives, such as punishments (in particular, material ones, since it is important for such workers not to "lag" behind others in terms of wages), threats of dismissal or demotion, comments, or reprimands.

Also, the workers with this motivation type can be stimulated with natural incentives, such as buying or renting a house, providing a car, or providing any services. Given the specifics of professional activity, paternalism is an important form of stimulation for such workers. This consists of "custody" or patronage through benefits, i.e., benefits, allowances, a developed system of personnel training, a system of advanced training, and increased wages for seniority.

The study of socio-psychological adaptation using the questionnaire of socio-psychological adaptation developed by K. Rogers and R. Diamond.
The percentage results according to the criteria of socio-psychological adaptation are presented in Figs. 5, 6, 7, 8, 9, 10.

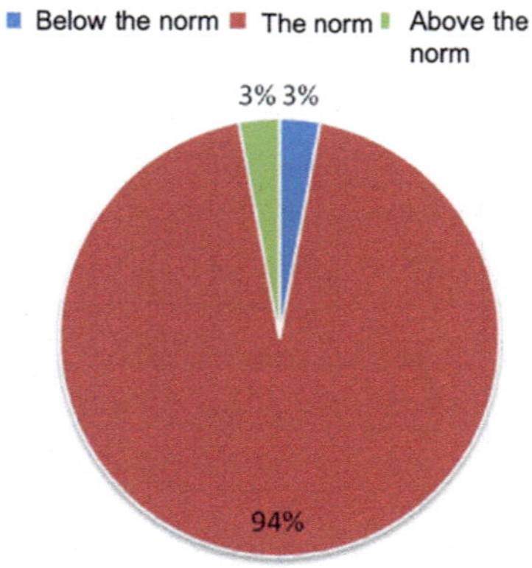

Fig. 5. The percentage results on the "Acceptance of Others" scale by Rogers-Diamond

According to the results of the methodology based on the criterion "acceptance of others" (Fig. 5), the majority of the respondents (94%) are within the normal limits, 3% of the respondents are below the normal limits, and 3% are above. Thus, 3% of the respondents have difficulty communicating; they neither understand nor accept other people. This percentage is small, which suggests that the vast majority of the respondents do not experience communication difficulties.

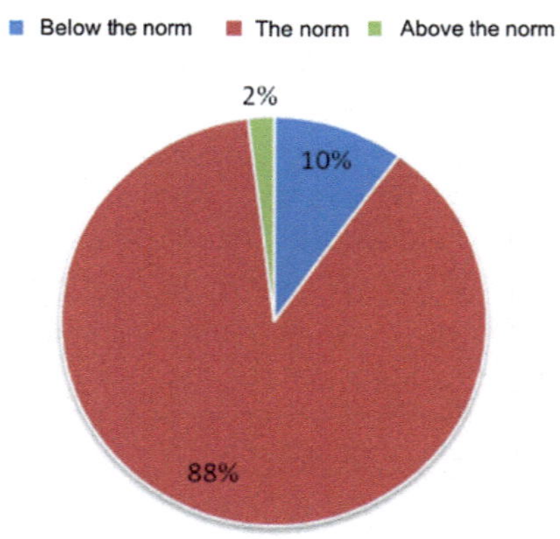

Fig. 6. The percentage results on the "Emotional Comfort" scale by Rogers-Diamond

According to the criterion "emotional comfort" (Fig. 6), 10% of the respondents have low results, 2% have high results, and most of the respondents (88%) have average results. The results show that 10% of the respondents have a high level of tension; they are characterized by sensitivity and irritability.

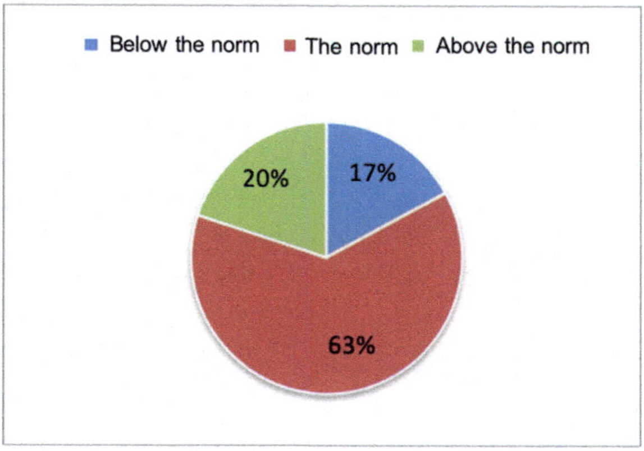

Fig. 7. The percentage results on the "Internal Control-External Control" scale by Rogers-Diamond

The criterion of socio-psychological adaptation, "voluntary self-control," corresponds to the "Internal Control-External Control" scale in the method of Rogers-Diamond (Fig. 7). The results of the study show that 17% of the respondents have difficulties with internal self-control, suggesting that they perform better under the supervision of colleagues and that independent work makes them difficult. Also, they may have low discipline, inability to control their behavior, and dependence on their mood. 20% of the respondents have high results by this criterion; 63% of the respondents have the average results.

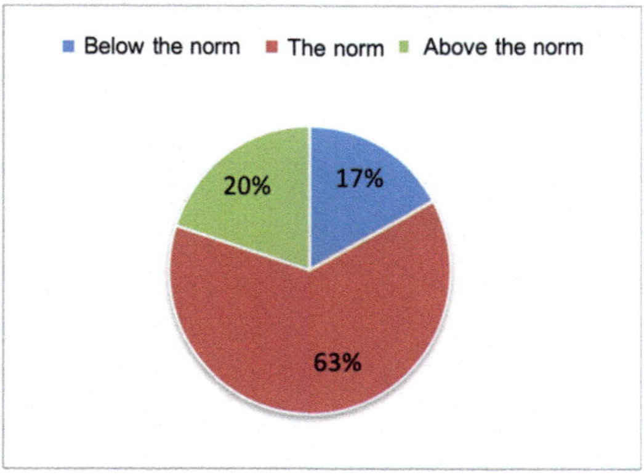

Fig. 8. The percentage results on the "Dominance-Bulletin" scale by Rogers-Diamond

The "Domination – Statement" scale in the Rogers-Diamond methodology corresponds to the criterion of socio-psychological adaptation of the "Autonomy and independence of behavior" that we have selected. On this basis, 17% of the respondents have low results, 20% high, and 63% average. Thus, 17% of the respondents are characterized by softness, pliability, helpfulness, humility, and dependence on others.

4 Discussion

In order to find out which of the criteria prevails in the group as a whole, the raw scores obtained by us in the course of the study were averaged and presented in the form of a diagram in Fig. 9.

In general, by almost all the criteria, the results of the respondents fit into the norms. The only deviation from the norm is the "Dominance" scale, which indicates a high degree of independence and autonomy of the behavior of most respondents.

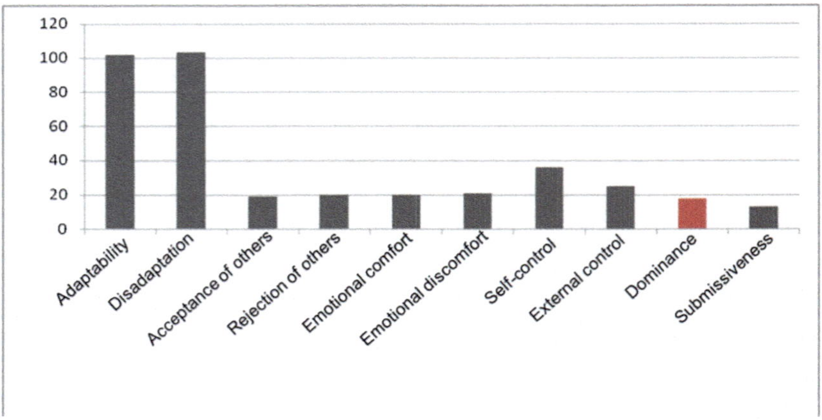

Fig. 9. The average scores on the Rogers-Diamond methodology scales.

Using the Rogers-Diamond methodology, one can also identify the general level of socio-psychological adaptation of the respondents. The results for this indicator are presented in Fig. 10.

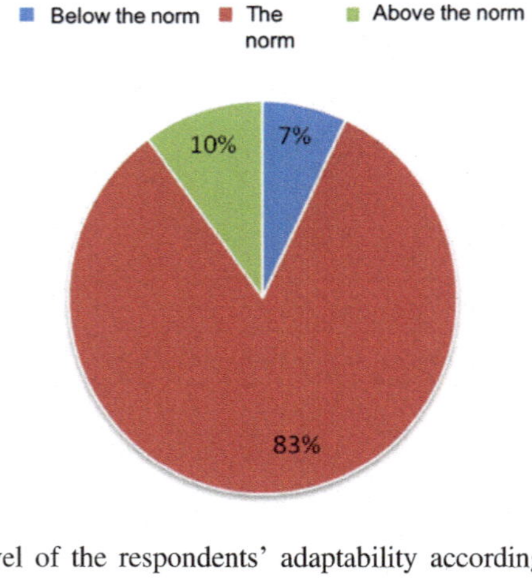

Fig. 10. The overall level of the respondents' adaptability according to the Rogers-Diamond methodology.

According to the indicator "Adaptability – Disadaptation," the results of the majority of respondents are within the normal limits (83%), 10% of the respondents have high rates, while 7% of the respondents have low indicators by this criterion.

According to the results of the study of socio-psychological adaptation according to the Rogers-Diamond methodology, in general, the majority of the team members are adapted to the professional activities (93%). The non-adapted respondents, as a rule, experience difficulties with self-control, are dependent on the colleagues and experience emotional discomfort.

To identify the relationship of labor motivation and socio-psychological adaptation (hereinafter SPA), a correlation analysis of all the data obtained during the study was carried out. For the analysis, we used the Pearson linear correlation coefficient.

Let us consider the results obtained for each type of labor motivation in detail:

1. The transforming motivation of the procedural orientation is negatively associated with the "non-acceptance of others" factor ($r = -0.327$, $p < 0.05$). This suggests that for the workers with this type of labor motivation, difficulties communicating with others do not arise; they are positively disposed to communication and quickly converge with people.
2. The communicative motivation of the resulting orientation is positively related to the factor of "job satisfaction" ($r = 0.343$, $p < 0.05$) and negatively related to the "submissiveness" factor ($r = -0.407$, $p < 0.05$). Such results indicate distrust, prudence in actions, cunning, independence, and autonomy in social behavior. The employees can communicate and are focused on active interactions with others and communication.
3. The communicative motivation of the process orientation is positively associated with the "statement" factors ($r = 0.367$, $p < 0.05$). It is negatively associated with the "emotional discomfort" ($r = -0.328$, $p < 0.05$) and "job satisfaction" factors ($r = -0.343$, $p < 0.05$). Such employees are flexible, accommodating to others, compliant, trusting, and dependent on others.
4. The utilitarian motivation of the process orientation is positively associated with the "submissiveness" factor ($r = 0.330$, $p < 0.05$). Such workers are group-dependent, shun rules, disorganized, and irresponsible.
5. The cooperative motivation of the process orientation did not reveal any significant relationships with the factors of the SPA.
6. The competitive motivation of the resulting orientation did not reveal a significant connection with the factors of the SPA.
7. The competitive motivation of the process orientation is positive with the factor "internal control" ($r = 0.313$, $p < 0.05$). Such employees are dependent on the opinions of others, are oriented towards social approval, are dependent on the group, and have a high level of self-control. Such workers feel the need not to be worse than others, so they will not spend unnecessary efforts on improving their professional skills.
8. The achieving motivation of the resulting orientation is negatively associated with the "rejection of others" factor ($r = -0.318$, $p < 0.05$). Such workers are dependent on the group, quickly converge with people, and strive to solve production problems together with others.

9. The achieving motivation of the process orientation did not reveal the significant relationships with the factors of SPA.
10. The instrumental motivation is negatively associated with the "non-acceptance of others" factor ($r = -0.338$, $p < 0.05$). Such employees easily establish social contacts; they are independent, proactive, intellectually developed, smart, and learn quickly. For them, work is a source of income, and they will strive to increase their professionalism in order to earn more. They can be quite effective employees, as long as they are satisfied with the level of earnings.
11. Professional motivation did not reveal significant relationships with SPA factors.
12. The patriotic motivation did not reveal significant relationships with the SPA factors.
13. The economic motivation is negatively associated with the factor of "non-acceptance of oneself" ($r = 0.326$, $p < 0.05$). Such employees are self-confident, active, entrepreneurial, independent, and prone to adventurism and leadership. Sellers with this motivation type take all the responsibility for the work performed on themselves, perform it with the maximum return. They are efficient workers but challenging to manage. Such people are most suitable for a managerial position.

5 Conclusion

Thus, the highest level of the SPA was shown by the workers with the resulting orientation of the communicative type of labor motivation. This is because communication is of particular importance in professional activities, which is why people with a focus on interacting with others most quickly adapt to such work.

The least adapted are workers with the resulting orientation of cooperative motivation. Employees with this type of motivation are most likely not to experience job satisfaction.

Our results can be applied for:

1. The development of a professional selection program for the position of a penal officer.
2. The development of the SPA program, taking into account the factors and criteria identified by us. In this case, it is necessary to diagnose the SPA level before taking up the position to identify employee reserves, as well as conduct diagnostics after passing the SPA program to identify the results.

References

Batishchev, V.N.: Psychological diagnosis of the competencies of police officers. Materials from the Professional education of employees of internal affairs bodies. Pedagogy and psychology of performance: state and prospects. Moscow University of the Ministry of Internal Affairs of the Russian Federation named after V. Ya. Kikotya, Moscow, Russia (2017)

Dmitrieva, M.A., Druzhilov, S.A.: Levels and criteria of professionalism: problems of the formation of a modern professional. Siberia Philos. Educ. Almanac of SB RAO **4**, 18–30 (2000)

Gerchikov, V.I.: Typological concept of labor motivation (part 1). Motiv. Remuneration **2**, 53–62 (2005)

Ilyin, E. P.: Motivation and motives. St. Petersburg, Russia, Peter (2000)

Kamyshanov, A.A.: Value orientations of cadets (students) of higher education institutions of the Ministry of Internal Affairs of the Russian Federation. Psychopedag. Law Enforc. **1**(7), 69–71 (1998)

Kamyshanov, A.A.: The motivation of the commander of the military unit of the Ministry of Internal Affairs of Russia in military service activities. Psychopedag. Law Enforc. **1**(15), 35–39 (2001)

Kamyshanov, A.A.: The problem of developing personal communication qualities during social and psychological training. In: Proceedings from Science for the Benefit of Humanity – 2016, pp. 44–50. Moscow State Regional University, Moscow (2016)

Kishikov, R.V.: Improving the leadership style of an aviation engineer of the Air Force (Ph.D. in psychology dissertation). Moscow, Russia (2001)

Kishikov, R.V: The psychological foundations of management, Moscow, Russia (2004)

Poruchkina, S.M.: Application of design technologies in vocational education. Train. Psychol. **5**, 61–67 (2017)

Putivtsev, P.V., Noss, I.N.: Problems of increasing the professional reliability of drug control officials. (Series: Psychological Sciences), vol. 3, pp. 118–123. Vestnik of Moscow State Regional University (2011)

Analyzing Key Tools for Digital Transformation of Educational and Scientific Organizations

Nataliia E. Ovchinnikova[1(✉)], Oxana P. Ovchinnikova[2], and Maxim M. Kharlamov[3]

[1] Business School SKOLKOVO, Moscow, Russia
n.e.ovchinnikova@mail.ru,
nataliia_ovchinnikova@skolkovo.ru
[2] RUDN University, Moscow, Russia
ovchinnikova_op@rudn.ru
[3] Yu. A. Gagarin Research & Test Cosmonaut Training Center, Star City, Russia
m.kharlamov@gctc.ru

Abstract. Digital technologies are actively developing currently, and educational organizations can no longer stand aside. Digital transformation has affected almost all spheres of life, from daily tasks and work processes up to complex medical interventions, etc. A new job market requires new level specialists, and it is impossible to prepare them unless radical changes in the personnel training are made. Educational digitalization, the ICT use in the educational process, and the creation of new campuses generation indicate the urgent need for the higher educational institutions to be digitally transformed. Digital transformation is not just the interactive whiteboards appliance or the electronic class-book usage. It implies a systematic change in key processes that could lead to the creation of a perfect new education system, which allows the preparation of people, able to learn and retrain without spending much time on this. This article analyzes the key tools for higher educational institutions digital transformation. The key ways of using the toolkit are highlighted, as well as its implementation effect. Further study stages are also presented.

Keywords: Digital transformation · Educational organizations · Digitalization · Artificial Intelligence · Internet of Things

1 Introduction

In our opinion, no university has completed the digital transformation process yet, which would fully use the digital technologies capabilities to improve not only the educational, but also the administrative process. In a world of rapidly changing technological solutions, scientific and educational organizations often lag behind, thereby widening the gap between market demand and supply on their part.

Despite the fact that the process known as digitalization affects the life of all spheres, non-governmental, educational and scientific organizations clearly remain on the sidelines (Dräger et al. 2014; Popkova et al. 2019). The most widespread

A. V. Bogoviz and Y. V. Ragulina (Eds.): ISCI 2019, LNNS 115, pp. 728–735, 2020.
https://doi.org/10.1007/978-3-030-40749-0_86

digitalization has happened in the information technologies field, with the smallest extent in the humanitarian sphere. The research and development processes are the same as in the medieval university – the same teacher, standing at the blackboard and giving a lecture, and there are the same textbooks (Hanna 2016). The only thing that has changed is the tools. There are projectors instead of a blackboard, as well as online books and various electronic devices instead of paper textbooks. However, this does not affect the organization of educational and administrative activities.

Every year it becomes more and more obvious that digital transformation will move to a more fundamental level and change the whole university life (Bowen 2013). The only question that remains open is whether the university would keep up with these changes?!

In addition to the changing context of consumer and labor markets, educational and scientific organizations have a strong influence on social such megatrends as education individualization, globalization, mobility, and lifelong learning. Education is becoming more accessible, the market for massive open online course is growing every year, and the education cost significantly differs from the university one. According to UNESCO, 13 million students studied online in 2015. The online courses market (Rieber 2017) creates a real competition to traditional education forms. The emergence of small private online courses (SPOC) opens up the access to the educational services market not only for educational and scientific giants, but also for individual educational courses developers.

Recorded lectures are also becoming popular, and the students can listen them to at the most comfortable time and place for them. Consequently, any educational organization can become a global distance platform for the advanced knowledge provision focused on different population segments in different parts of the world (Heflin et al. 2017).

Entering the international market, educational organizations need to engage in the struggle for foreign students, thereby creating competition to the world's leading universities. Educational and scientific organizations should work out the content in accordance with the global consumer and labor market demands, providing this content in the most convenient formats in English.

The digital transformation of the educational process organization and training courses formation is a mandatory stage at the present time. By creating modular online course designs, a university gives students the opportunity to form their own unique educational package and get exactly those skills they need (Scott et al. 2017). Along with online learning, there is a successful "blended learning" practice, which allows combining online and offline formats and shifts the teacher from being a lecturer to being a mentor (Horn and Fisher 2017).

New interaction formats between students and teachers, expanding learning opportunities and enhancing the applied direction of received knowledge, are applied during the digitalization of educational process. These interactions are gamification (Lameras et al. 2017), interactive videos (Barr 2018), and simulation models (Maurer and Engelmann 2018).

Digital transformation is also reflected in the tasks of academic staff and teachers. The lack of necessary skills for working with ICT and the slowness in mastering modern educational technologies create a gap between students and lecturers. This gap is easily neutralized by training staff and building a digital ecosystem that would facilitate the work of academic staff and increase the level of educational activities in the educational institution. In this article, we will look at the ways of conducting digital transformation in organizations and formulate a roadmap for digitization of key processes in educational and scientific organizations. In a rapidly changing world, the need for new competencies, related to the ability to use ICT in the educational process and the universities' "digititalization," is growing within the international politics framework, as well as by increasing the competition level for talent on the world stage (OECD 2018b). However, along with the importance of using ICT tools in the educational process, there are no guarantees of active students' participation in the educational process (Kirkwood 2009) and their high achievements (Tamim et al. 2011). That is why, it is important for universities to have such a pedagogical staff, combining both the public speaking skills and fluency in modern technology.

In this article, we will analyze the modern key tools for educational organizations digital transformation.

2 Materials and Methods

A detailed analysis of the "digital transformation" process in the context of educational organizations was conducted as a part of the study. For that purpose, we studied key works on this topic, analyzed trends in the field of modern digital technologies, and conducted a comparative analysis.

Key tools for educational organizations digital transformation were identified on the basis of the previously published scholarship.

3 Results

Currently, there is a wide variety of tools for achieving digital transformation of educational and scientific organizations. Most educational and scientific organizations limit their digital transformation process with such tools that do not make them more competitive in the market. Such tools include open online courses, LMS systems, the social networks use for educational purposes, among others. At this stage, considering advanced technologies in the field of digitalization and discussing possibilities of their application in educational and scientific organizations seem to be the most important. First of all, it is necessary because universities and research organizations train specialists for the future labor market, and they should be involved in the use of technologies and tools that future graduates would face in their jobs. Table 1 discusses the key tools of digital transformation, their possible application in educational and scientific organizations, as well it makes an estimate of the effect from implementing this digital transformation tool.

After analyzing the comparative study results of the key tools pf digital transformation of educational organizations, we can distinguish a certain challenges, arising when introducing tools to educational institutions (Aldowah et al. 2017):

- *The transition to open cloud technologies*

Table 1. Digital transformation tools for educational and scientific organizations.

Digital transformation tools	Mode of application	Implementation effect
Artificial Intelligence (AI)	Intelligent software systems, constantly learning and improving, can improve the educational process. Training through interactive programs using holograms and intelligent systems	In terms of attracting students, the AI use allows a round-the-clock support and an individual counseling, which affects the enrollment and retention rates In terms of the educational process, the AI helps students to select the necessary basic resources; it also makes the work with a teacher more fundamental and effective The AI allows students to receive counseling, rehearsal services, or other services based on previous and projected academic results The AI can use information from multiple campus systems to manage administrative decisions and tailor training programs to the employers' needs (McKenzie 2018)
Robotization	Using robots to perform routine and automatic tasks	AI can replace administrative staff. The saved funds would allow to reduce operating expenses. This will be reflected in training cost reduction, the opportunity to modernize facilities and develop new curricula (West 2019)
Internet of Things	Implementing a digital campus. Real-time data collection Monitoring, decision making, and process optimization. Progressive student assessment. Existing learning platforms integration	Digital campus will reduce crime rates, create an improved integrated infrastructure with real-time connections, provide access to data through sensors, improve working conditions, and reduce costs Real-time monitoring of student performance improves the level of those students completing their education (Aldowah et al. 2017; Bogoviz et al. 2019)
Augmented reality	Using augmented reality in the educational process immerses students in an environment in which they need to work or, conversely, return to the past for studying certain phenomena or	Augmented reality opens up new opportunities for the educational system to improve the training level of the highly professional specialists. At the same time, augmented reality reduces possible

(*continued*)

Table 1. (*continued*)

Digital transformation tools	Mode of application	Implementation effect
	circumstances. Information about the environment and its objects is superimposed on the real world	negative practice-oriented training consequences, which may be associated with a risk to life
Virtual reality	Applying educational situations for the visualization in the educational process Preparing students for jobs with VR technology. Attracting students by demonstrating the campus through VR technology	Increasing involvement into the educational process. The education level is enhanced by developing practical skills and preparing for real problems solving. Interest in an educational organization, which uses advanced technology, increases even at the stage of attracting students (Brereton 2019)
Digitalization	The inclusion, improvement and transformation of educational and research processes using digital technologies	The education digitalization will make it more scalable and affordable Lifelong learning provides access to education for people of different ages and social opportunities. The transition to digital education will make the educational process more adaptive and efficient The education personification through introducing digital technologies will allow to provide relevant feedback and flexibility in the formation of educational trajectory The digitalization of education will open new opportunities through mobile devices, cloud systems, and other applications The digital transformation of workplaces requires the specialists of a new level, constantly improving their knowledge and skills. The access to continuing education can be achieved through the digitalization of education (BCG 2019)

Many universities use hybrid cloud technologies, which are tied to corporate programs and are quite hard to use. Students prefer using open cloud technologies for their broadcast convenience and group work via the Internet. The universities' content lies in private cloud spaces, and access to the data can only be obtained through corporate networks.

– *The rapid development of educational technologies*

The widespread use of educational LMS systems (such as Moodle and Blackboard) creates a huge structured and non-structured data amount (audio, text, and video content). Students can access this educational content wherever they want. However, this serves as a certain challenge for universities, because there is a necessity in conducting lecture broadcasts in real-time as the database grows. Moreover, a necessity in constantly updating the educational content requires a different type of work and organization of students' engagement.

– *Competition with mobile applications*

Universities begin their way in the development of educational applications, the most advanced of which provide students with their own schedule, track performance, and an access to the educational content. However, it is important to understand that there are many convenient and modern formats in the mobile applications market that can distract and enthrall a student. Moreover, the university's mobile applications should be one step ahead, providing content that students cannot get from third-party developers.

– *Growing information technology costs*

The rise in the information technology cost continues to increase at the content level as well as at the application level. The ability to correctly allocate financial resources, assess the developing costs, using ICT in the educational process, is very important. Higher education should offer new formats for the infrastructure provision to implement the ICT in the educational process.

4 Discussion

The study results showed that there are many tools for conducting digital transformations in educational organizations. Despite all their diversity, most of them boil down to a few key goals:

- Improving the training level;
- Establishing a modern educational space;
- Optimization of the educational organizations' infrastructure;
- Increasing the competitiveness level of a particular educational organization.

The results of our comparative analysis demonstrate that there is a large number of positive effects from the digital transformation of educational organizations. However, there are challenges that entail transformation processes. The list of challenges can be expanded, new contexts can be added. However, it seems appropriate to consider these tools functioning within the transformation models framework.

The key limitations of our research lay in the fact that there are insufficient experiences by educational organizations in the implementation of digital transformation tools. Therefore, we (as researchers) face the inability to assess the level of theorization of the identified effects from implementing these tools.

5 Conclusions

The educational organizations digital transformation is an integral evolution process of the higher education institutions. New digital technologies offer unique opportunities that allow universities not only raising the education level, but also introducing significant changes in the educational infrastructure and the administrative organization of educational process. Using information technology, universities can solve such problems as forming sound financial policies, increasing the information resources availability, designing smart and safe campuses, and building curricula for the new generation.

The introduction of information technology in the educational process can increase students' interest in learning, provide an opportunity to get practice-oriented knowledge, and, most importantly, increase the learning speed.

The university does not exist separately from the whole world, and starting the digital transformation process stimulates changes not only within its structure, but also affects the surrounding ecosystem. Universities can not only introduce modern technologies into the educational process, but also these technologies serve as drivers for the formation of new social and technological contexts so as not to lose their relevance and usefulness.

The universities' digital transformation entails not only the changing of ideas about higher education, but also contributes to the formation of a new people generation, who have a fundamentally new outlook and a completely different position. Universities cease to be the knowledge transmission institutions, they become agents for training people with flexible skills and the ability to retrain under a new and changing environment, thereby providing themselves with work.

In further works, it seems appropriate to consider the digital transformation model of educational organizations, which allows using certain key elements, such as the artificial intelligence, the Internet of things, robotization, as well as the virtual and augmented reality.

References

Aldowah, H., Rehman, S.U., Ghazal, S., Umar, I.N.: Internet of Things in higher education: a study on future learning. J. Phys.: Conf. Ser. **892**, 012017 (2017). https://doi.org/10.1088/1742-6596/892/1/012017

Barr, M.: Student attitudes to games-based skills development: learning from video games in higher education. Comput. Hum. Behav. **80**, 283–294 (2018)

BCG: A bBlueprint for dDigital eEducation (2019). https://www.bcg.com/industries/public-sector/successfully-transitioning-digital-education.aspx

Bogoviz, A.V., Lobova, S.V., Ragulina, Y.V., Chernitsova, K.A., Shkodinsky, S.V.: Internet tools for development of knowledge economy: essence, tendencies, and perspectives. Adv. Intell. Syst. Comput. **726**, 39–45 (2019)

Bowen, W.G.: Higher Education in the Digital Age. Princeton University Press, Princeton (2013)

Brereton, E.: 4 ways colleges are embracing virtual reality (2019). https://edtechmagazine.com/higher/article/2018/05/4-ways-colleges-are-embracing-virtual-reality

Dräger, J., Friedrich, J.-D., Müller-Eiselt, R.: Digital as the new normal – how digitalization is changing higher education. Centrum für Hochschulentwicklung gGmbH., Gütersloh (2014)

Hanna, N.: Mastering Digital Transformation: Towards a Smarter Society, Economy, City and Nation. Emerald Publishing, Bingley (2016)

Heflin, H., Shewmaker, J., Nguyen, J.: Impact of mobile technology on student attitudes, engagement, and learning. Comput. Educ. **107**, 91–99 (2017)

Horn, M.B., Fisher, J.F.: New faces of blended learning. Educ. Leadersh. **74**, 59–63 (2017)

Kirkwood, A.: E-learning: you don't always get what you hope for. Technol. Pedag. Educ. **18**(2), 107–121 (2009). https://doi.org/10.1080/14759390902992576

Lameras, P., Arnab, S., Dunwell, I., Stewart, C., Clarke, S., Petridis, P.: Essential features of serious games design in higher education: linking learning attributes to game mechanics. Br. J. Educ. Technol. **48**, 972–994 (2017)

Maurer, H., Engelmann, C.: Aktives Lernen durch EU Simulationen: Eine kritische Auseinandersetzung mit Erfahrungen an der Universität Maastricht. In: Muno, W., Niemann, A.W., Guasti, P. (eds.). Europa spielerisch erlernen. PB. Springer, Wiesbaden (2018). https://doi.org/10.1007/978-3-658-17463-7_10

McKenzie, L.: Pushing the boundaries of learning with AI (2018). https://www.insidehighered.com/digital-learning/article/2018/09/26/academics-push-expand-use-ai-higher-ed-teaching-and-learning

OECD: The future of education and skills: Education 2030. OECD, Paris, France (2018b). http://www.oecd.org/education/2030/OECD%20Education%202030%20Position%20Paper.pdf

Popkova, E.G., Ragulina, Y.V., Bogoviz, A.V. (eds.): Industry 4.0: Industrial Revolution of the 21st Century. Springer, Berlin (2019)

Rieber, L.P.: Participation patterns in a massive open online course (MOOC) about statistics. Br. J. Educ. Technol. **48**, 1295–1304 (2017)

Scott, D., Ribeiro, J., Burns, A., Danyluk, P., Bodnaresko, S.: A Review of the Literature on Academic Writing Supports and Instructional Design Approaches within Blended and Online Learning Environments. University of Calgary Digital Repository, Calgary (2017)

Tamim, R.M., Bernard, R.M., Borokhovski, E., Abrami, P.C., Schmid, R.F.: What forty years of research says about the impact of technology on learning: AA second-order meta-analysis and validation study. Rev. Educ. Res. **81**(1), 4–28 (2011)

West, J.: How robots could affect the future of higher education (2019). https://www.qs.com/how-robots-affect-future-of-higher-education/

The Dual Education System of Germany in the Conditions of Migration Crisis

Anastasia Yu. Pleshakova(✉)

Ural State Mining University, Yekaterinburg, Russia
pleshakova.a@m.ursmu.ru

Abstract. The relevance of the my research is determined by the problems experienced by the European educational systems in connection with the increasing migration activity waves. Political and economic world crises, wars, and demographic instability entailed a threat to identity preservation by the national education systems. The objective of the paper is to examine the problem of the adaptation of migrants and refugees using the example of the German dual education system and efforts to resolve it by the federal government, the Federal Institute of Vocational Education, chambers of commerce, and industry and professional schools. The leading research methods were the scientific text analysis, expert interviews, and content analysis of empirical materials. As a result of the study, difficulties arising when migrants or refugees are involved in vocational education in the Federal Republic of Germany (FRG) were identified; necessary conditions for effectiveness and sustainability of the dual education system as the national identity product were determined. The study materials can be useful for the researchers of the vocational education system of those countries with a high migration activity to determine mechanisms for responding to globalization changes, as well as vocational education systems of the countries seeking to borrow the German dual education experience in order to create conditions for the effective functioning of a similar educational model.

Keywords: Globalization · Internalization · Dual education system · Migration crisis

1 Introduction

The globalization processes and internationalization of education all over the world are perceived as a hard fact and have consequences for the world education systems, including the dual educational system of the Federal Republic of Germany (Brown, Kirpal and Rauner 2007). The processes of globalization limit the desire of the educational system of any country for the uniqueness of its educational standards (Friedman 1994), as well as the opportunity to discuss the future vocational education of the country only within the national context. The global political crisis, wars, and the ensuing migration wave exacerbated the issue of preserving the identity of educational systems, especially under pressure from migrants and refugees (Brown 2017).

A. V. Bogoviz and Y. V. Ragulina (Eds.): ISCI 2019, LNNS 115, pp. 736–744, 2020.
https://doi.org/10.1007/978-3-030-40749-0_87

The ongoing internationalization of education, understood by us as an integrative process of introducing international educational standards into the educational process of universities based on (a) the activity approach (using active and interactive technologies for improving the teaching and research quality), (b) competence approach (the result of higher professional education represented by a set of competencies), (c) ethical approach (implementing intercultural communications based on traditions and innovations, the creation of the adapt conditions reducing the internationalization risks), (d) strategic approach (developing internationalization in management and education management, introducing international standards for measuring its quality) involves the introduction of elements of culture and traditions into any education system, including Germany and other countries (Pleshakova 2014). In this context, it is necessary to mention the impact of the global integration of labor, goods, and services markets on the structuring of the national vocational education system (Allen 2015) and such aspects as the integration of international and intercultural competencies (Padur and Zinke 2015), ensuring recognition of the professional qualifications obtained abroad (Ebbinghaus and Gei 2017).

The migration problems have become particularly relevant for Germany recently. It was the state that initiated the reception and social security of refugees from the Asian and African countries. The issue of attracting migrants and refugees to education and their adaptation in vocational schools in Germany was a challenge to global time and demanded an immediate solution. According to Ani Karliczek, the Minister of Education of the FRG, the most relevant topics in the vocational education development in Germany are the trends of demographic decline, increasing the attractiveness of vocational education, digitalization, and theoretical training and practice computerization. She paid particular attention to the involvement of migrants in labor, since the stability and effectiveness of the dual education system may be threatened (BIBB 2018).

2 Materials and Method

The main goal of the study is to comprehend the contemporary experience in solving problems related to the internationalization process of the dual education system in the FRG. A comparative analysis of the official data of the Federal Institute of Vocational Education, the Ministry of Education of Germany, the Ministry of Labor of Germany and the opinions of the direct participants of the dual education system (vocational schools employees, representatives of the chambers of commerce and enterprises), as well as experts who determine the conceptual development of the dual education system, allowed forming the estimates of globalization changes associated with the inclusion of migrants in the educational space. The purpose of the empirical study is to determine the conditions and ways to address the issues of migration policy in the vocational education system. The basis of the study was the German Federal Research Institute for the Study of Vocational Education, the Chamber of Commerce and Industry, and the vocational schools of Bavaria, Hesse, Berlin. A total of 127 experts participated in the study. The study did not take into account the differences in age and sex. The choice of our respondents occurred spontaneously. The method of expert survey was used to perform the empirical study tasks; its results were processed using the content analysis. The survey was anonymous. In addition to the interview, a questionnaire with open and closed questions, revealing such a problem as the inclusion and involvement of the migrants in a professional educational environment, was developed.

3 Results

Many German companies currently complain about serious difficulties in filling their training places; at the same time, young migrants do not find a place to study (Gei, Matthes 2016). Probably, selection procedures used by the companies create unfavorable conditions for young migrants. In particular, their school certificates are not recognized to the same extent as the certificates of non-migrants. This provision also applies to those refugees whose integration can only be successful due to the excellent quality of vocational training. The professional training of such people is a particularly severe problem in Germany these days, and its solution requires not only increasing the level of the recognition of migrants' education documents but also increasing the duration of teaching professional German language, as well as the measures for the potential applicants social and cultural adaptation.

For many respondents, the questions of globalization impacting the state of the dual educational system were associated with the actual problem of refugees and migrants adaptation. A high degree of influence of the migration crisis on the effectiveness of the dual education system was stated by 72% of the surveyed experts. Many experts noted the problems of cultural and moral nature.

Expert 11: *"...The issue of cultural adaptation of refugees is also acute. We try to educate them so that they are not late and follow the rules. We expect them to accept these requirements and accept our rules. Those students who follow the rules are successful..."*

Expert 21: *"...At school, German pupils dominate quantitatively and culturally, urging a minority of migrants and refugees to abide by the traditional school rules. In many cases, pupils turn to a teacher or principal if the set rules are violated..."*

Expert 8: *"... In many ways, the process of cultural adaptation depends on the school and the enterprise. The problem of attitudes towards the women's position is one of the most pressing since pupils among migrants or refugees are not ready to accept the study material or behavior criticism from a female teacher. The issue is resolved only through administrative and social pressure; in most cases, only a woman director or a woman employer can influence..."*

Expert 42: *"... When a migrant or refugee is outside the national community or the community is small, the adaptation process and the acceptance of the cultural norms of the indigenous population take place more actively. This process is easier in small cities, and more difficult in large cities and regional centers, in which the community creates a micro-society while preserving traditional values and the rules of behavior..."*

Expert 5: "*... assimilation and adoption of the norms and rules of behavior among people with common religious or cultural foundations are happening faster, for example, among Russian Germans...*"

Expert 28: "*... some schools have the classes for refugees where they study German and profession basics. The average age of students is 15-16. However, they do not have a working permit...*"

The answers of our experts testify the awareness of the problem of managing migration flows, the involvement of migrants in the professional environment, and confirm our assumption that the respondents consider the problem of the employment and adaptation of migrants through the globalization lens. Experts emphasize differences in the cultural code of indigenous Germans and migrants, offering various tools for overcoming them.

For the younger generation of Germans, cultural differences are not a particular problem. The processes of globalization and migration are more familiar to the younger generation, which makes it easier to adapt to other traditions, and creates conditions for the integration of cultures. However, experts are not inclined to consider the trend towards integration as positive and advocate the preservation of national identity. The detailed answers in the questionnaire to the question about national character traits that positively affect the learning process showed a clear connection between the reproduction of traditional national German character traits in the education process and the education results:

Expert 1: "*... Our clear education structure, reasonable prospect of applying knowledge in the profession for the student, the specific enterprise with which the contract is signed, and the workplace in the future demonstrate certainty, transparency, understandability, stability, and perspective. This requires each student to be diligent, responsible, to follow the rules and requirements of the teacher, and fosters clarity, accuracy, honesty, and devotion to the work or profession...*"

Expert 26: "*... Law and order are important for the Germans!*"

Expert 31: "*... There are rules with the basic rights and obligations, but the rules are not enough if one is not willing to follow these rules. Diligence and clarity are in the character of the Germans...*"

Expert 9: "*... The Germans are standing when the red light is on. We follow the rules!*"

Expert 16: "*... All participants in the learning process have their responsibilities, and we bring up that responsibilities are first and foremost!*"

In their answers, the experts emphasized the importance of such national qualities of the Germans as accuracy, punctuality, observance of rules and laws, and obligation. Possession of these character traits and familiarization with national values are one of the key conditions for the successful adaptation of migrants into the educational space and maintaining the effectiveness of the vocational education system.

Some answers on the question about the difficulties that respondents face during the implementation of the educational process within the dual education system reflect the awareness of the need for a new workforce and the expectation of actions from the federal authorities to legalize the involvement of migrants and refugees in the professional environment.

Expert 26: *"... At one of the meetings at the Chamber of Commerce and Industry, there was a group of independent artisans, school principals, agents of employment companies, and government officials. The issue related to the employment of migrants and refugees was discussed. It is essential for us! Many problems arise when applying for a job..."*

Expert 11: *"The profession of a baker is quite complex and physically difficult, not popular among the young generation in Germanym but extremely in demand in the market. Businesses want to attract migrants for baking and to enter into a contract with them for schooling. In turn, the Chamber of Commerce and Industry would like to help but may not always influence the state policy regarding migrants. All parties are waiting for a decision from the state."*

Identity preservation associated with the national mentality and character traits of the Germans is the primary condition for effectiveness of the dual education system, according to the experts surveyed. It was particularly evident in the issue of borrowing elements of the German vocational education by other countries directly related to the use of the dual education system by representatives of a different cultural code: 83% of the respondents believe that the introduction of the dual system cannot be successful in other countries, providing a list of arguments.

Expert 12: *"...The Chinese and Arabs are actively trying to borrow the German experience in the implementation of the dual system. However, to ensure effectiveness of the process, not only the desire of individual companies is needed, but the legislative framework that ensures the legitimacy and transparency of the entire learning process. Also, the financial side of the issue is important. When it comes to scholarships that companies must pay, the question arises, what is the benefit of the company? Moreover, there is nothing to answer here. Only costs are presented. The education cost per student is 10,000 EUR for large companies. Imagine that a company has 20-30 students! The companies must cover expenses, although a student does not produce anything during this period. Foreign companies do not understand what they should pay for?"*

In Germany, the state pays only for education. In other countries, students who come to work in a company after vocational schools most often need to adapt to the workplace and correct their professional knowledge and skills necessary for a particular company. That is why, students begin working immediately in Germany, and a company interested in hiring does not spend on the adaptation period. Even if a single company realizes the importance of investing in the educational process, the state support and the desire (or duty) of all the companies in a country to participate in the dual education are necessary. After all, infrastructure, school, and personnel are needed, and the state finances this purpose in Germany.

Expert 7: *"... A few years ago I was in the UK... teachers there say that they have no students sometimes. It all depends on the stability of the financial situation of the company. If there is money, the company sends students to school, ad vice versa. This is not a dual system..."*

Expert 22: *"... The dual system designates a strict framework for all participants in the educational process but defines a specific benefit, profit, and goal. The habit of following the rules and regulations is formed. However, students in other countries do not know their future; they remain in uncertainty and do not see the specific application of their knowledge. The state of uncertainty stimulates them, on the one hand, to strive for professional flexibility and mobility; and on the other hand, it forms such traits as a neglection of rules, non-binding, assiduity, and lack of punctuality."*

Expert 27: *"... If we compare how people work in Germany and how people work in other countries, then we must say that Germans always want to be accurate and consistent in their work, they are perfectionists, and this is inherent in their mentality. References to the goods manufactured in Germany must demonstrate the highest level of quality production. Therefore, professional training must be at a high level, and we are proud of our system ...".*

My analysis of expert opinions indicates various conditions for implementation: from legislative mechanisms ensuring the legality and legitimacy of the participants' actions in the educational process, from the compulsory financial support for educational institutions by the state and industrial enterprises to the value orientations of vocational school students. There is no optimism in their vision of the results of the introduction of dual education in other countries.

4 Discussion

The German vocational education and, in particular, the dual education system is seen as a model of success worldwide (Thomann and Mouillour 2012). In Germany, vocational training contributes significantly to the growth of the economy, employment, and social welfare of citizens (Sell and Schultheis 2014). This is achieved as a result of the great state, society, and business community work in creating an effective social partnership (Wossman 2005). Moreover, the dual education system has passed a long historical path from labor schools to modern vocational schools. It is supported by spiritual and moral attitudes of Protestantism and shaped the German professional image.

At present, the position of the titular German nation is quite difficult, since demographic indicators show a decline in the birth rate, and every fifth German resident is of a foreign origin (BIBB 2018). In 2014, a quarter of those who wanted to study in the vocational schools, representing a total of just under 560,000 people, were migrants (Beicht and Gei 2015). Migrant applicants were either born abroad and had foreign citizenship, or studied another language as their mother tongue. About two-thirds of potential migrant students were born in Germany, and about the same had German citizenship, almost two-thirds studied German as their mother tongue and passed through the German school system. However, this does not reduce tensions and generates an identity risk for the dual system. The studies conducted by the Federal Institute of Vocational Education in Germany show that migrants, even with diligence, are significantly less likely to succeed in vocational education than the native Germans (Beicht and Gei 2015).

For example, only 27% of young migrants received training based on enterprises in 2014, while the corresponding indicator for non-migrant young Germans was 42%. Despite some improvements in the situation, there has been practically no increase in the rate of vocational training at the enterprise base among migrants over the past ten years. It is known that the certificate obtained after graduation is important in finding a company to conclude a training contract. The higher the score, the higher the odds. However, even with the same average score on a school certificate, migrant candidates are less likely to succeed than non-migrants. The chances of migrants who have a high school diploma are currently only slightly higher than those of a lower secondary school diploma.

A similar situation is manifested when entering a university (Beicht and Gei 2015). Indeed, along with the quality of the graduate school certificate, other factors are affecting the migrant's ability to obtain a contract from the company, which draws attention to the social background of the future student, the school certificate level and takes into account the situation on the regional market for professional education. Thus, the probability of obtaining a training contract for a young migrant is significantly lower than that of a native resident, even if the same selection conditions apply. That is why the INA educational project (Integration through the Dual Education System) and the Rhine Chamber of Commerce and Industry deserves special attention (Imove 2018). "Trade and Industry Integration" is another program designed to encourage refugees to receive training. More than 1,000 young people went through vocational guidance to begin learning the craft (Imove 2018).

Despite the efforts undertaken by the Federal Ministry of Education of Germany, the number of participants cannot be compared with the actual number of refugees who have arrived in Germany over the past two years. Recent developments in the German migration policy have become a demonstration of the critical situation in the country, which is unable to cope with the waves of refugees and successfully integrate them into society. At the EU Summit held in June 2018, special attention was paid to the discussion of illegal immigration. Although official sources claim a decrease in the number of refugees in the EU, the majority of immigrants, statistically, rush to Germany. This problem has generated proposals for the creation of platforms for accommodating immigrants outside the EU and the introduction of country quotas for migrants. As a result, it was decided to establish transit centers with extraterritorial status on the border with Austria for illegal immigrants who would expect a decision on deportation back to the country of registration or the official asylum in Germany (Adler 2018).

5 Conclusion

The results obtained in the study identify the conditions for exporting the dual system to other countries and finding a different identity by migrants, which means adopting the traditional German values. This demonstrates the difficulty of blindly copying the experience of the vocational education system in Germany by other countries and the practical impossibility of assimilating migrants into the German socio-cultural space. It leads to the conclusion that the German dual education is a unique high-tech system with a high level of identity, which makes it difficult for the processes of globalization and integration to influence it. This largely determines the difficulty of involving migrants in the vocational educational environment, despite the growing need for labor. Therefore, attracting migrants to professional communities remains a nationally important issue for Germany.

The obtained results are of practical value in determining the prospects for borrowing some elements of the dual educational system by other countries. It is impossible to copy the whole system or transfer the dual education system model to the educational environment of another state since it is not provided with legislative and economic mechanisms for regulating the dual education system of the German sample. Moreover, the cultural code of each nation is individual. Therefore the education system, as a product of national identity, cannot be copied. The paper opens up the prospects for further studies on the identity of regional educational systems and unfavorable conditions of globalization.

Acknowledgment. This work was supported by the Russian Foundation of Basic Researches under the Grant 'The Prospects and Terms for the German Dual Education Elements Adopting to the RF Vocational Education' [18-013-00695A].

References

Adler, K.: Migrant crisis: EU summit divided amid migration row (2018). https://www.bbc.com/news/world-europe-44633606

Allen, M.: Brief history and description of the innovative national system in Germany (2015). www.researchgate.net/profile/Matthew_Allen6/publication/307938035_The_National_Innovation_System_in_Germany/links/5af164efaca272bf4255b7d6/The-National-Innovation-System-in-Germany.pdf?origin=publication_detail

Beicht, U., Gei, J.: Ausbildungschancen junger Migrantinnen und Migranten unterschiedlicher Herkunftsregionen. Aktuelle Situation 2014 und Entwicklung seit 2010 - Ergebnisse der BA/BIBB-Bewerberbefragungen], BIBB Report 3/2015 (2015). https://www.bibb.de/veroeffentlichungen/en/publication/show/id/7834

BIBB: Online documentation of the BIBB Congress (2018). https://kongress2018.bibb.de/en/?s

Brown, A.: Identity work and organizational identification. Int. J. Manag. Rev. **19**(3), 296–317 (2017)

Brown, A., Kirpal, S., Rauner, F.: Identities at Work. Springer, Netherlands (2007)

Ebbinghaus, M., Gei, J.: Duale Berufsausbildung junger Geflüchteter (2017). https://expertenmonitor.bibb.de/downloads/Ergebnisse_20170602.pdf

Ebner, Ch., Uhly, A.: Entstehung und Merkmale des dualen Ausbildungssystems (2016). https://www.bpb.de/gesellschaft/bildung/zukunft-bildung/228394/entstehung-und-merkmale/

Friedman, J.: Cultural Identity and Global Process. SAGE publications Ltd., Thousand Oaks (1994)

Gei, J., Matthes, S.: Refugees on their way into training – what support do they want? (2016). https://www.bibb.de/en/66615.php

Imove: 754 refugees in training (2018). https://www.imove-germany.de/cps/rde/xchg/imove_projekt_international/hs.xsl/news.htm?content-url=/cps/rde/xchg/imove_projekt_international/hs.xsl/754-refugees-in-training.htm

Padur, T., Zinke, G.: The digitalization of the world of work – perspectives and challenges facing vocational education and training 4.0 (2015). https://www.bibb.de/en/36985.php

Pleshakova, A.: Social and pedagogical terms of the high education internationalization process. The news of the Moscow state Agro-Engineering University named after V.P. Goryachkin, vol. 4, no. 64, pp. 55–58 (2014)

Sell, S., & Schultheis, K.: Hintergrund: Berufliche Bildunghttp (2014). https://www.bpb.de/politik/innenpolitik/arbeitsmarktpolitik/187848/hintergrund-berufliche-bildung

Thomann, B., Mouillour, I.: From a special case to a model that is in demand - what makes dual VET attractive abroad? (2012). https://www.bibb.de/en/22750.php

Wossmann, L.: Public-Private Partnerships in Schooling: Cross-Country Evidence on their Effectiveness in Providing Cognitive Skills. Program on Education Policy and Governance, Research Paper PEPG 05-09. Harvard University, Cambridge, MA

Quality Management of Pedagogical Personnel as a Driver for the Development of the Education System

Veronika V. Yankovskaya[1,2(✉)]

[1] Plekhanov Russian University of Economics them. Plekhanov G. V, Moscow, Russia
veronika28-2@mail.ru
[2] State University of Management (GUU), Moscow, Russia

Abstract. In this work, we conducted a study in the field of modifying the competencies of faculty and students, carried out in accordance with the *Concept of Modernization of Russian Education* and the signed agreements at the Bologna Process, during its deep integration into the international community with the assistance of the European Union. The conditions, nature, and problems of teacher quality management and the degree of humanization are considered. **Goals/objectives.** The goal is to discuss the results of a study of the parameters of a complex phenomenon; management, degree of humanization, conditions, and quality of work of "faculty at the Higher School." To achieve the designated goal, the following tasks were solved: (1) A comparative assessment of the criteria for working conditions of various concepts and the degree of their impact on the work of employees of the Higher School; (2) discussion of methods of managing labor processes in a higher educational institution that have a direct impact on the working conditions of the faculty; (3) the possibility of the impact of humanization on the degree of effectiveness and appropriateness of application for evaluating the above process, and the concept of humanization. **Methodology.** The traditional methodology of scientific research, that is, the formulation of a scientific hypothesis followed by argumentation, both in favor of confirmation and disavowal, was used. **Conclusion.** The article proposes an interaction algorithm through the modification of methods, techniques, approaches, and forms of management of the administration-faculty-department-teacher-student, which demonstrates the following. The management tools used make it possible to achieve a specific set of competencies required by the teacher and student during the period of integration of education, in order to ensure the breakthrough development of the individual and the highly skilled worker as creative persons. To increase the effectiveness of the educational process and ensure that the necessary set of student competencies is obtained, it is necessary to create conditions conducive to the disclosure of the teacher's creative abilities. This is possible if the quality management system is modified. The application of the concept of working conditions for teachers of the Higher School, the degree of humanization, which is determined by the factors of the external environment, can positively affect the management of the quality of work of the teaching staff and its professional competencies, in the context of increased digitalization of the economy. **Novelty**. The leading proposal is the recommendation of the author of the article on the need to reform the quality

A. V. Bogoviz and Y. V. Ragulina (Eds.): ISCI 2019, LNNS 115, pp. 745–753, 2020.
https://doi.org/10.1007/978-3-030-40749-0_88

management system of the labor management of the teaching staff, based on the principles of humanization, as well as the creation of a separate system of approved methods for improving the level and development of teachers' competencies at the initial stage of a career, depending on the specifics and profile of the institution.

Keywords: Modernization and integration in education · Bologna Process · Quality management · Humanization · Competencies

1 Introduction

When considering the working conditions of university professors, the degree of humanization of which is determined by the regulatory factors of the external environment, namely, the Russian Federation Codes (civil, labor, budget, administrative), laws of the Russian Federation regulating the activities of universities, one must acknowledge the following: the length of the working week, the number of working days of teachers per week, the duration of leave, the amount of teachers' remuneration, benefits. Contemporary reality leads to a need to change the paradigm of education and to improve the quality of it, to improve the working conditions of the faculty of tertiary institutions and the formation of a concept in education—to synergistic humanization. This concept refers to the use of synergistic and humanistic approaches in education at all levels, from the initial level of education to higher education.

Many elements are included in the whole system of this concept, namely the quality of work of the faculty, working conditions, quality management methods and the degree of humanization of labor, principles and signs, up to the synergies used in the university. Today, the humanization of the teacher's labor is extremely relevant. Under contemporary conditions, changes in accepted norms of ethics and moral values, ideology, culture, and society, in general, can lead to a crisis. Humanization of education involves the formation of moral principles in student/master and teacher relations, as well as among colleagues, the teaching staff, and the administration of the institution. The basis for the humanization of work should be caring for a teacher, employee, colleague, or subordinate, and an active desire to improve the level and quality of life. Realization is possible through capacity development and the full satisfaction of needs (Yankovskaya 2017). All this applies to students.

Humanization is nothing but a holistic system of worldviews that recognize a person as an individual, his rights to mercy, freedom, respect for the principles of equality, and justice, providing favorable conditions for realizing the individual's potential. Today, in the social sciences, these terms have gained significant meaning, since these concepts strictly determine the life of an individual in society. The modern period of time is characterized by a high degree of integration in education. This system of interaction exists also in the European Union. There is a close, mutually beneficial interaction between Russia and the West, almost all the subjects of the Russian Federation. An important task is to achieve an increase in the quality of specialists in accordance with the requirements set forth in the *Concept for the Modernization of Russian Education for the Period Until 2020* and international documents. Agreements signed at the Bologna Process emphasize the need for and importance of training highly competent specialists with a wide range of core competencies.

The accession of Russia to the Bologna Process strengthens the demand for harmonization of the work of the teaching staff. The proper effect of the work quality management of the teaching staff and its harmonization can only be achieved by using the strategy of quality and harmonization of work for teachers. This task, in the era of economic digitalization, is an integral and predominant element of universal human development. In the era of market relations, multiple aspects of the life of an individual have transformed. One of the most important among these aspects is education. Transformations that have occurred in higher education cannot be unambiguously assessed as positive or negative. Freedom of choice for both the applicant and the teacher has increased. In addition to state universities, non-state and municipal universities have appeared as options.

Following a unified teaching methodology is a thing of the past. Each of the teachers is free to choose their own methodology. Applicants choose between a budget and a commercial form of training, depending on multiple factors, including their own capabilities and abilities. The niche in the share of educational services provided has increased dramatically. Education has become more accessible. But this does not mean improving the quality of education, but rather the opposite (Maiorana et al. 2017). The market, being a customer for a specialist of a certain profile, with a certain set of skills and competencies, dictates its conditions. And so, the market orientates universities to the necessary level of training for a future specialist. A university is a mechanism for the reproduction of the field of science and a holistic system in education.

Accordingly, today, a radical transformation of the entire education system is extremely necessary and is dictated by time and conditions. It's about a new quality system. One of the important principles of managing the quality of the teaching staff during the modernization of the university will be a modification of the quality management system (QMS). QMS is the basis for organizing and ensuring a positive effect on the educational process. Today, comprehensive reform of the university is required and is not limited to only restructuring the organizational structure of management, which makes it possible to tighten administrative methods.

Deep modification, capable of providing the desired effect, should lie at its root, but this is impossible without a high degree of humanization of the work of the teaching staff and an ever-increasing quality of labor. In the author's opinion, the semantic meaning of quality management of the teacher's activities and the quality of the educational process lie in a single direction of action. This is in accordance with the chosen strategy outlined in the Concept of Higher School Reform and takes into account the existing values of society.

The external social and working conditions of teachers should include the remoteness of the place of residence from the educational buildings of the university, as well as the convenience of transport.

The degree of humanization of labor is also determined by the complexity of labor operations, as well as their sequence and relationship. For the teaching work, the educational process can be divided into the following elements (or, based on the terminology of the production process, operations): basic, auxiliary, and serving.

In our opinion, the humanity of the basic operations lies in the fact that computer classrooms have all the necessary equipment and working furniture, are prepared for the educational process, and comply with sanitary and hygienic standards of lighting, heating, insulation, and ventilation (Huggins and Thompson 2014).

2 Materials and Methods

In this article, empirically and by means of a survey and questionnaires, the conditions of successful work of the faculty were investigated, key management methods are described, and a study of the concept of humanization and the working conditions of higher education teachers is carried out (Yankovskaya 2017). In the future, this will allow further research into the subject of whether they are able to ensure the success of the educational process of the teaching staff. Three main aspects are working conditions, the degree of humanization, and the degree of effectiveness.

The degree of humanization of the teachers' work depends on the nature of the work and reflects the quality features of the work. It is categorized in a special group or type, according to the list of signs which are monotony, not attractiveness, fashion, prestige, creativity, complexity, efficiency, and strategic orientation (Goryainova et al. 2017). In our opinion, the opposites to the list of signs can be added: monotony-diversity, unattractiveness-attractiveness, fashionability-unpopularity, prestige-insignificance or non-authority, creativity-routine, complexity-no qualification or unpreparedness, efficiency-slowness, and strategic focus-ongoing activities (focus). The characteristics of the teacher's work (teaching staff) affecting the level of humanization are conditionality-unconditionality, target orientation-target non-orientation, and modernity-pastness.

Methods of managing labor processes in a university are divided into three components: organizational and administrative, economic, and socio-psychological.

So, the humanization of organizational and administrative methods is manifested in the degree of rationality in the organization of the research, consultative, and educational processes for each teacher in space and in time, as well as in the development of adequate working and leisure modes for teachers, means of current monitoring of activities, and reporting forms.

Economic management methods directly affect the degree of humanization of internal socio-economic conditions of work. Additionally, socio-psychological methods affect the degree of humanity in socio-psychological conditions of activity (Mainzer 2007).

The degree of humanization of the results of labor activity depends on the adequate humanity of all elements of labor, namely on the conditions, processes, and nature of labor, as well as on the methods used in managing all the processes that make up the teacher's labor. However, in our opinion, the degree of humanization of the listed elements of pedagogical, research, advisory, and educational work, as the integration work of a university teacher, does not have a direct proportional or adequate effect on the results of teachers' work in each type of activity.

3 Results

The realities of the modern period require a paradigm shift in existing education, a significant increase in quality, and improvement of the working conditions of the faculty of the Higher School, as well as the widespread introduction of concepts of education such as synergistic humanization.

The degree of humanization of teachers' work depends on the nature of this work, highlighting it in a special group or type according to a list of features. Methods of managing labor processes in a university are divided into three components: organizational and administrative, economic, and sociopsychological. From the author's point of view, the analysis of the degree of humanization and the quality of teacher work is the basis for reform processes, determining a strategic development alternative that is innovative for higher education. Today, an important task of the university is the preparation of bachelor and masters graduates that are in demand in the labor market, which is possible with a competent approach to teacher quality management. The contemporary period is a time of high rates and volumes of scientific and technical informatization, growth dynamics of technology and science, the use of convergent technologies, and the ever-increasing demand for high social and professional mobility (Némethy et al. 2014). Demand in new methods and forms of management, in the degree of humanization of labor, and in new approaches is growing. The volume and degree of infusion of the intellectual factor in almost all types of labor are increasing. This forces universities to modernize the methods, principles, forms, and approaches that are widely used in the implementation of the educational process. The demand for increasing the degree of humanization and modification of the teacher's quality management and educational process are the main tasks in the development and improvement of the dynamically changing world community, which leads to the formation of a new education paradigm.

4 Discussion

Today, dynamic reform processes are taking place at the highest level of education. There is an important and urgent demand for the use of strategic tasks in the educational process at all levels of management. This insufficiently high degree of return (efficiency and effectiveness) forces us to seek new methods and models, forms of teacher quality management, and approaches to teaching masters and bachelors and forms a new education paradigm in the university. In the Russian Federation, after the transition to a multi-stage system of higher education and refusal to train specialists, most students prefer to complete only the first stage of higher education, namely a bachelor's degree program. This situation is confirmed by the requirement of employers for higher education as a necessary stage of specialization due to the presence of a certain minimum of knowledge, skills, and competencies. At the same time, the demand for graduates with a master's degree in the labor market is much less. Although, for further innovative, intensive, balanced, and effective development of the economy, a layer of the so-called intellectual elite of the new generation needs to be formed. It is characterized by an extensive list of competencies, skills, and knowledge in the use of techniques, methods, tools, (mechanisms), and algorithms for highly productive and highly effective professional activities, often, not in one direction of training, but in several related areas. In addition, such a graduate must have humanitarian erudition, which makes it necessary to obtain a second higher or additional postgraduate education in those areas of knowledge that are not related and do not even have a common knowledge base and competencies.

This changes the work of teachers. In scientific literature, reasons are indicated that affect the intensity and degree of innovation of changes in the field of education, including higher education (Kukushkin and Yankovskaya 2017). So, contemporary society is distinguished by the presence of a new layer of educated employees, or personnel who have not only higher education, but are fluent in several foreign languages, information and computer technologies, modern methods and techniques of research, analysis and information processing, and are of highly intelligent personnel who respond to the requirements of the globalization of the economy. Further, after the signing of the *Bologna Declaration on Higher Education* in 1990, the structure of disciplines, requirements for the content of the educational process and its duration, and the level of competence of graduates, in accordance with the principles of developing the international educational space and creating open labor markets, have changed. These principles involve universities' mutual trust in the quality of higher education, the transparency of curricula and plans, the employers of other countries, and the competencies and qualifications obtained by university graduates in those countries (Bogoviz et al. 2019; Bogoviz et al. 2018; Yankovskaya 2016).

In addition, at the end of the 1970s, one of the reports to the Club of Rome revealed the problem of the inadequacy of the content and results of traditional education. Its principles and goals, compared to the requirements of the current stage of socio-economic development of most countries in Europe, Asia, and America, which have similar educational spaces, national policies in the field of higher education, and requirements for the quality of education, were found lacking (Maiorana et al. 2017).

This has given rise to the transformation of the educational process from a method of transferring knowledge from teacher to student to a way of developing a person based on subject-subject relations.

Another reason is the labor market requirements for the preparation of graduates in new areas, which arise almost every year. At the same time, the higher education system does not have time to respond to these requirements in a timely manner since it is reactive, functioning with a long delay in the development and adoption of new programs and curricula. The dynamics of changes in labor markets under the influence of crisis causes the need for changes in professional activity within three to five years (Goryainova et al. 2017). From this, we can conclude that it is almost impossible to prepare a graduate with a specific amount of knowledge in a certain direction. The concept of "profession," which traditionally is understood as the specialization of narrowly focused knowledge, is no longer relevant. The rate of accumulation of information, the formation of new directions in the development of science, and the emergence of various kinds of problems (called challenges) are growing so fast that the existing standard problem-solving algorithms can no longer provide adequate and effective management solutions. Only practical activity in a specific industry or field of activity will allow the graduate to receive those specific options for solving problems that cannot be foreseen in the learning process since the speed of updating curricula and plans and the emergence of new areas of preparation for bachelors' and masters' degrees is much lower.

Therefore, the tasks of preparing a graduate with professional and territorial mobility include the following: providing the necessary level of competence and skills for obtaining knowledge and ensuring its rational use in various situations based on individual target orientation, personal interests, and relevant aspirations for career growth on the basis of continuing self-education, self-organization and self-management throughout the entire labor activity.

It follows from this that all of the aforementioned reasons or challenges of higher education and the requirements for the student determine changes in the nature and content of the work of the teaching staff. In such conditions, the teacher must also become a student, and the speed of his self-training, self-education, and advanced training should be higher than that of students. We find a full reflection on the purpose and truth of the quality category in the works of Aristotle, Pestalozzi, Feuerbach, and Kant, the German classics of philosophy. We also find full reflection in the mechanistic philosophy of Hobbes, Descartes, and others. It should be noted that Hegel proposed a systematic approach to the concept of quality. Quality is a category from logic, and it can be attributed to the initial level of study. Hegel suggested in his works to search for the rational in man. This was his interpretation of education.

In K. Marx and V. I. Ulyanov, in Marxist philosophy, we can meet systematicity in the study of quality. In terms of quality, among the people of Russia, the worldview and the soul have always been the center of attention of thinkers of philosophy and science, sociology, and religion. Domestic philosophy paid considerable attention to the quality category, highlighting its value and consistency. Today, the theory and methodology of the education system are deeply studied in sociology and pedagogy, philosophy from the perspective of pedagogy, general philosophy, and other sciences. These are the problems and prospects of general vocational education; in part, we are talking about communities involved directly in the education process. As the market dictates new conditions and requirements for specialists, the modernization of education is underway, and requirements for the level of services provided and the quality of education are growing. In the scientific literature, there are many works devoted to the quality of the first stage and sometimes the second stage of education, but there is practically no systematic study on the quality of higher education. In general, the methodology of the prospects and problems of education are well represented in the works of domestic authors such as G. Shchedrovitsky, A. Leontyeva, S. I. Vershlovsky, etc.

We can read about the problems of education in higher education in the modern period and its prospects in the works of G. Motov, T. Barkhatova, G. Shilo, L. I. Ambaeva, etc.

Proceedings of A. Subetto, S. Shishov, V. I. Kachalov, and B. Prudkovsky, etc. are much closer and more specific to our problem raised in this article.

Without a doubt, each of the authors made a certain contribution to the study of the indicated problems. Various key categories of the education system, such as qualitative characteristics, structural characteristics, and functional characteristics have been studied, but from the point of view of theory and empirical observations. The need, dictated by the market and the employer, is a systematic study of the quality of higher education. This is the following contradiction: demand and offer.

V. Rozina, A. Kuszhanova, E. Gusinsky, and others dealt with issues of education, the study of its structure and content, as well as methods and connections with other sciences. On the role of education in the life of society, we can find it in the works of V. Kuptsov and T. Kuznetsova, A. Kochergina, G. Balykhin, etc.

The works of I. Gruenko and V. Sadovnichy, N. Borisova, L. Efremov, and others explore urgent problems and prospects in the formation of the contemporary period.

Studying the level of quality of the Higher School and the degree of humanization for the formation of his own view and model of studying this issue, the author focuses on the development of such authors as A. Subetto and V. Krichesky, B. Prudkovsky and B. Kolomiets, etc. Today, we can talk about quite high achievements in the used methods of quality management and the theoretical basis, improved methodology as well, which will create an effective quality management system; these issues are investigated in articles by Yu. Tatur, Yu. Adler, V. Versan, and several others. The philosophy of education is actively gaining momentum in domestic works, research, and abroad. These are the works of such well-known scientific researchers as P. Senge, P. Dalton, D. Norton, I. Mergen, D. Cotter, F. Johnson, R. Kaplan, P. Cardon, D. Collier, and L. Leslie, etc. Summarizing the experience gained in the studies of foreign authors, I would like to note that they adapted the theory and methodology of general quality management to the education system in a certain way.

5 Conclusion

The scientists listed in the article made a huge contribution to the study of the multiple aspects of education. Today, the theory and methodology of the primary research base on education are quite deeply studied in sociology and pedagogy, as, indeed, in other sciences. But too little attention is paid to the management of the quality of education and the work of the teacher. The needs of society on the part of the individual and the employer in the quality and level of higher education do not coincide with the offer on the education market. There are a complete contradiction and lack of elaboration of the issues of humanization of the teacher's work. The indicated points are a mismatch, along with highly complex and heterogeneous tasks in managing the quality of work of the teacher and education, humanization of labor, and the methods used today for successfully resolving them. Hence, the creation of the most humane conditions, process, nature of labor and management tools (methods) does not lead to a similar increase in the teacher's labor results.

Therefore, there is no direct proportional relationship between humanization and performance, which is one of the essential problems of the teachers' work. Even if all divisions of the university management system and education management bodies at the federal and subject levels of government in the Russian Federation embraced the idea of humanizing labor as an integrated system of aesthetic and ethical, psychological and moral, legislative, technical and economic measures to ensure the transformation of labor into the source of satisfaction of the subject with the results of labor and personal development (Yankovskaya 2017).

References

Bogoviz, A.V., Lobova, S.V., Alekseev, A.N., Yankovskaya, V.V., Shabarchina, I.V.: Transformation of the russian labor market as a result of development of internet technologies. In: Advances in Intelligent Systems and Computing, pp. 972–979. Springer, Switzerland AG (2019)

Bogoviz, A., Averin, A., Yankovskaya, V.V.: Gender analysis of local budgets in law enforcement and educational organizations of Russia. In: MATEC Web of Conferences, vol. 212, p. 09015 (2018)

Goryainova, L.V., Krishtal, I.S., Kuznetsova, O.D.: Financing of infrastructure in education: International experience of attracting private investments and opportunities for Russia to form a knowledge-driven economy. Eur. Res. Stud. J. **20**(2), 348–363 (2017)

Huggins, R., Thompson, P.: A network-based view of regional growth. J. Econ. Geogr. **14**(3), 511–545 (2014)

Kukushkin, S.N., Yankovskaya, V.V.: Current priorities of the object-process model of managing a firm and intangible asset. Econ. Bus. Banks **7**, 45–68 (2017)

Maiorana, F., Berry, M., Nelson, M., Lucarelli, C., Phillips, M., Mishra, S., Benassi, A.: International perspectives on CS teacher formation and professional development. In: Proceedings of the ITiCSE 2017: 2017 ACM Conference on Innovation and Technology in Computer Science Education, Bologna, Italy, pp. 236–237 (2017)

Némethy, K., Hegyesi, F., Osz, R., Kártyás, G., Gáti, J.: New paradigm in engineering changes higher education courses. In: Proceedings of the ICETA 2014: 12th IEEE International Conference on Emerging eLearning Technologies and Applications, Smokovec, Slovakia, pp. 121–125 (2014)

Yankovskaya, V.V.: Program and goal-oriented development of the municipality. Int. J. Adv. Stud. **6**(2), 137–151 (2016)

Yankovskaya, V.V.: Properties and the basic principle of the formation of professional and pedagogical potential. Econ. Entrepreneurship **11**(1), 741–745 (2017)

Mainzer, K.: Thinking in complexity: The computational dynamics of matter, mind, and mankind, 5th edn. Springer-Verlag, Berlin (2007)

Digital Competencies: Requirements for Information Technologies in the Framework "Management University-Industry-Science-Market"

Irina A. Rumyantseva, Tatiana Yu. Krotenko(✉), and Marina B. Zhernakova

State University of Management, Moscow, Russia
rumyantsevay@mail.ru, krotenkotatiana@rambler.ru, mzhernakova@yandex.ru

Abstract. Due to the growing need for better management in the application of digital tools, it becomes highly important to develop in managers those necessary competencies corresponding to the changing conditions. The process of economic digitalization is a challenge for education. The research argues that universities should respond to such challenges by developing new competencies not only among the specialists in information and communication technologies but also among their future managers. This process should facilitate interactions between universities, industries, and markets. The students of management specialties need to master both information retrieval and system-algorithmic thinking, and they even have to understand the very specific capabilities of programming languages. The authors also argue that future managers quite definitely recognize the need to master competencies in the field of information technology, which is confirmed by the research results obtained. Moreover, the authors show that some of the respondents already have specific knowledge in this area and their own experience in using complex information and communication tools. Both digital knowledge and skills necessary for future managers are identified by the authors while working with experts and students; they are additionally summarized as a list of competencies.

Keywords: Digital competencies · Information technology · Management university · Science · Future managers · Integration · Industry

1 Introduction

Digital technologies today define not only communication but also many types of social, cultural, educational, industrial, and individual activities—in other words, they define human life (Sacolick 2017; Udaltsova 2019). Thus, computing, multimedia, and information and communication technologies expand human capabilities, becoming, to a borrow a term from L. S. Vygotsky, full-fledged "cultural and historical tools".

Simultaneously with this trend, we are witnessing another one: the moving of education to the digitalization forefront and acting as a platform for introducing new

A. V. Bogoviz and Y. V. Ragulina (Eds.): ISCI 2019, LNNS 115, pp. 754–762, 2020.
https://doi.org/10.1007/978-3-030-40749-0_89

technologies. The digital coverage of the various forms, types, and levels of educational activities becomes comprehensive and irreversible (Antonov et al. 2019). The educational content is structured in a way that goes beyond acquiring visual, auditory, kinesthetic, and digital resolution (Gianelli 2018). In particular, a greater emphasis is placed not on transmitting information through the visual and auditory analyzers of students but on the transmission of thinking in the "*here and now*" mode (Khramtsova and Mayboroda 2019).

Science and practice put forward a whole range of acute problems discussed by information engineers, teachers, and psychologists. How do modern students' cognitive abilities and opportunities change? Do the newest information technologies create advantages for students? Or, are they creating an abundance of visual range, hypertextual presentation, content redundancy, spontaneity in the choice of material, and reduced the ability in subjects' cognition (Baskakova and Soboleva 2019)? How does the emergence of fundamentally new ways of production and reproduction of knowledge, types of thinking and communication, hybrid and artificial intelligence, and automation of life processes affect the intellectual and emotional spheres of man (Gianelli 2018)? How is the motivational structure of students transformed under the influence of new technologies (Antonov et al. 2019)?

These questions, together with the observations and analyses conducted by the authors of the social and educational activity of students at the State University of Management, as well as the desire to look at the problem in the context of "management university-industry-science-market", directed the research attention to confirming or disproving the following hypothesis: The students of managerial specialties, the future managers, quite clearly are aware of the need to master the competencies in the field of information technology for successful professional activities and self-actualization in the market conditions.

Working with this hypothesis suggests answers to such research questions as:

(1) What are the ideas of the expert community about the structure and composition of the competencies of a modern manager in a digitalized economy?
(2) What are the changes in the current demands of employers to applicants for managerial vacancies?
(3) What is the awareness of future managers of the need to form digital competencies for sustainable economic development?

2 Materials and Methods

In the conducted study, such methods as the individual expert survey in the form of an unstructured interview, the expert assessments analysis, the structured interview, and the survey on the questionnaire developed by the authors were applied consistently (Yadov 2003).

At the beginning of the work, an expert group was formed, which included (1) the leading IT software developer of a company; (2) the chief personnel officer of the company; (3) the head of the compensation and benefits department; (4) the head of the HR department; (5) the general director of an industrial enterprise; (6) the deputy director for IT of a fuel and energy company; (7) an auditor of a consulting company; (8) the head of the purchasing department of an aircraft manufacturing company; (9) the Professor of the Institute of Information Systems of the State University of Management; (10) and three associate professors of the management department of the State University of Management. The requirements for the expert group members were at least 5–8 years of experience in the IT-sphere, in the real economy, or in the higher education system. In fact, 12 representatives of the university-industry-science-market sphere took part in the study.

The interviewing was conducted in two stages. In the first stage, the interview was unstructured and contained a suggestion for the respondents to reflect in the following directions: (1) What are the additional competencies for the future managers in a digital economy? (2) What types of information technologies are necessary for the managers now, and what types require development in the near future? (3) How have the employers' requests changed in relation to the applicants?

After processing the experts' detailed answers, the authors selected seven main types of information technologies that the experts consider promising and necessary for further consideration. Further, a structured interview program was compiled for the selected types of information technologies. The experts' survey continued with the consideration of this program. The structured interview included questions about specific information tools included in the seven selected IT types. According to the authors, information technologies, once mastered, will allow future managers to form competencies necessary in the digital economy.

The types of information technologies identified by the experts in the process of unstructured interviews and applied to later stages of the study include the following:

- Programming languages;
- Databases;
- Corporate IT systems;
- Data analysis tools;
- Marketing tools for the promotion through social networks;
- Comprehensive tools for search, analysis, and forecast; and
- Machine-learning tools.

In the process of subsequent interviews, the experts termed the set of specific tools associated with each selected type of information technology.

The final stage of information gathering was a survey of students of management specialties at the State University of Management, identifying their conscious need for the study and application of specific information technologies. The questionnaire included seven types of information technologies (presented in Table 1), specific tools within each type, and an additional line for "Other", which allowed respondents to identify those types and tools that were not included in the questionnaire.

3 Results

3.1 The Results of the Expert Survey and Interview

The representatives of the expert group for the individual interviews noted the following. Digital skills not only enable the production and sharing of content but also create opportunities for solving enterprise problems, communication, functioning, and sustainable development. The accumulation of a critical mass of employees with developed digital competencies forms a competitive advantage for their organizations. By using technology platforms, the company's cost structure becomes more efficient. The high-quality products and services defined by high-tech solutions ensure an influx of investment.

In modern organizations, many situations require managers to have a predictive vision of the opportunities that can be provided by digital technology:

- Introducing electronic document management;
- Partial or complete automation of business processes;
- Implementing corporate IT systems;
- Developing the KPI system in an organization;
- All types of planning;
- Establishing quality systems;
- the Developing an organization business model;
- Facilitating changes by strengthening competitive advantages.

Six out of ten experts mentioned another phenomenon. The modern manager has to interact constantly and closely with such experts as frontend developers (create websites and web apps), web developers (frontend developers with backend knowledge), backend developers (work on sites and applications, responsible for the internal part of the project, hidden from the user's eyes), data analysts (collect, analyze, and visualize data, help in decision making), data engineers (create infrastructure so that data is stored and processed stably, then sent into analytical systems without fails), specialists in machine learning (teach models to predict user behavior and the complex system's response to changes), specialists in neural networks (create and train neural networks to solve applied problems), interface designers (design the appearance of sites, make them comfortable and beautiful), and product managers (responsible for the creation and development of digital products, know why, how, and what to do to solve the users' and businesses' problems).

Such communication, of course, requires high-quality engineering training of the manager and his knowledge of information technology. This fact is evidenced by the publication (Rudskoy et al. 2018; Chuchalin 2018).

As for employers' requests for digital competencies, which are added concerning the managerial ones, experts agree that managerial competencies are increasing at a rapid rate. It is impossible to imagine a so-called manager of a new type without professional knowledge and skills in IT. Even now, a manager needs experience in independent planning and the implementation of IT solutions, software architecture design (at least for the competent task setting for information specialists), and the compilation of catalogs of information services necessary for the production processes functioning (Piskunov and Glezman 2019). The head must, together with IT deputies and other specialists, form the information technology strategy of the enterprise, design information protection to ensure the economic security of the organization, and create a quality system (Solozhentsev 2018). They should be aware of the possibilities and features of the goods and services promotion on the Internet. They are obliged to work confidently with the ERP-systems, blockchain, business intelligence, Internet of Things (IoT), and Big Data. The experts' opinions are also confirmed by studies (Li et al. 2019; Lokuge et al. 2019).

3.2 The Survey on the Questionnaire Developed by the Authors

In the course of the study, 184 students of the State University of Management studying in the management direction in 11 groups were interviewed. The answers were located differently in random columns of the questionnaire, allowing identification of both the tools and the information technologies already used by the students and those only intended for future use. The questionnaire without details about specific tools (in order to save space) is given in Table 1.

Through the course of the processing, the students' selection of the information technologies that the students perceived as necessary for studying was obtained.

Table 1. The questionnaire used in our research.

№	Information technology types	Used IT types			Interesting and prospective IT applications	
		I already work with these technologies. I study them independently	I already work with these technologies. I study them in paid classes outside the SUM	I already work with these technologies. I study them on paid classes in the SUM	I assume that I need it	I am interested in these technologies, but not sure if I need them anytime soon
1	Programming languages					
2	Databases					
3	Corporate IT systems					
4	Data analysis tools					
5	Marketing tools for promotion through social networks					
6	Comprehensive tools for search, analysis, forecast					
7	Machine learning tools					
8	Other (name what, in any form)					

Source: Authors' study materials.

A generalized result of the survey processing is presented in Fig. 1.

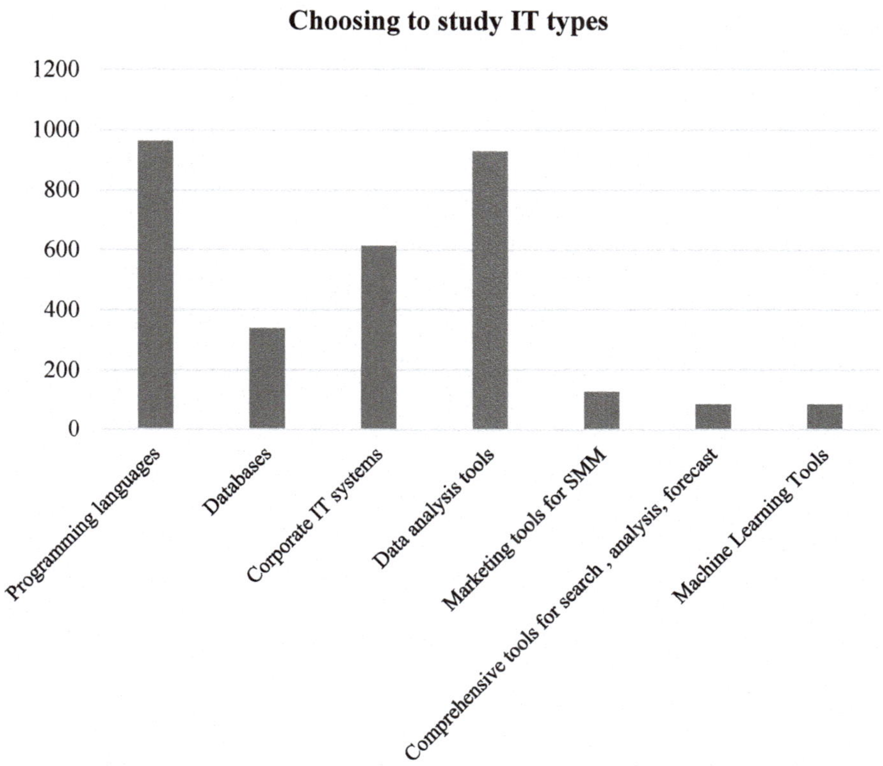

Fig. 1. The cumulative students' choice of the information technologies that the students perceived as necessary for studying

In the survey course, answers in the line "Other" were received in just six questionnaires. Moreover, either *Adobe* (the whole package) or *Photoshop* (specifically) was mentioned four times. In the other two cases, foreign languages and legal disciplines that were not related to IT were identified. On this basis, only seven main types of information technologies contained in the questionnaire are shown further in the tables and diagrams.

After the processing of the survey results, the most significant number of choices was received from the 1st IT type, programming languages, then the 4th type, various tools for data analysis, then the 3rd type, corporate IT systems, then the 2nd type, databases, and finally the 5th type, marketing tools to promote the product in social networks.

From group to group, the results are almost the same—only in one group, the 1st and 4th types changed places. The result is shown in Fig. 2.

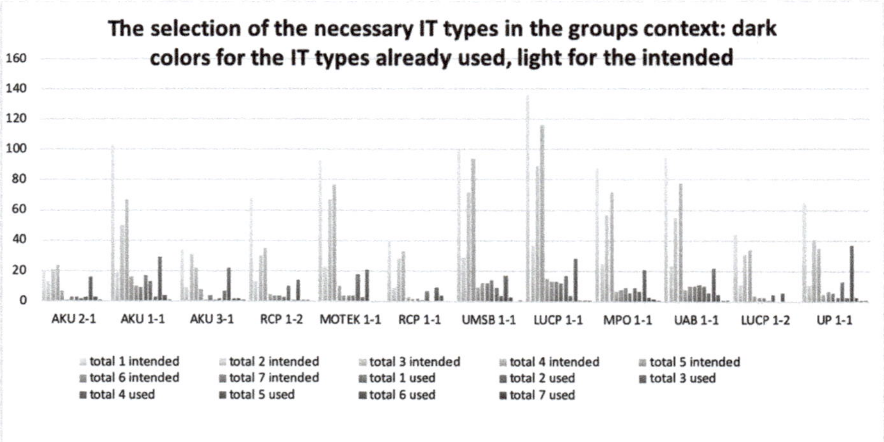

Fig. 2. The result of the selection by students of the necessary IT types in the groups' context.

The analysis of the obtained results allowed us to make the following conclusions. Students' views on the types of information technologies necessary for studying have a common structure; the most popular at the moment are programming languages, various data analysis tools, corporate IT systems, databases, and marketing tools for promoting the product on the Internet.

It should be noted that the types already used make up a smaller part than those expected for future use. Students choose the option, "I assume that I will need these technologies in my work", more often. That is, students are aware of not only the immediate need but also the tendency for further digitalization and the need to master digital competencies.

Thus, all the "digital" knowledge, skills, and abilities necessary for future managers, in addition to the typical managerial ones, can be summarized in the form of competencies:

- Advanced computer literacy, which includes not only information retrieval and data filtering but also programming languages;
- The use of digital data analysis tools, including various database activities;
- Solving various problems of automation of management processes using digital marketing tools and corporate systems;
- Digital content integration;
- Communication and interaction using the entire range of digital solutions;
- Data protection, privacy, and cybersecurity; and
- The creative use of various digital tools.

4 Discussion

The study hypothesis, consisting of the fact that future managers are quite clearly aware of the need to master the competencies in the information technology field, is confirmed by the results obtained.

The awareness of this need is extremely important for sustainable economic development, as digital transformation is becoming a critical factor for industrial development (Li et al. 2019; Piskunov et al. 2019). Furthermore, this is not just the popularization of the information and communication technologies in business through education. The digital competence of managers and all employees of the industrial companies provides a competitive business advantage. Knowledge, skills, and abilities in the application of digital devices, networks for information access, and communication applications make it possible to competently use, manage, and implement digital content in professional activities.

The consistent integration of the new competencies into modern educational practices is a necessity. However, integration should affect all learning formats, both full-time and distance, and support the learning process throughout life (Gianelli 2018). Under the conditions of the need to master information technologies, recognized by experts, students, and future managers, there is a discussion and understanding of this issue in the professional environment, in addition to a search for specific opportunities to form the required competencies.

5 Conclusion

Thus, the study, developed in the plane of "management university-industry-science-market," showed that, at present, future managers' awareness of the need to form competencies in the field of information technologies is changing significantly, and the need for them already has been formed. That is, in the conditions of the digital economy, there is a need to master new management competencies. This situation should meet the educational content and structure of the learning environment at the university.

Simplified, entertaining behavior of Internet users is replaced by well-thought-out strategies aimed at enriching educational and professional tools (Antonov et al. 2019).

There is another important feature of modern education that is impossible to ignore: young people become the initiators and main innovators. This is manifested not only in the natural desire of young people for novelty but also in the organic introduction of crowdsourcing interaction into education when the production of the idea, content, service, or product implies informal assistance and interest of all the group members. Prosumerism, which is not only consumption but also independent joint production of educational products, is more and more common in the university. The creation of non-commercial collective products and services is based on the partnership and the principle of transferring information and knowledge not from one source, but "from everyone to everyone". As a result, new opportunities emerge and must be consciously created for education, science, and industry integration to solve emerging problems and, as a result, ensure sustainable economic development as a whole.

Acknowledgments. The authors are grateful to the administration of the State University of Management for the opportunity to conduct the research, and to the experts and students for their active participation.

References

Antonov, V.G., Rumyantseva, I.A., Krotenko, T.Yu.: Assessment of the level and structure of motivation for studying undergraduate students as indicators of education quality. Perspect. Sci. Educ. **2**(38), 267–283 (2019)

Baskakova, M.E., Soboleva, I.V.: New facets of functional illiteracy in the digital economy. Educ. Issues **1**, 244–263 (2019)

Chuchalin, A.I.: Engineering education in the era of the industrial revolution and the digital economy. High. Educ. Russ. **10**, 47–62 (2018)

Gianelli, M.: E-learning in theory, practice, and research. Educ. Issues **4**, 81–98 (2018)

Khramtsova, N.G., Maiboroda, T.Yu.: Approaches to the development of digital competencies of law students. Perspect. Sci. Educ. **1**(37), 80–93 (2019)

Li, Y., Huang, J., Song, T.: Examining the business value of customer relationship management systems: IT usage and two-stage model perspectives. Inf. Manag. **3**(56), 392–402 (2019)

Lokuge, S., Sedera, D., Grover, V., Dongming, X.: Organizational readiness for digital innovation: development and empirical calibration of a construct. Inf. Manag. **3**(56), 445–461 (2019)

Piskunov, A.I., Glezman, L.V.: The development of industrial enterprises in the digital economy development. Creative Econ. **3**(13), 471–482 (2019). https://creativeconomy.ru/lib/40085

Rudskoy, A.I., Borovkov, A.I., Romanov, P.I., Kolosova, O.V.: General professional competence of a modern Russian engineer. High. Educ. Russ. **2**, 5–18 (2018)

Sacolick, I.: Driving digital: The leader's guide to business transformation through technology. AMACOM, New York (2017)

Solojentsev, E.D.: Ephemeral and digital safety and quality management in the economy. Risk Anal. Prob. **5**(15), 58–77 (2018)

Udaltsova, N.L.: Economic processes digitization in the context of the industrial revolution 4.0. Creative Econ. **1**(13), 49–62 (2019)

Yadov, V.A.: The Strategy of Sociological Research. Academkniga, Moscow (2003)

Scientific and Educational Cooperation in the Framework of the EAEU: Problems and Prospects

Oksana P. Ovchinnikova(✉), Tatiana V. Kokuytseva(✉), and Pavel A. Byshkov

Peoples' Friendship University (RUDN University), Moscow, Russia
{ovchinnikova-op, kokuytseva-tv, byshkov-pa}@rudn.ru

Abstract. The Eurasian Economic Union is a major integration project nowadays, but various problems arise in its implementation course, including in the economic field and the humanitarian cooperation sphere, which determines the relevance of the research. The paper focuses on identifying these problems and finding promising directions for the long-term development of the EAEU. The problems of the EAEU that hinder its effective development are analyzed with the use of classical scientific knowledge methods (analysis, deduction, and induction). The scientific novelty of the research is in the development of the author's proposals for creating conditions for sustainable science and education interaction. It is the one that can provide a solution to the key task of the EAEU, its innovative production and competitiveness modernization on the world market.

Keywords: Eurasian integration · Education · Science · Cooperation

1 Introduction

The Eurasian Economic Union (EAEU) is an ambitious integration project in the Eurasian space and, at the same time, a fairly young association (the operation's start was January 1, 2015). There have been several positive aspects received by countries from integration processes, including growth in mutual trade, the absence of customs duties, increased business activity, increased global competitiveness of member countries, etc. First of all, the EAEU's goal is economic development and humanitarian cooperation. Within the economic development framework, each country looks to get more than invest, and national interests also prevail in the implementation of economic interaction. The humanitarian sphere is based on the economy and, therefore, is mainly concerned with continuation in the fields of science, education, and culture.

One of the problems existing within the EAEU is that the member states produce approximately the same goods set competing in the domestic market of the EAEU without creating a developmental effect (Ovchinnikova and Ovchinnikova 2017). The same can be said about the scientific and educational fields—the training of scientific personnel and university graduates is carried out at approximately the same level in identical directions if we consider the mass segment of the educational sphere. At the

A. V. Bogoviz and Y. V. Ragulina (Eds.): ISCI 2019, LNNS 115, pp. 763–770, 2020.
https://doi.org/10.1007/978-3-030-40749-0_90

same time, technological development requires greater flexibility in the training system in order to occupy a competitive position in the global technology market.

However, education was not included in the list of service sectors for which the common EAEU market is being formed. Based on this, the development of scientific and educational cooperation within the EAEU framework is an important direction to follow.

2 Materials and Methods

In recent decades, cooperation in all spheres of human activity has been based on the partnership paradigm, which has replaced the notion of the "external imposition" in relation to goals, priorities, methods and interaction tools. One of the factors for such changes in world perception is the globalization and informatization of the world space (Van de Water et al. 2008).

Partners need to answer the main question: "What is the relationship in a partnership?" Furthermore, if this issue—at the level of individual universities or research organizations—is generally understood for the EAEU member states, it has a certain asymmetry with respect to resources, capabilities, academic and research work degrees, development priorities, etc. (Gutierrez 2008).

In the educational and research partnership field in foreign literature, much attention is given to the "responsibility of the parties," focusing on the need for the mutual management of interaction processes by the involved parties (Hoppers 1998).

After the EAEU Treaty came into force (Consultant Plus 2014), Russia, realizing the importance of a training system that met the requirements of the digital economy, initiated the creation of a mechanism for cooperation between the EAEU member countries in the field of education and science. The concept of the EAEU Network University (Eurasian Network University) was developed, and its main purpose was the formation of a modern training system adequate to the needs of the EAEU while promoting the formation of the latest technological and scientific agendas for the participating countries. However, this project did not find serious support from all the EAEU members.

At the Congress of Ministers of Education and Science of the countries participating in the EAEU, held on April 13, 2016, a memorandum of cooperation on educational and scientific-technological cooperation in the Eurasian region was signed. Kazakhstan did not sign this document. Thus, the need to formulate a common policy in the education and science field is not shared by individual participants in the integration process.

Therefore, the main task within the framework of integrating educational and scientific institutions can be solved with the help of participants in the educational and scientific process. Such an "initiative from below" can activate the process of forming a standard educational policy in the EAEU.

However, there also exist specific difficulties. Most Russian universities have a significant number of cooperation agreements with the universities of the EAEU member countries, but most of them are either non-working, working under a project, or sometimes initiative (for example, a double degree program).

The motivation for the active interaction of the educational institutions within the EAEU is the process of the internationalization of higher education and the requirements of the Ministry of Science and Higher Education of Russia on the need to attract international students. The students from the near abroad countries have the advantages, due to the knowledge of the Russian language, allowing them to join the educational process immediately. Scientific interaction is considered mainly as an opportunity to participate in joint projects of such funds as the RFBR, RNF. Often, "scientific cooperation" is the most interesting motivational mechanism directly for university scientists, which allows them to attract additional fundings for their research.

That is why it is necessary to study the causes of the weak interaction between partners and identify mechanisms for their activation.

3 Results

The interaction between educational and scientific organizations of the EAEU member countries faces several problems, both objective and subjective, which have a serious impact on the effectiveness of cooperation.

The main problem that prevents active interaction is "time." In the vast majority of Russian universities and many of those in the EAEU countries, a prerequisite for the teaching staff is the availability of indexed publications in international scientometric databases. The preparation of such articles requires a significant investment of time, which, combined with the training load, dramatically limits the time that employees can spend on implementing international partnerships.

The second problem is the difference in the level of scientific and educational potential. Almost thirty of the years after the collapse of the USSR were characterized by severe financial crises for all EAEU countries, which led to a reduction in funding for the scientific and educational sphere, a brain drain, and the development of the commercial component of educational policy. Altogether, this led to certain problems, including low labor productivity, lack of participation in global industry-specific discussions, increased dependence on external funding in individual countries, and the inability to form their research agenda.

Such an imbalance in resources can complicate the interaction significantly. Stronger universities want to see a reliable partner that can have a substantive dialogue and form a common research agenda.

Another challenge is the sustainability of partnerships. Even though the initiation of a partnership often begins with the possibility of implementing a project (for example, implementing a joint educational program), over time, the resources laid down for the project run out, and the cooperation itself gradually fades away. Besides, in the framework of the burden that is placed on the staff, it is difficult for universities to monitor and control cooperation on a systematic basis.

Thus, an effective partnership is a continual, dynamic process of cooperation between scientific and educational organizations, which brings mutual but not always equal benefits to the partners. As a rule, a partnership implies mutual trust, transparency, and respect for each other's cultural and working environment. The successful partnership develops over time and greatly benefits partners.

There are five steps to sustainable interactions.

The first step. When forming partnerships, it is necessary to focus on the mutual needs of future partners. In this context, the developmental goals of academic institutions, the possibilities of interaction and resources that can be allocated to this interaction, as well as the expected results, are considered. Educational and scientific partnerships are institutional cooperation that should be built into the overall structure of the educational institution. When considering a possible partnership, it is necessary to determine whether it is really necessary or perhaps limited to the implementation of a particular project.

The second step is the discussion of the partnership area. As a rule, a cooperation agreement is concluded, where the scientific and educational partnership areas are prescribed. However, this abstract approach does not meet the time requirements. Nowadays, it is essential to maximize the interaction, not just the documents signing time. Therefore, the discussion process includes the issues of interaction planning, determining the responsible circles for such interaction, and the allocation of the appropriate resources for interaction from both.

At this stage, it is necessary to decide whether the interaction will concern the implementation of joint educational programs, research projects, and programs of academic mobility, or affect the deeper processes associated with the increase in scientific and educational potential.

Partners should trust each other, as well as ensure the transparency of the activities and management approaches, taking into account the geographical distance between them. It is also necessary to determine the duration of the partnership.

The third step. When implementing partnerships, it is important to remember that educational and research institutions operate in different educational sociocultural environments. Therefore, a detailed environmental assessment of the partner is necessary. Such an approach neutralizes the overestimated expectations of partners and allows us to understand the basic capabilities of both institutions in order to ensure realistic deadlines for implementation of the project and implement joint activities. The reassessment of the potential or insufficient consideration of the specifics of planning and organizing activities at a partner university may significantly slow down the project implementation or even limit the partnership scope.

It is also necessary at the beginning of the implementation of partnership activities to clearly distinguish between the real duties and responsibilities of the parties. Roles and responsibilities in a partnership should be considered depending on the capabilities, resources, and available potential, as well as their development opportunities. It is important to accurately assess the resource capabilities of the partner and its ability to adapt role-playing installations as projects are implemented as the partnership part.

The fourth step is regular communication between the partners. Since a mobile environment characterizes the modern economy, regular communication between the partners will allow them to adjust the target settings and tactical objectives following changes in the external environment. The partnership should be aimed at solving the tasks of the EAEU by increasing the competitiveness of scientific and educational programs in the external market. However, technological changes may make adjustments in the projects' implementation. In order to neutralize possible temporary and material losses, it is necessary to create some advisory body that could regulate these moments at the personal communication level.

The fifth step is the joint strategic planning of partnership development. The educational and scientific partnership has its life cycle, which is characterized by different development stages. At these stages, new projects may arise, new participants may appear, and a number of projects may turn out to be unsuccessful and have to be closed. To avoid temporary, material, and other losses, it should be borne in mind that the strategic planning in the educational and scientific partnership context means:

- The coordination at the initial stage of clear strategic goals for the partnership. Although over time these goals can be adjusted, at the initial stage, the coordination of framework goals and objectives is necessary;
- The coordination of specific results that characterize each stated goal of the partnership;
- The development of an agreed-upon action plan within the framework of the identified goals and priorities. This plan should be detailed, indicating the terms and persons responsible for the implementation. Moreover, the deadlines for implementation should be realistic, so that there is no disruption to the plan by any party;
- A prerequisite is a separation of powers and responsibilities in the implementation of specific projects, as well as the development of a project management mechanism;
- A consolidation of these areas by a regulatory document defining the partnership's nature. However, within the individual projects, this may be an informal agreement, but the partners must have a certain trust amount in each other for this to happen.

It is also necessary to take into account changes in the external environment when forming a strategic partnership plan. The partnership must be adaptive and respond quickly to the changes that have occurred without losing the partnership as a whole.

The interest of university staff in implementing partnerships is an important element of the scientific and educational partnership. If the university headship is interested in the partnership, but the staff considers it as an additional burden, then an effective long-term partnership will fail.

Before entering into a partnership agreement, it is necessary to inform the team about the advantages and possible disadvantages of the agreement being signed, the stages of implementation, and the benefits for the scientific and educational process. Such a discussion will help to lay a solid foundation for the partnership.

4 Discussion

One of the EAEU primary tasks is an innovative joint modernization of production, which should be the basis for reforming economies, serve as a transition to scientific, technical, and technological cooperation. In the process of joint modernization, the formation of macroeconomic prospects for the innovative development of national economies should be ensured. Also, science-intensive and high-tech production should be a decisive factor in the economic development and enhancement of the international competitiveness of EAEU countries.

All the participating countries set themselves the task of enhancing international competitiveness, diversifying production, moving to an innovative development path, etc. At the same time, it is important that, in addition to goals and objectives, industry priority development directions coincide by approximately 70% (i.e., the list of industries that the EAEU countries have identified as priority for development) and that national industrial complexes intersect on target export sales markets, which may be India, China, and other fast-growing economies of the world (Rodionova et al. 2016).

The Decree of the Council of the Eurasian Economic Commission, October 18, 2016 No. 32 "On the Formation of Priority Eurasian Technological Platforms" determined that the technological platform is an innovation infrastructure object, which is designed to ensure the integration of science and business in order to concentrate intellectual and material resources on the priority areas of scientific and technological development of a particular industry. The world and Russian experience show that technology platforms are created primarily to ensure that the scientific activities' results become innovations, i.e., were introduced into production.

The EAEU has approved the creation of 12 technological platforms:

1. Space and geo-information technologies, products of global competitiveness;
2. Eurasian Biomedical Technology Platform;
3. Eurasian Supercomputer Technology Platform;
4. Photonics;
5. Eurasian LED technology platform;
6. Mining and solid minerals processing technology;
7. Environmental Development Technologies;
8. EurasiaBio;
9. Food and agro-industrial complex processing industry technologies, healthy food products;
10. Eurasian Agricultural Technology Platform;
11. Industrial technologies "Light industry";
12. Metallurgy technologies and new materials.

The main tasks of the Eurasian technology platforms in the document are: to identify the needs of the real economy sector for new technologies; to search and promote the priority scientific and technical projects development; to establish cooperation among Member States and support joint initiatives and joint projects; to indicate the barriers to the scientific and technological development of the Member States, and to develop recommendations for their correction; to assist in the improvement of documents on the cooperation of the Member States in the scientific, technical, and innovation fields; to popularize the achievements of the Member States in the scientific and technological development, as well as the achievements within the framework of Eurasian technology platforms; and to monitor the results of scientific, technical, and innovation cooperation.

However, the implementation of such ambitious tasks requires trained personnel. Therefore a prerequisite is to create a map of priority occupations for the EAEU member countries. This map, like a competency map, should be based on a forecast of personnel potential in the future until 2040. Considering that the training period of a qualified specialist is six years, on average, this forecasting horizon is optimal.

Further, the EAEU member states should develop a unified concept of scientific and educational cooperation, which should be based on the initiatives from the scientific and educational community, which, within the framework of the created scientific and educational clusters (partnerships), can present a real picture of the directions and mechanisms for implementing effective interaction in the EAEU countries without duplication of professions and areas of training.

5 Conclusion

Despite the diversity of educational cooperation in the EAEU countries, it is still an independent element and does not make a significant contribution to the integration processes. HSE scientists have put forward a proposal for the formation of a common Eurasian educational space (EAPA) (Eurasian Economic Integration 2019). However, the lack of harmonization of educational regulatory documents and the lack of a unified concept for the development of the EAEU's scientific and educational space does not currently allow us to implement this proposal fully. In this regard, the most effective way to implement partnership agreements between individual educational organizations based on the principles of the partnership described above, which are based on strategic development plans and a project-oriented approach to the implementation of the agreement, is seen as the most effective. Based on the forecast in staffing needs and clear differentiation of the roles in the project activities, partners will be able to create a working scientific and educational environment that will help to scale up the best practices to the EAEU level and create a unified scientific and educational space.

Acknowledgment. The study was carried out with the financial support of the Russian Foundation for Basic Research in the framework of the scientific project No. 19-010-00609 "Developing forecast scenarios for the socio-economic development of the EAEU on the medium and long-term perspective based on building simulation models for evaluating the integration potential of the EAEU and analyzing economic and geopolitical factors and challenges impact on the Union."

References

Consultant Plus: Treaty on the Eurasian Economic Union (Signed in Astana on 05/29/2014) (ed. on 03/15/2018) (2014). http://www.consultant.ru/document/cons_doc_LAW_163855/

Eurasian Economic Integration: Development prospects and strategic objectives for Russia. In: Proceedings from the XX April International Scientific Conference on the Development of the Economy and Society. HSE Publishing House, Moscow (2019)

Gutierrez, D.: Beyond disappointment: Transforming ideology and practice in North-South research partnerships. NORRAG News **41**, 19–22 (2008)

Hoppers, C.O.: The morning after midnight? Partnership and paradigms for development cooperation in the 21st century. NORRAG News **25**, 26–27 (1998)

Ministry of Education and Science of the Russian Federation: Memorandum on educational, scientific, and technological cooperation in the Eurasian region is signed (2016). https://минобрнауки.рф/новости/8152

Ovchinnikova, O.P., Ovchinnikova, N.E.: Financing higher education in developed countries and Russia: an analysis of current trends. Financ. Credit **23**(38), 2305–2316 (2017)
Rodionova, I., Kokuytseva, T., Semenov, A.: Features of migration processes in different world industries in the second half of the XX century. J. Appl. Econ. Sci. **11**(8), 1769–1780 (2016)
Van de Water, J., Green, M., Koch, K.: International Partnerships: Guidelines for Colleges and Universities. American Council on Education, Washington (2008)

The Functioning of Regional and Municipal Non-profit Organizations: Practice, Trends, and Prospects

Sergey P. Sazonov, Alexey S. Dmitriev(✉), and Alexander I. Chunakov

Volgograd State Technical University, Volgograd, Russia
sazonovsp@mail.ru, lesha_dmitriev@outlook.com, efp@vstu.ru

Abstract. Non-profit organizations take a direct part in the life of society, carrying out socially-oriented activities (including activities in such spheres as healthcare, education, cultural education, and others). The civil society institutions are developed by improving the work of the non-profit economy sector. Also, non-profit organizations can act as employers, which has a beneficial effect on the employment level. This paper discusses the functioning features of non-profit organizations in the Volgograd region, including the dynamics of their numbers in recent years, as well as specific features of the formation of their financial resources. The classification of the projects of socially-oriented non-profit organizations submitted for the presidential grant competition in 2019 is considered, and the conclusions about the need for the systematic development of the non-profit economy sector to improve the quality of life are drawn.

Keywords: Non-profit organizations · Presidential grants · Volgograd region · Financing · Social projects

1 Introduction

Non-profit organizations (NPOs) do not have profit as their main goal of activity and do not distribute the received fund among the founders. The mission of such an organization may have cultural, spiritual, educational, and other socially useful goals. Most of the higher educational institutions, schools, health care institutions, and sports sections (including state, i.e., budgetary) belong to non-profit organizations and are subdivided into institutions, foundations, autonomous non-profit organizations, and other legal forms. At present, the Russian non-profit sector, due to insufficient budget funding and underdeveloped cooperation with the private sector, can be called stagnant. As of 2017, the share of the non-profit economy sector in the GDP of the Russian Federation does not exceed 1%, and the percentage of the economically active population employed in NPOs barely reaches 1.1%. In developed countries, non-profit organizations can compete with the state in carrying out activities that have a social focus. Thus, the tertiary economic sector stimulates the development of the domestic policy of the

A. V. Bogoviz and Y. V. Ragulina (Eds.): ISCI 2019, LNNS 115, pp. 771–779, 2020.
https://doi.org/10.1007/978-3-030-40749-0_91

country, forcing the state to work continually on improving the quality of existing social projects.

2 Materials and Methods

This study used the methods of the economic and statistical structural and comparative analysis and a synthesis of the amounts of funds requested by non-profit organizations for the implementation of their projects.

3 Results

As of 2018, 10,638 non-profit organizations operated in the Volgograd region, 4,563 (42.9% of the total) are in Volgograd. 2,522 organizations are socially oriented. The dynamics of the number of non-profit organizations operating in the Volgograd region from 2016 to 2018 are given in Table 1.

Table 1. Non-profit organizations of the Volgograd region.

Organization form	2016	% of the total	2017	% of the total	2018	% of the total	Change 18/16, %
Consumer cooperatives	915	7.9	840	7.6	808	7.6	−11.7
Public organizations	2770	23.8	2732	24.6	2632	24.7	−5
Associations (unions)	452	3.9	453	4.1	414	3.9	−8.4
Funds	261	2.2	242	2.2	225	2.1	−13.8
Autonomous Non-Commercial Organizations	220	1.9	231	2.1	233	2.2	+5.9
Religious organizations	436	3.8	438	3.9	436	4.1	0
Institutions	5083	43.7	4775	43.0	4583	43.1	−9.8
- Private institutions	272	2.3	251	2.3	226	2.1	−16.9
- Budgetary institutions	1673	14.4	1388	12.5	1259	11.8	−24.7
- Autonomous Institutions	136	1.2	227	2.0	219	2.1	+61
- State institutions	3002	25.8	2909	26.2	2879	27.1	−4.1
Other forms	1488	12.8	1393	12.5	1307	12.3	−12.2
Total	11625	100	11104	100	10638	100.0	−8.5

Source: Developed by the authors based on the materials (Volgogradstat, n.d.).

Based on the tabular data, it can be noted that there is a trend toward a decrease in the total number of non-profit organizations in the Volgograd region. In 2018, the number amounted to 10,638, which is 8.5% fewer than the same indicator in 2016. The most significant reduction is observed in budgetary institutions: in 2018, the number decreased by almost a quarter (24.7%) compared to 2016. This is explained by difficulties in budget financing and the ongoing process of improving organizational and managerial decisions on the development of the national economy of the territories. At the same time, the number of autonomous institutions in 2018 increased by 61% compared with 2016, and autonomous non-profit organizations (ANPO)—by 5.9%. Figure 1 shows a diagram of non-profit organizations of the Volgograd region by the organizational and legal forms.

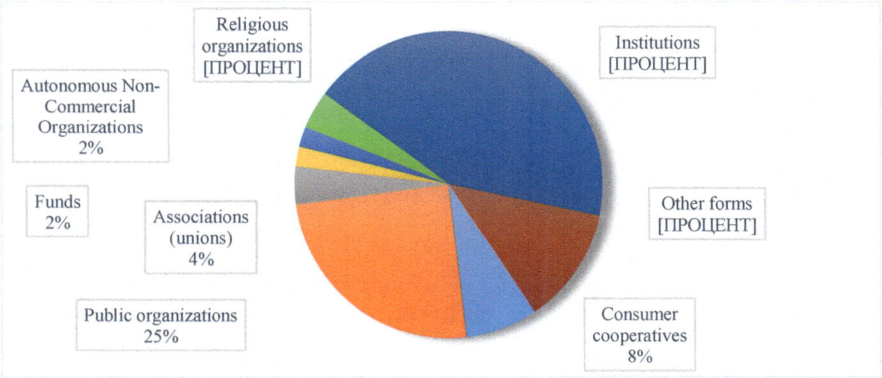

Fig. 1. The forms of incorporation of regional NCOs. *Source*: Developed by the authors based on the materials (Volgogradstat, n.d.).

Currently, the largest part of NCOs in the Volgograd region is represented by institutions (43.1% of the total number), followed by public organizations (24.7%). Consumer cooperatives occupy 7.6%, associations and religious organizations account for 4.1% and 3.9%, respectively. 2.2% are foundations, 2.1% are autonomous non-profit organizations. The remaining 12.3% includes organizations of other legal forms.

In the period from 2016–2018, 252 socially-oriented nonprofit organizations of the Volgograd region received support from the regional budget. The support consisted of the provision of financial and property benefits, such as full or partial rent payment of premises occupied by organizations and transfer to the ownership or operational management (in the institutions' case) of certain property items, as well as full or partial reimbursement of expenses for utility bills and communication services.

A small amount of private investments characterizes the nonprofit sector of the Volgograd region. The decrease in the number of potential investors in the Volgograd region is primarily associated with a decrease in the number of operating enterprises, and as a result, a decrease in budget revenues and a decrease in the number of job opportunities (Sazonov et al. 2014). Table 2 shows the analysis of the number of public organizations of the Volgograd region in the period 2017–2019.

Table 2. The number of public organizations of the Volgograd region.

Public organizations	2017	% of the total	2018	% of the total	2019	% of the total	Change 19/17, %
Political parties	62	2.3	58	2.2	50	2	−19.4
Trade union organizations	330	12.1	317	12	291	11.9	−11.8
Social movements	26	1	25	0.9	24	1	−7.7
Public funds	29	1.1	21	0.8	18	0.7	−37.9
Charity funds	130	4.8	128	4.9	118	4.8	−9.2
Unions of public associations	5	0.2	3	0.1	4	0.2	−20
Public institutions	11	0.4	11	0.4	10	0.4	−9.1
Territorial public self-government	1272	46.6	1199	45.6	1074	44	−15.6
Other public organizations	867	31.7	870	33.1	853	34.9	−1.6
Total	2732	100	2632	100	2442	100	−10.6

Source: Developed by the authors based on the materials (Volgogradstat, n.d.).

At the beginning of 2019, 2,442 public organizations were operating in the Volgograd region, 10.6% fewer than the same indicator in 2017. The territorial public self-government had the largest losses; their number in 2019 reached 1074, 15.6% less than in 2017 (the number of organizations decreased by 198). Figure 2 shows the dynamics of the number of public organizations in the Volgograd region from 2016–2019. The figure also shows the trend line of the number of public organizations in the Volgograd region.

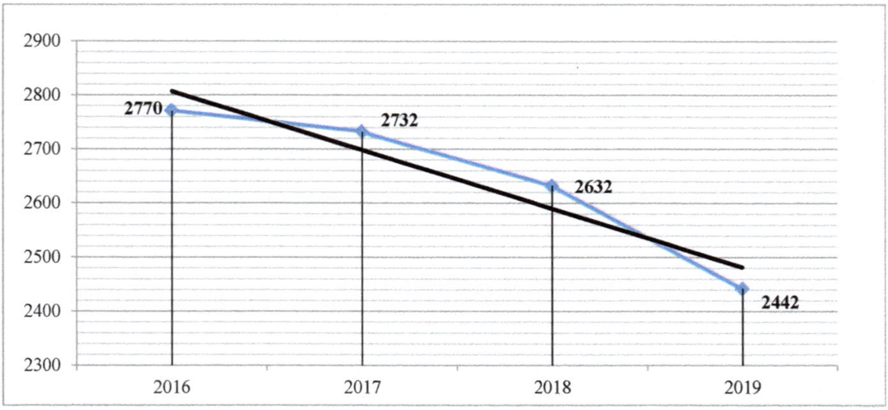

Fig. 2. The dynamics of the number of public organizations. *Source*: Developed by the authors based on the materials (Volgogradstat, n.d.).

The figure shows a tendency to decrease the number of public organizations operating in the region. In 2019, the total number of public organizations amounted to 88.2% of the 2016 level. This kind of dynamic is primarily due to the lack of equity for the development of organizations and low public demand for the services they provide. Non-profit organizations provide paid and free services according to their specialization. So, law firms conduct consultations for citizens on legal issues free of charge. Some private clinics include a free examination in their list of services. In 2018, 120 non-profit organizations of the Volgograd region received presidential grants for the development and implementation of the projects. The region, as in 2017, was among the top three entities in terms of the number of grants received. The grant support helped to reduce the cost and improved the quality of services provided, minimizing the cost of own funds by the organization. Table 3 presents the distribution by types of activities of non-profit organizations of the Volgograd region that received presidential grants in 2018.

Table 3. Non-profit organizations that received presidential grants in 2018.

Kind of activity	2018	% of the total
Environmental protection	7	5.8
The public diplomacy development	4	3.3
Health Activities	24	20
Protection of the citizens' rights and freedoms	7	5.8
Historical memory preservation	18	15
Social service	16	13.3
The civil society institutions development	5	4.2
Support for youth projects	6	5
Support for family, motherhood, and fatherhood	5	4.2
Science and education	8	6.7
Culture and art	16	13.3
Strengthening interethnic and interreligious harmony	4	3.3
Total	120	100

Source: Developed by the authors based on the materials (Presidential Grants Fund, n.d.).

Table 3 shows that the most significant number of presidential grants were granted to organizations operating in the spheres of healthcare (20% of the total number of organizations), historical memory preservation (15%), and social services for citizens and culture (13.3%). Also, the grant support was given to organizations operating in the fields of science (6.7%), environmental protection (5.8%), protection of rights and freedoms (5.8%), support for youth projects (5%), and others.

In 2019, 403 projects of nonprofit organizations of the Volgograd region were announced to participate in the first competition for being awarded presidential grants. The project data is presented in Table 4.

Table 4. The classification of the projects.

Activity area	Q-ty	Amount requested	Co-financing	Total amount
Social Services, Social Support	26	26 463 786	12 176 452	38 640 238
Identification and support of young talents in culture	1	443 313	364 576	807 889
Protection of the rights and freedoms of man and citizen	5	9 406 571	1 150 693	10 557 264
Protection of the citizens' health, promoting a healthy lifestyle	132	130 446 352	63 445 225	193 891 577
Environmental and animal welfare	23	18 508 207	9 674 962	28 183 169
Support for youth projects	20	25 113 691	9 739 813	35 853 630
Support for cultural and art projects	49	41 250 857	14 218 562	55 469 419
Support for projects in the field of science, education, enlightenment	29	41 055 271	15 569 399	56 624 670
Support for family, motherhood, fatherhood, and childhood	29	57 310 163	24 647 817	81 957 980
The development of civil society institutions	12	16 000 397	6 422 835	22 423 232
The public diplomacy development	6	24 133 862	33 930 228	58 064 090
Historical memory preservation	69	84 858 859	32 272 444	117 131 303
Strengthening interethnic and interreligious harmony	2	5 561 788	159 320	5 721 108
Total	403	481 553 243	223 772 326	705 325 569

Source: Developed by the authors based on the materials (Presidential Grants Fund, n.d.).

The total amount of the requested presidential grants is more than ₽ 481 million, while the most significant percentage of requests belongs to the organizations engaged in health and promoting a healthy lifestyle (the number of projects of such entities is 32.75% of the total), 27.1% of the total grant amounts requested. In the second place are the projects preserving the historical memory (17.1% of the total number). The third place is taken by the organizations that support motherhood, fatherhood, and childhood (7.2%); their total requests amounted to more than ₽ 57 million, which is 11.9% of the total amount of funds requested.

The ratio of the number of requested grants to the total amount of necessary financing of projects is 68.3%; the amount of co-financing funds is 31.7%. The only grant direction in which the number of co-financing funds exceeded the number of requested grants is for projects of public diplomacy development.

The student project, "Fifth semester," can be considered as an example. Its main purpose is to form a stable connection between students of various higher and secondary special institutions in Volgograd and the Volgograd region. As a result of these events and competitions, the organizers of the project plan to select a group of initiative students for further participation in the implementation of youth projects in the region.

The amount of the requested grant is ₽ 783,426 (80% of the project cost), and the indicated amount of co-financing is ₽ 195,500 (20% of the implementation cost). As a result, the total cost of the project is estimated at ₽ 978 926. The project is aimed primarily at young people, and its implementation can have a beneficial effect on the overall level of the development of all student projects in the region.

4 Discussion

The development of the non-profit sector of the economy of the Volgograd region is possible only if there is a growing public demand for goods and services sold by the organizations as part of the main goals of their activities. Thus, one of the main tasks in this direction is to increase public awareness regarding the activities of the tertiary sector and increase the degree of public approval (Oleinik and Korochkina 2014). In addition to the development of civil society institutions and the implementation of public benefit activities, non-profit organizations perform another important function – the creation of new jobs for citizens.

Table 5. The distribution of socially oriented non-profit organizations of the Volgograd region according to types of their activity.

Organizations engaged in:	2018	
	Number	%
Prevention of social orphans, support for motherhood and childhood	39	1.5
Improving the life quality of elderly people	58	2.3
Social adaptation of disabled people and their families	59	2.3
Social support and protection of citizens	154	6.1
Emergency preparedness	29	1.1
Assistance to victims of natural disasters, environmental, human-made or other disasters	41	1.6
Environmental protection and animal welfare	44	1.7
Protection and maintenance of objects and territories of historical, cult, cultural or environmental importance, and burial places	27	1.1
Providing legal assistance free of charge or on a preferential basis to citizens, legal education	106	4.2
Providing legal assistance free of charge or on a preferential basis to non-profit organizations	20	0.8
The prevention of socially dangerous behavior forms	29	1.1
Charity work	136	5.4
Promotion of charity	82	3.3
Volunteering activities	70	2.8
Activities in education, enlightenment, science, promotion of such activities	270	10.7

(*continued*)

Table 5. (*continued*)

Organizations engaged in:	2018	
	Number	%
Cultural and art activities, promotion of such activities	99	3.9
Activities in the field of healthcare, prevention, and protection of the citizens' health, promotion of a healthy lifestyle	101	4
Activities in the field of patriotic, including military-patriotic, spiritual and moral education of citizens	206	8,2
Activities in the field of improving the moral and psychological state of citizens and the spiritual development of the individual	50	2
Activities in the field of physical education and sports, the promotion of such activities	280	11.1
Development of interethnic cooperation, preservation, and protection of the identity, culture, language, and traditions	70	2.8
Support of socially significant youth initiatives, projects, children's and youth movement, children's and youth organizations	80	3.2
Promotion of the production and (or) distribution of social advertising	11	0.4
Anti-corruption activities, including promoting the formation of an active rejection towards the corruption	18	0.7
Conducting search work aimed at identifying unknown military graves and unburied remains of the Fatherland defenders	18	0.7
Participation in the prevention and (or) suppression of fires and emergency rescue operations	16	0.6
Social and cultural adaptation and integration of migrants	3	0.1
Measures for medical rehabilitation and social rehabilitation of persons engaged in illegal drug use	6	0.2
Other activities	400	15.9
The total number of organizations:	2522	100

Source: Developed by the authors based on the materials (Volgogradstat, n.d.).

According to the statistics, from November 2018 to January 2019, the unemployment rate in the Volgograd region amounted to 5.8%, which exceeds the all-Russian indicator of about 4.7%. The creation of favorable conditions for the functioning of non-profit organizations involves increasing the number of jobs for non-disabled citizens. Such measures can have a significant impact on increasing the employment level in the region.

Table 5 shows the distribution of socially-oriented nonprofit organizations of the Volgograd region by the activity type.

In 2018, the total number of socially-oriented nonprofit organizations in the Volgograd region amounted to 2,522; the organizations specializing in sports (280 organizations, 11.1%), educational activities (270 organizations, 10.7%) and organizations operating in the field of patriotic education of the population (206 organizations, 8.2%) have the most significant shares. 154 organizations specialize in social support and protection of citizens (6.1% of the total), 136 NCOs carry out charitable activities.

Also, the organizations operating in the region support people with disabilities, conduct measures to increase the legal literacy of the population, work in the healthcare sector, and do many other things. About 16% of the total amount consists of organizations whose activities do not correspond to the areas of work indicated in the table.

5 Conclusion

As a result of the study, it was found that in the conditions of the increasing influence of individual organizations, other NCOs exist with a constant lack of funding from the budget and own funds. For the development of the tertiary economy sector in Russia, and in particular in the Volgograd region, it is necessary to improve the regulatory framework governing the activities of nonprofit organizations, as well as to create favorable conditions for their cooperation with the private sector and the business community.

Also, it is necessary to increase the literacy of the population in matters of the activities of NCOs. There is a close relationship between the non-profit and the public sectors, which is expressed in expanding the sphere of influence of the organization in the presence of public demand for its services. An increase in the client base helps to strengthen the financial independence of a non-profit organization. By definition, a non-profit organization does not distribute profits among the founders but uses it solely as part of the goal. Based on this, most of the profit received by the NCOs from the entrepreneurial activity will go towards expanding the scope of the organization's activities, including diversifying the range of services and goods provided. Thus, public approval of the activities of a non-profit organization becomes the key to its financial well-being in the long term.

References

Oleinik, O.S., Korochkina, L.E.: Features of formation and use of financial and property resources of socially oriented non-profit organizations of the Volgograd region. Bull. VolSU (Ser. 3: Econ. Ecol.) **6**, 53–64 (2014). https://ges.jvolsu.com/index.php/ru/component/attachments/download/772

Presidential Grants Fund. (n.d.): Official website. https://президентскиегранты.рф

Sazonov, S.P., Gaponenko, Y.V., Weissbein, K.D.: The financial potential of the region as a component of economic potential and aspects that determine its efficiency. Bull. VolSU (Ser. 10: Innov.) **2**, 15–21 (2014). https://cyberleninka.ru/article/n/finansovyy-potentsial-regiona-kak-sostavlyayuschayaekonomicheskogo-potentsiala-i-aspekty-opredelyayuschie-ego-effektivnost/viewer

Volgogradstat. (n.d.): Official website. http://volgastat.gks.ru

The Concept of Humanization of Professional Activities of University Teachers as a Basis for Improving the Quality of Its Results

Veronika V. Yankovskaya[1,2,3](✉)

[1] State University of Management (GUU), Moscow, Russia
veronika28-2@mail.ru
[2] Higher School of Tariff Regulation, Moscow, Russia
[3] Plekhanov Russian University of Economics (PRUE), Moscow, Russia

Abstract. Research subject. In this paper, a study is made of the principles of humanization and their impact on the management of labor quality of higher education faculty in modern conditions. In the context of the digital economy, labor is transformed into creativity. This will especially affect the field of activity of scientists of the teaching staff of the Higher School. **Goals and objectives.** The goal is to discuss the results of the study of the parameters of the complex phenomenon "management of the work quality of teachers in the Higher School, based on the principles of humanization." To achieve the designated goal, the following tasks were completed: a comparative assessment of various concepts on the principles of humanization of labor in the digital economy; discussion of the content of the quality indicator "the degree of humanization of labor"; to assess the above process, the possibility and expediency of applying the generalizing characteristics (degree of humanization), by the indicators of which it is possible to judge the quality of work of the faculty and researchers. **Methodology.** The traditional methodology of scientific research is used, namely, the formulation of a scientific hypothesis with subsequent argumentation. **Conclusions.** In the scientific literature, there is the concept of the quality of work or the working life of teachers, which is interpreted as the degree of satisfaction of the teacher's personal needs and interests through the implementation of all types of labor activity, which is transformed into creative activity in the digital economy. It is obvious that the quality of work is directly related to self-actualization, self-development, self-education, and, therefore, to the degree of humanization of the teacher's work. In our opinion, the principles of humanization of labor proposed above allow us, if observed, to conclude the degree of humanization of the teacher's labor, and thus, it is possible to judge the quality of labor activity of the teaching staff. The same principles apply to the transformation of labor into creativity. **Scientific novelty** of the work consists in the development of the principles of humanization of labor with its transformation into creativity in a digital economy.

Keywords: Labor management · Humanization principles · Development index · Transformation of labor into creativity

A. V. Bogoviz and Y. V. Ragulina (Eds.): ISCI 2019, LNNS 115, pp. 780–786, 2020.
https://doi.org/10.1007/978-3-030-40749-0_92

1 Introduction

Humanization is the demand for social development. Today, an individual often asks himself what his role in social progress is. This is the need of society, which has reached a high stage of its development. At the end of the XX-beginning of the XXI centuries, Russia entered an era of cardinal transformations and changes. Humanist values are again at the forefront. Emphasis was on the formation of a civil society based on a creatively developed personality. Having lost a single geopolitical space in the early 1990s, abandoning the ideology that has developed over the past 80 years, we have practically lost the spirituality inherent in Russian people, such an important component of our life for more than a decade. In pursuit of Western and European permissiveness, freedom of personality development, we got the opposite effect, namely deep crisis phenomena, in almost all spheres of life. Over time, significant stabilization in our society has occurred, but, despite this, the principles of humanization are not fully reflected in the education system. In particular, we are talking about the highest level of education.

Humanization of education as a process of preparing a humanistic person comes from the recognition of the self-worth of the personality of not only the student, but also the teacher, his rights to free self-education, self-determination, self-development, based on the manifestation of his abilities, or, in other words, his potential.

Thus, the humanization of education is based on recognizing the humanity not only of the students but also of the teaching staff and their right to comprehensive development by creating the necessary conditions for training and work.

2 Materials and Method

The paper summarizes the theoretical and methodological approaches to the studied concept. The author conducted a study in the field of education, identifying the most promising trends of its further development (Yankovsckaya 2016).

In scientific literature, such principles of humanization of labor are considered as security, justice, self-actualization of personality, democracy and others (Hetmańska and Maciejewski 2017).

In our opinion, the basic principles of humanizing the work of university teachers should include (Yankovsckaya 2016):

- Ergonomics, which means the adaptability of working conditions to sanitary and hygienic norms and standards, the aesthetic perception of technical and other means that provide teachers with labor processes, including physical, informational and economic safety;
- Adequacy, which should be understood as the degree to which teachers meet the requirements for the conditions, processes, nature of work and management methods regarding, firstly, the resources of the state educational institution budget, and secondly, the results of the work of the teaching staff;

- Program and target orientation, which determines the existence of a set of strategic goals and programs for the development of the institution, including raising and applying the potential of teachers to achieve the main strategic long-term goal of the university in accordance with the needs of the external environment, namely clientele markets;
- Continuity, which means the ongoing, continuous, and annual implementation of specific activities by the university management system to improve conditions, processes, nature of work, and methods of managing faculty activities;
- Partisipativity, which is understood as a high degree of teacher participation in determining the nature of work; the organization of educational, research, consulting and educational activities; and improving management methods and working conditions;
- The systematic nature of requiring a higher education institution to be considered as an independent system, functioning in changing conditions of the external market environment and under the influence of education management systems at higher levels—federal, subject, district, and municipal—which leads to a university development strategy, changes in the processes, and nature of labor teachers;
- Complexity, which means the need to develop complexes of programs and action plans that are interconnected by the general strategic goals of the university, the department, and each teacher, according to the agreed-upon criteria for achieving these goals and the results of measures to improve the conditions, process, nature of work, and management methods;
- Efficiency, which implies planning and obtaining results of increasing the degree of humanization of labor of university teachers, comparing these results with previously achieved results, and obtaining effects for each element of humanization of labor and the whole system for improving the conditions, process, nature of work, and management methods, as well as calculating the effects of faculty performance;
- Transparency, which should be understood as the clarity of methods for assessing the potential, results and quality of work of each teacher; plans and results for increasing the degree of humanization of teachers; the availability of methods; and methods for determining the contribution of each teacher to the activities of the department, faculty and university;
- Comparability, a principle that is necessary to develop criteria for achieving goals and indicators for assessing the degree of the humanization of labor, which can be measured using qualitative and quantitative measurement methods and compared with retrospective ones in order to calculate effects and draw up further plans for improving humanization elements; and
- Proportionality, which means compiling and maintaining for a certain period the proportions established at the university between the costs and the results of the work of teachers and between the classroom and extracurricular workloads; the workload for teachers; and their contribution to educational, research, consulting and educational activities and salaries, bonuses and additional incentive payments.

Thus, the proposed principles of humanization of labor include, in essence, the few that are proposed in the scientific literature.

For example, the principles of proportionality, comparability and transparency ensure the fairness of the assessment of the results and the quality of work of teachers. The principle of participation is adequate to democracy in the development of goals and plans to increase the degree of humanization of teachers.

The principles of participation, program-oriented orientation and systematicity directly affect the self-actualization of the personality of the teacher as a teacher, scientific researcher, educator, consultant and developer of teaching materials. And finally, the safety of teachers is ensured through the introduction of the principles of ergonomics, adequacy and continuity.

In the scientific literature, there is the concept of the quality of work or the working life of teachers, which is interpreted as the degree of satisfaction of the personal needs and interests of the teacher through the implementation of all types of work.

3 Results

It should be noted that humanism is not a constant concept, the essence of which does not change over time. On the contrary, humanism is a system of views and opinions that changes under the influence of historical events; the evaluation criteria of a person's good; the degree to which the principles of equality, justice are applied, first of all, in creating adequate working conditions and their assessment and payment.

Humanity in labor, social and pedagogical activity in relation to each other is the desired norm of relations, ensuring the comprehensive development of the individual and the more complete satisfaction of his needs, regardless of social status. Thus, humanization is the dissemination in society and the affirmation in professional activities and life of ideas, attitudes, and affirmations permeated by humanism.

Consequently, humanism is the basis of humanization and can be seen as an increase in humanistic principles in society, the adoption of universal values that determine the cultural and moral development of a person, and his abilities in an aesthetic form, combined with humanity and gentleness.

In practice, humanism is manifested in the provision of real, material, scientific, technical, economic, political, cultural, and natural conditions for human life, as well as in adequate, objective social relations. The essence of humanism consists in the totality of objective and subjective relations to each person, in this study—to the teacher, as the highest independent individuality with a certain value. The manifestation of humanism in reality is carried out in the form of socially useful activities of the individual, a freely chosen form of labor that consciously benefits another person, group (collective), or society. Humanism can exist not only in the form of useful activity, but also in any manifestations of one person's concern for the good of another. Inhumane manifestations include factors such as life at the expense of other members of society; actions that threaten or reduce the well-being of others and reduce the quality of the environment; production of goods not having public utility; scientific research the results of which are not used for the benefit of the people and the individual and for the tasks of creation, but, on the contrary, for the destruction and decline in the well-being of society.

All these concepts relate directly to the relationship in the processes of pedagogical, scientific, educational, and consulting activities of teachers of higher educational institutions. However, the understanding of the term humanization for these types of activities is not unambiguous because, in our opinion, there are several types of humanization or humane relations, which are called levels of humanization in the scientific literature and are divided into subjective and objective levels. The latter, in turn, is divided into micro and macro levels; moreover, the micro level includes personal and business levels (Big Russian Encyclopedia 2007).

From the author's point of view, a change in production factors will occur in the new economy. The main factor of production will be knowledge, and labor and capital will become necessary conditions (Makarov and Khramov 2010; Kukushkin 2018). Moreover, labor is transformed into creativity (Inozemtsev 1998; Kukushkin 2018). As a result of the creative process in which teachers and scientists are engaged, new knowledge is created that is necessary not only to convey to students but also to turn it into a material product. This can be done with the help of information and communication technology, which is a key element of the digital economy.

Knowledge, as an economic good, has the following two features (Kukushkin 2018):

- Highly science-intensive and informative;
- Involves a consumer willingness to use/consume goods.

The transformation of labor into creativity will entail a change in the place and role of man in the organization. In his work, the American researcher E. Toffler notes, "[B] efore our eyes, a new type of organization person appears—this is a person who, despite his many contacts, does not join any organization. He wants to use his skills and creative energy to solve problems using the equipment of the organization, being inside the temporary associations created by her" (2004).

4 Discussion

The end of the twentieth century and the beginning of the twenty-first was marked by the appearance of the works of such authors as V. Lezhnikov, E. Pushkareva, R. Porfiryev, N. Bokachev, et al.; Lezhnikov in partial, viewed humanization as "integral humanism" (Pushkareva 2007). These authors devote their work to the rigid interweaving and interaction of the education system and philosophy with the idea of humanism as the basis for the development of society. T. Rubantsova believes that the emergence of such categories as humanism and the humanization of higher education and societies postmodern and modern are interdependent. O. Nedina notes in her works the multivariance of judgments and concepts of humanization, from the point of view of history, psychology, and philosophy. Klochkov focuses on such functions of humanization as socialization, ideology, and profiling. V. Koop highlights cultural and humanistic function. Recently, the number of research papers has increased significantly, but the degree of knowledge of humanization at the university is still insufficient. There is no systematization of this aspect; no in-depth study of the elements of humanization, its true purpose and role in the development of society, in general, and the education system, in particular.

5 Conclusion

Thus, we can conclude that there are two types of principles of humanization of labor. The first type includes those types that ensure the creation of the necessary working conditions for the faculty. The second type reflects the requirements for the teacher, who creates the conditions for increasing the degree of humanization of the educational, research, consulting, and educational process for students, colleagues, and the university management system. These two types of principles can be called system-wide: They are inherent in the whole society (Erzhanova 2009).

In addition, with regard to labor resources, the head of state expressed the meaning or purpose of the policy, which can be called labor. It consists of saving people (population), increasing human capital in Russia through the implementation of demographic programs and improving the environment, health and development of education and culture (Presidential Executive Office 2016). As a rule, the tasks of labor policy include the following: the need to attract talented young people and highly qualified specialists in social sectors, the possibility of increasing the wages of such specialists and the possibility of improving their working conditions.

Referring to the statement of academician D. Likhachev, V. Putin voiced the goal of education: to provide knowledge and educate a moral person on the basis that the moral foundation is the main factor that determines the economic, creative and state viability of society.

It is obvious that the quality of work is directly related to self-actualization, self-development, self-education and, therefore, the degree of humanization of the teachers' work (Gryazev and Arkhipova 2014).

If we draw an analogy between the quality of life as a human development index (HDI) and the quality of work of a teacher, then we first need to develop an integral or additive indicator that reflects the quality of working life and then determine the constituent indicators. Thus, for example, if the quality of life is determined by the life expectancy at birth, levels of income and education, then for the quality of work, one of the quality indicators may be the degree of humanization of labor (Johnson 2010).

In our opinion, the principles of the humanization of labor proposed above allow us to conclude that if they are observed, the degree of humanization of the teachers' work is the generalizing characteristic of the indicators with which we can judge the quality of the work of the teaching staff.

References

Gryazev, M.V., Arkhipova, S.A.: Motivation of labor activity of the academics of the university. Proc. Tula State Univ. **1**, 39–44 (2014)

Hetmańska, M., Maciejewski, A.: The application of a UHPLC system to study the formation of various chemical species by compounds undergoing efficient self-aggregation and to determine the homodimerization constants (KDM) with values in the high range of 106-1010 M-1. RSC Adv. **7**(71), 44843–44859 (2017). https://doi.org/10.1039/c7ra05051h

Inozemtsev, V.L.: Outside of an Economic Society. "Academia" - "Science", Moscow (1998)

Johnson, J.: The future of the social sciences and humanities in the science of complex systems. Eur. J. Soc. Sci. Res. **23**(2), 115–134 (2010)

Kukushkin, S.N.: Economics of the Information Society. SIC Impulse, Moscow (2018)

Presidential Executive Office: Message of the President of the Russian Federation to the Federal Assembly of the Russian Federation (2016). http://en.kremlin.ru/events/president/transcripts/messages/53379

Pushkareva, E.: The specifics of the philosophical analysis of the modern system of education and science. Philos. Educ. **1**, 60–65 (2007)

Toffler, E.: The Shock of the Future. "AST" Publishing House LLC, Moscow (2004)

Yankovsckaya, V.V.: Creation of permanent institutional monitoring mechanism of modern educational technology market significant for nanoindustry. IEJME: Math. Educ. **6**, 1545–1556 (2016)

Erzhanova, R. K.: Formation of professional and pedagogical potential. Karaganda, Kazakhstan: Karaganda State University (2009)

Makarov, A. M., Khramov, E.N. Harmonization of development of innovative systems as a factor of improvement of knowledge management system. Dig. Finance **19**(192), 15-20 (2010). https://elibrary.ru/item.asp?id=15322597

Author Index

A. V. Bogoviz and Y. V. Ragulina (Eds.): ISCI 2019, LNNS 115, pp. 787–789, 2020.
https://doi.org/10.1007/978-3-030-40749-0

Printed by Printforce, the Netherlands